Compound type	*Functional group*	*Simple example*	*Name ending*
Ester	$-C(=O)-O-C-$	$CH_3C(=O)-OCH_2CH_3$ Ethyl ethanoate (ethyl acetate)	*-oate*
Carboxylic acid	$-C(=O)-O-H$	$CH_3CH_2CH_2C(=O)-OH$ Butanoic acid	*-oic acid*
Carboxylic acid chloride	$-C(=O)-Cl$	$CH_3CH_2C(=O)-Cl$ Propanoyl chloride	*-yl chloride*
Amide	$-C(=O)-N<$	$CH_3C(=O)-NH_2$ Ethanamide	*-amide*
Amine	$>C-N<$	$CH_3CH_2NH_2$ Ethylamine	*-amine*
Nitrile	$-C\equiv N$	$CH_3C\equiv N$ Ethanenitrile (acetonitrile)	*-nitrile*
Nitro	$>C-\overset{+}{N}(=O)-O^-$	$CH_3CH_2NO_2$ Nitroethane	None
Organometallic	$>C-M$ M = metal	CH_3-Li Methyllithium	None

EDITION 2

Fundamentals of Organic Chemistry

Preface

As I stated in the first edition of this text, my goal has been to write a readable and effective teaching text that presents only those subjects needed for a one-semester or two-quarter course in organic chemistry but that keeps the important pedagogical tools commonly found in larger books. I have taken particular care to incorporate clear explanations of difficult subjects, effective artwork, repetition of important points, and varied end-of-chapter learning tools. The result, I believe, is a book that is easier to learn from than other short organic chemistry texts.

All of the features that made the first edition a success have been improved, and many new ones have been added.

The writing, already clear and accessible, has been further refined at the sentence level on every page.

The artwork has been redone with the use of computer-generated structures to ensure accuracy.

Text and reaction summaries have been added at the end of each chapter, and problem sets have been expanded.

Other changes include expanded treatments of resonance, isomerism, and enzyme structure and function.

Organization

The primary organization of this book is by functional group, beginning in Chapter 2 with alkanes and going on to more complex compounds. Within this primary organization, more emphasis is placed on explaining the fundamental mechanistic similarities of organic reactions than is common in other short texts. Chapter 11, "Carbonyl Alpha-Substitution and Condensation Reactions," for example, helps to remove the artificial lines between ketones and esters by showing how all carbonyl compounds undergo similar reactions.

Organic molecules and organic reactions are presented as early as possible. After a brief review of structure and bonding in Chapter 1, organic molecules and functional groups are introduced in Chapter 2, followed by an initial discussion of organic reactions in Chapter 3.

The Lead-off Reaction: Addition of HCl to Ethylene

An important yet simple polar reaction—the addition of HCl to ethylene—is used as the lead-off to illustrate general principles of organic reactions. This choice has the advantage of relative simplicity (no prior knowledge of stereochemistry and kinetics is required), yet it is also a broadly applicable reaction of a common functional group. Many students attach great importance to a text's lead-off reaction, and I believe that the choice of an important polar reaction serves to introduce students to functional-group chemistry better than a lead-off such as free-radical alkane chlorination.

Spectroscopy

Spectroscopy is treated as a tool, not as a specialized field of study. Infrared, ultraviolet, ^{13}C NMR, and ^{1}H NMR spectroscopies are all covered by showing the kind of information that can be derived from each and how each can be used to answer specific structural questions.

Nomenclature

The IUPAC system of nomenclature is used throughout. For the most part, this involves the use of systematic names, although a few IUPAC-approved nonsystematic names such as acetic acid, acetone, ethylene, and phenol are also employed. Since it's unlikely that these few common names will disappear from everyday use in the near future, it's probably best for students to learn them.

Coverage

The coverage in this book is up to date, reflecting important advances of the past decade. For example, ^{13}C NMR is introduced as a routine spectroscopic tool, equal in importance to ^{1}H NMR. Similarly, the chemistry of nucleic acids is covered, including a section on DNA sequencing by the Maxam-Gilbert method.

Interludes

Brief "interludes" are included at the end of each chapter. Meant to serve as short breathers between chapters, these interludes show interesting applications of organic chemistry to industrial and biological systems. They can be covered by the instructor or left for student reading.

Practice Problems

Each chapter contains many worked-out examples that illustrate how problems can be solved. Each practice problem and solution is then followed by a similar problem for the reader to solve. These worked-out examples are valuable because of their appearance in the text, but they are not meant to serve as a replacement for the accompanying *Study Guide and Solutions Manual.*

Pedagogy

In addition to the features mentioned above, every effort has been made to make this book as effective, clear, and readable as possible—in short, to make it easy to learn from.

Paragraphs usually start with summary sentences.

Transitions between paragraphs and between topics are smooth.

Extensive use is made of computer-generated, airbrushed art and carefully rendered stereochemical formulas.

Extensive cross-referencing to earlier material is used.

A second color is used to indicate the changes that occur during reactions.

More than 800 problems are included, both within the text and at the end of every chapter. These include both drill and thought problems.

Key terms are defined in the margin next to where they are first used.

An innovative vertical format is used to explain reaction mechanisms. The mechanisms are printed vertically, while explanations of the changes taking place in each step are printed next to the reaction arrow. This format allows the reader to see easily what is occurring at each step in a reaction without having to jump back and forth between the text and structures.

Study Guide and Solutions Manual

A carefully prepared *Study Guide and Solutions Manual* accompanies this text. Written by Susan McMurry, this companion volume answers all in-text and end-of-chapter problems and explains in detail how answers are obtained. In addition, many valuable supplemental materials are given, including: a list of study goals for each chapter, a glossary, a summary of name reactions, a summary of organic reaction mechanisms, a summary of the uses of important reagents, tables of spectroscopic information, a list of suggested readings, and a discussion on the rules of nomenclature for polyfunctional molecules.

Acknowledgments

I sincerely thank the many people whose help and suggestions were so valuable in the creation of this book. Foremost is my wife Susan who read, criticized, and improved all aspects of the text, and who authored the accompanying *Study Guide and Solutions Manual.* Among the reviewers providing thoughtful comments were R. A. Abramovitch, Clemson University; Orville L. Boge, Mesa State College; Gerald Caple, Northern Arizona University; George B. Clemans, Bowling Green State University; Donald Grant, University of Saskatchewan; George Kraus, Iowa State University; Gordon L. Lange, University of Guelph, Ontario, Canada; Anne G. Lenhert, Kansas State University; Barbara J. Mayer, California State University at Fresno; and Daniel O'Brien, Texas A & M University.

Special thanks are due Mary Douglas, Harvey Pantzis, Joan Marsh, and others of the Brooks/Cole staff for their usual fine work.

A Note for Students

We have similar goals. Yours is to learn organic chemistry; mine is to do everything possible to help you learn. It's going to require work on your part, but the following suggestions should prove helpful:

Don't read the text immediately. As you begin each new chapter, look it over first. Read the introductory paragraphs, find out what topics will be covered, and then turn to the end of the chapter and read the summary. You'll be in a much better position to understand new material if you first have a general idea of where you're heading. Once you've begun a chapter, read it several times. First read the chapter rapidly, making checks or comments in the margin next to important or difficult points; then return for an in-depth study.

Keep up with the material. Who's likely to do a better job—the runner who trains five miles per day for weeks before a race, or the one who suddenly trains twenty miles the day before the race? Organic chemistry is a subject that builds on previous knowledge. You have to keep up with the material on a daily basis.

Work the problems. There are no shortcuts here. Working problems is the only way to learn organic chemistry. The practice problems show you how to approach the material, the in-text problems provide immediate practice, and the end-of-chapter problems provide additional drill and some real challenges. Answers and explanations for all problems are given in the accompanying *Study Guide and Solutions Manual.*

Ask questions. Faculty members and teaching assistants are there to help you. Most of them will turn out to be extremely helpful and genuinely interested in seeing you learn.

Use molecular models. Organic chemistry is a three-dimensional science. Although this book uses many careful drawings to help you visualize molecules, there's no substitute for building a molecular model, turning it in your hands, and looking at it from different views.

Use the study guide. The *Study Guide and Solutions Manual* that accompanies this text gives complete solutions to all problems and provides a wealth of supplementary material. Included are a list of study goals for each chapter, outlines of each chapter, a large glossary, a summary of name reactions, a summary of methods for preparing functional groups, a summary of the uses of important reagents, tables of spectroscopic information, a list of suggested readings, and a discussion on the naming of polyfunctional compounds. Find out ahead of time what's there so that you'll know where to go when you need help.

Good luck. I sincerely hope you enjoy learning organic chemistry and that you come to see the logic and beauty of its structure. I would be glad to receive comments and suggestions from any who have learned from this book.

Brief Contents

Contents

6 STEREOCHEMISTRY 158

7 ALKYL HALIDES 186

8 ALCOHOLS, ETHERS, AND PHENOLS 219

9 ALDEHYDES AND KETONES: NUCLEOPHILIC ADDITION REACTIONS 252

CHAPTER

1 Structure and Bonding

What is organic chemistry? Why should you study it?

The answers to these questions are everywhere. Every living organism is composed of organic chemicals; the foods you eat and most medicines you take are organic chemicals; the wood, paper, plastics, and fibers that make modern life possible are organic chemicals. Anyone with a curiosity about life and living things must have a fundamental understanding of organic chemistry if their curiosity is to be satisfied.

The historical roots of organic chemistry can be traced to the mid-1700s when alchemists noticed unexplainable differences between compounds derived from living sources and those derived from minerals. Compounds from plants and animals were often difficult to isolate and purify. Even when pure, they were difficult to work with and tended to decompose more easily than compounds from minerals. The Swedish chemist Torbern Bergman was the first person to express this difference between "organic" and "inorganic" substances, and the term *organic chemistry* soon came to mean the chemistry of compounds from living organisms.

To many chemists of the time, the only explanation for the difference in behavior between organic and inorganic compounds was that organic compounds contained an undefinable "vital force" as a result of their origin in living sources. Further work showed, however, that organic compounds could be manipulated in the laboratory just like inorganic compounds. Friedrich Wöhler, for example, discovered in 1828 that it was possible to convert the "inorganic" salt ammonium cyanate into the already known "organic" substance urea:

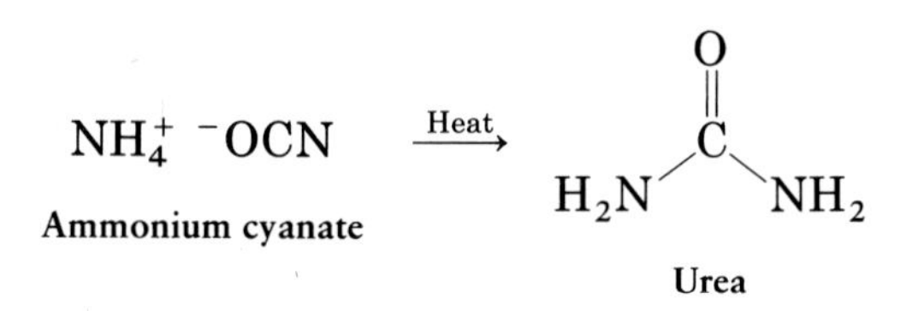

Ammonium cyanate

Urea

By the mid-1800s, the weight of evidence was against the vitalistic theory, and it had become clear that the same basic scientific principles are applicable to all compounds. The only unifying characteristic of organic compounds is that they all contain the element carbon.

organic chemistry the chemistry of the compounds of carbon

Organic chemistry, then, is the study of the compounds of carbon. But why is carbon special? What is it that sets carbon apart from all other elements in the periodic table? The answers to these questions derive from the unique ability of carbon atoms to bond together, forming rings and long chains. Carbon, alone of all elements, is able to form an immense diversity of compounds, from the simple to the staggeringly complex: from methane, containing one carbon atom, to DNA, which can contain tens of *billions.*

Nor are all organic compounds derived from living organisms. Modern chemists have become highly sophisticated in their ability to synthesize new organic compounds in the laboratory. Medicines, dyes, polymers, plastics, pesticides, and a host of other substances—all are prepared in the laboratory, and all are organic chemicals. Organic chemistry is a subject that touches the lives of everyone. Its study can be a fascinating undertaking.

1.1 ATOMIC STRUCTURE

Before beginning a study of organic chemistry, it's best to review some general ideas about atoms and bonds. Our present picture of atoms is that they consist of a dense, positively charged *nucleus* surrounded at a relatively large distance by negatively charged electrons (Figure 1.1). The nucleus consists of subatomic particles called *neutrons,* which are electrically neutral, and *protons,* which are positively charged. Though extremely small—about 10^{-14} to 10^{-15} meter (m) in diameter—the nucleus nevertheless contains essentially all the mass of the atom. Electrons have negligible mass and orbit the nucleus at a distance of approximately 10^{-10} m. Thus, the diameter of a typical atom is about 2×10^{-10} m, often called 2 *angstroms* (Å), where 1 Å = 10^{-10} m. To give you an idea of how small this is, a thin pencil line is about 3 *million* carbon atoms across.

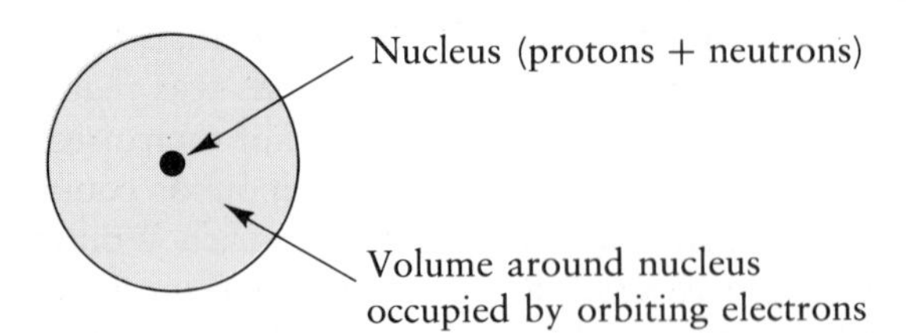

FIGURE 1.1 A schematic view of an atom. The dense, positively charged nucleus contains most of the atom's mass and is surrounded by negatively charged electrons.

atomic number the number of protons in an atom's nucleus

mass number the total number of protons and neutrons in an atom's nucleus

An atom is described by its **atomic number,** which is the number of protons in the atom's nucleus, and its **mass number,** which is the number of protons plus neutrons. All the atoms of a given element have the same atomic number—1 for hydrogen, 6 for carbon, 53 for iodine, and so on—but they can have different mass numbers depending on how many neutrons they contain. The *average* mass number

atomic weight the average mass of a large number of an element's atoms

of a great many atoms of an element is called the element's **atomic weight;** because it's an average, it's often not an integer (1.008 for hydrogen, 12.011 for carbon, 126.90 for iodine, and so on).

How are the electrons distributed in an atom? Because electrons are in constant motion around the nucleus, it's not possible to define their exact positions. It turns out, though, that electrons aren't completely free to move; different electrons are confined to different regions within the atom according to the amount of energy they have.

electron shell an imaginary layer around the nucleus occupied by electrons

Electrons can be thought of as belonging to different layers or **shells** at various distances from the nucleus. The farther a shell is from the nucleus, the more electrons it can hold and the greater the energies of those electrons. Thus, an atom's lowest-energy electrons occupy the first shell, which is nearest the nucleus and has a capacity of only two electrons. The second shell is farther from the nucleus and can hold eight electrons, the third shell is still farther from the nucleus and can hold eighteen electrons, and so on, as shown in Table 1.1.

TABLE 1.1 Distribution of electrons into shells

Number of shell	*Electron capacity of shell*
First	2
Second	8
Third	18
Fourth	32

orbital a specific region of space occupied by a given electron

Within each shell, electrons are further grouped by pairs into variously shaped regions of space called **orbitals.** It's helpful to think of an orbital as a kind of time-lapse photograph of an electron's movements around a nucleus. Such a photograph would show the orbital as a blurry cloud indicating where the electron has been. This electron cloud doesn't have a sharp boundary, but for practical purposes we can set the limits by saying that an orbital represents the space where an electron spends most (90–95%) of its time.

What do orbitals look like? There are four kinds of orbitals, denoted *s*, *p*, *d*, and *f*. Of the four, we'll be concerned only with *s* and *p* orbitals because most atoms found in living organisms use only these. The *s* orbitals have a spherical shape with the nucleus at the center, and the *p* orbitals have a dumbbell shape, as shown in Figure 1.2. Note that a given shell contains three different *p* orbitals, oriented in space so that each is perpendicular to the other two. They are denoted p_x, p_y, and p_z, depending on which coordinate axis they lie.

The different shells have different numbers and kinds of orbitals. The two electrons of the first shell occupy a single *s* orbital, designated 1*s*. The eight electrons of the second shell occupy one *s* orbital (designated 2*s*) and three *p* orbitals (each designated 2*p*). The eighteen electrons of the third shell occupy one *s* orbital (3*s*), three *p* orbitals (3*p*), and five *d* orbitals (3*d*). These electron distributions and their relative energy levels are indicated in Figure 1.3.

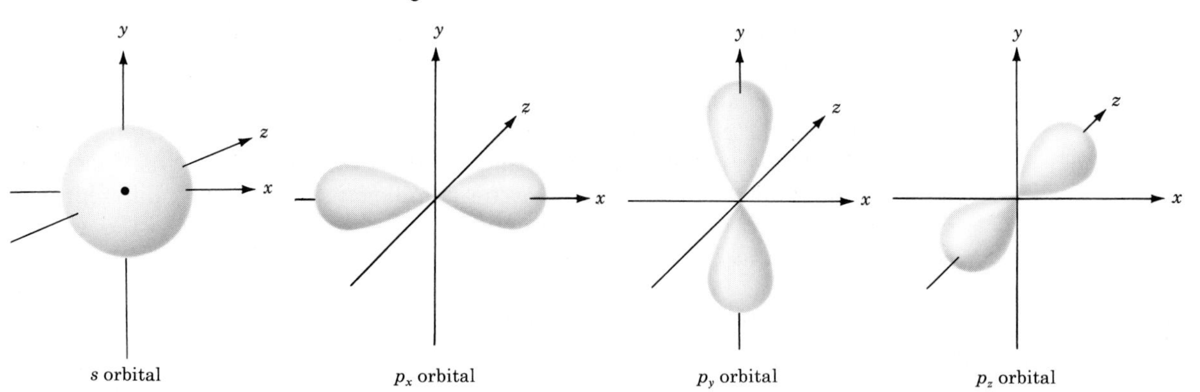

FIGURE 1.2 The shapes of *s* and *p* orbitals. There are three *p* orbitals, denoted p_x, p_y, and p_z.

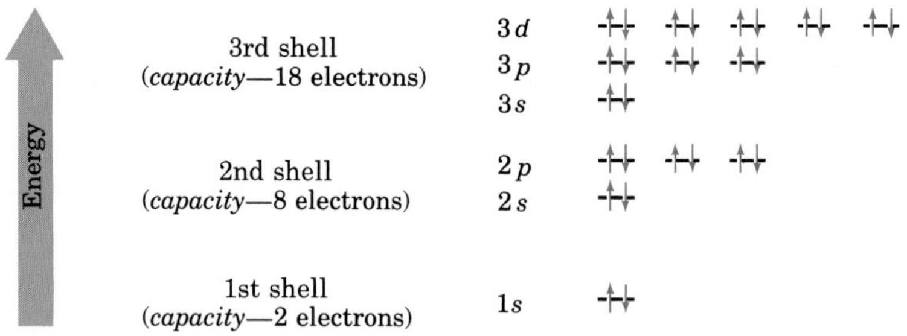

Energy

3rd shell
(*capacity*—18 electrons)
3*d* ↑↓ ↑↓ ↑↓ ↑↓ ↑↓
3*p* ↑↓ ↑↓ ↑↓
3*s* ↑↓

2nd shell
(*capacity*—8 electrons)
2*p* ↑↓ ↑↓ ↑↓
2*s* ↑↓

1st shell
(*capacity*—2 electrons)
1*s* ↑↓

FIGURE 1.3 The distribution of electrons in an atom. The first shell holds a maximum of two electrons in a 1*s* orbital; the second shell holds a maximum of eight electrons in one 2*s* and three 2*p* orbitals; the third shell holds a maximum of eighteen electrons in one 3*s*, three 3*p*, and five 3*d* orbitals; and so on. The two electrons in each orbital are represented by up and down arrows ↑↓.

1.2 ELECTRONIC CONFIGURATION OF ATOMS

The lowest-energy arrangement, or ground-state electronic configuration, of any atom is a description of the orbitals that the atom's electrons occupy. We can determine an atom's ground-state electronic configuration by following three rules:

1. The orbitals of lowest energy (those nearest the nucleus) are filled first.
2. Only two electrons can occupy the same orbital, and they must be of opposite spin.[1]

[1] Electrons can be thought of as spinning on an axis in much the same way that the earth spins. This spin can have two equal and opposite orientations, denoted as up ↑ and down ↓.

3. If two or more empty orbitals of equal energy are available, one electron is placed in each until all are half-full.

Some examples of how these rules are applied are shown in Table 1.2. Hydrogen, the lightest element, has only one electron, which must occupy the lowest-energy orbital. Thus, hydrogen has a 1*s* ground-state electronic configuration. Carbon has six electrons, and a ground-state electronic configuration $1s^2\ 2s^2\ 2p^2$ is arrived at by applying the three rules.[2]

TABLE 1.2 Ground-state electronic configurations of some elements

Element	*Atomic number*	*Configuration*	*Element*	*Atomic number*	*Configuration*
Hydrogen	1	1*s* ↑	Neon	10	2*p* ↑↓ ↑↓ ↑↓ 2*s* ↑↓ 1*s* ↑↓
Lithium	3	2*s* ↑ 1*s* ↑↓			
Carbon	6	2*p* ↑ ↑ — 2*s* ↑↓ 1*s* ↑↓	Chlorine	17	3*p* ↑↓ ↑↓ ↑ 3*s* ↑↓ 2*p* ↑↓ ↑↓ ↑↓ 2*s* ↑↓ 1*s* ↑↓

PRACTICE PROBLEM 1.1 Give the ground-state electronic configuration of nitrogen.

SOLUTION The periodic table on the rear inside cover shows that nitrogen has atomic number 7 and thus has seven electrons. Using Figure 1.3 to find the relative energy levels of orbitals and using the three rules to assign the seven electrons to orbitals, we find that the first two electrons go into the lowest-energy orbital ($1s^2$), the next two electrons go into the second-lowest-energy orbital ($2s^2$), and the remaining three electrons go into the three third-lowest-energy orbitals ($2p^3$). Thus, the complete configuration of nitrogen is $1s^2\ 2s^2\ 2p^3$.

PROBLEM 1.1 How many electrons does each of these elements have in its outermost electron shell?
(a) Potassium (b) Calcium (c) Aluminum

PROBLEM 1.2 Give the ground-state electronic configuration of these elements:
(a) Boron (b) Phosphorus (c) Oxygen (d) Argon

[2] A superscript is used here to represent the number of electrons at a particular energy level. For example, $1s^2$ indicates that there are two electrons in the 1*s* orbital. No superscript is used when there is only one electron in an orbital.

1.3 DEVELOPMENT OF CHEMICAL BONDING THEORY

By the mid-1800s, the new science of chemistry was developing rapidly, and chemists had begun to probe the forces holding compounds together. In 1858, August Kekulé and Archibald Couper independently proposed that, in all organic compounds, carbon always has four "affinity units." In modern terms, carbon is *tetravalent:* It always forms four bonds when it joins other elements to form stable compounds. Furthermore, said Kekulé, carbon atoms can bond to each other to form extended chains, and chains can double back on themselves to form rings of carbon atoms.

Although Kekulé was correct in describing the tetravalent nature of carbon, chemistry was still viewed in a two-dimensional way until 1874. In that year, Jacobus van't Hoff and Joseph Le Bel added a third dimension to our ideas about chemistry. They proposed that the four bonds of carbon are not randomly oriented but have a specific spatial orientation. Van't Hoff went even further and proposed that the four atoms to which a carbon atom is bonded sit at the corners of a tetrahedron, with carbon in the center. A representation of a tetrahedral carbon atom is shown in Figure 1.4. Note the conventions used to show three-dimensionality: Solid lines are assumed to be in the plane of the paper; heavy wedged lines come out of the plane of paper toward the viewer; and dashed lines recede into the plane. These representations will be used frequently in this text.

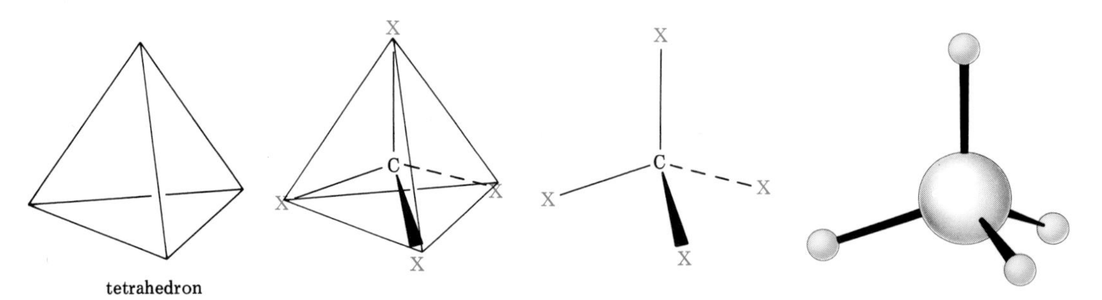

FIGURE 1.4 Van't Hoff's tetrahedral carbon atom. The solid lines are in the plane of the paper, the heavy wedged line comes out of the plane of the paper, and the broken line goes back into the plane.

PROBLEM 1.3 Draw chloromethane, CH_3Cl, using solid, wedged, and dashed lines to show its tetrahedral geometry.

1.4 THE NATURE OF CHEMICAL BONDS: IONIC BONDS

Why do atoms bond together, and how can bonds be described? The *why* question is relatively easy to answer: Atoms bond together because the product that results is more stable (has less energy) than the separate atoms themselves. Energy is always *released* when chemical bonds are formed. The *how* question is more difficult. To answer it, we need to know more about the properties of atoms.

valence shell the outermost electron shell of an atom

We know through observation that eight electrons (an electron octet) in the outermost shell (the **valence shell**) impart a special stability to the noble-gas elements in Group 8A: neon (2 + 8), argon (2 + 8 + 8), krypton (2 + 8 + 18 + 8). We also know that the chemistry of many elements with nearly noble-gas configurations is governed by a tendency to take on the stable noble-gas electronic makeup. For example, the alkali metals in Group 1A have a single *s* electron in their outer shells. By losing this electron, they can achieve a noble-gas configuration. Elements that tend to give up electrons are called **electropositive.**

electropositive an element with a tendency to donate electrons

Just as the alkali metals at the left of the periodic table have a tendency to form positive ions by losing an electron, the halogens (Group 7A elements) at the right of the periodic table have a tendency to form *negative* ions by *gaining* an electron. By so doing, the halogens achieve noble-gas configurations. Elements that tend to accept electrons are called **electronegative.**

electronegative an element with a tendency to withdraw electrons

The simplest kind of chemical bonding occurs between an electropositive element and an electronegative element. For example, when sodium metal (electropositive) reacts with chlorine gas (electronegative), sodium donates an electron to chlorine to form positively charged sodium ions and negatively charged chloride ions. The product, sodium chloride, is said to have **ionic bonding;** that is, the Na^+ and Cl^- ions are held together by an electrical attraction between their unlike charges. When a vast number of sodium atoms transfer electrons to an equal number of chlorine atoms, a visible crystal of sodium chloride results. As shown in Figure 1.5, each positively charged sodium ion in the NaCl crystal is surrounded by six negatively charged chloride ions and vice versa. This packing arrangement allows each ion to be stabilized by the attraction of unlike charges on its six nearest-neighbor ions while being as far as possible from like-charged ions.

ionic bond a bond between two atoms due to the electrostatic attraction of unlike charges

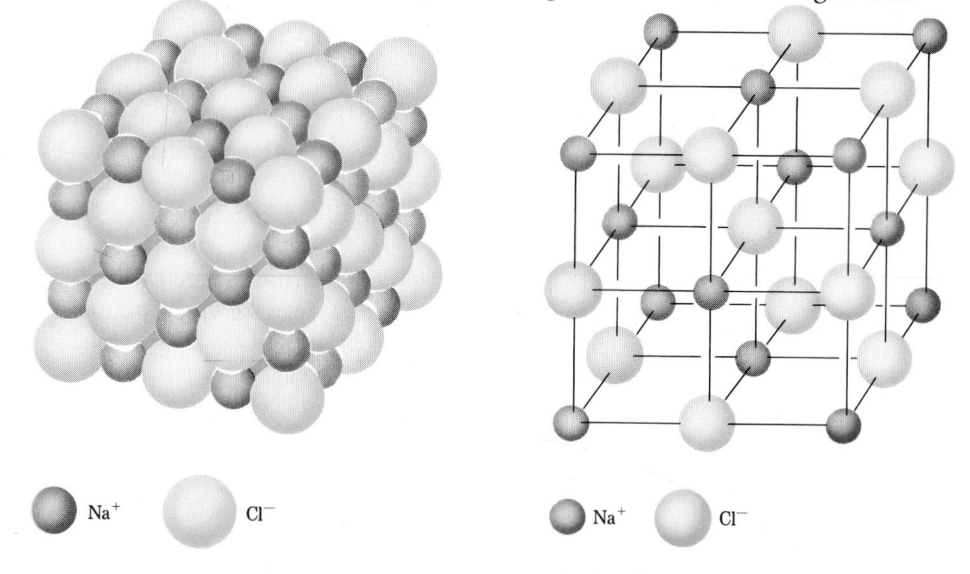

FIGURE 1.5 Ionic bonding in a sodium chloride crystal. Each sodium ion is surrounded by six chloride ions, and each chloride ion is surrounded by six sodium ions.

ionic solid
a crystal of positive and negative ions held together by ionic bonds

Because of the three-dimensional packing of ions in a sodium chloride crystal, we can't speak of specific ionic bonds between specific pairs of ions. That is, there's really no such thing as an individual NaCl "molecule." Rather, there are many ionic bonds between an ion and its nearest neighbors, and so we speak of the whole crystal as being an **ionic solid.**

PRACTICE PROBLEM 1.2 What charge would you expect a magnesium ion to have?

SOLUTION Because the periodic table indicates that magnesium is in Group 2A, it has two valence-shell electrons. Loss of these electrons results in the Mg^{2+} ion.

PROBLEM 1.4 How many valence-shell electrons do these elements have?
(a) Be (b) S (c) Br

PROBLEM 1.5 What charge would you expect the elements in Problem 1.4 to have on their ions?

PROBLEM 1.6 Judging from their position in the periodic table, which is more electronegative:
(a) Potassium or oxygen (b) Calcium or bromine

1.5 THE NATURE OF CHEMICAL BONDS: COVALENT BONDS

covalent bond
a bond formed by sharing electrons between two nuclei

molecule
a group of atoms joined together by covalent bonds

Lewis structure
a way of representing a molecule using dots to indicate an atom's outer-shell electrons

valence electrons
an atom's outer-shell electrons

We've just seen that elements at the far right (chlorine) and far left (sodium) of the periodic table form ionic bonds by gaining or losing an electron. How, though, do the elements in the middle of the periodic table form bonds? Let's look at methane, CH_4, the main constituent of natural gas, as an example. The bonding in methane is not ionic because it would be very difficult for carbon ($1s^2\ 2s^2\ 2p^2$) to either gain or lose *four* electrons to achieve a noble-gas configuration.[3] In fact, carbon bonds to other atoms, not by donating electrons, but by *sharing* them. Such shared-electron bonds, first proposed in 1916 by G. N. Lewis, are called **covalent bonds,** and the collection of atoms held together by covalent bonds is called a **molecule.** *The covalent bond is the most important bond in organic chemistry.*

A simple shorthand way of indicating covalent bonds in molecules is to use what are known as **Lewis structures.** In a Lewis structure, an atom's outer-shell electrons (**valence electrons**) are represented by dots. Thus, hydrogen has one dot ($1s$), carbon has four dots ($2s^2\ 2p^2$), oxygen has six dots ($2s^2\ 2p^4$), and so on. A stable molecule results when a noble-gas configuration with filled s and p valence-shell orbitals has been achieved for all atoms in the molecule, as in the following examples:

[3] The electronic configuration of carbon can be written either as $1s^2\ 2s^2\ 2p^2$ or as $1s^2\ 2s^2\ 2p_x\ 2p_y$. Both notations are correct, but the latter is more informative because it indicates that two of the three equivalent p orbitals are half filled.

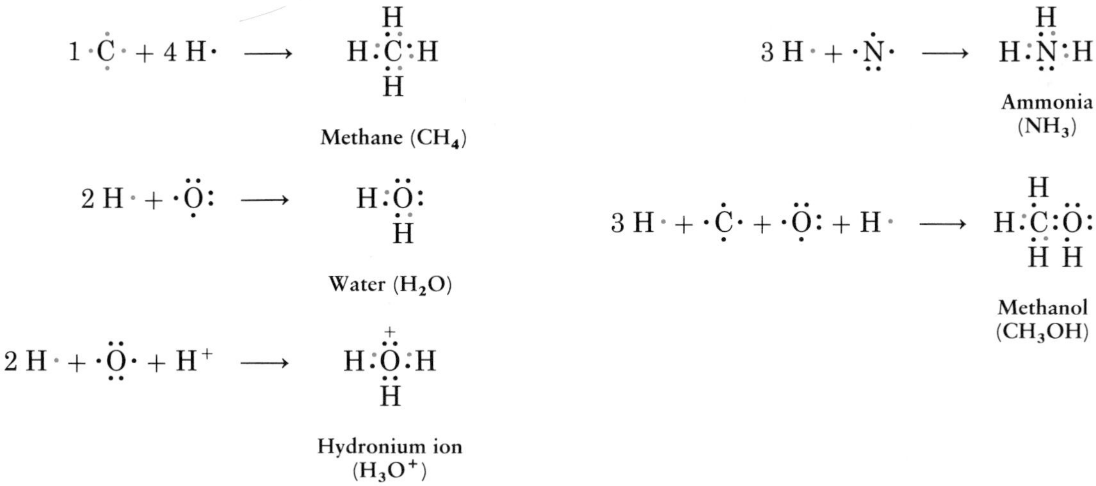

$1\cdot\dot{\text{C}}\cdot + 4\,\text{H}\cdot \longrightarrow$ H:C:H (with H above and below C)

Methane (CH_4)

$3\,\text{H}\cdot + \cdot\dot{\text{N}}\cdot \longrightarrow$ H:N:H (with H above N; lone pair below N)

Ammonia (NH_3)

$2\,\text{H}\cdot + \cdot\ddot{\text{O}}: \longrightarrow$ H:Ö: (with H below O)

Water (H_2O)

$3\,\text{H}\cdot + \cdot\dot{\text{C}}\cdot + \cdot\ddot{\text{O}}: + \text{H}\cdot \longrightarrow$ H:C:Ö: (with H above C; H below C and H below O)

Methanol (CH_3OH)

$2\,\text{H}\cdot + \cdot\ddot{\text{O}}\cdot + \text{H}^+ \longrightarrow$ H:Ö:H (+, with H below O)

Hydronium ion (H_3O^+)

The number of covalent bonds that an atom forms depends on how many valence-shell electrons it has. Atoms with one, two, or three valence-shell electrons form one, two, or three bonds, but atoms with four or more valence-shell electrons form as many bonds as they need electrons to fill the *s* and *p* levels of their valence shells. Thus, boron has three valence-shell electrons ($2s^2\ 2p^1$) and forms three bonds, as in BH_3; carbon ($2s^2\ 2p^2$) fills its valence shell and achieves the noble-gas configuration of neon ($2s^2\ 2p^6$) by forming four bonds as in CH_4; oxygen ($2s^2\ 2p^4$) achieves the neon configuration by forming two bonds, as in H_2O, and nitrogen ($2s^2\ 2p^3$) achieves the neon configuration by forming three bonds, as in NH_3.

Kekulé structure a representation of a molecule that indicates a covalent bond as a line between atoms

line-bond structure an alternative name for Kekulé structure

Lewis structures are valuable because they make electron bookkeeping possible and remind us of the number of valence electrons present. Simpler still is the use of **Kekulé structures,** also called **line-bond structures.** In a line-bond structure, the two electrons in a covalent bond are indicated simply by a line. Pairs of nonbonding valence electrons are often ignored when drawing line-bond structures, but we must remain aware of their existence. Some examples of these structures are shown in Table 1.3.

TABLE 1.3 Lewis and Kekulé structures of some simple molecules

Name	*Lewis structure*	*Kekulé structure*	*Name*	*Lewis structure*	*Kekulé structure*
Water (H_2O)	H:Ö:H	H—O—H	Methane (CH_4)	H H:C:H H	H | H—C—H | H
Ammonia (NH_3)	H H:N:H (lone pair on N)	H | H—N—H	Methanol (CH_3OH)	H H:C:Ö:H H	H | H—C—O—H | H

PRACTICE PROBLEM 1.3 How many hydrogen atoms does phosphorus bond to in forming phosphine, $PH_?$?

SOLUTION Because phosphorus is in Group 5A of the periodic table, it has five valence electrons. It needs to share three more electrons to make an octet and therefore bonds to three hydrogen atoms, giving PH_3.

PRACTICE PROBLEM 1.4 Draw a Lewis structure for chloromethane, CH_3Cl.

SOLUTION Hydrogen has one valence electron, carbon has four valence electrons, and chlorine has seven valence electrons. Thus, chloromethane is represented as

```
   H  ..
H:C:Cl:     Chloromethane
   H  ..
```

PROBLEM 1.7 What are the likely formulas of these molecules:

(a) $CCl_?$ (b) $AlH_?$ (c) $CH_?Cl_2$ (d) $SiF_?$

PROBLEM 1.8 Write both Lewis and line-bond structures for these molecules:

(a) $CHCl_3$, chloroform (b) H_2S, hydrogen sulfide
(c) CH_3NH_2, methylamine

PROBLEM 1.9 Which of these molecules would you expect to have covalent bonds and which ionic bonds?

(a) CH_4 (b) CH_2Cl_2 (c) LiI (d) KBr (e) $MgCl_2$ (f) Cl_2

PROBLEM 1.10 Write both a Lewis and a line-bond structure for ethane, C_2H_6.

1.6 FORMATION OF COVALENT BONDS

How are covalent bonds formed? The simplest way to picture the formation of a covalent bond is to imagine an *overlapping* of two atomic orbitals, each of which contains one electron. For example, we can picture the hydrogen molecule (H–H) by imagining what might happen if two hydrogen atoms, each with one electron in an atomic 1*s* orbital come together. As the two spherical atomic orbitals approach and combine, a new, egg-shaped H–H orbital results. The new orbital is filled by two electrons, one donated by each hydrogen:

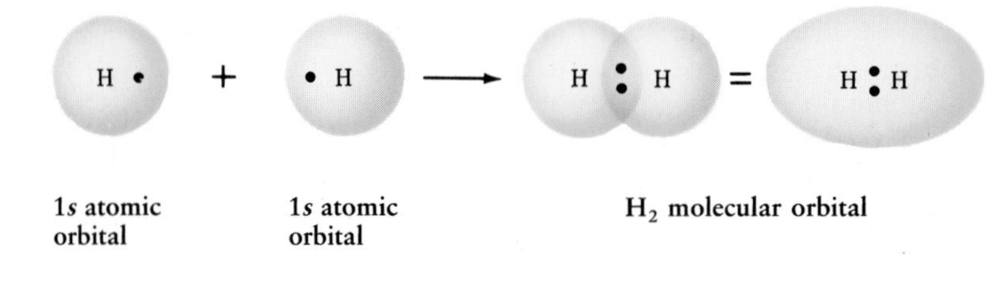

During the reaction $2H\cdot \rightarrow H_2$, 104 kcal/mol of energy is released.[4] Because the product H_2 molecule has 104 kcal/mol *less* energy than the starting $2H\cdot$, we say that the product is more stable than the starting material and that the new H–H bond has a **bond strength** of 104 kcal/mol. Looked at another way, we would have to put 104 kcal/mol of energy (heat) *into* the H–H bond to break the hydrogen molecule into two hydrogen atoms.

bond strength the amount of energy needed to break a covalent bond

How close are the two nuclei in the hydrogen molecule? If the two positively charged nuclei are too close together, they repel each other. Yet if they are too far apart, they won't be able to share the bonding electrons. Thus, there is an optimum distance between nuclei that leads to maximum bond stability. This optimum distance is called the **bond length.** In the hydrogen molecule, the bond length is 0.74 Å. Every covalent bond has both a characteristic bond strength and bond length.

bond length the optimum distance between atoms in a covalent bond

The orbital in the hydrogen molecule has the elongated egg shape that we might get by pressing two spheres together, and a cross-section cut by a plane through the middle of the H–H bond is a circle. In other words, the H–H bond is *cylindrically symmetrical,* as shown in Figure 1.6. Such bonds that have circular cross-sections and are formed by head-on overlap of two atomic orbitals are called **sigma (σ) bonds.**

sigma (σ) bond a covalent bond formed by head-on overlap of atomic orbitals

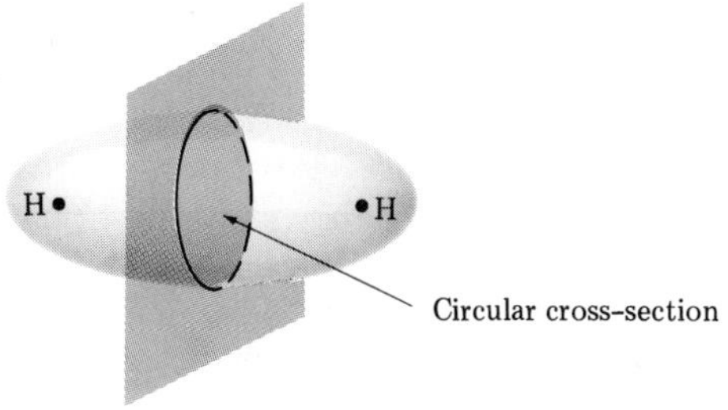

H• •H

Circular cross-section

FIGURE 1.6 Cylindrical symmetry of the H–H sigma bond

1.7 HYBRIDIZATION: THE FORMATION OF sp^3 ORBITALS

The bonding in the hydrogen molecule is fairly straightforward, but the situation becomes more complex when we turn to organic molecules with tetravalent carbon atoms. Let's start with the simplest case and consider methane, CH_4. Carbon has four electrons in its valence shell, two of which are paired in the 2*s* orbital and two of which are unpaired in different 2*p* orbitals. Thus, the electronic configuration of carbon is $1s^2\ 2s^2\ 2p_x\ 2p_y$.

The first problem is immediately apparent: Carbon needs to share four electrons to make four bonds, but only *two* of its valence electrons are unpaired. The $2s^2$ electrons are already paired with each other in the same orbital and thus can't be shared in bonds with other atoms. To get around this difficulty, carbon must

[4] Organic chemists still prefer to use kilocalories (kcal) as a measure of energy rather than kilojoules (kJ), the SI unit. The conversion factor is 1 kcal = 4.184 kJ.

excited-state configuration an electronic configuration higher in energy than the ground state

adopt an electronic configuration that's different from the ground-state configuration. By promoting one electron from the 2*s* orbital into the vacant $2p_z$ orbital, carbon can achieve an **excited-state configuration** $1s^2\ 2s\ 2p_x\ 2p_y\ 2p_z$.

	Ground-state carbon			Excited-state carbon
2*p*	↑ ↑ —	$\xrightarrow{\text{Energy}}$	2*p*	↑ ↑ ↑
2*s*	↑↓		2*s*	↑
1*s*	↑↓		1*s*	↑↓

Carbon has four unpaired electrons in the excited state and can now form four bonds to hydrogens. In Lewis structures:

$$\cdot\ddot{\mathrm{C}}\cdot \longrightarrow \cdot\dot{\underset{\cdot}{\mathrm{C}}}\cdot \xrightarrow{4\,\mathrm{H}\cdot} \mathrm{H}:\overset{\mathrm{H}}{\underset{\mathrm{H}}{\ddot{\underset{\cdot\cdot}{\mathrm{C}}}}}:\mathrm{H}$$

hybridization the mathematical combination of atomic orbitals to form new orbitals with different spatial properties

***sp*³ hybrid orbital** a hybrid orbital formed by combination of one *s* and three *p* atomic orbitals

What are the natures of the four C–H bonds in methane? Because excited-state carbon uses *two* kinds of C–H orbitals (2*s* and 2*p*) for bonding purposes, we might expect methane to have two kinds of C–H bonds. In fact, though, a large amount of evidence shows that all four C–H bonds in methane are identical. How can we explain this? The answer was provided by Linus Pauling in 1931. Pauling proposed that an *s* orbital and three *p* orbitals can combine or **hybridize** to form four equivalent atomic orbitals that are spatially oriented toward the four corners of a tetrahedron. These tetrahedral orbitals, shown in Figure 1.7, are called ***sp*³ hybrids**[5] because they arise from a combination of three *p* orbitals and one *s* orbital.

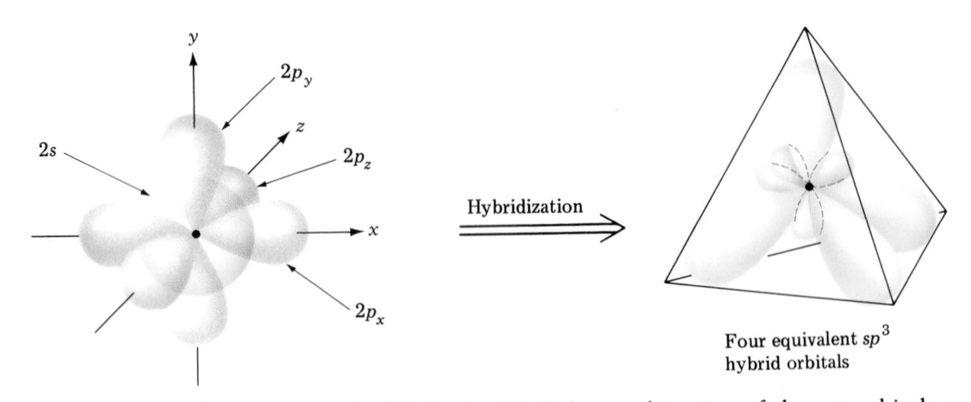

FIGURE 1.7 The formation of sp^3 hybrid orbitals by combination of three *p* orbitals and one *s* orbital

[5] Note that the superscript used to identify an sp^3 hybrid orbital tells how many of each type of atomic orbital combine in the hybrid; it doesn't tell how many electrons occupy that orbital.

The concept of hybridization explains how carbon forms four equivalent tetrahedral bonds but doesn't answer the question why it does so. Viewing an sp^3 hybrid orbital from the side suggests the answer. When an *s* orbital hybridizes with three *p* orbitals, the resultant hybrids are unsymmetrical about the nucleus. One of the two lobes of an sp^3 orbital is much larger than the other, as shown in Figure 1.8, and can therefore overlap better with another orbital. As a result, sp^3 hybrid orbitals form much stronger bonds than unhybridized *s* or *p* orbitals.

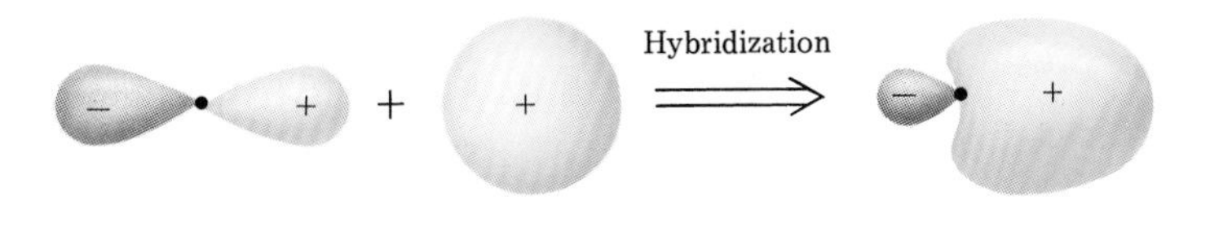

FIGURE 1.8 A side view of an sp^3 hybrid orbital, showing how it is strongly oriented in one direction

1.8 THE STRUCTURE OF METHANE

How can carbon use its hybrid orbitals to form covalent bonds? Just as we imagined the formation of an H–H bond by the overlap of two hydrogen 1*s* orbitals, we can imagine the formation of C–H bonds like those in methane by the overlap of a carbon sp^3 hybrid orbital with a hydrogen 1*s* orbital to give a strong C–H sigma bond (Figure 1.9).

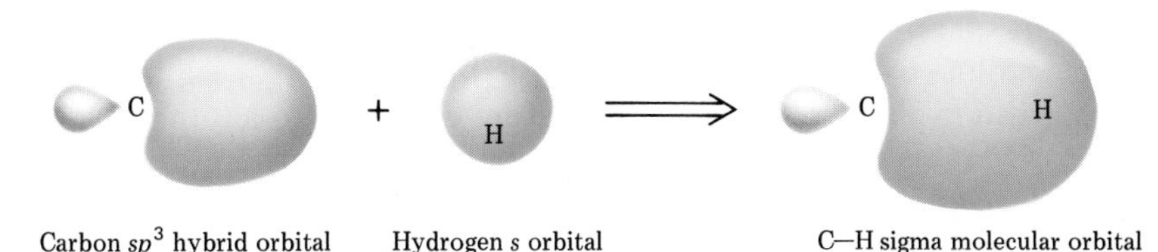

FIGURE 1.9 The formation of a C–H bond by overlap of a carbon sp^3 hybrid orbital with a hydrogen 1*s* orbital

When the four identical orbitals of an sp^3 hybridized carbon atom overlap with four hydrogen atoms, four identical C–H bonds are formed, and methane, CH_4, results. A C–H bond of methane has a strength of 104 kcal/mol and a length of 1.10 Å. Since the four bonds have a specific geometry, we can also define a third property called the **bond angle.** The angle formed by each H–C–H is exactly 109.5°, the so-called tetrahedral angle. Methane therefore has the structure shown in Figure 1.10.

bond angle
the angle formed by two adjacent bonds

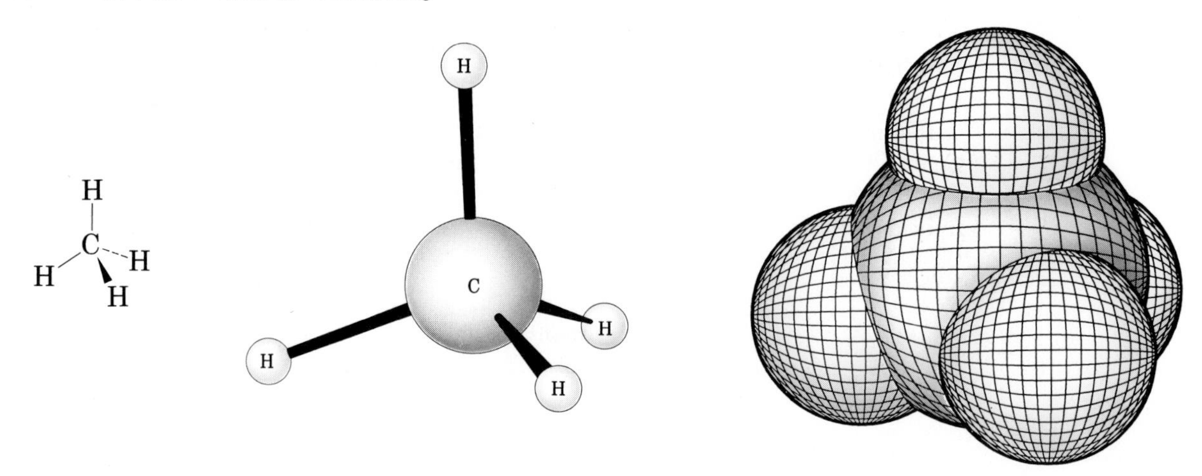

FIGURE 1.10 The structure of methane: The drawings are computer generated for accuracy.

PROBLEM 1.11 Draw a tetrahedral representation of tetrachloromethane, CCl_4, using the standard convention of solid, dashed, and wedged lines.

PROBLEM 1.12 Why do you think a C–H bond (1.09 Å) is longer than an H–H bond (0.74 Å)?

1.9 THE STRUCTURE OF ETHANE

A special characteristic of carbon is that it can form stable bonds to other carbon atoms. Exactly the same kind of hybridization that explains the methane structure also explains how one carbon atom can bond to another to form a chain. Ethane, C_2H_6, is the simplest molecule containing a carbon–carbon bond:

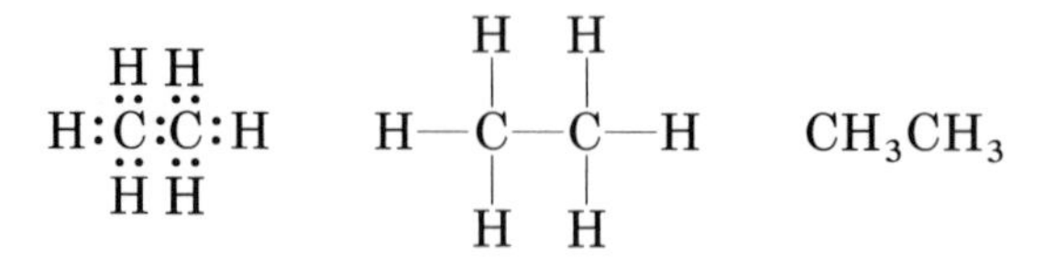

Some representations of ethane

We can picture the ethane molecule by imagining that the two carbon atoms bond to each other by head-on (sigma) overlap of an sp^3 hybrid orbital from each. The remaining three hybrid orbitals on each carbon form the six C–H bonds, as shown in Figure 1.11. The C–H bonds in ethane are similar to those in methane, though a bit weaker (98 kcal/mol for ethane versus 104 kcal/mol for methane).

The C–C bond is 1.54 Å long and has a strength of 88 kcal/mol. All the bond angles of ethane are very near the tetrahedral value, 109.5°.

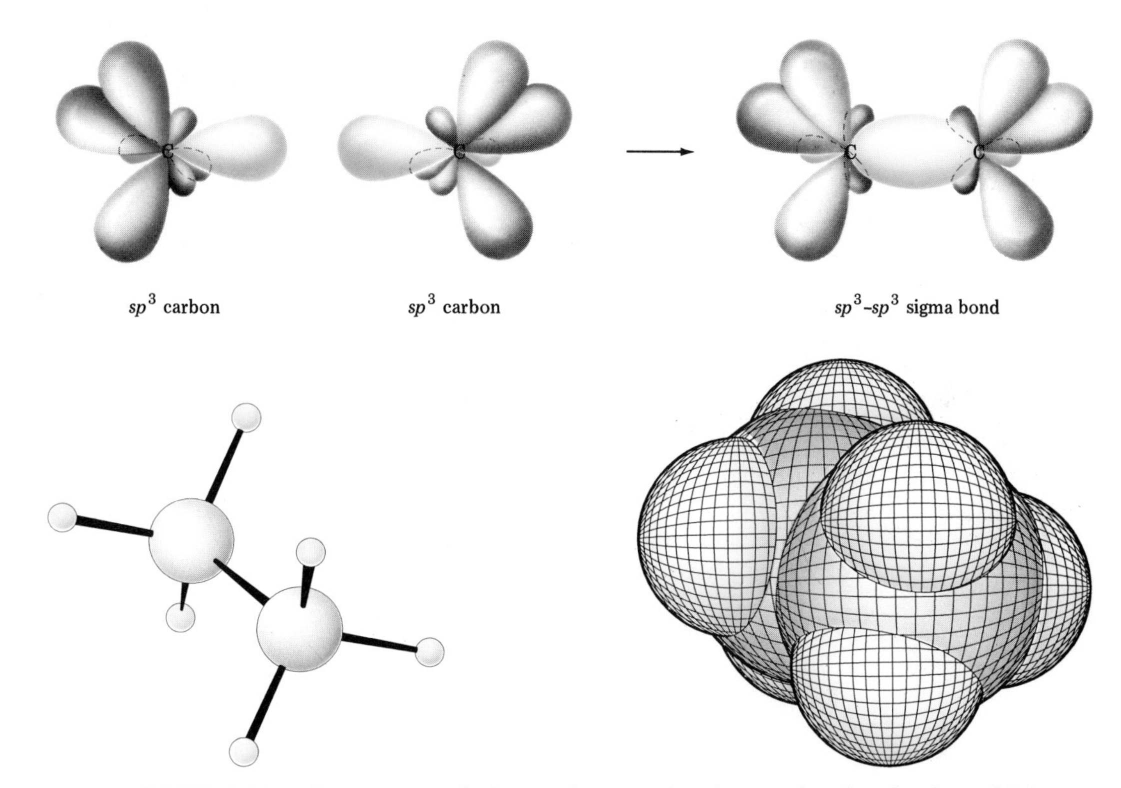

FIGURE 1.11 The structure of ethane: The central carbon–carbon bond is formed by overlap of two sp^3 hybrid orbitals.

PROBLEM 1.13 Draw a line-bond structure for propane, $CH_3CH_2CH_3$. Predict the value of each bond angle and indicate the overall shape of the molecule.

PROBLEM 1.14 Why can't an organic molecule have the formula C_2H_7?

1.10 HYBRIDIZATION: sp^2 ORBITALS AND THE STRUCTURE OF ETHYLENE

Although sp^3 hybridization is the most common electronic state of carbon, it's not the only possibility. Let's look at ethylene, C_2H_4, for example. It was recognized over 100 years ago that ethylene carbons can be tetravalent only if the two carbon atoms share *four* electrons and are linked by a *double* bond. How can we explain formation of the carbon–carbon double bond?

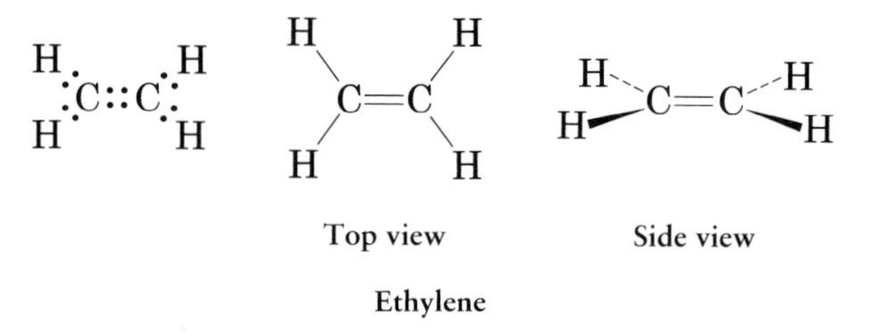

When we formed sp^3 hybrid orbitals to explain the bonding in methane, we first promoted an electron from the $2s$ orbital to a $2p$ orbital to form excited-state carbon with four unpaired electrons. We then combined all four singly occupied atomic orbitals to construct four equivalent sp^3 hybrids. Imagine instead that we combine the carbon $2s$ orbital with only two of the three available $2p$ orbitals. Three hybrid orbitals, which we call **sp^2 hybrids,** result, and one unhybridized $2p$ orbital remains unchanged. The three sp^2 orbitals lie in a plane at angles of 120° to each other, with the remaining p orbital perpendicular to the sp^2 plane, as shown in Figure 1.12.

sp^2 hybrid orbital a hybrid orbital formed by combination of one s and two p atomic orbitals

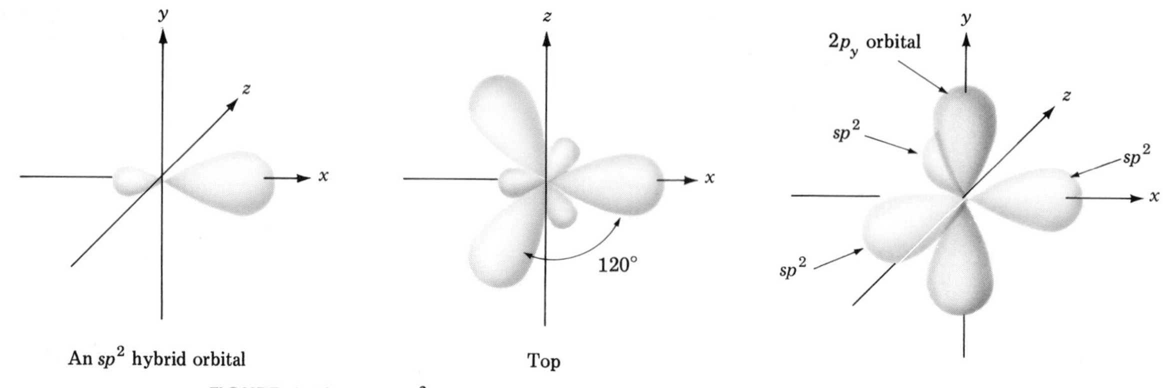

FIGURE 1.12 An sp^2-hybridized carbon atom

If we allow two sp^2-hybridized carbon atoms to approach each other, they can form a strong sigma bond by sp^2–sp^2 overlap. When this occurs, the unhybridized p orbitals on each carbon approach each other with the correct geometry for sideways rather than head-on overlap, leading to the formation of what is called a **pi (π) bond.** The combination of sp^2–sp^2 sigma overlap and $2p$–$2p$ pi overlap results in the net sharing of four electrons and the formation of a carbon–carbon double bond (Figure 1.13).

pi (π) bond a covalent bond formed by sideways overlap of two p orbitals

To complete the structure of ethylene, we need only allow four hydrogen atoms to sigma bond to the remaining four carbon sp^2 orbitals. The resultant ethylene molecule has a planar (flat) structure with H–C–H and H–C–C bond angles of approximately 120°. Each C–H bond has a length of 1.076 Å and a strength of 103 kcal/mol.

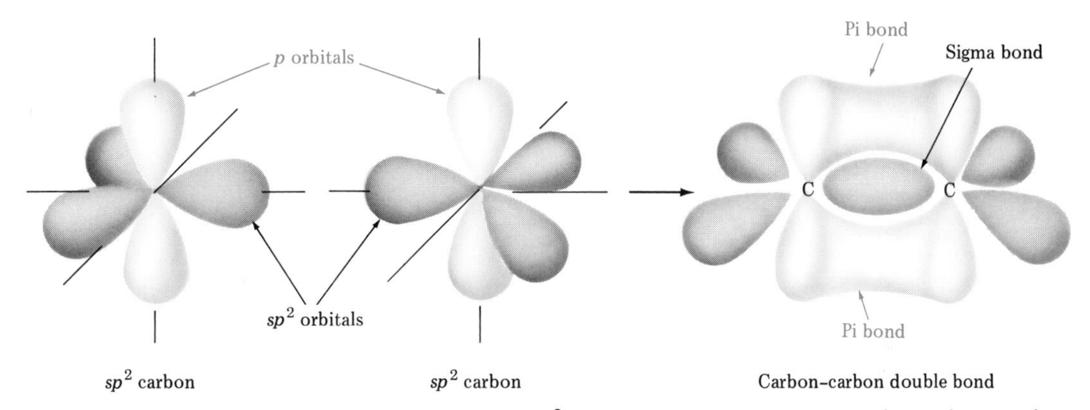

FIGURE 1.13 Orbital overlap of two sp^2-hybridized carbon atoms in the carbon–carbon double bond

We might expect the carbon–carbon double bond in ethylene to be both shorter and stronger than the ethane single bond because it results from the sharing of four electrons rather than two. This prediction has been verified, for ethylene has a carbon–carbon bond length of 1.33 Å and a bond strength of 152 kcal/mol, whereas ethane has values of 1.54 Å and 88 kcal/mol, respectively. The structure of ethylene is shown in Figure 1.14.

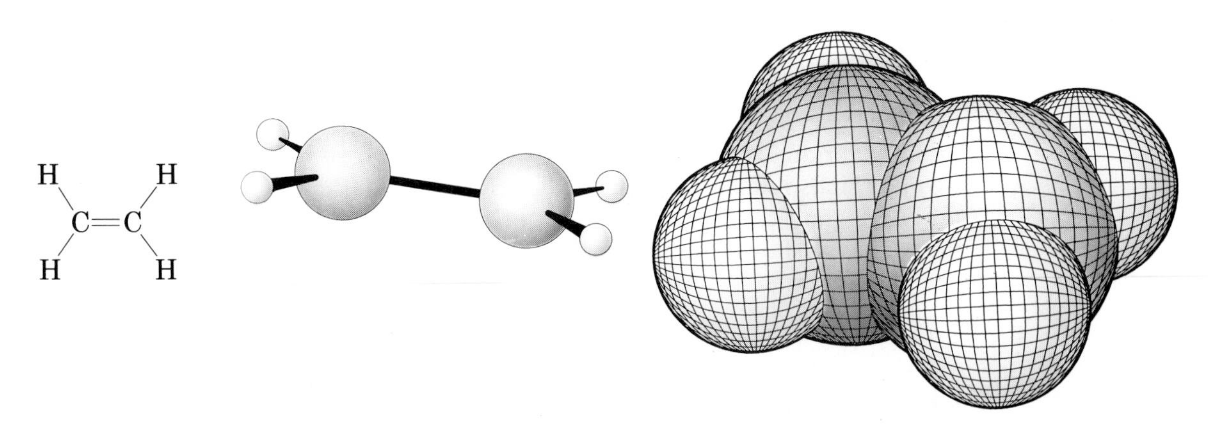

FIGURE 1.14 The structure of ethylene

PRACTICE PROBLEM 1.5 Formaldehyde, CH_2O, contains a carbon–*oxygen* double bond. Draw Lewis and line-bond structures of formaldehyde and indicate the hybridization of the carbon atom.

SOLUTION There is only one way that two hydrogens, one carbon, and one oxygen can combine:

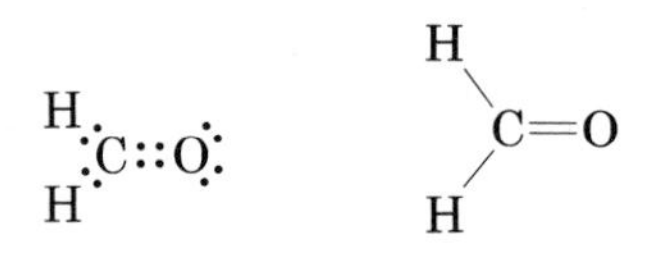

Like the carbon atoms in ethylene, the carbon atom in formaldehyde is sp^2 hybridized.

PROBLEM 1.15 Draw both a Lewis structure and a line-bond structure for acetaldehyde, CH_3CHO.

PROBLEM 1.16 Draw a line-bond structure for propene, $CH_3CH{=}CH_2$. Indicate the hybridization of each carbon and predict the value of each bond angle.

1.11 HYBRIDIZATION: *sp* ORBITALS AND THE STRUCTURE OF ACETYLENE

_sp_ hybrid orbital a hybrid orbital formed by combination of one *s* and one *p* atomic orbital

In addition to being able to form single and double bonds by sharing two and four electrons, carbon can also form *triple* bonds by sharing *six* electrons. To account for triple bonds like that in acetylene, C_2H_2, we have to construct yet a third kind of hybrid orbital: an ***sp* hybrid.**

H:C:::C:H   H—C≡C—H

Acetylene

Imagine that, instead of combining with two or three *p* orbitals, the carbon 2*s* orbital hybridizes with only a single *p* orbital. Two *sp* hybrid orbitals result, and two *p* orbitals remain unchanged. The two *sp* orbitals are linear, or 180° apart on the *x*-axis, and the remaining two *p* orbitals are perpendicular on the *y*-axis and the *z*-axis, as shown in Figure 1.15.

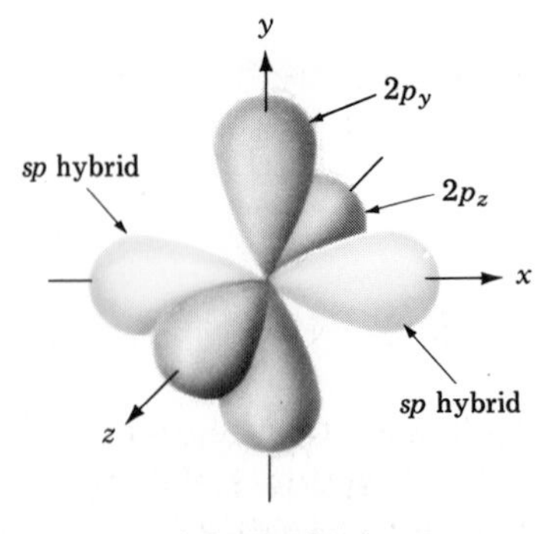

FIGURE 1.15 An *sp*-hybridized carbon atom

If we allow two *sp*-hybridized carbon atoms to approach each other, *sp* hybrid orbitals from each carbon overlap head-on to form a strong *sp–sp* sigma bond. In addition, the p_z orbitals from each carbon form a p_z–p_z pi bond by sideways overlap, and the p_y orbitals from each carbon overlap similarly to form a p_y–p_y pi bond. The overall effect is formation of one sigma bond and two pi bonds: a net carbon–carbon triple bond. The remaining *sp* hybrid orbitals form sigma bonds to hydrogen 1*s* orbitals to complete the acetylene molecule (Figure 1.16).

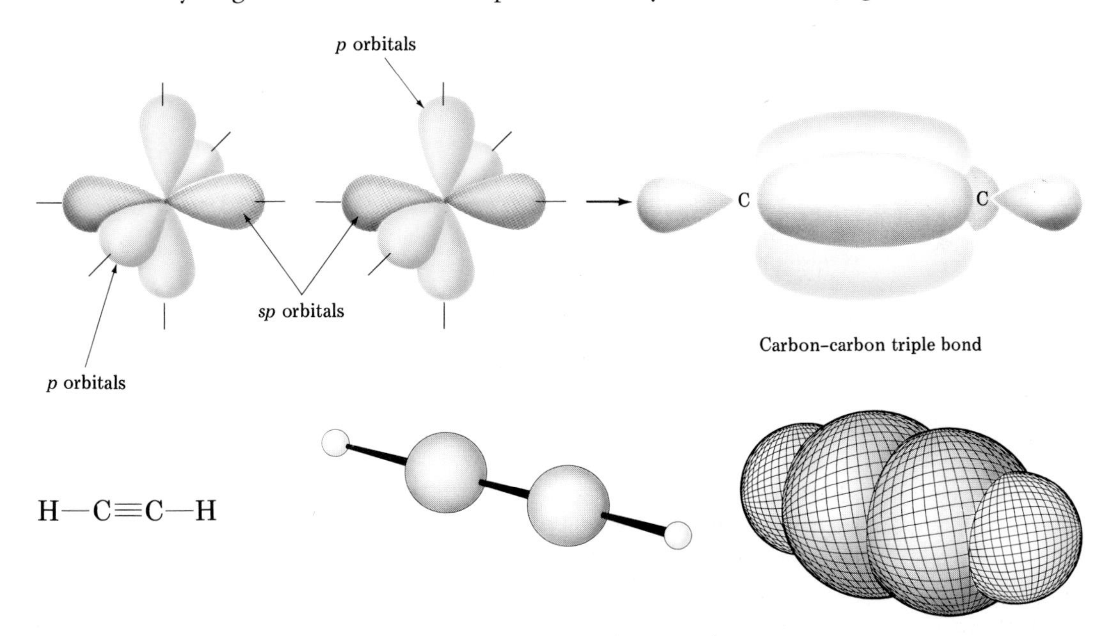

FIGURE 1.16 The carbon–carbon triple bond in acetylene

Because of *sp* hybridization, acetylene is a linear molecule with H–C–C bond angles of 180°. The C–H bond is 1.06 Å in length and has a strength of 125 kcal/mol. The C–C bond length is 1.20 Å, and its strength is 200 kcal/mol.

PROBLEM 1.17 Draw a line-bond structure for propyne, $CH_3C{\equiv}CH$. Indicate the hybridization of each carbon and predict a value for each bond angle.

PROBLEM 1.18 Draw a Lewis structure for carbon dioxide, CO_2.

PROBLEM 1.19 Draw a line-bond structure for 1,3-butadiene, $H_2C{=}CH{-}CH{=}CH_2$, indicate the hybridization of each carbon, and predict a value for each bond angle.

1.12 BOND POLARITY AND ELECTRONEGATIVITY

Up to this point, we've viewed chemical bonding in an either/or manner: A given bond is either ionic or covalent. It's more accurate, though, to look at bonding as a continuum of possibilities, from a perfectly covalent bond with a symmetrical electron distribution on the one hand, to a perfectly ionic bond on the other (Figure 1.17). For example, the carbon–carbon bond in ethane is electronically symmetrical and therefore perfectly covalent; the two bonding electrons are equally shared by the two equivalent carbon atoms. The bond in sodium chloride, by contrast, is purely ionic; an electron has been donated from sodium to chlorine to give Na^+ and Cl^- ions. Between these two extremes lie the great majority of chemical bonds, in which the bonding electrons are attracted *somewhat* more strongly by one atom than by the other. We call such bonds **polar covalent bonds.**

polar covalent bond
a covalent bond in which the electrons are attracted more strongly by one atom than by the other

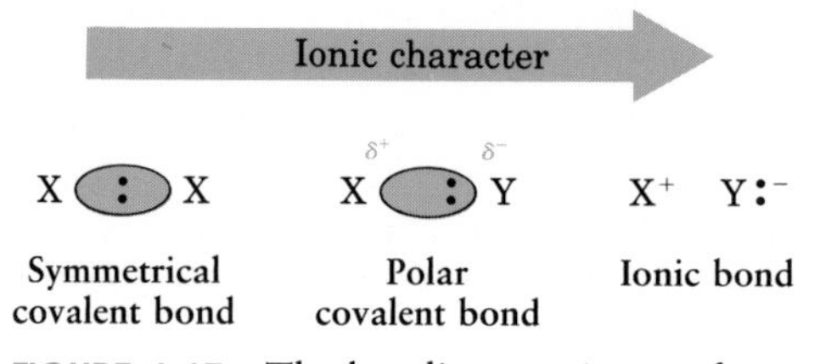

FIGURE 1.17 The bonding continuum from covalent to ionic bonds. The symbol δ (Greek delta) means *partial* charge, either positive (δ^+) or negative (δ^-).

electronegativity
the ability of an atom to attract electrons in a covalent bond

Bond polarity is due to **electronegativity,** the intrinsic ability of an atom to attract electrons in bonds. As shown in the electronegativity table (Table 1.4), carbon and hydrogen have similar electronegativities, and carbon–hydrogen bonds are therefore relatively nonpolar. Elements on the right side of the periodic table, such as oxygen, nitrogen, and chlorine, are more electronegative than carbon, however. When carbon bonds to one of these elements, the bond is polarized so that the bonding electrons are drawn more toward the electronegative atom than toward carbon. This unsymmetrical distribution of bonding electrons leaves carbon with a partial positive charge, denoted δ^+, and the electronegative atom with a partial negative charge, δ^- (δ is the Greek letter delta). For example, the C–Cl

TABLE 1.4 **Electronegativities of some common elements**

Group 1A	2A	3A	4A	5A	6A	7A
H 2.2						
Li 1.0	Be 1.6	B 2.0	C 2.5	N 3.0	O 3.4	F 4.0
Na 0.9	Mg 1.3	Al 1.6	Si 1.9	P 2.2	S 2.6	Cl 3.1
						Br 3.0
						I 2.6

bond in chloromethane is a polar covalent bond:

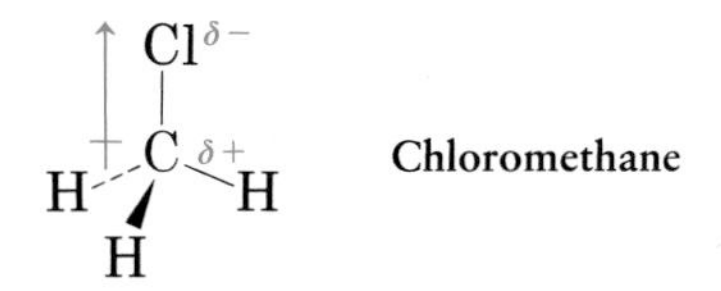

Chloromethane

The arrow ↛ is sometimes used to indicate the direction of polarity. By convention, *electrons move in the direction of the arrow.* The tail of the arrow is electron-poor (δ^+), and the head of the arrow is electron-rich (δ^-).

Metallic elements on the left side of the periodic table are less electronegative than carbon and attract electrons less strongly. When carbon bonds to one of these elements, the bond is polarized so that carbon bears a partial negative charge and the other atom bears a partial positive charge. Organometallic compounds like tetraethyllead, the "lead" in gasoline, provide good examples of this kind of polar bond.

$$\begin{array}{c} \overset{\delta^-}{CH_2}CH_3 \\ | \\ CH_3\overset{\delta^-}{CH_2}-\overset{\delta^+}{Pb}-\overset{\delta^-}{CH_2}CH_3 \\ | \\ \overset{\delta^-}{CH_2}CH_3 \end{array}$$

Tetraethyllead

inductive effect the ability of an atom to polarize a covalent bond by donating or withdrawing the bonding electrons

When we talk about an atom's ability to polarize a bond, we use the term **inductive effect.** Electropositive elements like lithium and sodium inductively *donate* electrons whereas electronegative elements like oxygen and chlorine inductively *withdraw* electrons. Inductive effects and bond polarity play major roles in chemical reactivity and will be encountered many times throughout this text.

PRACTICE PROBLEM 1.6 Predict the direction of polarization of an O–H bond in a water molecule.

SOLUTION Oxygen is more electronegative than hydrogen, according to Table 1.4, and therefore attracts electrons more strongly (δ^-):

$$\overset{\delta^+}{H}-\overset{\delta^-}{O}-\overset{\delta^+}{H}$$

PROBLEM 1.20 Which element in each of these pairs is more electronegative?

(a) Li or H (b) Be or Br (c) Cl or I

PROBLEM 1.21 Indicate the direction of expected polarity for each of the bonds shown:

(a) $Br-CH_3$ (b) H_2N-CH_3 (c) $Li-CH_3$ (d) $H-NH_2$
(e) $HO-CH_3$ (f) $BrMg-CH_3$ (g) $F-CH_3$

SUMMARY AND KEY WORDS

Organic chemistry is the study of carbon compounds. Although a division into inorganic and organic chemistry occurred historically, there's no scientific reason for the division.

Atoms are composed of a positively charged nucleus surrounded by negatively charged electrons that occupy specific regions of space called **orbitals.** Different orbitals have different energy levels and shapes. For example, *s* orbitals are spherical, and *p* orbitals are dumbbell shaped.

There are two fundamental kinds of chemical bonds: **ionic bonds** and **covalent bonds.** The ionic bonds commonly found in inorganic salts result from the electrical attraction of unlike charges. The covalent bonds found in organic molecules are formed by the sharing of an electron pair between atoms. Electron-sharing occurs when two atoms approach each other and their atomic orbitals overlap. Bonds that have a circular cross section and are formed by head-on overlap of atomic orbitals are called **sigma (σ) bonds.** Bonds formed by sideways overlap of two *p* orbitals are called **pi (π) bonds.**

To form bonds in organic compounds, carbon first **hybridizes** to an excited-state configuration. When forming only single bonds, carbon is sp^3 hybridized and has four equivalent sp^3 **hybrid orbitals** with tetrahedral geometry. When forming double bonds, carbon is sp^2 hybridized, has three equivalent sp^2 **orbitals** with planar geometry, and has one unhybridized *p* orbital. When forming triple bonds, carbon is *sp* hybridized, has two equivalent ***sp* orbitals** with linear geometry, and has two unhybridized *p* orbitals.

Most covalent bonds are **polar** because of unsymmetrical electron-sharing. For example, a carbon-chlorine bond is polar because chlorine is more **electronegative** than carbon and therefore attracts the bonding electrons more strongly. Carbon-metal bonds, however, are usually polarized in the opposite sense because carbon attracts electrons more strongly than most metals. Carbon-hydrogen bonds are relatively nonpolar.

Working Problems

Learning organic chemistry means knowing a large number of facts. Although careful reading and rereading of the text is important, reading alone isn't enough. In addition, you must be able to apply the information you read and be able to use your knowledge in new situations. Working problems gives you the opportunity to do this. There's no surer way to learn organic chemistry than by working problems.

Each chapter in this book provides many problems of different sorts. The in-chapter problems are placed for immediate reinforcement of ideas just learned. The end-of-chapter problems provide additional practice, but are of two types: The first problems tend to be the drill type, which provide an opportunity for you to practice your command of the fundamentals; later problems tend to be more challenging and thought provoking.

As you study organic chemistry, take the time to work the problems. Work those you can and ask for help with those you can't. If stumped by a particular exercise, check the answer book for an explanation that should help clarify the difficulty. Working problems takes effort, but the payoff in knowledge and understanding is immense.

ADDITIONAL PROBLEMS

1.22 How many valence (outer-shell) electrons does each of these atoms have?

(a) Oxygen (b) Magnesium (c) Fluorine

1.23 Give the ground-state electronic configuration of these elements: For example, carbon is $1s^2\ 2s^2\ 2p^2$.

(a) Lithium (b) Sodium (c) Aluminum (d) Sulfur

1.24 What are the likely formulas of these molecules?

(a) $AlCl_?$ (b) $CF_2Cl_?$ (c) $NI_?$

1.25 Which of these molecules would you expect to have covalent bonds and which ionic bonds?

(a) BeF_2 (b) SiH_4 (c) CBr_4

1.26 Write Lewis (electron dot) structures for these molecules:

(a) $H{-}C{\equiv}C{-}H$ (b) AlH_3 (c) CH_3OH (d) $H_2C{=}CHCl$

1.27 Write a Lewis (electron dot) structure for acetonitrile, $CH_3C{\equiv}N$. How many electrons does the nitrogen atom have in its valence shell? How many are used for bonding and how many are not used for bonding?

1.28 Fill in any unshared electrons that are missing in these line-bond structures:

(a) $CH_3{-}O{-}CH_3$

(b)
```
        O
        ‖
CH3 — C — CH3
```

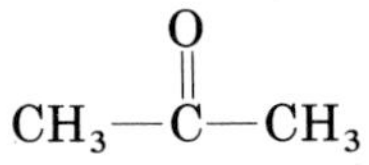

(c)
```
        O
        ‖
CH3 — C — NH2
```

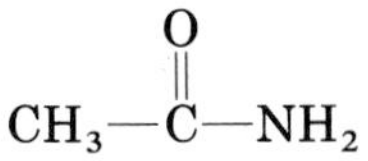

(d) CH_2ClF

1.29 There are two structures that correspond to the formula C_4H_{10}. Draw them.

1.30 Convert these molecular formulas into line-bond structures:

(a) C_3H_8 (b) C_3H_7Br (2 possibilities)
(c) C_3H_6 (2 possibilities) (d) C_2H_6O (2 possibilities)

1.31 Indicate the kind of hybridization (sp, sp^2, or sp^3) you would expect for each carbon atom in these molecules:

(a) Butane, $CH_3CH_2CH_2CH_3$ (b) 1-Butene, $CH_3CH_2CH{=}CH_2$
(c) Cyclobutene, (d) 1-Buten-3-yne, $H_2C{=}CH{-}C{\equiv}CH$

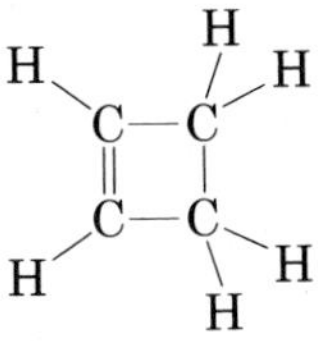

```
        H
 H      /  H
   \   /  /
    C — C
    ‖   |
    C — C
   /   \  \
 H      \  H
        H
```

1.32 What is the hybridization of each carbon atom in benzene? What shape would you expect benzene to have?

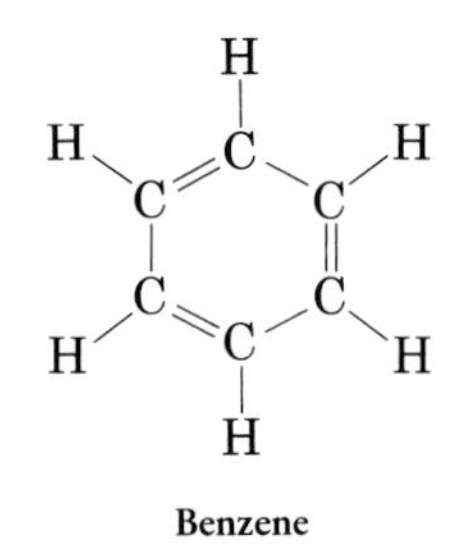

Benzene

1.33 What is the hybridization of each carbon atom in acetonitrile, $CH_3C{\equiv}N$?

1.34 Write Lewis (electron dot) structures for these molecules:

(a) $CH_3{-}Be{-}CH_3$ (b) $CH_3{-}P(CH_3){-}CH_3$ (c) $TiCl_4$

1.35 Draw line-bond structures for these covalent molecules:

(a) Br_2 (b) CH_3Cl (c) HF (d) CH_3CH_2OH

1.36 Indicate which of the bonds in the structures you drew for Problem 1.35 are polar. Indicate bond polarity by using the symbols δ^+ and δ^-.

1.37 Identify all the bonds in these molecules as either ionic or covalent:

(a) NaOH (b) HOH (c) CH_3OH (d) CH_3OCH_3 (e) FF

1.38 Sodium methoxide, $NaOCH_3$, contains both covalent and ionic bonds. Indicate which is which.

1.39 Use the electronegativity table (Table 1.4) to predict which bond in each of the following sets is more polar:

(a) $Cl{-}CH_3$ or $Cl{-}Cl$ (b) $H{-}CH_3$ or $H{-}Cl$
(c) $HO{-}CH_3$ or $(CH_3)_3Si{-}CH_3$

1.40 Indicate the direction of polarity for each bond in Problem 1.39.

1.41 Which atoms in these structures have unshared valence electrons? Draw in these unshared electrons.

(a) CH_3SH (b) $CH_3{-}N(CH_3){-}CH_3$ (c) CH_3CH_2Br

(d) $CH_3C({=}O){-}OH$ (e) $CH_3C({=}O){-}Cl$

1.42 Draw a three-dimensional representation of chloroform, $CHCl_3$, using the standard convention of solid, wedged, and dashed lines. Do the same for the oxygen-bearing carbon atom in ethanol, CH_3CH_2OH.

1.43 The ammonium ion, NH_4^+, has a geometry identical to that of methane, CH_4. What kind of hybridization do you think the nitrogen atom has? Explain.

1.44 Draw a three-dimensional representation of ethane, CH_3CH_3, using normal, dashed, and wedged lines for both carbons.

1.45 Indicate the kind of hybridization you would expect for each carbon atom in these molecules:

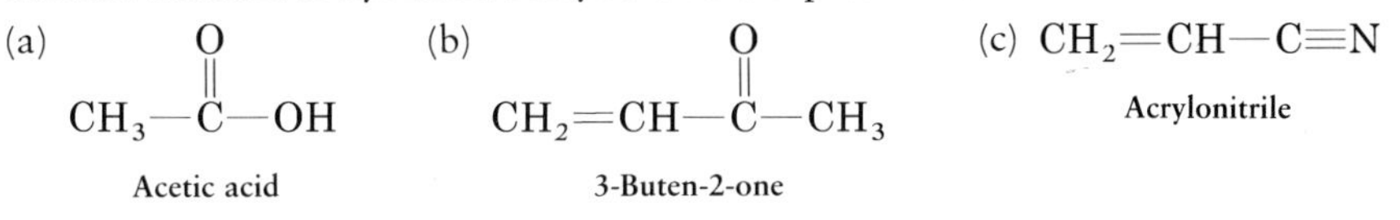

1.46 Use Table 1.4 to order the following molecules according to increasing positive character of the carbon atom:

$$CH_3F,\quad CH_3OH,\quad CH_3Li,\quad CH_3I,\quad CH_3CH_3,\quad CH_3NH_2$$

1.47 Allene is an unusual molecule that has the structure $H_2C{=}C{=}CH_2$. Draw an orbital picture of allene. What hybridization must the central carbon atom have since it has two double bonds? What shape do you predict for allene?

1.48 Although most stable organic compounds have tetravalent carbon atoms, species with trivalent carbon atoms also exist. *Carbocations* are one such class of compounds. If the positively charged carbon atom has planar geometry, what hybridization do you think it has? How many valence-shell electrons does the carbon have?

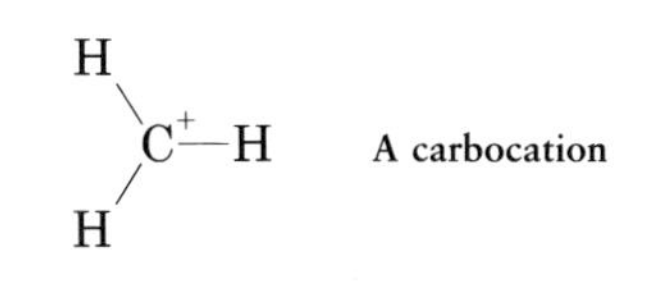

CHAPTER 2

The Nature of Organic Compounds: Alkanes

Believe it or not, there are nearly nine million known organic compounds. Fortunately, chemists have discovered over the years that these compounds can be classified into families according to their structural features and that the chemical reactivity of compounds in a given family is similar. Instead of nine million compounds with random reactivity, there are a few dozen families of compounds whose chemistry is reasonably predictable. Throughout this book, we'll study the chemistry of specific families of organic molecules, beginning in this chapter with a look at the simplest family, the *alkanes*.

2.1 FUNCTIONAL GROUPS

functional group a group of atoms within a molecule that has a characteristic chemical behavior

The structural features that allow us to classify compounds by reactivity are called functional groups. A **functional group** is a part of a larger molecule; it is composed of an atom or group of atoms that have a characteristic chemical behavior. Chemically, a given functional group behaves in nearly the same way in every molecule it's a part of. For example, one of the simplest functional groups is the carbon–carbon double bond. Since the electronic structure of the carbon–carbon double bond remains essentially the same in all molecules where it occurs, its chemical reactivity also remains the same. Ethylene, the simplest compound with a carbon–carbon double bond, undergoes reactions that are remarkably similar to those of α-pinene, a seemingly more complicated molecule (and major component of turpentine). Both, for example, react with bromine to give products in which a bromine atom has added to each of the double–bond carbons (Figure 2.1).

The example shown in Figure 2.1 is typical: *The chemistry of every organic molecule, regardless of size and complexity, is determined by the functional groups it contains.* Table 2.1 lists many of the common functional groups and gives simple examples of their occurrence. Look carefully at this table to see the many types of functional groups found in organic compounds. Some functional groups, such as alkenes, alkynes, and aromatic rings, have only carbon–carbon double or triple bonds; others have halogen; and still others have oxygen, nitrogen, or sulfur. Much of the chemistry you'll be studying in the remainder of this book is the chemistry of these functional groups.

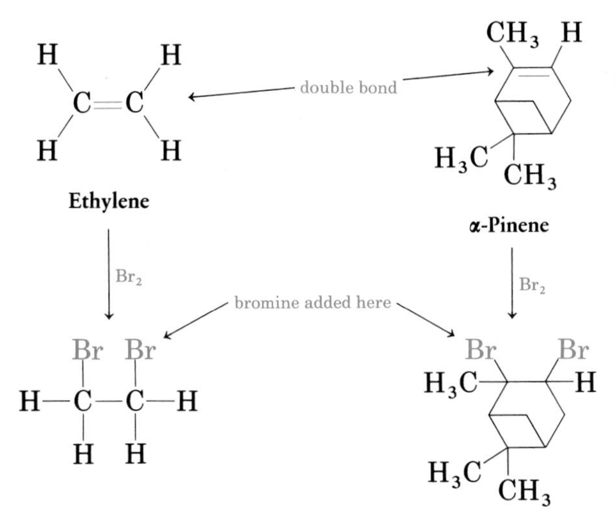

FIGURE 2.1 The reactions of ethylene and α-pinene with bromine. In both cases, bromine reacts with the C=C double-bond functional group in exactly the same way. The size and nature of the remainder of the molecule doesn't matter.

TABLE 2.1 Structures of some important functional groups

Family name	*Functional group structure*[a]	*Simple example*	*Name ending*
Alkane	Contains only C—H and C—C single bonds	CH_3CH_3 **Ethane**	-ane
Alkene	>C=C<	$H_2C{=}CH_2$ **Ethene (ethylene)**	-ene
Alkyne	—C≡C—	H—C≡C—H **Ethyne (acetylene)**	-yne
Arene	(six-membered ring of alternating C=C and C—C bonds, each C bearing one substituent bond)	C_6H_6 (ring structure) **Benzene**	None
Halide	—C—X (X = F, Cl, Br, I)	CH_3Cl **Chloromethane**	None
Alcohol	—C—O—H	CH_3OH **Methanol**	-ol

(*Continued*)

TABLE 2.1 (*Continued*)

Family name	*Functional group structure*[a]	*Simple example*	*Name ending*
Ether	$-\overset{\vert}{\underset{\vert}{C}}-O-\overset{\vert}{\underset{\vert}{C}}-$	CH_3OCH_3 Dimethyl ether	ether
Amine	$-\overset{\vert}{\underset{\vert}{C}}-\underset{\underset{H}{\vert}}{N}-H$, $-\overset{\vert}{\underset{\vert}{C}}-\underset{\vert}{N}-H$, $-\overset{\vert}{\underset{\vert}{C}}-\underset{\vert}{N}-$	CH_3NH_2 Methylamine	-amine
Nitrile	$-C\equiv N$	$CH_3C\equiv N$ Ethanenitrile (acetonitrile)	-nitrile
Thiol	$-\overset{\vert}{\underset{\vert}{C}}-S-H$	CH_3SH Methanethiol	-thiol
Carbonyl $\overset{O}{\overset{\Vert}{C}}$			
Aldehyde	$-\overset{O}{\overset{\Vert}{C}}-H$	$CH_3-\overset{O}{\overset{\Vert}{C}}-H$ Ethanal (acetaldehyde)	-al
Ketone	$-\overset{\vert}{\underset{\vert}{C}}-\overset{O}{\overset{\Vert}{C}}-\overset{\vert}{\underset{\vert}{C}}-$	$CH_3-\overset{O}{\overset{\Vert}{C}}-CH_3$ Propanone (acetone)	-one
Carboxylic acid	$-\overset{O}{\overset{\Vert}{C}}-OH$	$CH_3-\overset{O}{\overset{\Vert}{C}}-OH$ Ethanoic acid (acetic acid)	-oic acid
Ester	$-\overset{O}{\overset{\Vert}{C}}-O-\overset{\vert}{\underset{\vert}{C}}-$	$CH_3\overset{O}{\overset{\Vert}{C}}-O-CH_3$ Methyl ethanoate (methyl acetate)	-oate
Amide	$-\overset{O}{\overset{\Vert}{C}}-NH_2$, $-\overset{O}{\overset{\Vert}{C}}-\underset{\vert}{N}-H$, $-\overset{O}{\overset{\Vert}{C}}-\underset{\vert}{N}-$	$CH_3-\overset{O}{\overset{\Vert}{C}}-NH_2$ Ethanamide (acetamide)	-amide

[a] The bonds whose connections aren't specified are assumed to be attached to carbon or hydrogen atoms in the rest of the molecule.

It's a good idea at this point to familiarize yourself with the structures of the functional groups shown in Table 2.1 so that you'll recall them when you see them again. They can be grouped into several categories.

Functional Groups with Carbon–Carbon Multiple Bonds

Alkenes, alkynes, and arenes (aromatic compounds) all contain carbon–carbon multiple bonds. Alkenes have a double bond, alkynes have a triple bond, and arenes have three alternating double and single bonds in a six-membered ring of carbon atoms. Because of their structural similarities, these compounds also have some chemical similarities.

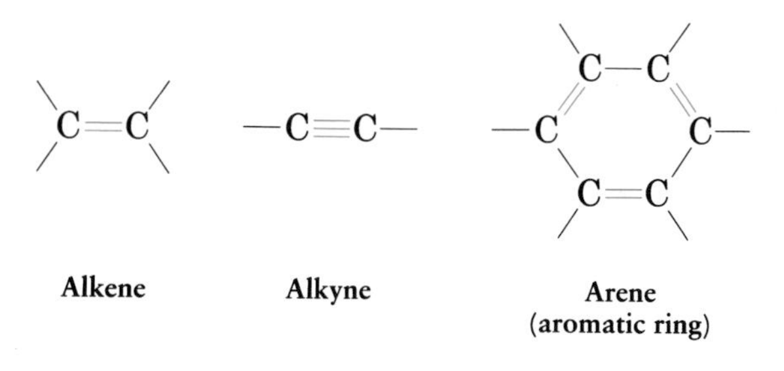

Functional Groups with Carbon Singly Bonded to an Electronegative Atom

Alkyl halides, alcohols, ethers, amines, and thiols all have a carbon atom singly bonded to an electronegative atom—a halogen, an oxygen, a nitrogen, or a sulfur. Alkyl halides have a carbon atom bonded to halogen (−X), alcohols have a carbon atom bonded to a hydroxyl (−OH) group, ethers have two carbon atoms bonded to the same oxygen, amines have a carbon atom bonded to a nitrogen, and thiols have a carbon atom bonded to an −SH group. In all cases, the bonds are polar, with the carbon atom bearing a slight positive charge (δ^+) and the electronegative atom bearing a slight negative charge (δ^-).

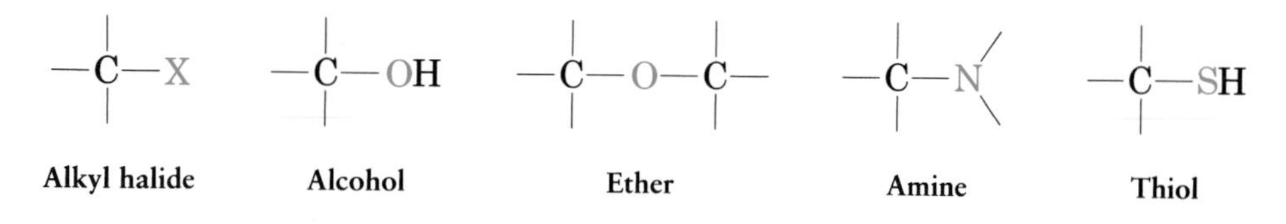

Functional Groups with a Carbon–Oxygen Double Bond (Carbonyl Groups)

carbonyl group the carbon–oxygen double bond, C=O

Note particularly in Table 2.1 the different families of compounds that contain the **carbonyl group, C=O** (pronounced car-bo-*neel*). Carbon–oxygen double bonds are present in some of the most important compounds in organic chemistry. These compounds are similar in many respects but differ depending on the identity of the atoms bonded to the carbonyl-group carbon. Aldehydes have one carbon and one hydrogen bonded to the C=O, ketones have two carbons bonded to the C=O, carboxylic acids have one carbon and one −OH group bonded to the C=O, esters have one carbon and one ether-like oxygen bonded to the C=O, and amides have one carbon and one amine-like nitrogen bonded to the C=O.

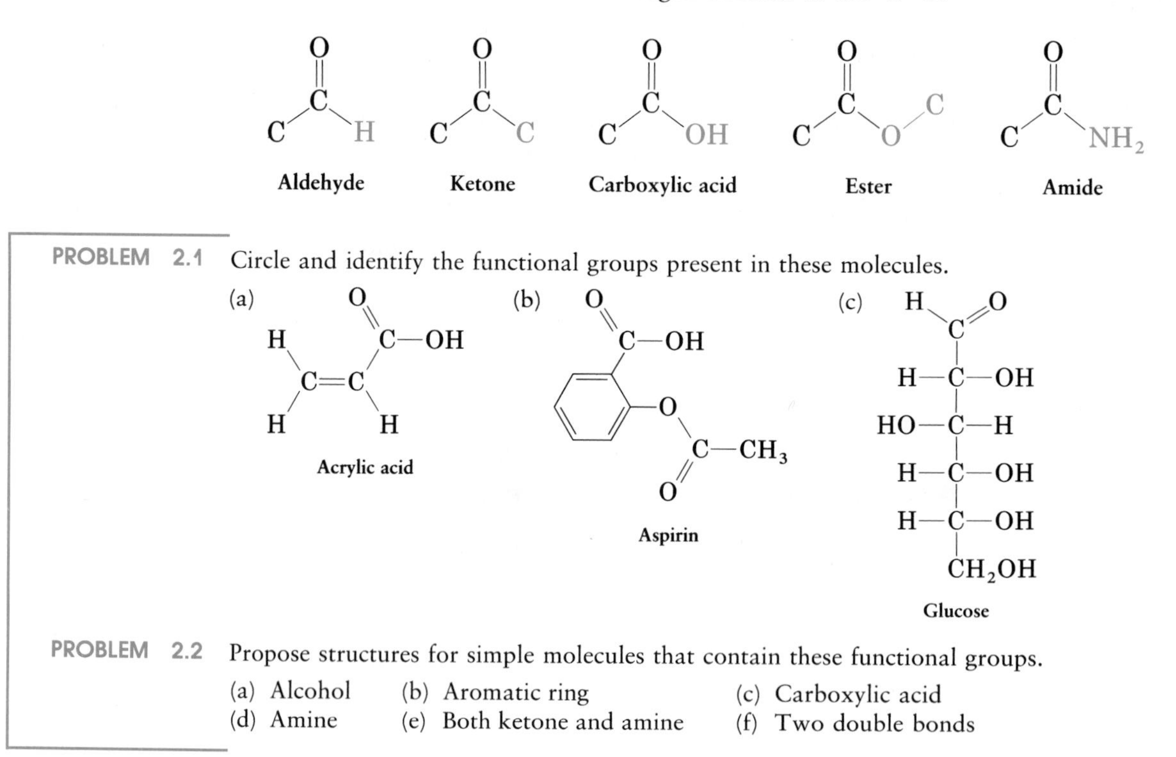

PROBLEM 2.1 Circle and identify the functional groups present in these molecules.

(a) Acrylic acid (b) Aspirin (c) Glucose

PROBLEM 2.2 Propose structures for simple molecules that contain these functional groups.

(a) Alcohol (b) Aromatic ring (c) Carboxylic acid
(d) Amine (e) Both ketone and amine (f) Two double bonds

2.2 ALKANES AND ALKYL GROUPS: ISOMERS

We saw in Section 1.9 that the carbon–carbon single bond in ethane results from sigma (head-on) overlap of carbon sp^3 orbitals. If we imagine joining three, four, five, or even more carbon atoms by carbon–carbon single bonds, we can generate the large family of molecules called alkanes.

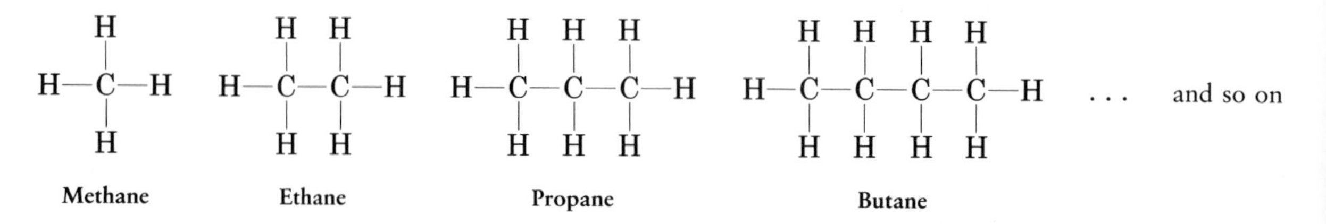

alkane
a compound of carbon and hydrogen that has only single bonds

Alkanes are often described as **saturated hydrocarbons—hydrocarbons** because they contain only carbon and hydrogen; **saturated** because they have only C–C and C–H single bonds and thus contain the maximum possible number of hydrogens per carbon. They have the general formula C_nH_{2n+2}, where n is any

hydrocarbon a compound that has only carbon and hydrogen

saturated a compound that has only single bonds

aliphatic an alternative word describing alkanes

integer. Alkanes are also occasionally referred to as **aliphatic** compounds (Greek *aleiphas*, "fat"). (We'll see later that animal fats do indeed contain long carbon chains similar to alkanes.)

Think for a moment about the ways that carbon and hydrogen can combine to make alkanes. With one carbon and four hydrogens, only one structure is possible: methane, CH_4. Similarly, there is only one possible combination of two carbons with six hydrogens (ethane, CH_3CH_3) and only one possible combination of three carbons with eight hydrogens (propane, $CH_3CH_2CH_3$). If larger numbers of carbons and hydrogens combine, however, more than one kind of molecule can form. For example, there are two ways that molecules with the formula C_4H_{10} can form: The four carbons can be in a row (butane), or they can branch (isobutane). Similarly, there are three ways in which C_5H_{12} molecules can form, and so on for larger alkanes.

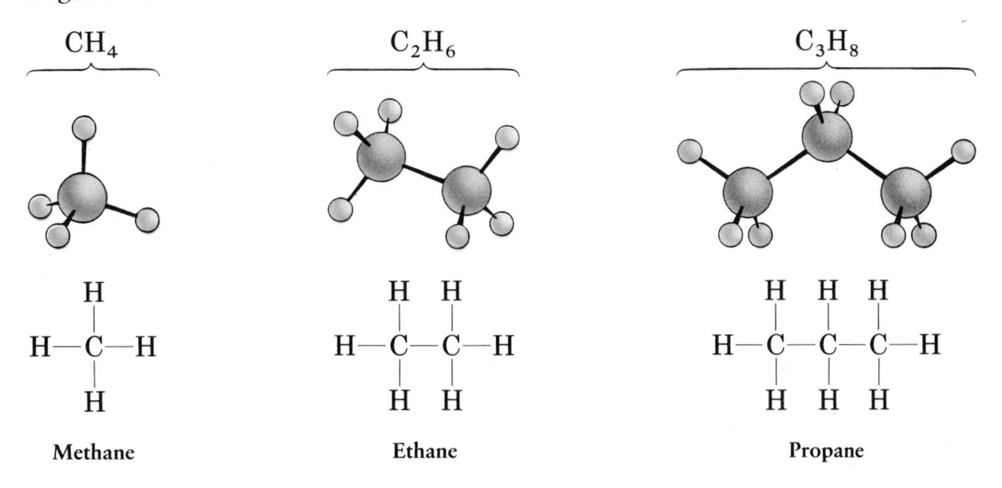

C_4H_{10}

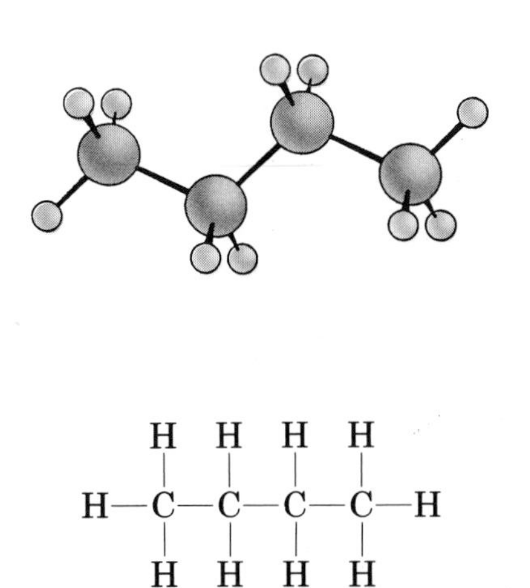

Butane

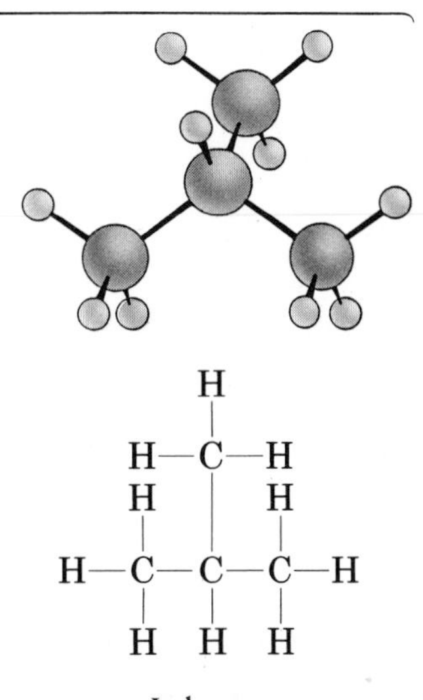

Isobutane
(2-methylpropane)

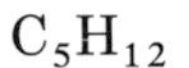

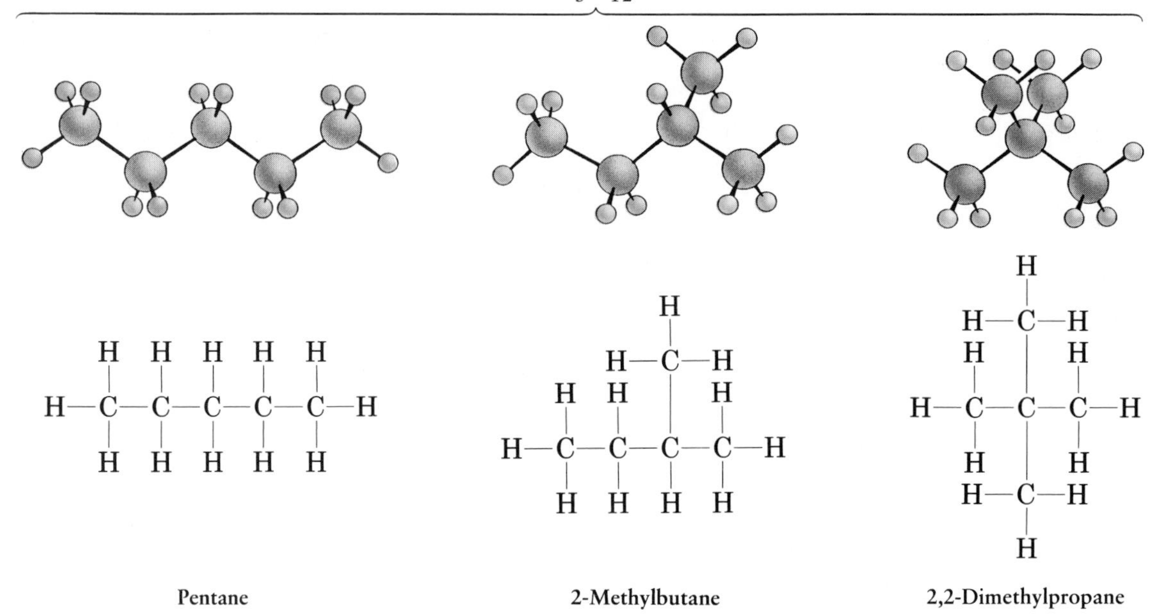

Pentane 2-Methylbutane 2,2-Dimethylpropane

Compounds like pentane, whose carbons are connected in a row, are called **straight-chain alkanes,** or **normal alkanes,** whereas compounds whose carbon chains branch, such as 2-methylpentane, are called **branched-chain alkanes.** Note that you can draw a line connecting all the carbons of a straight-chain alkane without retracing your path or lifting your pencil from the paper. For a branched-chain alkane, however, you either have to retrace your path or lift your pencil from the paper in order to draw a line connecting all the carbons.

Compounds like the two C_4H_{10} molecules that have the same formula but different structures are called *isomers* (*eye*-so-mers; Greek *isos* + *meros,* "made of the same parts"). **Isomers** are compounds that have the same numbers and kinds of atoms but differ in the way these atoms are arranged. Compounds like butane and isobutane, whose atoms are connected differently are called **constitutional isomers.** We'll see shortly that other kinds of isomerism are also possible, even among compounds whose atoms are connected in the same order. As Table 2.2 shows, the number of possible alkane isomers increases dramatically as the number of carbon atoms increases.

straight-chain alkane an alkane whose carbon atoms are connected in a row

normal alkane a straight-chain alkane

branched-chain alkane an alkane that contains a branching arrangement of carbon atoms in its chain

isomers compounds that have the same formula but different chemical structures

constitutional isomers isomers that have their atoms connected in a different order

TABLE 2.2 **Number of alkane isomers**

Formula	*Number of isomers*	*Formula*	*Number of isomers*
C_6H_{14}	5	$C_{10}H_{22}$	75
C_7H_{16}	9	$C_{15}H_{32}$	4,347
C_8H_{18}	18	$C_{20}H_{42}$	366,319
C_9H_{20}	35	$C_{30}H_{62}$	4,111,846,763

A given alkane can be arbitrarily drawn in many ways. For example, the straight-chain, four-carbon alkane called butane can be represented by any of the structures shown in Figure 2.2. These structures don't imply any particular three-dimensional geometry for butane; they only indicate the connections among its atoms. In practice, we soon tire of drawing all the bonds in a molecule and usually refer to butane by the shorthand **condensed structure,** $CH_3CH_2CH_2CH_3$, or even more simply as n-C_4H_{10}, where n signifies *normal,* straight-chain butane. In condensed structures like $CH_3CH_2CH_2CH_3$, the C–C and C–H bonds aren't usually shown but are "understood." If a carbon has three hydrogens bonded to it, we write CH_3; if a carbon has two hydrogens bonded to it, we write CH_2; and so on.

condensed structure a shorthand way of drawing structures in which bonds are understood rather than shown

$CH_3—CH_2—CH_2—CH_3$

$CH_3—CH_2—CH_2—CH_3$ (drawn with the terminal CH_3 above the last CH_2)

$CH_3—CH_2—CH_2—CH_3$ (drawn with $CH_2—CH_3$ above the second CH_2)

$CH_2—CH_2—CH_3$ with CH_3 below the first CH_2

$CH_3(CH_2)_2CH_3$

FIGURE 2.2 Some representations of butane, C_4H_{10}. The molecule is the same regardless of how it's drawn. These structures imply only that butane has a continuous chain of four carbon atoms.

Straight-chain alkanes are named according to the number of carbon atoms in the chain, as shown in Table 2.3. With the exception of the first four compounds—methane, ethane, propane, and butane—whose names have historical origins, the alkanes are named using Greek numbers, according to the number of carbons in the molecule. The suffix *-ane* is added to the end of each name to identify the molecule as an alkane.

TABLE 2.3 **Names of straight-chain alkanes**

Number of carbons (n)	*Name*	*Formula (C_nH_{2n+2})*	*Number of carbons (n)*	*Name*	*Formula (C_nH_{2n+2})*
1	Methane	CH_4	8	Octane	C_8H_{18}
2	Ethane	C_2H_6	9	Nonane	C_9H_{20}
3	Propane	C_3H_8	10	Decane	$C_{10}H_{22}$
4	Butane	C_4H_{10}	11	Undecane	$C_{11}H_{24}$
5	Pentane	C_5H_{12}	12	Dodecane	$C_{12}H_{26}$
6	Hexane	C_6H_{14}	20	Icosane	$C_{20}H_{42}$
7	Heptane	C_7H_{16}	30	Triacontane	$C_{30}H_{62}$

If one hydrogen atom is removed from an alkane, the part-structure that remains is called an **alkyl group.** Alkyl groups are named by replacing the *-ane* ending of the parent alkane with an *-yl* ending. For example, removal of a hydrogen atom from methane, CH_4, generates a *methyl group,* $-CH_3$, and removal of a hydrogen atom from ethane, CH_3CH_3, generates an *ethyl group,* $-CH_2CH_3$. Similarly, removal of a hydrogen atom from the end carbon of any n-alkane gives the series of n-alkyl groups shown in Table 2.4.

alkyl group the part-structure formed by removing a hydrogen from an alkane

TABLE 2.4 Some straight-chain alkyl groups

Alkane	*Name*	*Alkyl group*	*Name*
CH_4	Methane	$—CH_3$	Methyl
CH_3CH_3	Ethane	$—CH_2CH_3$	Ethyl
$CH_3CH_2CH_3$	Propane	$—CH_2CH_2CH_3$	Propyl
$CH_3CH_2CH_2CH_3$	Butane	$—CH_2CH_2CH_2CH_3$	Butyl
$CH_3CH_2CH_2CH_2CH_3$	Pentane	$—CH_2CH_2CH_2CH_2CH_3$	Pentyl

Just as *n*-alkyl groups are generated by removing a hydrogen from an end carbon, *branched* alkyl groups are generated by removing an internal hydrogen atom. Two 3-carbon alkyl groups and four 4-carbon alkyl groups are possible (Figure 2.3).

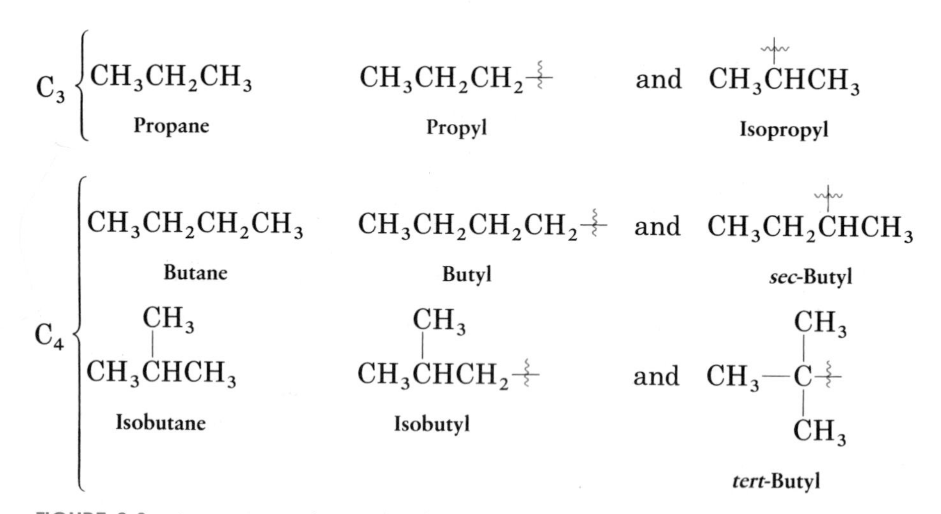

FIGURE 2.3 Generation of straight-chain and branched-chain alkyl groups from *n*-alkanes

One further word of explanation about naming alkyl groups: The prefixes used for the C_4 alkyl groups in Figure 2.3, *sec* (for secondary) and *tert* (for tertiary), refer to the degree of alkyl substitution at the carbon atom in question. There are four possible degrees of alkyl substitution for carbon, denoted 1°, 2°, 3°, and 4°:

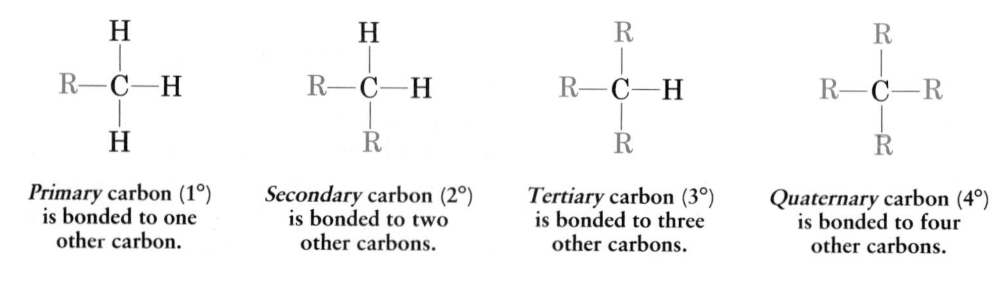

H
R—C—H
H
Primary carbon (1°) is bonded to one other carbon.
H
R—C—H
R
Secondary carbon (2°) is bonded to two other carbons.
R
R—C—H
R
Tertiary carbon (3°) is bonded to three other carbons.
R
R—C—R
R
Quaternary carbon (4°) is bonded to four other carbons.

R
the general symbol used for an organic part-structure

The symbol **R** is used here and throughout this text to represent a *generalized* alkyl group. The R group can be methyl, ethyl, or any of a multitude of other alkyl groups.

PRACTICE PROBLEM 2.1

Propose structures for two isomers of formula C_2H_6O.

SOLUTION We know that carbon forms four bonds, oxygen forms two, and hydrogen forms one. Putting the pieces together yields two isomeric structures:

2 –C–, 1 –O–, 6 H– give

$$\mathrm{H{-}\overset{H}{\underset{H}{C}}{-}\overset{H}{\underset{H}{C}}{-}O{-}H} \quad \text{and} \quad \mathrm{H{-}\overset{H}{\underset{H}{C}}{-}O{-}\overset{H}{\underset{H}{C}}{-}H}$$

PROBLEM 2.3 Draw structures of the five isomers of C_6H_{14}.

PROBLEM 2.4 Draw structures that meet these descriptions:

(a) Three isomers with the formula C_8H_{18}
(b) Two isomers with the formula $C_4H_8O_2$

PROBLEM 2.5 Draw the eight possible five-carbon alkyl groups (pentyl isomers).

PROBLEM 2.6 Draw alkanes that meet these descriptions:

(a) An alkane with two tertiary carbons
(b) An alkane that contains an isopropyl group
(c) An alkane that has one quaternary and one secondary carbon

PROBLEM 2.7 Identify the carbon atoms in these molecules as primary, secondary, tertiary, or quaternary:

(a) 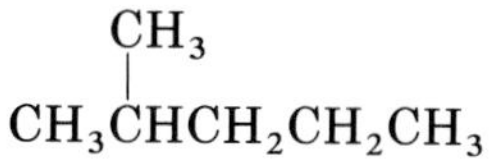 $\mathrm{CH_3\overset{\displaystyle CH_3}{\overset{|}{C}}HCH_2CH_2CH_3}$

(b) $\mathrm{CH_3CH_2\overset{\displaystyle CH_3CHCH_3}{\overset{|}{C}}HCH_2CH_3}$

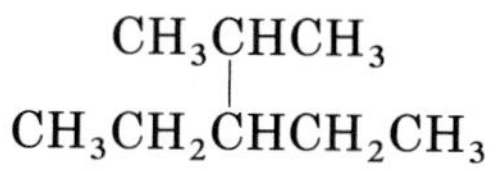

(c) $\mathrm{CH_3\overset{\displaystyle CH_3}{\overset{|}{C}}HCH_2\overset{\displaystyle CH_3}{\underset{\displaystyle CH_3}{\overset{|}{\underset{|}{C}}}}CH_3}$

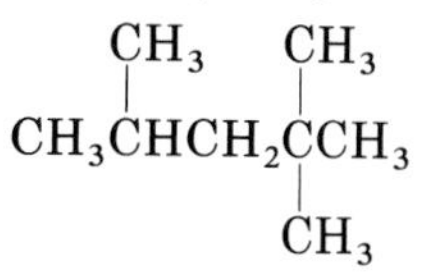

2.3 NAMING BRANCHED-CHAIN ALKANES

In earlier times when few pure organic chemicals were known, new compounds were named at the whim of their discoverer. Thus, urea (CH_4N_2O) is a crystalline substance isolated from urine, and barbituric acid is a tranquilizing agent named by its discoverer in honor of his friend Barbara. As the science of organic chemistry grew in the nineteenth century, however, so too did the number of compounds and the need for a rational and systematic method of naming them. The system of nomenclature (naming) most often used by organic chemists is that devised by the International Union of Pure and Applied Chemistry (IUPAC, usually spoken as *eye*-you-pac).

In the IUPAC system, a chemical name has three parts: prefix, parent, and suffix. The parent name specifies the overall size of the molecule by identifying the number of carbon atoms in its main chain; the suffix identifies the family that the molecule belongs to; and the prefix specifies the location of various substituents on the main chain:

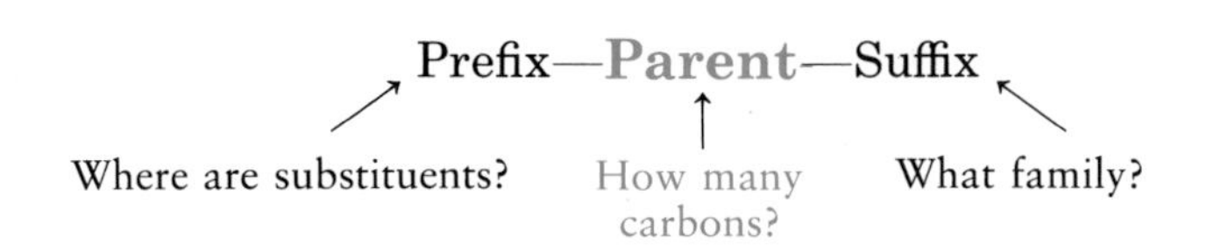

All but the most complex branched-chain alkanes can be named by following four steps:

Step 1. Find the parent hydrocarbon.

a. Find the longest continuous carbon chain in the molecule and use the name of that chain as the parent name. The longest chain may not always be obvious; you may have to "turn corners":

$$\begin{array}{l} \quad\quad\quad\quad\quad CH_2CH_3 \\ \quad\quad\quad\quad\quad | \\ CH_3CH_2CH_2CH{-}CH_3 \end{array}$$

named as a substituted *hexane*

b. If two different chains of equal length are present, choose the one with the larger number of branch points as the parent:

$$\begin{array}{l} \quad\;\; CH_3 \\ \quad\;\; | \\ CH_3CHCHCH_2CH_2CH_3 \\ \quad\quad\quad | \\ \quad\quad\quad CH_2CH_3 \end{array}$$

named as a hexane with *two* substituents

NOT

$$\begin{array}{l} \quad\;\; CH_3 \\ \quad\;\; | \\ CH_3CH{-}CHCH_2CH_2CH_3 \\ \quad\quad\quad\quad | \\ \quad\quad\quad\quad CH_2CH_3 \end{array}$$

as a hexane with *one* substituent

Step 2. Beginning at the end *nearer the first branch point*, number each carbon atom in the parent chain:

$$\begin{array}{l} \quad\quad\quad\;\; \overset{2}{C}H_2\overset{1}{C}H_3 \\ \quad\quad\quad\;\; | \\ CH_3{-}\underset{3}{C}H\underset{4}{C}H{-}CH_2CH_3 \\ \quad\quad\quad\quad\quad | \\ \quad\quad\quad\quad\quad \underset{5}{C}H_2\underset{6}{C}H_2\underset{7}{C}H_3 \end{array} \quad \text{NOT} \quad \begin{array}{l} \quad\quad\quad\;\; \overset{6}{C}H_2\overset{7}{C}H_3 \\ \quad\quad\quad\;\; | \\ CH_3{-}\underset{5}{C}H\underset{4}{C}H{-}CH_2CH_3 \\ \quad\quad\quad\quad\quad | \\ \quad\quad\quad\quad\quad \underset{3}{C}H_2\underset{2}{C}H_2\underset{1}{C}H_3 \end{array}$$

The first branch occurs at C3 in the proper order of numbering but at C4 in the improper order.

Step 3. Assign a number to each substituent according to its point of attachment on the parent chain. If there are two substituents on the same carbon, assign them both the same number. There must always be as many numbers in the name as there are substituents.

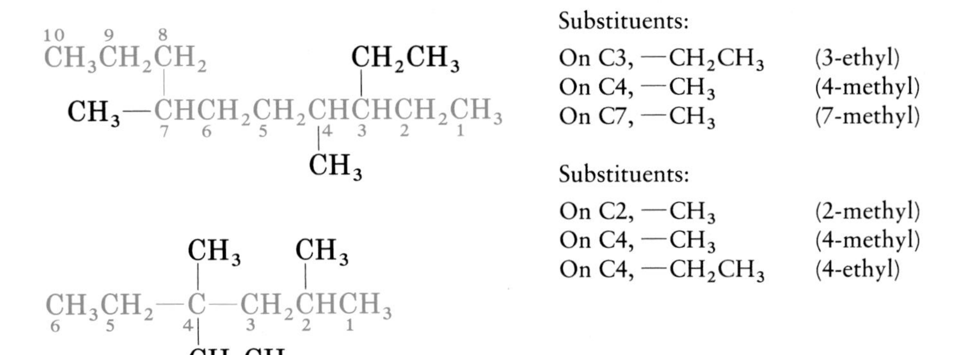

Substituents:
On C3, $-CH_2CH_3$ (3-ethyl)
On C4, $-CH_3$ (4-methyl)
On C7, $-CH_3$ (7-methyl)
Substituents:
On C2, $-CH_3$ (2-methyl)
On C4, $-CH_3$ (4-methyl)
On C4, $-CH_2CH_3$ (4-ethyl)

Step 4. Write the name as a single word, using hyphens to separate the various prefixes and commas to separate numbers. If two or more different side chains are present, cite them in alphabetical order. If two or more identical side chains are present, use one of the prefixes *di-*, *tri-*, *tetra-*, and so forth. Don't use these prefixes for alphabetizing, though.

$CH_3CH_2CH_2CH(CH_2CH_3)-CH_3$

3-Methylhexane

$CH_3CH(CH_3)CH(CH_2CH_3)CH_2CH_2CH_3$

3-Ethyl-2-methylhexane

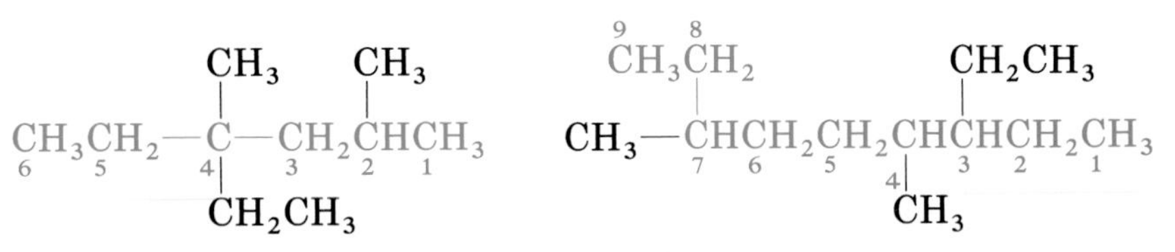

4-Ethyl-2,4-dimethylhexane

3-Ethyl-4,7-dimethylnonane

PRACTICE PROBLEM 2.2

What is the IUPAC name of this alkane?

$CH_3CH(CH_2CH_3)CH_2CH_2CH_2CH(CH_3)CH_3$

SOLUTION The molecule has a chain of eight carbons (octane) with two methyl substituents. Numbering from the end nearer the first methyl substituent indicates that the methyls are at C2 and C6, giving the name 2,6-dimethyloctane.

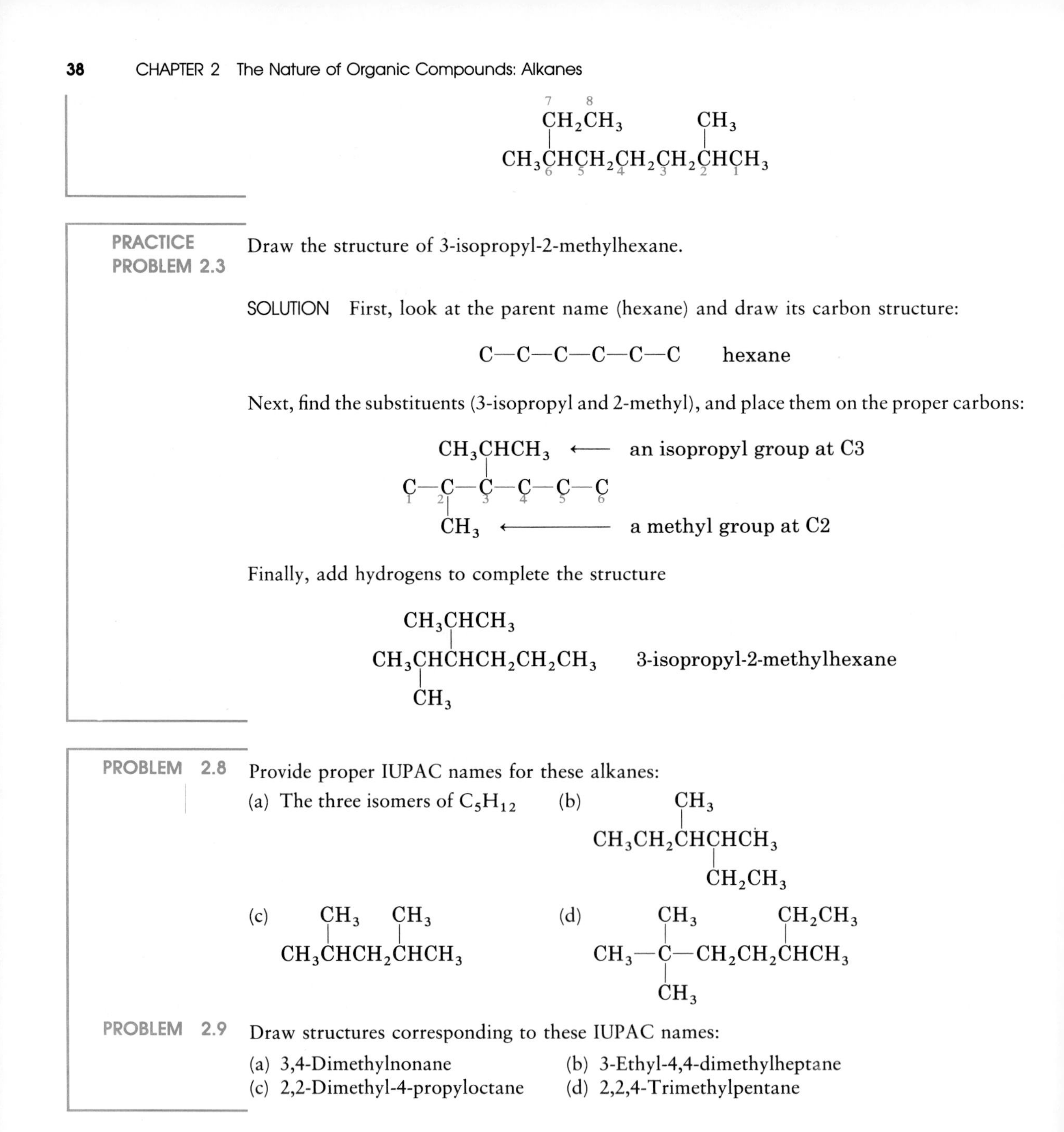

$$\begin{array}{l} \quad\ \ \overset{7\quad 8}{CH_2CH_3} \qquad\quad CH_3 \\ \quad\ \ | \qquad\qquad\qquad\quad | \\ CH_3\underset{6}{C}H\underset{5}{C}H_2\underset{4}{C}H_2\underset{3}{C}H_2\underset{2}{C}H\underset{1}{C}H_3 \end{array}$$

PRACTICE PROBLEM 2.3 Draw the structure of 3-isopropyl-2-methylhexane.

SOLUTION First, look at the parent name (hexane) and draw its carbon structure:

$$C—C—C—C—C—C \qquad \text{hexane}$$

Next, find the substituents (3-isopropyl and 2-methyl), and place them on the proper carbons:

$$\begin{array}{l} \qquad CH_3CHCH_3 \longleftarrow \text{an isopropyl group at C3} \\ \qquad\qquad\ | \\ \underset{1}{C}—\underset{2}{C}—\underset{3}{C}—\underset{4}{C}—\underset{5}{C}—\underset{6}{C} \\ \qquad | \\ \qquad CH_3 \longleftarrow \text{a methyl group at C2} \end{array}$$

Finally, add hydrogens to complete the structure

$$\begin{array}{l} \qquad CH_3CHCH_3 \\ \qquad\qquad\ | \\ CH_3CHCHCH_2CH_2CH_3 \qquad \text{3-isopropyl-2-methylhexane} \\ \qquad | \\ \qquad CH_3 \end{array}$$

PROBLEM 2.8 Provide proper IUPAC names for these alkanes:

(a) The three isomers of C_5H_{12}

(b)
$$\begin{array}{l} \qquad\qquad CH_3 \\ \qquad\qquad | \\ CH_3CH_2CHCHCH_3 \\ \qquad\qquad\quad | \\ \qquad\qquad\quad CH_2CH_3 \end{array}$$

(c)
$$\begin{array}{l} \quad CH_3 \quad CH_3 \\ \quad | \qquad\quad | \\ CH_3CHCH_2CHCH_3 \end{array}$$

(d)
$$\begin{array}{l} \qquad\quad CH_3 \qquad\qquad CH_2CH_3 \\ \qquad\quad | \qquad\qquad\qquad | \\ CH_3—C—CH_2CH_2CHCH_3 \\ \qquad\quad | \\ \qquad\quad CH_3 \end{array}$$

PROBLEM 2.9 Draw structures corresponding to these IUPAC names:

(a) 3,4-Dimethylnonane
(b) 3-Ethyl-4,4-dimethylheptane
(c) 2,2-Dimethyl-4-propyloctane
(d) 2,2,4-Trimethylpentane

2.4 PROPERTIES OF ALKANES

paraffin an alternative name for alkanes

Alkanes are sometimes referred to as **paraffins** (Latin *parum affinis*, "slight affinity"). This term for alkanes aptly describes their behavior: Alkanes show little chemical affinity for other molecules and are chemically inert to most laboratory reagents. They do, however, react with oxygen and with chlorine under appropriate conditions.

Reaction of alkanes with oxygen occurs during combustion in an engine or furnace when the alkane is used as a fuel. Carbon dioxide and water are formed as products, and a large amount of heat is released. For example, methane (natural gas) reacts with oxygen according to the equation:

$$CH_4 + 2\,O_2 \longrightarrow CO_2 + 2\,H_2O + 213\text{ kcal/mol}$$

Reaction of alkanes with chlorine occurs when a mixture of the two reagents is irradiated with ultraviolet light (denoted $h\nu$, where ν is the Greek letter *nu*). Depending on the relative amounts of the two starting materials (reactants) and on the time allowed for reaction, a sequential substitution of the alkane hydrogen atoms by chlorine occurs, leading to chlorinated products. Methane reacts with chlorine to yield a mixture of chloromethane (CH_3Cl), dichloromethane (CH_2Cl_2), trichloromethane ($CHCl_3$), and tetrachloromethane (CCl_4):

$$CH_4 + Cl_2 \xrightarrow{\text{light }(h\nu)} CH_3Cl + CH_2Cl_2 + CHCl_3 + CCl_4 + HCl$$

We'll see how this chlorination reaction occurs when we take up the chemistry of alkyl halides in Chapter 7.

Alkanes show regular increases in both boiling point and melting point as molecular weight increases. Average carbon–carbon bond parameters are nearly the same in all alkanes, with bond lengths of 1.54 ± 0.01 Å and bond strengths of 85 ± 3 kcal/mol. Carbon–hydrogen bond parameters are also nearly constant at 1.09 ± 0.01 Å and 95 ± 3 kcal/mol.

2.5 CONFORMATIONS OF ETHANE

We saw earlier that methane has a tetrahedral structure and that carbon–carbon bonds in alkanes result from sigma overlap of two tetrahedral carbon sp^3 orbitals. Let's now look into the three-dimensional consequences of such bonding. What are the spatial relationships between the hydrogens on one carbon and the hydrogens on the other?

We know that sigma bonds result from the head-on overlap of two atomic orbitals and that a cross section through a sigma bond is circular. As a consequence of this circular symmetry, *rotation* is possible around carbon–carbon single bonds. Orbital overlap in the C–C bond is exactly the same for all geometric arrangements of the hydrogens (Figure 2.4).

The different arrangements of atoms caused by rotation around a single bond are called **conformations,** and a specific conformation is called a **conformer** (*confor*mational iso*mer*). Unlike constitutional isomers (Section 2.2), though, different conformers can't be isolated because they interconvert too rapidly.

conformation the specific arrangement of atoms, assuming there is no rotation around a single bond

conformers conformational isomers that differ only in rotation around a single bond

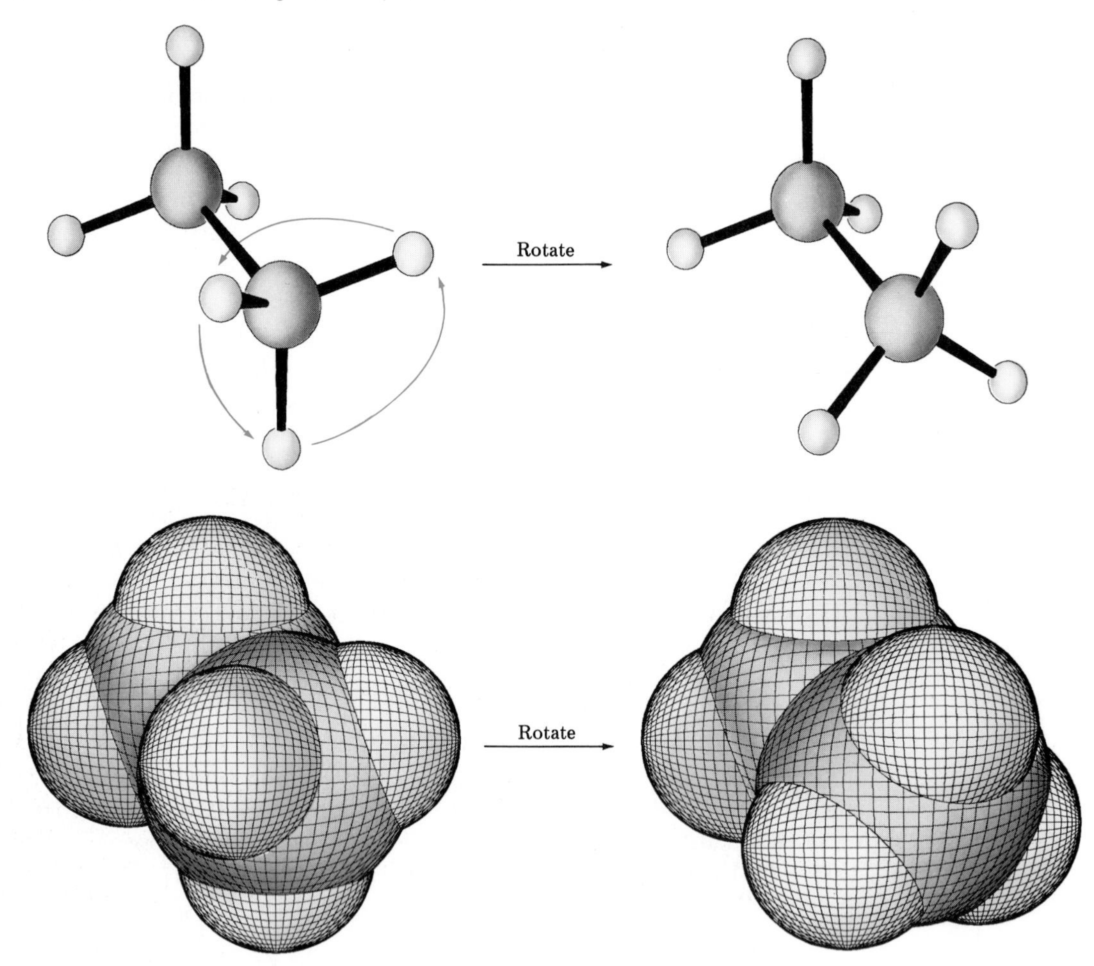

FIGURE 2.4 Two conformations of ethane. Rotation around the carbon–carbon single bond interconverts the forms.

sawhorse representation
a way of viewing a molecule's spatial arrangement by looking at a given carbon–carbon bond from an oblique angle

Newman projection
a way of viewing a molecule's spatial arrangement by looking end-on at a carbon–carbon bond

Chemists represent conformational isomers in two ways, as shown in Figure 2.5. **Sawhorse representations** view the carbon–carbon bond from an oblique angle and indicate spatial relationships by showing all the C–H bonds. **Newman projections** view the carbon–carbon bond end-on and represent the two carbon atoms by a circle. Bonds attached to the front carbon are represented by lines to the center of the circle, and bonds attached to the rear carbon are represented by lines to the edge of the circle.

In spite of what we've just said about sigma-bond symmetry, we don't observe perfectly free rotation in ethane. Experiments show that there is a slight (2.9 kcal/mol) barrier to rotation and that some conformations are more stable than others. The lowest-energy (most stable) conformation is the one in which all six carbon–hydrogen bonds are as far away from each other as possible (**staggered** when viewed end-on in a Newman projection). The highest-energy (least stable)

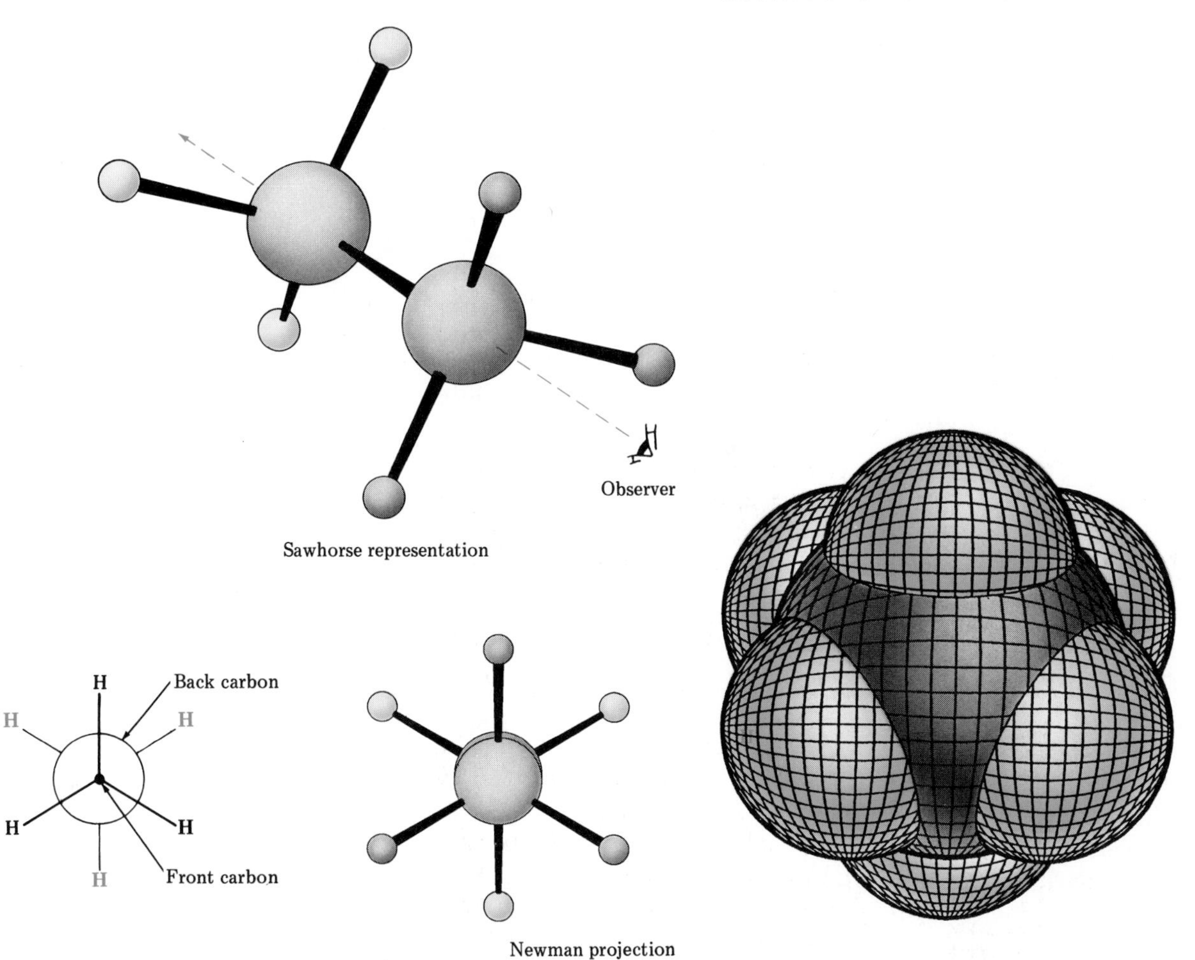

FIGURE 2.5 A sawhorse representation and a Newman projection of ethane

conformation is the one in which the six carbon–hydrogen bonds are as close as possible (**eclipsed** in a Newman projection). At any given instant, about 99% of ethane molecules have an approximately staggered conformation, and only about 1% are near the eclipsed conformation.

staggered
the conformation around a carbon–carbon single bond that places all attached atoms as far from each other as possible

eclipsed
the conformation around a carbon–carbon single bond that places attached atoms as close as possible

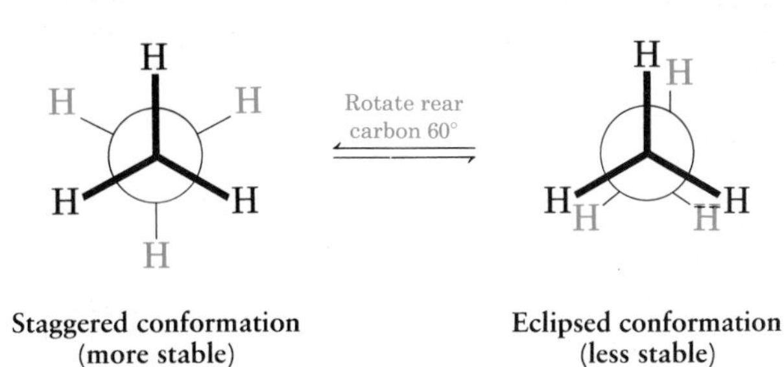

What's true for ethane is also true for propane, butane, and all higher alkanes. The most favored conformation for any alkane is the one in which all bonds have staggered arrangements (Figure 2.6).

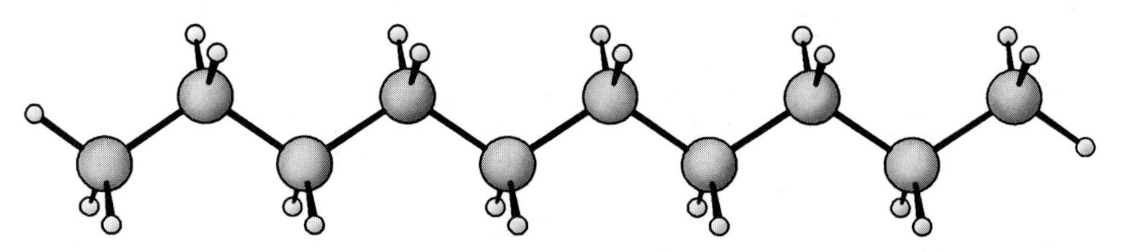

FIGURE 2.6 The most stable conformation of any alkane is that in which the bonds on adjacent carbons are staggered and the carbon chain is fully extended, as in this computer-generated structure of decane.

PROBLEM 2.10 Make a graph of energy versus amount of bond rotation for ethane. Place energy on the vertical (y) axis and the amount of bond rotation on the horizontal (x) axis, using a staggered conformation as the 0° starting point.

PROBLEM 2.11 Sight along a carbon–carbon bond of propane and draw a Newman projection of the most stable conformation. Draw a Newman projection of the least stable conformation.

PROBLEM 2.12 Looking along the C2–C3 bond of butane, there are two different staggered conformations and two different eclipsed conformations. Draw them.

PROBLEM 2.13 Which of the butane conformations you drew in Problem 2.12 looks the most stable?

2.6 DRAWING CHEMICAL STRUCTURES

In the structures we've been using, a line between atoms represents the two electrons in a covalent bond. Most chemists find themselves drawing many structures each day, however, and it would soon become awkward if every bond and every atom had to be indicated. Chemists have therefore devised a shorthand way of drawing structures that greatly simplifies matters, particularly for the cyclic compounds that we'll see shortly.

The rules for this shorthand are simple:

1. Carbon atoms aren't usually shown. Instead, a carbon atom is assumed to be at the intersection of two lines (bonds) and at the open end of each line. Occasionally, a carbon atom might be indicated for emphasis or for clarity.
2. Hydrogen atoms bonded to carbon aren't shown. Since carbon always has a valence of four, we mentally supply the correct number of hydrogen atoms for each carbon.
3. All atoms other than carbon and hydrogen *are* shown.

The following examples indicate how these rules are applied in specific cases.

Butane, C_4H_{10} $CH_3CH_2CH_2CH_3$

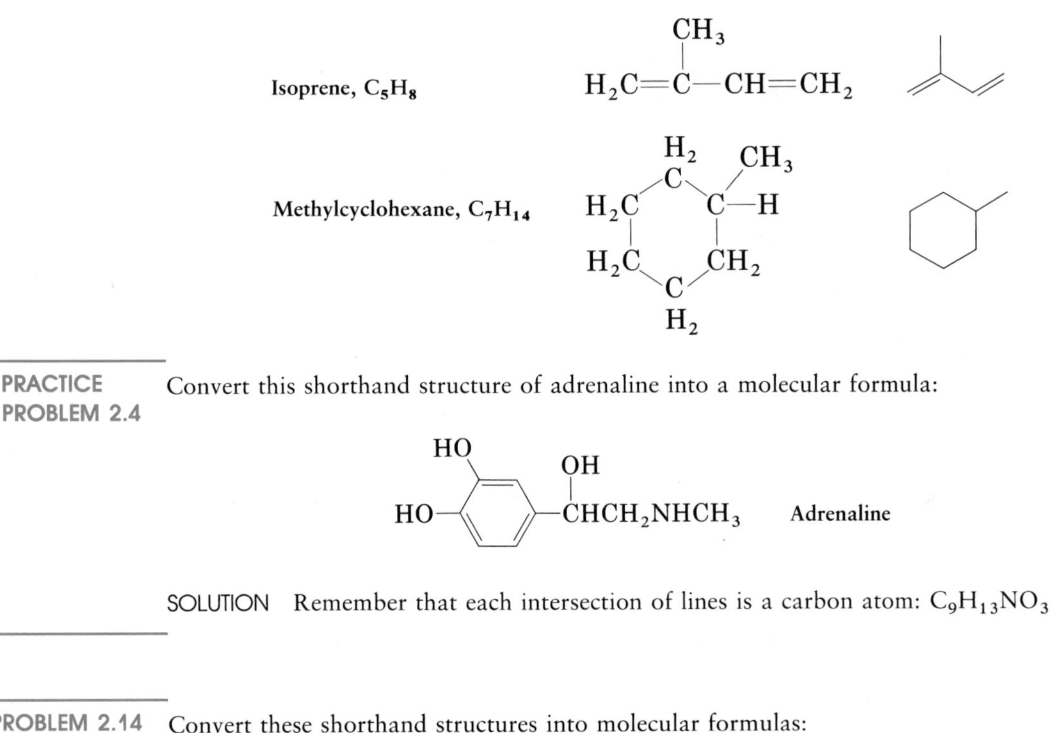

Isoprene, C_5H_8

Methylcyclohexane, C_7H_{14}

PRACTICE PROBLEM 2.4 Convert this shorthand structure of adrenaline into a molecular formula:

SOLUTION Remember that each intersection of lines is a carbon atom: $C_9H_{13}NO_3$

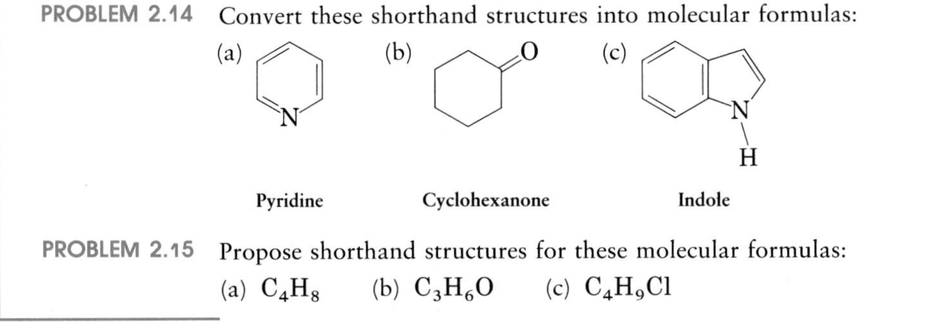

PROBLEM 2.14 Convert these shorthand structures into molecular formulas:

PROBLEM 2.15 Propose shorthand structures for these molecular formulas:

(a) C_4H_8 (b) C_3H_6O (c) C_4H_9Cl

2.7 CYCLOALKANES: CIS–TRANS ISOMERISM

cycloalkane an alkane with a ring of carbons

alicyclic an alternative word for cycloalkane

We've only discussed open-chain alkanes up to now, but compounds that contain *rings* of carbon atoms are also well known. Such compounds are called **cycloalkanes** or **alicyclic** (*ali*phatic *cyclic*) compounds. Since cycloalkanes consist of rings of $-CH_2-$ units, they have the general formula $(CH_2)_n$ or C_nH_{2n} as in the following examples:

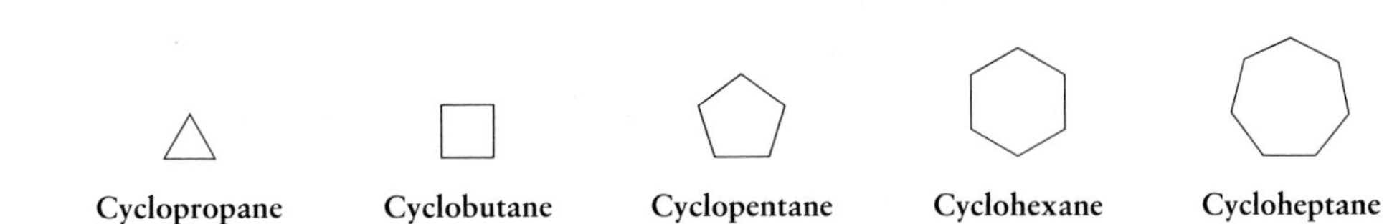

Cyclopropane Cyclobutane Cyclopentane Cyclohexane Cycloheptane

In most respects, the chemistry of cycloalkanes mimics that of open-chain (acyclic) alkanes. Both classes of compounds are relatively nonpolar and are chemically inert to most reagents. There are, however, some important differences.

One difference is that cycloalkanes have less conformational freedom than their open-chain counterparts. Although open-chain alkanes have nearly free rotation around their carbon–carbon single bonds, cycloalkanes have much less freedom of rotation around bonds and are therefore more constrained in their geometry. For example, cyclopropane is constrained by geometry to be a flat (planar) molecule with a rigid structure. No rotation around a carbon–carbon bond in cyclopropane is possible without breaking open the ring.

Because of their cyclic structure, cycloalkanes have two sides, a "top" side and a "bottom" side. Isomerism is therefore possible in substituted cycloalkanes. For example, there are two 1,2-dimethylcyclopropane isomers, one with the two methyl groups on the same side of the ring and the other isomer with them on opposite sides. Both isomers are stable compounds that can't be interconverted without breaking chemical bonds.

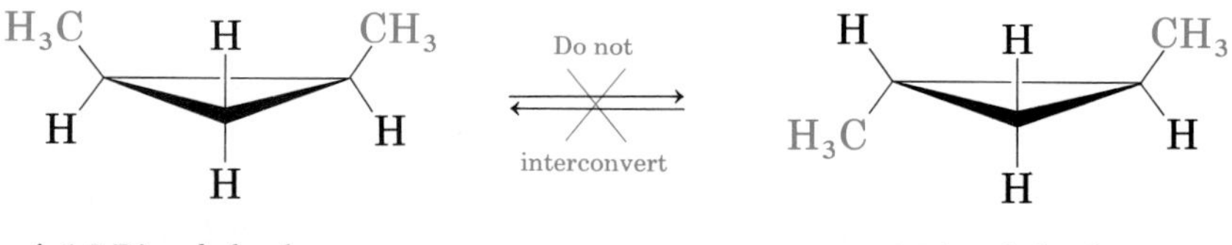

cis-1,2-Dimethylcyclopropane
(methyl groups on same side of ring)

trans-1,2-Dimethylcyclopropane
(methyl groups on opposite sides of ring)

stereoisomers isomers that have their atoms connected in the same order but with a different three-dimensional arrangement

cis-trans isomers stereoisomers that differ in having substituent groups attached to the same side (cis) or opposite sides (trans) of the molecule

Unlike the constitutional isomers, butane and isobutane (Section 2.2), which have a different order of connection between their atoms, the two 1,2-dimethylcyclopropanes have the *same* order of connection. They differ, however, in the spatial orientation of their atoms. Compounds that have their atoms connected in the same order but differ in three-dimensional orientation are called **stereoisomers.** The 1,2-dimethylcyclopropanes are special kinds of stereoisomers called **cis–trans isomers;** the prefixes *cis-* (Latin, "on the same side") and *trans-* (Latin, "across") are used to distinguish between them.

PROBLEM 2.16 Draw both cis and trans isomers of 1,2-dibromocyclobutane.

2.8 NAMING CYCLOALKANES

Substituted cycloalkanes are named by rules similar to those used for open-chain alkanes. For most compounds, there are only two steps:

1. Count the number of carbon atoms in the ring and add the prefix *cyclo-* to the name of the corresponding alkane. If a substituent is present on the ring, the compound should be named as an alkyl-substituted cycloalkane rather than as a cycloalkyl-substituted alkane.

$-CH_3$ Methylcyclopentane

2. Start at a point of attachment and number the substituents on the ring so as to arrive at the lowest sum. If two or more different substituents are present, number them in order of their alphabetical priority.

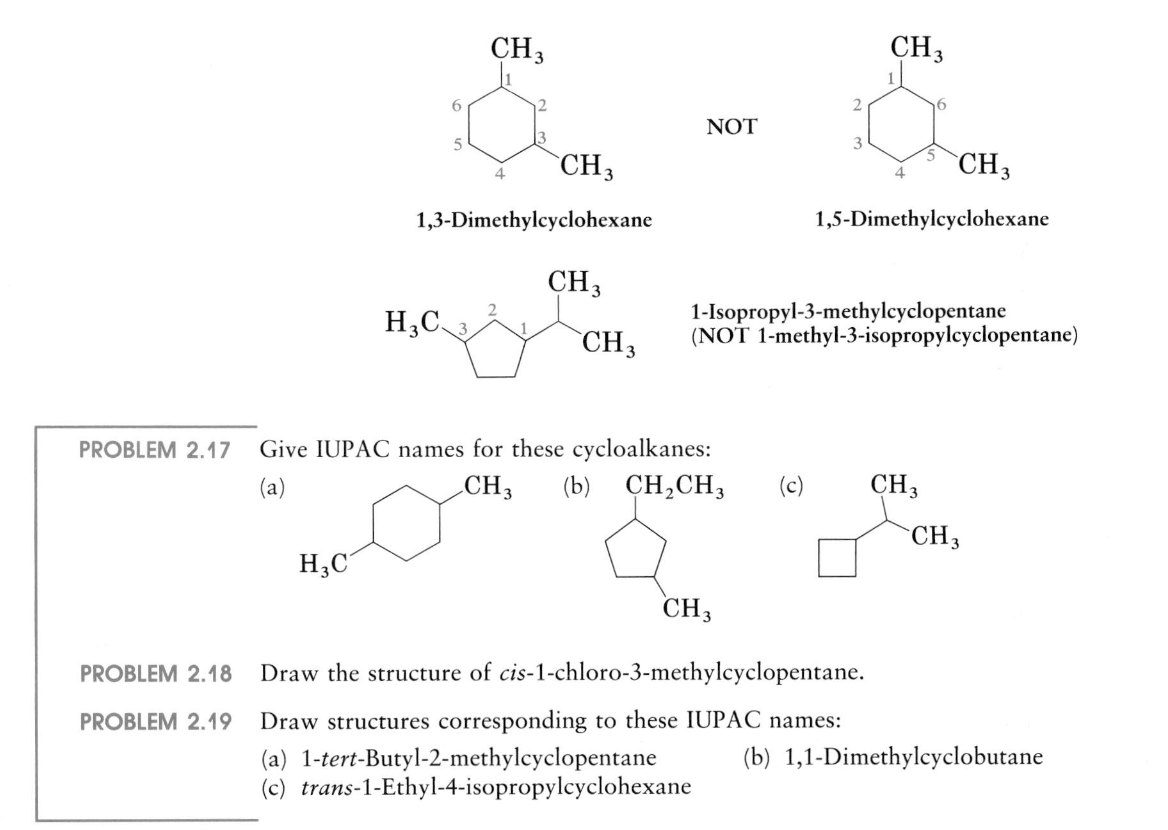

PROBLEM 2.17 Give IUPAC names for these cycloalkanes:

PROBLEM 2.18 Draw the structure of *cis*-1-chloro-3-methylcyclopentane.

PROBLEM 2.19 Draw structures corresponding to these IUPAC names:

(a) 1-*tert*-Butyl-2-methylcyclopentane
(b) 1,1-Dimethylcyclobutane
(c) *trans*-1-Ethyl-4-isopropylcyclohexane

2.9 CONFORMATIONS OF SOME COMMON CYCLOALKANES

In the early days of organic chemistry, cycloalkanes provoked a good deal of consternation among chemists. The problem was that if carbon prefers to have bond angles of 109°, how is it possible for cyclopropane and cyclobutane to exist? After all, cyclopropane must have a triangular shape with bond angles near 60°, and cyclobutane must have a square or rectangular shape with bond angles near 90°. Nonetheless, these compounds *do* exist and are perfectly stable.

Let's look at the most common cycloalkanes.

Cyclopropane

Cyclopropane is a flat, symmetrical molecule with C–C–C bond angles of 60°, as indicated in Figure 2.7. The three carbons form an equilateral triangle, with the six hydrogens protruding above and below the plane of the carbons. All six of the C–H bonds have an eclipsed, rather than staggered, arrangement with their neighbors.

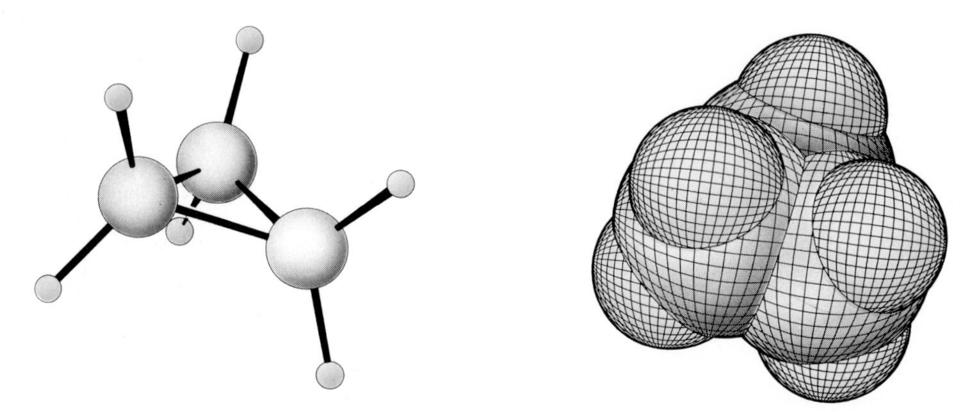

FIGURE 2.7 Computer-generated models of cyclopropane

angle strain the strain in a molecule caused by expanding or compressing a bond angle away from its normal value

The simplest way to account for the distortion of the C–C–C bond angles from their normal value of 109° to a value of 60° in cyclopropane is to think of cyclopropane as having *bent bonds* (Figure 2.8). In a normal alkane, maximum bonding efficiency is achieved when two atoms are located so that their overlapping orbitals point directly toward each other. In cyclopropane, however, the orbitals can't point directly toward each other; they must instead overlap at a slight angle. The results of this poor overlap are that cyclopropane carbon–carbon bonds are weaker than normal alkane bonds due to what is called **angle strain** and that cyclopropane is therefore more reactive than normal alkanes.

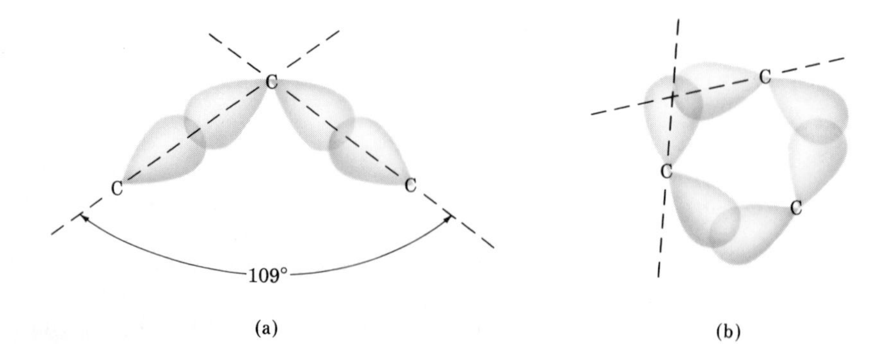

FIGURE 2.8 An orbital view of cyclopropane. (a) Normal C–C bonds have good overlap of orbitals; (b) cyclopropane bent bonds have poor overlap of orbitals.

Cyclobutane and Cyclopentane

Cyclobutane and cyclopentane are slightly puckered rather than flat, as indicated in Figure 2.9. This puckering makes the C–C–C bond angles a bit smaller than they would otherwise be and increases the angle strain, but it relieves the eclipsing interactions of adjacent C–H bonds that would occur in a flat ring.

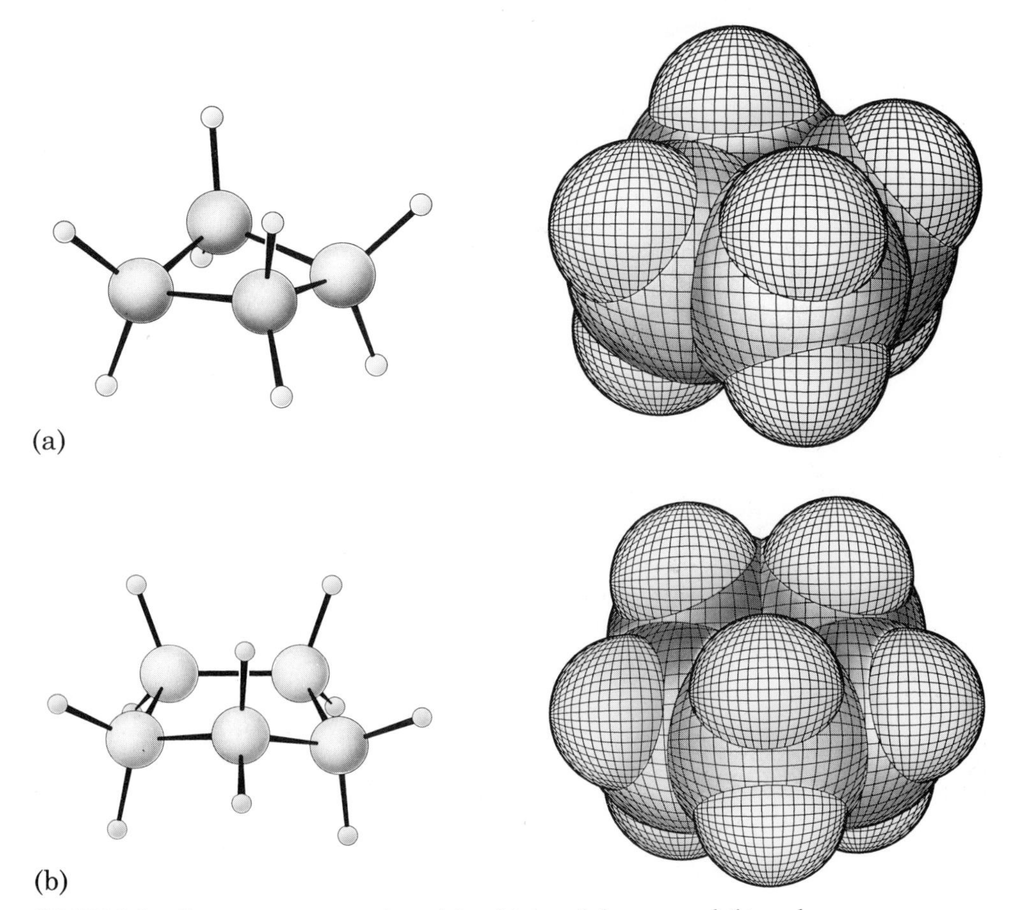

FIGURE 2.9 Computer-generated models of (a) cyclobutane and (b) cyclopentane

Cyclohexane

chair conformation the energetically favored conformation of cyclohexane in which all bond angles are 109° and all bonds have a staggered arrangement

Six-membered (cyclohexane) rings are the most important of all cycloalkanes because of their wide occurrence in nature. A large number of compounds, including steroids and many other pharmaceutical agents, have cyclohexane rings.

Cyclohexane is not flat. Rather, it is puckered into a three-dimensional arrangement called a **chair conformation,** in which the C–C–C bond angles have the ideal 109° tetrahedral value (Figure 2.10). In addition to having ideal bond angles and being free of angle strain, chair cyclohexane is also free of all C–H eclipsing interactions: All neighboring C–H bonds are perfectly staggered.

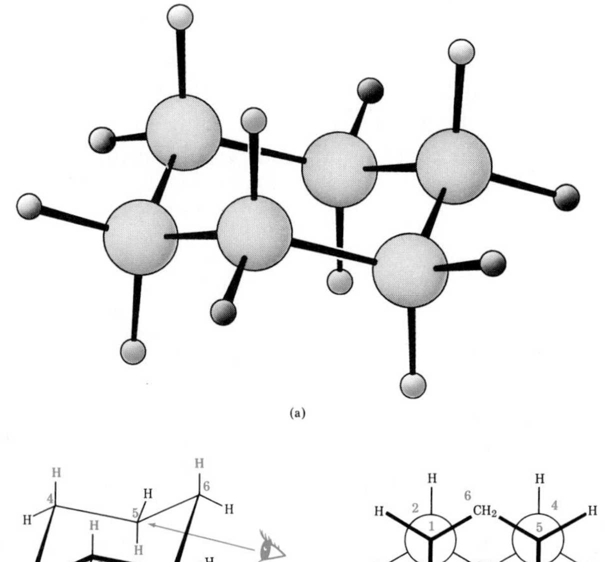

(a)

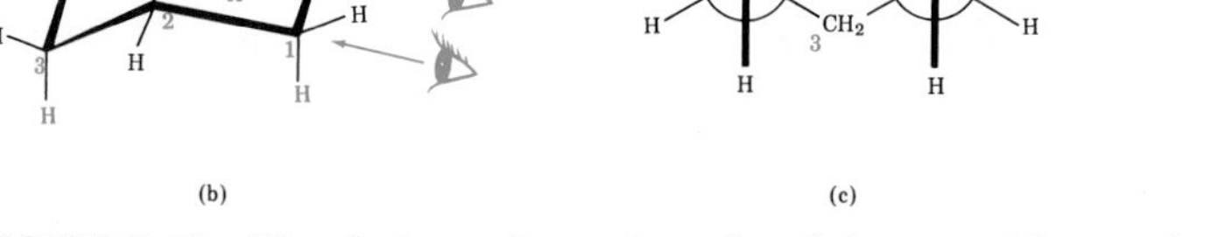

(b) (c)

FIGURE 2.10 The chair conformation of cyclohexane. This conformation has no eclipsing between neighboring C–H bonds, and all bond angles are 109°.

Chair conformations are drawn by following three steps:

1. Draw two parallel lines, slanted downward and slightly offset from each other. These lines show that four of the cyclohexane carbon atoms lie in a plane.
2. Place the topmost carbon atom above and to the right of the plane of the other four and connect the bonds.
3. Place the bottommost carbon atom below and to the left of the plane of the middle four and connect the bonds. Note that the bonds to the bottommost carbon atom are parallel to the bonds to the topmost carbon.

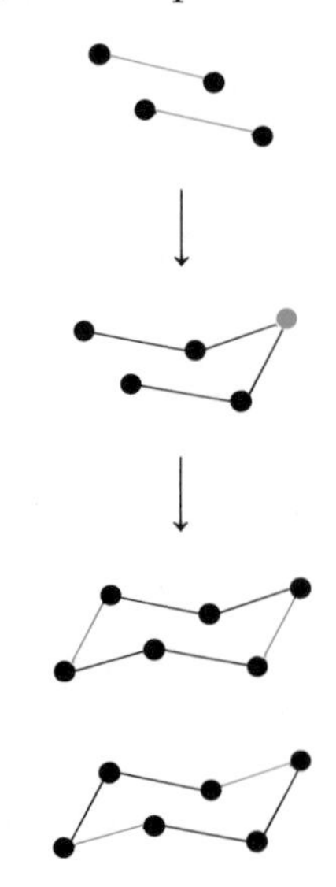

It's important to remember when viewing cyclohexane that the lower bond is in front, and the upper bond is in back. If this convention is not defined, an optical illusion can make it appear that the reverse is true.

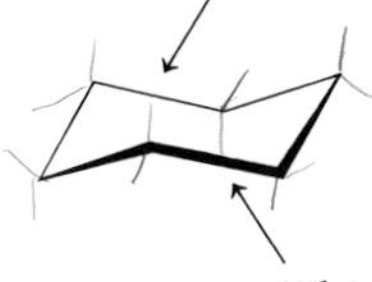

2.10 AXIAL AND EQUATORIAL BONDS IN CYCLOHEXANE

The chair conformation of cyclohexane leads to many important consequences. One such consequence is that there are two kinds of positions for hydrogen atoms on the ring: *axial* positions and *equatorial* positions (Figure 2.11).

As indicated in Figure 2.11 on page 50, cyclohexane has six axial hydrogens, which are perpendicular to the ring (parallel to the ring *axis*), and six equatorial hydrogens, which are more-or-less in the rough plane of the ring (around the ring *equator*). Each carbon atom is bonded to one axial hydrogen and one equatorial hydrogen. Axial and equatorial bonds can be drawn in the following way:

axial bond
the bonds on chair cyclohexane that are roughly perpendicular to the plane of the ring

1. **Axial bonds:** All six axial bonds (one on each carbon) are parallel and alternate between top and bottom faces.

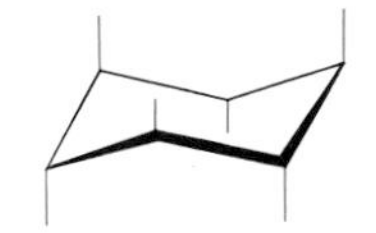

equatorial bond
the bonds on chair cyclohexane that are roughly parallel to the plane of the ring

2. **Equatorial bonds:** The six equatorial bonds (one on each carbon) are represented by three sets of parallel lines. Each set is also parallel to two ring bonds. Equatorial bonds alternate between top and bottom faces in proceeding around the ring.

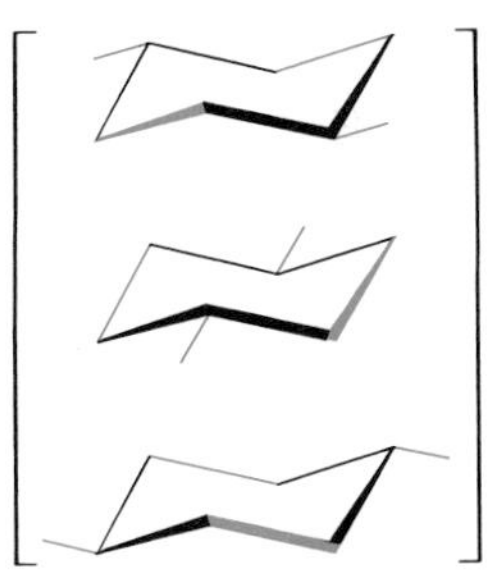

PROBLEM 2.20 Draw two structures for methylcyclohexane, one with the methyl group axial and one with the methyl group equatorial.

2.11 CONFORMATIONAL MOBILITY OF CYCLOHEXANE

Because chair cyclohexane has two kinds of positions, axial and equatorial, we might expect to be able to isolate two isomeric forms of a monosubstituted cyclohexane. In fact, this expectation is wrong: There is only one methylcyclohexane, one bromocyclohexane, and so forth because cyclohexane is *conformationally mobile*. The two chair cyclohexanes can readily interconvert, with the result that

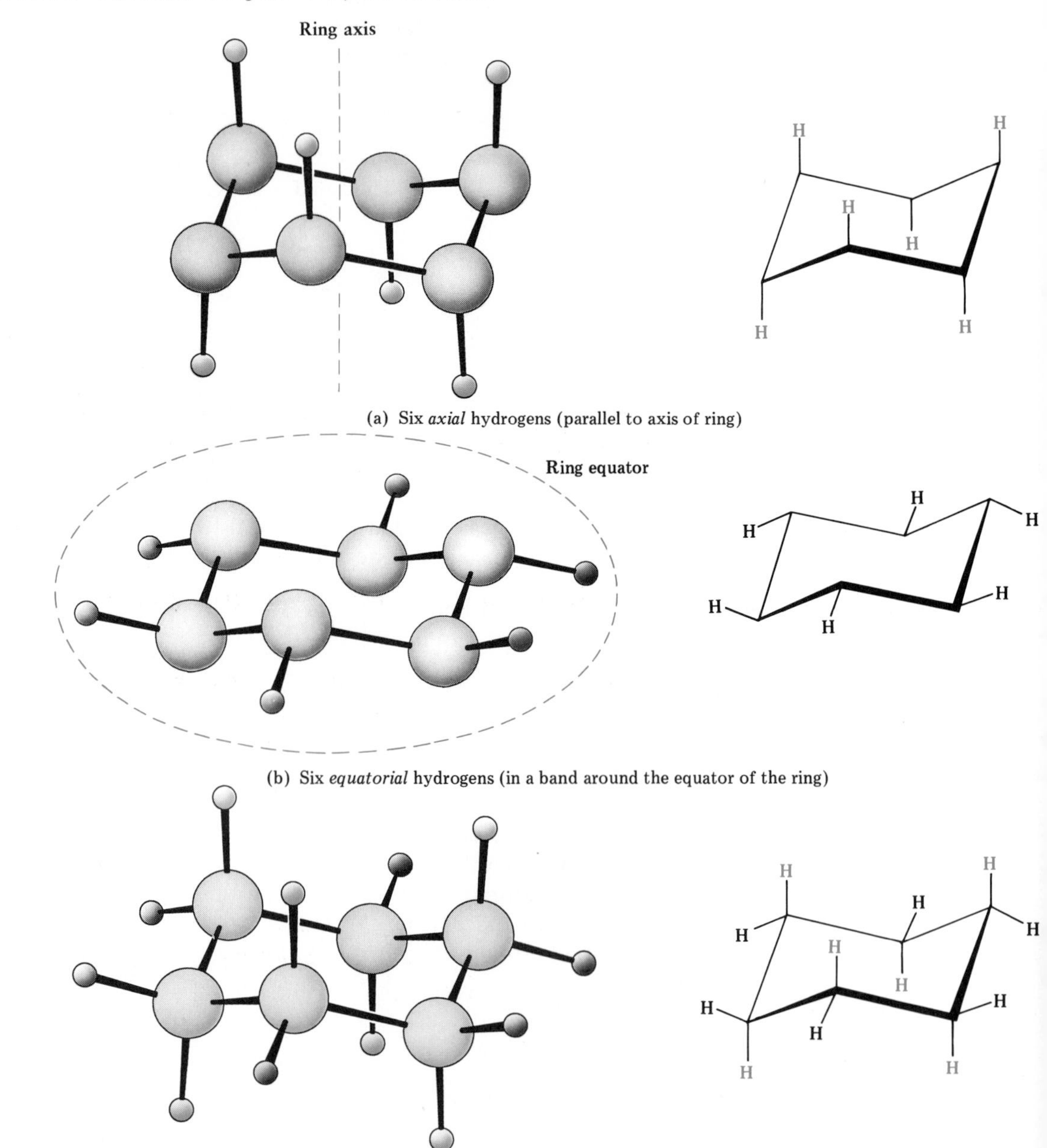

(a) Six *axial* hydrogens (parallel to axis of ring)

(b) Six *equatorial* hydrogens (in a band around the equator of the ring)

(c) Chair cyclohexane with all its hydrogen atoms

FIGURE 2.11 Axial and equatorial hydrogen atoms in cyclohexane

axial and equatorial positions are interchanged. This interconversion of chair conformations, usually referred to as a *ring flip*, is shown in Figure 2.12.

We can mentally ring-flip a chair cyclohexane by holding the middle four carbon atoms in place while folding the two ends in opposite directions. The net result of carrying out a ring-flip is that axial and equatorial positions interconvert. An axial substituent in one chair form becomes an equatorial substituent in the

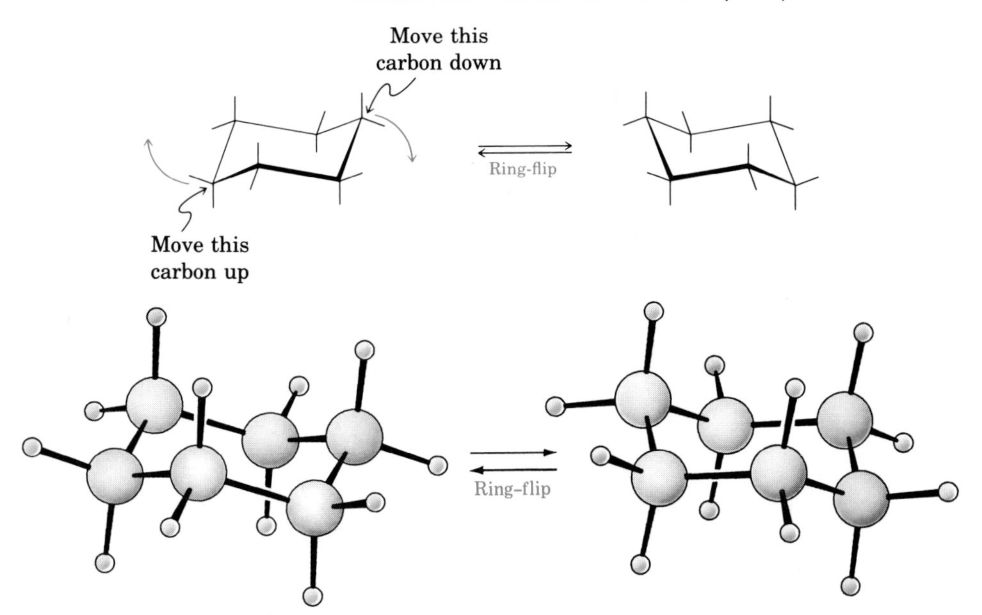

FIGURE 2.12 Cyclohexane ring-flips interconvert axial and equatorial positions

ring-flipped chair form, and vice versa. For example, axial methylcyclohexane becomes equatorial methylcyclohexane after ring-flip. Since this interconversion occurs rapidly at room temperature, we can isolate only an interconverting mixture rather than distinct axial and equatorial isomers.

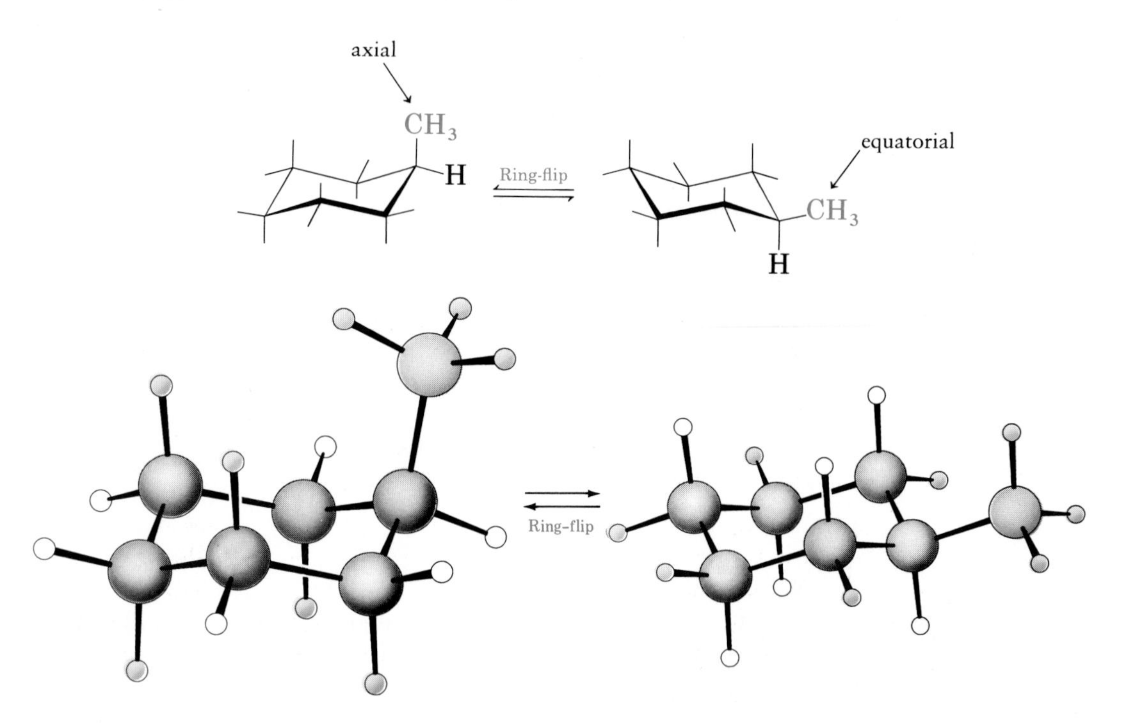

1,3-diaxial interaction the unfavorable spatial interaction between an axial substituent and a hydrogen atom three carbons away in a substituted chair cyclohexane

steric strain the strain in a molecule caused by atoms that try to occupy the same space

Even though axial and equatorial methylcyclohexanes interconvert rapidly, they aren't equally stable. At any instant about 95% of methylcyclohexane molecules have their methyl group equatorial and only 5% have their methyl group axial. Equatorial methylcyclohexane is unstrained, but axial methylcyclohexane has an unfavorable spatial interaction between the methyl group on carbon 1 and the axial hydrogen atoms on carbons 3 and 5 (Figure 2.13). This so-called **1,3-diaxial interaction** introduces 1.8 kcal/mol of **steric** (spatial) **strain** into the molecule because the axial methyl group and the nearby axial hydrogen are too close together and are trying to occupy the same space.

What's true for methylcyclohexane is also true for all other monosubstituted cyclohexanes: A substituent is always more stable in an equatorial position than in an axial position. As you might expect, the amount of steric strain increases as the size of the axial substituent group increases.

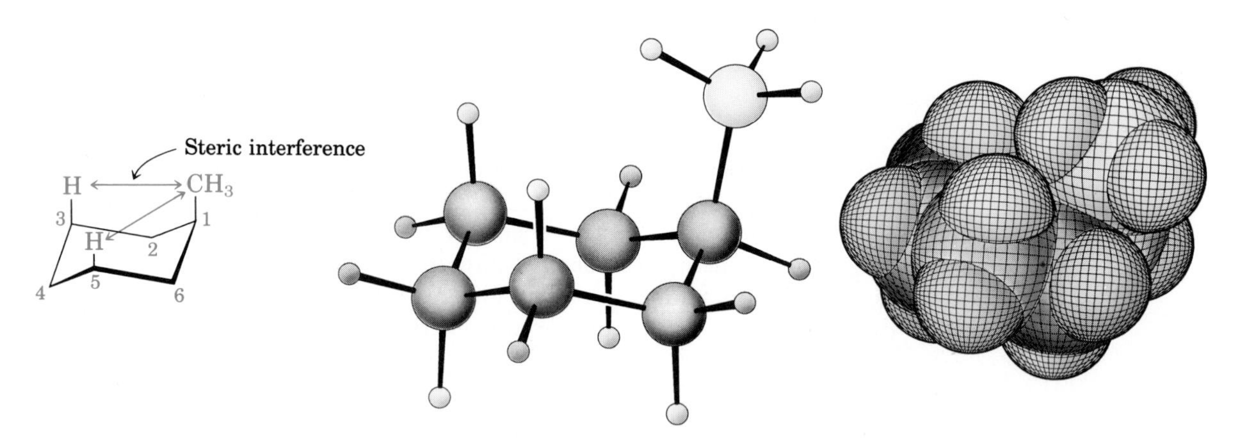

FIGURE 2.13 1,3-Diaxial steric interactions in axial methylcyclohexane

PRACTICE PROBLEM 2.5 Draw 1,1-dimethylcyclohexane, indicating whether each methyl group is axial or equatorial.

SOLUTION First draw a chair cyclohexane ring and then put two methyl groups on the same carbon. The methyl group in the rough plane of the ring must be equatorial and the other (above or below the ring) must be axial.

Axial methyl group
CH_3
CH_3
Equatorial methyl group

PROBLEM 2.21 Draw two different chair conformations of bromocyclohexane showing all hydrogen atoms. Label all positions as axial or equatorial. Which of the two conformations do you think is more stable?

PROBLEM 2.22 Explain why a cis-1,2-disubstituted cyclohexane such as *cis*-1,2-dichlorocyclohexane must have one group axial and one equatorial.

PROBLEM 2.23 Explain why a trans-1,2-disubstituted cyclohexane must either have both groups axial or both equatorial.

INTERLUDE

Petroleum

Many alkanes occur naturally in the plant and animal world. For example, the waxy coating on cabbage leaves contains nonacosane ($C_{29}H_{60}$), and the wood oil of the Jeffrey pine common to the Sierra Nevada mountains contains heptane (C_7H_{16}). By far the major sources of alkanes, however, are the world's natural gas ana petroleum deposits. Laid down eons ago, these natural deposits are thought to be derived from the decomposition of organic matter, primarily of marine origin.

Natural gas consists chiefly of methane but also contains ethane, propane, and butane. Petroleum is a highly complex mixture of hydrocarbons that must be refined before it can be used. Refining begins by fractional distillation of crude petroleum into three principal cuts: straight-run gasoline (boiling point, bp 30–200°C), kerosene (bp 175–300°C), and gas oil (bp 275–400°C). Finally, distillation under reduced pressure yields lubricating oils and waxes and leaves an undistillable tarry residue of asphalt.

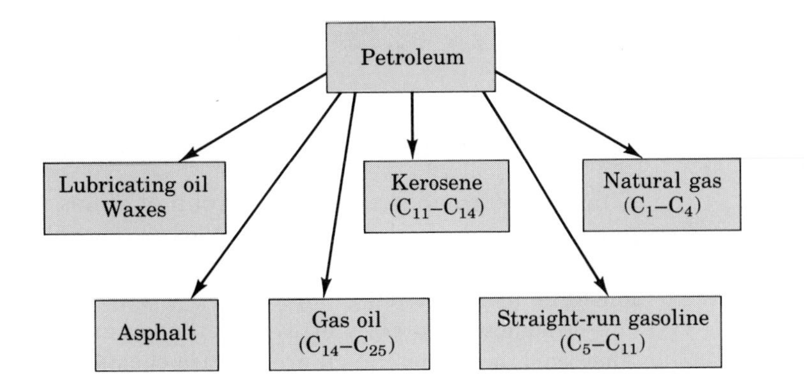

It turns out that straight-run gasoline is a rather poor fuel because of the phenomenon of *engine knock*. Thus, the simple distillation of petroleum into fractions is just the beginning of the process by which automobile fuel is made. In the ordinary automobile engine, a piston draws a mixture of fuel and air into a cylinder on its downward stroke and compresses the mixture on its upward stroke. Just before the end of the compression, a spark plug ignites the fuel and smooth combustion occurs, pushing the piston downward. Not all fuels burn equally well, though. When poor fuels are used, combustion can be initiated in an uncontrolled manner by a hot surface in the cylinder before the spark plug fires. This *preignition,* detected as an engine knock, can destroy the engine by putting irregular forces on the crankshaft.

The *octane number* of a fuel is the measure by which its antiknock properties are judged. It has long been known that straight-chain alkanes are far more prone to induce engine knock than are branched-chain compounds. Heptane, a particularly bad fuel, is assigned a base value of 0 octane number; 2,2,4-trimethylpentane (commonly known as isooctane) has a rating of 100.

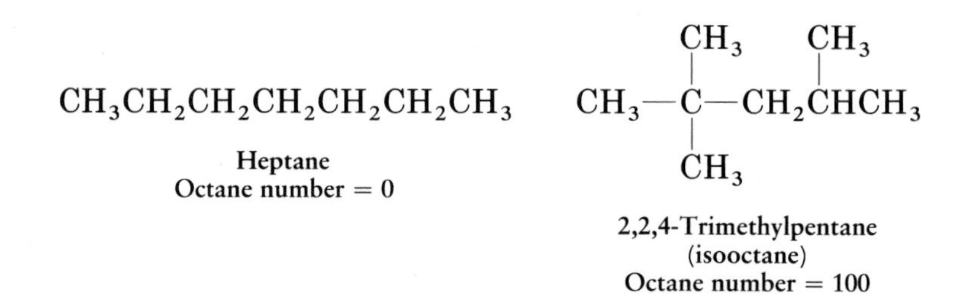

Because straight-run gasoline has a high percentage of unbranched alkanes, it is a poor fuel. Petroleum chemists, however, have devised sophisticated methods for producing better fuels. One of these methods, *catalytic cracking,* involves taking the kerosene cut (C_{11}–C_{14}) and "cracking" it into smaller molecules at high temperature on a silica–alumina catalyst. The major products of cracking are light hydrocarbons in the C_3–C_5 range. These small hydrocarbons are then catalytically recombined to yield C_7–C_{10} branched-chain molecules that are perfectly suited for use as high-octane fuels.

SUMMARY AND KEY WORDS

A **functional group** is an atom or group of atoms within a larger molecule. Functional groups have the same characteristic chemical reactivities in all molecules where they occur. The chemical reactions of an organic molecule are largely determined by its functional groups.

Alkanes are a class of hydrocarbons having the general formula C_nH_{2n+2}. They contain no functional groups, are chemically rather inert, and can be either straight-chain or branched. Isomerism is possible for all but the simplest alkanes. Compounds that have the same chemical formula but different structures are called **isomers.** Compounds such as butane and isobutane, which have the same formula but differ in the order of connection of their atoms, are called **constitutional isomers.** Alkanes can be named by a series of **IUPAC rules of nomenclature.**

As a result of their symmetry, rotation is possible about carbon–carbon single bonds. Alkanes can therefore adopt any of a large number of rapidly interconverting **conformations. Newman projections** allow us to visualize the spatial consequences of bond rotation by sighting directly along a carbon–carbon bond axis. **Staggered conformations** are more stable than **eclipsed conformations.**

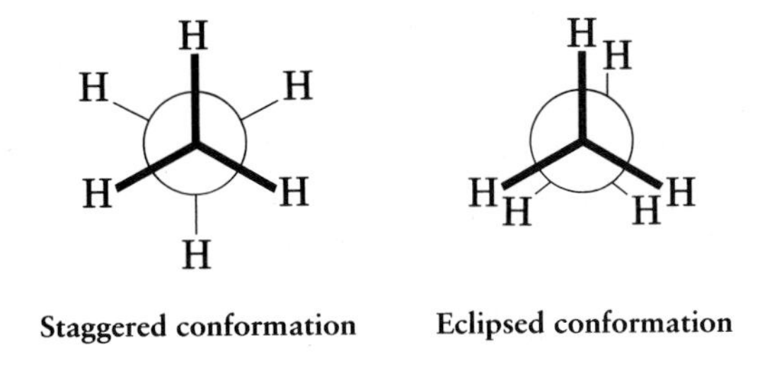

Cycloalkanes, which contain rings of carbon atoms, have the general formula C_nH_{2n}. Conformational mobility (rotation) is reduced in cycloalkanes, and complete rotation around carbon–carbon bonds is not possible. Disubstituted cycloalkanes can therefore exist as **cis–trans isomers.** In the cis isomer both substituents are on the same side of the ring whereas in the trans isomer the substituents are on opposite sides of the ring. Cyclohexanes are the most important of all rings because of their wide occurrence in nature. Cyclohexane is strain-free by virtue of its puckered **chair conformation,** in which all bond angles are 109° and all neighboring C–H bonds are staggered. Chair cyclohexane has two kinds of bonds, axial and equatorial. **Axial bonds** are directed up and down, parallel to the ring axis; **equatorial bonds** lie in a belt around the ring equator. Chair cyclohexanes are conformationally mobile and can undergo a **ring-flip** that interconverts axial and equatorial positions. Substituents on the ring are more stable in the equatorial than in the axial position.

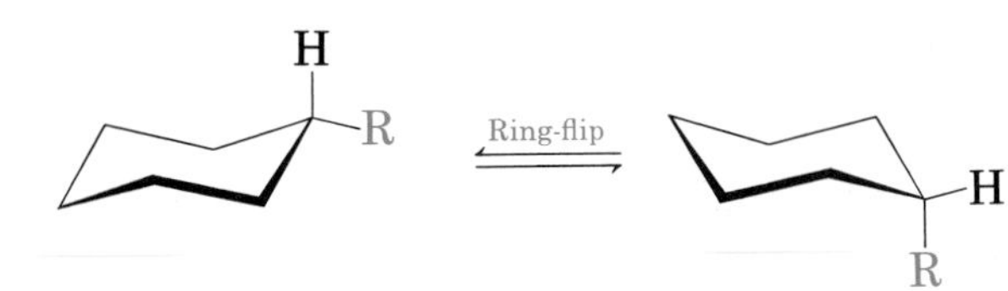

R equatorial, more stable　　　R axial, less stable

ADDITIONAL PROBLEMS

2.24 Locate and identify the functional groups in these molecules:

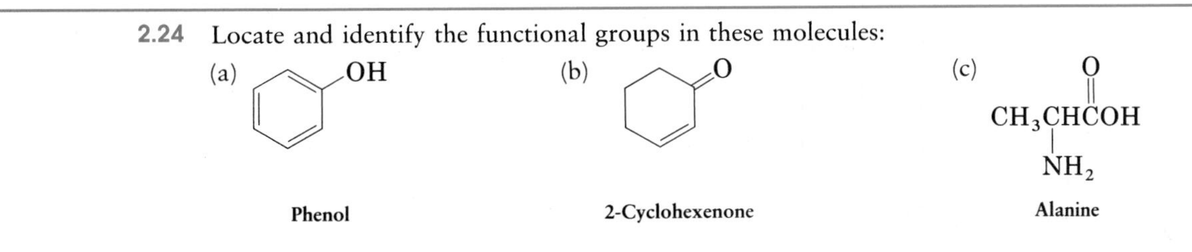

Phenol　　　2-Cyclohexenone　　　Alanine

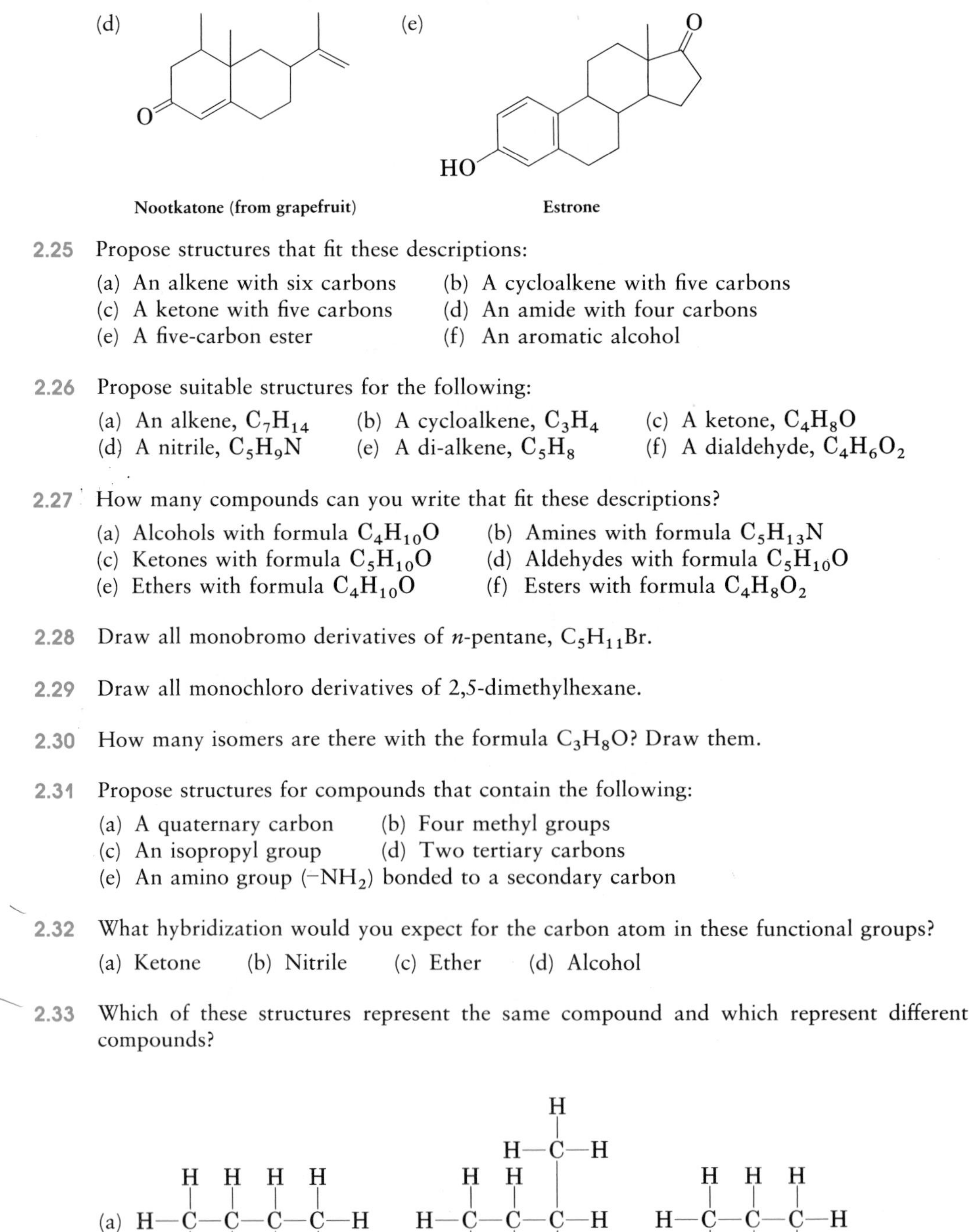

Nootkatone (from grapefruit)

Estrone

2.25 Propose structures that fit these descriptions:

(a) An alkene with six carbons
(b) A cycloalkene with five carbons
(c) A ketone with five carbons
(d) An amide with four carbons
(e) A five-carbon ester
(f) An aromatic alcohol

2.26 Propose suitable structures for the following:

(a) An alkene, C_7H_{14}
(b) A cycloalkene, C_3H_4
(c) A ketone, C_4H_8O
(d) A nitrile, C_5H_9N
(e) A di-alkene, C_5H_8
(f) A dialdehyde, $C_4H_6O_2$

2.27 How many compounds can you write that fit these descriptions?

(a) Alcohols with formula $C_4H_{10}O$
(b) Amines with formula $C_5H_{13}N$
(c) Ketones with formula $C_5H_{10}O$
(d) Aldehydes with formula $C_5H_{10}O$
(e) Ethers with formula $C_4H_{10}O$
(f) Esters with formula $C_4H_8O_2$

2.28 Draw all monobromo derivatives of *n*-pentane, $C_5H_{11}Br$.

2.29 Draw all monochloro derivatives of 2,5-dimethylhexane.

2.30 How many isomers are there with the formula C_3H_8O? Draw them.

2.31 Propose structures for compounds that contain the following:

(a) A quaternary carbon
(b) Four methyl groups
(c) An isopropyl group
(d) Two tertiary carbons
(e) An amino group ($-NH_2$) bonded to a secondary carbon

2.32 What hybridization would you expect for the carbon atom in these functional groups?

(a) Ketone (b) Nitrile (c) Ether (d) Alcohol

2.33 Which of these structures represent the same compound and which represent different compounds?

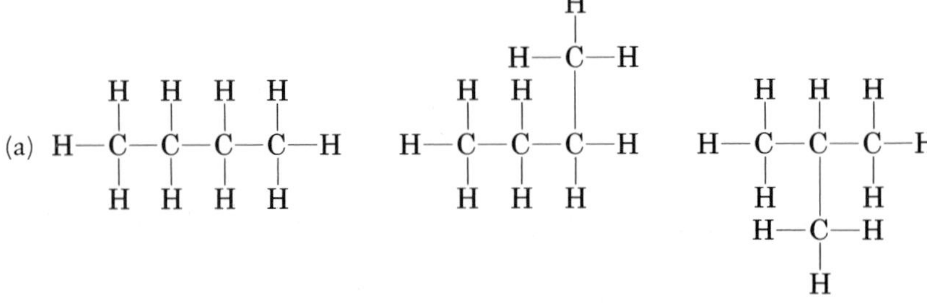

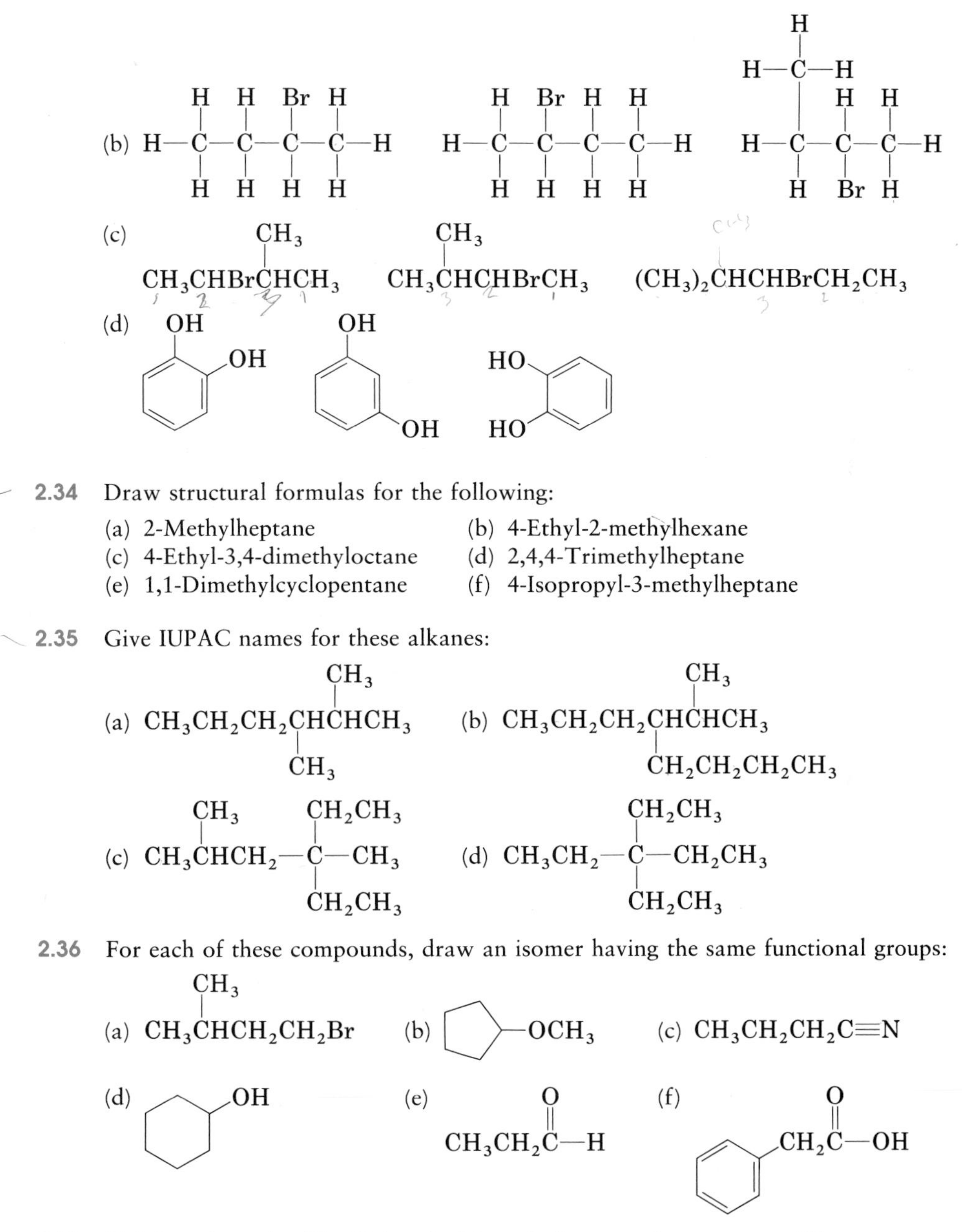

(b) [structural formulas of bromobutanes]

(c) $CH_3CHBrCH(CH_3)CH_3$ $CH_3CH(CH_3)CHBrCH_3$ $(CH_3)_2CHCHBrCH_2CH_3$

(d) [structures of dihydroxybenzenes]

2.34 Draw structural formulas for the following:

(a) 2-Methylheptane
(b) 4-Ethyl-2-methylhexane
(c) 4-Ethyl-3,4-dimethyloctane
(d) 2,4,4-Trimethylheptane
(e) 1,1-Dimethylcyclopentane
(f) 4-Isopropyl-3-methylheptane

2.35 Give IUPAC names for these alkanes:

(a) $CH_3CH_2CH_2CH(CH_3)CH(CH_3)CH_3$
(b) $CH_3CH_2CH_2CH(CH_2CH_2CH_2CH_3)CH(CH_3)CH_3$
(c) $CH_3CH(CH_3)CH_2—C(CH_2CH_3)_2—CH_3$
(d) $CH_3CH_2—C(CH_2CH_3)_2—CH_2CH_3$

2.36 For each of these compounds, draw an isomer having the same functional groups:

(a) $CH_3CH(CH_3)CH_2CH_2Br$
(b) cyclopentyl—OCH_3
(c) $CH_3CH_2CH_2C{\equiv}N$
(d) cyclohexyl—OH
(e) $CH_3CH_2C(=O)—H$
(f) phenyl—$CH_2C(=O)—OH$

2.37 Sighting along the C2–C3 bond of 2-methylbutane, there are two different staggered conformations. Draw them both in Newman projections and say which you think is the more stable. Explain your answer.

2.38 Sighting along the C2–C3 bond of 2-methylbutane (Problem 2.37), there are also two possible eclipsed conformations. Draw them both in Newman projection. Which do you think is lower in energy? Explain.

2.39 *cis*-1-*tert*-Butyl-4-methylcyclohexane exists almost exclusively in the conformation shown. What does this tell you about the relative size of a *tert*-butyl substituent and a methyl substituent?

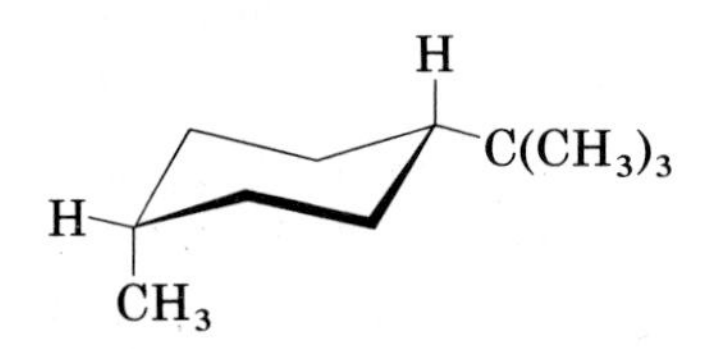

Cis-1-*tert*-butyl-4-methylcyclohexane

2.40 Supply proper IUPAC names for the following:

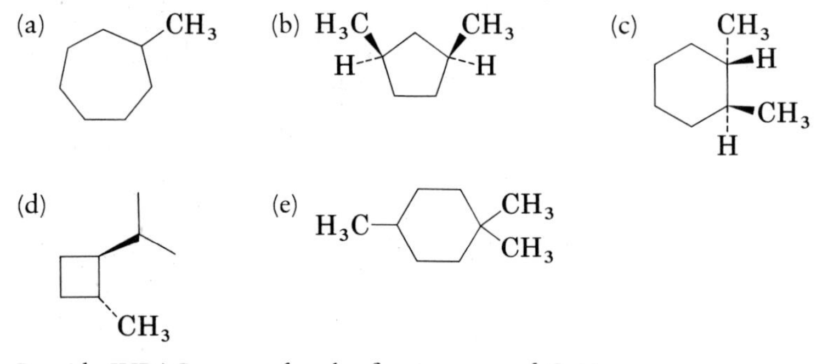

2.41 Provide IUPAC names for the five isomers of C_6H_{14}.

2.42 Draw structures for the nine isomers of C_7H_{16}.

2.43 Propose structures and give correct IUPAC names for the following:
(a) A dimethyloctane (b) A diethyldimethylhexane
(c) A cycloalkane with three methyl groups

2.44 The barrier to rotation about the C–C bond in bromoethane is 3.6 kcal/mol. If each hydrogen–hydrogen interaction in the eclipsed conformation is responsible for 0.9 kcal/mol, how much is the H–Br eclipsing interaction responsible for?

2.45 Malic acid, $C_4H_6O_5$, has been isolated from apples. Because malic acid reacts with two equivalents of base, it can be formulated as a di-carboxylic acid.
(a) Draw at least five possible structures for malic acid.
(b) If malic acid is also a secondary alcohol (has an –OH group attached to a secondary carbon), what is its structure?

2.46 Cyclopropane was first prepared by reaction of 1,3-dibromopropane with sodium.
(a) Formulate the reaction.
(b) What product might this reaction give? What geometry would you expect for the product?

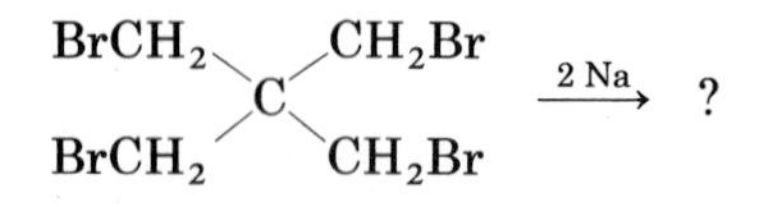

2.47 Draw *trans*-1,2-dimethylcyclohexane in its most stable chair conformation. Are the methyl groups axial or equatorial?

2.48 Draw *cis*-1,2-dimethylcyclohexane in its most stable chair conformation. Are the methyl groups axial or equatorial? Which do you think is more stable, *cis*-1,2-dimethylcyclohexane or *trans*-1,2-dimethylcyclohexane (Problem 2.47)?

2.49 Which do you think is more stable, *cis*-1,3-dimethylcyclohexane or *trans*-1,3-dimethylcyclohexane? Draw chair conformations of both and explain your answer.

2.50 *N*-Methylpiperidine is known to have the conformation shown. What does this tell you about the relative steric requirements of a methyl group versus an electron lone pair?

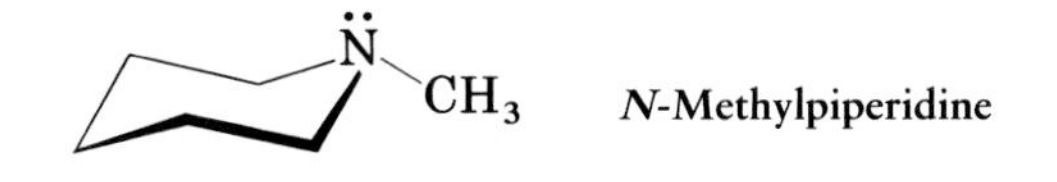

2.51 Glucose contains a six-membered ring in which all of the substituents are equatorial. Draw glucose in its more stable chair conformation.

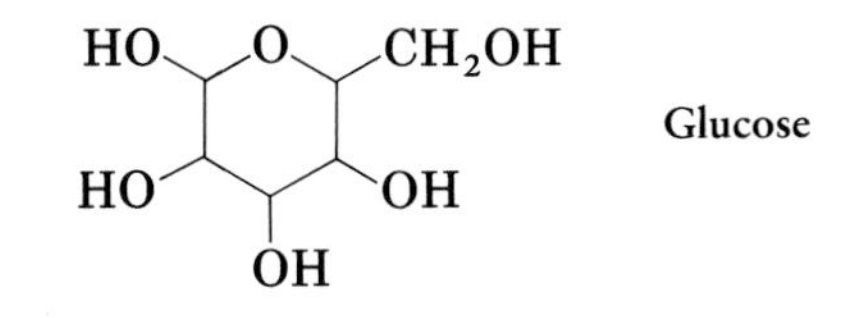

CHAPTER 3

Alkenes: The Nature of Organic Reactions

alkene
a hydrocarbon with one or more carbon–carbon double bonds

Alkenes are hydrocarbons that contain a carbon–carbon double bond functional group. They occur abundantly in nature, and many have important biological roles. For example, ethylene is a plant hormone that induces ripening in fruit, and α-pinene is the major constituent of turpentine.

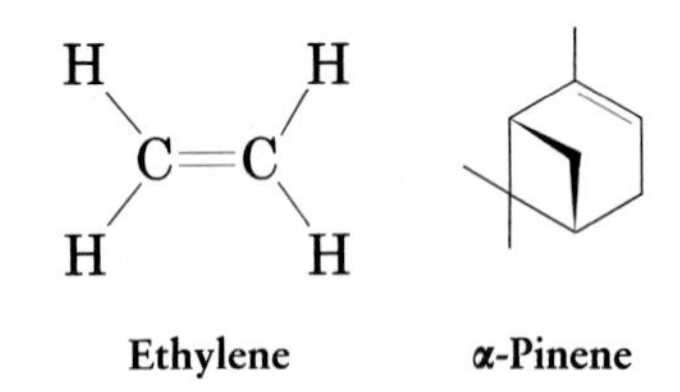

Ethylene **α-Pinene**

We'll see in this chapter how and why alkenes behave the way they do, and we'll develop some general ideas about organic chemical reactivity that can be applied to all molecules.

3.1 NAMING ALKENES

unsaturated
containing one or more double or triple bonds

Because of their double bond, alkenes have fewer hydrogens per carbon than related alkanes and are therefore referred to as **unsaturated.** Ethylene, for example, has the formula C_2H_4 whereas ethane has the formula C_2H_6.

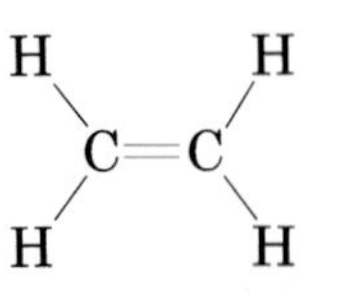

Ethylene, C_2H_4
(fewer hydrogens: *unsaturated*)

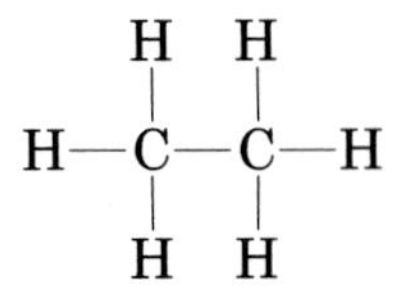

Ethane, C_2H_6
(more hydrogens: *saturated*)

Alkenes are named according to a series of rules similar to those used for naming alkanes, with the suffix *-ene* used in place of *-ane* to identify the family. There are three steps:

Step 1. Name the parent hydrocarbon. Find the longest carbon chain that contains the double bond and name the compound accordingly, using the suffix *-ene*.

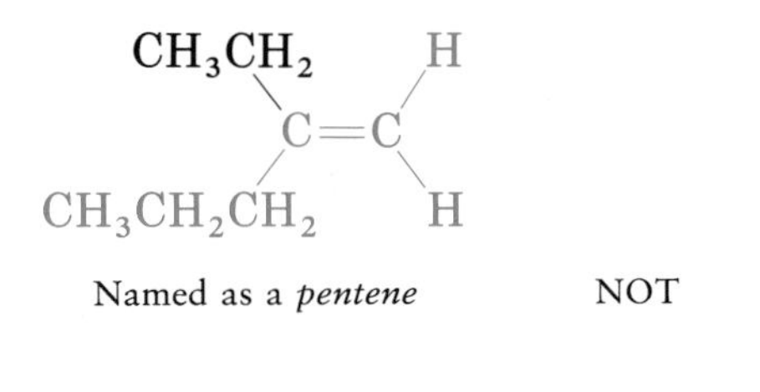

Named as a *pentene*

NOT

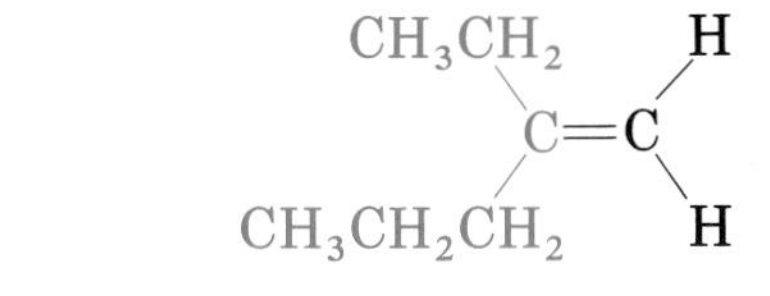

as a hexene, because the double bond is not contained in the six-carbon chain.

Step 2. Number the carbon atoms in the chain, beginning at the end nearer the double bond. If the double bond is equidistant from the two ends, begin numbering at the end nearer the first branch point:

$$\underset{6}{CH_3}\underset{5}{CH_2}\underset{4}{CH_2}\underset{3}{CH}{=}\underset{2}{CH}\underset{1}{CH_3} \qquad \underset{1}{CH_3}\underset{2}{\overset{CH_3}{\overset{|}{C}}H}\underset{3}{CH}{=}\underset{4}{CH}\underset{5}{CH_2}\underset{6}{CH_3}$$

Step 3. Write the full name, numbering the substituents according to their position in the chain and listing them alphabetically. Indicate the position of the double bond by giving the number of the *first* alkene carbon. If more than one double bond is present, give the position of each and use one of the suffixes *-diene, -triene,* and so on.

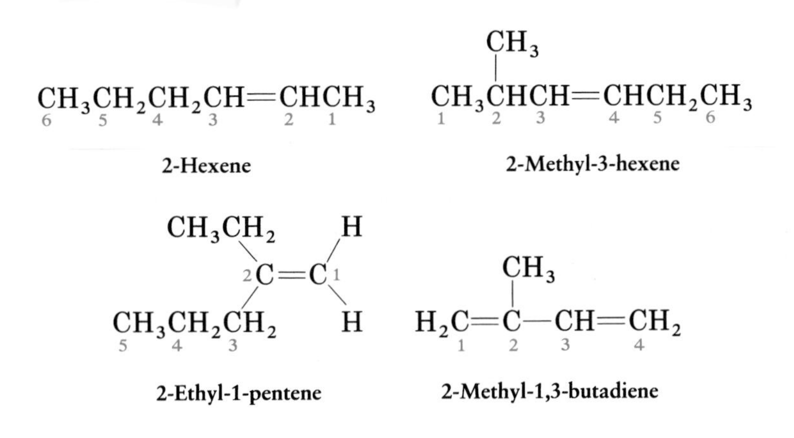

2-Hexene **2-Methyl-3-hexene**

2-Ethyl-1-pentene **2-Methyl-1,3-butadiene**

Cycloalkenes are named in a similar way, but because there is no chain end to begin from, we number the cycloalkene so that the double bond is between C1 and C2 and the first substituent has as low a number as possible:

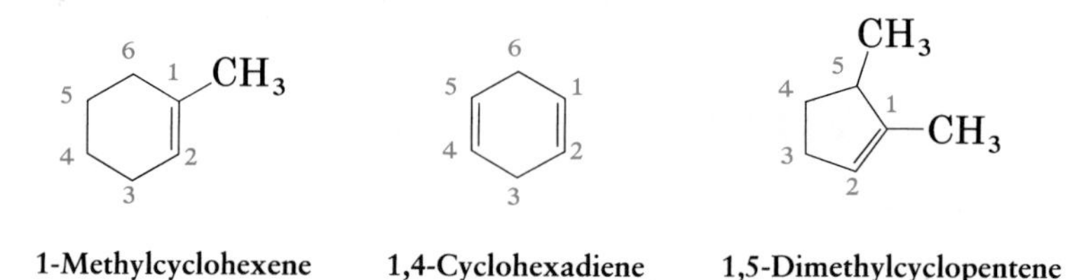

For historical reasons, there are a few alkenes whose names don't conform to the rules. For example, the alkene corresponding to ethane should be called *ethene*, but the name *ethylene* has been used for so long that it is accepted by IUPAC. Table 3.1 lists some other common names.

TABLE 3.1 Common names of some alkenes

Compound	*Systematic name*	*Common name*
$H_2C{=}CH_2$	Ethene	Ethylene
$CH_3CH{=}CH_2$	Propene	Propylene
$CH_3C(CH_3){=}CH_2$	2-Methylpropene	Isobutylene
$H_2C{=}C(CH_3){-}CH{=}CH_2$	2-Methyl-1,3-butadiene	Isoprene

PRACTICE PROBLEM 3.1

What is the IUPAC name of this alkene:

$$CH_3{-}C(CH_3)_2{-}CH_2CH_2CH{=}C(CH_3)CH_3$$

SOLUTION First, find the longest chain containing the double bond. In this case, it's a heptene.

Next, number the chain, beginning at the end nearer the double bond, and identify the substituents at each position. In this case, there are methyl groups at C2 and C6 (two).

$$\underset{7}{CH_3}{-}\underset{6}{C}(CH_3)_2{-}\underset{5}{C}H_2\underset{4}{C}H_2\underset{3}{C}H{=}\underset{2}{C}(CH_3)\underset{1}{C}H_3$$

The full name is 2,6,6-trimethyl-2-heptene.

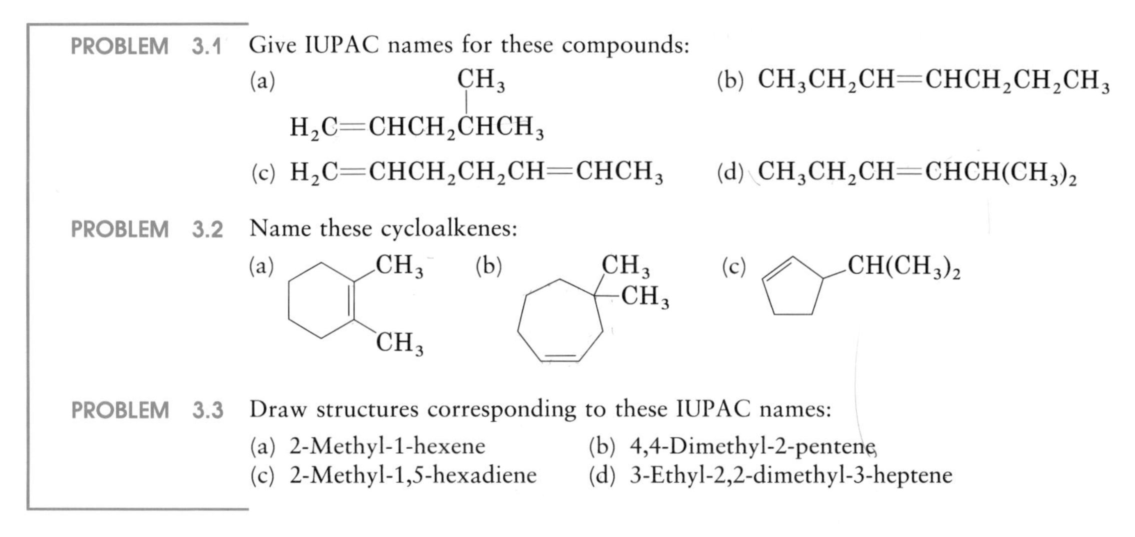

PROBLEM 3.1 Give IUPAC names for these compounds:

(a) $H_2C{=}CHCH_2CH(CH_3)CH_3$ (b) $CH_3CH_2CH{=}CHCH_2CH_2CH_3$

(c) $H_2C{=}CHCH_2CH_2CH{=}CHCH_3$ (d) $CH_3CH_2CH{=}CHCH(CH_3)_2$

PROBLEM 3.2 Name these cycloalkenes:

PROBLEM 3.3 Draw structures corresponding to these IUPAC names:

(a) 2-Methyl-1-hexene (b) 4,4-Dimethyl-2-pentene

(c) 2-Methyl-1,5-hexadiene (d) 3-Ethyl-2,2-dimethyl-3-heptene

3.2 ELECTRONIC STRUCTURE OF ALKENES

We saw in Section 1.10 that the carbon atoms in a double bond are sp^2 hybridized and have three equivalent orbitals that lie in a plane at angles of 120° to one another. The fourth carbon orbital is an unhybridized *p* orbital, which is perpendicular to the sp^2 plane. When two such carbon atoms approach each other, they form two kinds of bonds: a sigma bond, formed by head-on overlap of sp^2 orbitals, and a pi bond, formed by sideways overlap of *p* orbitals. The doubly bonded carbons and the four atoms attached to them lie in a plane, with bond angles of approximately 120° (Figure 3.1).

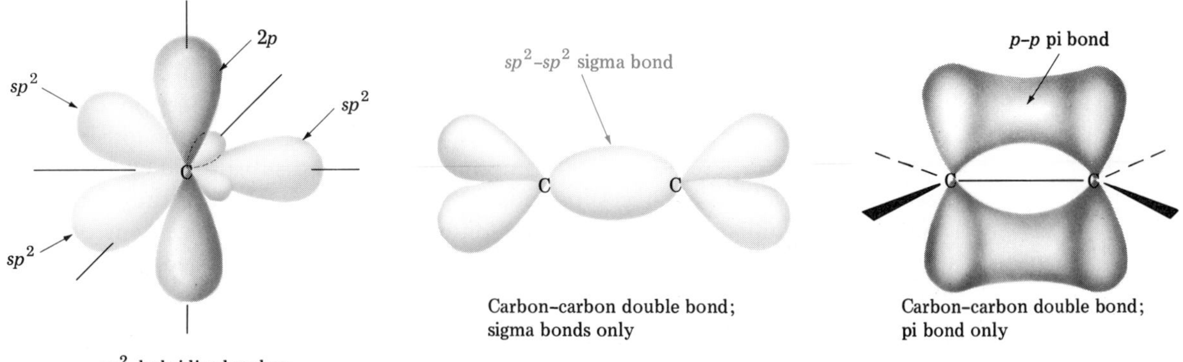

FIGURE 3.1 An orbital picture of the carbon–carbon double bond

We know from Section 2.5 that free rotation is possible around single bonds, and that open-chain alkanes like ethane and propane therefore have many rapidly interconverting conformations. The same is not true for double bonds, however. No rotation can take place around carbon–carbon double bonds because doing so would break the pi part of the bond (Figure 3.2). In fact, the energy barrier to rotation around a double bond is as large as the strength of the pi bond itself, an estimated 65 kcal/mol. (Recall that the rotation barrier for a single bond is only about 2.9 kcal/mol.)

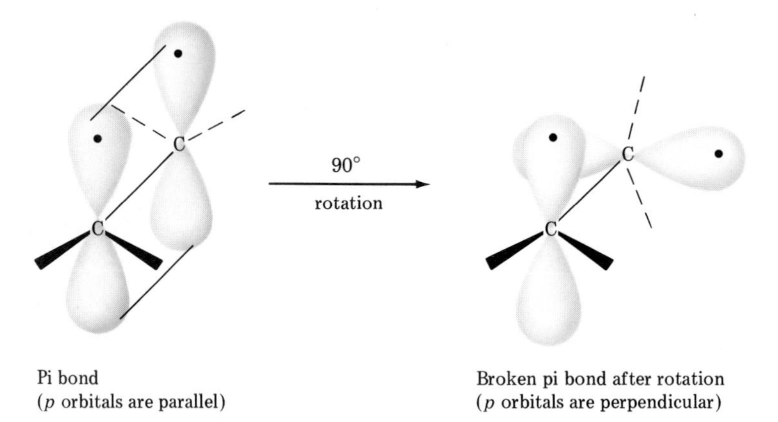

FIGURE 3.2 Breaking the pi bond is necessary for rotation around a carbon–carbon double bond to take place

3.3 CIS–TRANS ISOMERS

The lack of rotation around carbon–carbon double bonds is of more than just theoretical interest; it also has chemical consequences. Imagine the situation for a disubstituted alkene like 2-butene. (*Disubstituted* means that two substituents other than hydrogen are linked to the double-bond carbons.) In 2-butene, the two methyl groups can be either on the same side of the double bond or on opposite sides (Figure 3.3), a situation reminiscent of substituted cycloalkanes (Section 2.7).

Since bond rotation can't occur, the two 2-butenes don't spontaneously interconvert; they are different chemical compounds. As with disubstituted cycloalkanes (Section 2.7), we call such compounds *cis–trans isomers* because they have the same formula and overall skeleton but differ in the spatial arrangement of their atoms. The isomer with both substituents on the same side is called *cis*-2-butene, and the isomer with substituents on opposite sides is *trans*-2-butene.

Cis–trans isomerism is not limited to disubstituted alkenes: It can occur whenever both of the double-bond carbons are attached to two different groups. If one of the double-bond carbons is attached to two identical groups, however, then cis–trans isomerism is not possible (Figure 3.4).

Although the cis–trans interconversion of alkene isomers doesn't occur spontaneously, it can be made to happen by treating the alkene with a strong-acid catalyst. If we interconvert *cis*-2-butene with *trans*-2-butene and allow them to

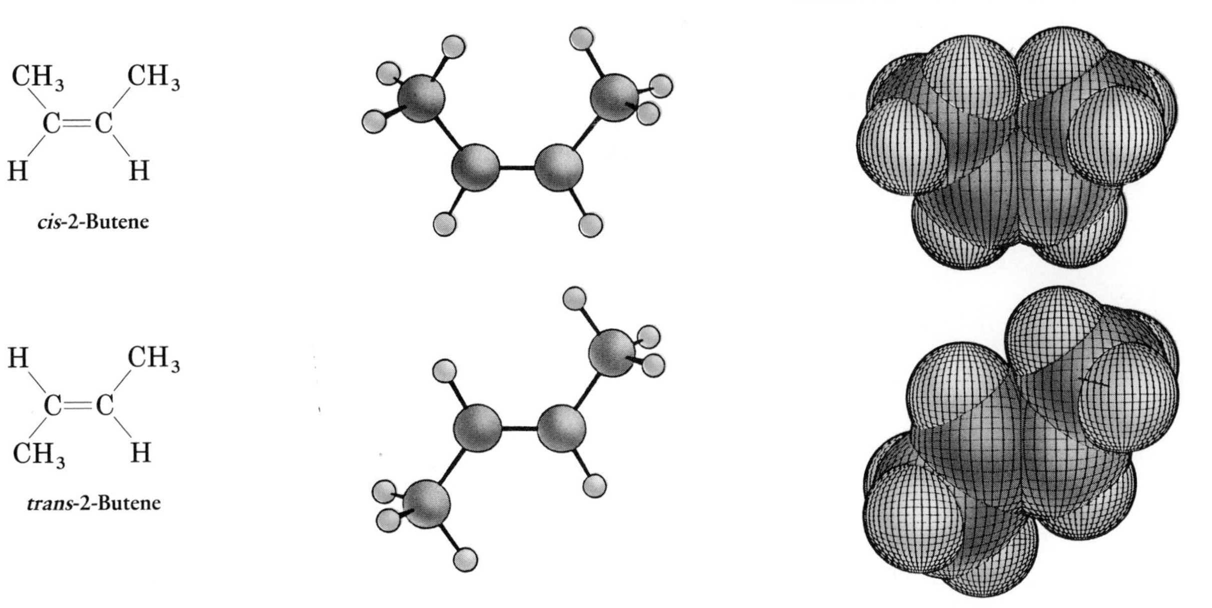

FIGURE 3.3 Cis and trans isomers of 2-butene. The cis isomer has the substituent methyl groups on the same side of the double bond, and the trans isomer has the methyl groups on opposite sides of the double bond.

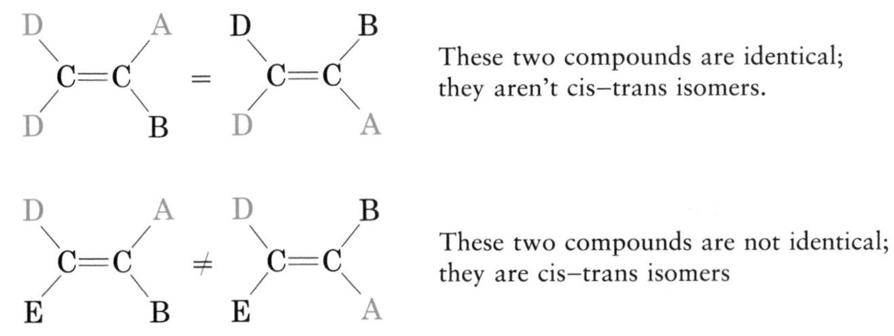

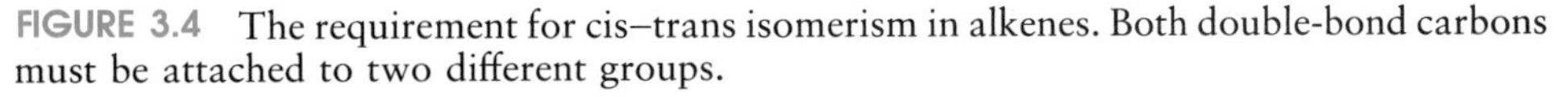
FIGURE 3.4 The requirement for cis–trans isomerism in alkenes. Both double-bond carbons must be attached to two different groups.

reach equilibrium, we find that they aren't of equal stability. At equilibrium, the trans isomer is more favored than the cis isomer by a ratio of 76% trans to 24% cis.

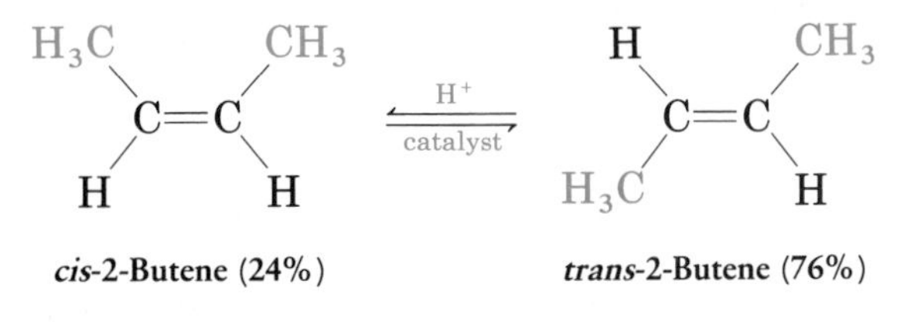

Cis alkenes are less stable than their trans isomers because of spatial interference between the bulky substituents on the same side of the double bond. This is the same kind of interference, or *steric strain,* that we saw in axial methylcyclohexane (Section 2.11).

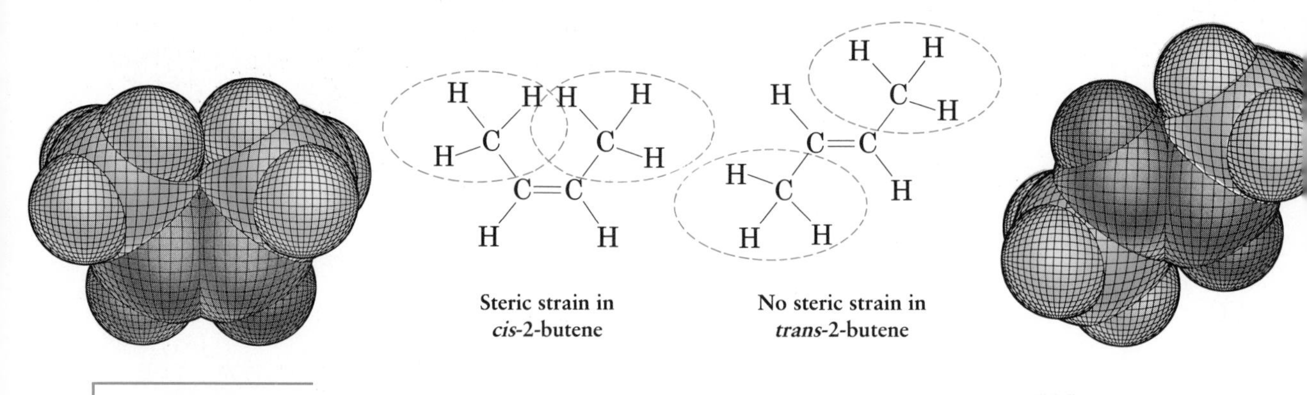

Steric strain in *cis*-2-butene

No steric strain in *trans*-2-butene

PRACTICE PROBLEM 3.2 Draw the cis and trans isomers of 5-chloro-2-pentene.

SOLUTION 5-Chloro-2-pentene is $ClCH_2CH_2CH{=}CHCH_3$. The chloroethyl and methyl groups are on the same side of the double bond in one isomer and on opposite sides in the other isomer.

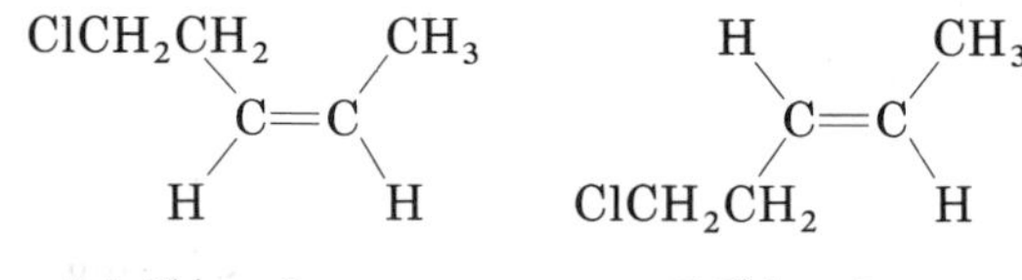

***cis*-5-Chloro-2-pentene** ***trans*-5-Chloro-2-pentene**

PROBLEM 3.4 Which of these compounds can exist as pairs of cis–trans isomers? Draw each cis–trans pair.

(a) $CH_3CH{=}CH_2$ (b) $(CH_3)_2C{=}CHCH_3$ (c) $ClCH{=}CHCl$
(d) $CH_3CH_2CH{=}CHCH_3$ (e) $CH_3CH_2CH{=}CBrCH_3$ (f) 3-Methyl-3-heptene

PROBLEM 3.5 Which is more stable, *cis*-2-methyl-3-hexene or *trans*-2-methyl-3-hexene?

PROBLEM 3.6 How can you account for the observation that cyclohexene does not show cis–trans isomerism?

3.4 SEQUENCE RULES: THE *E,Z* DESIGNATION

In the discussion of isomerism in the 2-butenes, we used the terms *cis* and *trans* to specify alkenes whose two substituents were on the same side and opposite sides of a double bond, respectively. This cis–trans naming system is unambiguous for disubstituted alkenes, but how do we denote the geometry of *tri*substituted and *tetra*substituted double bonds? (*Trisubstituted* means three substituents other than hydrogen on the double bond; *tetrasubstituted* means four substituents other than hydrogen on the double bond.)

sequence rules a series of rules for assigning priorities to groups so that double-bond geometry can be specified

The answer is provided by the *E,Z* system of nomenclature, which uses a system of **sequence rules** to assign priorities to the groups on the double-bond carbons. Considering each of the double-bond carbons separately, we use the sequence rules to decide which of the two groups on each carbon is higher in priority. If the higher-priority groups are on the same side of the double bond, the alkene is designated *Z* (for the German *zusammen*, "together"). If the higher-priority groups are on opposite sides, the alkene is designated *E* (for the German *entgegen*, "opposite"). One way to remember which is which is to remember that *Z* = groups on ze zame zide (*E* = the other one). These assignments are shown in Figure 3.5.

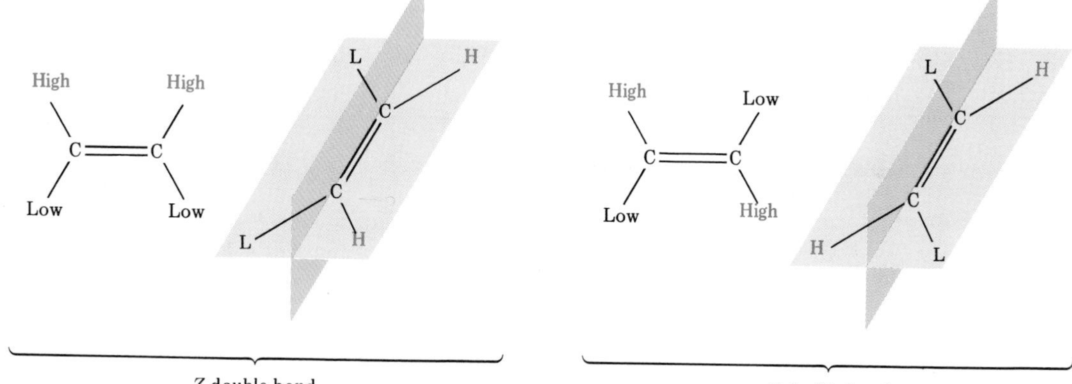

FIGURE 3.5 The *E,Z* system of nomenclature

The sequence rules used in assigning priorities are as follows:

Sequence rule 1 Look at the atoms directly attached to each of the double-bond carbons and rank them in order of decreasing atomic number. That is, an atom with a high atomic number like Cl receives higher priority than an atom with a low atomic number like H. Thus, the atoms that we commonly find attached to a double-bond carbon are assigned priorities as follows:

$$\overset{35}{\mathrm{Br}} > \overset{17}{\mathrm{Cl}} > \overset{8}{\mathrm{O}} > \overset{7}{\mathrm{N}} > \overset{6}{\mathrm{C}} > \overset{1}{\mathrm{H}}$$

For example:

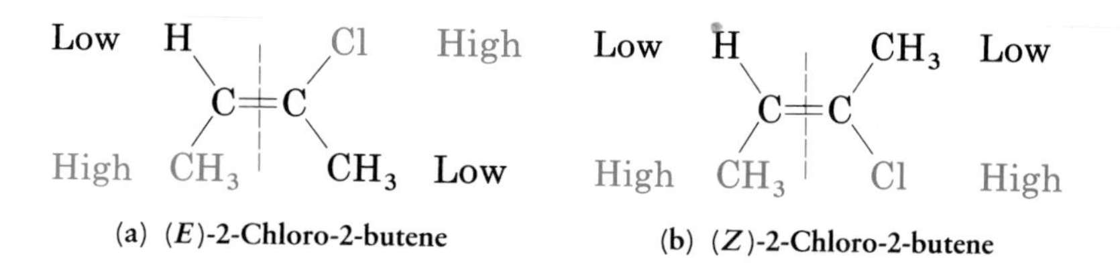

(a) (*E*)-2-Chloro-2-butene (b) (*Z*)-2-Chloro-2-butene

Because chlorine has a higher atomic number than carbon, it receives higher priority than a methyl (CH_3) group. Methyl receives higher priority than hydrogen, however, and isomer (a) is therefore assigned *E* geometry (high-priority groups on opposite sides of the double bond). Isomer (b) has *Z* geometry (high-priority groups on ze zame zide of the double bond).

Sequence rule 2 If a decision can't be reached by ranking the first atoms in the substituents (rule 1), look at the second, third, or fourth atoms away from the double-bond carbons until a difference is found. Thus, an ethyl substituent, $-CH_2CH_3$, and a methyl substituent, $-CH_3$, are equivalent by rule 1 since both have carbon as the first atom. But by rule 2, ethyl receives higher priority than methyl because the *second* atoms are one carbon (and two hydrogens) rather than three hydrogens. Look at the following examples to see how this rule is applied:

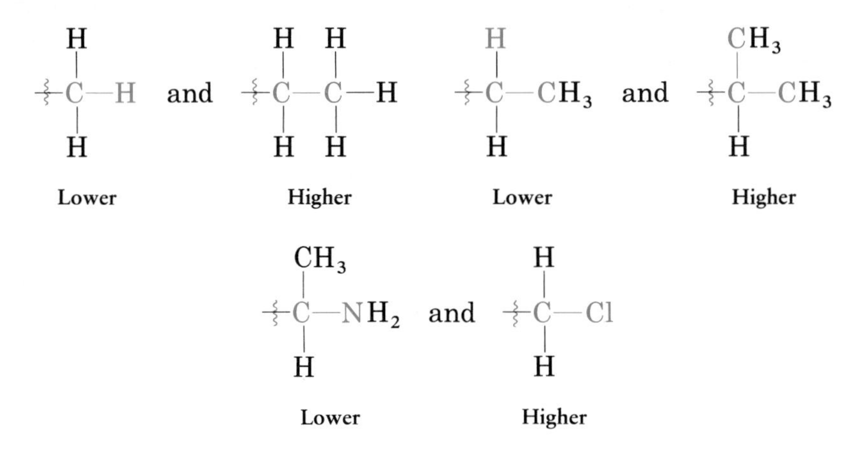

Sequence rule 3 Multiple-bonded atoms are considered to be equivalent to the same number of single-bonded atoms. For example, an aldehyde substituent ($-CH{=}O$), which has a carbon atom *doubly* bonded to *one* oxygen, is considered equivalent to a substituent having a carbon atom *singly* bonded to *two* oxygens.

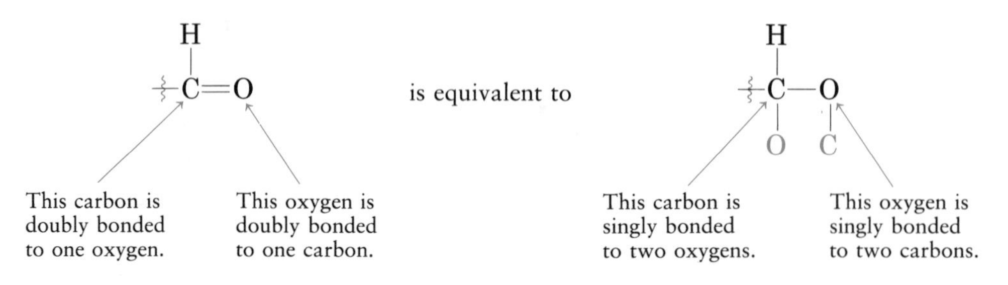

In the following examples, the indicated pairs are equivalent:

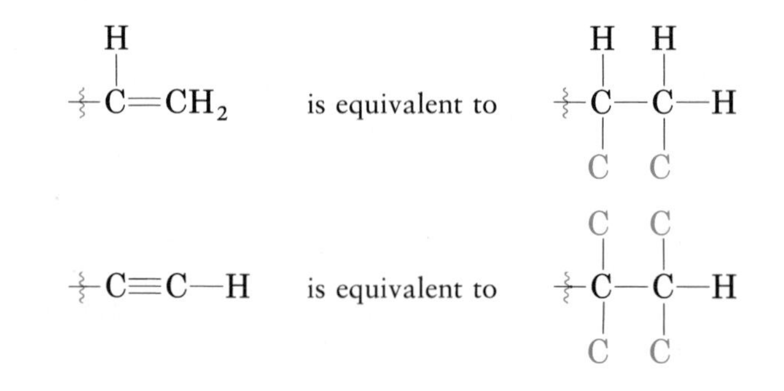

PRACTICE PROBLEM 3.3 Assign *E* or *Z* configuration to the double bond in this compound:

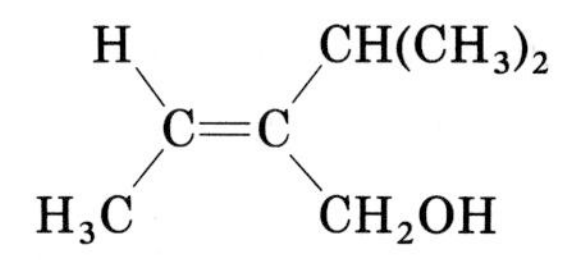

SOLUTION Look at the two double-bond carbons individually. The left-hand carbon has two substituents, $-H$ and $-CH_3$, of which $-CH_3$ receives higher priority by rule 1. The right-hand carbon also has two substituents, $-CH(CH_3)_2$ and $-CH_2OH$, but rule 1 does not allow a priority assignment to be made since both groups have carbon as their first atom. By rule 2, however, $-CH_2OH$ receives higher priority than $-CH(CH_3)_2$, since $-CH_2OH$ has an *oxygen* (and two hydrogens) as its second atom whereas $-CH(CH_3)_2$ has two *carbons* (and one hydrogen) as its second atom. Thus, the two high-priority groups are on the same side of the double bond, and we assign *Z* configuration.

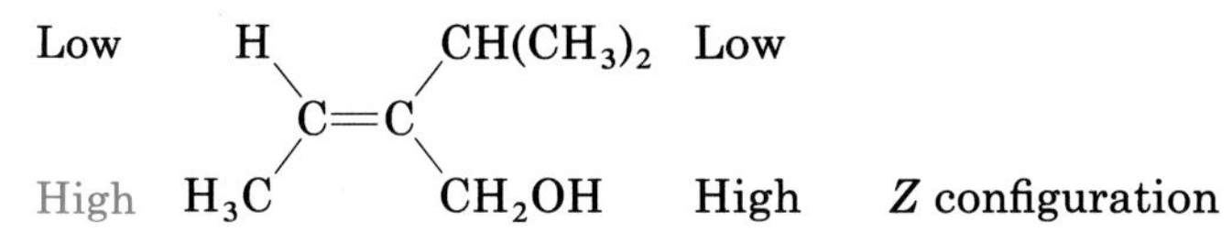

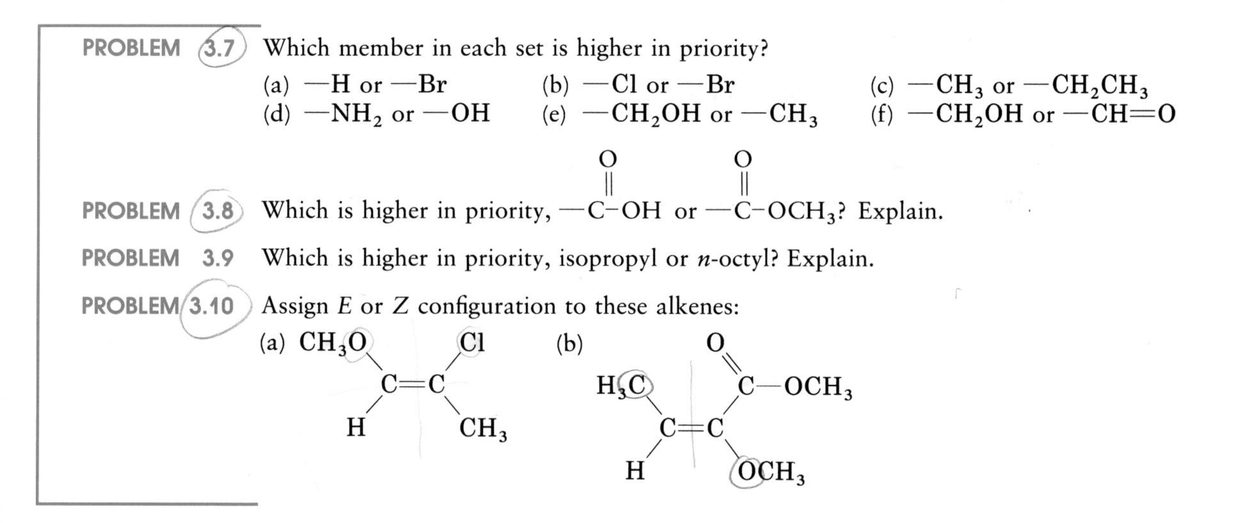

PROBLEM 3.7 Which member in each set is higher in priority?

(a) $-H$ or $-Br$ (b) $-Cl$ or $-Br$ (c) $-CH_3$ or $-CH_2CH_3$
(d) $-NH_2$ or $-OH$ (e) $-CH_2OH$ or $-CH_3$ (f) $-CH_2OH$ or $-CH{=}O$

PROBLEM 3.8 Which is higher in priority, $-\overset{O}{\overset{\|}{C}}-OH$ or $-\overset{O}{\overset{\|}{C}}-OCH_3$? Explain.

PROBLEM 3.9 Which is higher in priority, isopropyl or *n*-octyl? Explain.

PROBLEM 3.10 Assign *E* or *Z* configuration to these alkenes:

3.5 KINDS OF ORGANIC REACTIONS

Now that we know something about alkenes and the double-bond functional group, it's time to learn about their chemical reactivity. As an introduction, though, we'll first look at some of the basic principles that underlie all organic reactions. In particular, we'll develop some general notions about why compounds react the way they do and see some methods that have been developed to help understand how reactions take place.

Organic chemical reactions can be organized in two ways: by *what kinds* of reactions occur and by *how* reactions occur. We'll begin by looking at the kinds of reactions that take place. There are four particularly important kinds of organic reactions: additions, eliminations, substitutions, and rearrangements.

addition reaction the reaction that occurs when two reagents combine together to form a single new product with no atoms left over

Addition reactions occur when two reactants add together to form a single new product with no atoms "left over." We can generalize the process as

$$A + B \longrightarrow C$$

These two reactants add together . . . . . . to give this single product.

As an example of an important addition reaction that we'll be studying soon, alkenes react with HCl to yield alkyl chlorides:

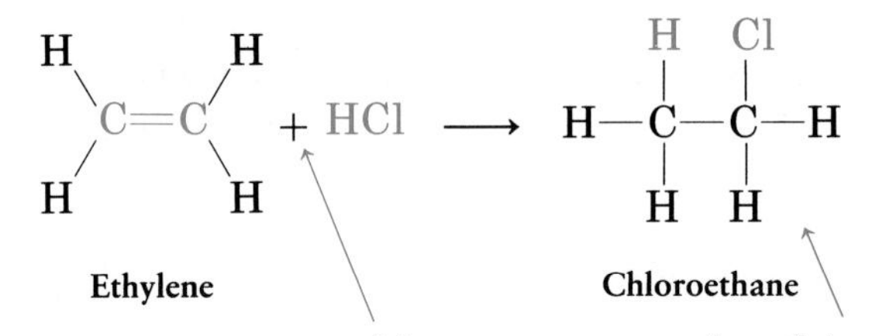

These two reactants add . . . . . . to give this product.

elimination reaction the reaction that occurs when a single reactant splits apart into two products

Elimination reactions are the opposite of addition reactions. They occur when a single reactant splits into two products, a process we can generalize as

$$A \longrightarrow B + C$$

This one reactant . . . . . . splits apart to give these two products.

As an example of an important elimination reaction, alkyl halides split apart into an acid and an alkene when treated with base:

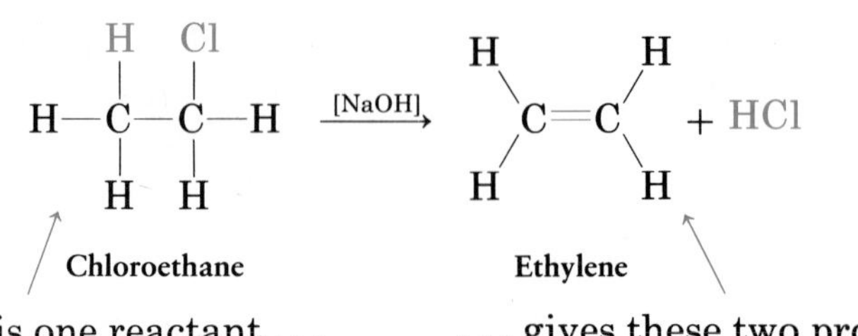

substitution reaction the reaction that occurs when two reactants exchange parts to give two products

Substitution reactions occur when two reactants exchange parts to give two new products, a process we can generalize as

$$A{-}B + C{-}D \longrightarrow A{-}C + B{-}D$$

These two reactants . . . . . . exchange parts to give these two new products.

As an example of a substitution reaction, we saw in Section 2.4 that alkanes react with chlorine gas in the presence of ultraviolet light to yield alkyl chlorides. A $-Cl$ group from chlorine replaces (substitutes for) the $-H$ group of the alkane, and two new products result:

$$\underset{\text{Methane}}{\text{H}-\overset{\displaystyle \text{H}}{\underset{\displaystyle \text{H}}{\text{C}}}-\text{H}} + \text{Cl}-\text{Cl} \longrightarrow \underset{\text{Chloromethane}}{\text{H}-\overset{\displaystyle \text{H}}{\underset{\displaystyle \text{H}}{\text{C}}}-\text{Cl}} + \text{H}-\text{Cl}$$

These two reactants . . . . . . give these two products.

rearrangement reaction the reaction that occurs when a single reactant undergoes a reorganization of bonds and atoms to give a single isomeric product

Rearrangement reactions occur when a single reactant undergoes a reorganization of bonds and atoms to yield a single isomeric product, a process we can generalize as

$$A \longrightarrow B$$

This single reactant . . . . . . gives this isomeric product.

As an example of a rearrangement reaction, we saw in Section 3.3 that 1-butene can be converted into its isomer 2-butene by treatment with an acid catalyst:

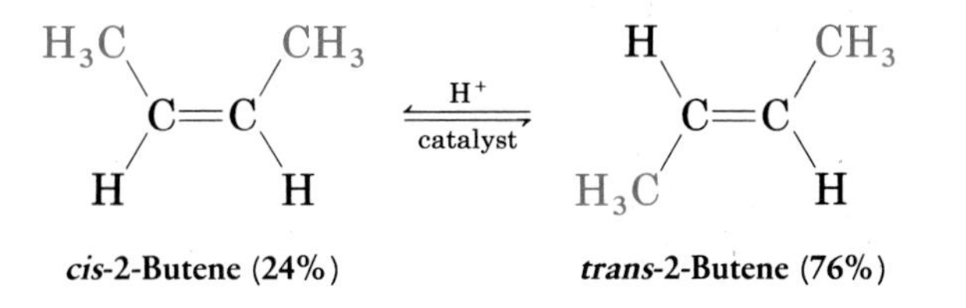

cis-2-Butene (24%) *trans*-2-Butene (76%)

PROBLEM 3.11 Classify these reactions as additions, eliminations, substitutions, or rearrangements.

(a) $CH_3Br + KOH \longrightarrow CH_3OH + KBr$
(b) $CH_3CH_2OH \longrightarrow H_2C{=}CH_2 + H_2O$
(c) $H_2C{=}CH_2 + H_2 \longrightarrow CH_3CH_3$

3.6 HOW REACTIONS OCCUR: MECHANISMS

reaction mechanism a description of the details by which a reaction occurs

Having looked at the kinds of reactions that take place, let's now see how reactions occur. An overall description of how a reaction occurs is called a **reaction mechanism.** A mechanism describes in detail exactly what takes place at each stage of a chemical transformation. It describes which bonds are broken and in what order, which bonds are formed and in what order, and what the relative rate of each step is.

All chemical reactions involve bond breaking and bond making. When two starting materials come together, react, and yield products, certain chemical bonds in the starting materials are broken, and new bonds are formed to make the products. Fundamentally, there are only two ways in which a covalent two-electron bond can break: an electronically *symmetrical* way such that one electron remains with each product fragment or an electronically *unsymmetrical* way such that both electrons remain with one product fragment, leaving the other fragment with an empty orbital. The symmetrical cleavage is called a **homolytic** process, and the unsymmetrical cleavage is called a **heterolytic** process:

homolytic electronically symmetrical breaking of a covalent bond to yield two radicals

heterolytic electronically unsymmetrical breaking of a covalent bond to yield an anion and a cation

$$\mathrm{A{:}B} \longrightarrow \mathrm{A\cdot} + \mathrm{\cdot B} \quad \text{Radical bond-breaking (homolytic)}$$

$$\mathrm{A{:}B} \longrightarrow \mathrm{A^+} + \mathrm{{:}B^-} \quad \text{Polar bond-breaking (heterolytic)}$$

Conversely, there are two ways in which a covalent two-electron bond can form: an electronically symmetrical (**homogenic**) way, when one electron is donated to the new bond by each reactant, or an electronically unsymmetrical (**heterogenic**) way, when both bonding electrons are donated to the new bond by one reactant:

homogenic electronically symmetrical formation of a covalent bond by combination of two radicals

heterogenic electronically symmetrical formation of a covalent bond by combination of an anion and a cation

$$\mathrm{A\cdot} + \mathrm{\cdot B} \longrightarrow \mathrm{A{:}B} \quad \text{Radical bond-making (homogenic)}$$

$$\mathrm{A^+} + \mathrm{{:}B^-} \longrightarrow \mathrm{A{:}B} \quad \text{Polar bond-making (heterogenic)}$$

Processes that involve symmetrical bond breaking and making are called **radical reactions.** A **radical** is a chemical species that contains an odd number of valence electrons and thus has an orbital with only one electron. Processes that involve unsymmetrical bond breaking and making are called **polar reactions.** Polar reactions always involve species that contain an even number of valence electrons and have only electron pairs in their orbitals. Polar processes are the more common reaction type, and a large part of this book is devoted to their description.

radical reaction a reaction that involves electronically symmetrical bond making and bond breaking

radical a species that contains an odd number of electrons

polar reaction a reaction that involves electronically unsymmetrical bond making and bond breaking

organometallic containing a carbon–metal bond

To see how polar reactions occur, we need to recall our discussion of polar covalent bonds and to look more deeply into the effects of bond polarity on organic molecules. We saw in Section 1.12 that certain bonds in a molecule, particularly the bonds in functional groups, often have an unsymmetrical distribution of electrons and are therefore polar. When a carbon atom bonds to an electronegative atom such as chlorine or oxygen, the bond is polarized in such a way that the carbon bears a partial positive charge (δ^+) and the electronegative atom bears a partial negative charge (δ^-). Conversely, when carbon bonds to an atom that is less electronegative than itself, the opposite polarity results. Such is the case with most carbon–metal (**organometallic**) bonds:

Where Y = O, N, Cl, Br, I **Where M = a metal such as Mg or Li**

What does bond polarity mean with respect to chemical reactions? Because species with unlike charges attract each other, *the fundamental characteristic of all polar reactions is that the electron-rich sites in one molecule react with the electron-poor sites in another molecule.* Covalent bonds form in a polar reaction when the electron-rich reactant donates a *pair* of electrons to the electron-poor reactant; conversely, covalent bonds break in polar reactions when one of the two product fragments leaves with the electron *pair.*

nucleophile an electron-rich reagent that can donate an electron pair to an electrophile in a polar reaction

electrophile an electron-poor reagent that can accept an electron pair from a nucleophile in a polar reaction

Chemists usually indicate the electron movement that occurs during a polar reaction by curved arrows. By convention, a curved arrow means that an electron pair moves from the tail to the head of the arrow during the reaction. In referring to polar reactions, chemists have coined the words *nucleophile* and *electrophile.* A **nucleophile** is a reagent that is "nucleus loving"; it has an electron-rich site and forms a bond by donating an electron pair to an electron-poor site. An **electrophile,** by contrast, is "electron-loving"; it has an electron-poor site and forms a bond by accepting an electron pair from a nucleophile.

The curved arrow shows that electrons are moving from $:B^-$ to A^+.

$$A^+ + :B^- \longrightarrow A:B$$

Electrophile (electron-poor) — Nucleophile (electron-rich)

PRACTICE PROBLEM 3.4 What is the direction of bond polarity in the amine functional group, $C{-}NH_2$?

SOLUTION According to the electronegativity table (Table 1.4), nitrogen is more electronegative than carbon. Thus an amine is polarized with carbon as δ^+ and nitrogen as δ^-.

PROBLEM 3.12 What is the direction of bond polarity in these functional groups?

(a) Ketone (b) Alkyl chloride (c) Alcohol (d) Alkyllithium

PROBLEM 3.13 Identify the functional groups and show the direction of bond polarity in each of these molecules.

(a) Acetone, $CH_3\overset{\overset{O}{\|}}{C}CH_3$ (b) Chloroethane, CH_3CH_2Cl

(c) Methanethiol, CH_3SH (d) Tetraethyllead, $(CH_3CH_2)_4Pb$ (the "lead" in gasoline)

PROBLEM 3.14 Which of the following would you expect to behave as electrophiles and which as nucleophiles? Explain.

(a) H^+ (b) $H\ddot{\underset{..}{O}}:^-$ (c) Br^+

(d) $:NH_3$ (e) $H{-}C{\equiv}C{-}H$ (f) CO_2

3.7 AN EXAMPLE OF A POLAR REACTION: ADDITION OF HCl TO ETHYLENE

Let's look in detail at a typical polar reaction, the reaction of ethylene with HCl. When ethylene is treated with hydrogen chloride at room temperature, chloroethane is produced. Overall, the reaction can be formulated as follows:

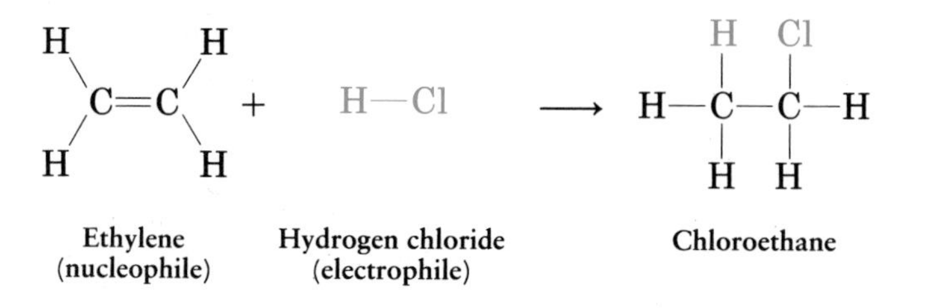

electrophilic addition the addition of an electrophile to an unsaturated acceptor, usually an alkene

This reaction, an example of a general polar reaction type known as an **electrophilic addition,** can be understood in terms of the general concepts just discussed. We'll begin by looking at the natures of the two reactants.

What do we know about ethylene? We know from Sections 1.10 and 3.2 that a carbon–carbon double bond results from orbital overlap of two sp^2-hybridized carbon atoms: The sigma part of the double bond results from sp^2–sp^2 overlap, and the pi part results from p–p overlap.

What kind of chemical reactivity might we expect of carbon–carbon double bonds? We know that alkanes are rather inert because all of their valence electrons are tied up in strong, nonpolar, carbon–carbon and carbon–hydrogen bonds. Furthermore, alkane bonding electrons are inaccessible to external reagents because they are localized in sigma orbitals between atoms.

The situation for ethylene and other alkenes is quite different. For one thing, double bonds have greater electron density than single bonds: four electrons in a double bond versus only two electrons in a single bond. Equally important, the electrons in the pi bond are accessible to external reagents because they are located above and below the plane of the double bond rather than between the nuclei (Figure 3.6).

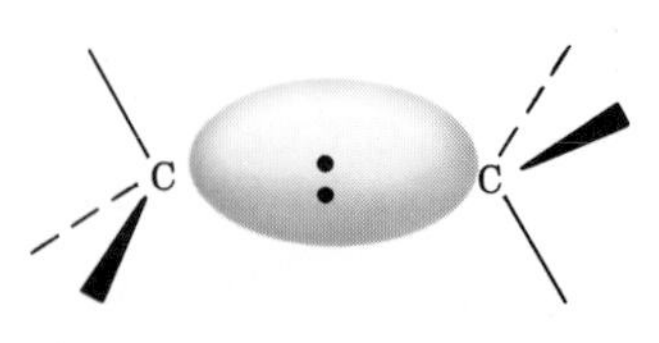

Carbon–carbon sigma bond: strong; inaccessible bonding electrons

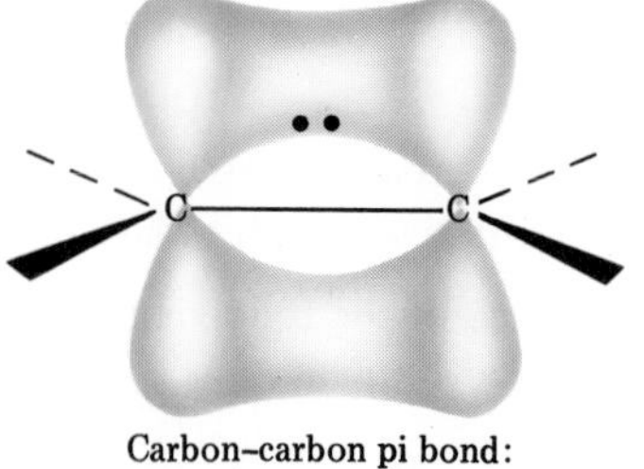

Carbon–carbon pi bond: weak; accessible electrons

FIGURE 3.6 A comparison of carbon–carbon single and double bonds. A double bond is both more electron-rich and more accessible to external reagents than a single bond.

Both electron richness and electron accessibility lead us to predict high reactivity for carbon–carbon double bonds. In the terminology of polar reactions used earlier, we might predict that carbon–carbon double bonds should behave as nucleophiles. That is, the chemistry of alkenes should involve reaction of the electron-rich double bond with electron-poor reagents. This is exactly what we find: The most important reaction of alkenes is their reaction with electrophiles.

What about HCl? As a strong acid, HCl is a powerful proton (H^+) donor. Since a proton is positively charged and electron-poor, it is a good electrophile. Thus, the reaction between H^+ and ethylene is a typical electrophile–nucleophile combination, characteristic of all polar reactions.

3.8 THE MECHANISM OF AN ORGANIC REACTION: ADDITION OF HCl TO ETHYLENE

We can view the electrophilic addition reaction between ethylene and HCl as proceeding by the mechanism shown in Figure 3.7. The reaction takes place in two steps, beginning with an attack on the electrophile, H^+, by the electron pair from the nucleophilic ethylene pi bond. Two electrons from the pi bond form a new sigma bond between the entering hydrogen and one of the ethylene carbons, as shown by tracing the path of the curved arrow in Figure 3.7. (Remember: A curved arrow is used to indicate how electrons move in a polar reaction. In this case, the

The electrophile H^+ is attacked by the pi electrons of the double bond, and a new C–H sigma bond is formed. This leaves the other carbon atom with a + charge and a vacant p orbital.

Cl$^-$ donates an electron pair to the positively charged carbon atom, forming a C–Cl sigma bond and yielding the neutral addition product.

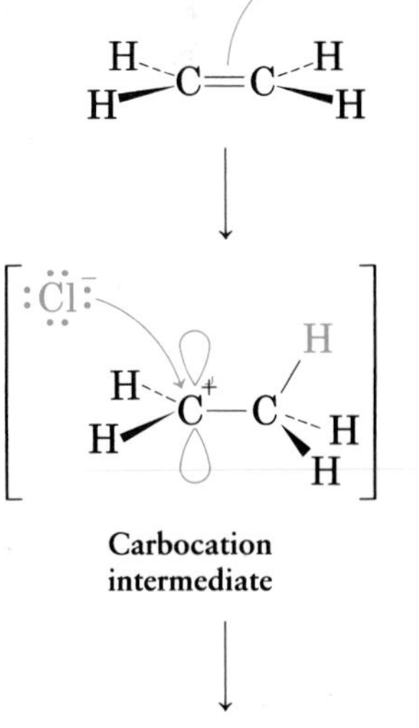

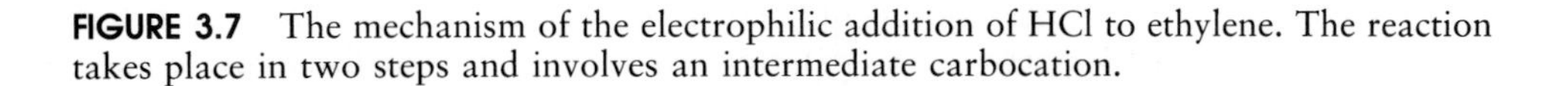
FIGURE 3.7 The mechanism of the electrophilic addition of HCl to ethylene. The reaction takes place in two steps and involves an intermediate carbocation.

electrons move away from the carbon–carbon pi bond to form a new bond with the incoming H^+.) The other ethylene carbon atom, having lost its share of the pi electrons, is now trivalent and is left with a vacant *p* orbital. Since the double-bond pi electrons were used in the formation of the new C–H bond, the trivalent carbon has only six valence electrons and therefore carries a positive charge. In the second step, this positively charged species, a carbon-cation or **carbocation,** is itself an electrophile that can accept an electron pair from the nucleophilic chloride anion to form a C–Cl bond, yielding the neutral addition product.

carbocation a species that has a positively charged, trivalent carbon atom

PRACTICE PROBLEM 3.5 What product would you expect from reaction of HCl with cyclohexene?

SOLUTION HCl should add to the double-bond functional group in cyclohexene in exactly the same way it adds to ethylene, yielding an addition product.

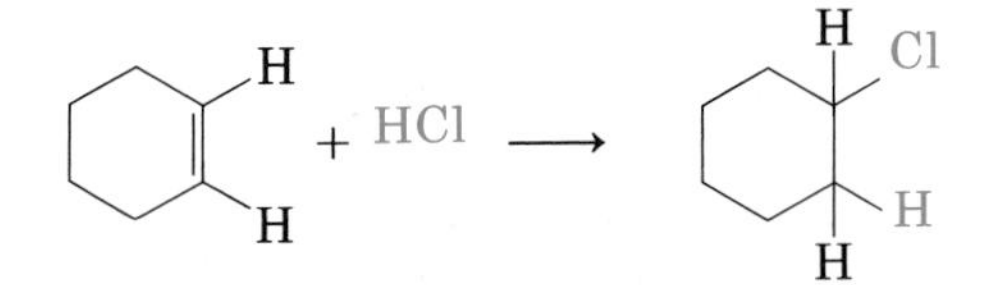

PROBLEM 3.15 Reaction of HCl with 2-methylpropene yields 2-chloro-2-methylpropane. Formulate the mechanism of the reaction. What is the structure of the carbocation formed during the reaction?

$$(CH_3)_2C{=}CH_2 + HCl \longrightarrow (CH_3)_3C{-}Cl$$

PROBLEM 3.16 Reaction of HCl with 2-pentene yields a mixture of two addition products. Write the reaction and show the two products.

3.9 DESCRIBING A REACTION: RATES AND EQUILIBRIA

Every chemical reaction can go in two directions. Starting materials can react to give products, and products can revert to starting materials. We usually express the resultant chemical equilibrium by an equation in which K_{eq}, the **equilibrium constant,** is equal to the concentration of products, divided by the concentration of starting materials. For the reaction,

equilibrium constant a value that expresses the extent to which a given reaction takes place at equilibrium

$$A + B \rightleftharpoons C + D$$

we have

$$K_{eq} = \frac{[\text{Products}]}{[\text{Reactants}]} = \frac{[C][D]}{[A][B]}$$

The equilibrium constant tells us the position of the equilibrium, that is, which side of the reaction arrow is energetically favored. If K_{eq} is large, then the product concentrations [C] and [D] are larger than the reactant concentrations [A] and [B], and the reaction proceeds as written from left to right. Conversely, if K_{eq} is small, the reaction does not take place as written but instead goes from right to left.

What the equilibrium equation does not tell us is the rate of the reaction: how fast the equilibrium is established. Some reactions are extremely slow even though they have highly favorable equilibrium constants. For example, gasoline is stable when stored because its reaction rate with oxygen is slow under normal circumstances. Under the proper reaction conditions, however (contact with a lighted match, for example), gasoline reacts rapidly with oxygen and undergoes complete conversion to water and carbon dioxide. Rates (*how fast* a reaction occurs) and equilibria (*how much* a reaction occurs) are two entirely different things.

Rate ⟶ Is reaction fast or slow?

Equilibrium ⟶ In what direction does reaction proceed?

exothermic a favorable reaction that gives off energy (heat)

endothermic an unfavorable reaction that absorbs energy (heat)

What determines whether a reaction takes place? For a reaction to have a favorable equilibrium constant, the energy level of the products must be lower than the energy level of the reactants. In other words, energy (heat) must be given off. Such reactions are said to be **exothermic** (from the Greek *exo*, "outside," and *therme*, "heat"). Heat is produced during exothermic reactions. If the energy level of the products is higher than the energy level of the reactants, then the equilibrium constant for the reaction is unfavorable, and heat must be added to make the reaction take place. Such reactions are said to be **endothermic** (Greek *endon*, "within").

A good analogy for the relationship between energy and chemical reactivity (stability) is that of a rock poised near the top of a hill. The rock, in its unstable position, has stored the energy that was required to get it up there. When it rolls downhill, it releases its energy until it reaches a stable, low-energy position at the bottom of the hill. In the same way, the energy level in a chemical reaction goes downhill as the energy stored in the chemical bonds of a reactant is released and a more stable product is formed (Figure 3.8).

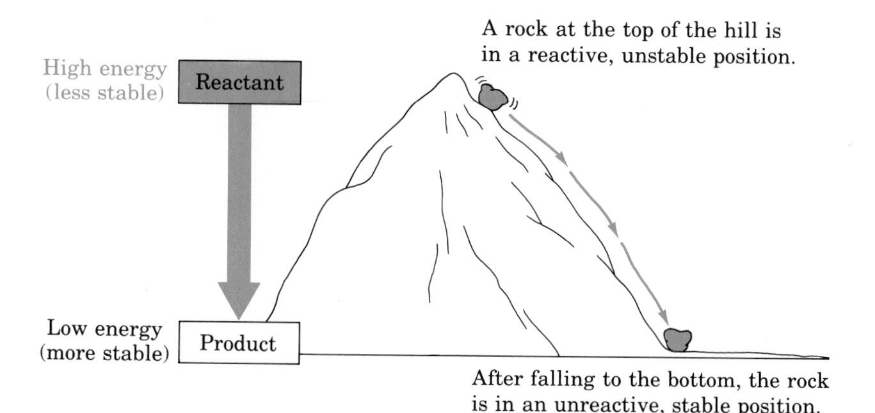

FIGURE 3.8 The relationship between energy and stability. Like a rock near the top of a hill, high-energy substances are unstable. They release their energy by dropping downhill to form low-energy, stable products.

heat of reaction (ΔH) the amount of heat released or absorbed in a reaction

The exact amount of energy either released in an exothermic reaction or absorbed in an endothermic reaction is called the **heat of reaction, ΔH** (spoken as delta-H). By convention, ΔH has a negative value in an exothermic reaction since heat is released, and a positive value in an endothermic reaction since heat is absorbed. ΔH is a direct measure of the difference in energy between products and starting materials. As such, the size of ΔH determines the size of the equilibrium constant K_{eq}. Favorable reactions with large K_{eq}'s are highly exothermic and have negative heats of reaction whereas unfavorable reactions with small K_{eq}'s are endothermic and have positive heats of reaction.

$$A + B \rightleftharpoons C + D$$

$$K_{eq} = \frac{[C][D]}{[A][B]} \qquad \begin{array}{l} \text{Exothermic if } K_{eq} > 1\text{; negative value of } \Delta H \\ \text{Endothermic if } K_{eq} < 1\text{; positive value of } \Delta H \end{array}$$

PRACTICE PROBLEM 3.6 Which reaction is more favorable, one with $\Delta H = -15$ kcal/mol or one with $\Delta H = +15$ kcal/mol?

SOLUTION According to convention, reactions with negative ΔH are exothermic and thus are favorable, but reactions with positive ΔH are endothermic and unfavorable.

PROBLEM 3.17 Which reaction is more exothermic, one with $\Delta H = -10$ kcal/mol or one with $\Delta H = +10$ kcal/mol?

PROBLEM 3.18 Which reaction is more exothermic, one with $K_{eq} = 100$ or one with $K_{eq} = 0.001$?

3.10 DESCRIBING A REACTION: REACTION ENERGY DIAGRAMS AND TRANSITION STATES

For a reaction to take place, reactant molecules must collide, and reorganization of atoms and bonds must occur. Let's look again at the addition reaction between ethylene and HCl:

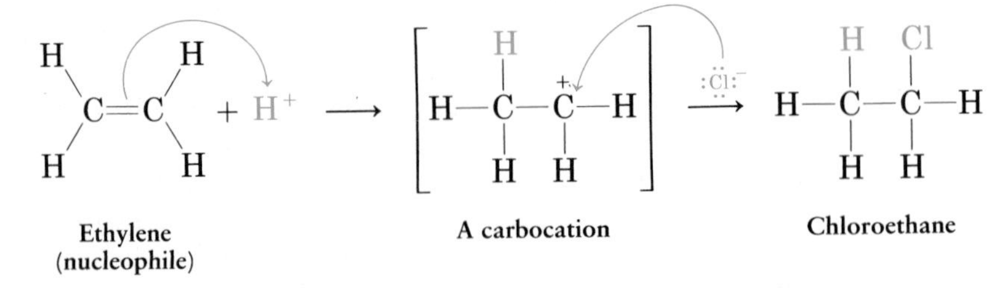

As the reaction proceeds, ethylene and HCl approach each other, the pi bond breaks, a new carbon–hydrogen bond forms in the first step, and a new carbon–chlorine bond forms in the second step.

Over the years, chemists have developed a method for depicting the energy changes that occur during a reaction using *reaction energy diagrams* of the sort shown in Figure 3.9. The vertical axis of the diagram represents the total energy of all reactants, and the horizontal axis represents the progress of the reaction from beginning (left) to end (right). Let's take a careful look at the reaction, one step at a time, and see how the addition of HCl to ethylene can be described on a reaction energy diagram.

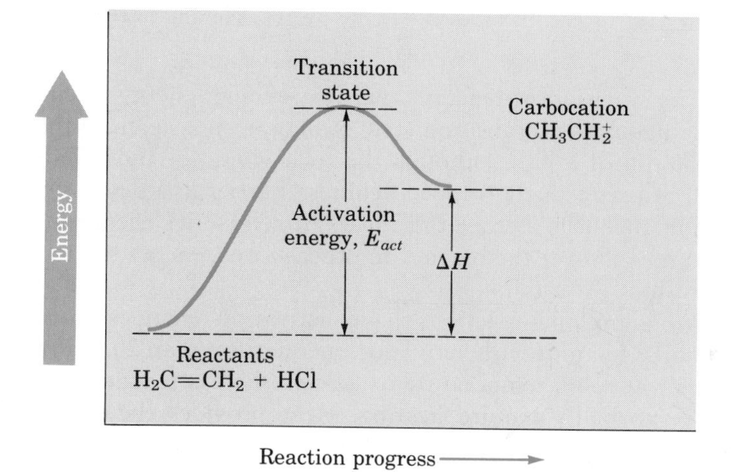

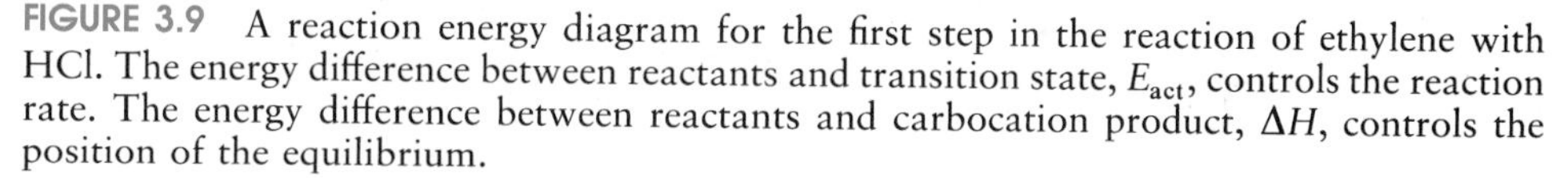
FIGURE 3.9 A reaction energy diagram for the first step in the reaction of ethylene with HCl. The energy difference between reactants and transition state, E_{act}, controls the reaction rate. The energy difference between reactants and carbocation product, ΔH, controls the position of the equilibrium.

At the beginning of the reaction, ethylene and HCl have the total amount of energy indicated by the reactant level on the left side of the diagram. As the two molecules crowd together, their electron clouds repel each other, causing the energy level to rise. If the collision has occurred with sufficient force and proper orientation, the reactants continue to approach each other, despite the repulsion, until the new carbon–hydrogen bond starts to form. At some point, a structure of maximum energy is reached, a structure we call the **transition state.**

transition state a hypothetical structure of maximum energy formed during the course of a reaction

The transition state represents the highest-energy structure involved in this step of the reaction and can't be isolated. Nevertheless, we can imagine the transition state to be a kind of activated complex of the two reactants in which the carbon–carbon pi bond is partially broken and the new carbon–hydrogen bond is partially formed (Figure 3.10).

activation energy E_{act} the energy difference between starting material and transition state

The energy difference between reactants and transition state, called the **activation energy, E_{act},** measures how rapidly the reaction occurs. A large activation energy corresponds to a large energy difference between reactants and transition state, and results in a slow reaction because few of the reacting molecules collide with enough energy to climb the high barrier. A small activation energy, however, results in a rapid reaction because almost all reacting molecules are energetic enough to climb to the transition state.

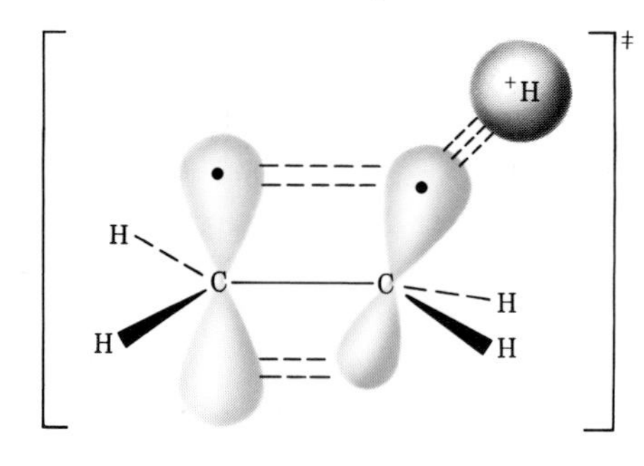

FIGURE 3.10 A hypothetical transition-state structure for the first step of the reaction of ethylene with HCl. The C–C pi bond is just beginning to break, and the C–H bond is just beginning to form.

The situation of reactants needing enough energy to climb the barrier from starting material to transition state is similar once again to the situation of a rock near the top of a hill. Although the rock would be more stable at the bottom of the hill, it is effectively trapped behind a barrier in a depression and is not able to fall spontaneously. Before the rock can release its energy in a fall, it has to be shoved up and over the barrier. In other words, energy has to be put into the rock to activate it for a fall.

Most organic reactions have activation energies in the range of 10–35 kcal/mol. Reactions with activation energies less than 20 kcal/mol take place spontaneously at room temperature or below whereas reactions with higher activation energies normally require heating. Heat provides the energy necessary for the reactants to climb the activation barrier.

Once the high-energy transition state has been reached, the reaction proceeds to the carbocation product. Energy is released as the new C–H bond forms fully, and the curve on the reaction energy diagram therefore turns downward until it reaches a minimum. This minimum point represents the energy level of the carbocation product of the first step. The energy change, ΔH, between starting materials and carbocation product is simply the difference between the two levels on the diagram.[1] Since the carbocation is less stable than the starting alkene, the first step is endothermic, and energy is absorbed.

PROBLEM 3.19 Which reaction is faster: one with $E_{act} = 15$ kcal/mol or one with $E_{act} = 20$ kcal/mol? Is it possible to predict which of the two has the larger K_{eq}?

3.11 DESCRIBING A REACTION: INTERMEDIATES

How can we describe the carbocation structure formed in the first step of the reaction of ethylene with HCl? The carbocation is clearly different from the starting materials, yet it isn't a transition state and it isn't a final product.

[1] Strictly speaking, it's not correct to say that the energy difference between starting materials and products is due entirely to the heat of the reaction, ΔH. The energy difference is actually defined as the Gibbs free energy (ΔG), which is equal to the heat of reaction (ΔH) minus an entropy contribution ΔS: ($\Delta G = \Delta H - T\,\Delta S$). Normally, though, the entropy contribution is small, and we make the simplifying assumption that ΔG and ΔH are approximately equal.

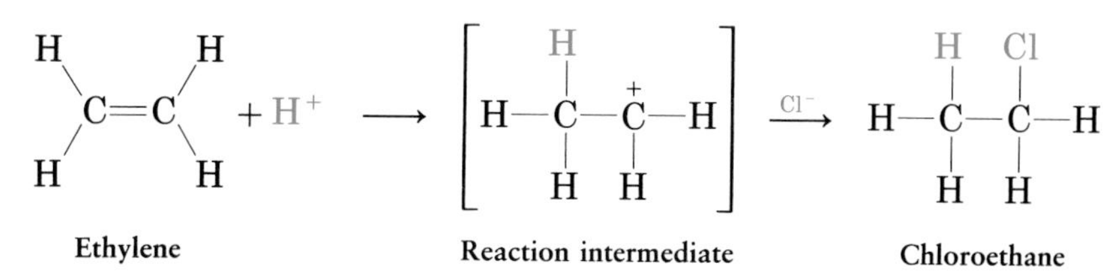

intermediate a species that is formed during the course of a multistep reaction but is not the final product. Intermediates lie at minima in reaction energy diagrams

We call the carbocation, which is formed briefly during the course of the multistep reaction, a reaction **intermediate.** As soon as the intermediate is formed in the first step by reaction of ethylene with H^+, it reacts with Cl^- in a second step to give the final product, chloroethane. This second step has its own activation energy E_{act}, its own transition state, and its own energy change ΔH. We can view the second transition state as an activated complex between the electrophilic carbocation intermediate and nucleophilic chloride anion, a complex in which the new C–Cl bond is just starting to form.

A complete energy diagram for the overall reaction of ethylene with HCl can be constructed as in Figure 3.11. In essence, we draw diagrams for each of the individual steps and join them in the middle so that the product of step 1 (the carbocation) serves as the starting material for step 2. As indicated in Figure 3.11, the reaction intermediate lies at an energy minimum between steps 1 and 2. Since the energy level of this intermediate is higher than the level of either the starting material (ethylene + HCl) or the product (chloroethane), the intermediate is highly reactive and can't be isolated. It is, however, more stable than either of the two transition states that surround it.

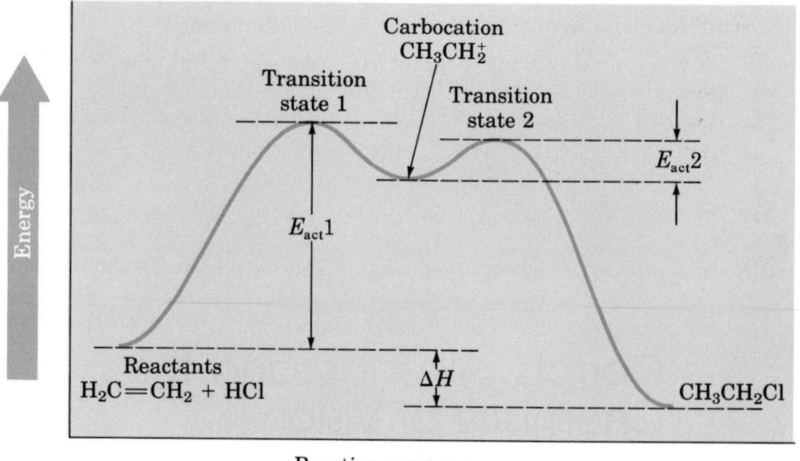

Reaction progress 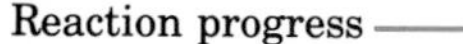

FIGURE 3.11 Overall reaction energy diagram for the reaction of ethylene with HCl. Two steps are involved, each with its own transition state. The energy minimum between the two steps represents the reaction intermediate.

Each step in a multistep process can be considered separately. Each step has its own E_{act} (rate) and its own ΔH (energy change). The overall ΔH of the reaction, however, is the energy difference between initial reactants (far left) and final products (far right). This is always true regardless of the shape of the reaction energy curve. Note, for example, that the energy diagram for the reaction of HCl with ethylene in Figure 3.11 shows the energy level of the final product to be lower than the energy level of the starting material. Thus, the overall reaction is exothermic.

PRACTICE PROBLEM 3.7 Sketch a reaction energy diagram for a one-step reaction that is very fast and highly exothermic.

SOLUTION A very fast reaction has a small E_{act}, and a highly exothermic reaction has a large negative ΔH. Thus, the diagram will look like this one:

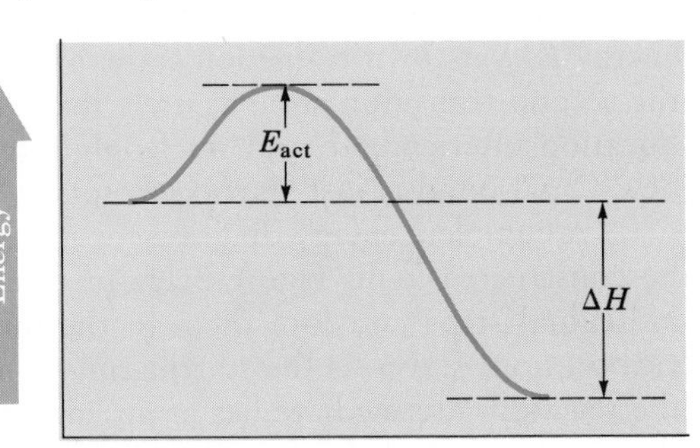

PROBLEM 3.20 Sketch reaction energy diagrams to represent the following situations and label the parts of the diagram corresponding to starting material, product, transition state, intermediate (if any), activation energy, and ΔH.

(a) An exothermic reaction that takes place in one step.
(b) An endothermic reaction that takes place in one step.

PROBLEM 3.21 Draw a reaction energy diagram for a two-step reaction with an endothermic first step and an exothermic second step. Label the intermediate.

INTERLUDE

Carrots, Alkenes, and the Chemistry of Vision

Folk medicine has long maintained that eating carrots improves night vision. Although that's probably not true for healthy adults on a proper diet, there's no question that the chemistry of carrots and the chemistry of vision are related. Alkenes play a role in both.

Carrots are rich in β-carotene, a purple-orange alkene that is an excellent dietary source of vitamin A. β-Carotene is converted to vitamin A in the liver, where enzymes first cut the molecule in half and then change the geometry of the C11–C12 double bond to produce 11-*cis*-retinal, the light-sensitive pigment on which the visual systems of all living things are based.

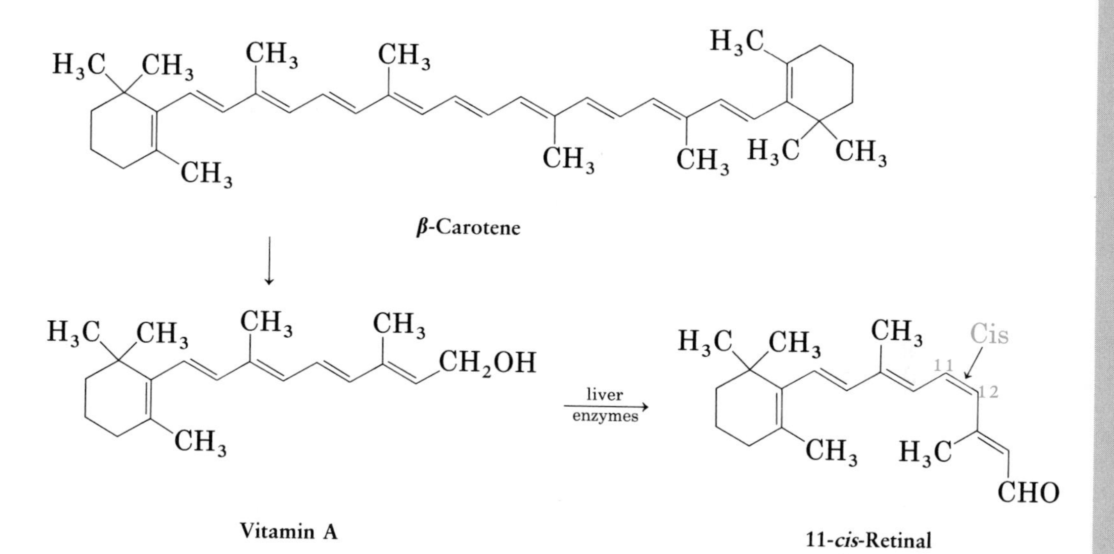

β-Carotene

Vitamin A

11-*cis*-Retinal

The retina of the eye contains two types of light-sensitive receptor cells, *rod* cells and *cone* cells. Rod cells are primarily responsible for seeing in dim light, whereas cone cells are responsible for seeing in bright light and for the perception of colors. In the rod cells of the eye, 11-*cis*-retinal is converted into *rhodopsin,* a light-sensitive substance formed from the protein *opsin* and 11-*cis*-retinal. When light strikes the rod cell, isomerization of the C11–C12 double bond occurs, and 11-*trans*-rhodopsin, also called metarhodopsin II, is produced. This cis–trans isomerization of rhodopsin is accompanied by a change in molecular geometry, which in turn causes a nerve impulse to be sent to the brain where it is perceived as vision.

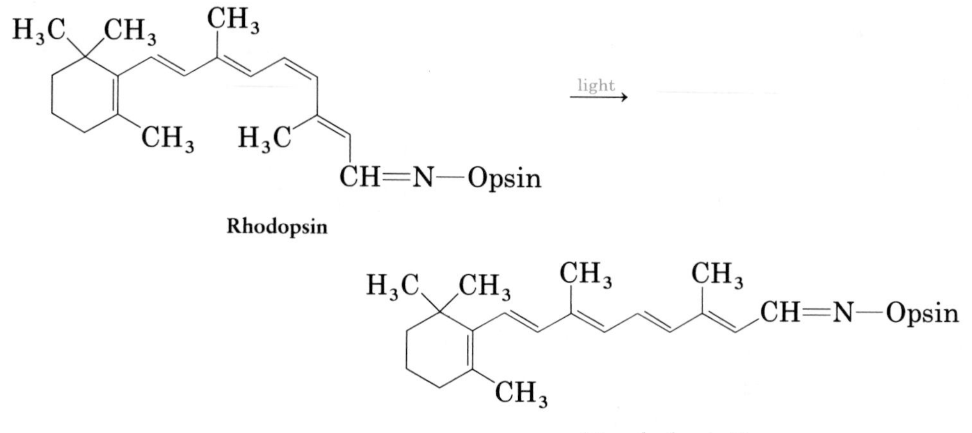

Rhodopsin

Metarhodopsin II

Metarhodopsin II is then recycled into rhodopsin by a multistep sequence involving cleavage into all-*trans*-retinal, conversion to vitamin A, cis–trans isomerization to 11-*cis*-vitamin A, and conversion back to 11-*cis*-retinal.

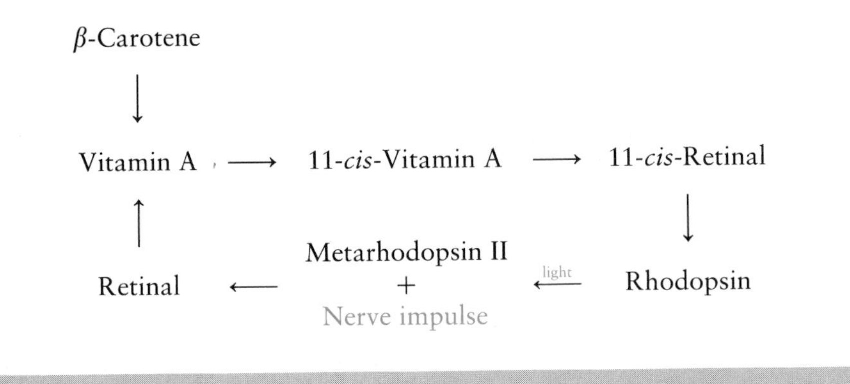

SUMMARY AND KEY WORDS

Alkenes are hydrocarbons that contain carbon–carbon double bonds. A double bond consists of two parts: a **sigma bond** formed by head-on overlap of two sp^2 orbitals and a **pi bond** formed by sideways overlap of two p orbitals. The bond strength of an alkene double bond is greater than that of a carbon–carbon single bond, with the strength of the pi part estimated to be 64 kcal/mol.

Rotation around the double bond is restricted, and substituted alkenes can therefore exist as **cis–trans isomers.** The geometry of a double bond can be described as either ***Z*** (*zusammen*) or ***E*** (*entgegen*) by application of a series of sequence rules.

All organic reactions involve bond making and bond breaking. Fundamentally, covalent two-electron bonds can break or form in only two ways: Bonds can break in an electronically symmetrical (**homolytic**) way such that each product retains one electron or in an electronically unsymmetrical (**heterolytic**) way such that one product retains both electrons, leaving the other product with a vacant valence orbital. Conversely, bonds can form in an electronically symmetrical (**homogenic**) way if each of two reactants donates one electron or in an electronically unsymmetrical (**heterogenic**) way if one reactant donates two electrons. Electronically symmetrical bond making and bond breaking occur in **radical** reactions whereas electronically unsymmetrical bond making and bond breaking occur in **polar** reactions:

$$A\cdot + \cdot B \longrightarrow A{:}B \qquad \textbf{Radical reactions}$$
$$A^+ + {:}B^- \longrightarrow A{:}B \qquad \textbf{Polar reactions}$$

The energy changes that take place during a reaction can be described by **rates** (how fast a reaction occurs) and **equilibria** (to what extent the reaction occurs). The equilibrium position of a reaction is determined by ΔH, the energy change that takes place during the reaction. If the reaction is **exothermic,** energy is given off, and the reaction has a favorable equilibrium constant. If the reaction is **endo-**

thermic, however, energy is absorbed, and the reaction has an unfavorable equilibrium constant. Reactions can be described pictorially by **reaction energy diagrams,** which follow the course of a reaction from starting material through transition state to product.

Every reaction proceeds through a **transition state,** which is the highest energy point reached. Transition-state structures can't be isolated because they are unstable, but we can imagine them to be activated complexes between starting materials, in which old bonds are beginning to break and new bonds are beginning to form. The amount of energy needed by starting materials to reach the high-energy transition state is the **activation energy, E_{act}.** The larger the magnitude of the activation energy, the slower the reaction.

Many reactions, such as the addition of HCl to ethylene, take place in more than one step and involve the formation of **intermediates.** A reaction intermediate is a structure that is formed during the course of a multistep reaction and that lies in an energy minimum between two transition states. Intermediates are more stable than transition states but are often too reactive to be isolated.

ADDITIONAL PROBLEMS

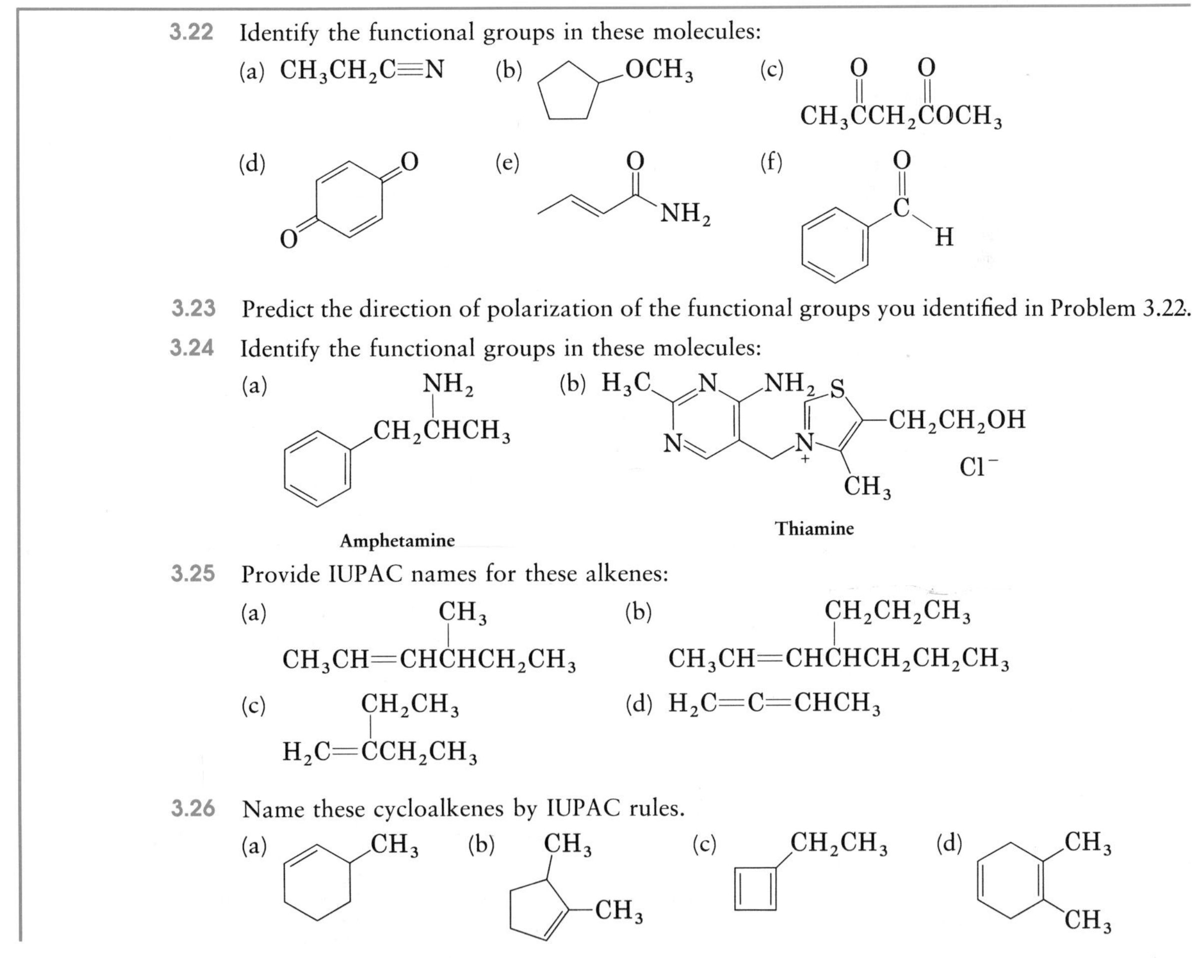

3.27 Draw structures corresponding to these IUPAC names.

(a) 3-Propyl-2-heptene (b) 2,4-Dimethyl-2-hexene
(c) 1,5-Octadiene (d) 4-Methyl-1,3-pentadiene
(e) *cis*-4,4-Dimethyl-2-hexene (f) (*E*)-3-Methyl-3-heptene

3.28 Draw the structures of these cycloalkenes.

(a) *cis*-4,5-Dimethylcyclohexene (b) 3,3,4,4-Tetramethylcyclobutene

3.29 These names are incorrect. Draw each molecule and give its correct name.

(a) 1-Methyl-2-cyclopentene (b) 1-Methyl-1-pentene
(c) 6-Ethylcycloheptene (d) 3-Methyl-2-ethylcyclohexene

3.30 Neglecting cis–trans isomers, there are five possible isomers of formula C_4H_8. Draw and name them.

3.31 Which of the molecules you drew in Problem 3.30 show cis–trans isomerism? Draw and name their cis–trans isomers.

3.32 Draw four possible structures for each of these formulas.

(a) C_6H_{10} (b) C_8H_8O (c) $C_7H_{10}Cl_2$

3.33 How can you explain the fact that cyclohexene does not show cis–trans isomerism but cyclodecene does?

3.34 Rank the following sets of substituents in order of priority according to the sequence rules.

(a) $—CH_3$, $—Br$, $—H$, $—I$
(b) $—OH$, $—OCH_3$, $—H$, $—COOH$
(c) $—CH_3$, $—COOH$, $—CH_2OH$, $—CHO$
(d) $—CH_3$, $—CH{=}CH_2$, $—CH_2CH_3$, $—CH(CH_3)_2$

3.35 Assign *E* or *Z* configuration to these alkenes.

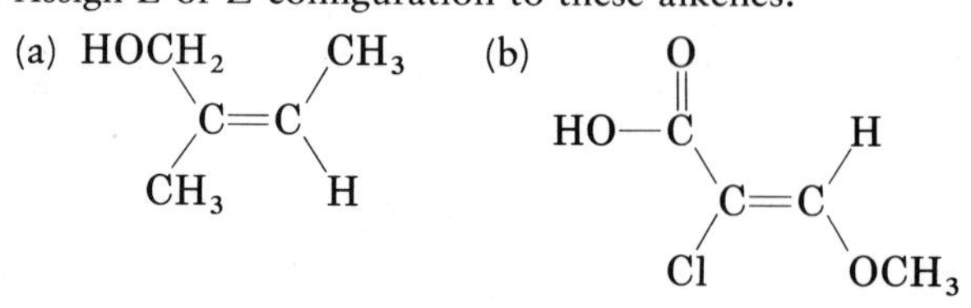

3.36 Draw and name the five possible C_5H_{10} alkene isomers. Ignore cis–trans isomers.

3.37 Menthene, a hydrocarbon found in mint plants, has the IUPAC name 1-isopropyl-4-methylcyclohexene. What is the structure of menthene?

3.38 Name these cycloalkenes by IUPAC rules.

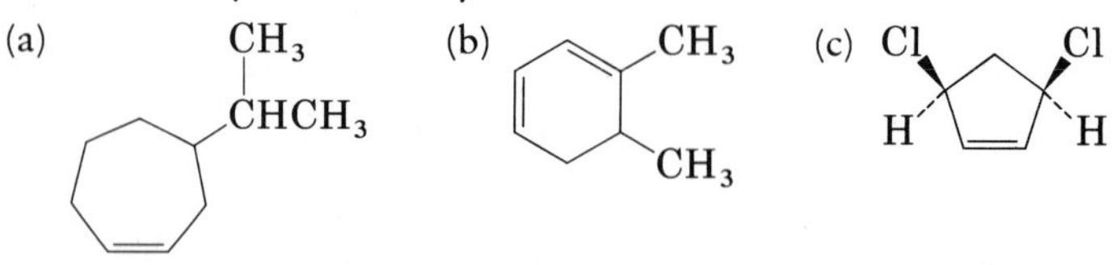

3.39 Classify these reagents as either electrophiles or nucleophiles.

(a) Zn^{2+} (b) $CH_3\ddot{N}H_2$ (c) $CH_3—C(=\ddot{O}:)—\ddot{O}:^-$ (d) $H\ddot{S}:^-$

3.40 α-Farnesene is a constituent of the natural waxy coating found on apples. What is its IUPAC name?

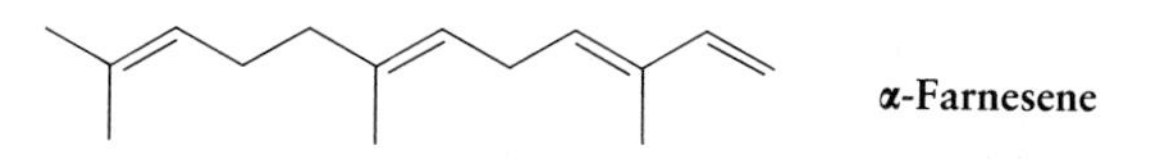

3.41 Indicate E or Z configuration for each of the double bonds in α-farnesene (Problem 3.40).

3.42 Define these terms.

(a) Polar reaction
(b) Radical reaction
(c) Functional group
(d) Reaction intermediate

3.43 Give an example of each of the following.

(a) An electrophile
(b) A nucleophile
(c) An oxygen-containing functional group

3.44 If a reaction has $K_{eq} = 0.001$, is it likely to be exothermic or endothermic? Explain.

3.45 If a reaction has $E_{act} = 5$ kcal/mol, is it likely to be fast or slow? Explain.

3.46 If a reaction has $\Delta H = 12$ kcal/mol, is it exothermic or endothermic. Is it likely to be fast or slow? Explain.

3.47 Draw a reaction energy diagram for a two-step exothermic reaction whose first step is faster than its second step. Label the parts of the diagram corresponding to reactants, products, transition state, activation energies, and overall ΔH.

3.48 Draw a reaction energy diagram for a two-step reaction whose second step is faster than its first step.

3.49 Draw a reaction energy diagram for a reaction with $K_{eq} = 1$.

3.50 Describe the difference between a transition state and a reaction intermediate.

3.51 Consider the reaction energy diagram shown here and answer the following questions.

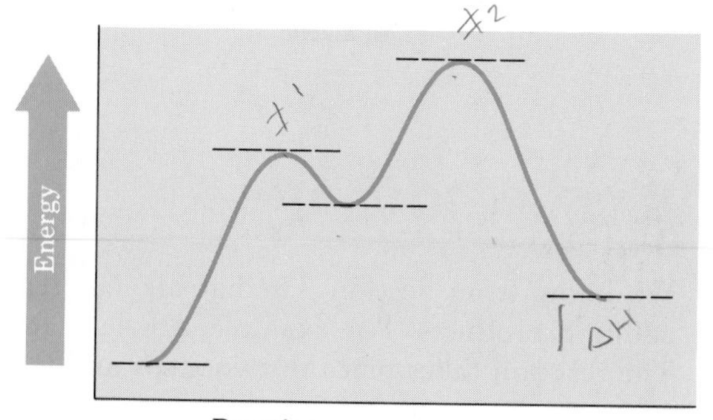

(a) Indicate the overall ΔH for the reaction. Is it positive or negative?
(b) How many steps are involved in the reaction?
(c) Which step is faster (has the lower activation energy)?
(d) How many transition states are there? Label them.

CHAPTER 4

Alkenes and Alkynes

We saw in the previous chapter how organic reactions can be classified, and we developed some general ideas about how reactions can be described. In this chapter, we'll apply those general ideas to a systematic study of the alkene and alkyne families of compounds. In particular, we'll see that the most important reaction of these two functional groups is the addition of various reagents X–Y to yield saturated products:

$$>C{=}C< \; + \; X{-}Y \longrightarrow -\overset{X}{\underset{|}{\overset{|}{C}}}-\overset{Y}{\underset{|}{\overset{|}{C}}}-$$

An alkene **An addition product**

4.1 ADDITION OF HX TO ALKENES

We know from Section 3.8 that alkenes react with HCl to yield alkyl chloride addition products. For example, ethylene reacts with HCl to give chloroethane. The reaction takes place in two steps and involves a carbocation intermediate:

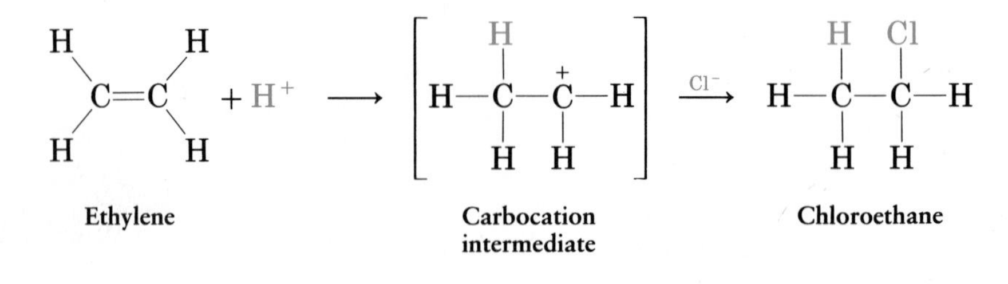

Ethylene **Carbocation intermediate** **Chloroethane**

The addition of halogen acids HX to alkenes is a general reaction that allows chemists to prepare a variety of products. Thus, HCl, HBr, and HI all add to alkenes:[1]

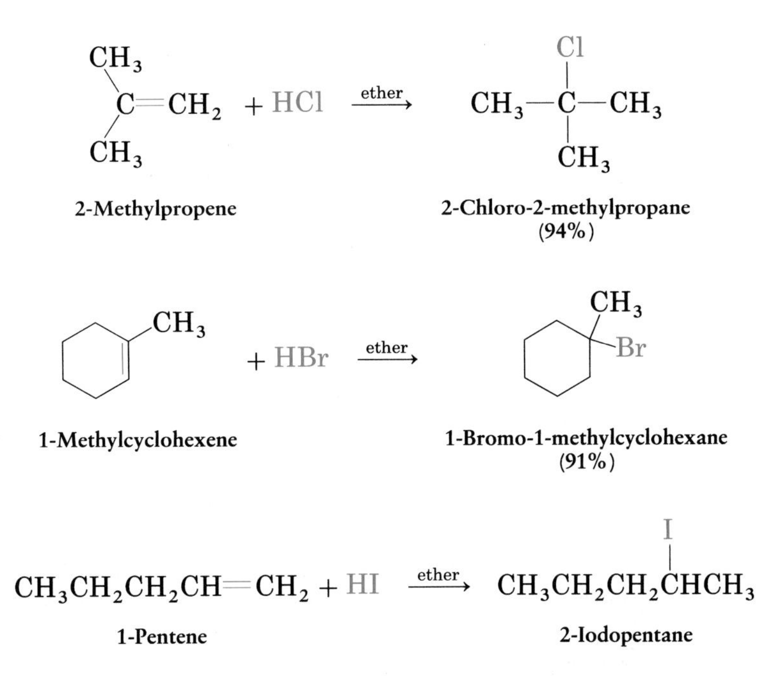

[1] Organic reaction equations can be written in different ways to emphasize different points. For example, the reaction of ethylene with HCl might be written in the format A + B → C to emphasize that *both* reaction partners are equally important for the purposes of the discussion. The reaction solvent and notes about other reaction conditions such as temperature are usually written either above or below the reaction arrow.

solvent

$$H_2C{=}CH_2 + HCl \xrightarrow[25°C]{\text{ether}} CH_3CH_2Cl$$

Alternatively, we might choose to write the same reaction in the format

$$A \xrightarrow{B} C$$

to emphasize that reagent A is the organic starting material whose chemistry is of greater interest. Reagent B is then placed above the reaction arrow, together with notes about solvent and reaction conditions. For example:

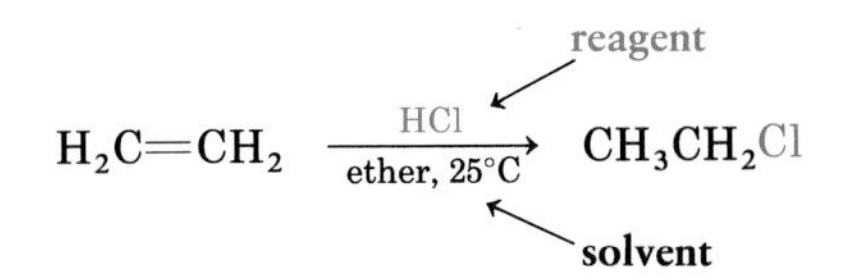

Both reaction formats are frequently used in chemistry, and you sometimes have to look at the overall transformation to see what the different roles of the chemicals shown next to the reaction arrows are.

4.2 ORIENTATION OF ALKENE ADDITION REACTIONS: MARKOVNIKOV'S RULE

regiospecific describing the orientation of an addition reaction that occurs on an unsymmetrical substrate and that leads to a single product

Look carefully at the reactions in the previous section. In every case, an unsymmetrically substituted alkene has given a *single* addition product rather than the mixture that might have been expected. For example, 2-methylpropene might have added HCl to give 1-chloro-2-methylpropane, but it didn't; it gave only 2-chloro-2-methylpropane. We say that reactions are **regiospecific** (*ree*-jee-oh-specific) when only one of the two possible directions of addition is observed.

A regiospecific reaction:

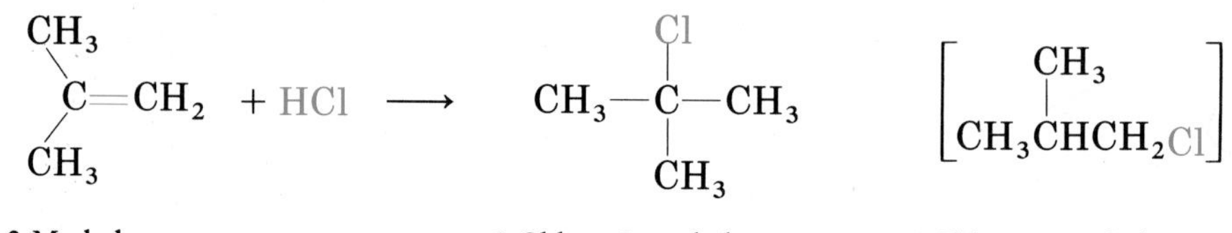

2-Methylpropene — 2-Chloro-2-methylpropane (sole product) — 1-Chloro-2-methylpropane (not formed)

Markovnikov's rule a rule for predicting the orientation of alkene electrophilic addition reactions

From an examination of many such reactions, the Russian chemist Vladimir Markovnikov proposed in 1905 what has come to be known as **Markovnikov's rule:** *In the addition of HX to an alkene, the H attaches to the carbon that has fewer alkyl substituents, and the X attaches to the carbon that has more alkyl substituents.*

2 alkyl groups on this carbon → $(CH_3)_2C{=}CH_2$ ← no alkyl groups on this carbon

$$(CH_3)_2C{=}CH_2 + HCl \xrightarrow{\text{ether}} CH_3—\overset{Cl}{\underset{CH_3}{C}}—CH_3$$

2-Methylpropene — 2-Chloro-2-methylpropane

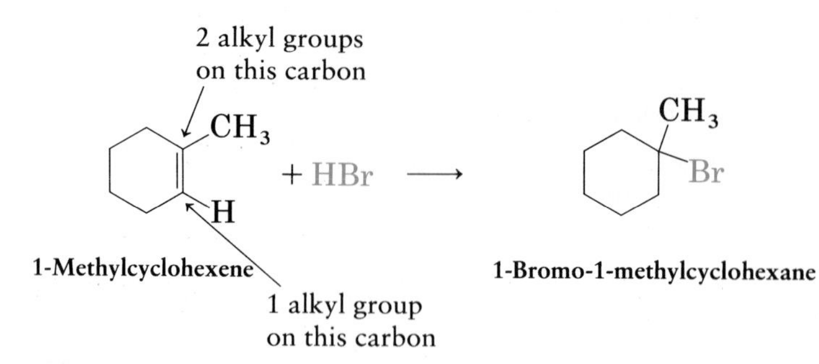

1-Methylcyclohexene — 1-Bromo-1-methylcyclohexane

When both ends of the double bond have the same degree of substitution, however, a mixture of addition products results:

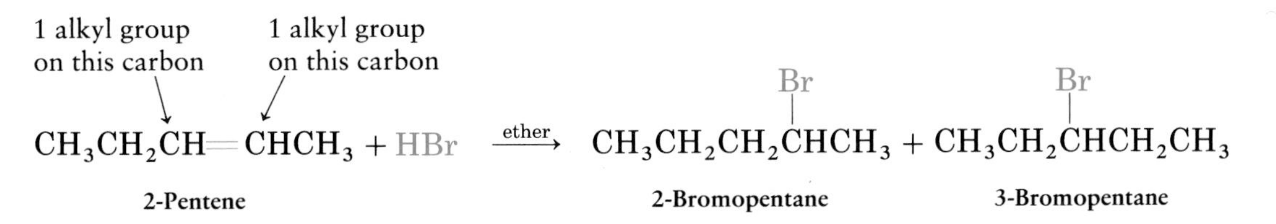

Since carbocations are involved as intermediates in these reactions (Section 3.11), another way to express Markovnikov's rule is to say that, in the addition of HX to alkenes, the more highly substituted carbocation intermediate is formed in preference to the less highly substituted one. For example, addition of H^+ to 2-methylpropene yields the intermediate tertiary carbocation rather than the primary carbocation. Why should this be?

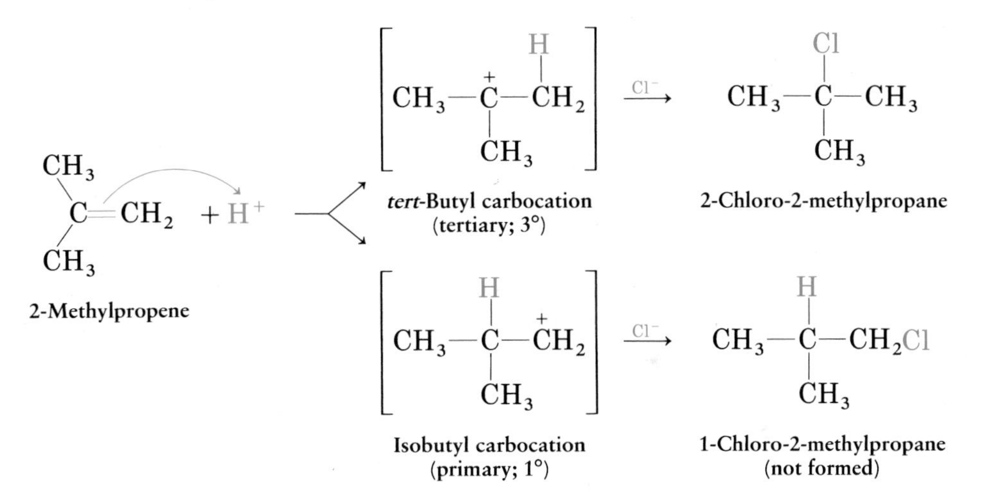

PRACTICE PROBLEM 4.1 What product would you expect from reaction of HCl with 1-ethylcyclopentene?

$$\text{1-Ethylcyclopentene } (CH_2CH_3 \text{ on ring}) + HCl \longrightarrow ?$$

SOLUTION Markovnikov's rule predicts that the hydrogen will add to the double-bond carbon that has one alkyl group (C2 on the ring), and the chlorine will add to the double-bond carbon that has two alkyl groups (C1 on the ring). The expected product is 1-chloro-1-ethylcyclopentane.

2 alkyl groups on this carbon

$$\text{1-Ethylcyclopentene } (CH_2CH_3) + HCl \longrightarrow \text{1-Chloro-1-ethylcyclopentane } (CH_2CH_3, Cl)$$

1 alkyl group on this carbon

1-Chloro-1-ethylcyclopentane

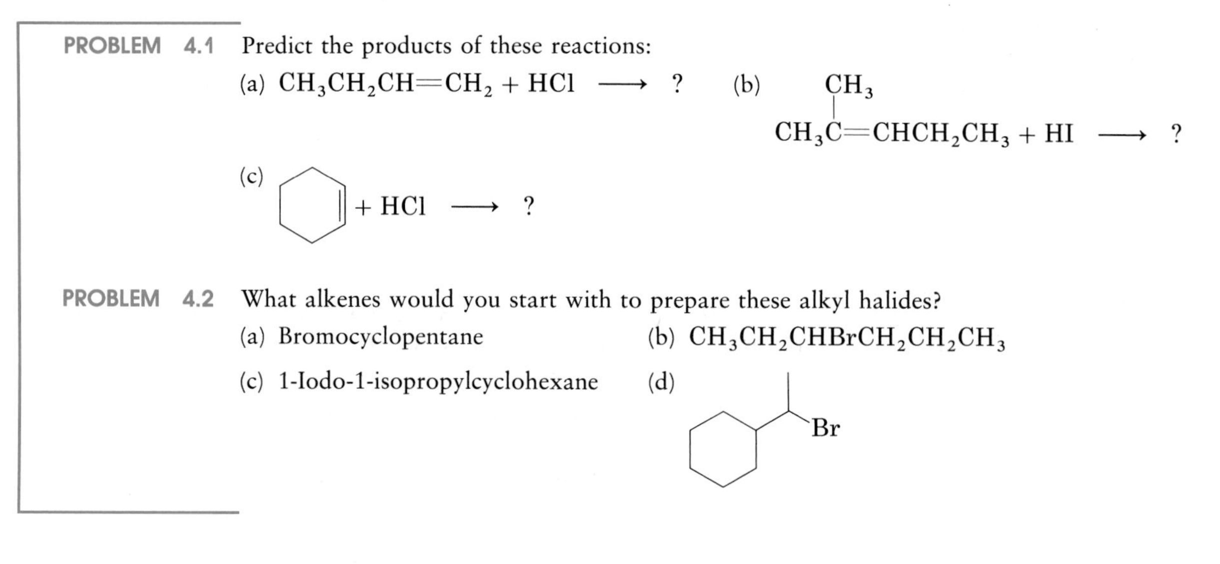

PROBLEM 4.1 Predict the products of these reactions:

(a) $CH_3CH_2CH{=}CH_2 + HCl \longrightarrow$?

(b) $CH_3C(CH_3){=}CHCH_2CH_3 + HI \longrightarrow$?

(c) Cyclohexene $+ HCl \longrightarrow$?

PROBLEM 4.2 What alkenes would you start with to prepare these alkyl halides?

(a) Bromocyclopentane

(b) $CH_3CH_2CHBrCH_2CH_2CH_3$

(c) 1-Iodo-1-isopropylcyclohexane

(d) (1-bromoethyl)cyclohexane structure, Br

4.3 CARBOCATION STRUCTURE AND STABILITY

To understand why Markovnikov's rule works, we need to learn more about the structure and stability of substituted carbocations. Regarding structure, evidence has shown that carbocations are *planar*. The positively charged carbon atom is sp^2 hybridized, and the three substituents are oriented to the corners of an equilateral triangle (Figure 4.1). Since there are only six electrons in the carbon valence shell, and since all six are used in the three sigma bonds, the p orbital extending above and below the plane is vacant.

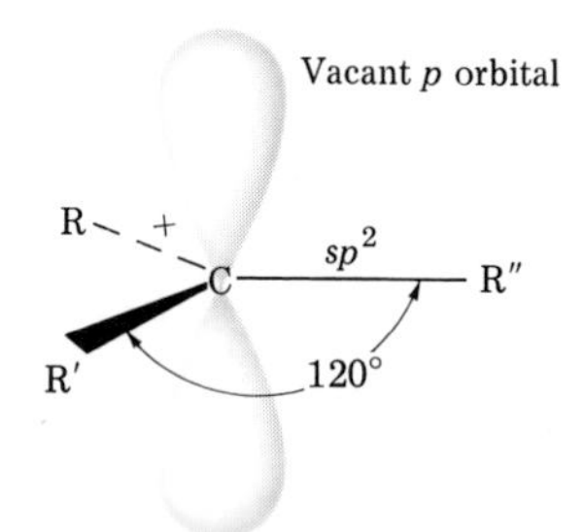

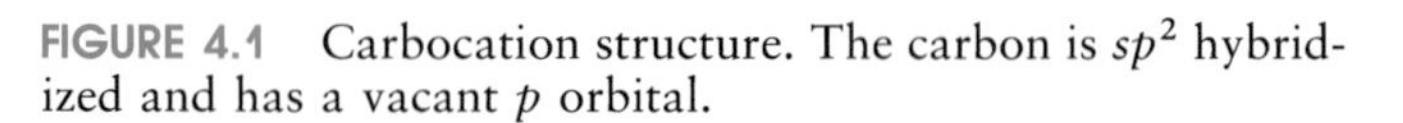

FIGURE 4.1 Carbocation structure. The carbon is sp^2 hybridized and has a vacant p orbital.

Regarding stability, measurements show that carbocation stability increases with increasing alkyl substitution. More highly substituted carbocations are more stable than less highly substituted ones because alkyl groups tend to donate electrons to the positively charged carbon atom. The more alkyl groups there are, the more electron donation there is and the more stable the carbocation.

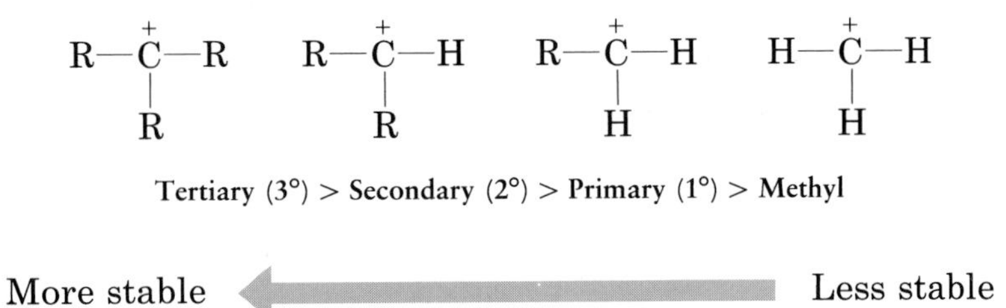

With the above information, we can now explain Markovnikov's rule. In the reaction of 2-methylpropene with HCl, for example, the intermediate carbocation might have either *three* alkyl substituents (a tertiary cation, 3°) or *one* alkyl substituent (a primary cation, 1°). Since the tertiary cation is more stable than the primary one, it's the tertiary cation that forms as the reaction intermediate, thus leading to the observed tertiary alkyl chloride product.

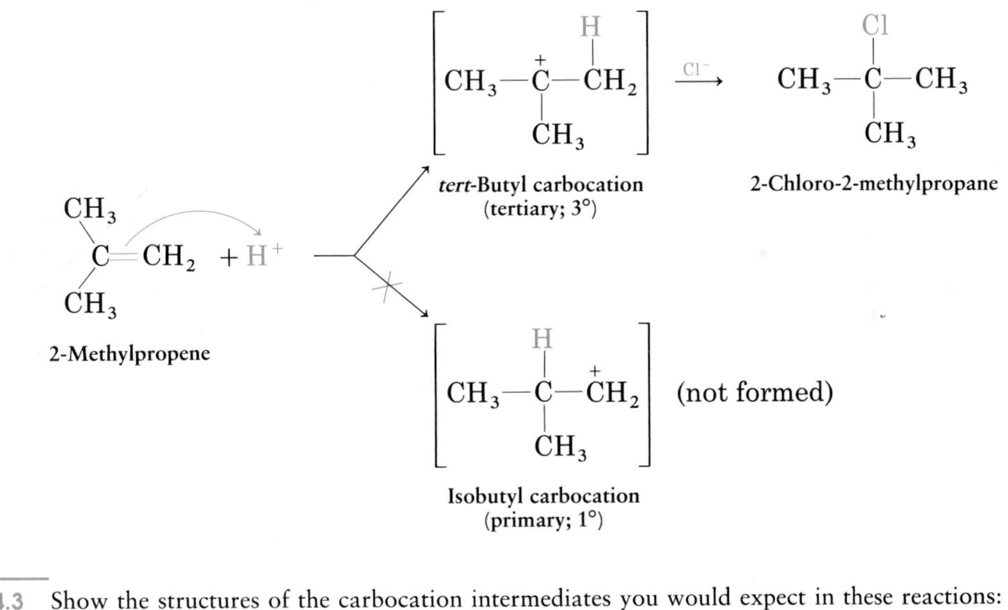

PROBLEM 4.3 Show the structures of the carbocation intermediates you would expect in these reactions:

(a) $CH_3CH_2C(CH_3)=CHCH(CH_3)CH_3 + HBr \longrightarrow ?$

(b) (cyclopentylidene)$=CHCH_3 + HI \longrightarrow ?$

4.4 HYDRATION OF ALKENES

Water can be added to simple alkenes like ethylene and 2-methylpropene to yield alcohols, ROH. Industrially, more than 300,000 tons of ethanol are produced each

hydration the addition of water to a substrate, usually an alkene

year in the United States by this **hydration** method:

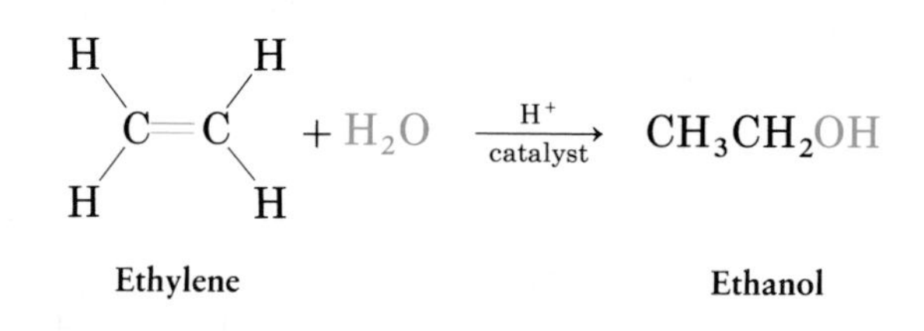

The hydration of an alkene takes place on reaction with aqueous acid by a mechanism similar to that of HX addition. Thus, reaction of the alkene double bond with H^+ yields a carbocation intermediate that then reacts with water as nucleophile to yield a protonated alcohol (ROH_2^+) product. Loss of H^+ from the protonated alcohol gives the neutral alcohol and regenerates the acid catalyst (Figure 4.2). The addition of water to an unsymmetrical alkene follows Markovnikov's rule, just as addition of HX does, giving the more highly substituted alcohol as product.

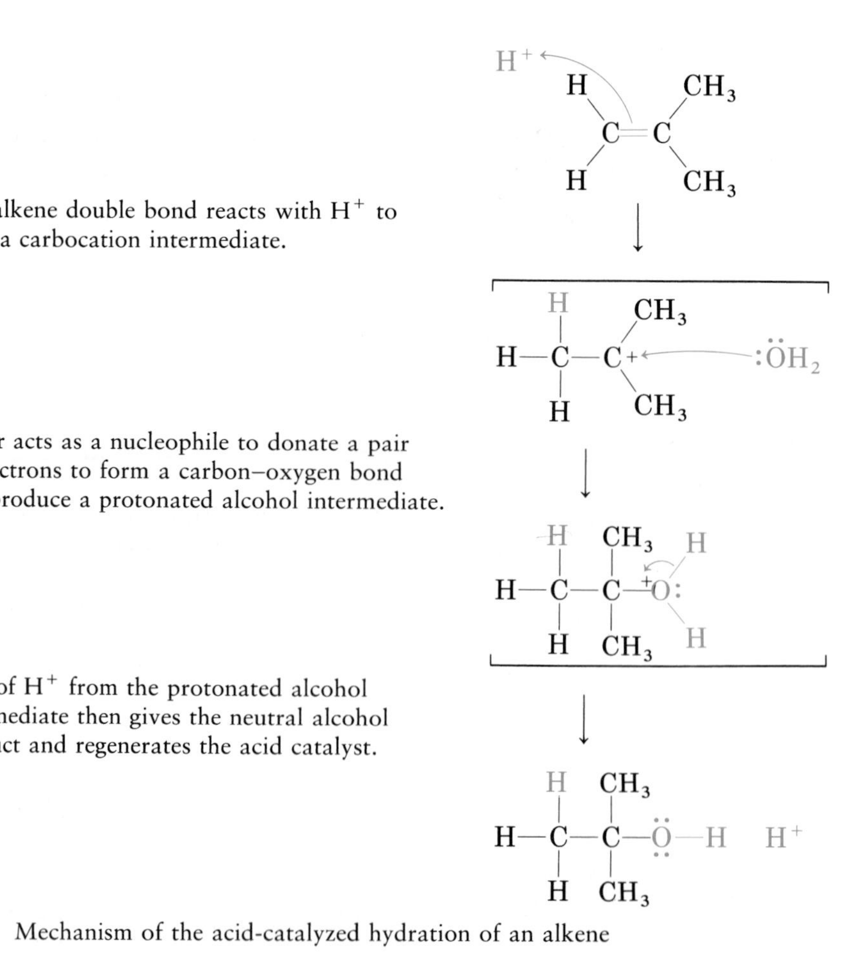

The alkene double bond reacts with H+ to yield a carbocation intermediate.
Water acts as a nucleophile to donate a pair of electrons to form a carbon–oxygen bond and produce a protonated alcohol intermediate.
Loss of H+ from the protonated alcohol intermediate then gives the neutral alcohol product and regenerates the acid catalyst.
H+
H
CH3
C=C
H
CH3
:ÖH2
H+

FIGURE 4.2 Mechanism of the acid-catalyzed hydration of an alkene

Unfortunately, the reaction conditions required for hydration are so severe that molecules are sometimes destroyed by the high temperatures and strongly acidic conditions. For example, the hydration of ethylene to produce ethanol requires a sulfuric acid catalyst and reaction temperatures of up to 250°C.

PRACTICE PROBLEM 4.2 What product would you expect from addition of water to methylenecyclopentane?

$$\text{(cyclopentane)}{=}CH_2 + H_2O \longrightarrow ?$$

Methylenecyclopentane

SOLUTION According to Markovnikov's rule, H^+ adds to the carbon that already has more hydrogens (the CH_2 carbon), and $-OH$ adds to the carbon that has fewer hydrogens (the ring carbon). Thus, the product will be a tertiary alcohol.

$$\text{(cyclopentane)}{=}CH_2 + H_2O \longrightarrow \text{1-methylcyclopentanol (ring C bearing OH and } CH_3\text{)}$$

PROBLEM 4.4 What product would you expect to obtain from addition of water to these alkenes?

(a) $CH_3CH_2C(CH_3){=}CHCH_2CH_3$ (b) 1-Methylcyclopentene
(c) 2,5-Dimethyl-2-heptene

PROBLEM 4.5 What alkenes do you suppose these alcohols were made from?

(a) $CH_3CH_2CH(OH)CH_3$

(b) $CH_3CH_2-C(OH)(CH_3)-CH_2CH_3$

(c) cyclohexane ring bearing OH and CH_3 on one carbon and CH_3 on the adjacent carbon

4.5 ADDITION OF HALOGENS TO ALKENES

Many other reagents besides HX and H_2O add to alkenes. Bromine and chlorine are particularly effective, and their reaction with alkenes provides a general method of synthesis of 1,2-dihaloalkanes. More than 5 million tons of 1,2-dichloroethane (also called ethylene dichloride) are synthesized each year in the chemical industry by addition of Cl_2 to ethylene. The product is used both as a solvent and as starting material for the synthesis of poly(vinyl chloride), PVC.

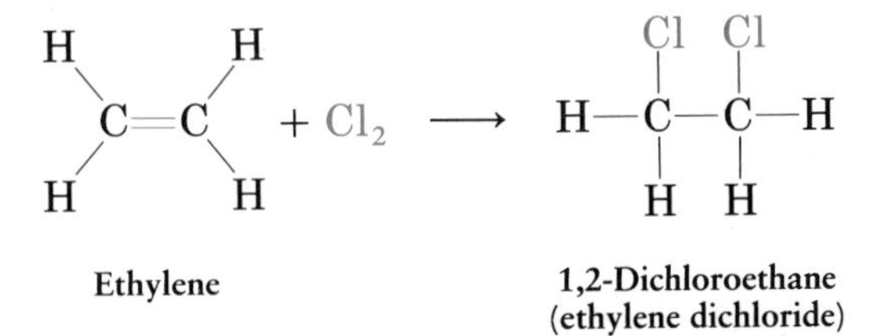

Ethylene **1,2-Dichloroethane (ethylene dichloride)**

Addition of bromine also serves as a simple and rapid laboratory test for the presence of a carbon–carbon double bond in a molecule of unknown structure. A sample of unknown structure is dissolved in tetrachloromethane, CCl_4, and several drops of bromine are added. Immediate disappearance of the reddish bromine color signals a positive test, indicating that the sample is an alkene.

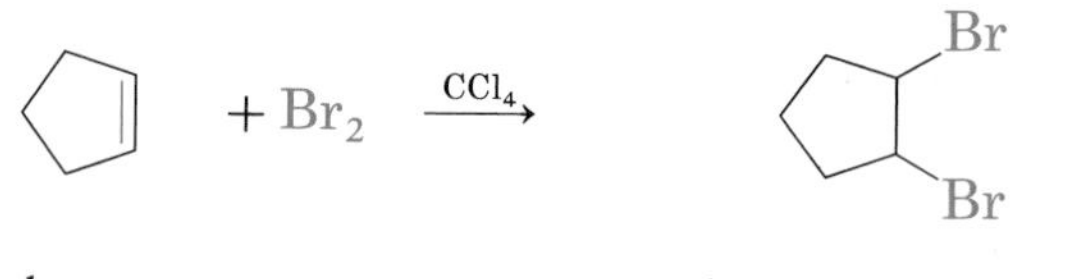

Cyclopentene 1,2-Dibromocyclopentane (95%)

Bromine and chlorine react with alkenes by the pathway shown in Figure 4.3. The pi-electron pair of the alkene attacks the Br_2 molecule, displacing Br^-. The net result is that electrophilic Br^+ adds to the alkene in much the same way that H^+ does, yielding an intermediate carbocation that immediately reacts further with Br^- to give the dibromo addition product.

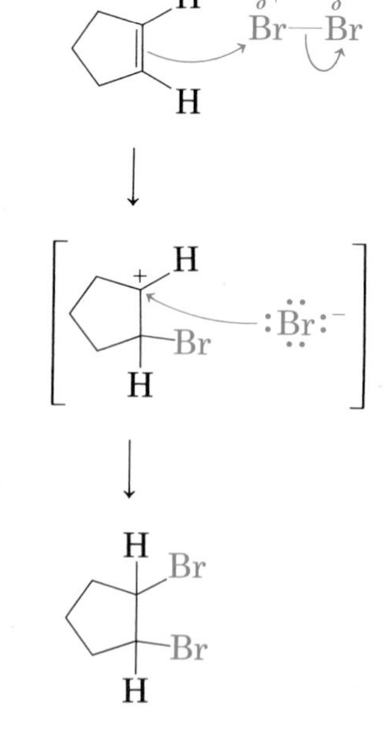

The electron pair from the double bond attacks the polarized bromine, forming a C–Br bond and causing the Br–Br bond to break. Bromide ion departs with both electrons from the former Br–Br bond.

Bromide ion uses an electron pair to attack the carbocation intermediate, forming a C–Br bond and giving the neutral addition product.

FIGURE 4.3 Addition of bromine to cyclopentene

The mechanism of halogen addition to alkenes shown in Figure 4.3 looks reasonable, but it's not completely consistent with known facts. In particular, the mechanism doesn't explain the *stereochemistry* of halogen addition. That is, the mechanism doesn't explain what product stereoisomers (Section 2.7) are formed in the reaction.

Let's look again at the reaction of Br_2 with cyclopentene and assume that Br^+ adds from the bottom face to form the cation intermediate shown in Figure 4.4. (The addition could just as well occur from the top face, but we'll consider only one possibility for simplicity.) Since this carbocation intermediate is planar and

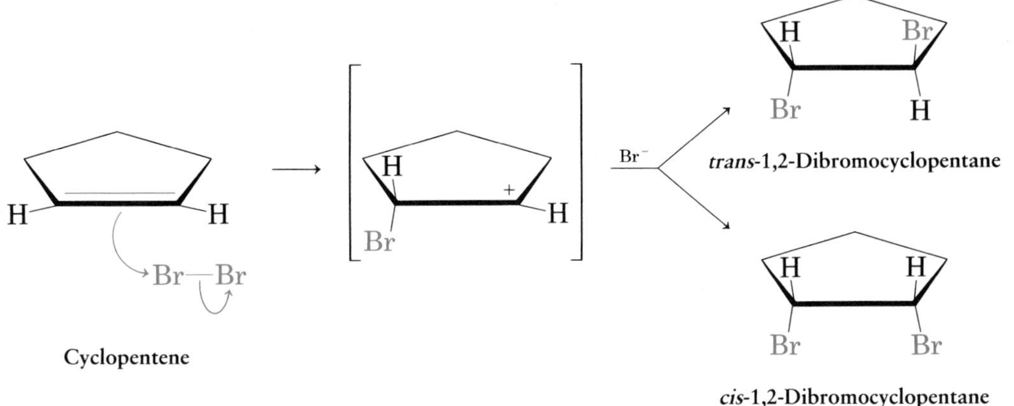

FIGURE 4.4 Stereochemistry of the addition of bromine to cyclopentene. Only the trans product is formed.

sp^2 hybridized, it could be attacked by bromide ion in the second step of the reaction from either the top or the bottom side. Thus, a mixture of products might result, in which the two bromine atoms are either on the same side of the ring (cis) or on opposite sides (trans). We find, however, that only *trans*-1,2-dibromocyclopentane is produced: The two bromine atoms add to opposite faces of the double bond, a result described by saying that the reaction occurs with **anti stereochemistry.** (*Anti* means that the two bromines that have added came from opposite sides of the molecule approximately 180° apart.)

anti stereochemistry Referring to a reaction in which both top and bottom sides of a reactant are involved

bromonium ion a species with a positively charged, divalent bromine atom

The stereochemistry of bromine addition is best explained by imagining that the reaction intermediate is not a true carbocation. Instead, the intermediate is a *bromonium ion,* formed by the overlap of the vacant carbocation p orbital with a lone pair of electrons on the neighboring bromine atom (Figure 4.5). (A **bromonium ion** is a species that contains a positively charged, divalent bromine, R_2Br^+.) Since the bromine atom shields one face of the molecule, reaction with bromide ion in the second step can occur only from the opposite, more accessible face to give the anti product.

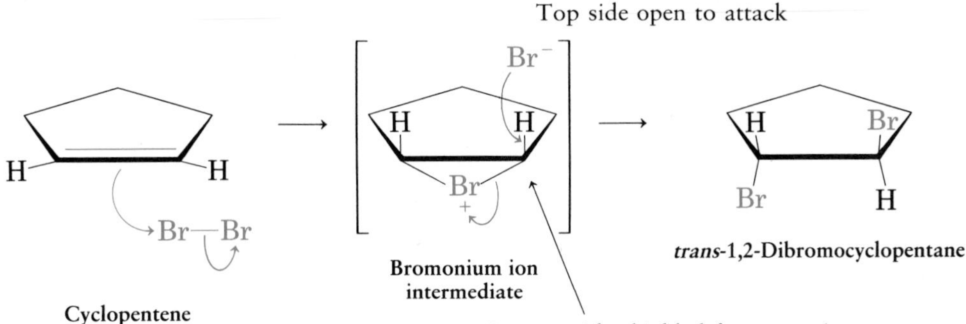

FIGURE 4.5 Formation of a bromonium-ion intermediate by addition of Br^+ to an alkene

PROBLEM 4.6 What product would you expect to obtain from addition of Br_2 to 1,2-dimethylcyclohexene? Show the stereochemistry of the product.

PROBLEM 4.7 Show the structure of the intermediate bromonium ion formed in Problem 4.6.

4.6 HYDROGENATION OF ALKENES

hydrogenation the addition of H_2 to a molecule, usually an alkene

reduction the addition of hydrogen to a molecule or the removal of oxygen from it

Addition of hydrogen to the double bond occurs when alkenes are exposed to an atmosphere of hydrogen gas in the presence of a catalyst. We describe the result by saying that the double bond has been **hydrogenated,** or **reduced.** (The word *reduction* in organic chemistry refers to the addition of hydrogen or removal of oxygen from a molecule.) For most alkene hydrogenations, either palladium or platinum (as PtO_2) is used as the catalyst.

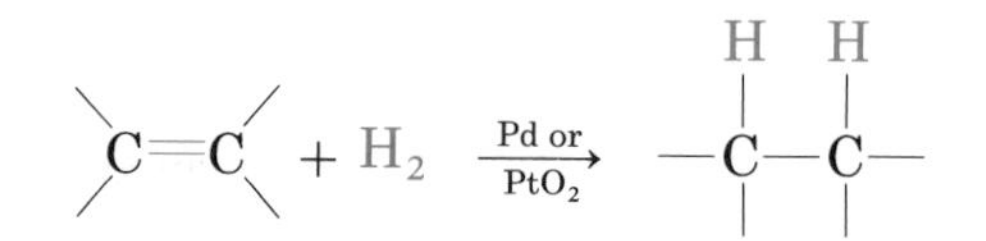

syn stereochemistry referring to a reaction in which only one side of a reactant is involved

Catalytic hydrogenation of alkenes is unlike most other organic reactions in that it is a heterogeneous process, rather than a homogeneous one. That is, the hydrogenation reaction occurs on the surface of solid catalyst particles rather than in solution. The reaction occurs with **syn stereochemistry** (the opposite of *anti*), meaning that both hydrogens add to the double bond from the same side.

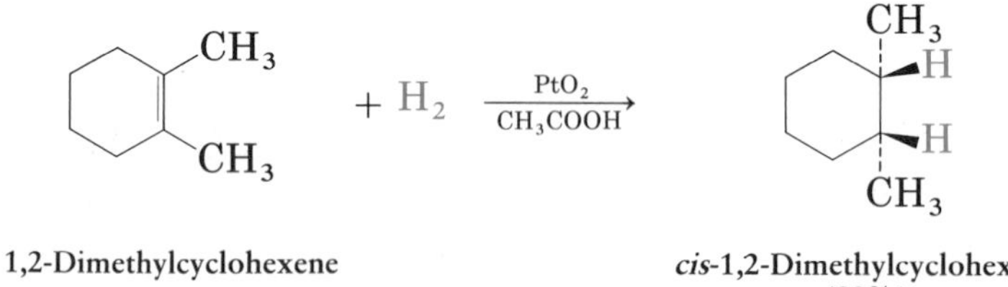

In addition to its usefulness in the laboratory, alkene hydrogenation is a reaction of great commercial value. In the food industry, unsaturated vegetable oils are catalytically hydrogenated on a vast scale to produce the saturated fats used in margarine.

PROBLEM 4.8 What product would you expect to obtain from catalytic hydrogenation of these alkenes?
(a) $(CH_3)_2C{=}CHCH_2CH_3$ (b) 3,3-Dimethylcyclopentene

4.7 OXIDATION OF ALKENES

hydroxylation the addition of –OH groups to a molecule

Hydroxylation of an alkene—the addition of a hydroxyl group to each of the alkene carbons—can be carried out by treatment of the alkene with potassium permanganate, $KMnO_4$, in basic solution. Since oxygen is added to the alkene

oxidation the addition of oxygen to a molecule or the removal of hydrogen from it

diol a dialcohol

during the reaction, we call this an **oxidation.** The reaction occurs with syn stereochemistry and yields a cis 1,2-dialcohol (**diol**) product. For example, cyclohexene gives *cis*-1,2-cyclohexanediol in 37% yield.

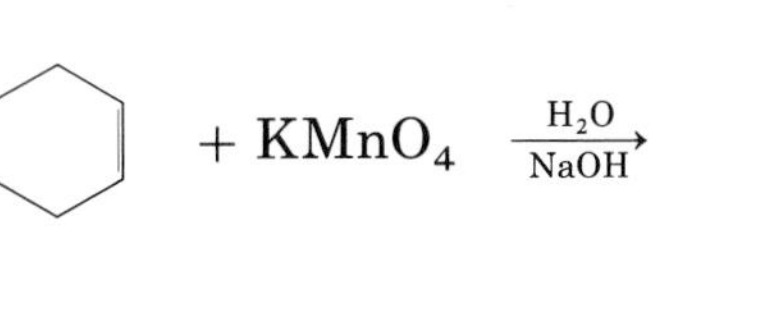

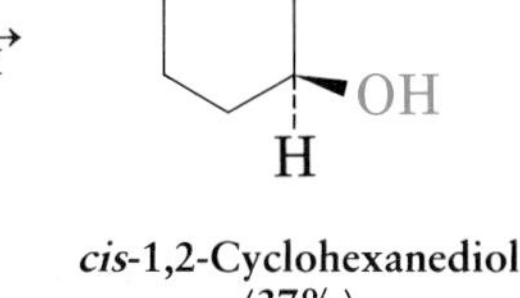

Cyclohexene **cis-1,2-Cyclohexanediol** (37%)

If the reaction of the alkene with $KMnO_4$ is carried out in either neutral or acidic solution, cleavage of the double bond occurs, giving carbonyl-containing products in moderate yield. If the double bond is tetrasubstituted, the two carbonyl-containing products are ketones; if a hydrogen is present on the double bond, one of the carbonyl-containing products is a carboxylic acid; and if two hydrogens are present on one carbon, CO_2 is formed:

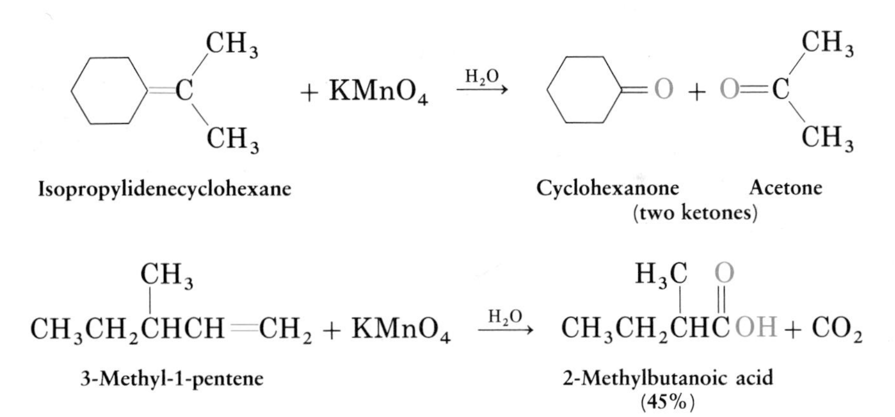

ozonide the addition product of ozone and an alkene

An alternative method for oxidatively cleaving carbon–carbon double bonds is to treat an alkene with ozone, O_3. Conveniently prepared by passing a stream of oxygen through a high-voltage electrical discharge, ozone adds rapidly to alkenes at low temperature to yield **ozonides.**

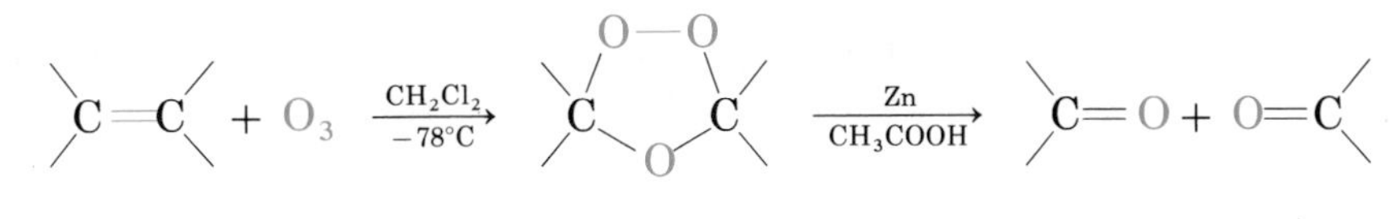

An ozonide

Since they're sometimes explosive, ozonides aren't usually isolated. Instead, they are treated with a reducing agent such as zinc metal in acetic acid to convert them to carbonyl compounds. The net result of the ozonolysis–zinc-reduction sequence is that the carbon–carbon double bond is cleaved, and oxygen becomes doubly bonded to each of the original alkene carbons. If a tetrasubstituted double bond is ozonized, two ketones result; if a trisubstituted double bond is ozonized, one ketone and one aldehyde result; and so on.

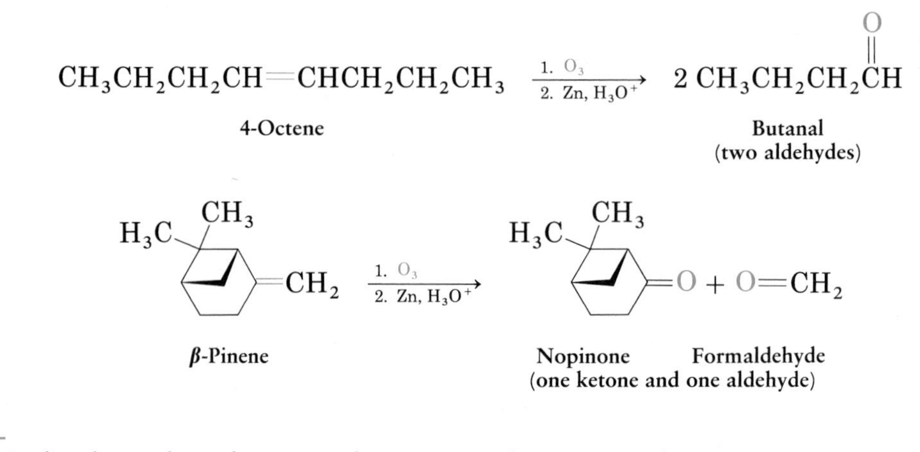

PRACTICE PROBLEM 4.3 Predict the product of reaction of 2-pentene with aqueous acidic $KMnO_4$.

SOLUTION Reaction of acidic $KMnO_4$ with an alkene yields carbonyl-containing products in which the double bond is broken and the two fragments have C=O in place of the original alkene C=C. If a hydrogen is present on the double bond, a carboxylic acid is produced. Thus, 2-pentene gives the following reaction:

$$\underset{\text{2-Pentene}}{CH_3CH_2CH{=}CHCH_3} + KMnO_4 \xrightarrow{H_2O} \underset{\text{Propanoic acid}}{CH_3CH_2\overset{O}{\overset{\|}{C}}OH} + \underset{\text{Acetic acid}}{HO\overset{O}{\overset{\|}{C}}CH_3}$$

PRACTICE PROBLEM 4.4 What alkene gives a mixture of acetone and propanal on ozonolysis followed by reduction with zinc?

$$? \xrightarrow[\text{2. Zn, } H_3O^+]{\text{1. } O_3} CH_3\overset{O}{\overset{\|}{C}}CH_3 + CH_3CH_2\overset{O}{\overset{\|}{C}}H$$

SOLUTION To find out what starting alkene gives the ozonolysis products shown, simply remove the oxygen atoms from the two products and rejoin the carbon fragments with a double bond:

$$\underset{\text{2-Methyl-2-pentene}}{CH_3\overset{CH_3}{\overset{|}{C}}{=}CHCH_2CH_3} \xrightarrow[\text{2. Zn, } H_3O^+]{\text{1. } O_3} \underset{\text{Acetone}}{CH_3\overset{O}{\overset{\|}{C}}CH_3} + \underset{\text{Propanal}}{CH_3CH_2\overset{O}{\overset{\|}{C}}H}$$

PROBLEM 4.9 Predict the product of the reaction of 1,2-dimethylcyclohexene with the following:

(a) Aqueous acidic $KMnO_4$ (b) Ozone, followed by zinc

PROBLEM 4.10 Propose structures for alkenes that yield these products on ozonolysis–reduction:

(a) $(CH_3)_2C{=}O + CH_2{=}O$ (b) 2 equiv. $CH_3CH_2CH{=}O$

(c) Cyclohexanone + $CH_3\overset{\displaystyle O}{\overset{\|}{C}}{-}H$

4.8 ALKENE POLYMERS

polymer
a large molecule built up by repetitive bonding of smaller units

monomer
a small building block from which polymers are made

No other group of synthetic organic compounds has had as great an impact on our day-to-day living as the synthetic polymers. A **polymer** is a large molecule built up by repetitive bonding together of many smaller units, called **monomers.** As we'll see in later chapters, nature makes wide use of biological polymers. For example, cellulose is a polymer built of repeating sugar units; proteins are polymers built of repeating amino acid units; and nucleic acids are polymers built of repeating nucleotide units. Although synthetic polymers are chemically much simpler than biopolymers, there is an immense diversity to the structures and properties of synthetic polymers, depending on the nature of the monomers and on the reaction conditions used for polymerization.

Radical Polymerization of Alkenes

Many simple alkenes undergo rapid polymerization when treated with a small amount of a radical catalyst. For example, ethylene yields polyethylene. Ethylene polymerization is usually carried out at high pressure (1000–3000 atm) and high temperature (100–250°C) with a radical catalyst like benzoyl peroxide. The resultant polymer may have anywhere from a few hundred to a few thousand monomer units incorporated into the chain.

$$\underset{\text{Ethylene}}{H_2C{=}CH_2} \xrightarrow[\text{peroxide}]{\text{benzoyl}} \underset{\text{A segment of polyethylene}}{\sim CH_2CH_2{-}CH_2CH_2{-}CH_2CH_2{-}CH_2CH_2{-}CH_2CH_2 \sim}$$

Radical polymerizations of alkenes involve three kinds of steps: initiation steps, propagation steps, and termination steps. *Initiation* occurs when small amounts of radicals are generated by the catalyst (step 1). For example, when benzoyl peroxide is used as initiator, the oxygen–oxygen bond is broken on heating to yield benzoyloxy radicals. One of these radicals adds to the double bond of an ethylene molecule to generate a new carbon radical (step 2), and the polymerization is off and running. Note that this radical addition step results in formation of a bond between the initiator and the ethylene molecule in which one electron has been contributed by each partner. The remaining electron from the ethylene pi bond remains on carbon as the new radical site.

Initiation

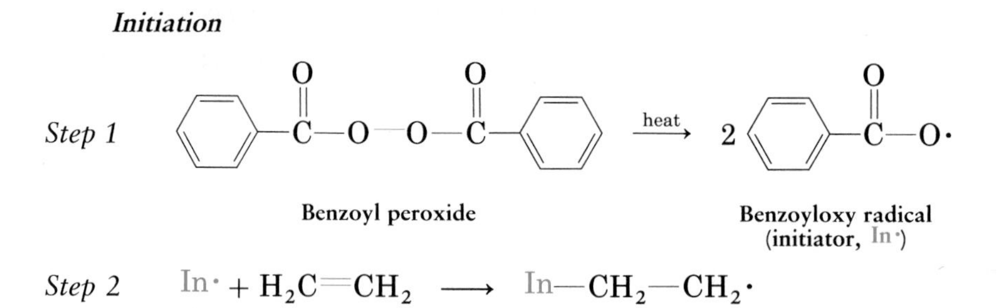

Step 2 $\quad \text{In}\cdot + H_2C{=}CH_2 \longrightarrow \text{In}{-}CH_2{-}CH_2\cdot$

Propagation of the reaction occurs when the carbon radical adds to another ethylene molecule (step 3). Repetition of step 3 for hundreds or thousands of times builds the polymer chain.

Propagation

Step 3 $\quad \text{In}{-}CH_2{-}CH_2\cdot + H_2C{=}CH_2$

$$\longrightarrow \text{In}{-}CH_2{-}CH_2{-}CH_2{-}CH_2\cdot$$

$$\xrightarrow[\text{times}]{\text{repeat many}} \text{In}{-}(CH_2CH_2)_n CH_2CH_2\cdot$$

Eventually, the polymer chain is *terminated* by reactions that consume the radical. For example, combination of two chains by chance meeting (step 4) is a possible chain-terminating reaction.

Termination

Step 4 $\quad 2\,R{-}CH_2CH_2\cdot \longrightarrow R{-}CH_2CH_2{-}CH_2CH_2{-}R$

Polymerization of Substituted Ethylenes

vinyl monomer a simple substituted ethylene used to make polymers

Many substituted ethylenes (**vinyl monomers**) undergo radical-initiated polymerization to yield polymers with substituent groups (denoted by a circled S) regularly spaced along the polymer backbone.

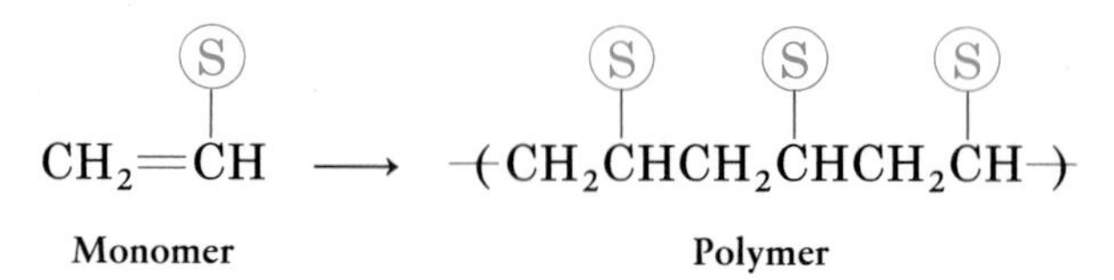

Table 4.1 shows some of the more important vinyl monomers and lists the industrial uses of the different polymers that result.

TABLE 4.1 **Some alkene polymers and their uses**

Monomer name	*Formula*	*Trade or common names of polymer*	*Uses*
Ethylene	$H_2C{=}CH_2$	Polyethylene	Packaging, bottles, cable insulation, films and sheets
Propene (propylene)	$H_2C{=}CHCH_3$	Polypropylene	Automotive moldings, rope, carpet fibers
Chloroethylene (vinyl chloride)	$H_2C{=}CHCl$	Poly(vinyl chloride), Tedlar	Insulation, films, pipes
Styrene	$H_2C{=}CHC_6H_5$	Polystyrene, Styron	Foam and molded articles
Tetrafluoroethylene	$F_2C{=}CF_2$	Teflon	Valves and gaskets, coatings
Acrylonitrile	$H_2C{=}CHCN$	Orlon, Acrilan	Fibers
Methyl methacrylate	$H_2C{=}C(CH_3)CO_2CH_3$	Plexiglas, Lucite	Molded articles, paints
Vinyl acetate	$H_2C{=}CHOCOCH_3$	Poly(vinyl acetate)	Paints, adhesives
Vinyl alcohol	"$H_2C{=}CHOH$"	Poly(vinyl alcohol)	Fibers, adhesives

PRACTICE PROBLEM 4.5 Show the structure of poly(vinyl chloride) by drawing several repeating units. Vinyl chloride is $H_2C{=}CHCl$.

SOLUTION The general structure of poly(vinyl chloride) is

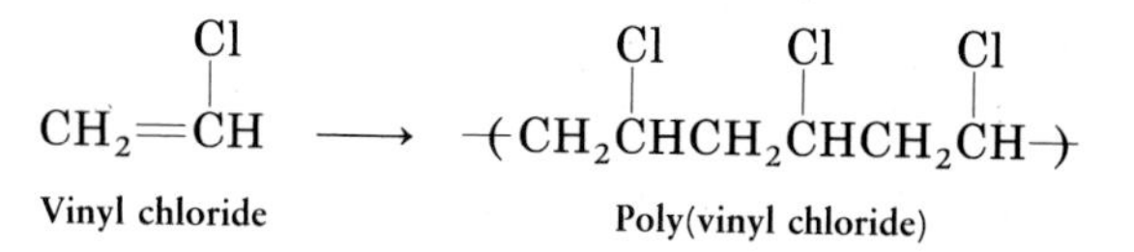

Vinyl chloride

Poly(vinyl chloride)

PROBLEM 4.11 Show the structure of polypropylene by drawing several repeating units. Propylene is $CH_3CH{=}CH_2$.

4.9 PREPARATION OF ALKENES: ELIMINATION REACTIONS

Just as addition reactions account for most of the chemistry that alkenes undergo, *elimination reactions* account for most of the ways used to prepare alkenes. Additions and eliminations are, in many respects, two sides of the same coin:

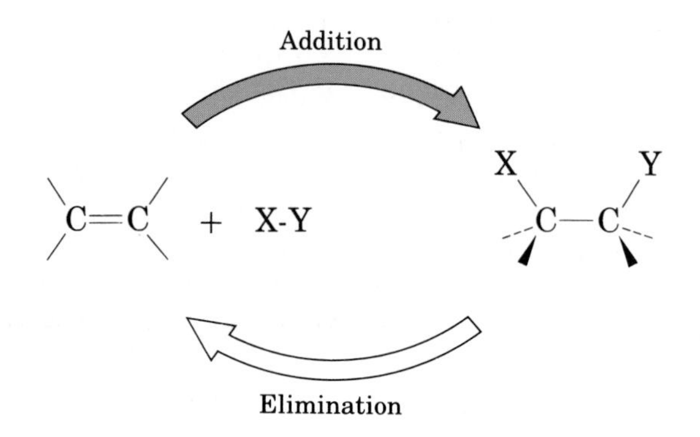

dehydrohalogenation the elimination of HX from an alkyl halide to yield an alkene

dehydration the loss of water from an alcohol to yield an alkene

Let's look briefly at two elimination reactions, the **dehydrohalogenation** of an alkyl halide (elimination of HX) and the **dehydration** of an alcohol (elimination of water, H_2O). We'll return for a closer look at how these reactions take place in Chapter 7.

Elimination of HX from Alkyl Halides: Dehydrohalogenation

Alkyl halides can be synthesized by addition of HX to alkenes. Conversely, alkenes can be synthesized by elimination of HX from alkyl halides. Dehydrohalogenation is usually effected by treating the alkyl halide with a strong base. Thus, bromocyclohexane yields cyclohexene when treated with potassium hydroxide in alcohol solution:

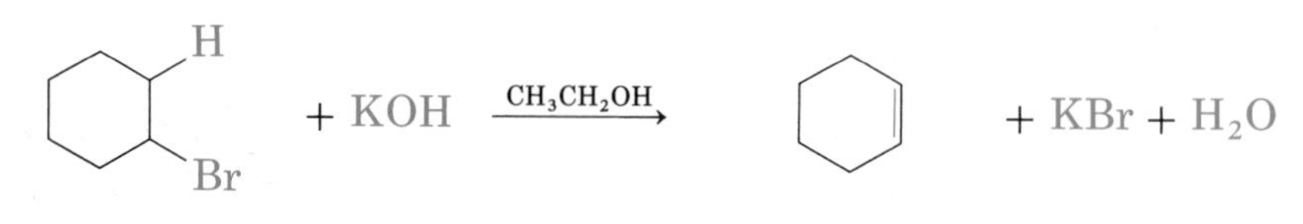

Bromocyclohexane **Cyclohexene** (81%)

Elimination reactions are somewhat more complex than addition reactions because of the regiochemistry problem: what products will result from dehydrohalogenation of unsymmetrical halides? In fact, elimination reactions almost always give mixtures of alkene products. The best we can usually do is to predict which product will be major.

According to a rule formulated by the Russian chemist Alexander Zaitsev[2], base-induced elimination reactions generally give the more highly substituted alkene product. For example, if 2-bromobutane is treated with sodium ethoxide in ethanol, Zaitsev's rule predicts that 2-butene (disubstituted; two alkyl-group substituents on

[2] Also spelled Saytzeff, according to the German pronunciation.

the double-bond carbons) should predominate over 1-butene (monosubstituted; one alkyl-group substituent on the double-bond carbons). This is exactly what is found.

$$CH_3CH_2\overset{\overset{Br}{|}}{C}HCH_3 \xrightarrow[CH_3CH_2OH]{KOH} CH_3CH{=}CHCH_3 + CH_3CH_2CH{=}CH_2$$

2-Bromobutane **2-Butene (81%)** **1-Butene (19%)**

PRACTICE PROBLEM 4.6 What product would you expect from reaction of 1-chloro-1-methylcyclohexane with KOH?

SOLUTION Treatment of an alkyl halide with a strong base like KOH causes dehydrohalogenation and yields an alkene. To find the products in a specific case, draw the structure of the starting material and locate the hydrogen atoms on each neighboring carbon. Then generate the potential alkene products by removing HX in as many ways as possible. The major product will be the one that has the most highly substituted double bond:

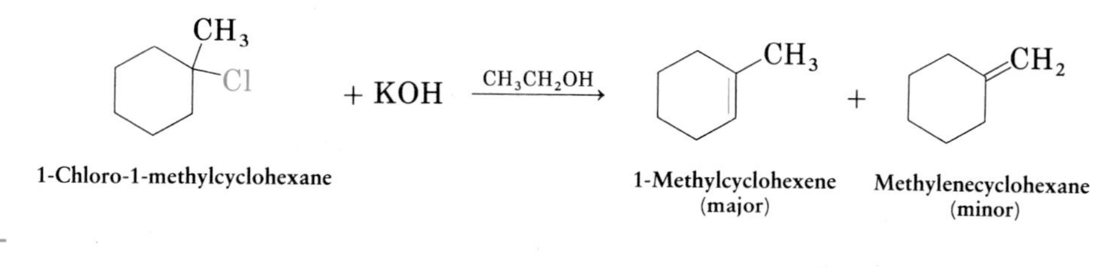

PROBLEM 4.12 What products would you expect from the reaction of 2-bromo-2-methylbutane with KOH? Which will be major?

PROBLEM 4.13 What alkyl halide starting materials might these alkenes have come from?

(a) $CH_3\overset{\overset{CH_3}{|}}{C}HCH_2CH_2\overset{\overset{CH_3}{|}}{C}HCH{=}CH_2$

(b) 3,4-dimethylcyclopentene structure with two CH_3 groups

Elimination of H_2O from Alcohols: Dehydration

The dehydration of alcohols is one of the most useful methods of alkene synthesis, and many ways of carrying out the reaction have been devised. A method that works particularly well for tertiary alcohols is acid-catalyzed dehydration. For example, when 1-methylcyclohexanol is treated with aqueous sulfuric acid, dehydration occurs to yield 1-methylcyclohexene:

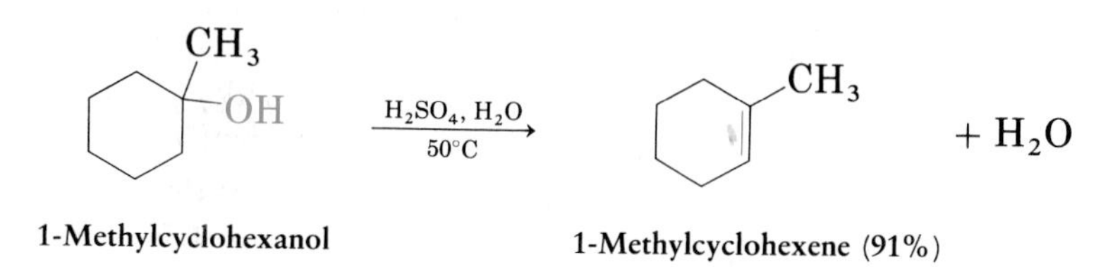

Acid-catalyzed dehydrations usually follow Zaitsev's rule and yield the more highly substituted alkene as major product. Thus, 2-methyl-2-butanol gives primarily 2-methyl-2-butene (trisubstituted) rather than 2-methyl-1- butene (disubstituted):

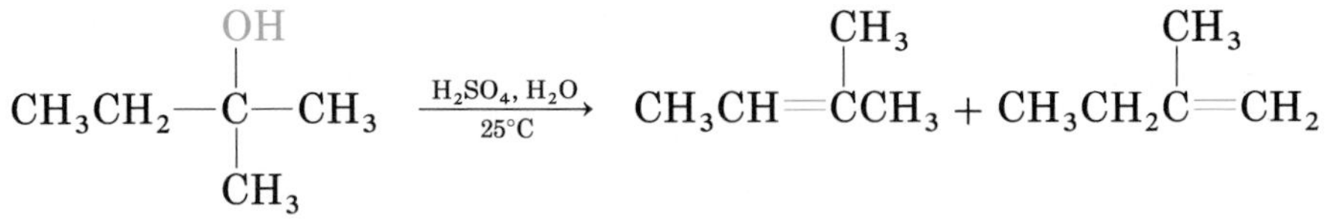

$$CH_3CH_2-\overset{OH}{\underset{CH_3}{C}}-CH_3 \xrightarrow[25°C]{H_2SO_4,\ H_2O} CH_3CH{=}\overset{CH_3}{C}CH_3 + CH_3CH_2\overset{CH_3}{C}{=}CH_2$$

2-Methyl-2-butanol — 2-Methyl-2-butene (major) — 2-Methyl-1-butene (minor)

PRACTICE PROBLEM 4.7 Predict the major product of this reaction:

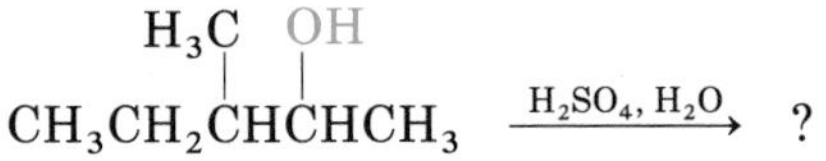

$$CH_3CH_2\overset{H_3C}{C}H\overset{OH}{C}HCH_3 \xrightarrow{H_2SO_4,\ H_2O} ?$$

SOLUTION Treatment of an alcohol with acid leads to dehydration and formation of the more highly substituted alkene product (Zaitsev's rule). Thus, dehydration of 3-methyl-2-pentanol should yield 3-methyl-2-pentene as the major product rather than 3-methyl-1-pentene:

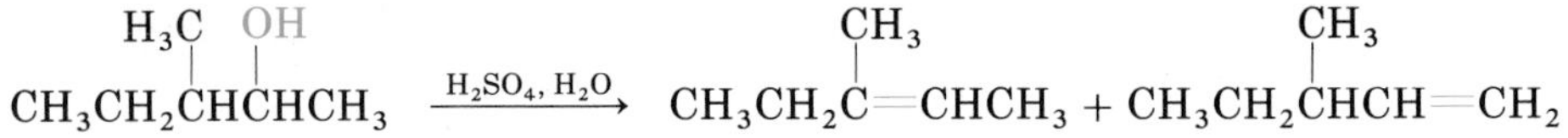

$$CH_3CH_2\overset{H_3C}{C}H\overset{OH}{C}HCH_3 \xrightarrow{H_2SO_4,\ H_2O} CH_3CH_2\overset{CH_3}{C}{=}CHCH_3 + CH_3CH_2\overset{CH_3}{C}HCH{=}CH_2$$

3-Methyl-2-pentanol — 3-Methyl-2-pentene (major) — 3-Methyl-1-pentene (minor)

PROBLEM 4.14 Predict the products you would expect from these reactions. Indicate the major product in each case.

(a) 2-Bromo-2-methylpentane + KOH $\longrightarrow$?

(b) $CH_3\overset{H_3C}{C}H-\overset{OH}{\underset{CH_3}{C}}-CH_2CH_3 \xrightarrow{H_2SO_4} ?$

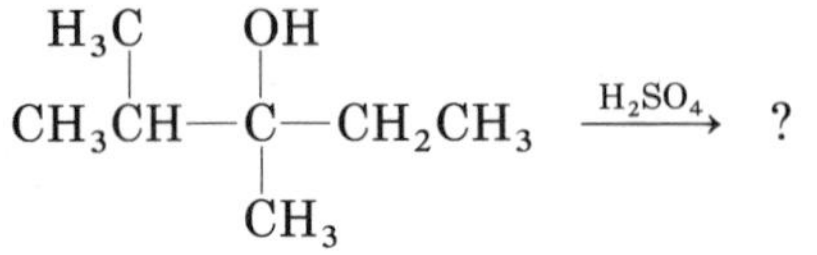

PROBLEM 4.15 What alcohols might these alkenes have come from?

(a) [cyclohexene ring bearing CH_3 and CH_3 substituents] (b) $CH_3CH_2CH{=}CHCH_2CH_2CH_3$

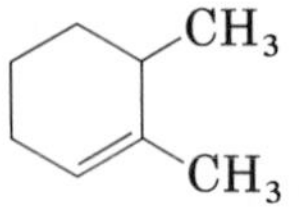

4.10 CONJUGATED DIENES

conjugation alternating single and double bonds in a molecule

Double bonds that alternate with single bonds are said to be **conjugated.** Thus, 1,3-butadiene is a **conjugated diene** whereas 1,4-pentadiene is a nonconjugated diene with isolated double bonds.

conjugated diene a diene whose two double bonds are separated by a single bond

$H_2C{=}CH{-}CH{=}CH_2$

1,3-Butadiene

A conjugated diene with alternating single and double bonds

$H_2C{=}CH{-}CH_2{-}CH{=}CH_2$

1,4-Pentadiene

A nonconjugated diene with nonalternating single and double bonds

What's so special about conjugated dienes that we need to look at them separately? The orbital view of 1,3-butadiene shown in Figure 4.6 provides a clue to the answer: *There is an electronic interaction between the two double bonds of a conjugated diene* because of *p*-orbital overlap across the central single bond. This interaction of *p* orbitals across a single bond gives conjugated dienes some unusual properties.

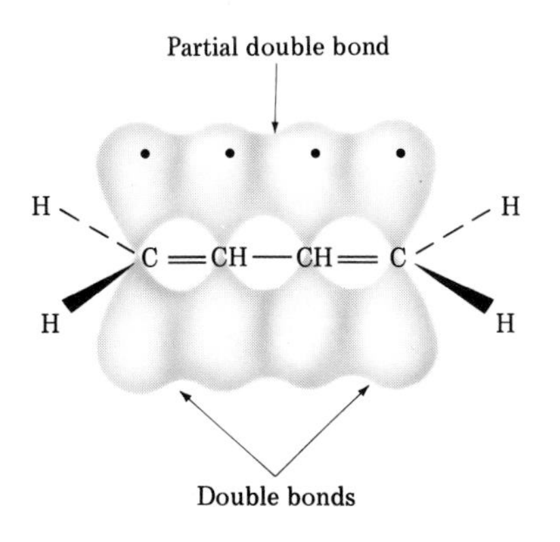

FIGURE 4.6 An orbital view of 1,3-butadiene. Each of the four carbon atoms has a *p* orbital, allowing for an electronic interaction across the C2–C3 single bond.

Although much of the chemistry of conjugated dienes and isolated alkenes is similar, there's a striking difference in their addition reactions with electrophiles like HX and X_2. When HX adds to an isolated alkene, Markovnikov's rule usually predicts the formation of a single product. When HX adds to a conjugated diene, though, mixtures of products are usually obtained. For example, reaction of HBr with 1,3-butadiene yields two products:

$$\underset{\text{1,3-Butadiene}}{CH_2{=}CH{-}CH{=}CH_2} + HBr \longrightarrow \underset{\substack{\text{3-Bromo-1-butene (71\%)}\\\text{(1,2-addition)}}}{\overset{4}{C}H_2{=}\overset{3}{C}H{-}\overset{2}{C}H(Br){-}\overset{1}{C}H_2(H)} + \underset{\substack{\text{1-Bromo-2-butene (29\%)}\\\text{(1,4-addition)}}}{\overset{4}{C}H_2(Br){-}\overset{3}{C}H{=}\overset{2}{C}H{-}\overset{1}{C}H_2(H)}$$

1,4-addition the addition of an electrophile to carbons 1 and 4 of a conjugated diene

allylic next to a double bond

3-Bromo-1-butene (a secondary bromide) is the normal product of Markovnikov addition, but 1-bromo-2-butene (a primary bromide) is unexpected. The double bond in this product has moved to a position between carbons 2 and 3, and H–Br has added to carbons 1 and 4. How can we account for the formation of this **1,4-addition** product?

The answer is that an *allylic carbocation* is involved as an intermediate in the reaction (**allylic** means next to a double bond). When H^+ adds to an electron-rich pi bond of 1,3-butadiene, two carbocation intermediates are possible: a primary nonallylic carbocation and a secondary allylic carbocation. Allylic carbocations are very stable and therefore form in preference to less stable, nonallylic carbocations.

$$CH_2{=}CH{-}CH{=}CH_2 + H^+ \longrightarrow CH_2{=}CH{-}\overset{+}{C}H{-}CH_3$$

1,3-Butadiene → Secondary allylic carbocation

$$CH_2{=}CH{-}CH{=}CH_2 + H^+ \not\longrightarrow CH_2{=}CH{-}CH_2{-}\overset{+}{C}H_2$$

Primary nonallylic carbocation (not formed)

4.11 STABILITY OF ALLYLIC CARBOCATIONS: RESONANCE

Why are allylic carbocations stable? To get an idea of the reason, look at the orbital picture of an allylic carbocation in Figure 4.7. The positively charged carbon atom has a vacant *p* orbital that can overlap the *p* orbitals of the neighboring double bond.

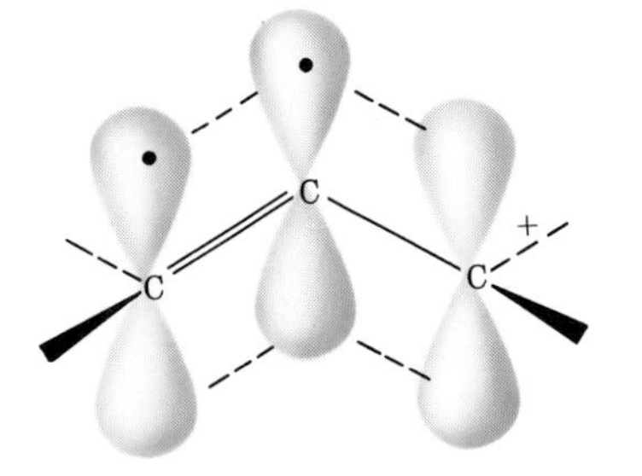

FIGURE 4.7 An orbital picture of an allylic carbocation. The vacant *p* orbital on the positively charged carbon can overlap the double-bond *p* orbitals.

From a *p*-orbital point of view, an allylic carbocation is symmetrical. All three carbon atoms are sp^2 hybridized, and each has a *p* orbital. Thus, the *p* orbital on the central carbon can overlap equally well with *p* orbitals on *either* of the two neighboring carbons. The two electrons are free to move about and spread out over the entire three-orbital array, as indicated in Figure 4.7.

One consequence of this orbital picture is that there are two ways to draw an allylic carbocation. We can draw it with the vacant orbital on the left and the double bond on the right, or we can draw it with the vacant orbital on the right and the double bond on the left. *Neither structure is completely correct: The true structure of the allylic carbocation is somewhere in between the two.*

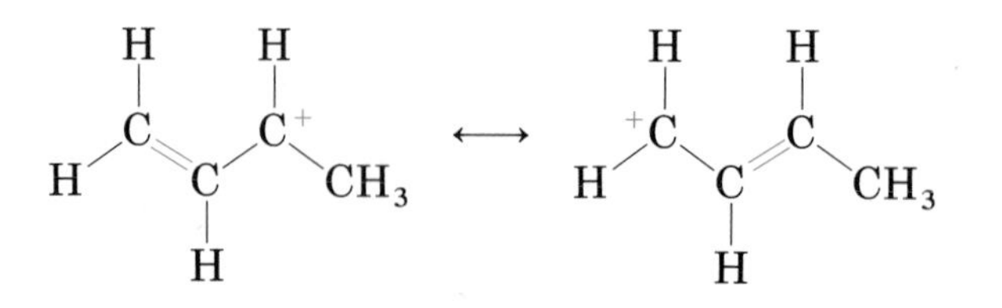

Two resonance forms of an allylic carbocation

resonance forms two representations of a molecule that differ only in where the bonding electrons are placed

The two individual structures are called **resonance forms,** and their special relationship is indicated by the double-headed arrow between them. The only difference between the resonance forms is the position of the bonding electrons. The nuclei don't move but occupy exactly the same places in both resonance forms.

The best way to think about resonance is to realize that a species like an allylic carbocation is no different from any other organic substance. An allylic carbocation doesn't jump back and forth between two resonance forms, spending part of its time looking like one and the rest of its time looking like the other; rather, it has a single, unchanging structure that we call a **resonance hybrid.** (A useful analogy is to think of a resonance hybrid as being like a mutt, or mixed-breed dog. Just as a dog that's a mixture of dachshund and German shepherd doesn't change back and forth from one to the other, a resonance hybrid doesn't change back and forth.)

resonance hybrid the true structure of a molecule described by different resonance forms

The difficulty in understanding resonance hybrids is visual, because we can't draw an accurate single picture of a resonance hybrid by using familiar kinds of structures. The line-bond structures that serve so well to represent most organic molecules just don't work well for resonance hybrids like allylic carbocations. We might try to represent the allylic carbocation by using a dotted line to indicate that the two C–C bonds are equivalent and that each is approximately $1\frac{1}{2}$ bonds, but such a drawing really doesn't help much and won't be used again in this book.

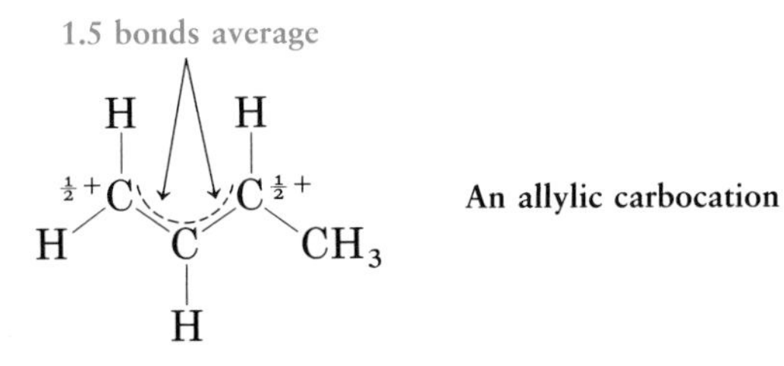

An allylic carbocation

One of the most important postulates of resonance theory is that *the greater the number of possible resonance forms, the greater the stability of the compound.* Since an allylic carbocation is a resonance hybrid of two line-bond structures, it's therefore more stable than a normal carbocation. This stability is due to the fact that the pi electrons can be spread out (*delocalized*) over an extended *p*-orbital network rather than centered on only one site.

In addition to affecting stability, the resonance picture of an allylic carbocation also has chemical consequences. When the allylic carbocation produced by protonation of 1,3-butadiene reacts with bromide ion to complete the addition reaction, attack can occur at either C1 or C3 because both share the positive charge. The result is a mixture of 1,2- and 1,4-addition products:

$$CH_2{=}CH{-}CH{=}CH_2$$

$$\downarrow H^+$$

$$\left[\overset{+}{C}H_2{-}CH{=}CH{-}CH_3 \longleftrightarrow CH_2{=}CH{-}\overset{+}{C}H{-}CH_3\right]$$

$$\downarrow Br^-$$

$$\underset{\text{1,4-Addition}}{\overset{Br}{\overset{|}{C}}H_2{-}CH{=}CH{-}CH_3} + \underset{\text{1,2-Addition}}{CH_2{=}CH{-}\overset{Br}{\overset{|}{C}}H{-}CH_3}$$

PROBLEM 4.16 1,3-Butadiene reacts with Br_2 to yield a mixture of 1,2- and 1,4-addition products. Show the structure of each.

4.12 DRAWING AND INTERPRETING RESONANCE FORMS

Resonance is an extremely useful concept for explaining a variety of phenomena. In inorganic chemistry, for example, the carbonate ion CO_3^{2-} is known to have identical bond lengths for its three C–O bonds. Although there is no single line-bond structure that can account for this equality of C–O bonds, resonance theory accounts for it nicely. The carbonate ion is simply a resonance hybrid of three resonance forms. The three oxygens share the pi electrons and the negative charges equally:

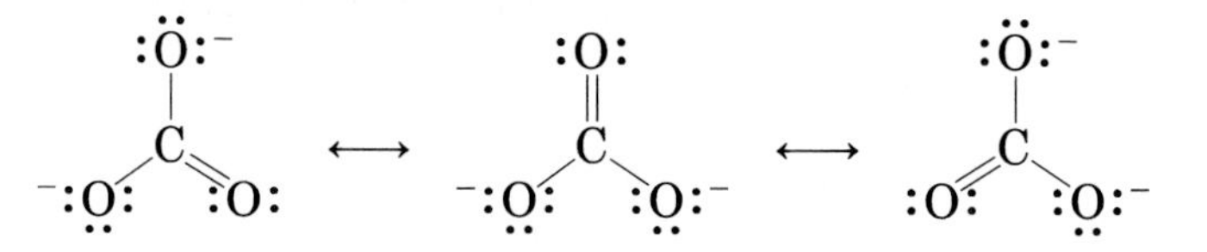

As an example from organic chemistry, we'll see in the next chapter that the six C–C bonds in aromatic compounds like benzene are equivalent because benzene is a resonance hybrid of two forms. Each form has alternating single and double bonds, and neither form is correct by itself. The true benzene structure is a hybrid of the two forms.

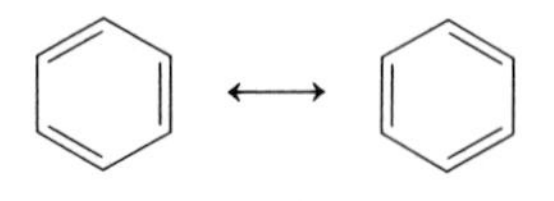

Two resonance forms of benzene

When first dealing with resonance theory, it's often useful to have a set of guidelines that describe how to draw and interpret resonance forms. The following five rules should prove helpful:

Rule 1. *Resonance forms are imaginary, not real.* The real structure is a composite hybrid of the different forms. Substances like the allylic carbocation, the carbonate ion, and benzene are no different from any other substance in having single, unchanging structures. The only difference is in the way they must be represented on paper.

Rule 2. *Resonance forms differ from each other only in the placement of the pi electrons.* Neither the position nor the hybridization of atoms changes from one resonance form to another. In benzene, for example, the pi electrons in the double bonds move, but the six carbon atoms remain in place:

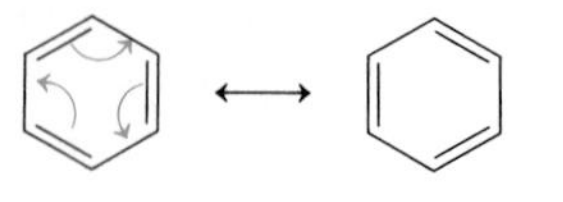

By contrast, two structures like 1,3-cyclohexadiene and 1,4-cyclohexadiene are *not* resonance structures because their hydrogen atoms don't occupy the same positions. Instead, the two dienes are constitutional isomers:

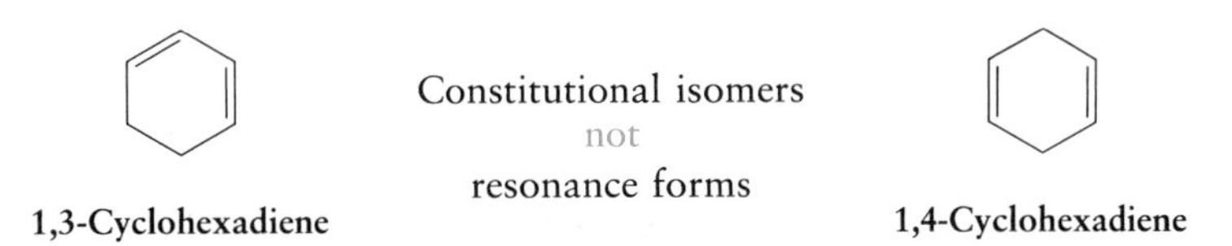

Rule 3. *Different resonance forms of a substance don't have to be equivalent.* For example, the allylic carbocation obtained by reaction of 1,3-butadiene with H^+ is unsymmetrical. One end of the delocalized pi-electron system has a methyl substituent, and the other end is unsubstituted. Even though the two resonance forms aren't equivalent, they both contribute to the overall resonance hybrid.

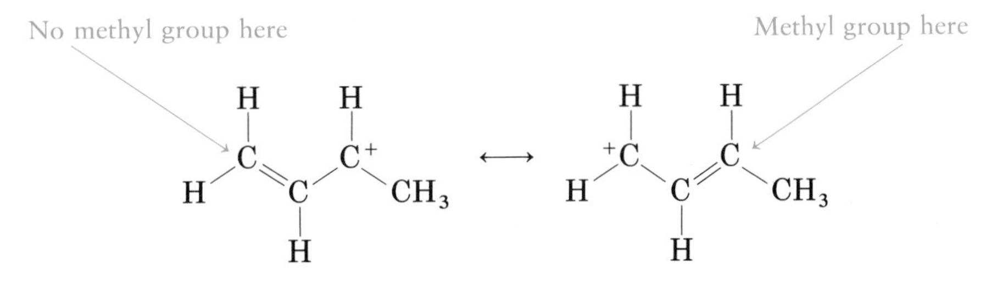

In general, when two resonance forms are not equivalent, the actual structure of the resonance hybrid is closer to the more stable form than to the less stable form. Thus, we might expect the butenyl carbocation to look a bit more like a secondary carbocation than like a primary one.

Rule 4. *All resonance forms must obey normal rules of valency.* Resonance forms are like any other structure: The octet rule still holds. For example, one of the following structures for the carbonate ion is not a valid resonance form because the carbon atom has five bonds and ten electrons:

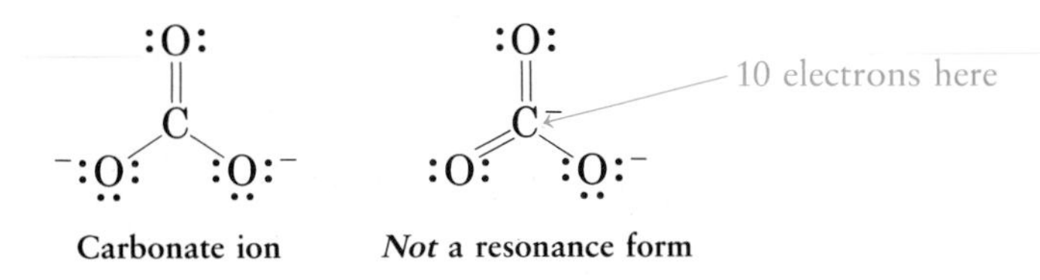

Rule 5. *The resonance hybrid is more stable than any single resonance form.* In other words, resonance leads to stability. The greater the number of resonance forms possible, the more stable the substance. We've already seen, for example, that an allylic carbocation is more stable than a normal carbocation. In a similar manner, we'll see in the next chapter that a benzene ring is more stable than a cyclic alkene.

PRACTICE PROBLEM 4.8 Use resonance structures to explain why the two C–O bonds of sodium formate are equivalent.

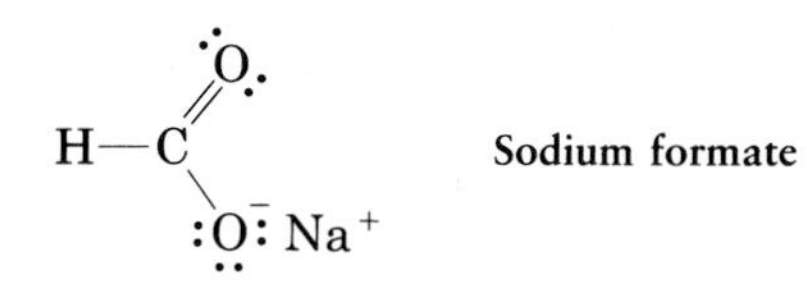

SOLUTION The formate anion is a resonance hybrid of two equivalent resonance forms. The two resonance forms can be drawn by showing the double bond either to the top oxygen or to the bottom oxygen. Only the positions of the electrons are different in the two structures.

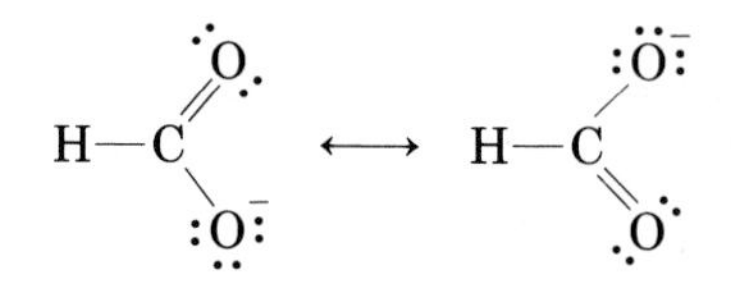

PROBLEM 4.17 Give the structure of all possible monoadducts of HCl and 1,3-pentadiene.

PROBLEM 4.18 Look at the possible carbocation intermediates produced during addition of HCl to 1,3-pentadiene (Problem 4.17) and predict which is the most stable.

PROBLEM 4.19 Draw as many resonance structures as you can for these species:

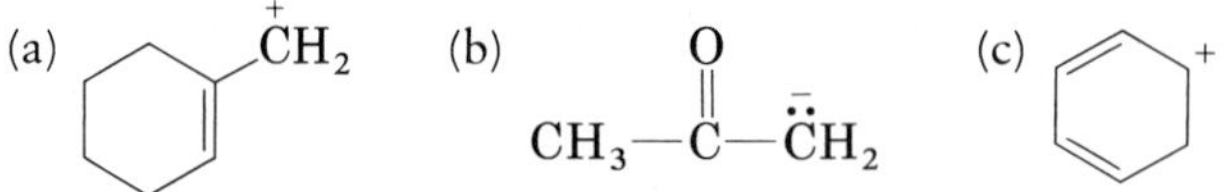

4.13 ALKYNES

alkyne a hydrocarbon that has a carbon–carbon triple bond

Alkynes are hydrocarbons that contain a carbon–carbon triple bond. Since two pairs of hydrogens must be removed from an alkane, C_nH_{2n+2}, to generate a triple bond, the general formula for an alkyne is C_nH_{2n-2}.

As we saw in Section 1.11, a carbon–carbon triple bond results from the overlap of two *sp*-hybridized carbon atoms. The two *sp*-hybrid orbitals of carbon lie at an angle of 180° to each other along an axis that is perpendicular to the axes of the two unhybridized $2p_y$ and $2p_z$ orbitals. When two such *sp*-hybridized carbons approach each other for bonding, the geometry is perfect for the formation of one *sp–sp* sigma bond and two *p–p* pi bonds—a net triple bond (Figure 4.8). The two remaining *sp* orbitals form bonds to other atoms at an angle of 180° from the carbon–carbon sigma bond. For example, acetylene, H–C≡C–H, is a linear molecule with H–C–C bond angles of 180°.

Alkynes follow closely the general rules of hydrocarbon nomenclature already discussed for alkanes (Section 2.3) and alkenes (Section 3.1). The suffix *-yne* is used

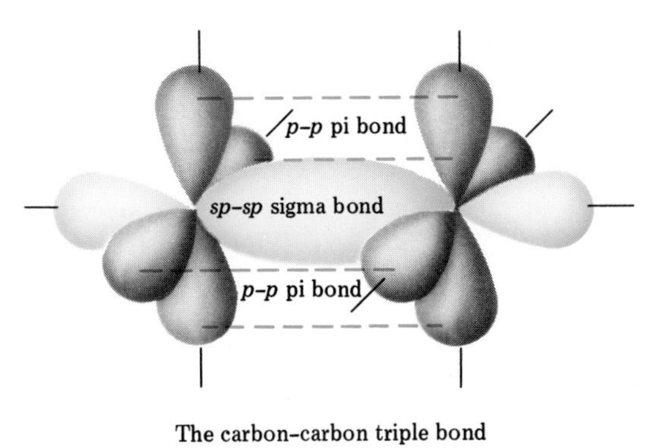

FIGURE 4.8 The electronic structure of a carbon–carbon triple bond

in the base hydrocarbon name to denote an alkyne, and the position of the triple bond is indicated by its number in the chain. Numbering always begins at the chain end nearer the triple bond so that the triple bond receives as low a number as possible.

$$\underset{7}{CH_3}\underset{6}{CH_2}\underset{5}{\overset{\overset{\displaystyle CH_3}{|}}{C}H}\underset{4}{CH_2}\underset{3}{C}{\equiv}\underset{2}{C}\underset{1}{CH_3}$$

Begin numbering carbons at the end nearer the triple bond

5-Methyl-2-heptyne

Compounds containing both double and triple bonds are called *enynes,* not *ynenes*. Numbering of the hydrocarbon chain always starts from the end nearer the first multiple bond, but if there's a choice in numbering, double bonds receive lower numbers than triple bonds. For example:

$$\underset{1}{CH_3}\underset{2}{CH}{=}\underset{3}{C}H\underset{4}{CH_2}\underset{5}{CH_2}\underset{6}{C}{\equiv}\underset{7}{C}\underset{8}{CH_3}$$

2-Octen-6-yne (not 6-octen-2-yne)

PROBLEM 4.20 Provide IUPAC names for these compounds:

(a) $CH_3CH_2C{\equiv}CCH_2CH(CH_3)_2$ (b) $HC{\equiv}CC(CH_3)_3$
(c) $CH_3CH(CH_3)CH_2C{\equiv}CCH_3$ (d) $CH_3CH{=}CHCH_2C{\equiv}CCH_3$

4.14 REACTIONS OF ALKYNES: ADDITION OF H_2, HX, AND X_2

Based on their structural similarity, we might expect alkynes and alkenes to show chemical similarities also. As a general rule, this prediction is true: Alkynes react in much the same way that alkenes do.

Addition of H_2 to Alkynes

Alkynes are easily converted into alkanes by reduction with two molar equivalents of hydrogen over a palladium catalyst.

$$\underset{\textbf{4-Octyne}}{CH_3CH_2CH_2C{\equiv}CCH_2CH_2CH_3} + 2\,H_2 \xrightarrow[\text{catalyst}]{\text{Pd}} \underset{\textbf{Octane (95\%)}}{CH_3CH_2CH_2CH_2CH_2CH_2CH_2CH_3}$$

The catalytic hydrogenation of an alkyne to yield an alkane proceeds through an intermediate alkene, and the reaction can be stopped at the alkene stage if the proper catalyst is used. The catalyst most often used for this purpose is the Lindlar catalyst, a specially prepared form of palladium metal. Because hydrogenation occurs with syn stereochemistry, alkynes are catalytically reduced to give cis alkenes. For example:

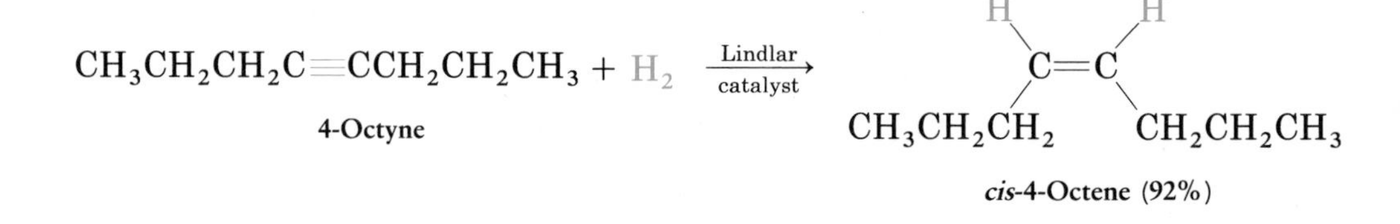

Another method for the reduction of alkynes to alkenes employs lithium metal in liquid ammonia solvent. Remarkably, lithium metal dissolves in pure liquid ammonia solvent at −33°C to produce a deep blue solution. When an alkyne is added to this blue solution, reduction of the triple bond occurs. This method is complementary to the Lindlar reduction, since it yields trans alkenes rather than cis alkenes:

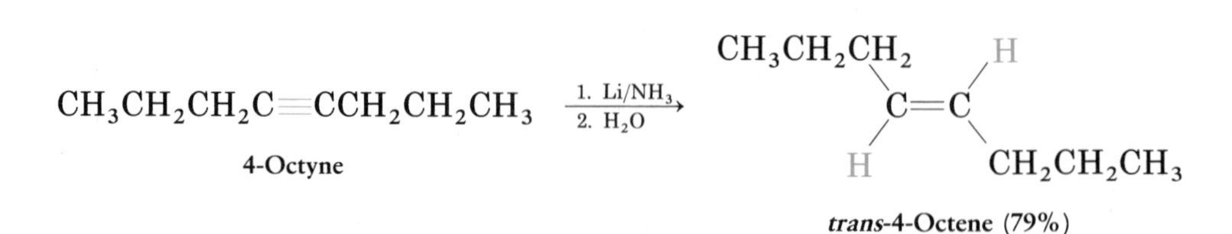

Addition of HX to Alkynes

Alkynes give the expected addition products with HCl, HBr, and HI. Although the reactions can usually be stopped after addition of 1 molar equivalent of HX to yield a haloalkene, an excess of reagent leads to formation of the dihalide product. As the following examples indicate, the regiochemistry of addition to monosubstituted alkynes follows Markovnikov's rule: The H atom adds to the terminal carbon of the triple bond, and the X atom adds to the internal, more highly substituted, carbon:

$$CH_3CH_2CH_2CH_2C{\equiv}CH + HBr \longrightarrow CH_3CH_2CH_2CH_2C(Br){=}CH_2$$

1-Hexyne 2-Bromo-1-hexene

Addition of X_2 to Alkynes

Bromine and chlorine add to alkynes to give addition products with trans stereochemistry:

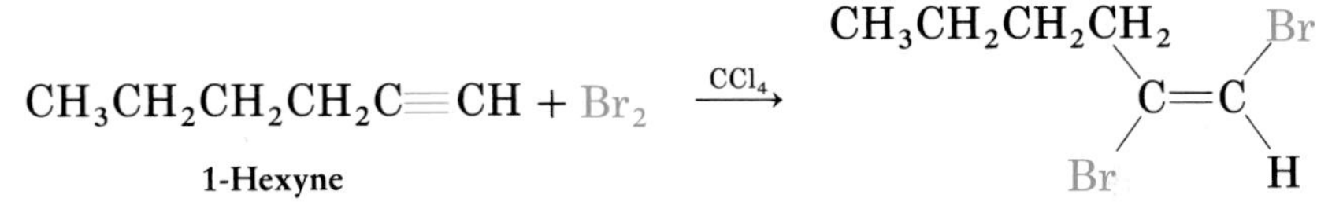

$$CH_3CH_2CH_2CH_2C{\equiv}CH + Br_2 \xrightarrow{CCl_4} (CH_3CH_2CH_2CH_2)(Br)C{=}C(Br)(H)$$

1-Hexyne

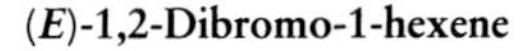

(*E*)-1,2-Dibromo-1-hexene

PROBLEM 4.21 What products would you expect from these reactions?

(a) $CH_3CH_2CH_2C{\equiv}CH + 1\ \text{equiv}\ Cl_2$

(b) $CH_3CH_2CH_2C{\equiv}CCH_2CH_3 + 1\ \text{equiv}\ HBr$

(c) $CH_3CH(CH_3)CH_2C{\equiv}CCH_2CH_3 + H_2 \xrightarrow[\text{catalyst}]{\text{Lindlar}}\ ?$

4.15 ADDITION OF WATER TO ALKYNES

Addition of water takes place when an alkyne is treated with aqueous sulfuric acid in the presence of mercuric sulfate catalyst:

$$CH_3CH_2CH_2C{\equiv}CH + H_2O \xrightarrow[HgSO_4]{H_2SO_4} \left[CH_3CH_2CH_2C(OH){=}CH_2\right] \longrightarrow CH_3CH_2CH_2C({=}O)CH_3$$

1-Pentyne An enol 2-Pentanone (78%)

Markovnikov regiochemistry is found for the hydration reaction, with the H attaching to the less substituted carbon and the OH attaching to the more substituted carbon. Interestingly, though, the expected alkenyl alcohol or *enol* (*ene* = alkene; *ol* = alcohol) is not isolated. Instead, this intermediate enol rearranges to a more stable isomer, a ketone ($R_2C{=}O$). It turns out that enols and ketones rapidly interconvert—a process called **tautomerism.** Tautomers, special kinds of isomers that are readily interconvertible through a rapid equilibration, will be studied in more detail in Section 11.1. With few exceptions, the tautomeric equilibrium heavily favors the ketone; enols are almost never isolated.

tautomerism a word used to describe two rapidly interconverting constitutional isomers

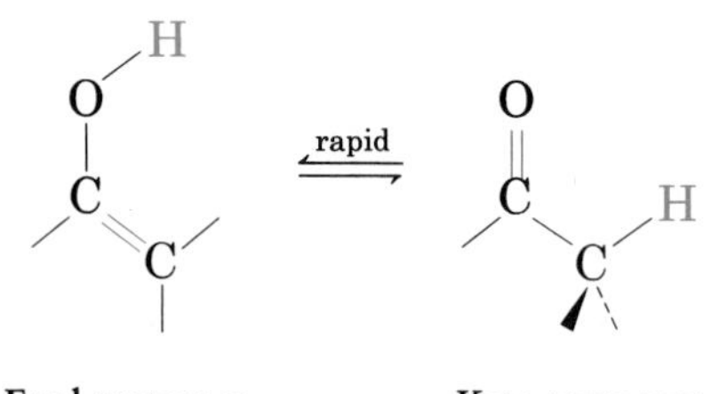

Enol tautomer (less favored) **Keto tautomer** (more favored)

A mixture of both possible ketones results when an internal alkyne (R−C≡C−R′) is hydrated, but only a single product is formed from reaction of a terminal alkyne (R−C≡CH).

$$CH_3CH_2C{\equiv}CCH_3 + H_2O \xrightarrow[HgSO_4]{H_2SO_4} CH_3CH_2\overset{O}{\overset{\|}{C}}CH_2CH_3 + CH_3CH_2CH_2\overset{O}{\overset{\|}{C}}CH_3$$

2-Pentyne (an internal alkyne) — **3-Pentanone** — **2-Pentanone**

$$CH_3CH_2CH_2C{\equiv}CH + H_2O \xrightarrow[HgSO_4]{H_2SO_4} CH_3CH_2CH_2\overset{O}{\overset{\|}{C}}CH_3$$

1-Pentyne (a terminal alkyne) — **2-Pentanone**

PRACTICE PROBLEM 4.9 What product would you obtain by hydration of 4-methyl-1-hexyne?

SOLUTION Addition of water to 4-methyl-1-hexyne according to Markovnikov's rule should yield a product with the OH group attached to C2 rather than to C1. This enol then isomerizes to yield a ketone:

$$CH_3CH_2\overset{CH_3}{\overset{|}{C}}HCH_2C{\equiv}CH + H_2O \xrightarrow[HgSO_4]{H_2SO_4} \left[CH_3CH_2\overset{CH_3}{\overset{|}{C}}HCH_2\overset{OH}{\overset{|}{C}}{=}CH_2\right]$$

4-Methyl-1-hexyne

$$\longrightarrow CH_3CH_2\overset{CH_3}{\overset{|}{C}}HCH_2\overset{O}{\overset{\|}{C}}CH_3$$

4-Methyl-2-hexanone

PROBLEM 4.22 What product would you obtain by hydration of 4-octyne?

PROBLEM 4.23 What alkynes would you start with to prepare these ketones by a hydration reaction?

(a) $CH_3CH_2CH_2\overset{O}{\overset{\|}{C}}CH_3$ (b) $CH_3CH_2CH_2\overset{O}{\overset{\|}{C}}CH_2CH_3$

INTERLUDE

Natural Rubber

Rubber—a most unusual name for a most unusual substance—is a naturally occurring alkene polymer produced by more than 400 different plants. The major source, however, is the so-called rubber tree, *Hevea brasiliensis,* from which the crude material is harvested as it drips from a slice made through the bark. The name *rubber* was coined by Joseph Priestley, the discoverer of oxygen and early researcher of rubber chemistry, for the simple reason that one of its early uses was to rub out pencil marks on paper.

Unlike polyethylene and other simple alkene polymers, natural rubber is a polymer of a conjugated diene, isoprene, or 2-methyl-1,3-butadiene. The polymerization takes place by 1,4-addition (Section 4.10) of each isoprene monomer unit to the growing chain, leading to formation of a polymer that still contains double bonds spaced regularly at four-carbon intervals. As the following structure shows, these double bonds have *Z* stereochemistry.

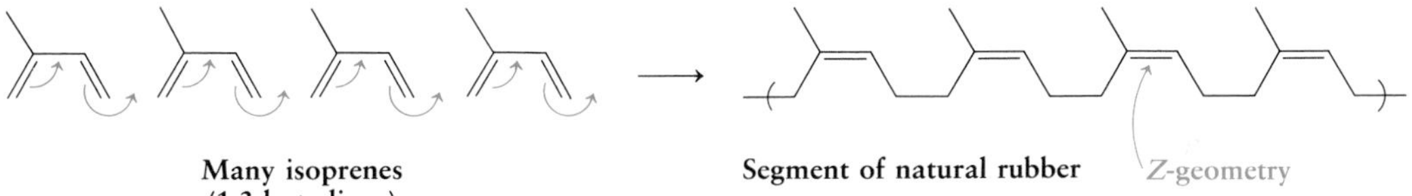

Many isoprenes (1,3-butadiene) → Segment of natural rubber

Z-geometry

Crude rubber (latex) is collected from the tree as an aqueous dispersion that is washed, dried, and coagulated by warming in air to give a polymer with chains that average about 5000 monomer units in length and have molecular weights of 200,000 to 500,000. This crude coagulate is too soft and tacky to be useful until it is hardened by heating with elemental sulfur, a process called *vulcanization*. By mechanisms that are still not fully understood, vulcanization cross-links the rubber chains by forming carbon–sulfur bonds between them, thereby hardening and stiffening the polymer. The exact degree of hardening can be varied, yielding material soft enough for automobile tires or hard enough for bowling balls (*ebonite*).

The remarkable ability of rubber to stretch and then contract to its original shape is due to the irregular shapes of the polymer chains caused by the double bonds. These double bonds introduce bends and kinks into the polymer chains, thereby preventing neighboring chains from nestling together into tightly packed, semicrystalline regions. When stretched, the randomly coiled chains straighten out and orient along the direction of the pull but are kept from sliding over each other by the cross-links. When the stretch is released, the polymer reverts to its original random state (Figure 4.9).

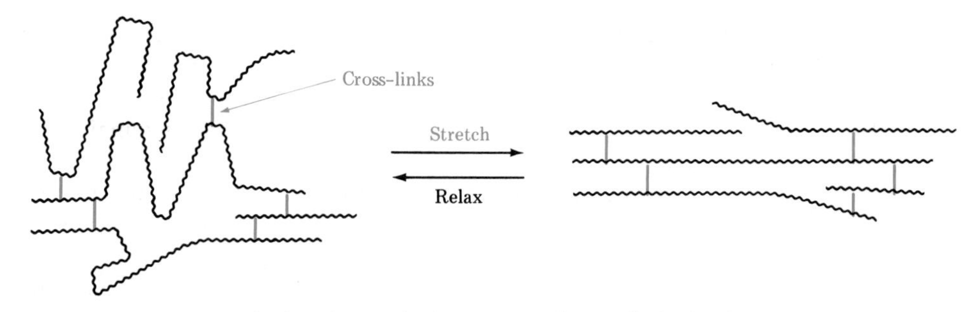

FIGURE 4.9 Unstretched and stretched sections of cross-linked rubber chains

SUMMARY AND KEY WORDS

The chemistry of alkenes is dominated by **addition reactions** of electrophiles. When HX reacts with an alkene, **Markovnikov's rule** predicts that the hydrogen will add to the carbon that has fewer alkyl substituents, and the X group will add to the carbon that has more alkyl substituents. For example:

$$(H_3C)_2C{=}CH_2 + HCl \longrightarrow H_3C{-}C(Cl)(CH_3){-}CH_3$$

Many other electrophiles besides HX add to alkenes. Thus, bromine and chlorine add to give **1,2-dihalide** addition products having **anti stereochemistry.** Addition of water takes place on reaction of the alkene with aqueous acid. Hydrogen can be added to alkenes by reaction in the presence of a metal catalyst such as platinum or palladium.

Oxidation of alkenes is carried out using potassium permanganate, $KMnO_4$. Under basic conditions, $KMnO_4$ reacts with alkenes to yield cis 1,2-diols. Under neutral or acidic conditions, however, $KMnO_4$ cleaves double bonds to yield carbonyl-containing products. Double-bond cleavage can also be effected by reaction of the alkene with ozone, followed by treatment with zinc in acetic acid.

Alkenes are prepared from alkyl halides and alcohols by **elimination reactions.** Treatment of an alkyl halide with strong base effects **dehydrohalogenation,** and treatment of an alcohol with acid effects **dehydration.** These elimination reactions usually give a mixture of alkene products in which the more highly substituted alkene predominates (**Zaitsev's rule**).

Conjugated dienes like 1,3-butadiene contain alternating single and double bonds. Conjugated dienes undergo **1,4-addition** of electrophiles through the formation of a resonance-stabilized allylic carbocation intermediate. No single line-bond representation can depict the true structure of an allylic carbocation. Rather, the true structure is a **resonance hybrid** somewhere intermediate between two contributing resonance forms. The only difference between two **resonance structures** is in the location of bonding electrons: The nuclei remain in the same places in both structures.

Many simple alkenes undergo **polymerization** when treated with a radical catalyst. **Polymers** are large molecules built up by the repetitive bonding together of many small **monomer** units.

Alkynes are hydrocarbons that contain carbon–carbon triple bonds. Much of the chemistry of alkynes is similar to that of alkenes. For example, alkynes react with one equivalent of HBr and HCl to yield **vinylic** halides, and with one equivalent of Br_2 and Cl_2 to yield 1,2-dihalides. Alkynes can also be hydrated by reaction with aqueous sulfuric acid in the presence of mercuric sulfate catalyst. The reaction leads initially to an intermediate enol that immediately isomerizes to a ketone. Alkynes can also be hydrogenated. Reduction over the Lindlar catalyst yields cis alkenes whereas reduction with lithium metal in liquid ammonia yields the trans alkene.

SUMMARY OF REACTIONS

1. Addition reactions of alkenes

(a) Addition of HX, where X = Cl, Br, or I (Sections 4.1 and 4.2)

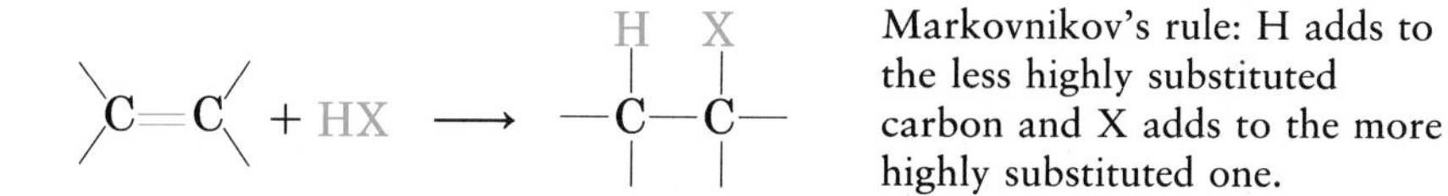

Markovnikov's rule: H adds to the less highly substituted carbon and X adds to the more highly substituted one.

(b) Addition of H_2O (Section 4.4)

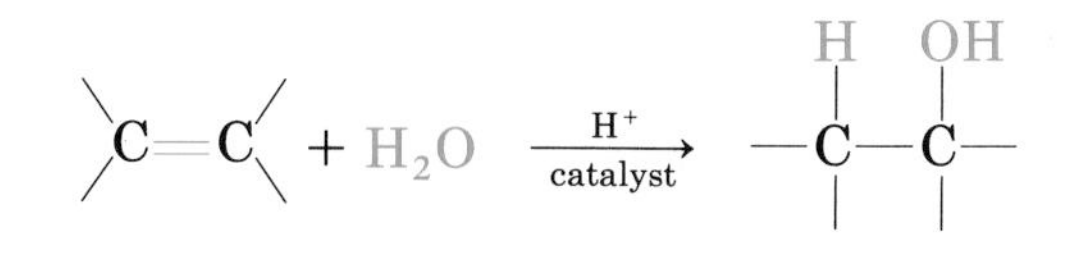

Markovnikov's rule: H adds to the less highly substituted carbon and OH adds to the more highly substituted one.

(c) Addition of X_2, where X = Cl, Br (Section 4.5)

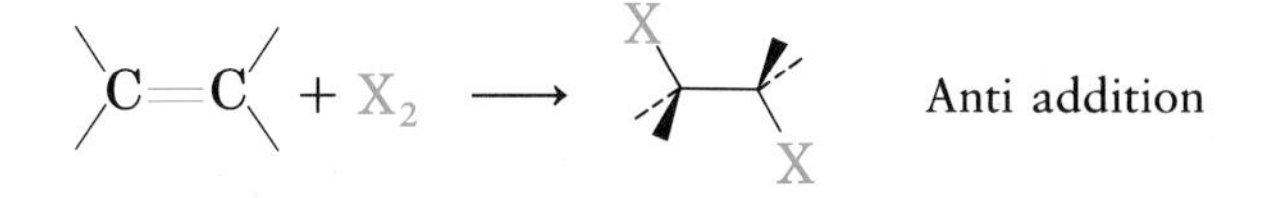

Anti addition

(d) Addition of H_2 (Hydrogenation; Section 4.6)

C=C + H_2 —(Pd or PtO_2 catalyst)→ H H

Syn addition

(e) Hydroxylation (Section 4.7)

$$\mathrm{>C{=}C<} \xrightarrow[\text{NaOH, } H_2O]{KMnO_4} \text{(OH OH, syn diol)}$$

Syn addition

2. Oxidative cleavage of alkenes with ozone (Section 4.7)

$$\mathrm{>C{=}C<} \xrightarrow[\text{2. Zn, } H_3O^+]{\text{1. } O_3} \mathrm{>C{=}O + O{=}C<}$$

3. Radical-induced polymerization of alkenes (Section 4.8)

$$n\ H_2C{=}CH_2 \xrightarrow[\text{initiator}]{\text{radical}} \mathrm{-(CH_2CH_2)_n-}$$

4. Synthesis of alkenes by elimination reactions

(a) Dehydrohalogenation of alkyl halides (Section 4.9)

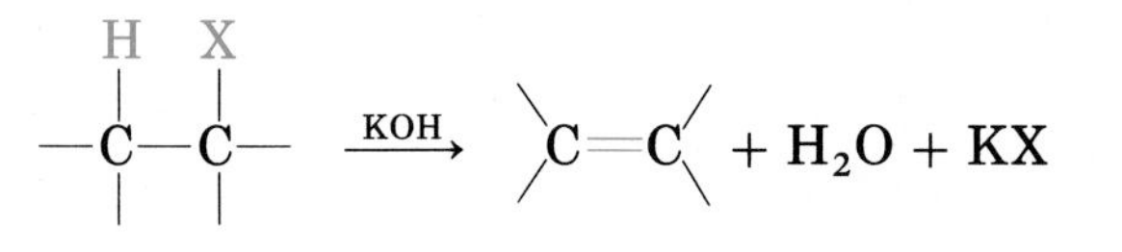

Zaitsev's rule: Major product formed is the alkene with the more highly substituted double bond.

(b) Dehydration of alcohols (Section 4.9)

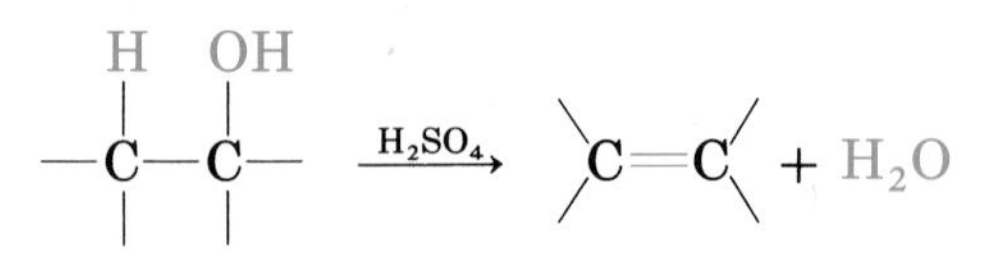

Zaitsev's rule: Major product formed is the alkene with the more highly substituted double bond.

5. Addition reactions of alkynes

(a) Addition of H_2 (hydrogenation; Section 4.14)

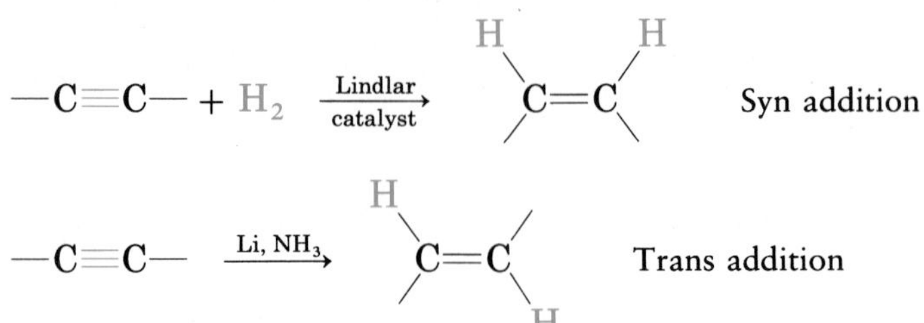

(b) Addition of HX, where X = Cl, Br, I (Section 4.14)

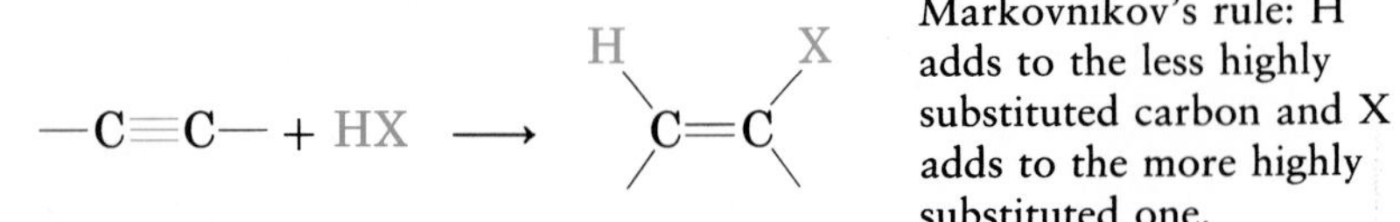

Markovnikov's rule: H adds to the less highly substituted carbon and X adds to the more highly substituted one.

(c) Addition of X_2, where X = Cl, Br (Section 4.14)

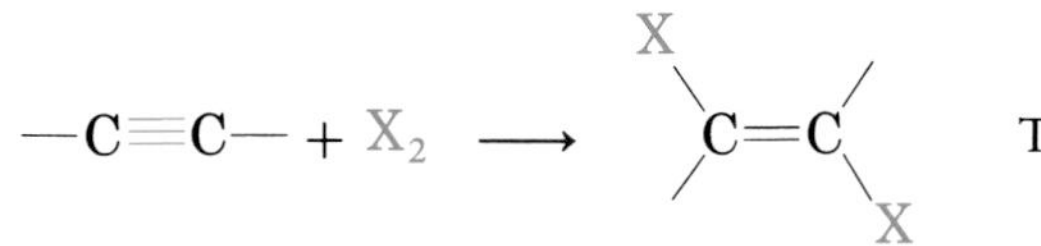

(d) Addition of H_2O to yield ketones (Section 4.15)

$$—C{\equiv}C— + H_2O \xrightarrow[HgSO_4]{H_2SO_4} \left[\overset{OH}{\diagdown} C{=}C \overset{H}{\diagup} \right] \longrightarrow \overset{O}{\underset{}{\|}}C—C\overset{H}{\diagup}$$

ADDITIONAL PROBLEMS

4.24 Provide IUPAC names for these compounds:

(a) $CH_3CH{=}CHC(CH_3){=}CHCH_3$ (with CH_3 on the fourth carbon)

(b) $CH_3CH{=}CHCH(CH_2CH_2CH_3)CH_2C{\equiv}CH$

(c) $CH_2{=}C{=}C(CH_3)CH_3$

(d) $HC{\equiv}CCH_2C{\equiv}CCH(CH_3)CH_3$

4.25 Draw structures corresponding to these IUPAC names:

(a) 3-Ethyl-1-heptyne (b) 3,5-Dimethyl-4-hexen-1-yne
(c) 1,5-Heptadiyne (d) 1-Methyl-1,3-cyclopentadiene

4.26 Draw three possible structures for each of these formulas:

(a) C_6H_8 (b) C_6H_8O

4.27 Name these alkynes according to IUPAC rules:

(a) $CH_3CH_2C{\equiv}CCH_2CH_2CH_3$ (b) $CH_3CH_2C{\equiv}CC(CH_3)_3$
(c) $CH_3C{\equiv}CCH_2C{\equiv}CCH_2CH_3$ (d) $H_2C{=}CHCH{=}CHC{\equiv}CH$

4.28 Draw structures corresponding to these IUPAC names:

(a) 3-Heptyne (b) 3,3-Dimethyl-4-octyne
(c) 3,4-Dimethylcyclodecyne (d) 2,2,5,5-Tetramethyl-3-hexyne

4.29 Draw and name all of the possible pentyne isomers, C_5H_8.

4.30 Draw and name the six possible diene isomers of formula C_5H_8. Which of the six are conjugated dienes?

4.31 Predict the products of these reactions. Indicate regiochemistry where relevant. (The aromatic ring is inert to all of the indicated reagents.)

$C_6H_5CH{=}CH_2$

Styrene

(a) Styrene + H_2 $\xrightarrow{Pd}$?

(b) Styrene + Br_2 $\longrightarrow$?

(c) Styrene + HBr $\longrightarrow$?

(d) Styrene + $KMnO_4$ $\xrightarrow{NaOH,\ H_2O}$?

4.32 Using an oxidative cleavage reaction, explain how you would distinguish between these two isomeric cyclohexadienes:

4.33 Formulate the reaction of cyclohexene with Br_2, showing the reaction intermediate and the final product with correct stereochemistry.

4.34 What products would you expect to obtain from reaction of 1,3-cyclohexadiene with each of the following?

(a) 1 mol Br_2 in CCl_4 (b) O_3, followed by Zn (c) 1 mol HCl
(d) 1 mol DCl (D = deuterium) (e) H_2 over a Pd catalyst

4.35 Draw the structure of a hydrocarbon that reacts with only 1 mol equiv. of hydrogen on catalytic hydrogenation and that gives only pentanal, $CH_3CH_2CH_2CH_2CHO$, on treatment with ozone. Write the reactions involved.

4.36 Give the structure of an alkene that yields the following keto aldehyde on reaction with ozone, followed by treatment with Zn/H_3O^+.

$$? \xrightarrow[\text{2. Zn, } H_3O^+]{\text{1. } O_3} \overset{O}{\overset{\|}{H C}}CH_2CH_2CH_2CH_2\overset{O}{\overset{\|}{C}}CH_3$$

4.37 What alkenes would you hydrate to obtain these alcohols?

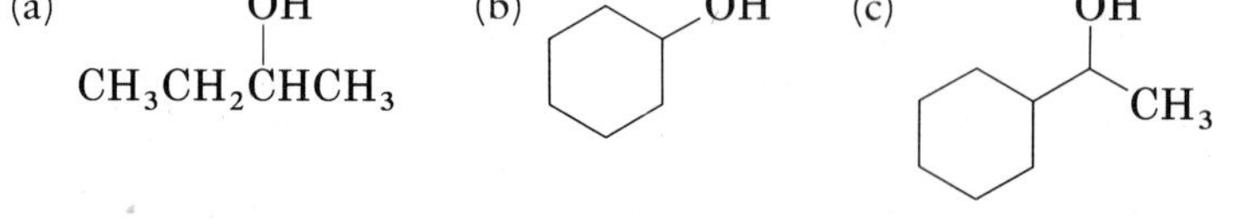

4.38 What alkynes would you hydrate to obtain these products?

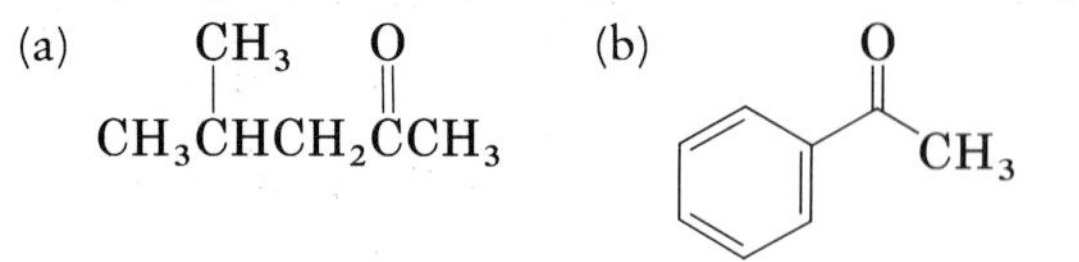

4.39 Draw the structure of a hydrocarbon that reacts with 2 mol equiv. of hydrogen on catalytic hydrogenation and that gives only butanedial, $OHCCH_2CH_2CHO$, on reaction with ozone.

4.40 Predict the products of these reactions:

$$CH_3CH_2CH_2CH_2C{\equiv}CH$$

1-Hexyne

(a) $\xrightarrow{\text{1 equiv HBr}}$?

(b) $\xrightarrow{\text{1 equiv } Cl_2}$?

(c) $\xrightarrow{H_2\text{, Lindlar catalyst}}$?

4.41 Predict the products of these reactions:

$CH_3CH_2CH_2CH_2C{\equiv}CCH_2CH_2CH_2CH_3$

5-Decyne

(a) $\xrightarrow{H_2,\ \text{Lindlar catalyst}}$

(b) $\xrightarrow{Li,\ NH_3}$

(c) $\xrightarrow{2\ \text{equiv.}\ Br_2}$

(d) $\xrightarrow{H_2O,\ H_2SO_4,\ HgSO_4}$

4.42 Acrylonitrile, $H_2C{=}CHC{\equiv}N$, contains a carbon–carbon double bond and a carbon–nitrogen triple bond. Sketch the orbitals involved in the bonding in acrylonitrile and indicate the hybridization of the carbons. Is acrylonitrile conjugated?

4.43 Using 1-butyne as the only organic starting material, along with any inorganic reagents needed, how would you synthesize these compounds? More than one step may be needed.

(a) Butane
(b) 1,1,2,2-Tetrachlorobutane
(c) 2-Bromobutane
(d) 2-Butanone ($CH_3CH_2COCH_3$)

4.44 Give the structure of an alkene that provides only acetone, $(CH_3)_2C{=}O$, on reaction with ozone.

4.45 Compound A has the formula C_8H_8. It reacts rapidly with acidic $KMnO_4$ but reacts with only 1 equiv of H_2 over a palladium catalyst. On hydrogenation under conditions that reduce aromatic rings, A reacts with 4 equiv of H_2, and hydrocarbon B, C_8H_{16} is produced. The reaction of A with $KMnO_4$ gives CO_2 and a carboxylic acid C, $C_7H_6O_2$. What are the structures of A, B, and C? Write all of the reactions.

4.46 Draw a reaction energy diagram for the addition of HBr to 1-pentene. Let one curve on your diagram show the formation of 1-bromopentane product and another curve on the same diagram show the formation of 2-bromopentane product. Label the position for all reactants, intermediates, and products.

4.47 Make sketches of what you imagine the transition-state structures to look like in the reaction of HBr with 1-pentene (Problem 4.43).

4.48 Methylenecyclohexane, on treatment with strong acid, isomerizes to yield 1-methylcyclohexene:

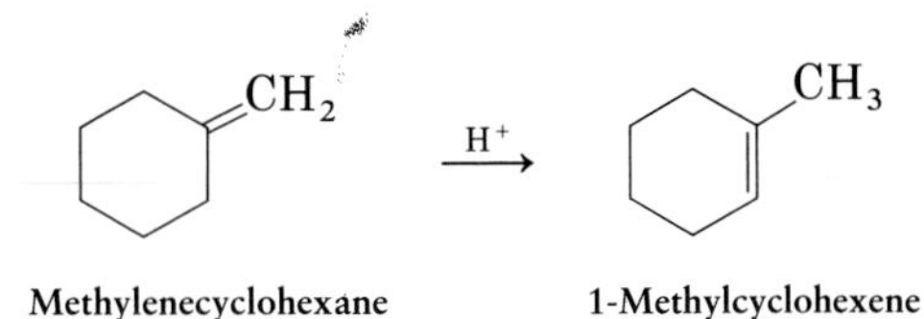

Propose a mechanism by which this reaction might occur.

CHAPTER

5 Aromatic Compounds

In the early days of organic chemistry, the word *aromatic* was used to describe fragrant substances such as benzaldehyde (from cherries, peaches, and almonds), toluene (from tolu balsam), and benzene (from coal distillate). It was soon realized, however, that the substances grouped as aromatic differ from most other organic compounds in their chemical behavior.

aromatic referring to the class of compounds that contain a benzene-like six-membered ring with three double bonds

Today, we use the word **aromatic** to refer to the class of compounds that contain benzene-like, six-membered rings with three double bonds. Many important compounds are aromatic in part, including the steroidal hormone estrone and the tranquilizer diazepam (Valium). We'll see in this chapter how aromatic substances behave and why they're different from the alkanes, alkenes, and alkynes we've studied up to this point.

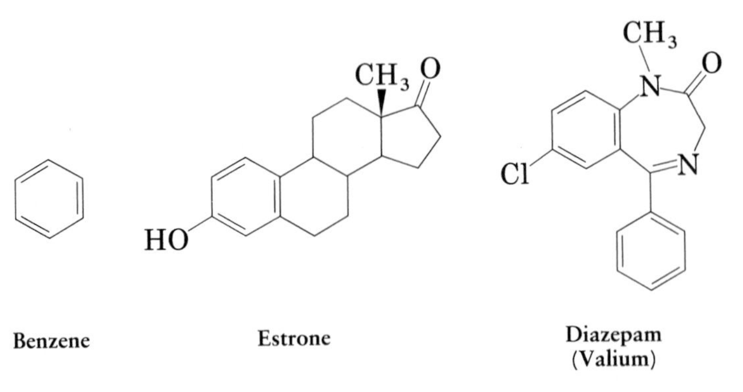

Benzene   Estrone   Diazepam (Valium)

5.1 NAMING AROMATIC COMPOUNDS

Aromatic substances, more than any other class of organic compounds, have acquired a large number of nonsystematic names. Although the use of such names is discouraged, IUPAC rules allow for the more common ones shown in Table 5.1 to be retained. Thus, methylbenzene is commonly known as toluene, hydroxybenzene as phenol, aminobenzene as aniline, and so on.

TABLE 5.1 Common names of some aromatic compounds

Structure	*Name*	*Structure*	*Name*
CH_3	Toluene (bp 110°C)	CHO	Benzaldehyde (bp 178°C)
OH	Phenol (mp 43°C)	COOH	Benzoic acid (mp 122°C)
NH_2	Aniline (bp 184°C)	CN	Benzonitrile (bp 191°C)
O, CH_3	Acetophenone (mp 21°C)	CH_3, CH_3	*ortho*-Xylene (bp 144°C)

phenyl the C_6H_5- group

benzyl the $C_6H_5CH_2-$ group

Monosubstituted benzene derivatives are systematically named in the same manner as other hydrocarbons, with *-benzene* as the parent name. Thus, C_6H_5Br is bromobenzene and $C_6H_5CH_2CH_3$ is ethylbenzene. The name **phenyl** (*fen*-nil) is used for the $-C_6H_5$ unit when the benzene ring is considered a substituent group, and the name **benzyl** is used for the $C_6H_5CH_2-$ alkyl group.

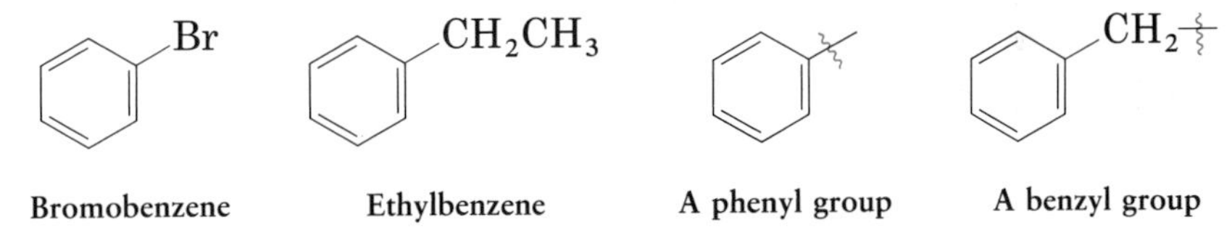

Disubstituted benzenes are named using one of the prefixes *ortho-* (*o*), *meta-*(*m*), or *para-* (*p*). An ortho-disubstituted benzene has its two substituents in a 1,2-relationship on the ring; a meta-disubstituted benzene has its two substituents in a 1,3 relationship; and a para-disubstituted benzene has its substituents in a 1,4-relationship.

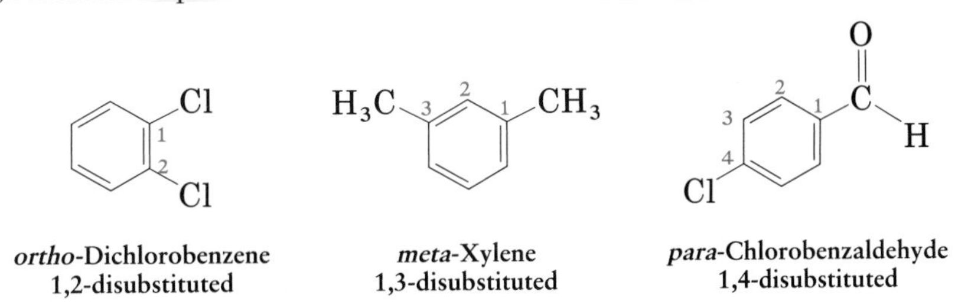

Benzenes with more than two substituents are named by numbering the position of each substituent on the ring in such a way that the lowest possible numbers are used. The substituents are listed alphabetically when writing the name.

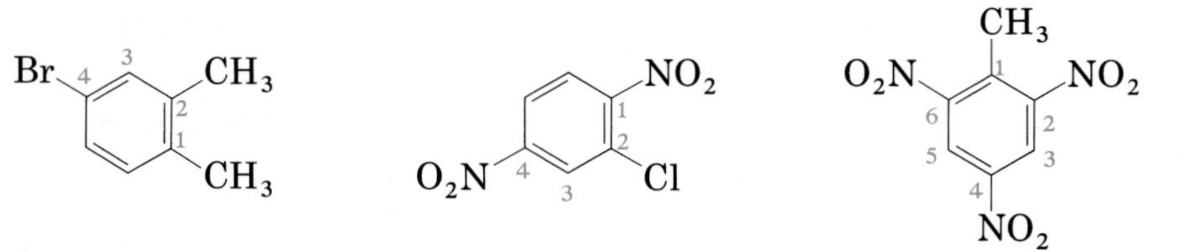

4-Bromo-1,2-dimethylbenzene   2-Chloro-1,4-dinitrobenzene   2,4,6-Trinitrotoluene (TNT)

In the third example shown, note that *-toluene* is used as the parent name rather than *-benzene*. Any of the monosubstituted aromatic compounds shown in Table 5.1 can serve as a parent name, with the principal substituent ($-CH_3$ in toluene, for example) assumed to be on carbon 1. The following two examples further illustrate this practice.

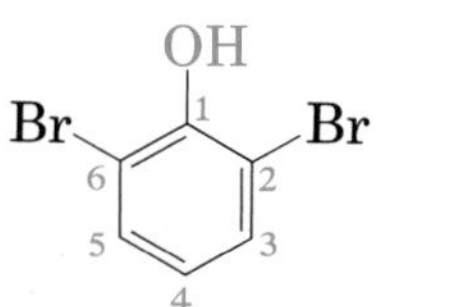

2,6-Dibromo*phenol*   *m*-Chloro*benzoic acid*

PRACTICE PROBLEM 5.1 What is the IUPAC name of this compound?

Cl, NO$_2$

SOLUTION Because the nitro group ($-NO_2$) and chloro group are on carbons 1 and 3, they have a meta relationship. Citing the two substituents in alphabetical order gives the IUPAC name *m*-chloronitrobenzene.

PROBLEM 5.1 Tell whether these compounds are ortho, meta, or para substituted:

(a) Cl, CH$_3$ (b) Br, NO$_2$ (c) SO$_3$H, OH

PROBLEM 5.2 Give IUPAC names for these compounds:

(a) Cl, Br (b) CH$_3$, CH$_2$CHCH$_3$ (c) Br, NH$_2$

PROBLEM 5.3 Draw structures corresponding to these IUPAC names:

(a) *p*-Bromochlorobenzene (b) *p*-Bromotoluene
(c) *m*-Chloroaniline (d) 1-Chloro-3,5-dimethylbenzene

5.2 STRUCTURE OF BENZENE: THE KEKULÉ PROPOSAL

By the mid-1800s, benzene was known to have the molecular formula C_6H_6, and its chemistry was being actively explored. It was known that although benzene is relatively unreactive toward most reagents that attack alkenes, it reacts with bromine in the presence of iron to give the substitution product C_6H_5Br rather than the possible *addition* product $C_6H_6Br_2$. Furthermore, only one monobromo substitution product was known; no isomers had been prepared.

$$C_6H_6 + Br_2 \xrightarrow{Fe} C_6H_5Br + HBr \qquad \left[C_6H_6Br_2 \right]$$

Benzene — **Bromobenzene (substitution product)** — **(addition product; not formed)**

On the basis of these and other results, August Kekulé proposed in 1865 that benzene consists of a ring of carbon atoms and can be formulated as 1,3,5-cyclohexatriene. Kekulé reasoned that this structure would readily account for the isolation of only a single monobromo substitution product, since all six carbon atoms and all six hydrogens in 1,3,5-cyclohexatriene are equivalent.

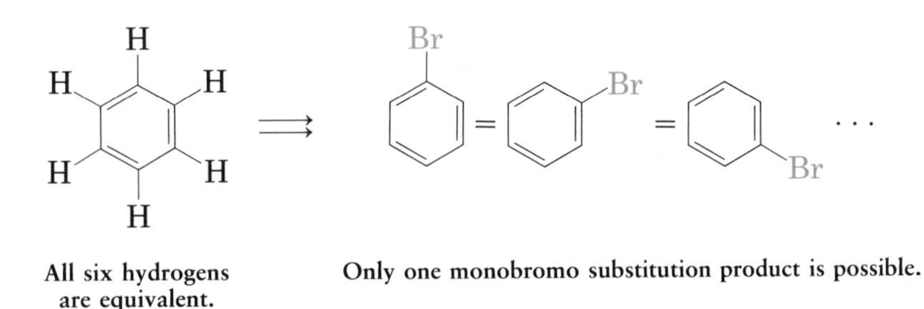

All six hydrogens are equivalent.

Only one monobromo substitution product is possible.

Kekulé's proposal was widely criticized by other chemists of the day. Although it satisfactorily accounts for the correct number of monosubstituted benzene isomers, the proposal doesn't answer the critical questions of why benzene is unreactive compared with other alkenes and why benzene gives a substitution product rather than an addition product on reaction with bromine.

PROBLEM 5.4 How many dibromobenzene derivatives are possible according to Kekulé's theory? Draw them.

5.3 STABILITY OF BENZENE

The unusual chemical stability of benzene was a great puzzle to early chemists. Although its formula, C_6H_6, indicates that several double bonds must be present, benzene shows none of the behavior characteristic of alkenes. For example, alkenes readily react with $KMnO_4$ to give 1,2-diols; they react with aqueous acid to give alcohols; and they react with gaseous HCl to give saturated chloroalkanes. Benzene does none of these things. *Benzene does not undergo electrophilic addition reactions* (Figure 5.1).

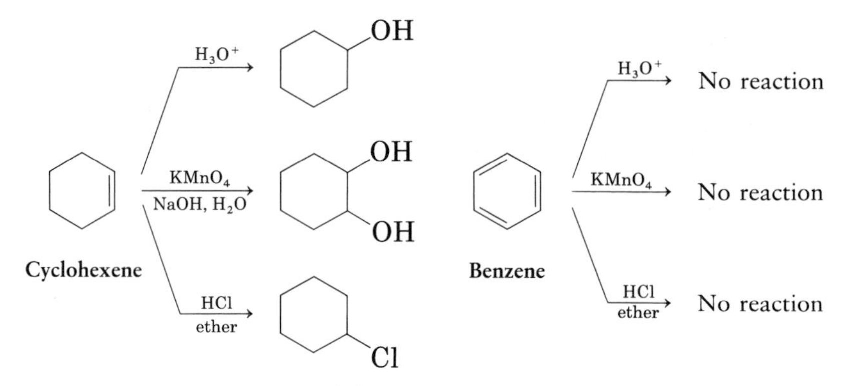

FIGURE 5.1 A comparison of the reactivity of cyclohexene and benzene

Further evidence for the unusual nature of benzene comes from studies indicating that all carbon–carbon bonds in benzene have the same length, intermediate between a normal single and a normal double bond. Most carbon–carbon single bonds have lengths near 1.54 Å, and most carbon–carbon double bonds are about 1.34 Å long. All carbon-carbon bonds in benzene, however, are 1.39 Å long (Figure 5.2).

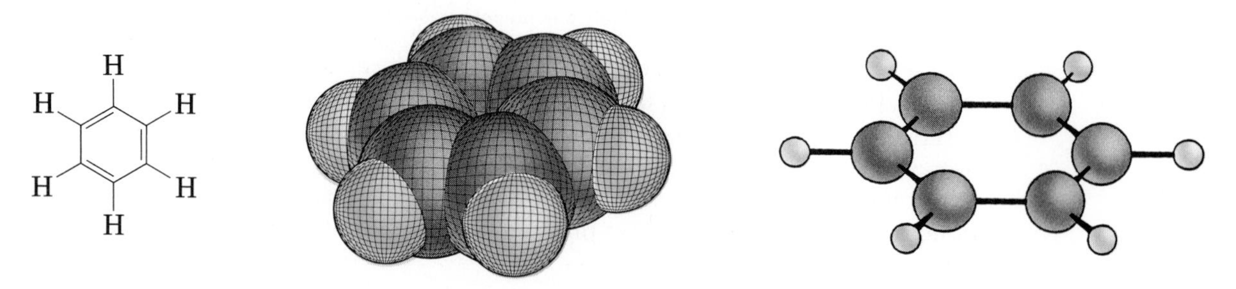

FIGURE 5.2 The structure of benzene. All six carbon–carbon bonds are identical.

5.4 STRUCTURE OF BENZENE: THE RESONANCE PROPOSAL

How can we account for benzene's properties, and how can we best represent its structure? To answer these questions, we need to look again at the concept of *resonance*. We saw in Section 4.11 that an allylic carbocation is best described as a resonance hybrid of two contributing resonance forms. Neither form is correct by itself; the true structure of an allylic carbocation can't be drawn using a single line-bond structure because it is intermediate between the two resonance forms:

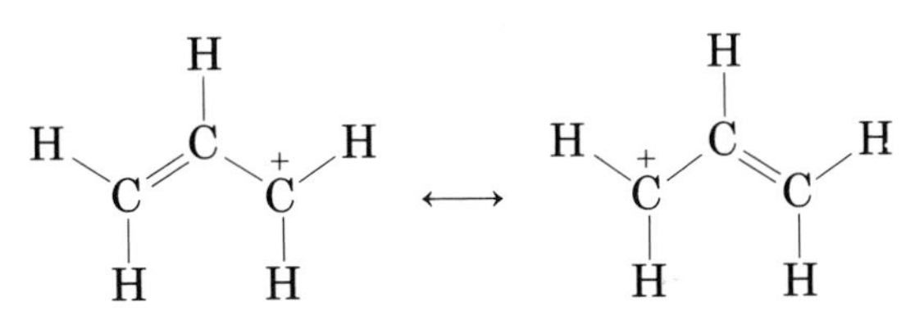

In the same way, resonance theory says that benzene can't be described satisfactorily by a single line-bond structure. Rather, benzene is a resonance hybrid of two equivalent structures. Benzene doesn't oscillate back and forth between two extremes; its true structure is somewhere between the two and is impossible to draw with our usual conventions (Figure 5.3).

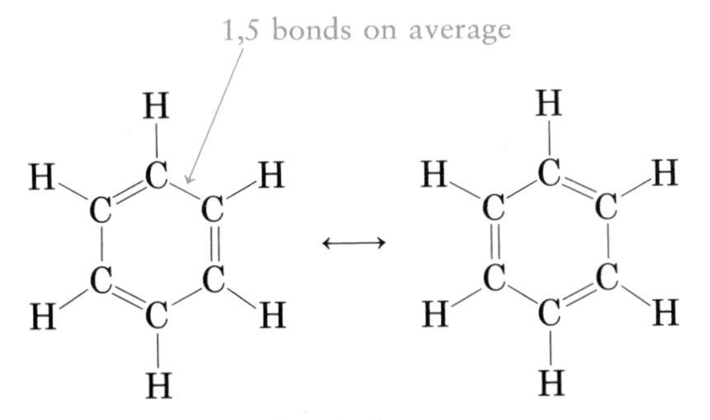

FIGURE 5.3 Two equivalent resonance structures of benzene. Each carbon–carbon connection is an average of 1.5 bonds, midway between a single and a double bond.

An orbital view of benzene shows the situation more clearly, emphasizing the *cyclic conjugation* of the benzene molecule and the equivalence of the six carbon–carbon bonds. Benzene is a flat, symmetrical molecule in the shape of a regular hexagon. All C–C–C bond angles are 120°; each carbon atom is sp^2 hybridized; and each carbon has a *p* orbital perpendicular to the plane of the six-membered ring. Since all six *p* orbitals are equivalent, it's impossible to define three localized alkene pi bonds in which a given *p* orbital overlaps only one neighboring *p* orbital. Rather, each *p* orbital overlaps equally well with both neighboring *p* orbitals, leading to a structure for benzene in which the pi electrons are completely delocalized around the ring in two doughnut-shaped clouds (Figure 5.4).

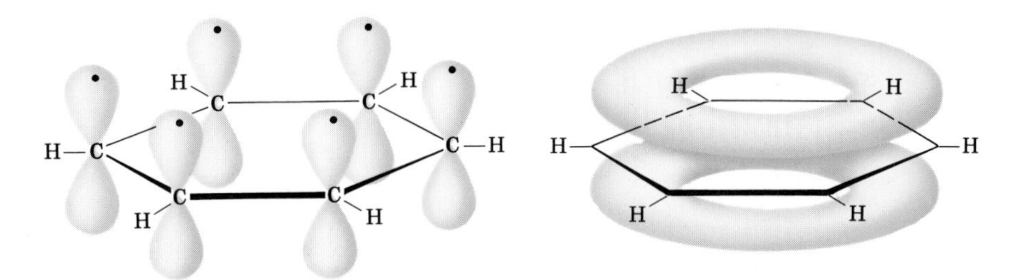

FIGURE 5.4 An orbital picture of benzene

We can now see why benzene is unusually stable. According to resonance theory, the more resonance forms a substance has, the more stable it is. Benzene, with two resonance forms of equal energy, is thus highly stabilized.

PROBLEM 5.5 How can you use resonance theory to account for the fact that there is only one known *o*-dibromobenzene rather than the two isomers that Kekulé's theory would suggest?

5.5 CHEMISTRY OF BENZENE: ELECTROPHILIC AROMATIC SUBSTITUTION

electrophilic aromatic substitution reaction in which an electrophile substitutes for a hydrogen atom on a benzene ring

The most important reaction of aromatic compounds is **electrophilic aromatic substitution.** That is, an electron-poor reagent (an electrophile, E^+) reacts with an aromatic ring and substitutes for one of the ring hydrogens.

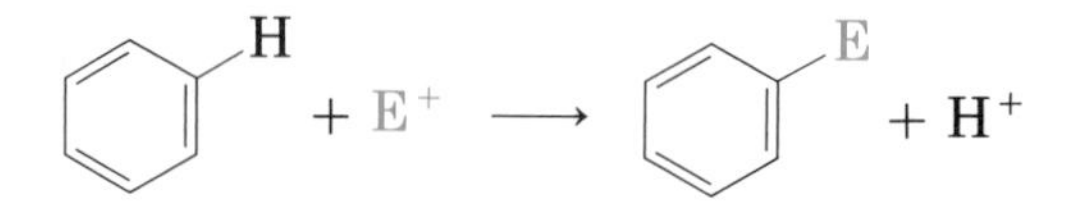

Many different substituents can be introduced onto the aromatic ring by electrophilic substitution reactions. By choosing the proper reagents, we can **halogenate** the aromatic ring (substitute a halogen: −F, −Cl, −Br, or −I), **nitrate** it (substitute a nitro group: $-NO_2$), **sulfonate** it (substitute a sulfonic acid group: $-SO_3H$), or **alkylate** it (substitute an alkyl group: −R). Starting with only a few simple materials, we can prepare many thousands of substituted aromatic compounds (Figure 5.5).

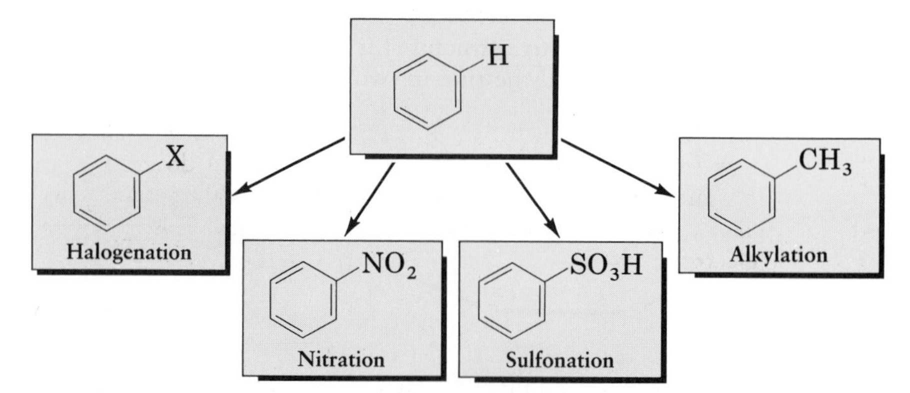

FIGURE 5.5 Some electrophilic aromatic substitution reactions

All of these reactions (and many more) take place by a similar mechanism. Let's begin a study of this fundamental reaction type by looking at one reaction in detail, the bromination of benzene.

PROBLEM 5.6 There are three products that might form on bromination of toluene. Draw and name them.

5.6 BROMINATION OF BENZENE

Benzene reacts with bromine in the presence of $FeBr_3$ as catalyst to yield the substitution product, bromobenzene:

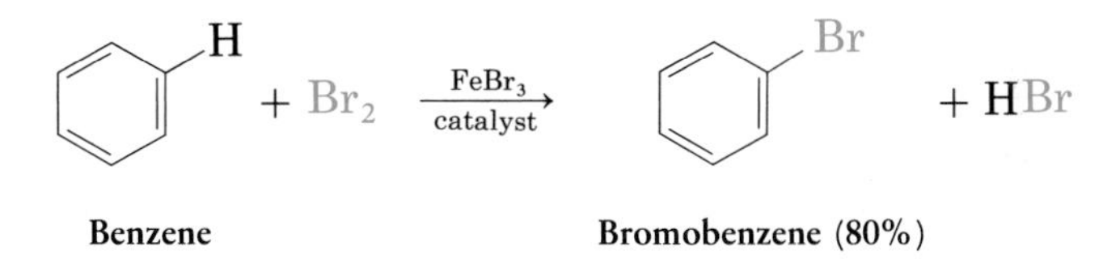

Before seeing how this electrophilic *substitution* reaction occurs, let's briefly recall what we learned about electrophilic *additions* to alkenes in Sections 3.8–3.11. When an electrophile such as H^+ adds to an alkene, it approaches the *p* orbitals of the double bond and forms a bond to one carbon, leaving a positive charge on the other carbon. The carbocation intermediate is then attacked by a nucleophile such as chloride ion to yield the addition product (Figure 5.6).

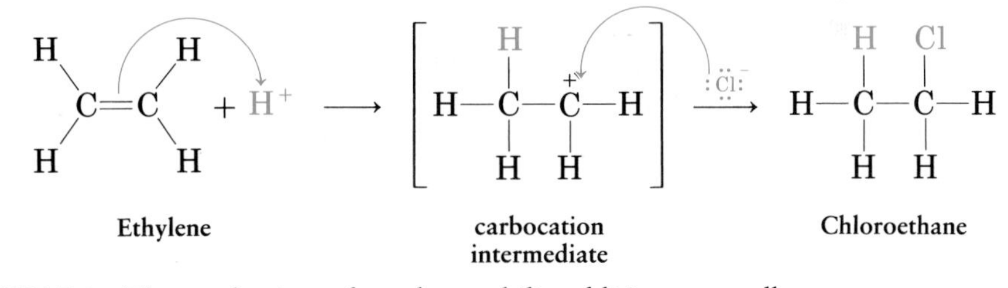

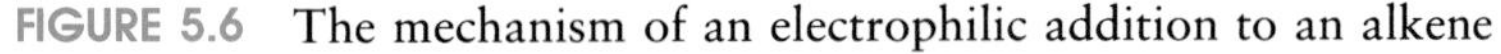
FIGURE 5.6 The mechanism of an electrophilic addition to an alkene

An electrophilic aromatic substitution reaction begins in a similar way, but there are a number of differences. One difference is that aromatic rings are much less reactive than alkenes toward electrophiles. For example, bromine in CCl_4 solution reacts instantly with most alkenes but does not react with benzene. For bromination of benzene to take place, a catalyst such as $FeBr_3$ is needed. The catalyst acts by reacting with bromine to form $FeBr_4^-$ and Br^+, a highly reactive electrophile:

$$FeBr_3 + Br_2 \longrightarrow FeBr_4^- + Br^+$$

The electrophilic Br^+ then reacts with the electron-rich (nucleophilic) benzene ring to yield a nonaromatic carbocation intermediate. This carbocation is allylic (recall the allyl cation, Section 4.11) and can be drawn in three resonance forms:

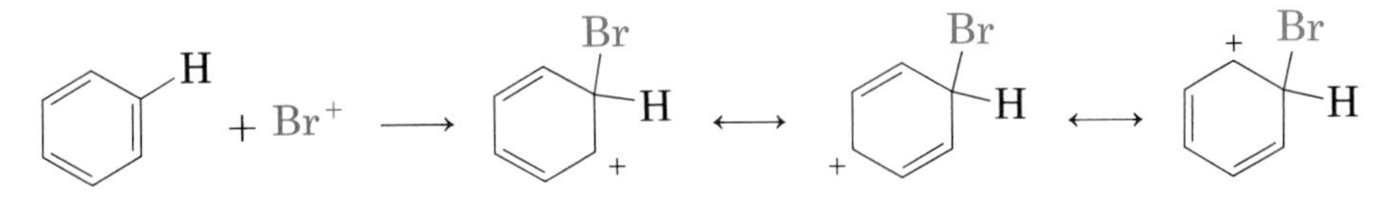

Although stable by comparison with nonallylic carbocations, the intermediate in electrophilic aromatic substitution is nevertheless much less stable than the starting benzene ring. Thus, reaction of an electrophile with a benzene ring has a relatively high activation energy and is therefore rather slow. Figure 5.7 gives reaction energy diagrams that compare the reaction of an electrophile E^+ with an alkene and with benzene. The benzene reaction is slower (that is, has a higher E_{act}) because the starting material is so stable.

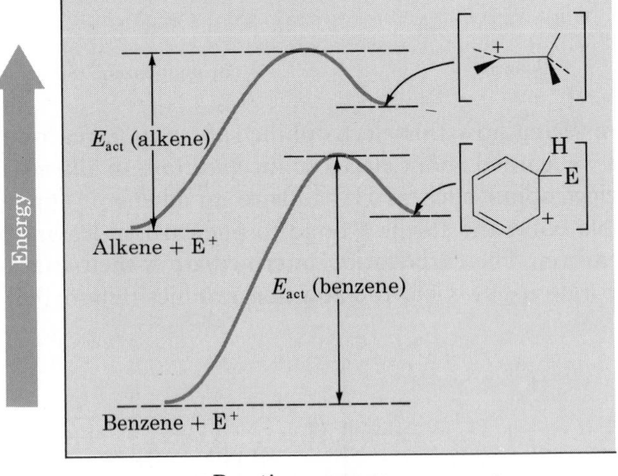

FIGURE 5.7 A comparison of the reactions of an electrophile with an alkene and with benzene: E_{act} (alkene) $\ll$ E_{act} (benzene)

A second difference between alkene addition and aromatic substitution reactions occurs after the electrophile has added to the benzene ring to give the carbocation intermediate. Although it would presumably be possible for a nucleophile such as bromide ion to react with the carbocation intermediate to yield the addition product dibromocyclohexadiene, this is not observed. Instead, the bromide ion abstracts H^+ to yield the neutral aromatic substitution product plus HBr. The net effect of reaction of Br_2 with benzene is the substitution of H^+ by Br^+ by the overall mechanism shown in Figure 5.8.

Why does the reaction of bromine with benzene take a different course than its reaction with an alkene? The answer is simple: If *addition* of bromine to benzene occurred, the stability of the aromatic ring would be lost, and the reaction would be endothermic. When *substitution* occurs, though, the stability of the aromatic ring is retained. A reaction energy diagram for the overall process is shown in Figure 5.9.

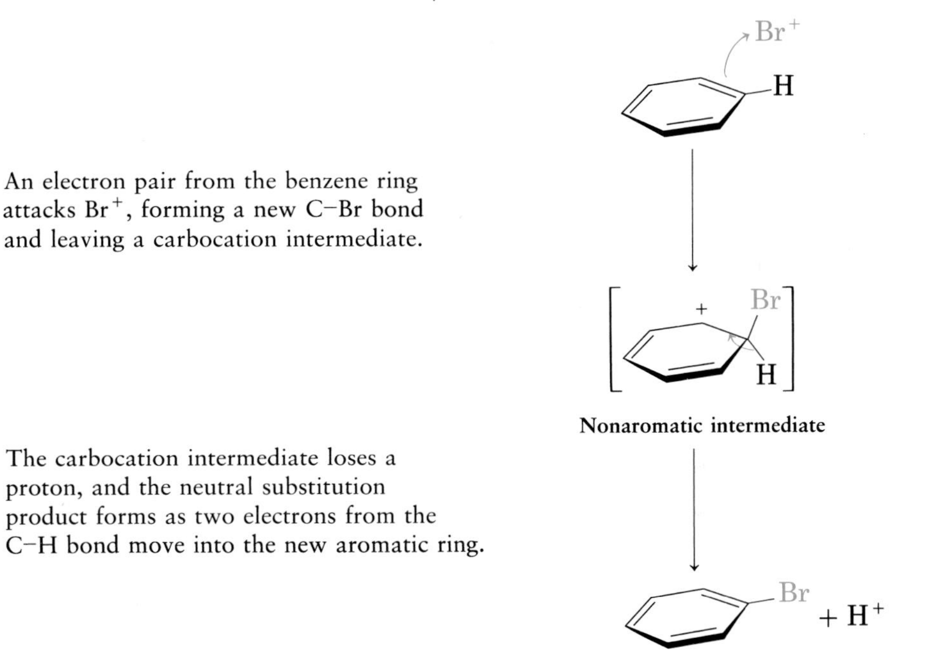

An electron pair from the benzene ring attacks Br^+, forming a new C–Br bond and leaving a carbocation intermediate.
Br^+
H
+
Br
H
Nonaromatic intermediate
The carbocation intermediate loses a proton, and the neutral substitution product forms as two electrons from the C–H bond move into the new aromatic ring.
Br
$+ H^+$

FIGURE 5.8 The mechanism of the electrophilic bromination of benzene

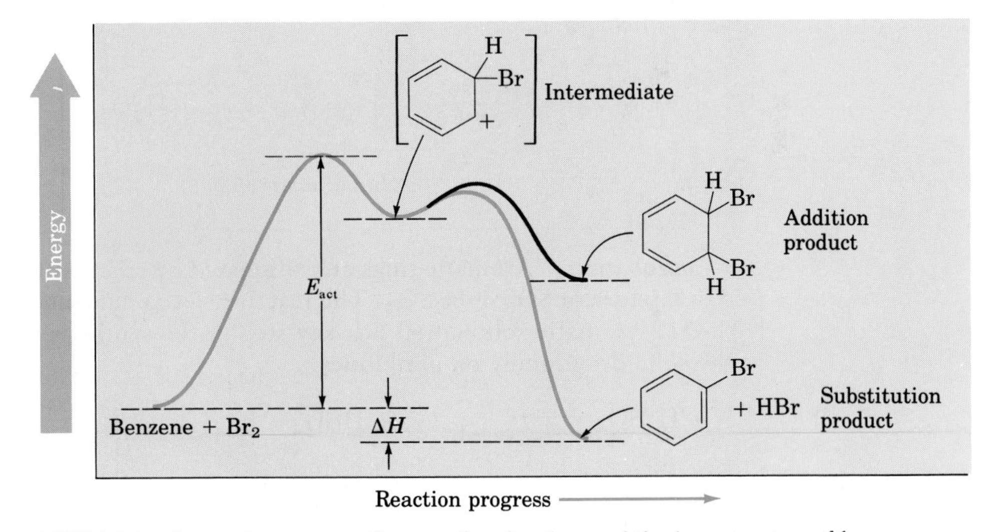

FIGURE 5.9 A reaction energy diagram for the electrophilic bromination of benzene

5.7 OTHER ELECTROPHILIC AROMATIC SUBSTITUTION REACTIONS

Many electrophilic aromatic substitutions besides bromination occur by the same general mechanism. Let's look briefly at some of these other reactions.

Chlorination. Aromatic rings react with chlorine in the presence of $FeCl_3$ catalyst to yield chlorobenzenes. This kind of reaction is used in the synthesis of numerous pharmaceutical agents, including the tranquilizer Librium:

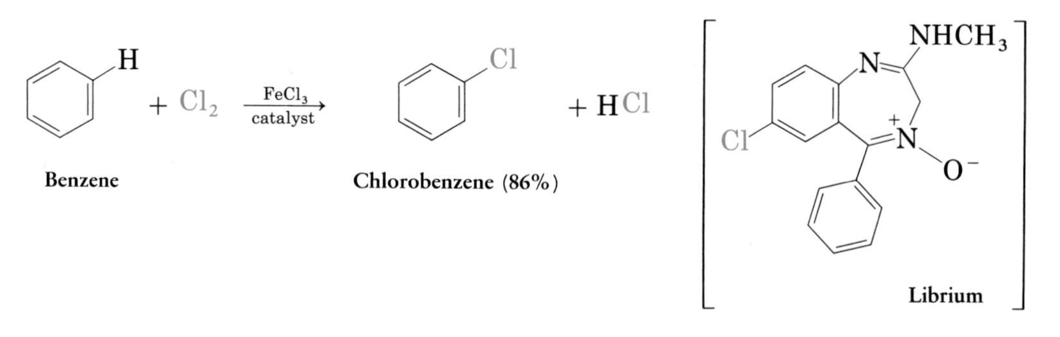

Nitration. Aromatic rings are nitrated by reaction with a mixture of concentrated nitric and sulfuric acids. The electrophile in this reaction is the nitronium ion, NO_2^+, which reacts with benzene in much the same way as does Br^+. Nitration of aromatic rings is a key step in the synthesis of explosives such as TNT (2,4,6-trinitrotoluene), dyes, and many pharmaceutical agents.

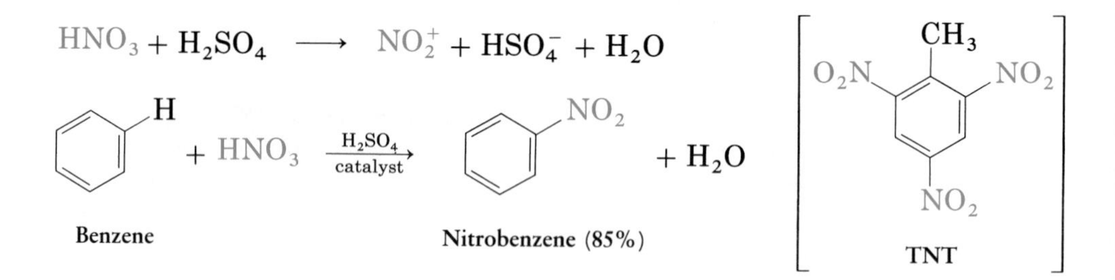

Sulfonation. Aromatic rings are sulfonated by reaction with fuming sulfuric acid, a mixture of $SO_3 + H_2SO_4$. The reactive electrophile under these conditions is HSO_3^+. Aromatic sulfonation is a key step in the synthesis of such compounds as the sulfa-drug family on antibiotics:

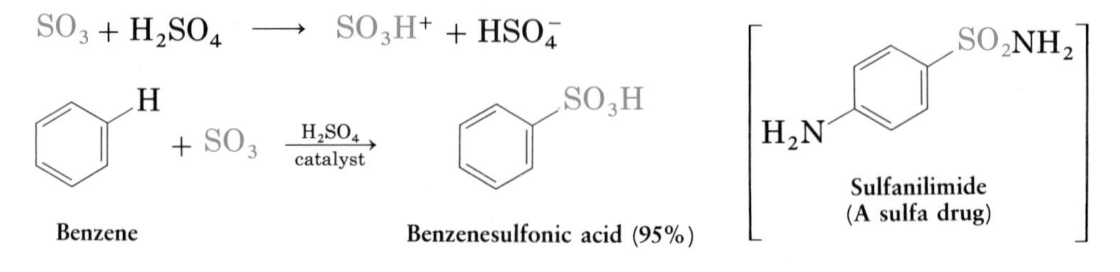

PRACTICE PROBLEM 5.2 Show the mechanism of the reaction of benzene with fuming sulfuric acid to yield benzenesulfonic acid.

SOLUTION The electrophile in sulfonation reactions is HSO_3^+, and the reaction occurs by the same two-step process common to all electrophilic aromatic substitutions:

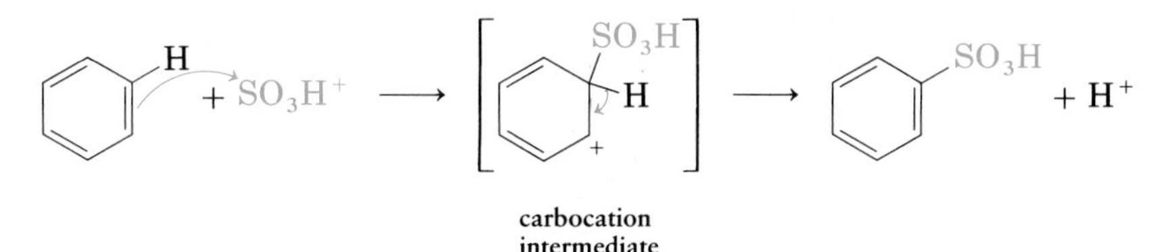

PROBLEM 5.7 Show the mechanism of the reaction of benzene with nitric acid and sulfuric acid to yield nitrobenzene.

PROBLEM 5.8 Chlorination of *o*-xylene (dimethylbenzene) yields a mixture of two products, but chlorination of *p*-xylene yields a single product. Explain.

PROBLEM 5.9 How many products might be formed on chlorination of *m*-xylene?

PROBLEM 5.10 How can you account for the fact that deuterium slowly replaces hydrogen in the aromatic ring when benzene is treated with D_2SO_4.

5.8 THE FRIEDEL–CRAFTS ALKYLATION AND ACYLATION REACTIONS

Charles Friedel and James Crafts reported in 1877 that benzene rings are *alkylated* by reaction with an alkyl chloride in the presence of aluminum chloride catalyst. That is, an alkyl group is attached to an aromatic ring. For example, benzene reacts with 2-chloropropane in the presence of $AlCl_3$ to yield isopropylbenzene (also called cumene).

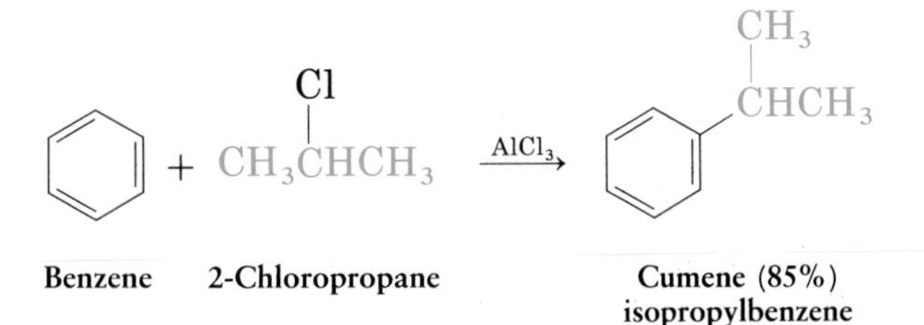

Friedel–Crafts alkylation reaction the introduction of an alkyl group onto a benzene ring by electrophilic substitution

The **Friedel–Crafts alkylation reaction** is an aromatic substitution in which the aromatic ring attacks a carbocation electrophile. Loss of a proton from the intermediate then yields the alkylated aromatic ring. The carbocation is usually generated by reaction of an alkyl chloride with aluminum chloride catalyst. It's thought that the $AlCl_3$ catalyst acts by helping the alkyl chloride to ionize, in much the same way that $FeCl_3$ catalyzes aromatic-ring chlorinations by helping Cl_2 to ionize (Section 5.7). The overall Friedel–Crafts mechanism for the synthesis of isopropylbenzene is shown in Figure 5.10.

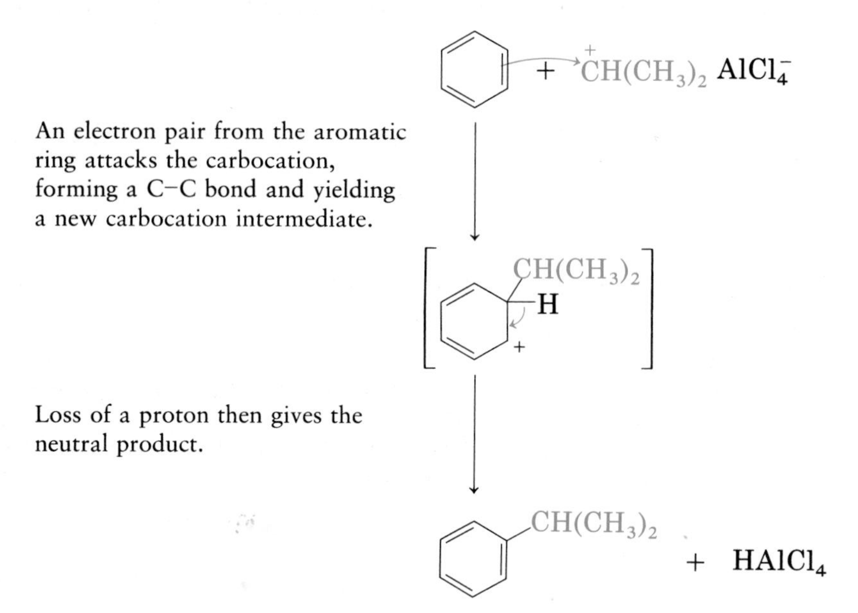

$(CH_3)_2CHCl + AlCl_3 \longrightarrow (CH_3)_2CH^+ AlCl_4^-$
$+ \overset{+}{C}H(CH_3)_2 AlCl_4^-$
An electron pair from the aromatic ring attacks the carbocation, forming a C–C bond and yielding a new carbocation intermediate.
$CH(CH_3)_2$
H
+
Loss of a proton then gives the neutral product.
$CH(CH_3)_2$
$+ HAlCl_4$

FIGURE 5.10 Mechanism of the Friedel–Crafts alkylation reaction in the synthesis of isopropylbenzene

Although extremely useful, the Friedel–Crafts alkylation reaction has several important limitations. For example, only *alkyl* halides can be used; aryl halides like chlorobenzene don't react. In addition, Friedel–Crafts reactions don't succeed on aromatic rings that are already substituted by the groups $-NO_2$, $-C{\equiv}N$, $-SO_3H$, or $-COR$. Such aromatic rings are much less reactive than benzene for reasons we'll discuss in Section 5.9.

Friedel–Crafts acylation reaction the introduction of an acyl group onto a benzene ring by electrophilic substitution

acyl group a name for the $-\overset{O}{\overset{\|}{C}}-R$ group

Closely related to the Friedel–Crafts alkylation reaction, is the **Friedel–Crafts acylation** (a-sil-*a*-tion) **reaction.** When an aromatic compound is allowed to react with a carboxylic acid chloride, RCOCl, in the presence of $AlCl_3$, an **acyl** (*a*-sil) **group,** $-COR$, is introduced onto the ring. For example, reaction of benzene with acetyl chloride yields the ketone, acetophenone.

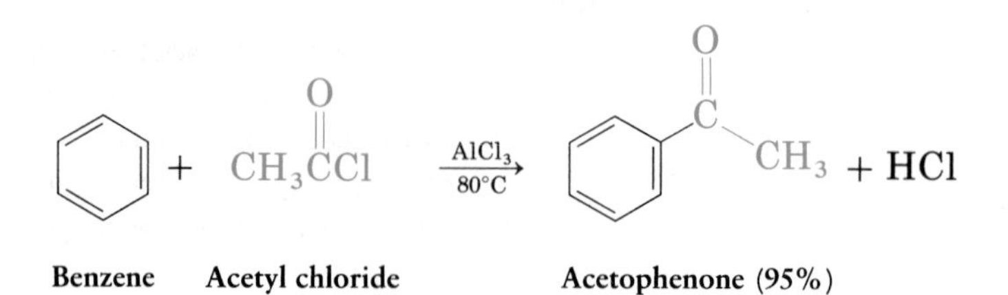

PROBLEM 5.11 What products would you expect to obtain from the reaction of these compounds with chloroethane and $AlCl_3$?

(a) Benzene (b) *p*-Xylene

PROBLEM 5.12 What products would you expect to obtain from the reaction of benzene with these reagents:
(a) $(CH_3)_3CCl$, $AlCl_3$ (b) CH_3CH_2COCl, $AlCl_3$

5.9 REACTIVITY IN ELECTROPHILIC AROMATIC SUBSTITUTION

Only one monosubstitution product can result when electrophilic substitution occurs on benzene. But what would happen if we were to carry out an electrophilic substitution reaction on a ring that's already substituted? Substituents already present on an aromatic ring have two effects:

1. Substituents affect the *reactivity* of the aromatic ring. Some substituents make the ring more reactive than benzene, and some make it less reactive.
2. Substituents affect the *orientation* of the reaction. Three possible disubstituted products can result: ortho, meta, and para. These products aren't formed in random ratios, though; instead, the nature of the substituent already present on the benzene ring determines the position of the second substitution.

Let's look at these two effects more closely.

Classification of Substituents

Substituents can be classified into two groups: those that activate the aromatic ring for further electrophilic substitution and those that deactivate it. Rings that contain an activating substituent are more reactive than benzene whereas those that contain a deactivating substituent are less reactive than benzene. Table 5.2 lists some groups in both categories.

TABLE 5.2 Activating and deactivating substituents for electrophilic aromatic substitution

Activating groups		*Deactivating groups*	
Strongly activating ↑		Weakly deactivating ↓	$-F$, $-Cl$, $-Br$, $-I$
	$-NH_2$, $-NHR$, $-NR_2$		$-C(=O)-H$
	$-OH$, $-OR$		$-C(=O)-OH$, $-C(=O)-OR$
	$-CH_3$ (alkyl)		$-C(=O)-R$
	$-C_6H_5$ (phenyl)		$-C{\equiv}N$
Weakly activating			$-NO_2$
		Strongly deactivating	$-\overset{+}{N}R_3$

The common feature of all substituents within a category is that all activating groups donate electrons to the ring and all deactivating groups withdraw electrons from the ring. An aromatic ring with an electron-donating substituent is more electron-rich (more nucleophilic) than benzene and therefore more reactive toward electrophiles. An aromatic ring with an electron-withdrawing substituent is less electron-rich (less nucleophilic) than benzene and therefore less reactive toward electrophiles.

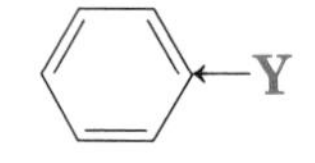

Y is an electron-donor; ring is electron-rich and more reactive than benzene

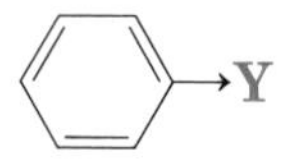

Y is an electron-acceptor; ring is electron-poor and less reactive than benzene

Resonance and Inductive Effects

inductive effects donation or withdrawal of electrons through sigma bonds or through space

A substituent can donate or withdraw electrons from the aromatic ring in two ways, by *inductive effects* and by *resonance effects*. **Inductive effects** are due to the intrinsic electronegativity of atoms (Section 1.12) and thus to the bond polarity in functional groups. These effects operate by donating or withdrawing electrons through sigma bonds. For example, halogens, carbonyl groups, cyano groups, and nitro groups deactivate an aromatic ring by inductively withdrawing electrons from the ring through the sigma bond linking the substituent to the ring:

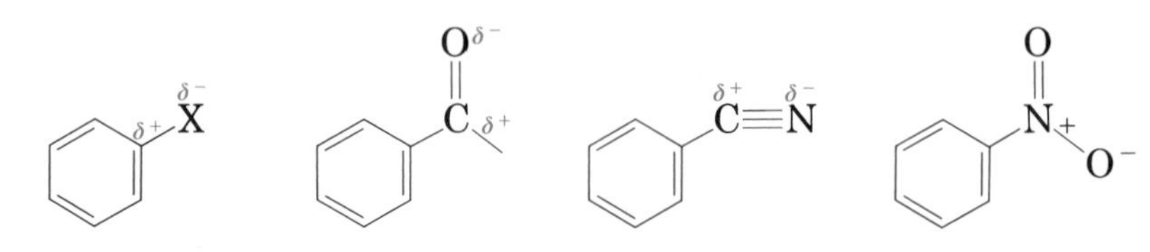

The groups attached to the aromatic rings are inductively electron-withdrawing because of the polarity of their bonds.

Alkyl groups are weakly electron-donating and therefore activate the ring. The reasons for this are complex, though, and aren't fully understood.

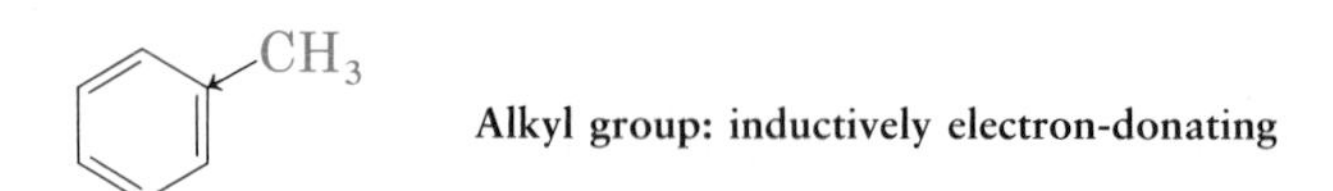

resonance effects donation or withdrawal of electrons through pi bonds

Resonance effects operate by donating or withdrawing electrons through pi bonds by overlap of an orbital on the substituent with a *p* orbital of the aromatic ring. For example, we can draw resonance structures of benzaldehyde showing how the carbonyl group deactivates an aromatic ring by allowing aromatic-ring pi electrons to move onto the substituent. In so doing, a positive charge is left in the ring, thereby deactivating it toward reaction with an electrophile.

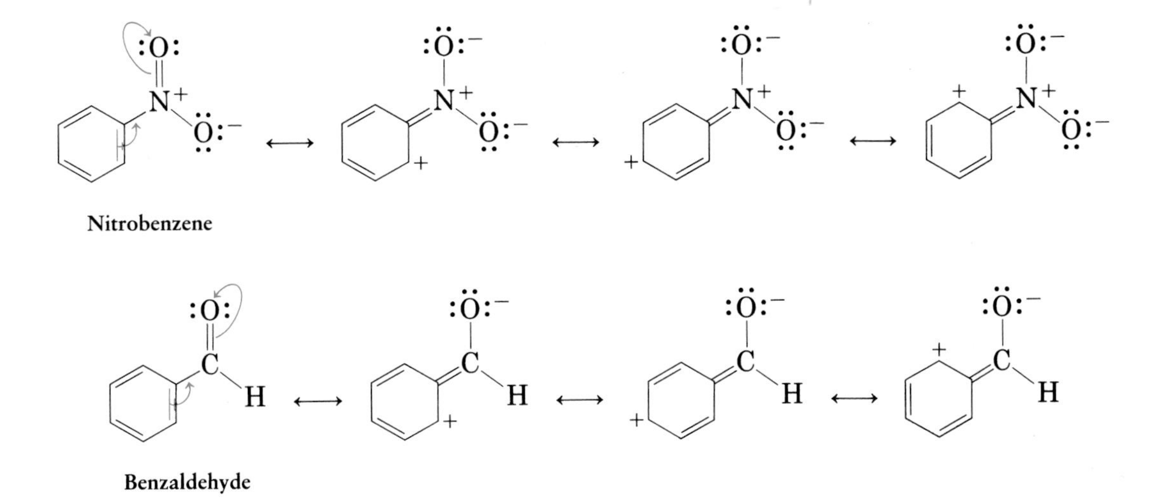

Note that substituents with a resonance-deactivating effect have the general structure −X=Y, where the Y atom is more electronegative than X:

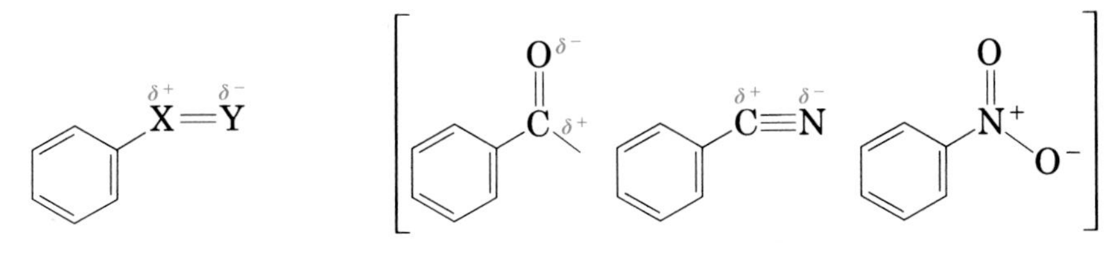

Deactivated rings have this general structure.

Conversely, substituents such as hydroxyl and amino activate the aromatic ring by resonance effects that donate pi electrons from the substituents to the ring. As the following resonance structures indicate, this electron donation places a negative charge in the ring, making the ring more reactive toward electrophiles.

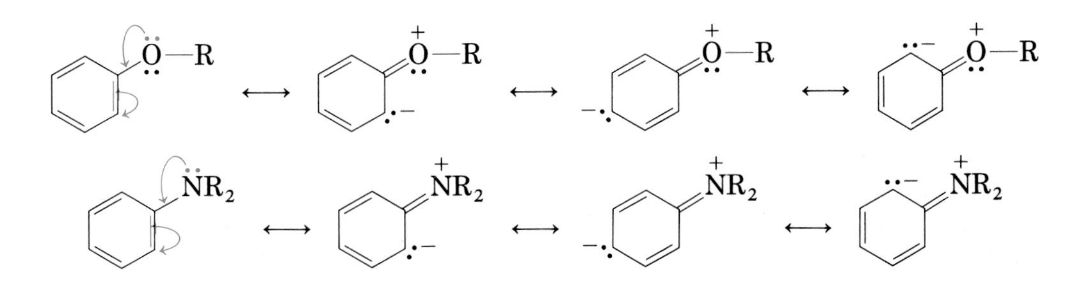

Note that substituents with a resonance-activating effect have the general structure $-\ddot{Y}$, in which the atom Y has a lone pair of electrons it can donate to the ring:

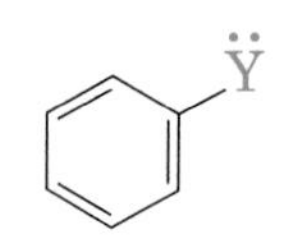

Activated rings have
this general structure.

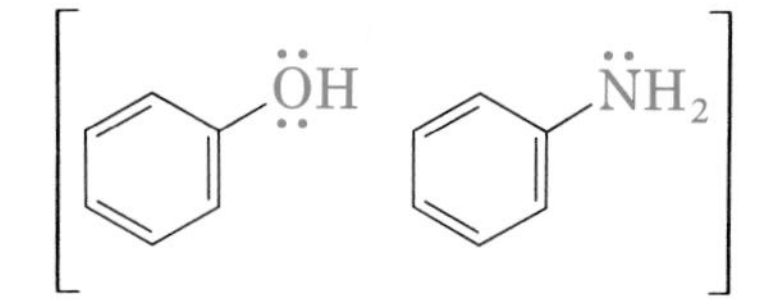

It probably comes as a surprise to learn that hydroxyl, methoxyl, and amino groups activate the ring. After all, both oxygen and nitrogen are highly electronegative and might be expected to deactivate the ring inductively. We find, however, that the resonance electron-donation effect through pi bonds far outweighs the inductive electron-withdrawal effect through sigma bonds for these substituents. Thus, there is a net activating influence.

PRACTICE PROBLEM 5.3 Which would you expect to react faster in an electrophilic substitution reaction, chlorobenzene or ethylbenzene? Explain.

SOLUTION According to Table 5.2, a chloro substituent is electron-withdrawing and deactivating whereas an alkyl group is electron-donating and activating. Thus, ethylbenzene is more reactive than chlorobenzene.

PROBLEM 5.13 Use Table 5.2 to rank the compounds in each of the following groups in order of their reactivity to electrophilic substitution:

(a) Nitrobenzene, phenol (hydroxybenzene), toluene
(b) Phenol, benzene, chlorobenzene, benzoic acid
(c) Benzene, bromobenzene, benzaldehyde, aniline (aminobenzene)

PROBLEM 5.14 Draw resonance structures to show how a methoxyl group ($-OCH_3$) activates an aromatic ring toward electrophilic substitution.

PROBLEM 5.15 Draw resonance structures to show how an acetyl group

$$CH_3\overset{\overset{\displaystyle O}{\|}}{C}-$$

deactivates an aromatic ring toward electrophilic substitution.

5.10 ORIENTATION OF REACTIONS ON SUBSTITUTED AROMATIC RINGS

In addition to affecting the reactivity of an aromatic ring, a substituent can also direct the position of electrophilic substitution. For example, a methyl substituent shows a strong ortho- and para-directing effect. Nitration of toluene yields predominately *o*-nitrotoluene (63%) and *p*-nitrotoluene (34%), along with only 3% of the meta isomer.

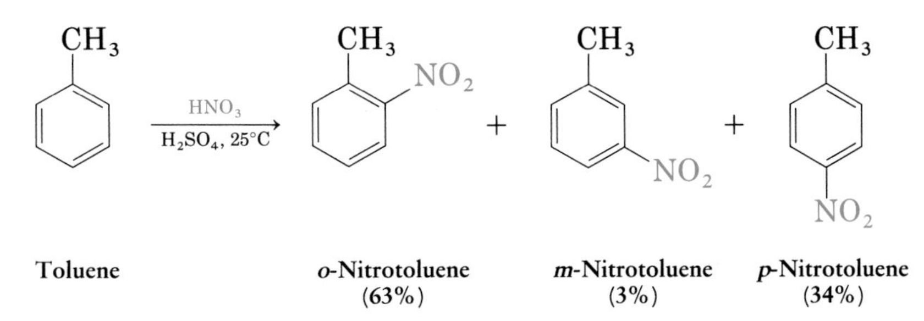

On the other hand, a cyano substituent shows a strong meta-directing effect. Nitration of benzonitrile (cyanobenzene) yields 81% *m*-nitrobenzonitrile, along with only 17% of the ortho isomer and 2% of the para isomer.

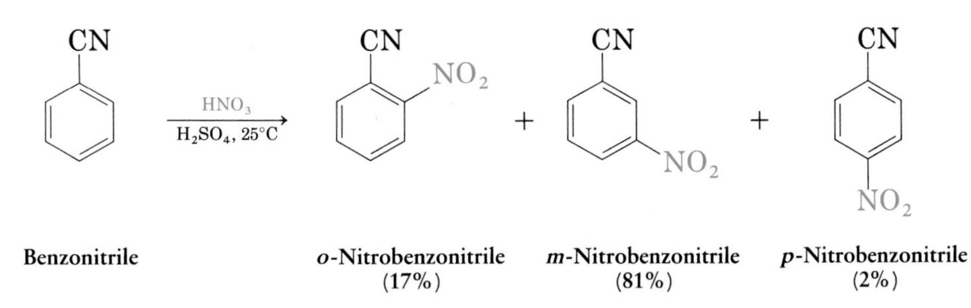

Table 5.3 lists some of the groups in ortho, para-directing and in meta-directing categories. Note that ortho, para directors can be either activating or deactivating whereas all meta directors are deactivating.

TABLE 5.3 Classification of directing effects of common substituents

Ortho- and para-directing activators	*Ortho- and para-directing deactivators*	*Meta-directing deactivators*
$-NH_2$, $-NHR$, $-NR_2$	$-F$:	$-C(=O)-H$
$-OH$, $-OR$	$-Cl$:	
$-CH_3$ (alkyl)	$-Br$:	$-C(=O)-OH$, $-C(=O)-OR$
	$-I$:	$-C(=O)-R$
		$-C\equiv N$
		$-NO_2$
		$-\overset{+}{N}R_3$

PROBLEM 5.16 Use Table 5.3 to predict the major monosubstitution products of these reactions:

(a) Nitration of bromobenzene
(b) Bromination of nitrobenzene
(c) Chlorination of phenol
(d) Bromination of aniline

Ortho and Para Directors

Let's look at the nitration of toluene as an example of how ortho and para directors work. In the first step, attack on the nitronium-ion electrophile (NO_2^+) can occur either ortho, meta, or para to the methyl group, giving the carbocation intermediates shown in Figure 5.11. The ortho and para intermediates are more stable than the meta intermediate because there is a resonance form that places the positive charge directly on the methyl-substituted carbon, where it can best be stabilized by the methyl group's electron-donating inductive effect.

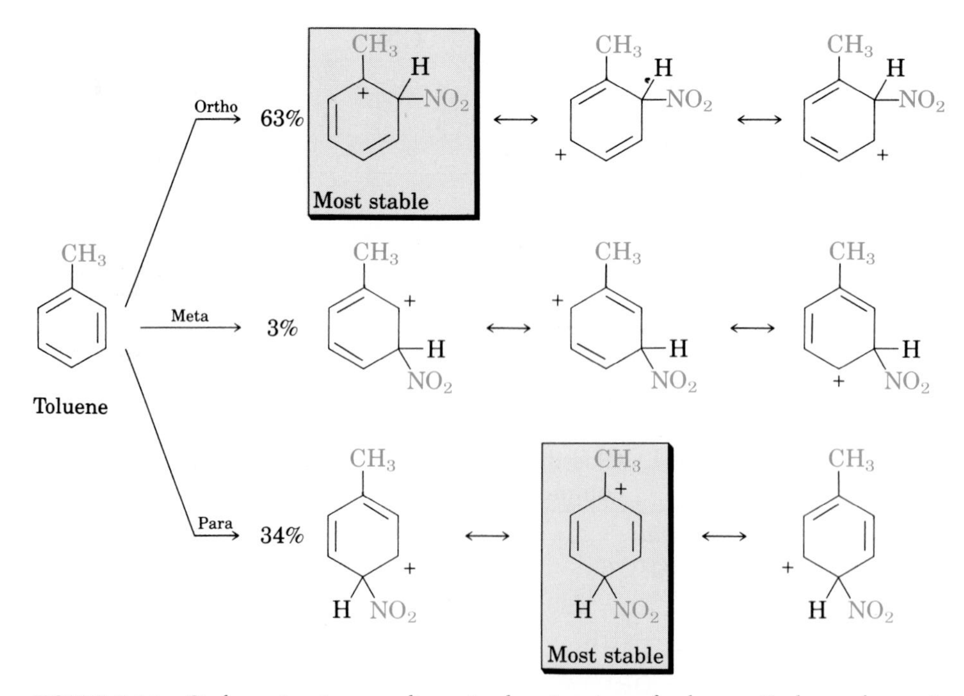

FIGURE 5.11 Carbocation intermediates in the nitration of toluene. Ortho and para intermediates are more stable than the meta intermediate.

A similar line of reasoning explains why hydroxyl and amino groups (and their derivatives) are ortho- and para-directing. These groups exert their electron-donating, activating influence through a strong resonance effect at the ortho and para positions. For example, when phenol is nitrated, the three carbocation intermediates shown in Figure 5.12 are possible. Only in the intermediates from ortho and para attack is the positive charge stabilized by resonance electron donation from the oxygen atom.

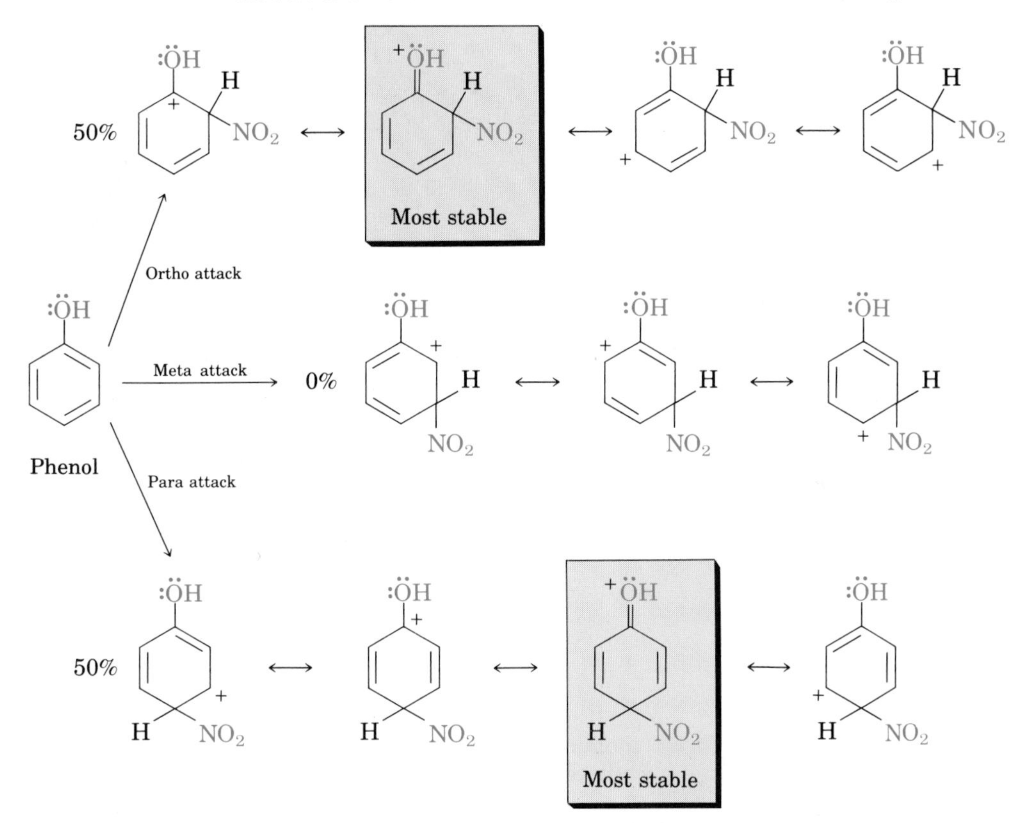

FIGURE 5.12 Intermediates in the nitration of phenol. The ortho and para intermediates are more stable than the meta intermediate because of resonance electron-donation from oxygen.

In general, any substituent that has a lone pair of electrons on the atom bound to the aromatic ring has an electron-donating resonance effect and is thus an ortho and para director:

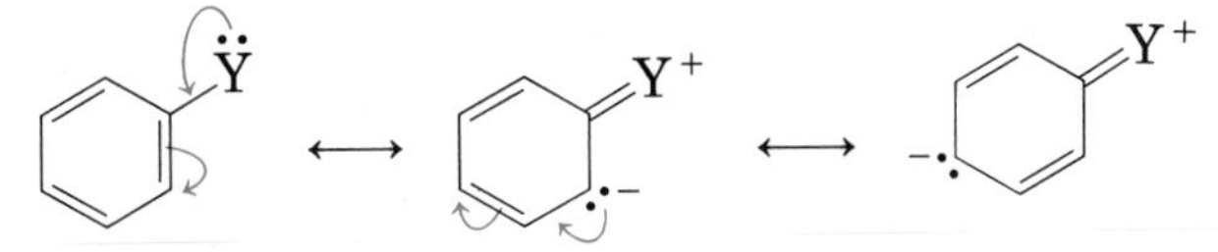

Meta Directors

We can explain the influence of meta directors by using the same kinds of arguments we used for ortho and para directors. For example, let's look at the chlorination of benzaldehyde, shown in Figure 5.13. Of the three possible intermediates, the carbocations produced by reaction at ortho and para positions are least stable. In both ortho and para intermediates, the unfavorable resonance forms indicated in Figure 5.13 place the positive charge directly on the carbon that bears the deactivating aldehyde group, where it is disfavored by a repulsive interaction with the positively polarized carbon atom of the carbonyl group. Hence the meta intermediate is most favored.

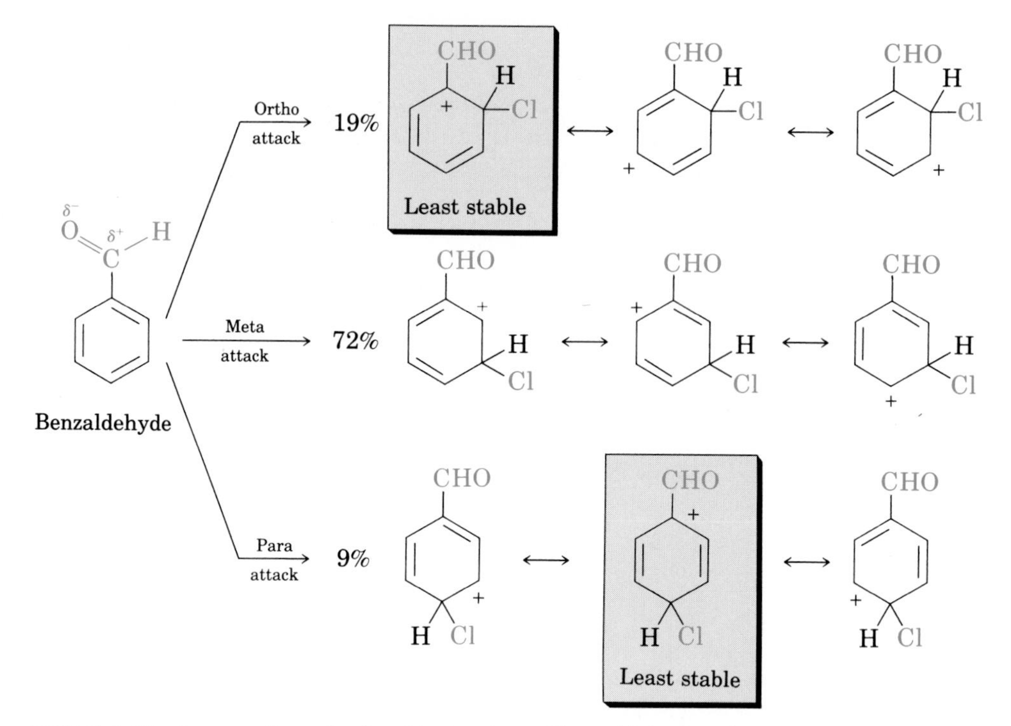

FIGURE 5.13 Intermediates in the chlorination of benzaldehyde. The ortho and para intermediates are less stable than the meta intermediate.

PROBLEM 5.17 What products would you expect from mononitration of these compounds:

(a) Nitrobenzene (b) Bromobenzene (c) Toluene
(d) Benzoic acid (e) *p*-Xylene

PROBLEM 5.18 Which would you expect to be more reactive toward electrophilic substitution, toluene or (trifluoromethyl)benzene? Explain your answer.

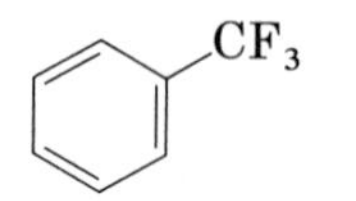

(Trifluoromethyl)benzene

PROBLEM 5.19 The nitroso group, $-N{=}O$, is an ortho and para director. Draw resonance structures of the three possible intermediates from electrophilic attack on nitrosobenzene, and explain why the ortho and para intermediates are favored over meta.

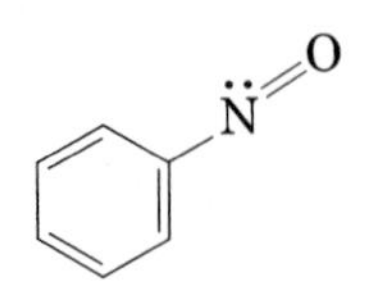

Nitrosobenzene

5.11 TRISUBSTITUTED BENZENES: ADDITIVITY OF EFFECTS

Further electrophilic substitution of a disubstituted benzene is governed by the same orientation effects just discussed. The only difference is that now we have to consider the additive effects of two directing groups. Three rules are sufficient:

1. If the directing effects of the groups reinforce each other, there's no problem. For example, in *p*-nitrotoluene, both the methyl group and the nitro group direct further substitution to the same position (ortho to the methyl = meta to the nitro), and a single product is thus formed by electrophilic substitution.

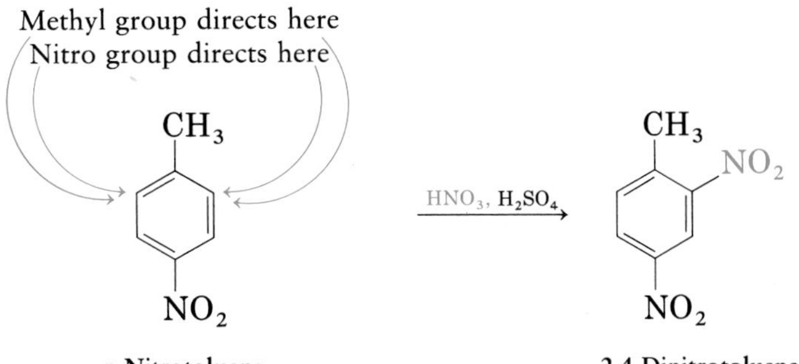

p-Nitrotoluene 2,4-Dinitrotoluene

2. If the directing effects of the two groups oppose each other, the more powerful activating group has the dominant influence. For example, bromination of *p*-methylphenol yields mostly 2-bromo-4-methylphenol because hydroxyl is a more powerful activator than methyl.

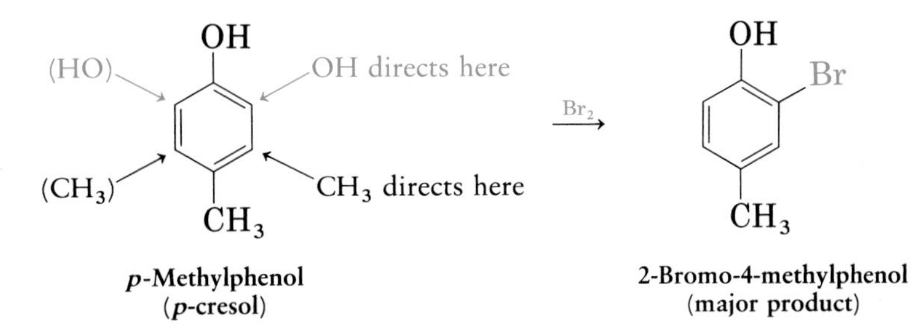

p-Methylphenol (*p*-cresol) 2-Bromo-4-methylphenol (major product)

3. Further substitution rarely occurs between the two groups in a meta-disubstituted compound because this site is too crowded for reaction to occur easily.

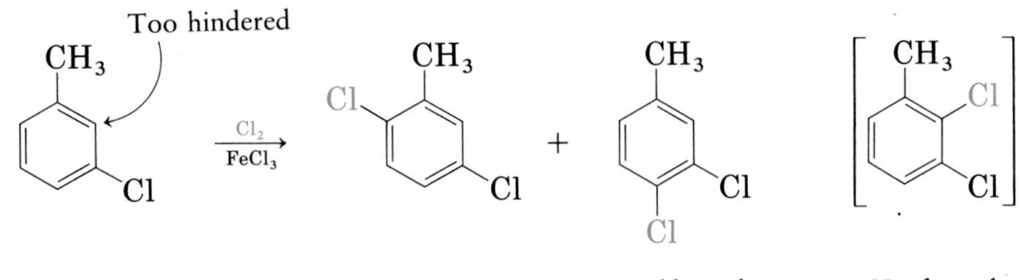

m-Chlorotoluene 2,5-Dichlorotoluene 3,4-Dichlorotoluene Not formed

PRACTICE PROBLEM 5.4 What product would you expect from bromination of *p*-methylbenzoic acid?

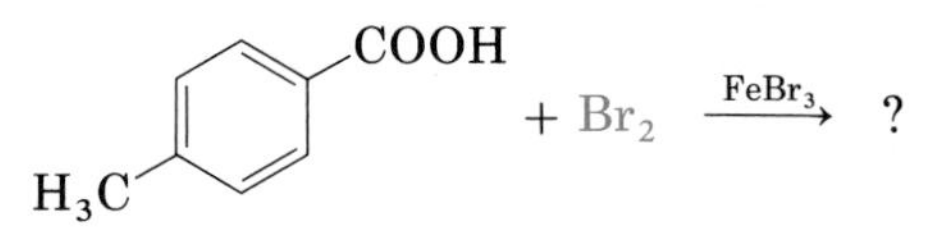

SOLUTION The carboxyl group (−COOH) is a meta director, and the methyl group is an ortho and para director. Both groups therefore direct bromination to the position next to the methyl group, yielding 3-bromo-4-methylbenzoic acid.

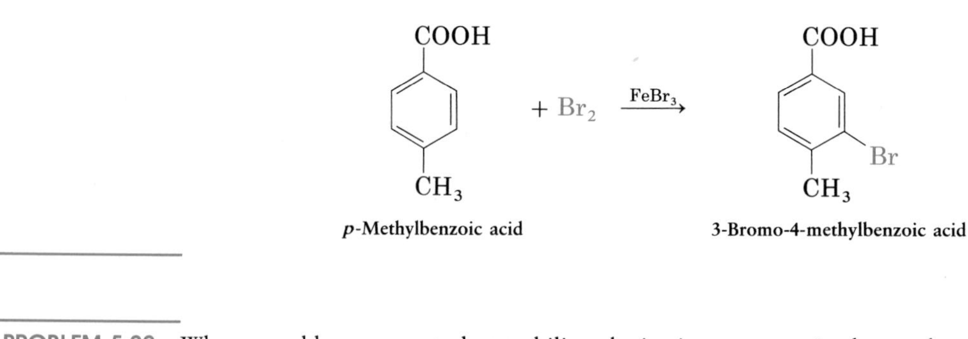

p-Methylbenzoic acid 3-Bromo-4-methylbenzoic acid

PROBLEM 5.20 Where would you expect electrophilic substitution to occur in these substances?

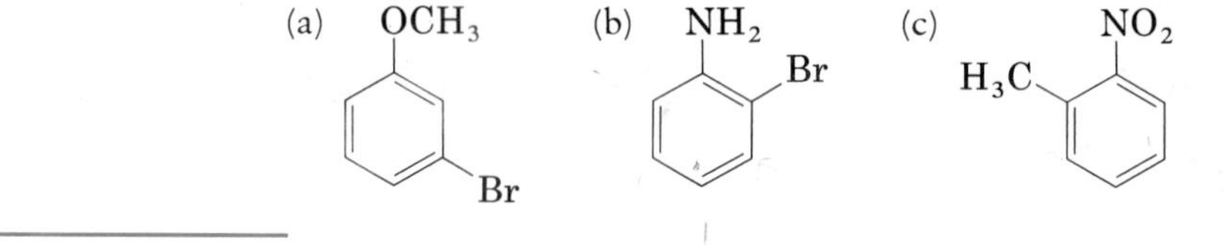

5.12 OXIDATION AND REDUCTION OF AROMATIC COMPOUNDS

The benzene ring, despite its unsaturation, is normally inert to strong oxidizing agents such as potassium permanganate. (Recall that this reagent cleaves alkene carbon–carbon double bonds; Section 4.7.) Alkyl groups attached to the aromatic ring are readily attacked by these reagents, however, and are converted into carboxyl groups (−COOH). For example, butylbenzene is oxidized by $KMnO_4$ in high yield to give benzoic acid.

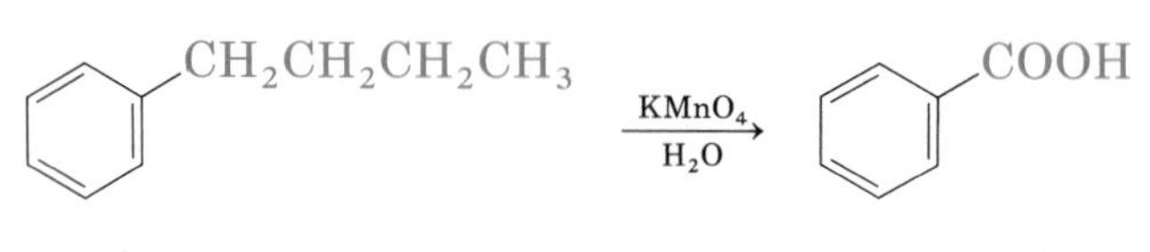

Butylbenzene **Benzoic acid (85%)**

The exact mechanism of this side-chain oxidation reaction isn't fully understood but probably involves attack on the side-chain C–H bonds at the position next to the aromatic ring (the **benzylic** position) to give radical intermediates.

benzylic
the position next to an aromatic ring

As well as being inert to oxidation, aromatic rings are also inert to reduction under standard alkene hydrogenation conditions. Only if high temperatures and pressures are used does reduction of aromatic rings occur. For example, *o*-dimethylbenzene gives 1,2-dimethylcyclohexane.

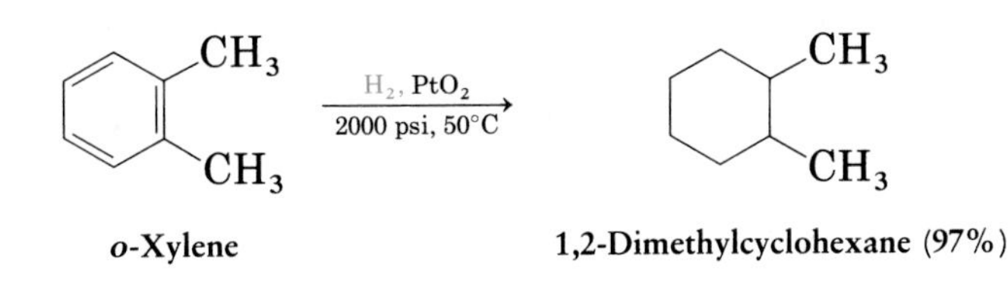

PROBLEM 5.21 What aromatic products would you expect to obtain from the $KMnO_4$ oxidation of these substances?

(a) *m*-Chloroethylbenzene (b) 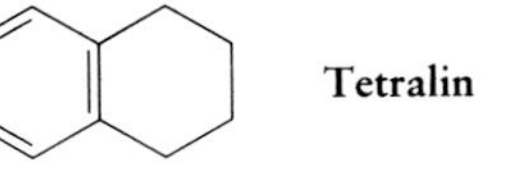 Tetralin

5.13 POLYCYCLIC AROMATIC HYDROCARBONS

The concept of aromaticity—the unusual chemical stability that arises in cyclic conjugated molecules like benzene—can be extended beyond simple monocyclic compounds to include **polycyclic aromatic compounds.** Naphthalene, with two benzene-like rings fused together, and anthracene, with three fused rings, are two of the simplest polycyclic aromatic molecules.

polycyclic aromatic compound
a molecule that has two or more fused benzene rings

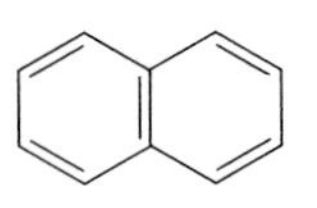
Naphthalene

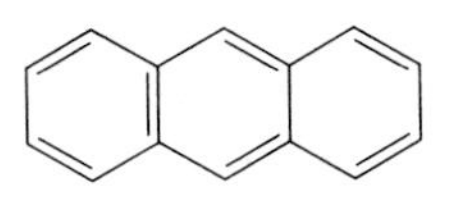
Anthracene

Naphthalene and other polycyclic aromatic hydrocarbons show many of the chemical properties we associate with aromaticity. Both, for example, react with electrophilic reagents such as bromine to give substitution products rather than double-bond addition products.

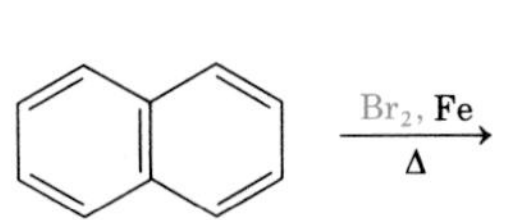

Naphthalene

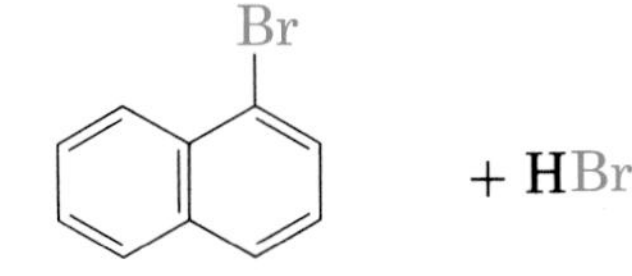

1-Bromonaphthalene (75%)

We'll see in Chapter 12 that compounds like pyridine and pyrrole are also aromatic even though they don't contain benzene rings.

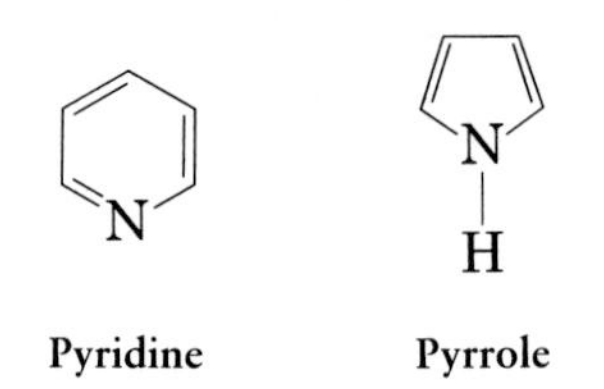

Pyridine **Pyrrole**

PROBLEM 5.22 There are three resonance structures of naphthalene, of which only one is shown. Draw the other two.

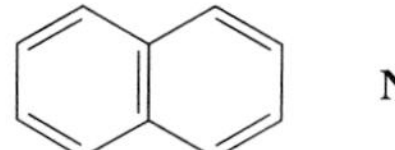

Naphthalene

5.14 ORGANIC SYNTHESIS

The laboratory synthesis of organic molecules from simple precursors is carried out for many reasons. In the pharmaceutical industry, new organic molecules are designed and synthesized for evaluation as medicines. In the chemical industry, syntheses are often undertaken to devise more economical routes to known compounds. In this book, too, we'll sometimes devise syntheses of complex molecules from simpler precursors, but our purpose is simply to learn organic chemistry: Attempts at synthesis make us approach chemical problems in a logical way, drawing on our knowledge and organizing that knowledge into workable plans.

The only real secret to planning an organic synthesis is to *work backward.* Look at the final product and ask yourself, "What is the immediate precursor of that product?" Having found an immediate precursor, proceed backward again, one step at a time, until a suitable starting material is found. Let's try some examples.

PRACTICE PROBLEM 5.5 Synthesize *m*-chloronitrobenzene starting with benzene.

SOLUTION Ask "What is an immediate precursor of *m*-chloronitrobenzene?"

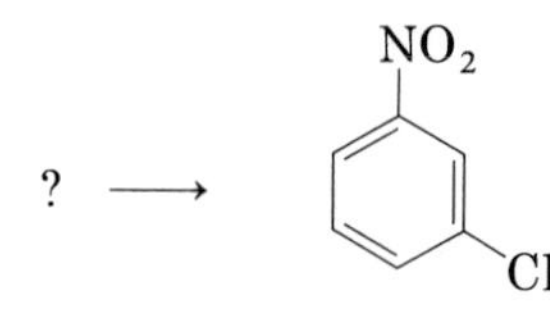

***m*-Chloronitrobenzene**

There are only two substituents on the ring, a chloro group, which is ortho-para-directing, and a nitro group, which is meta-directing. We can't nitrate chlorobenzene because the wrong isomers (*o*- and *p*-chloronitrobenzenes) would result, but chlorination of nitrobenzene should give the desired product.

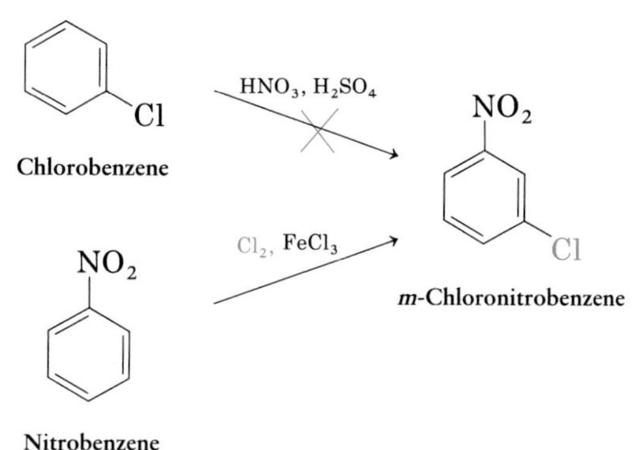

"What is an immediate precursor of nitrobenzene?" Benzene, which can be nitrated. Thus, in two steps, we've solved the problem.

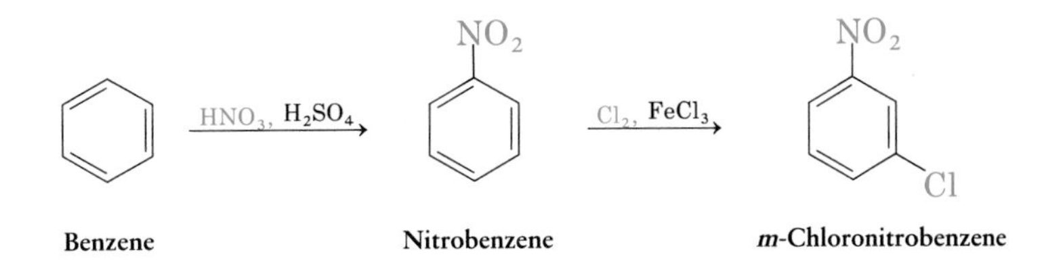

PRACTICE PROBLEM 5.6

Synthesize *p*-bromobenzoic acid starting from benzene.

SOLUTION Ask "What is an immediate precursor of *p*-bromobenzoic acid?"

$$? \longrightarrow \text{Br—C}_6\text{H}_4\text{—COOH}$$

p-Bromobenzoic acid

There are only two substituents on the ring, a carboxyl group (–COOH), which is meta-directing, and a bromine, which is ortho- and para-directing. We can't brominate benzoic acid because the wrong isomer (*m*-bromobenzoic acid) would be produced. We know, however, that oxidation of alkylbenzene side chains yields benzoic acids. An immediate precursor of our target molecule might therefore be *p*-bromotoluene.

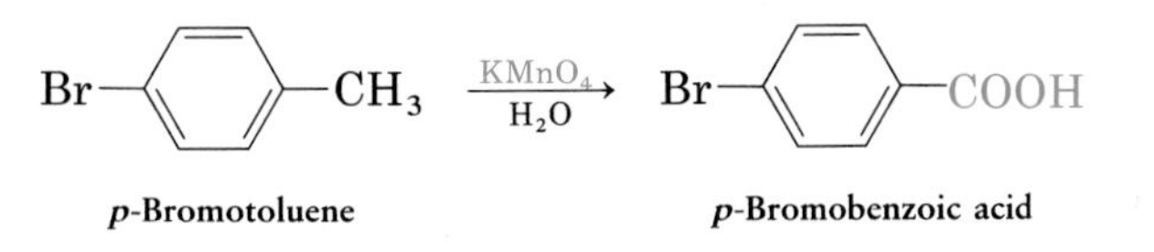

"What is an immediate precursor of *p*-bromotoluene?" Perhaps toluene because the methyl group would direct bromination to the ortho and para positions, and we could then separate isomers. Alternatively, bromobenzene might be an immediate precursor because we could carry out a Friedel–Crafts alkylation and obtain para product. Both answers are satisfactory.

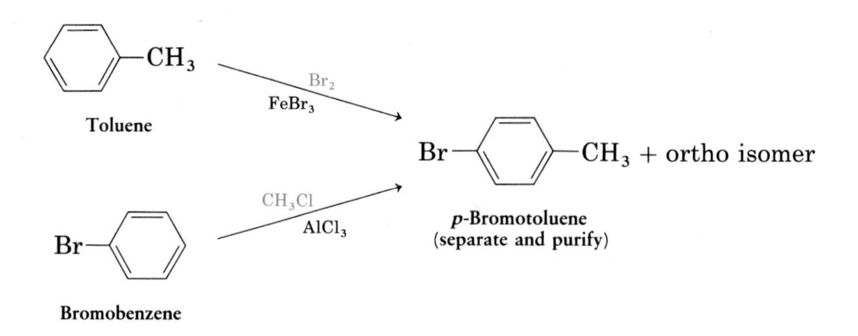

"What is an immediate precursor of toluene?" Benzene, which can be methylated in a Friedel–Crafts reaction.

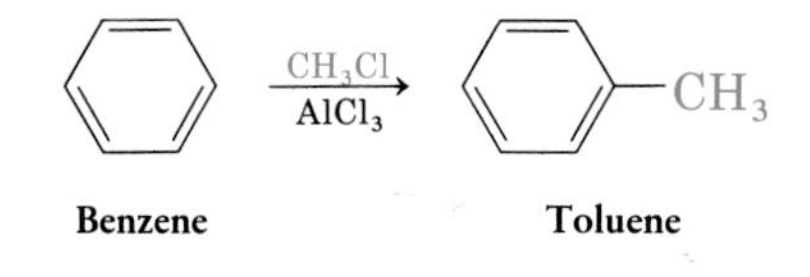

Alternatively, what is an immediate precursor of bromobenzene? Benzene, which can be brominated.

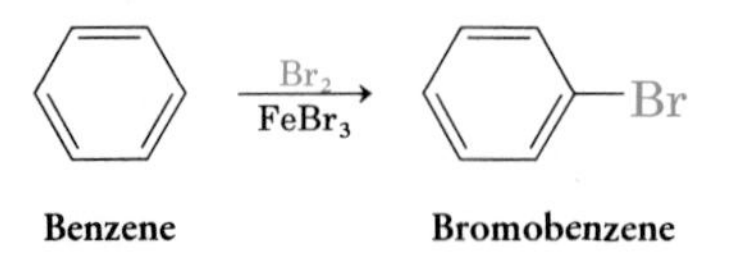

Our backward synthetic (*retrosynthetic*) analysis has provided two valid routes from benzene to *p*-bromobenzoic acid (Figure 5.14).

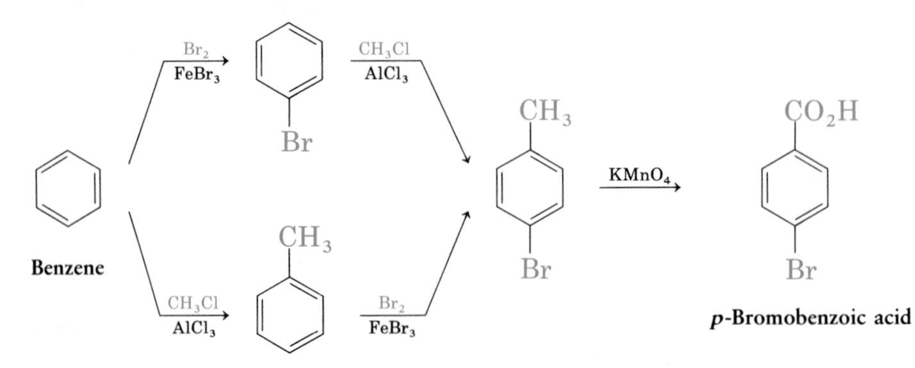

FIGURE 5.14 Two routes for the synthesis of *p*-bromobenzoic acid from benzene

PROBLEM 5.23 Propose syntheses of these substances, starting with benzene.
(a) *p*-Methylacetophenone (b) *p*-Chloronitrobenzene

PROBLEM 5.24 Synthesize these substances starting with benzene.
(a) *o*-Bromotoluene (b) 2-Bromo-1,4-dimethylbenzene

PROBLEM 5.25 How would you prepare *m*-chlorobenzoic acid from benzene?

INTERLUDE

Polycyclic Aromatic Hydrocarbons and Cancer

In addition to naphthalene and anthracene, there are a great many more-complex *polycyclic aromatic hydrocarbons* (PAHs). Benzo[a]pyrene, for example, contains five benzene rings joined together, and ordinary graphite (the "lead" in pencils) consists of essentially infinite, two-dimensional sheets of benzene rings.

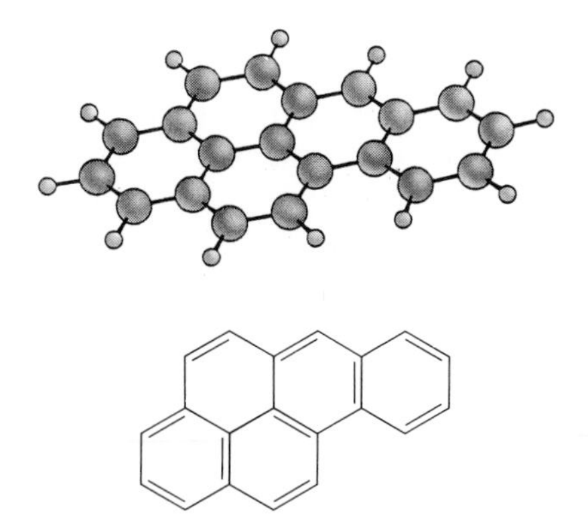

Benzo[a]pyrene

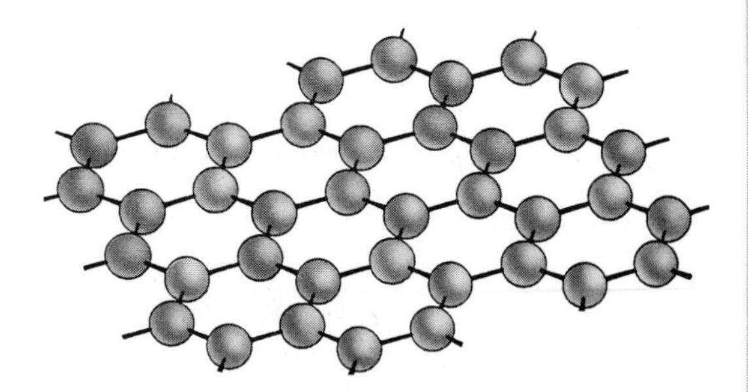

A graphite segment

Benzo[a]pyrene is a particularly important and well-studied PAH because it is one of the cancer-causing (*carcinogenic*) substances found in chimney soot and cigarette smoke. Exposure to a tiny amount is sufficient to induce a skin tumor in susceptible mice.

Recent studies have given a clear picture of how these PAHs cause tumors. After a PAH is absorbed by eating or inhaling, the body attempts to rid itself of the foreign substance by converting it into a water-soluble metabolite that can be excreted. In the case of benzo[a]pyrene, oxidation in the liver converts it into an oxygenated product called a *diol-epoxide*. Unfortunately, this metabolite is able to bind to cellular DNA, causing mutations and cancers.

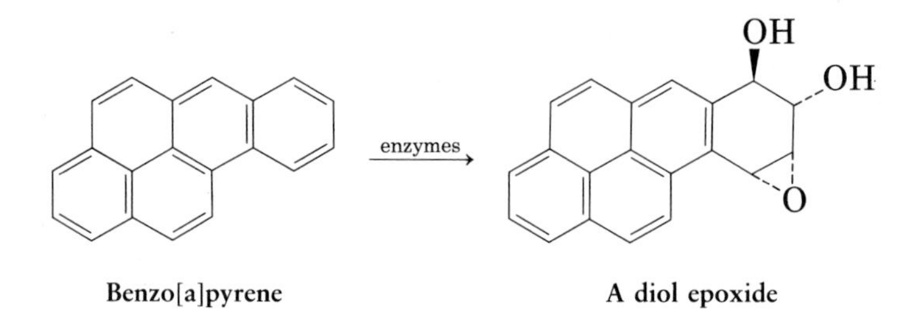

Benzo[a]pyrene **A diol epoxide**

Even benzene itself can cause certain types of cancers, particularly leukemia, on prolonged exposure. Breathing the fumes of volatile aromatic hydrocarbons should therefore be avoided.

SUMMARY AND KEY WORDS

The word **aromatic** refers to the class of compounds structurally related to benzene. Aromatic compounds are named according to IUPAC rules, with disubstituted benzenes referred to as either **ortho** (1,2-disubstituted), **meta** (1,3-disubstituted), or **para** (1,4-disubstituted).

Benzene is described in resonance terms as a resonance hybrid of two equivalent line-bond structures. Neither structure is correct by itself; the true structure of benzene is intermediate between the two.

Electrophilic aromatic substitution is the single most important reaction of aromatic compounds. In this two-step polar reaction, the pi electrons of the aromatic ring first attack the electrophile to yield a resonance-stabilized carbocation intermediate. Loss of H^+ from this intermediate regenerates the stable aromatic ring and gives a substituted product. Bromination, chlorination, iodination, nitration, sulfonation, Friedel–Crafts alkylation, and Friedel–Crafts acylation can all be carried out with the proper choice of reagent. **Friedel–Crafts alkylation** is a particularly useful reaction for preparing a variety of alkylbenzenes but is limited by the facts that only alkyl halides can be used and that strongly deactivated rings don't react.

Substituents on the benzene ring affect both the reactivity of the ring toward further substitution and the orientation of further substitution. We can classify substituents either as **activators** or **deactivators,** and either as **ortho and para directors** or as **meta directors.** Substituent effects are due to an interplay of resonance and inductive effects. **Resonance effects** are transmitted by pi-orbital overlap whereas **inductive effects** are transmitted by sigma bonds.

The side chains of alkylbenzenes have unique reactivity because of the neighboring aromatic ring. Thus, an alkyl group attached to the aromatic ring can be degraded to a carboxyl group (–COOH) by oxidation with aqueous $KMnO_4$.

SUMMARY OF REACTIONS

1. Electrophilic aromatic substitution

   (a) Bromination (Section 5.6)

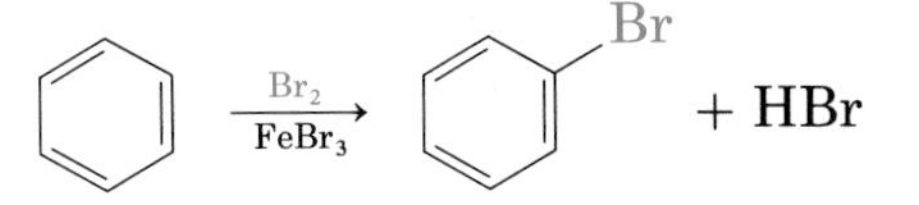

   (b) Chlorination (Section 5.7)

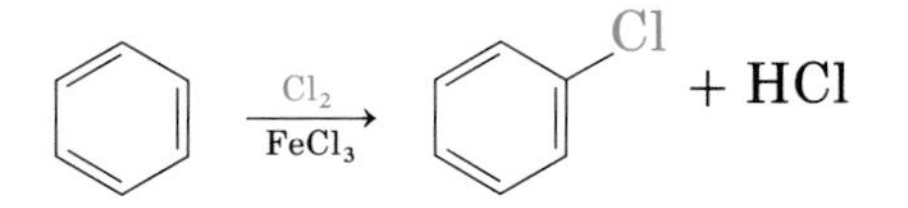

   (c) Nitration (Section 5.7)

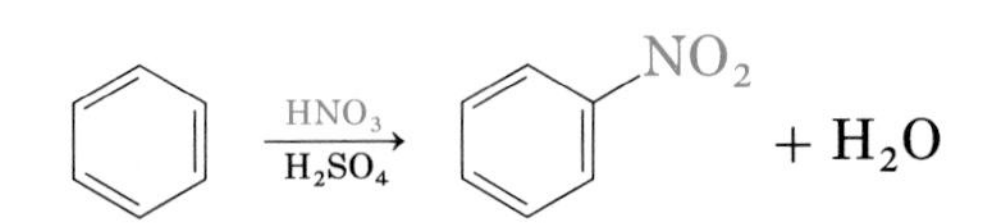

   (d) Sulfonation (Section 5.7)

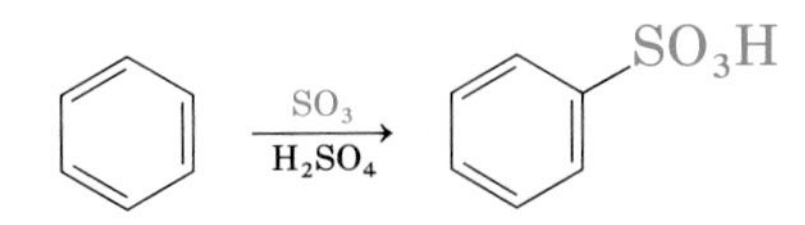

   (e) Friedel–Crafts alkylation (Section 5.8)

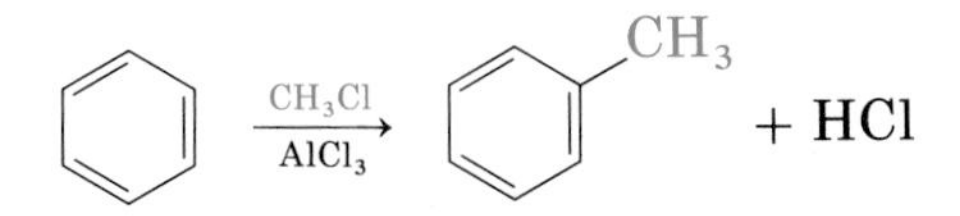

   (f) Friedel–Crafts acylation (Section 5.8)

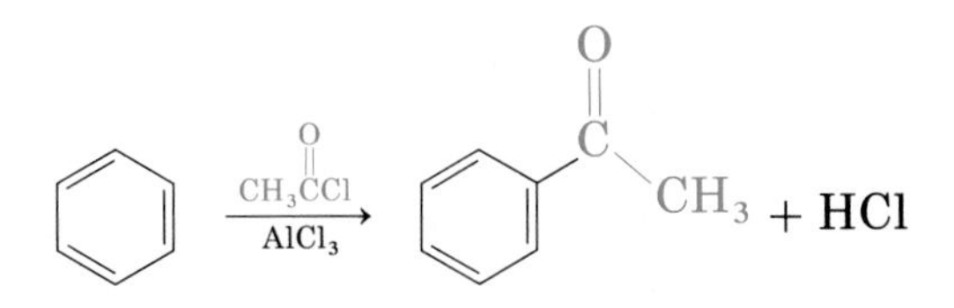

2. Oxidation of aromatic side chains (Section 5.12)

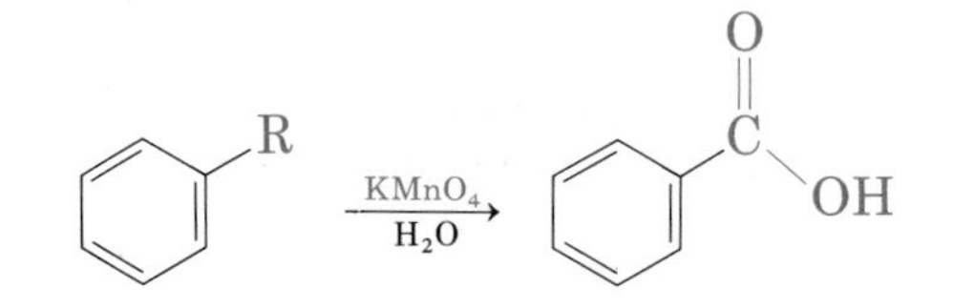

3. Hydrogenation of aromatic rings (Section 5.12)

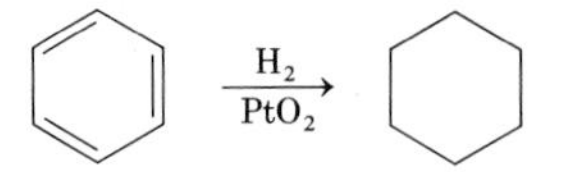

ADDITIONAL PROBLEMS

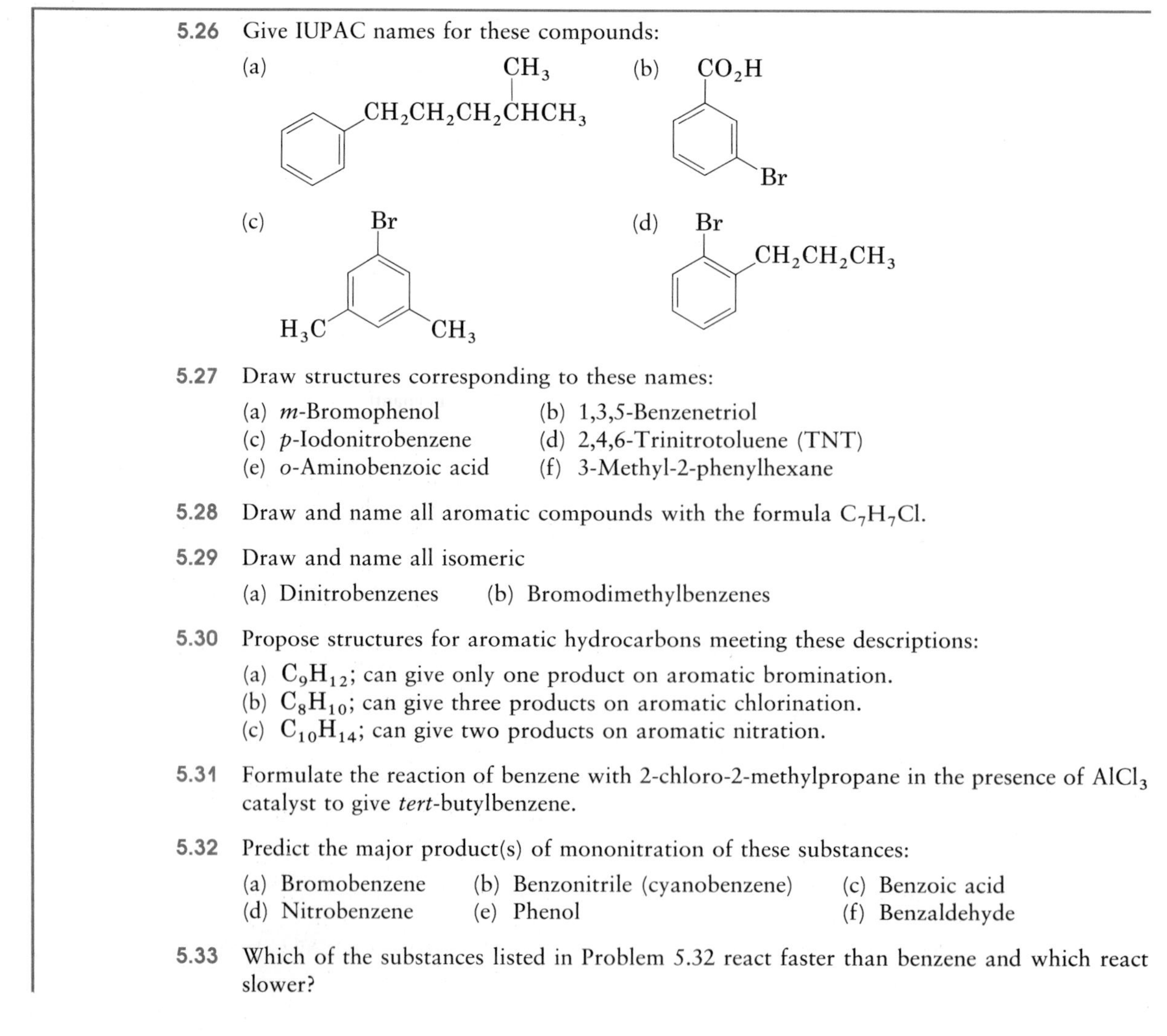

5.26 Give IUPAC names for these compounds:

(a) $CH_2CH_2CH_2CH(CH_3)CH_3$ on benzene (b) CO_2H and Br on benzene

(c) Br, H_3C, CH_3 on benzene (d) Br and $CH_2CH_2CH_3$ on benzene

5.27 Draw structures corresponding to these names:

(a) *m*-Bromophenol (b) 1,3,5-Benzenetriol
(c) *p*-Iodonitrobenzene (d) 2,4,6-Trinitrotoluene (TNT)
(e) *o*-Aminobenzoic acid (f) 3-Methyl-2-phenylhexane

5.28 Draw and name all aromatic compounds with the formula C_7H_7Cl.

5.29 Draw and name all isomeric

(a) Dinitrobenzenes (b) Bromodimethylbenzenes

5.30 Propose structures for aromatic hydrocarbons meeting these descriptions:

(a) C_9H_{12}; can give only one product on aromatic bromination.
(b) C_8H_{10}; can give three products on aromatic chlorination.
(c) $C_{10}H_{14}$; can give two products on aromatic nitration.

5.31 Formulate the reaction of benzene with 2-chloro-2-methylpropane in the presence of $AlCl_3$ catalyst to give *tert*-butylbenzene.

5.32 Predict the major product(s) of mononitration of these substances:

(a) Bromobenzene (b) Benzonitrile (cyanobenzene) (c) Benzoic acid
(d) Nitrobenzene (e) Phenol (f) Benzaldehyde

5.33 Which of the substances listed in Problem 5.32 react faster than benzene and which react slower?

5.34 Rank the compounds in each group according to their reactivity toward electrophilic substitution:

(a) Chlorobenzene, *o*-dichlorobenzene, benzene
(b) *p*-Bromonitrobenzene, nitrobenzene, phenol
(c) Fluorobenzene, benzaldehyde, *o*-dimethylbenzene

5.35 Show in detail the steps involved in the Friedel–Crafts reaction of benzene with CH_3Cl.

5.36 Name and draw the structure(s) of the major product(s) of electrophilic chlorination of these substances:

(a) *m*-Nitrophenol (b) *o*-Dimethylbenzene
(c) *p*-Nitrobenzoic acid (d) 2,4-Dibromophenol

5.37 Predict the major product(s) you would expect to obtain from sulfonation of these substances:

(a) Bromobenzene (b) *m*-Bromophenol (c) 2,4-Dichloronitrobenzene
(d) 2,4-Dichlorophenol (e) 2,5-Dibromotoluene

5.38 What is the structure of the compound, C_8H_9Br, that gives *p*-bromobenzoic acid on oxidation with $KMnO_4$?

5.39 Draw the four resonance structures of anthracene.

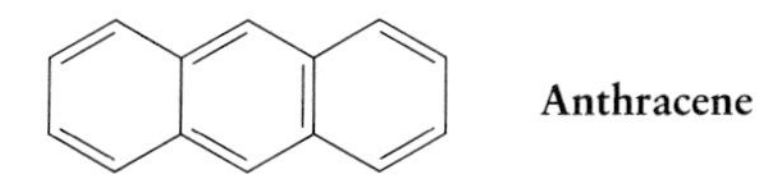

Anthracene

5.40 Draw the five resonance structures of phenanthrene.

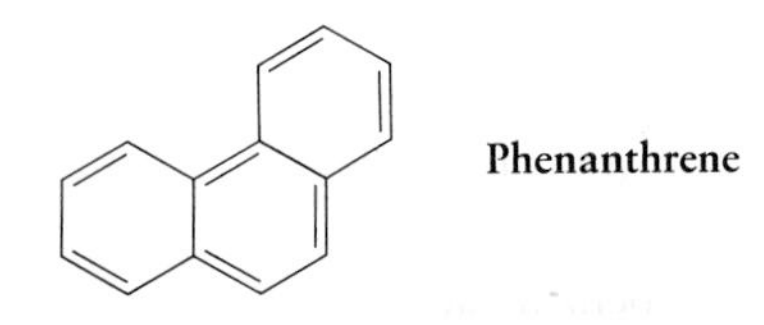

Phenanthrene

5.41 Suggest a reason for the observation that bromination of biphenyl occurs at ortho and para positions rather than at meta. Use resonance structures of the carbocation intermediates to explain your answer.

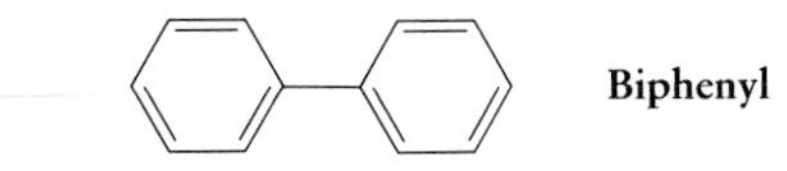

Biphenyl

5.42 In light of your answer to Problem 5.41, at what position and on which ring would you expect nitration of 4-bromobiphenyl to occur?

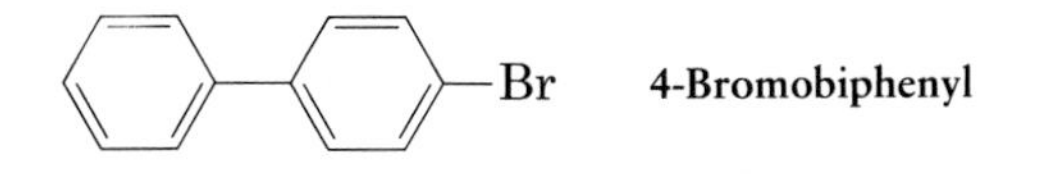

4-Bromobiphenyl

5.43 Starting with benzene, how would you synthesize these substances? Assume that you can separate ortho and para isomers if necessary.

(a) *m*-Bromobenzenesulfonic acid (b) *o*-Chlorobenzenesulfonic acid
(c) *p*-Chlorotoluene

5.44 Starting with either benzene or toluene, how would you synthesize these substances? Ortho and para isomers can be separated if necessary.

(a) 2-Bromo-4-nitrotoluene (b) 1,3,5-Trinitrobenzene

5.45 Starting from any aromatic hydrocarbon necessary, how would you synthesize these substances? Ortho and para isomers can be separated if necessary.

(a) 2,4,6-Trinitrobenzoic acid (b) *p-tert*-Butylbenzoic acid

5.46 Draw resonance structures of the intermediate carbocations and account for the fact that naphthalene undergoes electrophilic aromatic substitution at C1 rather than C2.

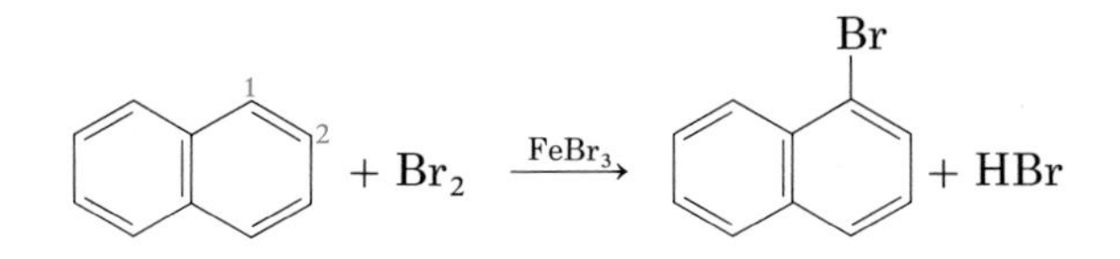

5.47 We said in Section 4.11 that allylic carbocations are stabilized by resonance. Draw resonance structures to account for a similar stabilization of benzylic carbocations.

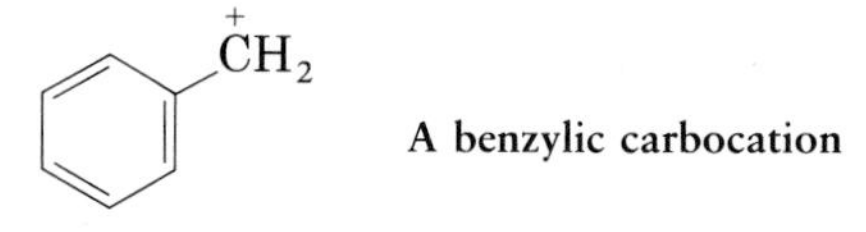

A benzylic carbocation

5.48 In light of your answer to Problem 5.47, which of the two possible addition products would you expect from reaction of HBr with octalin?

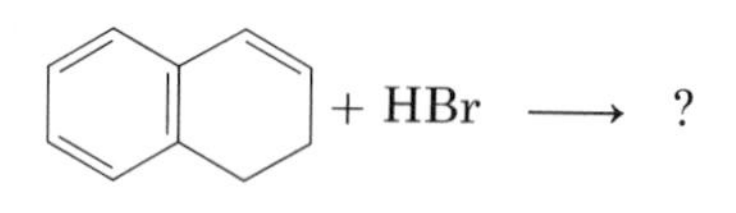

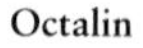
Octalin

5.49 Starting with benzene, how would you synthesize the following?

(a) 2,4-Dinitrobenzoic acid (b) 3,5-Dinitrobenzoic acid

5.50 Pyridine is a cyclic nitrogen-containing organic compound that shows many of the properties associated with aromaticity. For example, pyridine undergoes electrophilic substitution reactions. Draw an orbital picture of pyridine and account for its aromatic properties.

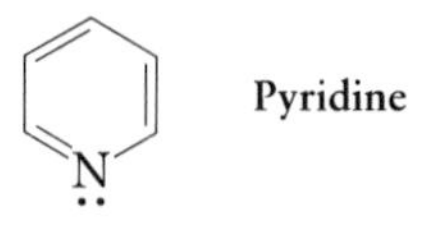

Pyridine

5.51 Would you expect the trimethylammonium group to be an activating or deactivating substituent? Explain your answer.

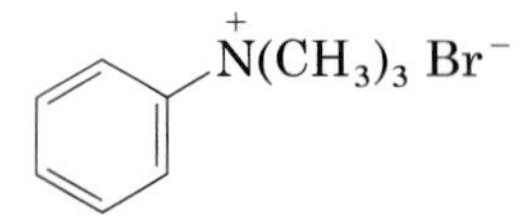

Phenyltrimethylammonium bromide

5.52 Starting with toluene, how would you synthesize the three nitrobenzoic acids?

5.53 Carbocations generated by reaction of an alkene with a strong acid catalyst can react with aromatic rings in a Friedel–Crafts reaction. Propose a mechanism to account for the industrial synthesis of the food preservative BHT from *p*-cresol and 2-methylpropene.

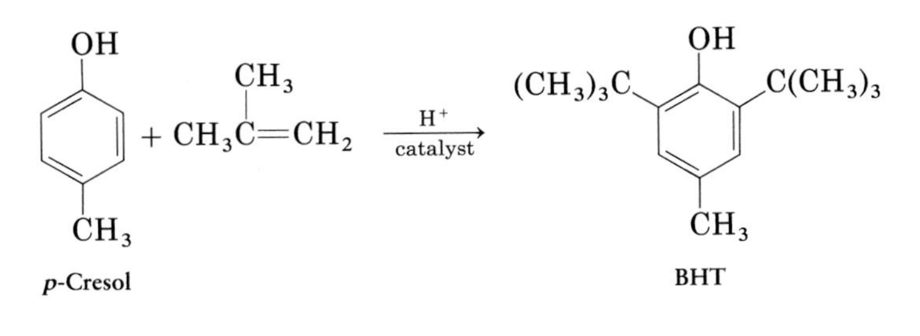

CHAPTER

6 Stereochemistry

stereochemistry the branch of chemistry concerned with three-dimensional consequences of structure

Up to this point, we've been concerned only with the general nature of chemical reactions and with the specific chemistry of hydrocarbon functional groups. Although we took a brief look at constitutional isomers of alkanes in Section 2.2 and cis–trans stereoisomers of cycloalkanes in Section 2.7, we've given little thought to the chemical consequences of the spatial arrangements of atoms in molecules. It's now time to look more deeply into these consequences. **Stereochemistry** is the branch of chemistry concerned with the three-dimensional nature of molecules.

6.1 STEREOCHEMISTRY AND THE TETRAHEDRAL CARBON

Are you right-handed or left-handed? Although most of us don't often think about it, handedness plays a surprisingly large role in our daily activities. Musical instruments like oboes and clarinets have a handedness to them; the last available softball glove always fits the wrong hand; and left-handed people write in a "funny" way. The fundamental reason for these difficulties is that our hands aren't identical, they're mirror images. When you hold your right hand up to a mirror, the reflection looks like a left hand. Try it.

Handedness also plays a large role in organic chemistry. To see why, look at the molecules shown in Figure 6.1. On the left of Figure 6.1 are three molecules and on the right are their images reflected in a mirror. The CH_3X and CH_2XY molecules are identical to their mirror images and thus are not handed. If you make molecular models of each molecule and of its mirror image, you find that you can superimpose one on the other. Unlike the CH_3X and CH_2XY molecules, the CHXYZ molecule isn't identical to its mirror image. You can't superimpose a model of the molecule on a model of its mirror image for the same reason that you can't superimpose a left hand on a right hand; they simply aren't identical. You might superimpose *two* of the substituents, X and Y for example, but H and Z would be reversed. If the H and Z substituents were superimposed, X and Y would be reversed.

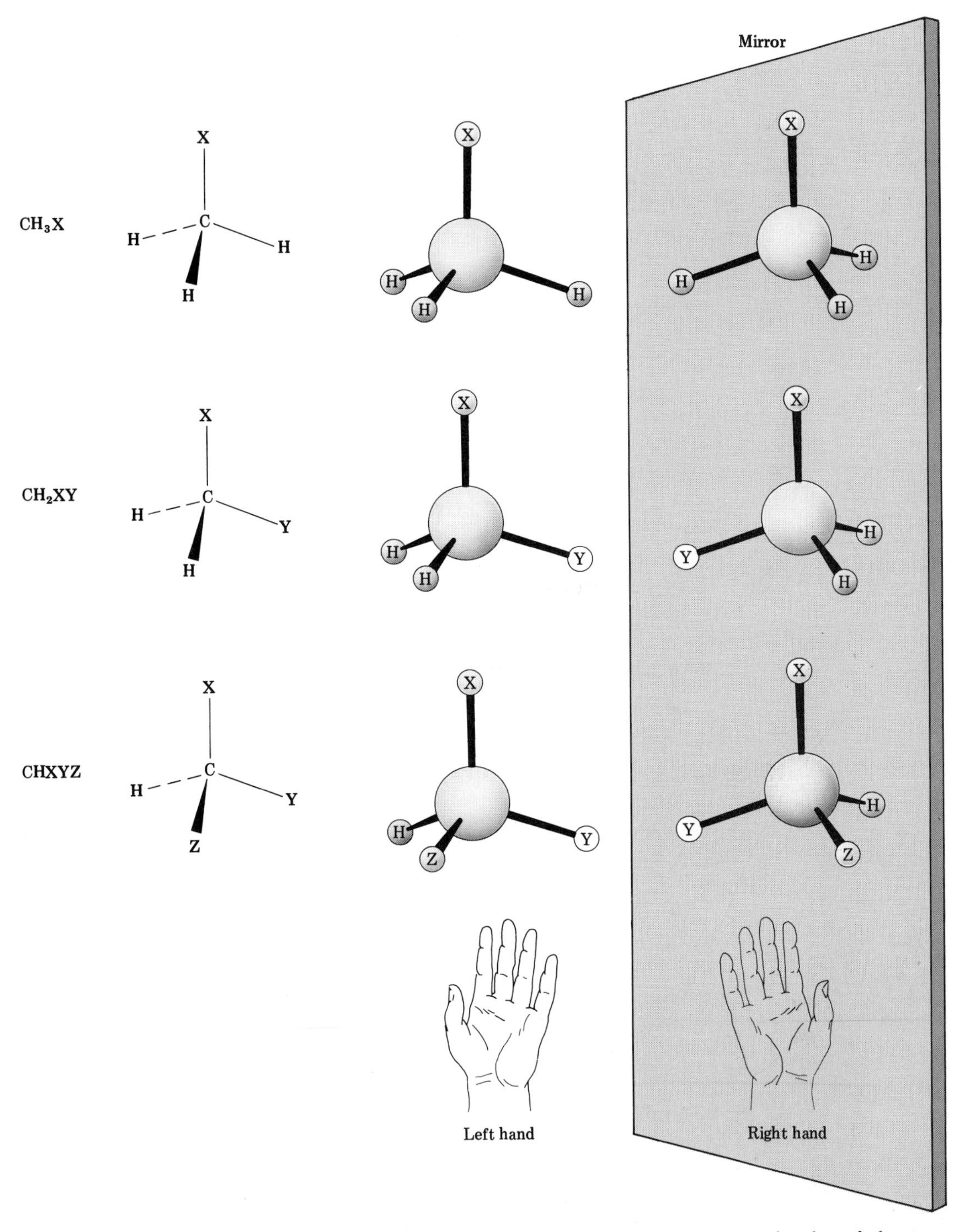

FIGURE 6.1 Three tetrahedral carbons and their mirror images. Molecules of the type CH_3X and CH_2XY are superimposable on their mirror images, but a molecule of the type CHXYZ is not superimposable on its mirror image for the same reason that a left hand is not superimposable on a right hand.

enantiomers stereoisomers that have a mirror-image relationship

Molecules that are nonsuperimposable mirror images of each other are a special kind of stereoisomers called **enantiomers** (e-*nan*-tee-o-mer; Greek *enantio,* "opposite"). Enantiomers are related to each other as a right hand is related to a left hand. Handedness (enantiomerism) can result whenever a tetrahedral carbon is bonded to four different substituents (one need not be H). For example, lactic acid (2-hydroxypropanoic acid) can exist in either right- or left-handed form because there are four different groups ($-H$, $-OH$, $-CH_3$, $-COOH$) attached to the central carbon atom:

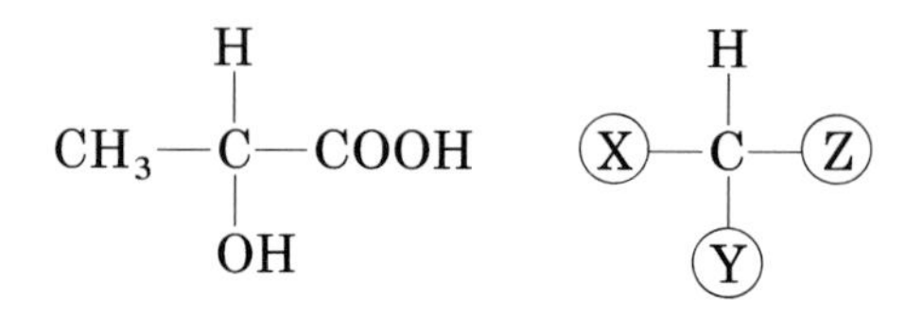

Lactic acid; a molecule of general formula CHXYZ

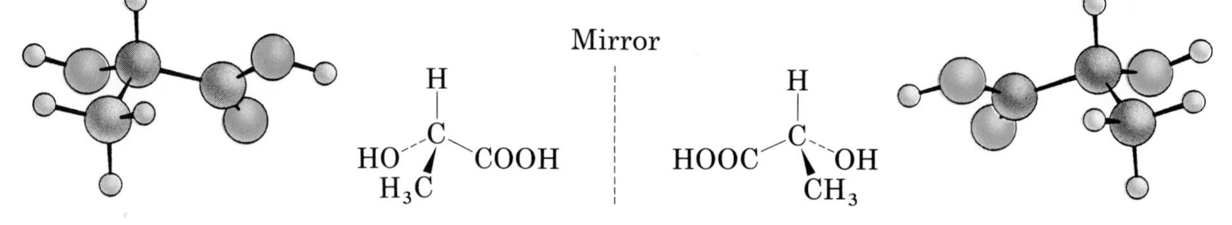

No matter how hard you try, you can't superimpose a molecule of "right-handed" lactic acid on top of a molecule of "left-handed" lactic acid; the two molecules aren't identical, as shown in Figure 6.2.

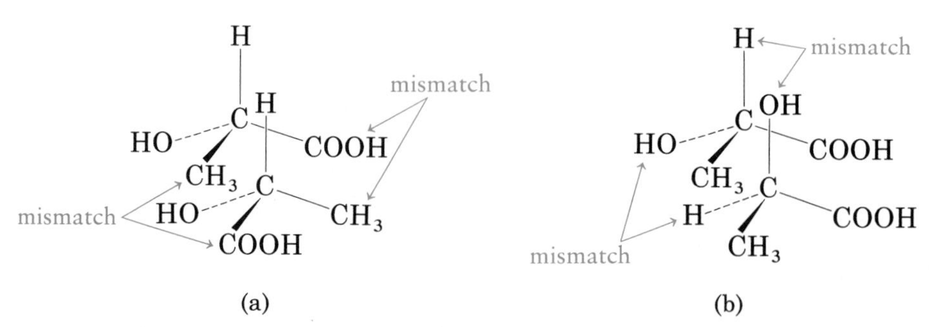

FIGURE 6.2 Attempts at superimposing the mirror-image forms of lactic acid: (a) When the $-H$ and $-OH$ substituents match up, the $-COOH$ and $-CH_3$ substituents don't; (b) when $-COOH$ and $-CH_3$ match up, $-H$ and $-OH$ don't. Regardless of how the molecules are oriented, they aren't identical.

6.2 CHIRALITY

chiral having handedness

Compounds that aren't the same as their mirror images and thus exist as a pair of enantiomers are said to be **chiral** (*ky*-ral; Greek *cheir,* "hand"). We can't take a chiral molecule and its mirror image (enantiomer) and place one on top of the other so that all atoms coincide.

plane of symmetry an imaginary plane that bisects an object or molecule so that one half is a mirror image of the other half

How can we predict whether a certain compound is or is not chiral? *A compound is not chiral if it contains a plane of symmetry.* A **plane of symmetry** is an imaginary plane that bisects an object (or molecule) so that one half of the object is an exact mirror image of the other half. For example, a flask has a plane of symmetry. If we were to cut the flask in half, one half would be an exact mirror image of the second half. A hand, however, doesn't have a plane of symmetry. One half of a hand is not a mirror image of the other half (Figure 6.3).

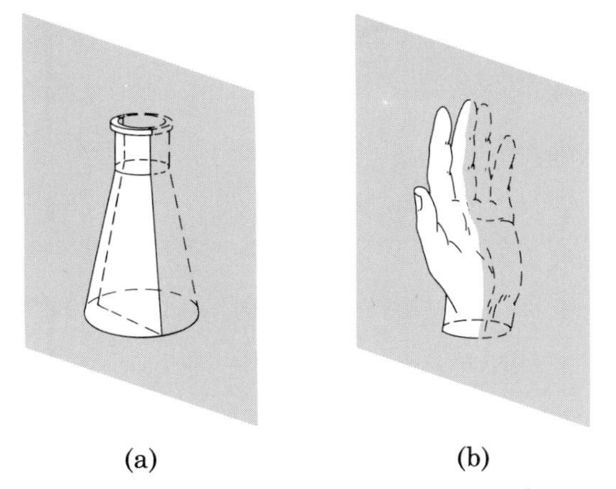

FIGURE 6.3 The meaning of symmetry planes. An object like the flask (a) has a plane of symmetry passing through it, making the right and left halves mirror images. An object like a hand (b) doesn't have a symmetry plane. The right half of a hand is not a mirror image of the left half.

achiral not having handedness

Molecules that have planes of symmetry *must* be superimposable on their mirror images and hence must be nonchiral or **achiral** (a-*ky*-ral). Thus, hydroxyacetic acid, $HOCH_2COOH$, contains a plane of symmetry and is achiral whereas lactic acid $CH_3CH(OH)COOH$ has no plane of symmetry and is chiral (Figure 6.4).

chiral center an atom in a molecule that is a local center of chirality

The most common (though not the only) cause of chirality in organic molecules is the presence of a carbon atom bonded to four different groups, for example, the central carbon atom in lactic acid. Such carbons are referred to as **chiral centers.** Detecting chiral centers in a complex molecule takes practice because it's not always immediately apparent that four different groups are bonded to a given carbon: The differences don't necessarily appear right next to the chiral center. For example, 5-bromodecane is a chiral molecule because four different groups are bonded to C5, the chiral center (marked by an asterisk):

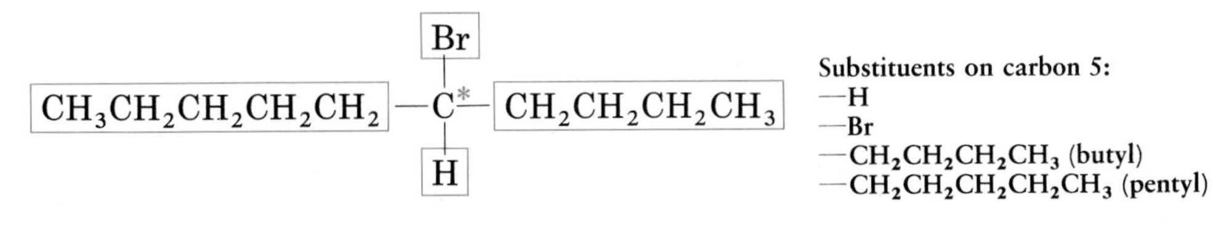

5-Bromodecane (chiral)

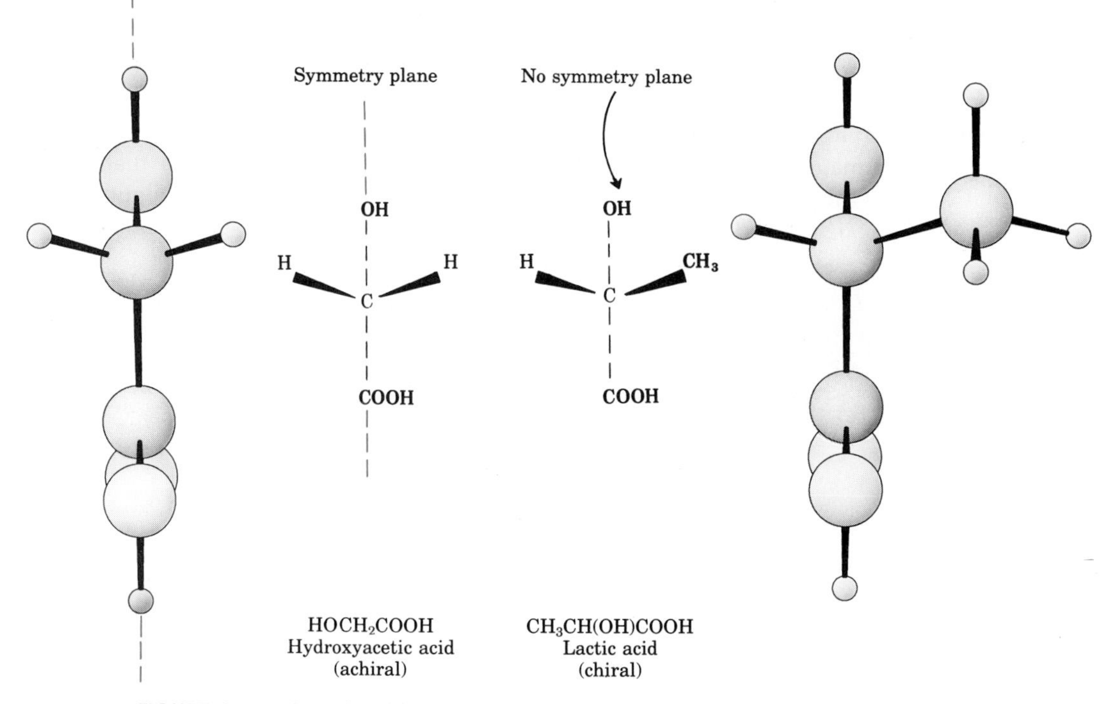

FIGURE 6.4 The achiral hydroxyacetic acid molecule versus the chiral lactic acid molecule. Hydroxyacetic acid has a plane of symmetry that makes one side of the molecule a mirror image of the other side. Lactic acid, however, has no such symmetry plane.

A butyl substituent is very similar to a pentyl substituent, but it isn't identical. The difference isn't apparent until we look four carbons away from the chiral center, but there's still a difference.

In the examples of chiral molecules that follow, check for yourself that the labeled centers are indeed chiral. (When checking for chiral centers, it's helpful to realize that $-CH_2-$, $-CH_3$, C=C, and C=O carbons *can't* be chiral because they have at least two identical bonds.)

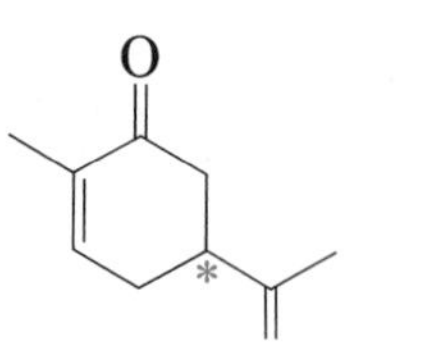

Carvone
(from spearmint oil)

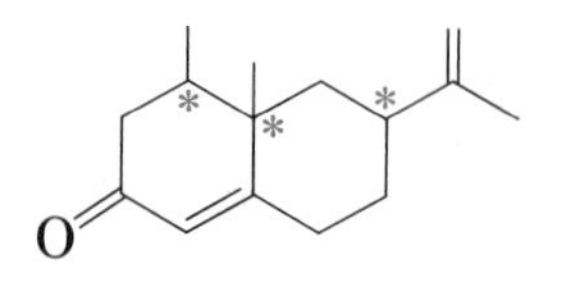

Nootkatone
(from grapefruit oil)

PRACTICE PROBLEM 6.1 Draw the structure of a chiral alcohol.

SOLUTION An alcohol is a compound that contains the $-OH$ functional group. To make an alcohol chiral, we need to have four different groups bonded to a single carbon atom, say $-H$, $-OH$, $-CH_3$, and $-CH_2CH_3$:

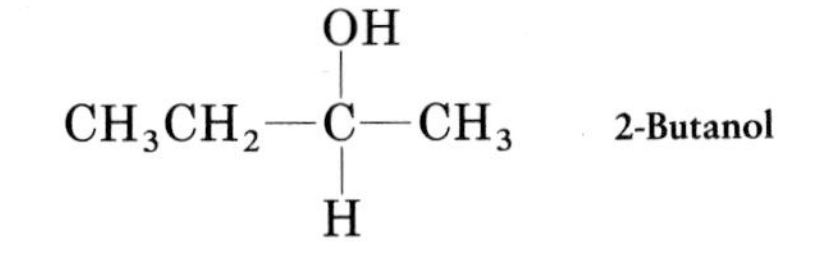

2-Butanol

PRACTICE PROBLEM 6.2 Is 3-methylhexane chiral?

SOLUTION Draw the structure of 3-methylhexane and cross out all the $-CH_2-$ and $-CH_3$ carbons because they can't be chiral. Then look closely at any carbon that remains to see if it's bonded to four different groups. Since C3 is bonded to $-H$, $-CH_3$, $-CH_2CH_3$, and $-CH_2CH_2CH_3$, the molecule is chiral.

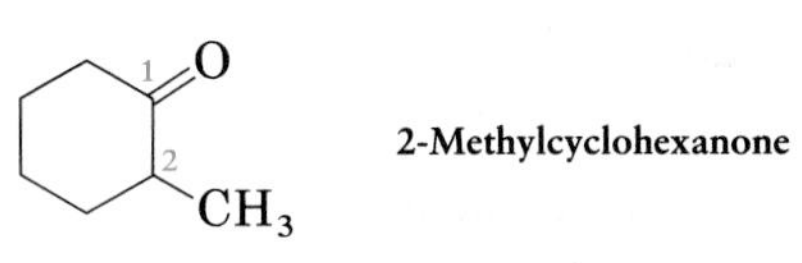

3-Methylhexane (chiral)

PRACTICE PROBLEM 6.3 Is 2-methylcyclohexanone chiral?

2-Methylcyclohexanone

SOLUTION Ignoring the $-CH_3$ carbon, the four $-CH_2-$ carbons in the ring, and the C=O carbon, look carefully at C2. Carbon 2 is bonded to four different groups: a $-CH_3$ group, an $-H$ atom, a $-C{=}O$ carbon in the ring, and a $-CH_2-$ ring carbon. Thus, 2-methylcyclohexanone is chiral.

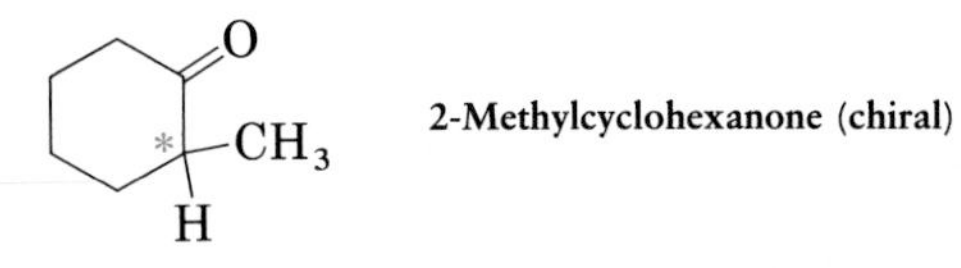

2-Methylcyclohexanone (chiral)

PROBLEM 6.1 Which of these objects are chiral (handed)?

(a) Bean stalk (b) Screwdriver (c) Screw (d) Shoe

PROBLEM 6.2 Which of these compounds are chiral?

(a) 3-Bromopentane
(b) 1,3-Dibromopentane
(c) 3-Methyl-1-hexene
(d) *cis*-1,4-Dimethylcyclohexane

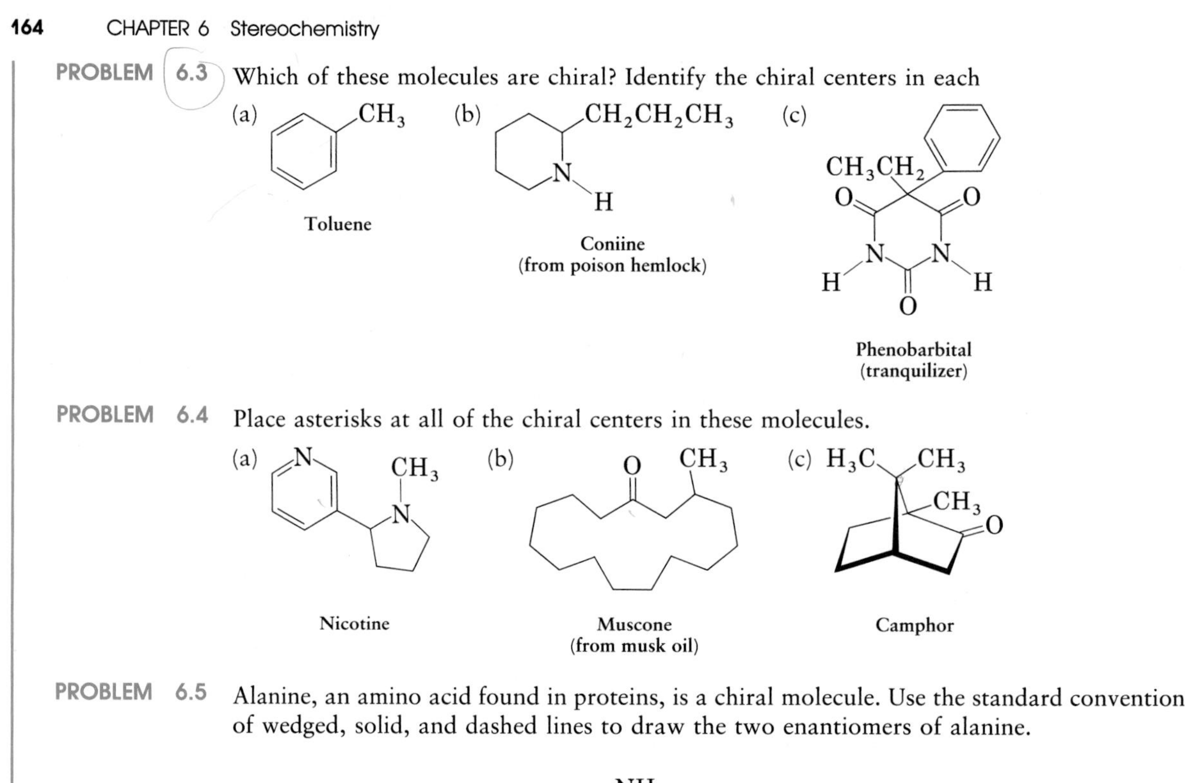

PROBLEM 6.3 Which of these molecules are chiral? Identify the chiral centers in each

PROBLEM 6.4 Place asterisks at all of the chiral centers in these molecules.

PROBLEM 6.5 Alanine, an amino acid found in proteins, is a chiral molecule. Use the standard convention of wedged, solid, and dashed lines to draw the two enantiomers of alanine.

$$\underset{}{CH_3\overset{\overset{\displaystyle NH_2}{|}}{C}HCOOH} \quad \text{Alanine}$$

6.3 OPTICAL ACTIVITY

The study of stereochemistry has its origins in the work of the French scientist, Jean Baptiste Biot in the early nineteenth century. Biot, a physicist, was investigating the nature of plane-polarized light. A beam of ordinary light consists of electromagnetic waves that oscillate in an infinite number of planes at right angles to the direction of light travel. When a beam of ordinary light passes through a device called a *polarizer,* though, only the light waves oscillating in a single plane get through (hence the name *plane-polarized light*). Light waves in all other planes are blocked out. The polarization process is represented in Figure 6.5.

Biot made the remarkable observation that, when a beam of plane-polarized light passes through solutions of certain organic molecules such as sugar or camphor, the plane of polarization is *rotated.* Not all organic molecules exhibit this property, but those that do are said to be **optically active.**

optical activity the property of a molecule that can rotate plane-polarized light

The amount of rotation can be measured with an instrument called a polarimeter, represented schematically in Figure 6.6. Optically active organic molecules are placed in a sample tube; plane-polarized light is then passed through the tube, and rotation of the plane occurs. The light then goes through a second polarizer known as the analyzer. By rotating the analyzer until light passes through it, we

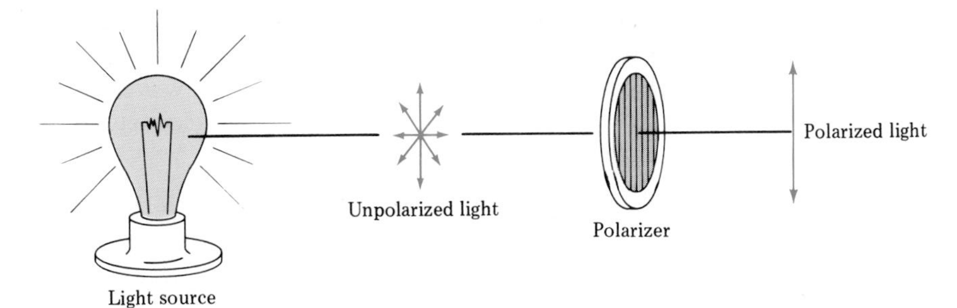

FIGURE 6.5 The nature of plane-polarized light. Only electromagnetic waves that oscillate in a single plane pass through the polarizer.

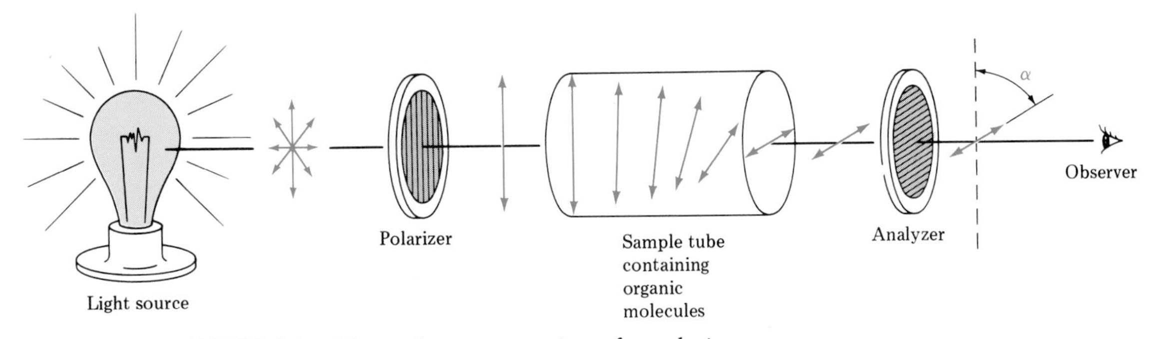

FIGURE 6.6 Schematic representation of a polarimeter

can find the new plane of polarization and can tell to what extent rotation has occurred. The amount of rotation observed is denoted α (Greek alpha) and is expressed in degrees.

In addition to determining the extent of rotation, we can also find out its direction. From the vantage point of the observer looking at the analyzer, some optically active molecules rotate plane-polarized light to the left (counterclockwise) and are said to be **levorotatory.** Other molecules rotate light to the right (clockwise) and are said to be **dextrorotatory.** By convention, rotation to the left is given a minus sign (−), and rotation to the right is given a plus sign (+). For example, (−)-morphine is levorotatory and (+)-sucrose is dextrorotatory.

levorotatory rotating plane-polarized light in a counterclockwise (left-handed) direction

dextrorotatory rotating plane-polarized light in a clockwise (right-handed) direction

6.4 SPECIFIC ROTATION

The amount of rotation observed in a polarimetry experiment depends both on the structure of the specific molecule and on the number of molecules encountered by the light beam. The more molecules the light encounters, the greater the observed rotation. Thus, the amount of rotation observed depends both on sample concentration and on sample path length. If we double the concentration of sample in a tube, the observed rotation is doubled; similarly, if we keep concentration constant but double the length of the sample tube, the observed rotation is doubled.

specific rotation the amount that an optically active compound rotates plane-polarized light under standard conditions

To express optical-rotation data in a meaningful way, we must choose standard conditions. By convention, the **specific rotation** $[\alpha]_D$ of a compound is defined as the observed rotation when light of 5896 Å wavelength (the yellow sodium D line) is used with a sample path length of 1 decimeter (1 dm = 10 cm) and a sample concentration of 1 g/mL.

Specific rotation:

$$[\alpha]_D = \frac{\text{observed rotation (deg)}}{\text{path length (dm)} \times \text{concentration (g/mL)}} = \frac{\alpha}{l \times C}$$

When optical-rotation data are expressed in this standard way, the specific rotation $[\alpha]_D$ is a physical constant that is characteristic of each optically active compound. Some examples are listed in Table 6.1.

TABLE 6.1 **Specific rotations of some organic molecules**

Compound	*$[\alpha]_D$ (degrees)*	*Compound*	*$[\alpha]_D$ (degrees)*
Camphor	+44.26	Penicillin V	+223
Morphine	−132	Monosodium glutamate	+25.5
Sucrose	+66.47	Benzene	0
Cholesterol	−31.5	Hexane	0

PRACTICE PROBLEM 6.4 A 1.2 g sample of cocaine, $[\alpha]_D = -16°$, was dissolved in 7.5 mL chloroform and placed in a sample tube having a path length of 5.0 cm. What was the observed rotation?

SOLUTION Observed rotation α is equal to specific rotation $[\alpha]_D$ times sample concentration C times path length l:

$$\alpha = [\alpha]_D \times C \times l$$

where $[\alpha]_D = -16°$; $l = 5.0$ cm $= 0.50$ dm; $C = 1.2$ g/7.5 mL $= 0.16$ g/mL. Thus,

$$\alpha = -16° \times 0.50 \times 0.16 = -1.3°$$

PROBLEM 6.6 Is cocaine (Practice Problem 6.4) dextrorotatory or levorotatory?

PROBLEM 6.7 A 1.50 g sample of coniine, the toxic extract of poison hemlock, was dissolved in 10.0 mL ethanol and placed in a sample tube with a path length of 5.00 cm. The observed rotation at the sodium D line was +1.21°. Calculate the specific rotation $[\alpha]_D$ for coniine.

6.5 PASTEUR'S DISCOVERY OF ENANTIOMERS

Little was done after Biot's discovery of optical activity until Louis Pasteur entered the picture in 1848. Pasteur was working on crystalline salts of tartaric acid derived from wine when he made a surprising observation. When he recrystallized a concentrated solution of sodium ammonium tartrate below 28°C, two distinct kinds of crystals precipitated. Furthermore, the two kinds of crystals were *mirror images* of each other. They were related to each other in exactly the same way that a right hand is related to a left hand.

Working carefully with a pair of tweezers, Pasteur was able to separate the crystals into two piles, one of "right-handed" crystals and one of "left-handed" crystals like those shown in Figure 6.7. Although the original sample (a 50:50 mixture of right and left) was optically inactive, *solutions of the crystals in each of the sorted piles were optically active,* and their specific rotations were equal in amount but opposite in sign.

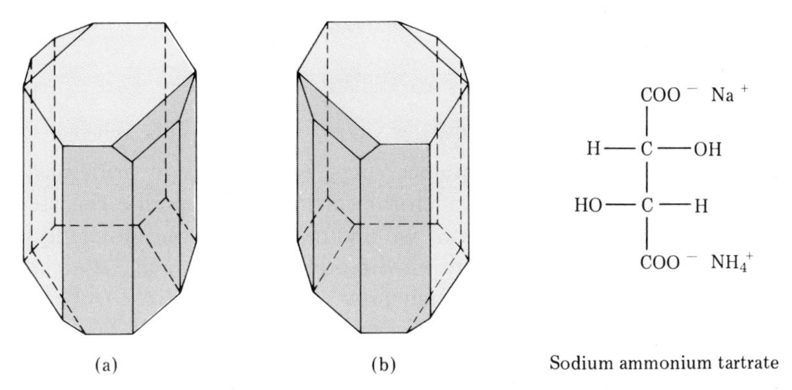

FIGURE 6.7 Crystals of sodium ammonium tartrate. One of the crystals is dextrorotatory in solution, and the other is levorotatory. The drawings are taken from Pasteur's original sketches.

Pasteur was far ahead of his time. Although the structural theory of Kekulé had not yet been proposed, Pasteur explained his results by speaking of the molecules themselves, saying "It cannot be a subject of doubt that [in the *dextro* tartaric acid] there exists an asymmetric arrangement having a nonsuperimposable image. It is no less certain that the atoms of the *levo* acid possess precisely the inverse asymmetric arrangement." Pasteur's vision was extraordinary, for it was not until 25 years later that his theories regarding the asymmetric carbon atom were confirmed.

optical isomerism an alternative name for enantiomerism

Today, we would describe Pasteur's work by saying that he had discovered the phenomenon of enantiomerism or **optical isomerism.** Enantiomers (also called optical isomers) have identical physical properties, such as melting points and boiling points, but differ in the direction in which they rotate plane-polarized light.

6.6 SEQUENCE RULES FOR SPECIFICATION OF CONFIGURATION

configuration the exact three-dimensional arrangement of atoms in space

Although drawings provide pictorial representations of stereochemistry, they're difficult to translate into words. Thus, a verbal procedure for specifying the exact three-dimensional arrangement (the **configuration**) of substituents around a chiral center is also necessary. The method used employs the same sequence rules that we used for the specification of alkene stereochemistry (Z versus E) in Section 3.4. Let's briefly review these sequence rules and see how they're used to specify the configuration of a chiral center. For a more thorough review, though, you should refer back to Section 3.4.

1. Look at the four atoms directly attached to the chiral center and assign priorities in order of decreasing atomic number. The atom with highest atomic number is ranked first; the atom with lowest atomic number is ranked fourth.
2. If a decision about priority can't be reached by applying rule 1, compare atomic numbers of the second atoms in each substituent, continuing on as necessary through the third or fourth atoms outward until the first point of difference is reached.
3. Multiple-bonded atoms are considered as if they were an equivalent number of single-bonded atoms.

Following these sequence rules, we can assign priorities to the four substituent groups attached to a chiral carbon. To describe the stereochemical configuration around that carbon, we mentally orient the molecule so that the group of lowest priority (4) is pointing directly away from us. We then look at the three remaining substituents, which now appear to radiate toward us from the chiral center like the spokes on a steering wheel. If a curved arrow, drawn from highest to second-highest to third-highest substituent (1 → 2 → 3) is clockwise, we say that the chiral center has the R configuration (Latin *rectus,* "right"). If a curved arrow from 1 → 2 → 3 is counterclockwise, the chiral center has the S configuration (Latin *sinister,* "left"). To remember these assignments, think of a car's steering wheel when making a *R*ight (clockwise) turn.

For example, let's look at (+)-lactic acid:

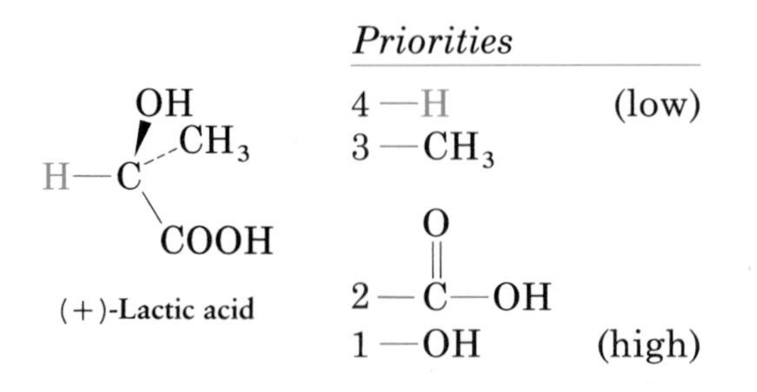

Sequence rule 1 says that –OH is first priority (1) and –H is fourth priority (4), but it doesn't distinguish between $-CH_3$ and –COOH since both groups have carbon as their first atom. Sequence rule 2, however, says that –COOH is higher in priority than $-CH_3$ because oxygen outranks hydrogen (the second atom in each group).

We next orient the molecule so that the fourth-priority group (–H) points away from us. Since the direction of the arrow 1 → 2 → 3 is counterclockwise (left turn of the steering wheel), we assign the S configuration to (+)-lactic acid (Figure 6.8). Applying the same procedure to (−)-lactic acid should (and does) lead to the opposite assignment. Try it for yourself.

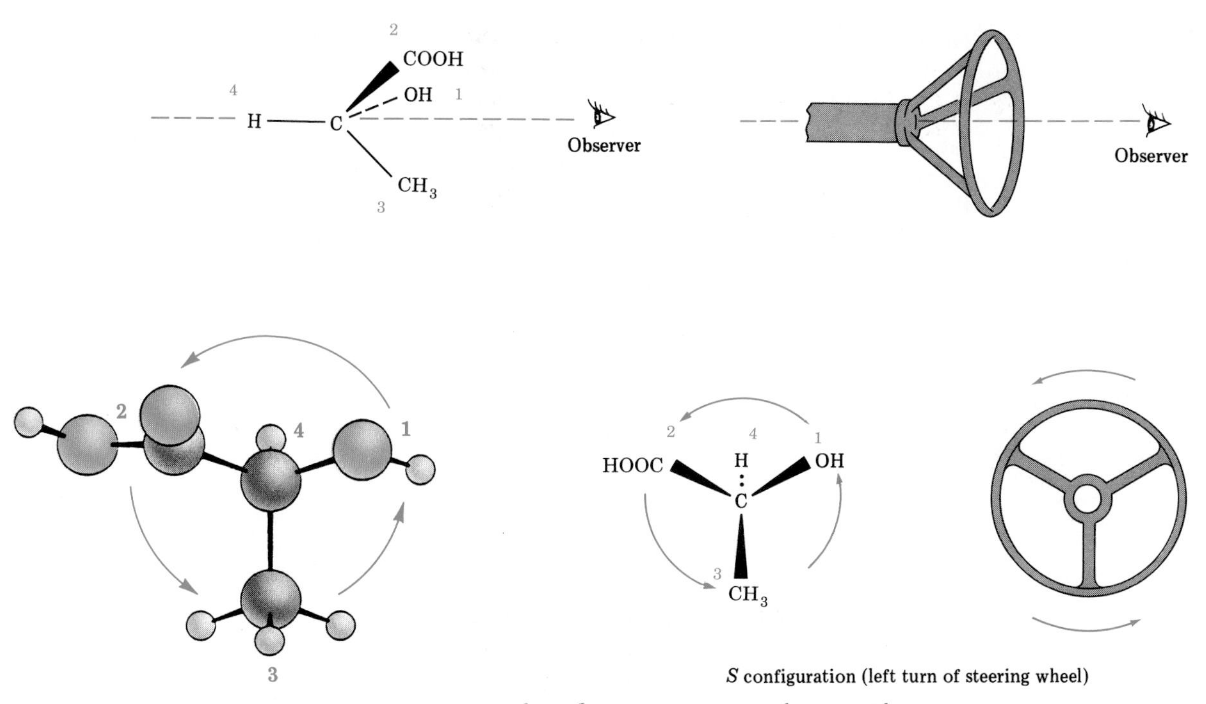

FIGURE 6.8 Assignment of configuration to *S*-(+)-lactic acid

As another example, look at (−)-glyceraldehyde, which has the *S* configuration shown in Figure 6.9. Note that the sign of optical rotation isn't related to the *R*,*S* designation. *S*-Lactic acid happens to be dextrorotatory (+), and *S*-glyceraldehyde happens to be levorotatory (−), but there's no simple correlation between of the direction of rotation and *R*,*S* configuration.

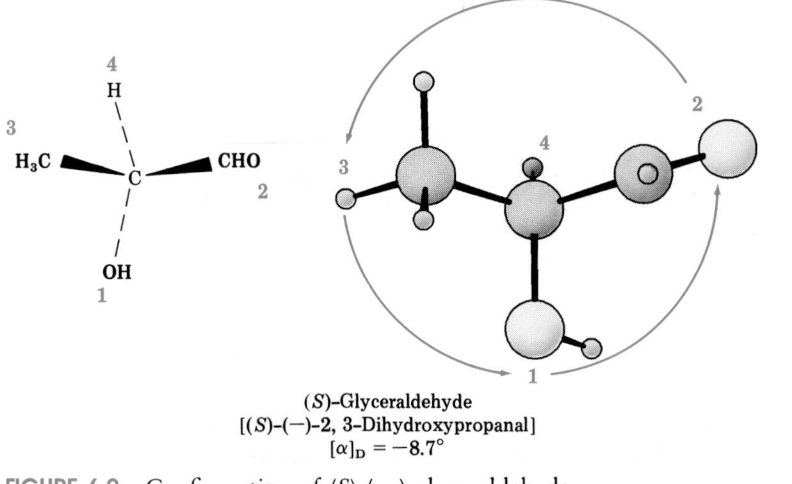

FIGURE 6.9 Configuration of (*S*)-(−)-glyceraldehyde

PRACTICE PROBLEM 6.5 Draw a tetrahedral representation of *R*-2-chlorobutane.

SOLUTION The four substituents bonded to the chiral carbon of *R*-2-chlorobutane can be assigned the following priorities: (1) $-Cl$, (2) $-CH_2CH_3$, (3) $-CH_3$, (4) $-H$. To draw a tetrahedral representation of the molecule, first orient the low-priority $-H$ group toward the rear and imagine that the other three groups are coming out of the page toward you. Place the remaining three substituents in order so that the direction of travel from 1 → 2 → 3 is clockwise (right turn), and then tilt the molecule to bring the rear hydrogen into view.

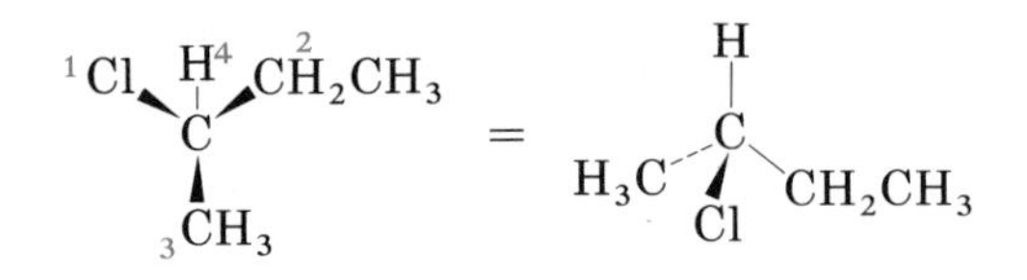

Using molecular models is a great help in working problems of this sort.

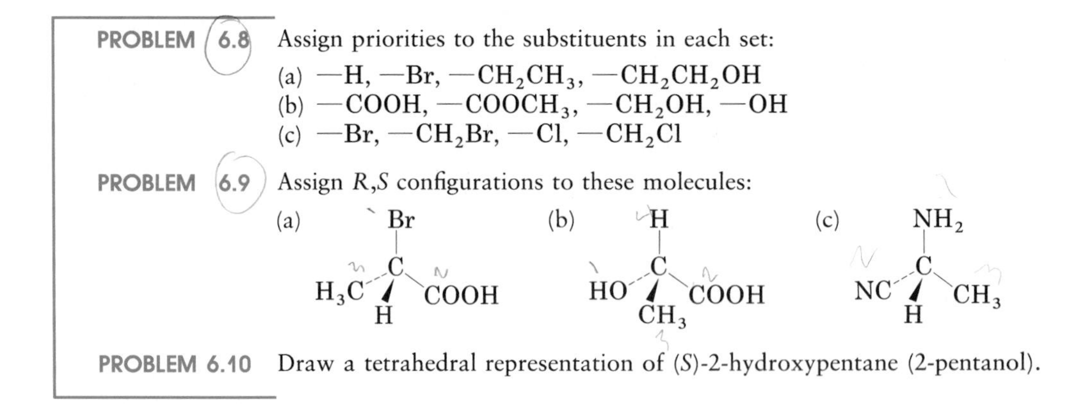

PROBLEM 6.8 Assign priorities to the substituents in each set:

(a) $-H$, $-Br$, $-CH_2CH_3$, $-CH_2CH_2OH$
(b) $-COOH$, $-COOCH_3$, $-CH_2OH$, $-OH$
(c) $-Br$, $-CH_2Br$, $-Cl$, $-CH_2Cl$

PROBLEM 6.9 Assign *R*,*S* configurations to these molecules:

(a) (b) (c)

PROBLEM 6.10 Draw a tetrahedral representation of (*S*)-2-hydroxypentane (2-pentanol).

6.7 DIASTEREOMERS

Molecules like lactic acid and glyceraldehyde are relatively simple to deal with because each has only one chiral center and only two enantiomeric forms. The situation becomes more complex, however, for molecules that have more than one chiral center.

Let's take the amino acid threonine (2-amino-3-hydroxybutanoic acid) as an example. Since threonine has two chiral centers (C2 and C3), there are four possible stereoisomers, which are shown in Figure 6.10. (Check for yourself that the *R*,*S* configurations are correct as shown.)

The four threonine stereoisomers can be classified into two mirror-image pairs of enantiomers. The 2*R*,3*R* stereoisomer is the mirror image of 2*S*,3*S*, and the 2*R*,3*S* stereoisomer is the mirror image of 2*S*,3*R*. But what's the relationship between any two configurations that aren't mirror images? What, for example, is the relationship between the 2*R*,3*R* compound and the 2*R*,3*S* compound? These

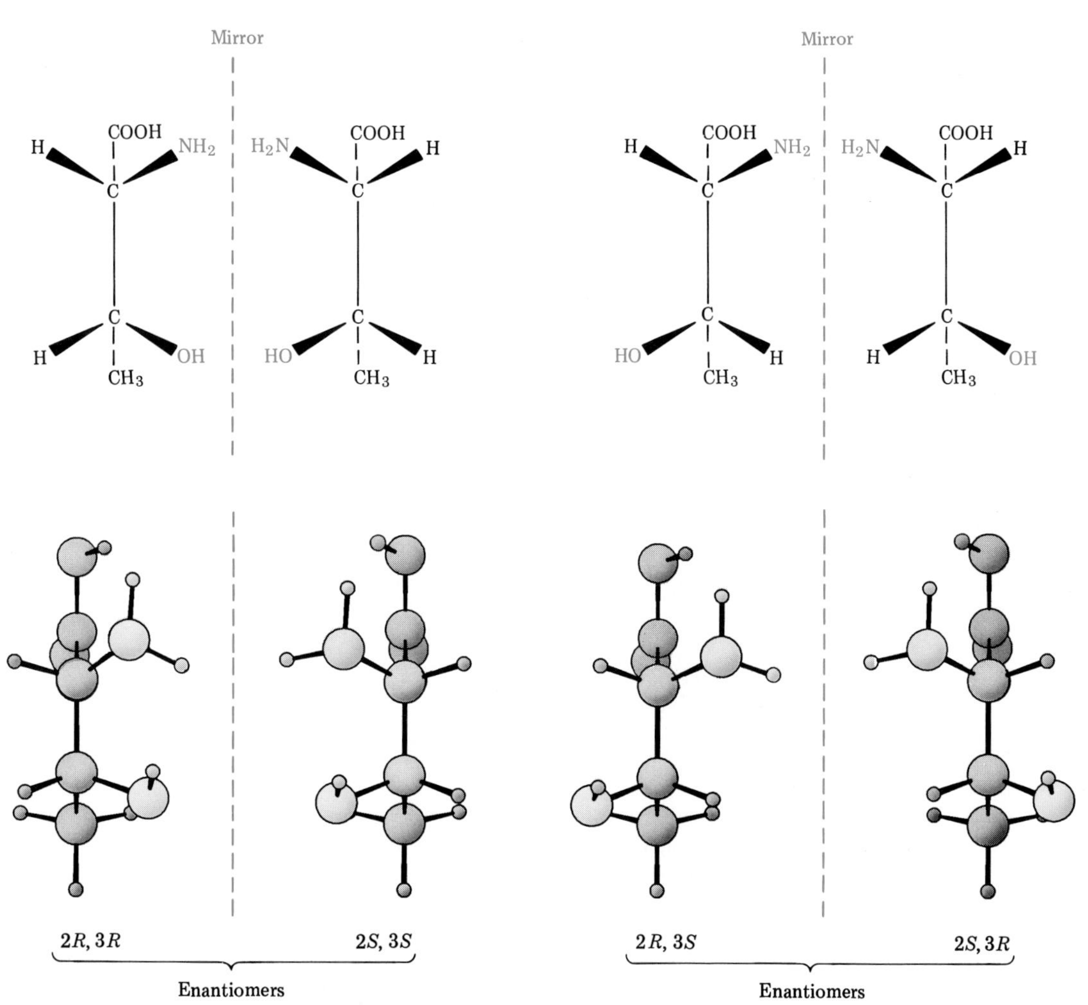

FIGURE 6.10 Four stereoisomers of 2-amino-3-hydroxybutanoic acid

two compounds are stereoisomers, yet they aren't superimposable, and they aren't enantiomers.

diastereomer stereoisomers that aren't mirror images

To describe such a relationship, we need a new term, **diastereomers,** which are stereoisomers that aren't mirror images of each other. Since we used the right-hand–left-hand analogy to describe the relationship between two enantiomers, we might extend the analogy by saying that diastereomers have a hand–foot relationship. Hands and feet look *similar,* but they aren't identical and aren't mirror images. The same is true of diastereomers: They're similar but not identical and not mirror images.

A full description of the four threonine stereoisomers is given in Table 6.2. Enantiomers must have opposite (mirror image) configurations around *all* chiral centers; diastereomers have the same configuration around at least one chiral center but have opposite configurations around the others.

TABLE 6.2 Relationships among the four threonine stereoisomers

Stereoisomer	*Enantiomeric with*	*Diastereomeric with*
$2R,3R$	$2S,3S$	$2R,3S$, $2S,3R$
$2S,3S$	$2R,3R$	$2R,3S$, $2S,3R$
$2R,3S$	$2S,3R$	$2R,3R$, $2S,3S$
$2S,3R$	$2R,3S$	$2R,3R$, $2S,3S$

Of the four possible threonine stereoisomers, only the $2S,3R$ isomer, $[\alpha]_D = -28.3°$, occurs naturally and is an essential human nutrient. Most biologically important molecules are chiral, and usually only one stereoisomer is found in nature.

PROBLEM 6.11 Assign *R,S* configurations to each chiral center in these molecules.

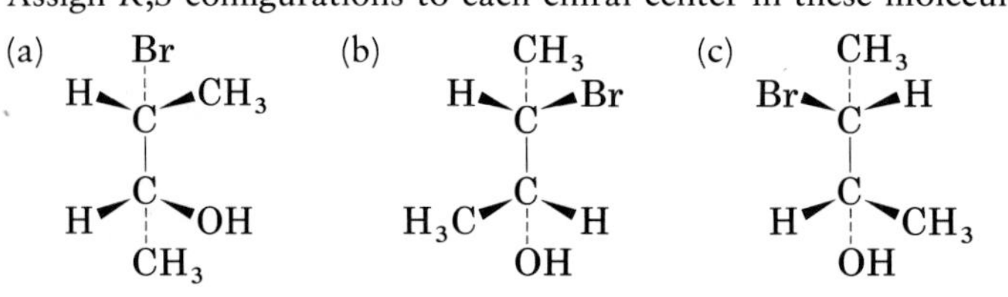

PROBLEM 6.12 Which of the compounds in Problem 6.11 are enantiomers and which are diastereomers?

PROBLEM 6.13 Chloramphenicol is a powerful antibiotic isolated from the *Streptomyces venezuelae* bacterium. It is active against a broad spectrum of bacterial infections and is particularly valuable against typhoid fever. Assign *R,S* configurations to the chiral centers in chloramphenicol.

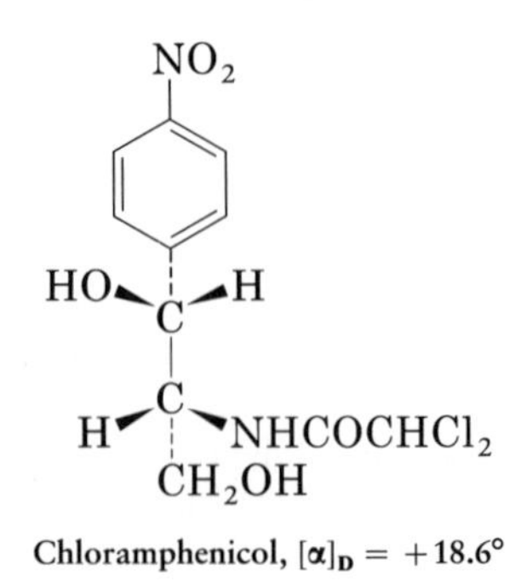

Chloramphenicol, $[\alpha]_D = +18.6°$

6.8 MESO COMPOUNDS

Let's look at one more example of a compound with two chiral centers: tartaric acid. We're already acquainted with tartaric acid because of its role in Pasteur's discovery of optical activity, and we can now draw the four stereoisomers:

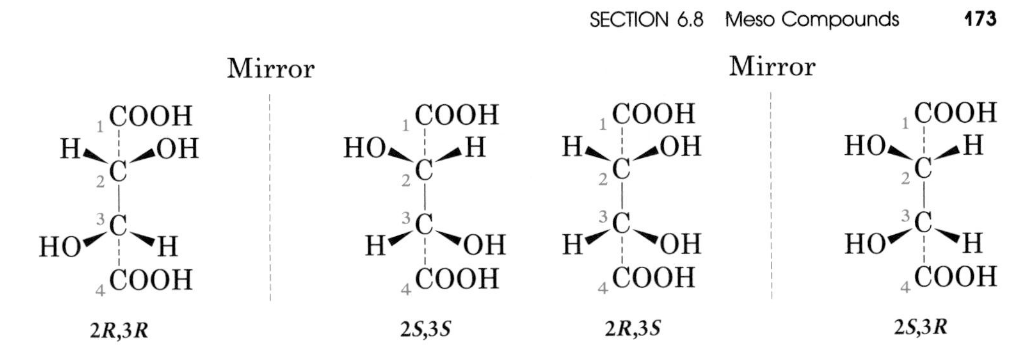

The mirror-image 2*R*,3*R* and 2*S*,3*S* structures are nonsuperimposable and therefore represent an enantiomeric pair. A close look, however, reveals that the 2*R*,3*S* and 2*S*,3*R* structures are *identical,* as we can see by rotating one structure 180°:

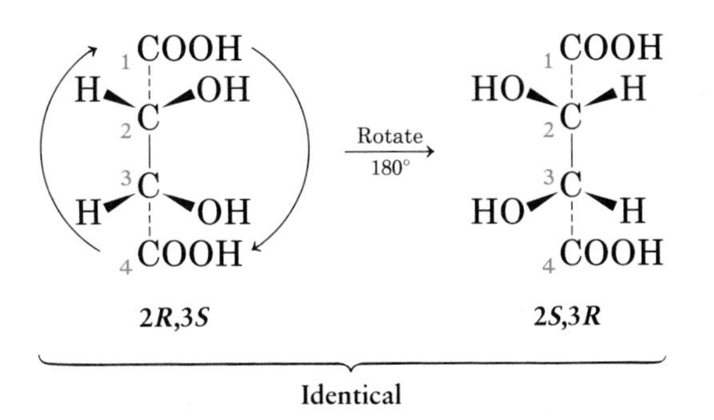

The identity of the 2*R*,3*S* and 2*S*,3*R* structures is due to the fact that the molecule has a plane of symmetry. The symmetry plane cuts through the C2–C3 bond, making one half of the molecule a mirror image of the other half (Figure 6.11).

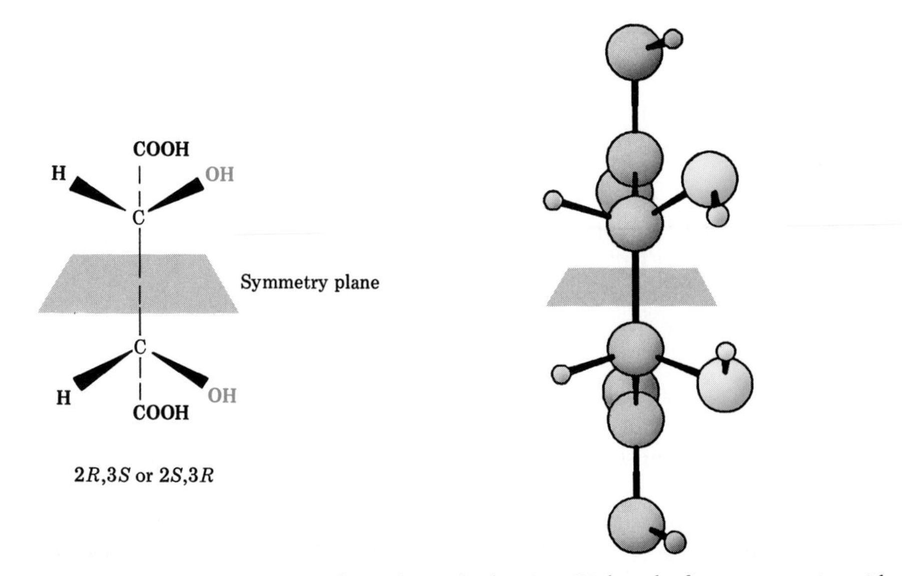

FIGURE 6.11 A symmetry plane through the C2–C3 bond of *meso*-tartaric acid

meso compounds compounds that have chiral centers but are achiral overall because of a symmetry plane

Because of the plane of symmetry, the tartaric acid stereoisomer shown in Figure 6.11 must be achiral, despite the fact that it has two chiral centers. Compounds that are achiral by virtue of a symmetry plane, yet contain chiral centers, are called **meso compounds** (*me*-zo). Thus, tartaric acid exists in only three stereoisomeric configurations: two enantiomers and one meso form.

PRACTICE PROBLEM 6.6 Does *cis*-1,2-dimethylcyclobutane have any chiral centers? Is it a chiral molecule?

SOLUTION Looking at the structure of *cis*-1,2-dimethylcyclobutane, we see that both of the methyl-bearing ring carbons (C1 and C2) are chiral. Overall, though, the compound is achiral because there's a symmetry plane bisecting the ring between C1 and C2. Thus, the molecule is a meso compound.

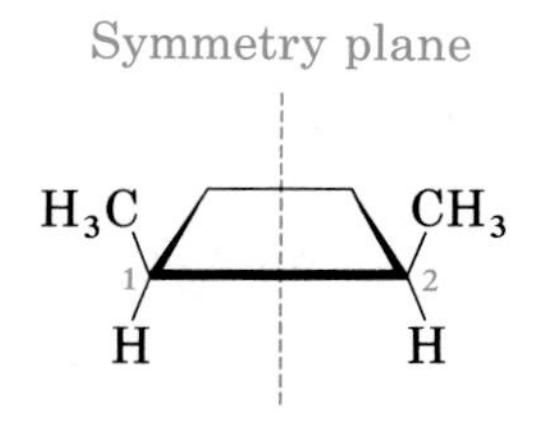

PROBLEM 6.14 Which of these substance can exist in a meso form?

(a) 2,3-Dibromobutane (b) 2,3-Dibromopentane (c) 2,4-Dibromopentane

PROBLEM 6.15 Which of these structures represent meso compounds?

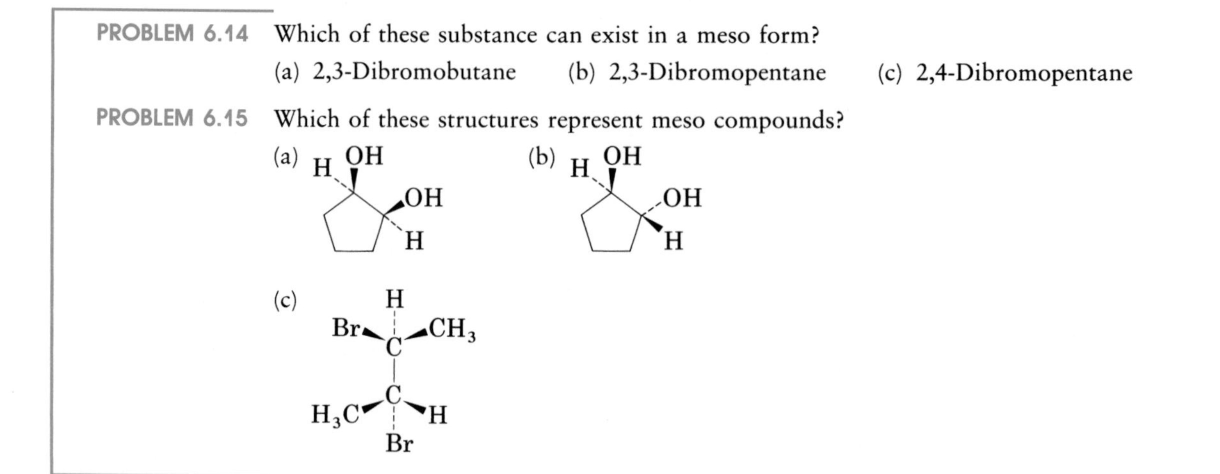

6.9 MOLECULES WITH MORE THAN TWO CHIRAL CENTERS

We've seen how a single chiral center gives rise to two stereoisomers (one pair of enantiomers) and how two chiral centers give rise to a maximum of four stereoisomers (two pairs of enantiomers). In general, a molecule with n chiral centers gives rise to a maximum of 2^n stereoisomers (2^{n-1} pairs of enantiomers). For example, cholesterol has eight chiral centers. Thus, $2^8 = 256$ stereoisomers, or 128 pairs of enantiomers, are possible in principle, though many would be too strained to exist. Only one, however, is produced in nature.

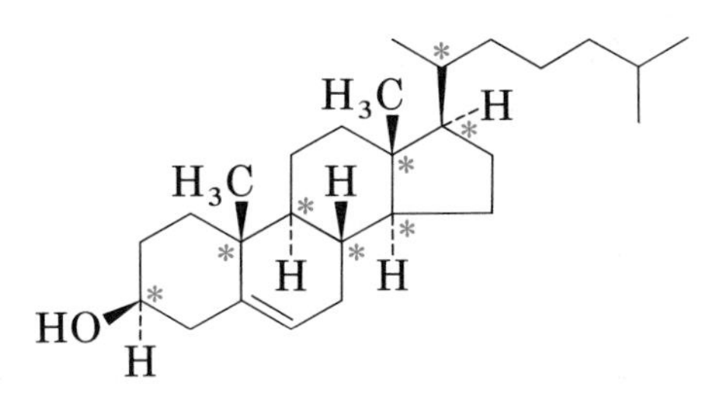

Cholesterol (eight chiral centers)

PROBLEM 6.16 Nandrolone is an anabolic steroid used illegally by some athletes to build muscle mass. How many chiral centers does nandrolone have? How many stereoisomers of nandrolone are possible in principle?

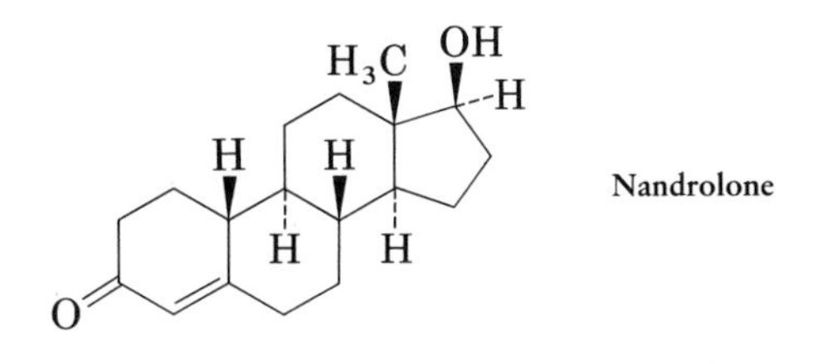

Nandrolone

6.10 RACEMIC MIXTURES

To conclude our discussion of stereoisomerism, let's return for a last look at Pasteur's pioneering work. Pasteur took an optically inactive form of a tartaric acid salt and found that he could crystallize two optically active forms from it. The two optically active forms had the 2*R*,3*R* and 2*S*,3*S* configurations, but what was the optically inactive form he started with? It couldn't have been *meso*-tartaric acid because *meso*-tartaric acid is a different chemical compound and can't interconvert with the two chiral enantiomers without breaking and re-forming chemical bonds.

racemic mixture a 50:50 mixture of enantiomers

resolve separate a racemic mixture into its pure component enantiomers

The answer is that Pasteur started with a 50:50 mixture of the two chiral tartaric acid enantiomers. Such a mixture is called a **racemic** (ray-*see*-mic) **mixture,** or **racemate.** Racemic mixtures, often denoted by the symbol (±), must show zero optical rotation because they contain equal amounts of (+) and (−) forms. The (+) rotation from one enantiomer exactly cancels the (−) rotation from the other. Through good fortune, Pasteur was able to separate, or **resolve,** (±)-tartaric acid into its (+) and (−) enantiomers. Unfortunately, the fractional-crystallization technique he used doesn't work for most racemic mixtures, and so other methods are required. We'll see a better method in Section 12.4.

6.11 PHYSICAL PROPERTIES OF STEREOISOMERS

If seemingly simple compounds like tartaric acid can exist in different stereoisomeric configurations, the question arises whether the different stereoisomers have different physical properties. The answer is yes, they do.

Some physical properties of the three stereoisomers of tartaric acid and of the racemic mixture are shown in Table 6.3. As indicated, the (+) and (−) enantiomers have identical melting points, solubilities, and densities. They differ only in the sign of their rotation of plane-polarized light. The meso isomer, by contrast, is diastereomeric with the (+) and (−) forms. As such, it's a different compound altogether and has different physical properties.

TABLE 6.3 Some properties of the stereoisomers of tartaric acid

Stereoisomer	*Melting point (°C)*	*$[\alpha]_D$ (degrees)*	*Density (g/cm^3)*	*Solubility at 20°C (g/100 mL H_2O)*
(+)	168–170	+12	1.7598	139.0
(−)	168–170	−12	1.7598	139.0
meso	146–148	0	1.6660	125.0
(±)	206	0	1.7880	20.6

The racemic mixture is different still. For reasons beyond our present scope, racemates act as though they were pure compounds, different from either enantiomer. Thus, the physical properties of racemic tartaric acid differ from those of the two enantiomers and from those of the meso form.

6.12 A BRIEF REVIEW OF ISOMERISM

We've seen several kinds of isomers in the past few chapters, and it's a good idea at this point to see how they relate to one another. As noted earlier, isomers are compounds that have the same chemical formula but different structures. There are two fundamental types of isomerism, both of which we've now encountered: constitutional isomerism and stereoisomerism.

Constitutional isomers (Section 2.2) are compounds whose atoms are connected differently. Among the kinds of constitutional isomers we've seen are skeletal, functional, and positional isomers.

Constitutional isomers—Different connections among atoms:

Skeletal isomers
(different carbon skeletons)

$$\begin{array}{c} CH_3 \\ | \\ CH_3\text{–}CH\text{–}CH_3 \end{array} \quad \text{and} \quad CH_3\text{–}CH_2\text{–}CH_2\text{–}CH_3$$

2-Methylpropane **Butane**

Functional-group isomers (different functional groups)

CH_3-CH_2-OH and CH_3-O-CH_3

Ethanol — Dimethyl ether

Positional isomers (different positions of functional group)

$CH_3-CH(OH)-CH_3$ and $CH_3-CH_2-CH_2-OH$

2-Propanol — 1-Propanol

Stereoisomers (Section 2.7) are compounds whose atoms are connected in the same way but with a different geometry. Among the kinds of stereoisomers we've seen are enantiomers, diastereomers, and cis–trans isomers (both in alkenes and in cycloalkanes). To be accurate, though, cis–trans isomers are really just another kind of diastereomer because they meet the definition of non-mirror-image stereoisomers:

Stereoisomers—Same connections among atoms but different geometry:

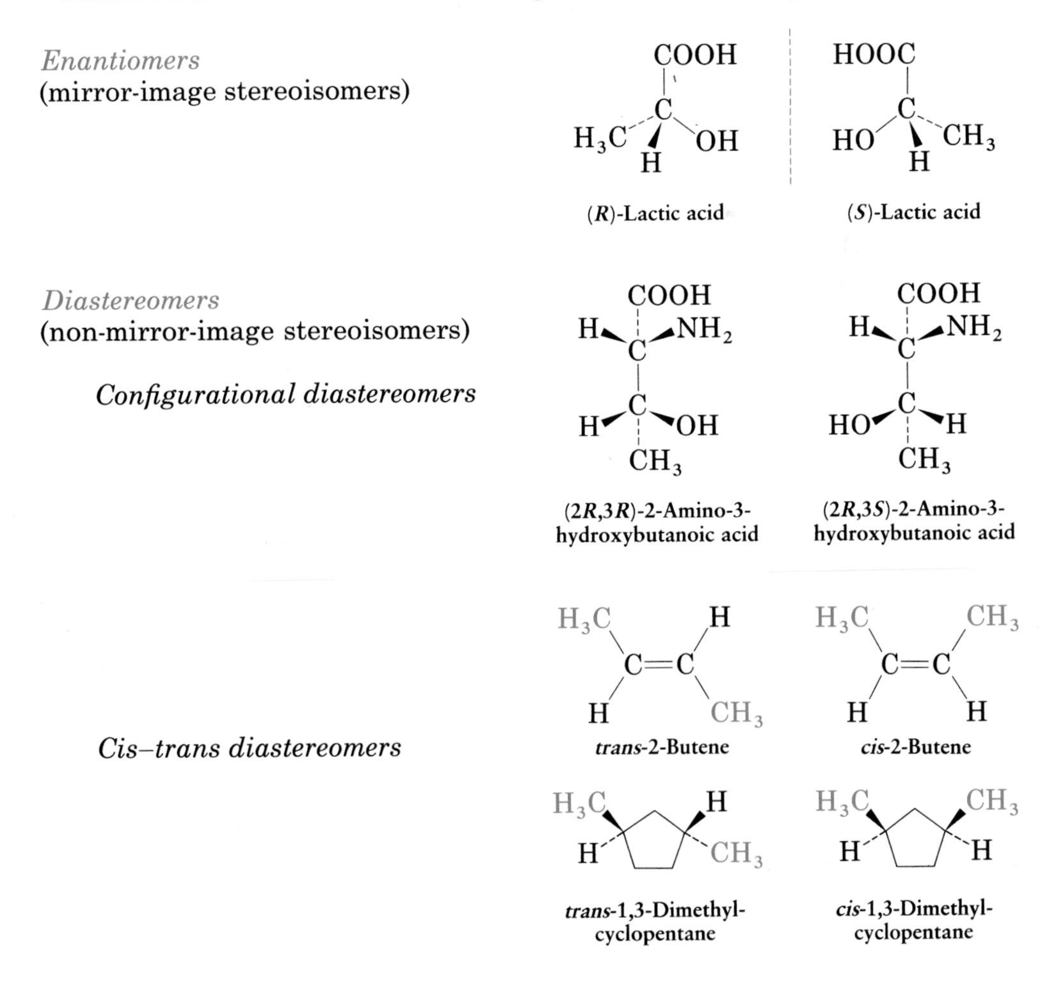

PROBLEM 6.17 Tell what kinds of isomers the following pairs are.

(a) *S*-5-Chloro-2-hexene and chlorocyclohexane
(b) 2*R*,3*R*-Dibromopentane and 2*S*,3*R*-dibromopentane

6.13 STEREOCHEMISTRY OF REACTIONS: ADDITION OF HBr TO ALKENES

Many organic reactions, including some that we've studied, yield products with chiral centers. For example, addition of HBr to 1-butene yields 2-bromobutane, a chiral molecule. What predictions can we make about the stereochemistry of this chiral product? The answer is that 2-bromobutane is produced as a racemic mixture of *R* and *S* enantiomers.

$$CH_3CH_2CH{=}CH_2 + HBr \xrightarrow{\text{ether}} CH_3CH_2\overset{*}{C}H(Br)CH_3$$

1-Butene (achiral) → **2-Bromobutane** (chiral but racemic)

To understand why a racemic product results, let's consider what happens during the reaction. 1-Butene is first protonated to yield an intermediate secondary carbocation. This ion is sp^2 hybridized and therefore has a plane of symmetry, making it achiral. As a result, it can be attacked by bromide ion equally well from either the top or the bottom (Figure 6.12). Attack from the top leads to (*S*)-2-bromobutane, and attack from the bottom leads to (*R*)-2-bromobutane. Since both pathways occur with equal probability, a racemic product mixture results.

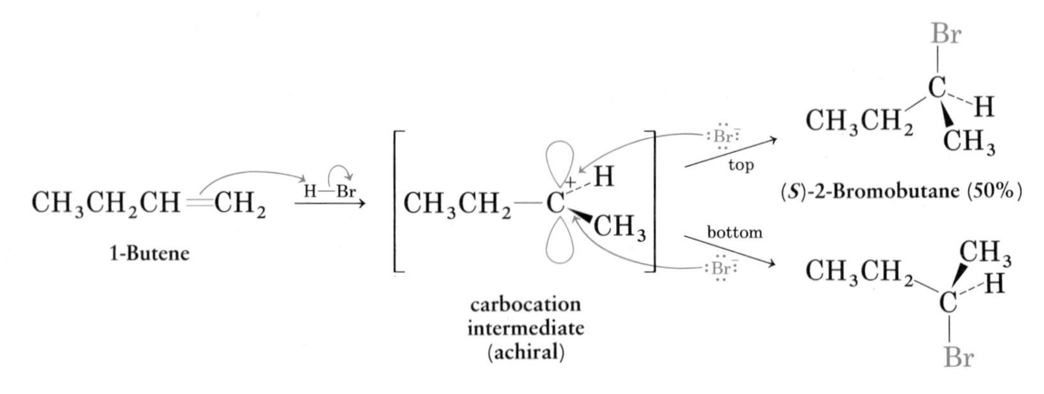

FIGURE 6.12 Stereochemistry of the addition of HBr to 1-butene. The intermediate carbocation is attacked equally well from both top and bottom sides, giving rise to a racemic mixture of products that is 50% *R* and 50% *S*.

What's true for the reaction of 1-butene with HBr is also true for all other reactions: *Reaction of achiral starting materials always leads to optically inactive products.* Optically active products can't be produced from optically inactive intermediates or starting materials.

INTERLUDE

Chirality in Nature

Just as the different stereoisomers of a substance have different physical properties, it's usually the case that stereoisomers have different chemical and biological properties as well. For example, (+)-lactic acid is rapidly converted into pyruvic acid by the enzyme lactic acid dehydrogenase, but (−)-lactic acid is unaffected by the enzyme.

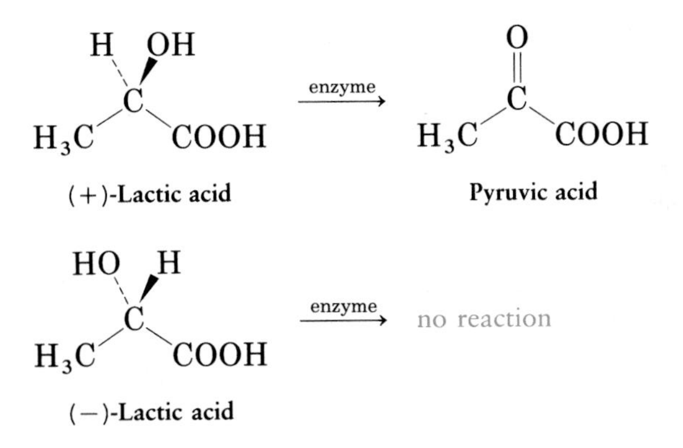

A remarkable example of how a simple change in chirality can affect the biological properties of a molecule is found in the amino acid, dopa. Dopa, more properly named 2-amino-3-(3,4-dihydroxyphenyl)propanoic acid, has a single chiral center and thus exists in two stereoisomeric forms. Although the dextrorotatory enantiomer, D-dopa, has no physiological effect on humans, the levorotatory enantiomer, L-dopa, has dramatic activity against Parkinsonism, a chronic disease of the central nervous system.

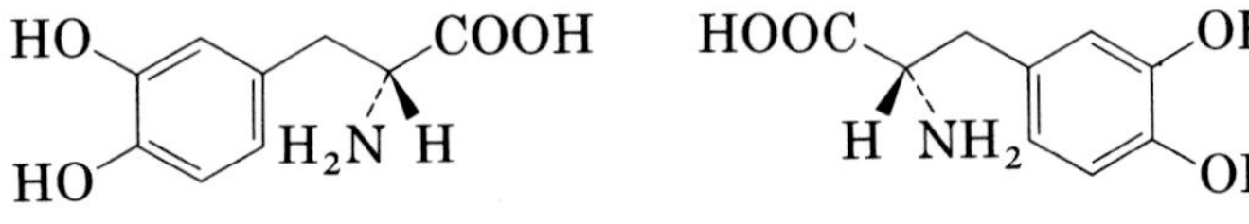

D-Dopa
(no biological effect)

L-Dopa
(anti-Parkinsonism agent)

Why do different stereoisomers have such widely different biological properties? To exert its biological effect, a chiral molecule must fit into a chiral receptor at the target site, much as a hand fits into a glove. Just as a right hand can fit only into a right-hand glove, so a particular stereoisomer can fit only into a receptor with the complementary shape. Any other stereoisomer is a misfit like a right hand in a left-hand glove. For example, we might imagine that L-dopa fits perfectly into a receptor site but that its L-enantiomer is unable to fit (Figure 6.13).

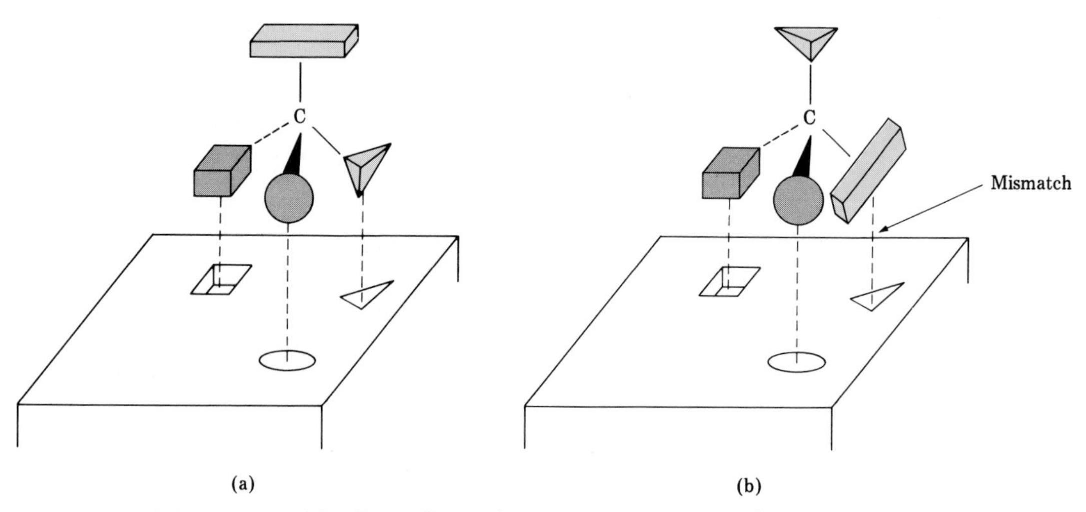

FIGURE 6.13 (a) L-Dopa fits easily into a receptor site, where it exerts its biological effect, but (b) D-dopa can't fit into the same receptor and therefore has no effect.

SUMMARY AND KEY WORDS

When a beam of **plane-polarized** light is passed through a solution of certain organic molecules, the plane of polarization is rotated. Compounds that exhibit this behavior are called **optically active.** Optical activity is due to the asymmetric structures of the molecules.

An object that is not superimposable on its mirror image is said to be **chiral,** meaning "handed." For example, a glove is chiral, but a coffee cup is **achiral.** A chiral object is one that does not contain a **plane of symmetry.** The usual cause of chirality in organic molecules is the presence of a tetrahedral carbon atom bonded to four different groups. Chiral compounds can exist as a pair of mirror-image stereoisomers called **enantiomers.** Enantiomers are identical in their physical properties except for the direction in which they rotate plane-polarized light; they're related to each other as a right hand is related to a left hand.

The stereochemical **configuration** of chiral carbon centers is specified as either *R* (*rectus*) or *S* (*sinister*). **Sequence rules** are used to assign priorities to the four substituents on the chiral carbon, and then the molecule is oriented so that the lowest-priority group points directly away from the viewer. If the direction of a curved arrow drawn from highest to second-highest to third-highest priority groups is clockwise, the configuration is labeled *R*. If the direction of the arrow is counterclockwise, the configuration is labeled *S*.

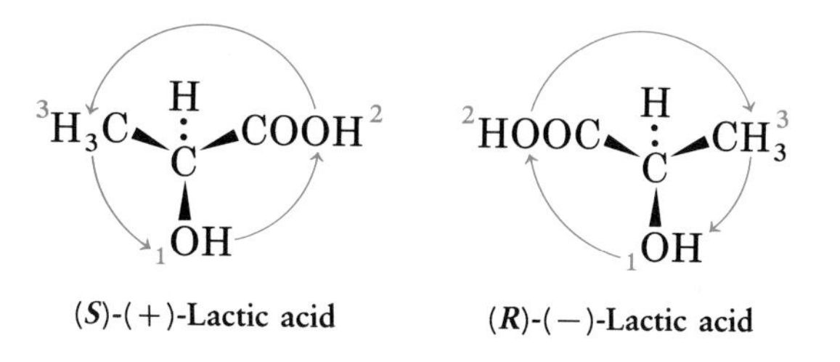

(*S*)-(+)-Lactic acid (*R*)-(−)-Lactic acid

Some molecules possess more than one chiral center. Enantiomers have opposite configurations at all chiral centers whereas **diastereomers** have the same configuration in at least one center and opposite configurations at the others. **Meso compounds** contain chiral centers but are achiral overall because they contain a plane of symmetry. **Racemates** are 50:50 mixtures of (+) and (−) enantiomers. Racemic mixtures and individual diastereomers differ in both their physical properties and their biological properties.

ADDITIONAL PROBLEMS

6.18 Cholic acid, the major steroid found in bile, was observed to have a specific rotation of +2.22° when a 3.0-g sample was dissolved in 5.0 mL alcohol in a sample tube with a 1.0-cm path length. Calculate $[\alpha]_D$ for cholic acid.

6.19 Polarimeters are quite sensitive and can measure rotations to the thousandth of a degree, an important point when only small amounts of sample are available. For example, when 7.0 mg of ecdysone, an insect hormone that controls molting in the silkworm moth, was dissolved in 1.0 mL chloroform in a cell with a 2.0-cm path length, an observed rotation of +0.087° was found. Calculate $[\alpha]_D$ for ecdysone.

6.20 Define these terms:

(a) Chirality (b) Chiral center (c) Diastereomer
(d) Racemate (e) Meso compound (f) Enantiomer

6.21 Which of these objects are chiral?

(a) A basketball (b) A wine glass (c) An ear
(d) A snowflake (e) A coin (f) Scissors

6.22 Which of these compounds are chiral?

(a) 2,4-Dimethylheptane (b) 5-Ethyl-3,3-dimethylheptane
(c) *cis*-1,3-Dimethylcyclohexane

6.23 Penicillin V is a broad-spectrum antibiotic that contains three chiral centers. Identify them with asterisks.

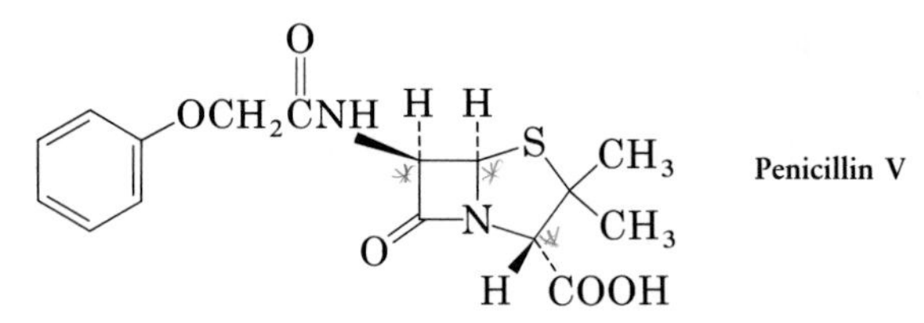

6.24 Draw chiral molecules that meet these descriptions:

(a) A chloroalkane, $C_5H_{11}Cl$ (b) An alcohol, $C_6H_{14}O$
(c) An alkene, C_6H_{12} (d) An alkane, C_8H_{18}

6.25 Which of these compounds are chiral? Label all chiral centers.

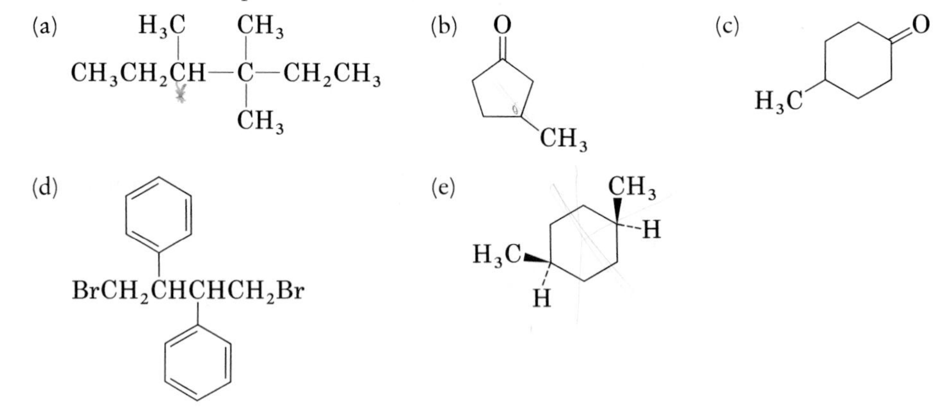

6.26 There are eight alcohols with the formula $C_5H_{12}O$. Draw them and tell which are chiral.

6.27 Propose structures for compounds that meet these descriptions:

(a) A chiral alcohol with four carbons
(b) A chiral carboxylic acid
(c) A compound with two chiral centers

6.28 Assign priorities to the substituents in each of these sets:

(a) $-H$, $-OH$, $-OCH_3$, $-CH_3$
(b) $-Br$, $-CH_3$, $-CH_2Br$, $-Cl$
(c) $-CH{=}CH_2$, $-CH(CH_3)_2$, $-C(CH_3)_3$, $-CH_2CH_3$
(d) $-COOCH_3$, $-COCH_3$, $-CH_2OCH_3$, $-OCH_3$

6.29 One enantiomer of lactic acid is shown below. Which one is it, *R* or *S*? Draw its mirror image in the standard tetrahedral representation.

6.30 Draw tetrahedral representations of both enantiomers of the amino acid serine. Tell which of your structures is *S* and which is *R*.

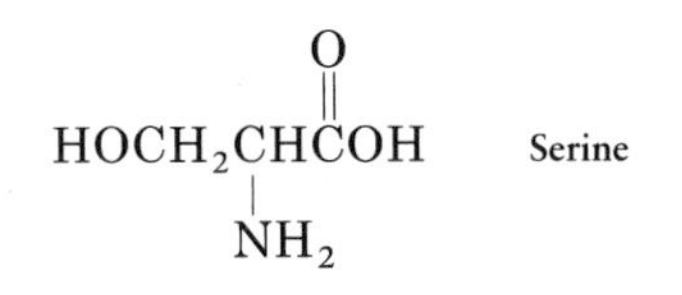

6.31 If naturally occurring *S*-serine has $[\alpha]_D = -6.83°$, what specific rotation do you expect for *R*-serine?

6.32 Assign *R* or *S* configuration to the chiral centers in these molecules:

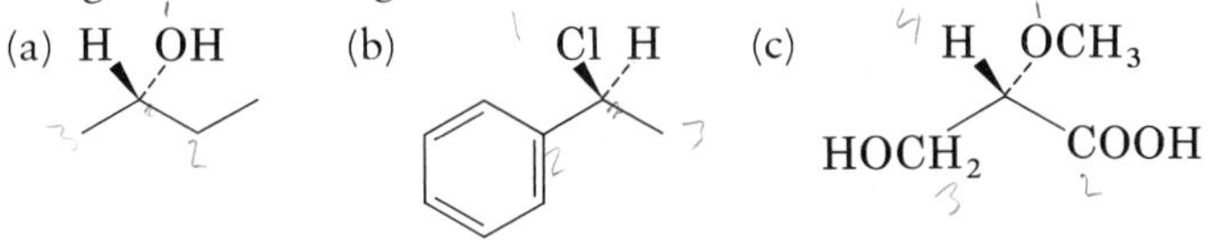

6.33 What is the relationship between the specific rotations of 2*R*,3*R*-dihydroxypentane and 2*S*,3*S*-dihydroxypentane? Between 2*R*,3*S*-dihydroxypentane and 2*R*,3*R*-dihydroxypentane?

6.34 What is the stereochemical configuration of the enantiomer of 2*S*,4*R*-dibromooctane?

6.35 What are the stereochemical configurations of the two diastereomers of 2*S*,4*R*-dibromooctane?

6.36 Draw examples of the following:

(a) A meso compound with the formula C_8H_{18}

(b) A compound with two chiral centers, one *R* and the other *S*.

6.37 Draw a tetrahedral representation of (*S*)-2-butanol, $CH_3CH_2CH(OH)CH_3$.

6.38 Tell whether this Newman projection of 2-chlorobutane is *R* or *S*.

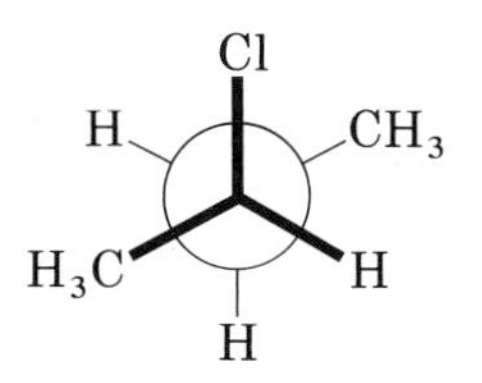

6.39 Draw a Newman projection that is enantiomeric with the one shown in Problem 6.38.

6.40 Draw a Newman projection of *meso*-tartaric acid.

6.41 Draw Newman projections of 2*R*,3*R*- and 2*S*,3*S*-tartaric acid and compare them to the projection you drew in Problem 6.40 for the meso form.

6.42 The sugar glucose has four chiral centers. How many stereoisomers of glucose are possible?

6.43 Draw a tetrahedral representation of (*R*)-3-chloro-1-pentene.

6.44 Draw all of the stereoisomers of 1,2-dimethylcyclopentane. Assign *R*,*S* configurations to the chiral centers in all isomers, and indicate which stereoisomers are chiral and which, if any, are meso.

6.45 Assign *R* or *S* configuration to each chiral center in these molecules:

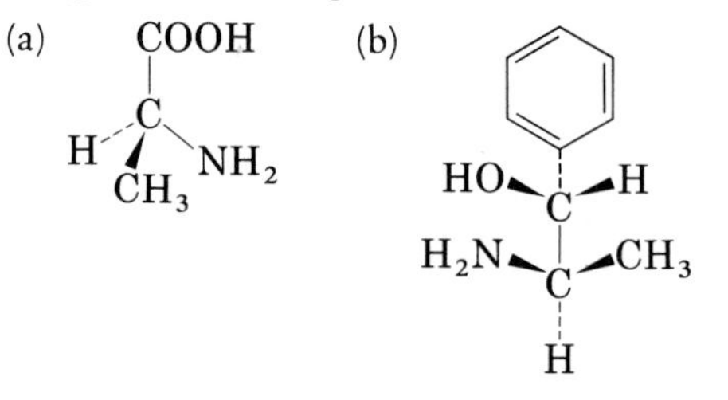

6.46 Hydroxylation of *cis*-2-butene with $KMnO_4$ yields 2,3-butanediol. What is the stereochemistry of the product? (Review Section 4.7.)

6.47 Answer Problem 6.46 for *trans*-2-butene.

6.48 How many stereoisomers of 2,4-dibromo-3-chloropentane are there? Draw them and indicate which are optically active.

6.49 Alkenes undergo reaction with peroxycarboxylic acids (RCO_3H) to give compounds called *epoxides*. For example, *cis*-2-butene gives 2,3-epoxybutane:

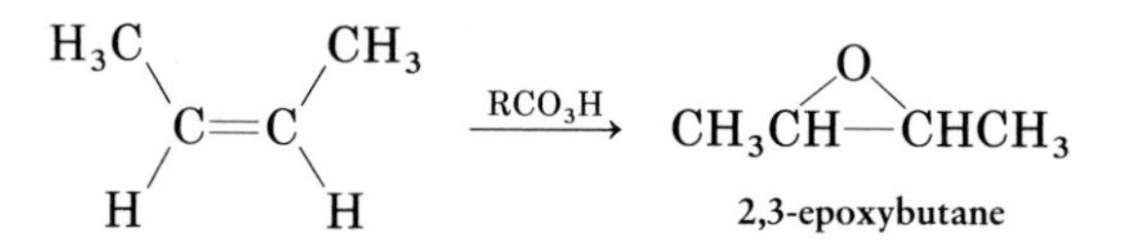

Assuming that both C–O bonds form from the same side of the molecule (syn stereochemistry), show the stereochemistry of the product. Is the epoxide chiral? How many chiral centers does it have? How would you describe the product stereochemically?

6.50 Answer Problem 6.49 assuming that the epoxidation was carried out on *trans*-2-butene.

6.51 Ribose, an essential part of ribonucleic acid (RNA) has the following structure:

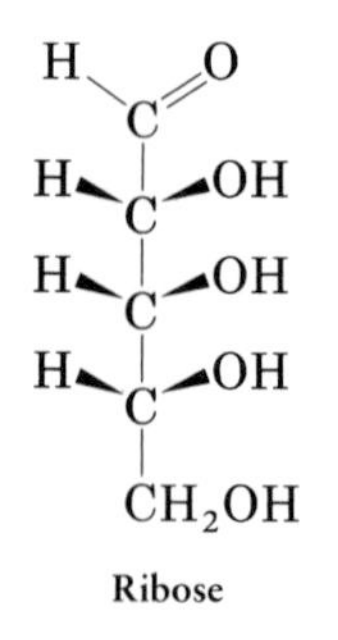

Ribose

How many chiral centers does ribose have? Identify them with asterisks. How many stereoisomers of ribose are there?

6.52 Draw the structure of the enantiomer (mirror image) of ribose (Problem 6.51).

6.53 Draw the structure of a diastereomer of ribose (Problem 6.51).

6.54 On catalytic hydrogenation over a platinum catalyst, ribose (Problem 6.51) is converted into ribitol. Is ribitol optically active or inactive? Explain.

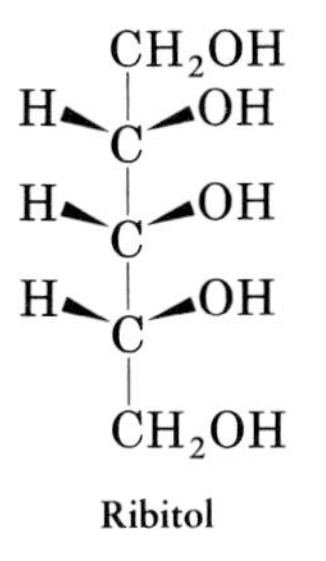

Ribitol

6.55 Draw the two enantiomers of the amino acid cysteine, $HSCH_2CH(NH_2)COOH$, and identify each as *R* or *S*.

6.56 Draw the structure of (*R*)-2-methylcyclohexanone.

6.57 Compound **A**, C_7H_{12}, was found to be optically active. On catalytic reduction over a palladium catalyst, two equivalents of hydrogen were absorbed, yielding compound **B**, C_7H_{16}. On cleavage with ozone, two fragments were obtained. One fragment was identified as acetic acid, CH_3COOH, and the other fragment, **C**, was found to be an optically active carboxylic acid. Formulate the reactions, and propose structures for **A**, **B**, and **C**.

6.58 *Allenes* are compounds with adjacent carbon–carbon double bonds. Even though they do not contain chiral carbon atoms, many allenes are chiral. For example, mycomycin, an antibiotic isolated from the bacterium *Nocardia acidophilus*, is chiral and has $[\alpha]_D = -130°$. Can you explain why mycomycin is chiral? Making a molecular model should be helpful.

$$HC{\equiv}C{-}C{\equiv}C{-}CH{=}C{=}CH{-}CH{=}CH{-}CH{=}CH{-}CH_2COOH$$

Mycomycin (an allene)

Alkyl Halides

It would be difficult to study organic chemistry for long without becoming aware of the importance of halo-substituted alkanes. Among their many uses, alkyl halides are employed as industrial solvents, as inhaled anesthetics in hospitals, as insecticides, and as refrigerants.

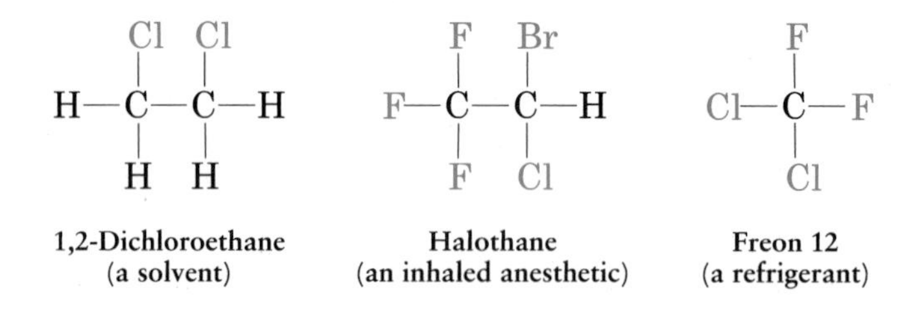

1,2-Dichloroethane (a solvent) **Halothane** (an inhaled anesthetic) **Freon 12** (a refrigerant)

7.1 NAMING ALKYL HALIDES

Alkyl halides are named in the same way that alkanes are (Section 2.3) by considering the halogen as a substituent on the parent alkane chain. There are three rules:

1. Find and name the parent chain. As in naming alkanes, select the longest chain as the parent. If a multiple bond is present, the parent chain must contain it.
2. Number the carbons of the chain beginning at the end nearer the first substituent, regardless of whether it's alkyl or halo. Assign each substituent a number according to its position on the chain.

$$\underset{1}{CH_3}\underset{2}{\overset{\overset{CH_3}{|}}{C}H}\underset{3}{CH_2}\underset{\underset{CH_3}{|}}{\underset{4}{C}H}\underset{5}{\overset{\overset{Br}{|}}{C}H}\underset{6}{CH_2}\underset{7}{CH_3}$$

5-Bromo-2,4-dimethylheptane

$$\underset{1}{CH_3}\underset{2}{\overset{\overset{Br}{|}}{C}H}\underset{3}{CH_2}\underset{\underset{CH_3}{|}}{\underset{4}{C}H}\underset{5}{\overset{\overset{CH_3}{|}}{C}H}\underset{6}{CH_2}\underset{7}{CH_3}$$

2-Bromo-4,5-dimethylheptane

3. Write the name, listing all substituents in alphabetical order and using one of the prefixes *di-*, *tri-*, and so forth if more than one of the same halogen is present.

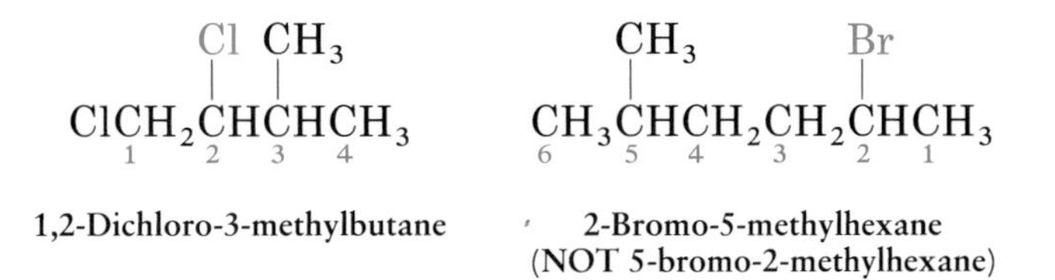

In addition to their systematic names, many simple alkyl halides are also named by identifying first the alkyl group and then the halogen. For example, CH_3I can be called either iodomethane or methyl iodide.

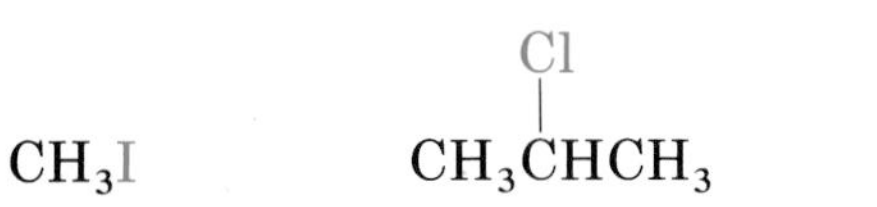

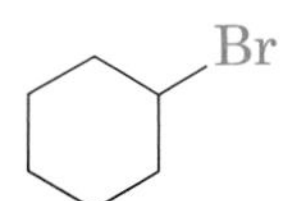

Iodomethane (or methyl iodide) — 2-Chloropropane (or isopropyl chloride) — Bromocyclohexane (or cyclohexyl bromide)

PROBLEM 7.1 Give the IUPAC names of these alkyl halides:

(a) $CH_3CH_2CHBrCH_3$
(b) $CH_3CH_2CHClCH(CH_3)_2$
(c) $(CH_3)_2CHCH_2CH_2Cl$
(d) $(CH_3)_2CClCH_2CH_2Cl$
(e) $BrCH_2CH_2CH_2CH_2Cl$
(f) $CH_3CHBrCH_2CH_2CH_2Cl$

PROBLEM 7.2 Draw structures corresponding to these names:

(a) 2-Chloro-3,3-dimethylhexane
(b) 3,3-Dichloro-2-methylhexane
(c) 3-Bromo-3-ethylpentane
(d) 2-Bromo-5-chloro-3-methylhexane

7.2 PREPARATION OF ALKYL HALIDES: RADICAL CHLORINATION OF ALKANES

We've already seen several methods of alkyl halide preparation, including the addition reaction of HX and X_2 with alkenes (Sections 4.1 and 4.5).

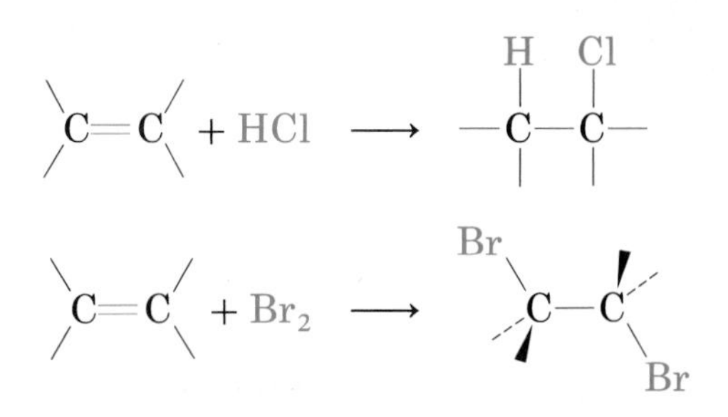

Another method of alkyl halide synthesis is the reaction of alkanes with chlorine or bromine. Although inert to most reagents, alkanes react readily with chlorine in the presence of ultraviolet light ($h\nu$) to give chlorinated alkane products. For example, methane reacts with chlorine gas to give chloromethane.

$$\underset{\text{Methane}}{CH_4} + Cl_2 \xrightarrow{h\nu} \underset{\text{Chloromethane}}{CH_3Cl} + HCl$$

radical substitution reaction a substitution reaction that takes place via a radical mechanism

The chlorination of methane is a typical **radical substitution reaction** rather than a polar reaction of the sort we've been studying until now. Recall from Section 3.6 that radical reactions involve odd-electron reagents. Bonds are formed in radical reactions when each partner donates one electron to the new bond and are broken when each fragment leaves with one electron.

$$A\cdot + \cdot B \longrightarrow A{:}B \qquad \text{Radical bond-making}$$
$$A{:}B \longrightarrow A\cdot + \cdot B \qquad \text{Radical bond-breaking}$$

Radical substitution reactions normally require three kinds of steps: an *initiation* step, *propagation* steps, and *termination* steps. As its name implies, the initiation step starts the reaction by producing reactive radicals. In the present case, the relatively weak chlorine–chlorine bond is broken by irradiation with ultraviolet light to give two chlorine radicals.

Once chlorine radicals have been produced in small amounts, reaction of Cl_2 with methane occurs by a sequence of two propagation steps. In the first propagation step, a chlorine radical abstracts a hydrogen atom from methane to produce HCl and a methyl radical ($\cdot CH_3$). In the second propagation step, the methyl radical abstracts a chlorine atom from Cl_2 to yield chloromethane and a new chlorine radical, which then cycles back to the first propagation step, making the overall process a *chain reaction*. Once the reaction has been initiated, it becomes a self-sustaining cycle of endlessly repeating propagation steps 1 and 2.

Occasionally, two radicals collide and combine to form a stable product. When this kind of termination step happens, the reaction cycle is interrupted and the chain is ended. The overall mechanism for radical chlorination of methane is shown in Figure 7.1.

Though interesting from a mechanistic point of view, alkane chlorination is not a generally useful method of alkyl halide synthesis because mixtures of products usually result. Chlorination of methane doesn't stop cleanly at the monochlorinated stage but continues on, giving a mixture of dichloro, trichloro, and even tetrachloro products that must be separated.

$$CH_4 + Cl_2 \xrightarrow{h\nu} CH_3Cl + CH_2Cl_2 + CHCl_3 + CCl_4 + HCl$$

The situation is even worse for chlorination of alkanes that have more than one kind of hydrogen. Chlorination of butane gives two monochlorinated products

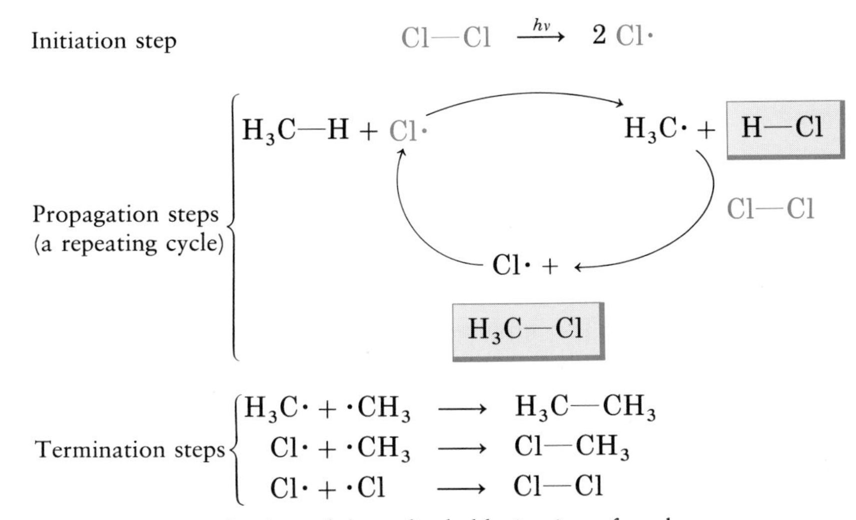

FIGURE 7.1 Mechanism of the radical chlorination of methane

as well as several dichlorobutanes, trichlorobutanes, and so on. Of the monochloro products, 30% is 1-chlorobutane, and 70% is 2-chlorobutane:

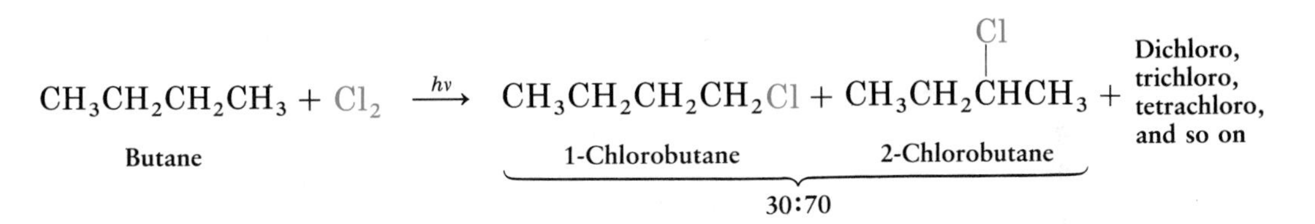

PRACTICE PROBLEM 7.1 Draw the monochloro products you might get from radical chlorination of 2-methylbutane.

SOLUTION Draw the structure of the starting material and begin systematically replacing each kind of hydrogen by chlorine. In this example, there are four possibilities.

$$CH_3-CH_2-\underset{H}{\overset{CH_3}{C}}-CH_3 \longrightarrow CH_3CH_2\overset{CH_3}{CH}CH_2Cl + CH_3CH_2-\underset{Cl}{\overset{CH_3}{C}}-CH_3 + CH_3CHCl\overset{CH_3}{CH}CH_3 + CH_2ClCH_2\overset{CH_3}{CH}CH_3$$

2-Methylbutane

PROBLEM 7.3 Draw and name all monochloro products you would expect to obtain from radical chlorination of 3-methylpentane. Which, if any, are chiral?

PROBLEM 7.4 Radical chlorination of pentane is a poor way to prepare 1-chloropentane, but radical chlorination of 2,2-dimethylpropane is a good way to prepare 1-chloro-2,2-dimethylpropane. Explain.

7.3 ALKYL HALIDES FROM ALCOHOLS

The most valuable method for the preparation of alkyl halides is their synthesis from alcohols. The simplest method for converting an alcohol to an alkyl halide involves treating the alcohol with HX:

$$R{-}OH + HX \longrightarrow R{-}X + H_2O$$

where X = Cl, Br, or I. For reasons to be discussed in Section 7.8, the reaction works best when applied to tertiary alcohols. Primary and secondary alcohols react much more slowly.

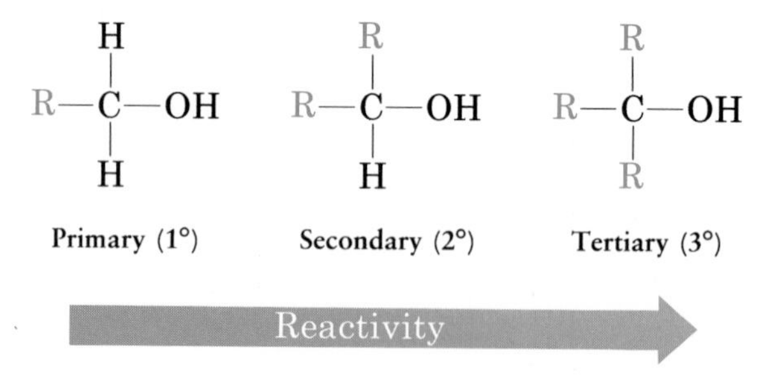

Primary and secondary alcohols are usually converted into alkyl halides by treatment with either thionyl chloride ($SOCl_2$) or phosphorus tribromide (PBr_3). These reactions normally take place under fairly mild conditions, and yields are usually high.

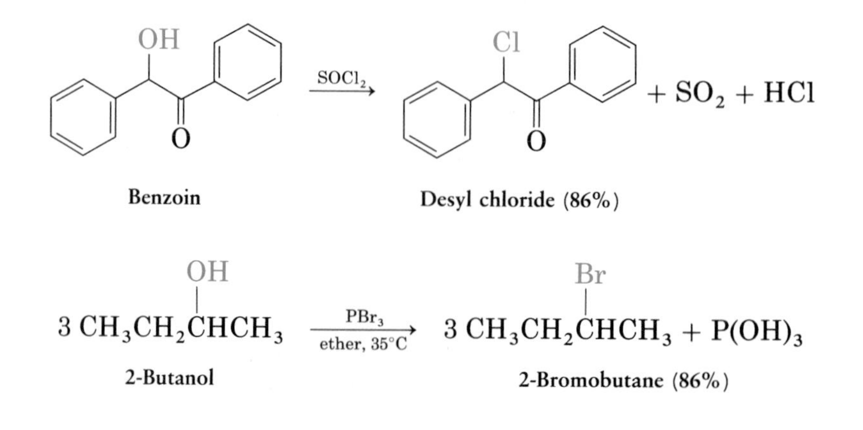

PRACTICE PROBLEM 7.2 Predict the product of this reaction:

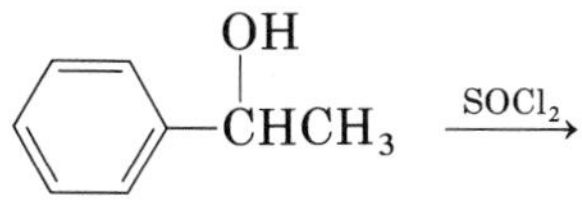

SOLUTION Alcohols yield alkyl chlorides on treatment with $SOCl_2$:

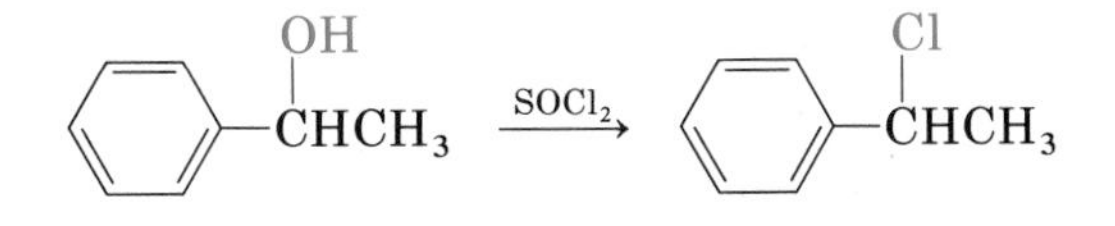

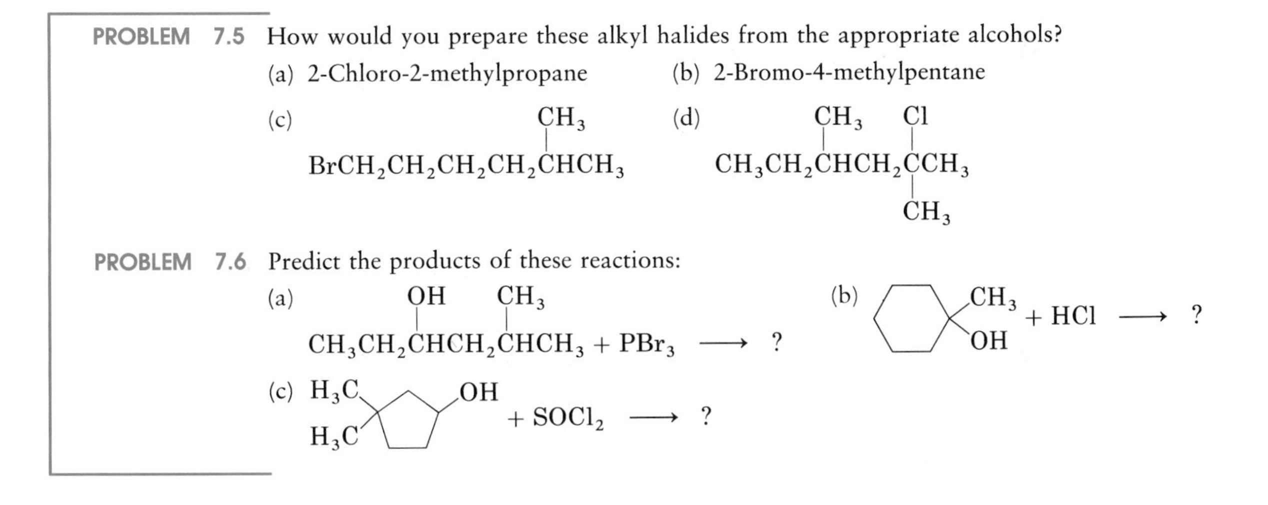

PROBLEM 7.5 How would you prepare these alkyl halides from the appropriate alcohols?

(a) 2-Chloro-2-methylpropane (b) 2-Bromo-4-methylpentane

(c) $BrCH_2CH_2CH_2CH_2CH(CH_3)CH_3$ (d) $CH_3CH_2CH(CH_3)CH_2C(Cl)(CH_3)CH_3$

PROBLEM 7.6 Predict the products of these reactions:

(a) $CH_3CH_2CH(OH)CH_2CH(CH_3)CH_3 + PBr_3 \longrightarrow ?$

(b) 1-methylcyclohexanol $+ HCl \longrightarrow ?$

(c) 3,3-dimethylcyclopentanol $+ SOCl_2 \longrightarrow ?$

7.4 REACTIONS OF ALKYL HALIDES: GRIGNARD REAGENTS

Grignard reagent an organomagnesium halide, RMgX

Alkyl halides react with magnesium metal in ether solvent to yield organomagnesium halides, called **Grignard reagents** after their discoverer, Victor Grignard. Grignard reagents contain a carbon–metal bond and thus are examples of *organometallic compounds.*

$$R{-}X + Mg \xrightarrow{\text{ether}} R{-}Mg{-}X$$

where R = 1°, 2°, or 3° alkyl or aryl
X = Cl, Br, or I

For example:

$$CH_3CH_2CH_2Br + Mg \xrightarrow{\text{ether}} CH_3CH_2CH_2{-}Mg{-}Br$$

1-Bromopropane **Propylmagnesium bromide**

Grignard reagents are extraordinarily useful and versatile compounds. As you might expect from the discussion of electronegativity and bond polarity in Section 1.12, a carbon–magnesium bond is strongly polarized, making the organic part strongly nucleophilic.

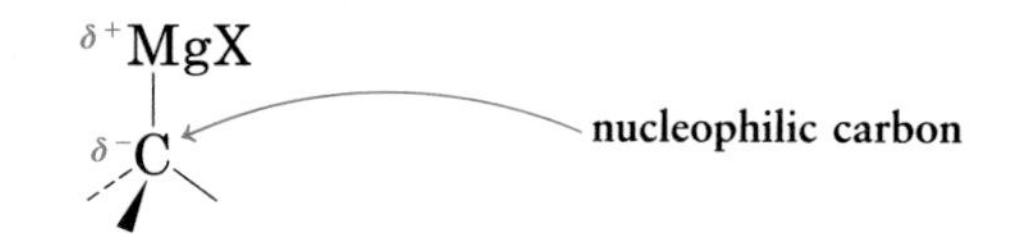

Because of their nucleophilic character, Grignard reagents react with a wide variety of electrophiles. For example, they react with acids such as HCl or H_2O to yield hydrocarbons. The overall sequence, R–X → R–MgX → R–H, is a useful method for reducing organic halides to yield alkanes:

$$\underset{\text{Alkyl halide}}{\text{R—X}} \xrightarrow{\text{Mg}} \underset{\text{Grignard reagent}}{\text{R—Mg—X}} \xrightarrow{H_2O} \underset{\text{Alkane}}{\text{R—H}} + \text{HOMgX}$$

For example,

$$\underset{\text{1-Bromodecane}}{CH_3(CH_2)_8CH_2Br} \xrightarrow[2.\ H_2O]{1.\ Mg} \underset{\text{Decane (85\%)}}{CH_3(CH_2)_8CH_3}$$

PRACTICE PROBLEM 7.3 By using several reactions in sequence, you can accomplish transformations that can't be done in a single step. How would you prepare the alkane methylcyclohexane from the alcohol 1-methylcyclohexanol?

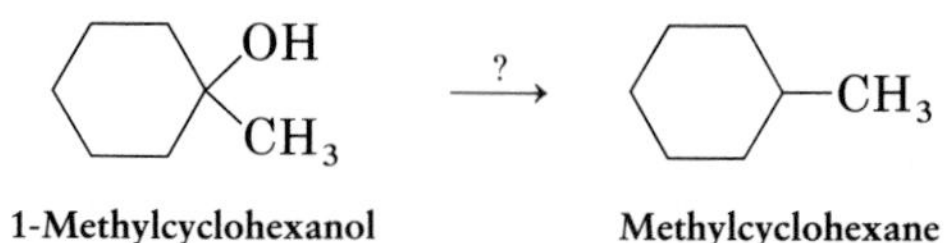

SOLUTION We know that alcohols can be converted into alkyl halides and that alkyl halides can be converted into alkanes. Carrying out the two reactions sequentially thus converts 1-methylcyclohexanol into methylcyclohexane.

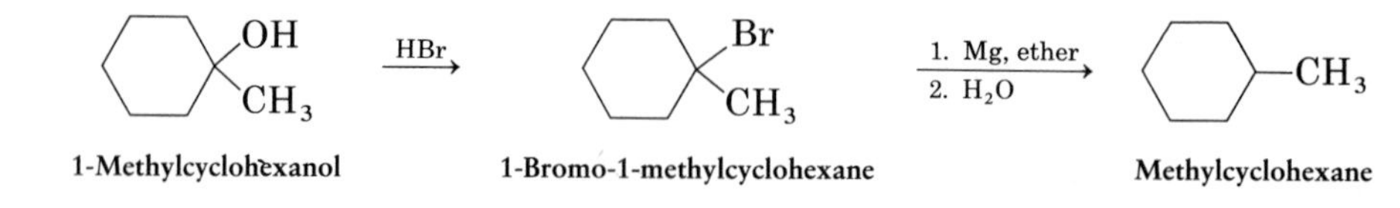

PROBLEM 7.7 An advantage to preparing alkanes from Grignard reagents is that deuterium (D, the isotope of hydrogen with atomic weight 2) can be introduced into a specific site in a molecule. How could you convert 2-bromobutane into 2-deuteriobutane?

$$CH_3\overset{\overset{\displaystyle Br}{|}}{C}HCH_2CH_3 \xrightarrow{?} CH_3\overset{\overset{\displaystyle D}{|}}{C}HCH_2CH_3$$

PROBLEM 7.8 Show how you could convert 4-methyl-1-pentanol into 2-methylpentane.

$$\begin{array}{c} CH_3 \\ | \\ CH_3CHCH_2CH_2CH_2OH \end{array} \quad \textbf{4-Methyl-1-pentanol}$$

7.5 NUCLEOPHILIC SUBSTITUTION REACTIONS: THE DISCOVERY

In 1896, the German chemist Paul Walden reported a remarkable discovery. He found that (+)- and (−)-malic acids can be interconverted by a series of simple reactions. When Walden treated (−)-malic acid with PCl_5, he isolated (+)-chlorosuccinic acid. This, on reaction with wet silver oxide, gave (+)-malic acid. Similarly, reaction of (+)-malic acid with PCl_5 gave (−)-chlorosuccinic acid, which was converted into (−)-malic acid when treated with wet silver oxide. The full cycle of reactions reported by Walden is shown in Figure 7.2.

$$\begin{array}{ccc} \begin{array}{c} OH \\ | \\ HO_2CCH_2CHCO_2H \end{array} & \xrightarrow[\text{ether}]{PCl_5} & \begin{array}{c} Cl \\ | \\ HO_2CCH_2CHCO_2H \end{array} \\ \textbf{(−)-Malic acid} & & \textbf{(+)-Chlorosuccinic acid} \\ [\alpha]_D = -2.3^\circ & & \\ \uparrow Ag_2O,\ H_2O & & \downarrow Ag_2O,\ H_2O \\ \begin{array}{c} Cl \\ | \\ HO_2CCH_2CHCO_2H \end{array} & \xleftarrow[\text{ether}]{PCl_5} & \begin{array}{c} OH \\ | \\ HO_2CCH_2CHCO_2H \end{array} \\ \textbf{(−)-Chlorosuccinic acid} & & \textbf{(+)-Malic acid} \\ & & [\alpha]_D = +2.3^\circ \end{array}$$

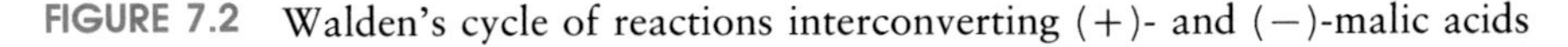

FIGURE 7.2 Walden's cycle of reactions interconverting (+)- and (−)-malic acids

At the time, the results were astonishing. The eminent chemist Emil Fischer called Walden's discovery "the most remarkable observation made in the field of optical activity since the fundamental observations of Pasteur." Since (−)-malic acid was being converted into (+)-malic acid, some reactions in the cycle must have occurred with a change in the configuration of the chiral center. But which ones, and how?

nucleophilic substitution reaction a substitution reaction in which one nucleophile replaces another

Today we refer to the transformations taking place in Walden's cycle as **nucleophilic substitution reactions** because each step involves the substitution of one nucleophile (chloride ion, Cl^-, or hydroxide ion, HO^-) for another. Nucleophilic substitution reactions are one of the most important general reaction types in organic chemistry.

7.6 KINDS OF NUCLEOPHILIC SUBSTITUTION REACTIONS

Following the work of Walden, a series of investigations was undertaken during the 1920s and 1930s to clarify the mechanism of nucleophilic substitution reactions and to find out how inversions of configuration occur. We now know that there are two major ways by which nucleophilic substitutions can occur, named the *S_N1 mechanism* and the *S_N2 mechanism*. In both cases, the "S_N" part of the name stands for "substitution, nucleophilic." What the 1 and the 2 stand for will become clear soon.

leaving group the group that is replaced in a nucleophilic substitution reaction

Regardless of mechanism, the overall change during all nucleophilic substitution reactions is the same: A nucleophile (Nu:) reacts with a substrate R–X and substitutes for X: (the **leaving group**) to yield the product R–Nu. The nucleophile can be either neutral (Nu:) or negatively charged ($Nu:^-$). If it's neutral, then the product is positively charged to maintain charge conservation; if it's negatively charged, the product is neutral.

$$Nu: + R{-}X \longrightarrow R{-}\overset{+}{Nu} + X:^-$$
$$Nu:^- + R{-}X \longrightarrow R{-}Nu + X:^-$$

Because of the wide scope of nucleophilic substitution reactions, many products can be prepared from alkyl halides. Table 7.1 lists some of the possibilities.

TABLE 7.1 **Some nucleophilic substitution reactions on bromomethane:**

$$Nu:^- + CH_3Br \longrightarrow Nu{-}CH_3 + :Br^-$$

Attacking nucleophile		*Product*	
Formula	*Name*	*Formula*	*Name*
$H:^-$	Hydride	CH_4	Methane
$CH_3\ddot{S}:^-$	Methanethiolate	CH_3SCH_3	Dimethyl sulfide
$H\ddot{S}:^-$	Hydrosulfide	$HSCH_3$	Methane thiol
$:N{\equiv}C:^-$	Cyanide	$N{\equiv}CCH_3$	Acetonitrile
$:\ddot{I}:^-$	Iodide	ICH_3	Iodomethane
$H\ddot{O}:^-$	Hydroxide	$HOCH_3$	Methanol
$CH_3\ddot{O}:^-$	Methoxide	CH_3OCH_3	Dimethyl ether
$:\ddot{N}{=}N{=}\ddot{N}:^-$	Azide	N_3CH_3	Azidomethane
$:\ddot{Cl}:^-$	Chloride	$ClCH_3$	Chloromethane
$CH_3CO_2:^-$	Acetate	$CH_3CO_2CH_3$	Methyl acetate
$H_3N:$	Ammonia	$H_3\overset{+}{N}CH_3\ Br^-$	Methylammonium bromide
$(CH_3)_3N:$	Trimethylamine	$(CH_3)_3\overset{+}{N}CH_3\ Br^-$	Tetramethylammonium bromide

PRACTICE PROBLEM 7.4 What is the product from reaction of 1-chloropropane with sodium hydroxide?

SOLUTION Write the two starting materials and identify the nucleophile, in this instance hydroxide ion. Then replace the –Cl group by –OH and write the complete equation.

$$\underset{\text{1-Chloropropane}}{CH_3CH_2CH_2Cl} + Na^+ \ ^-OH \longrightarrow \underset{\text{1-Propanol}}{CH_3CH_2CH_2OH} + Na^+ \ ^-Cl$$

PRACTICE PROBLEM 7.5 How would you prepare 1-propanethiol, $CH_3CH_2CH_2SH$, using a nucleophilic substitution reaction?

SOLUTION Since the product contains an –SH group, it could be prepared by reaction of ^-SH (hydrosulfide ion) on 1-bromopropane:

$$\underset{\text{1-Bromopropane}}{CH_3CH_2CH_2Br} + Na^+ \ ^-SH \longrightarrow \underset{\text{1-Propanethiol}}{CH_3CH_2CH_2SH} + Na^+ \ ^-Br$$

PROBLEM 7.9 What products would you expect to obtain from these reactions?

(a) $CH_3CH_2CHBrCH_3 + LiI \longrightarrow ?$ (b) $(CH_3)_2CHCH_2Cl + HS^- \longrightarrow ?$

(c) $C_6H_5\text{–}CH_2Br + NaCN \longrightarrow ?$

PROBLEM 7.10 How could you prepare these substances by using nucleophilic substitution reactions?

(a) $CH_3CH_2CH_2CH_2OH$ (b) $(CH_3)_2CHCH_2CH_2N_3$

7.7 THE S_N2 REACTION

The mechanism of the S_N2 reaction is shown in Figure 7.3. The reaction takes place in a single step without intermediates when the entering nucleophile attacks the substrate from a direction 180° away from the leaving group. As the nucleophile comes in on one side of the molecule, the leaving group departs from the other side.

We can picture the S_N2 reaction as occurring when an electron pair on the nucleophile $Nu:^-$ forces out the leaving group $X:^-$ with the electron pair from the C–X bond. This takes place through a transition state in which the new Nu–C bond is partially forming at the same time that the old C–X bond is partially breaking and in which the negative charge is shared by both the incoming nucleophile and the outgoing leaving group.

The nucleophile Nu:$^-$ uses its lone-pair electrons to attack the alkyl halide 180° away from the halogen. This leads to a transition state with a partially formed C–Nu bond and a partially broken C–Y bond.

The stereochemistry at carbon is inverted as the C–Nu bond forms fully and the halide departs with the electron pair from the original C–Y bond.

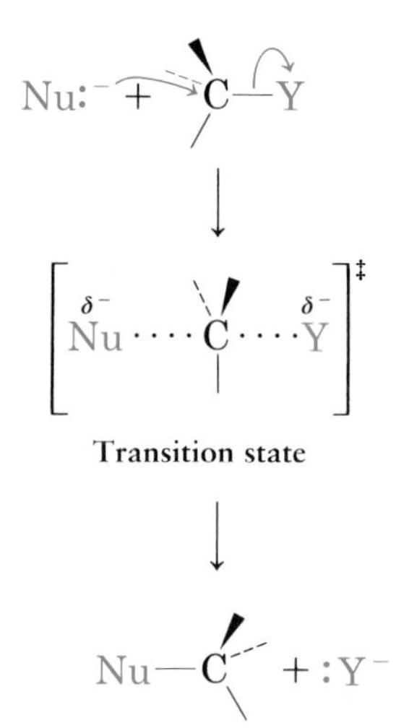

FIGURE 7.3 The mechanism of the S_N2 reaction

Let's see what evidence there is for this mechanism and what the chemical consequences are.

Rates of S_N2 Reactions

reaction rate the exact speed of a reaction under defined conditions

Chemists often speak about a reaction being fast or slow. The exact speed at which a reaction occurs is called the **reaction rate** and is a quantity that can often be measured. The determination of reaction rates and of how those rates depend on reagent concentrations is a powerful tool for probing reaction mechanisms. As an example, let's look at the effect of reagent concentrations on the rate of the S_N2 reaction of hydroxide ion with bromomethane to yield methanol:

$$HO:^- + CH_3\text{—}Br: \longrightarrow HO\text{—}CH_3 + :Br:^-$$

The S_N2 reaction of bromomethane with hydroxide ion takes place when substrate and nucleophile collide and react in a single step. At a given concentration of reagents, the reaction takes place at a certain rate. If we double the concentration of hydroxide ion, the frequency of encounter between the two reagents is also doubled, and we therefore find that the reaction rate doubles. Similarly, if we double the concentration of bromomethane, the reaction rate doubles. Thus, the derivation of the "2" in S_N2: S_N2 reactions are said to be **bimolecular** because two molecules, alkyl halide and nucleophile, are involved in the step whose rate is measured.

bimolecular describing a step that involves two molecules

PROBLEM 7.11 What effects would these changes have on the rate of reaction between iodomethane and sodium acetate?

(a) The CH_3I concentration is tripled.
(b) Both CH_3I and Na^+ $^-OOCCH_3$ concentrations are doubled.

Stereochemistry of S_N2 Reactions

Look carefully at the mechanism of the S_N2 reaction shown in Figure 7.3. As the incoming nucleophile attacks the substrate and begins pushing out the leaving group on the opposite side, the stereochemistry of the molecule becomes *inverted*. For example, treatment of *S*-2-bromobutane with hydroxide ion yields *R*-2-butanol:

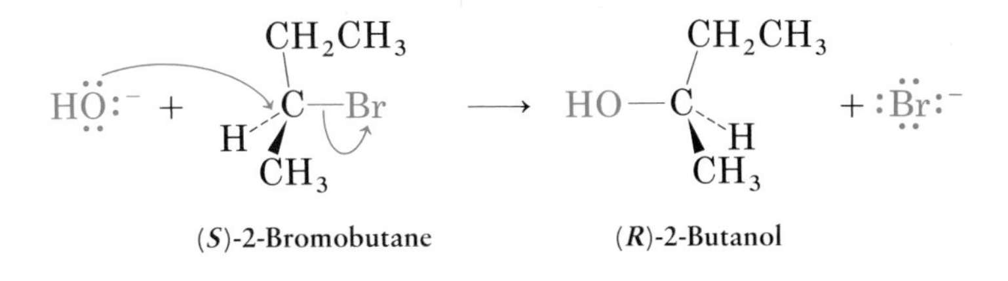

The inversion that takes place during an S_N2 reaction is similar to what happens when an umbrella turns inside out in the wind (Figure 7.4)

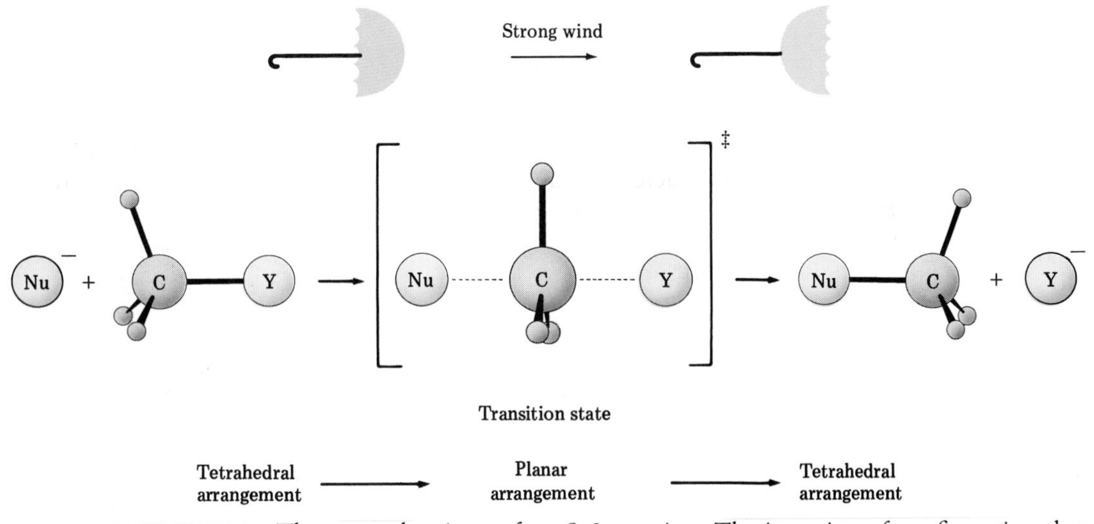

FIGURE 7.4 The stereochemistry of an S_N2 reaction. The inversion of configuration that occurs is like the inversion of an umbrella in a strong wind.

PRACTICE PROBLEM 7.6 What product would you expect to obtain from reaction of (*R*)-2-iodooctane with sodium cyanide, $Na^+ \; ^-CN$?

SOLUTION Table 7.1 shows that cyanide ion is a good nucleophile in the S_N2 reaction. We therefore expect it to displace iodide ion from (*R*)-2-iodooctane, with inversion of configuration to yield (*S*)-2-methyloctanenitrile.

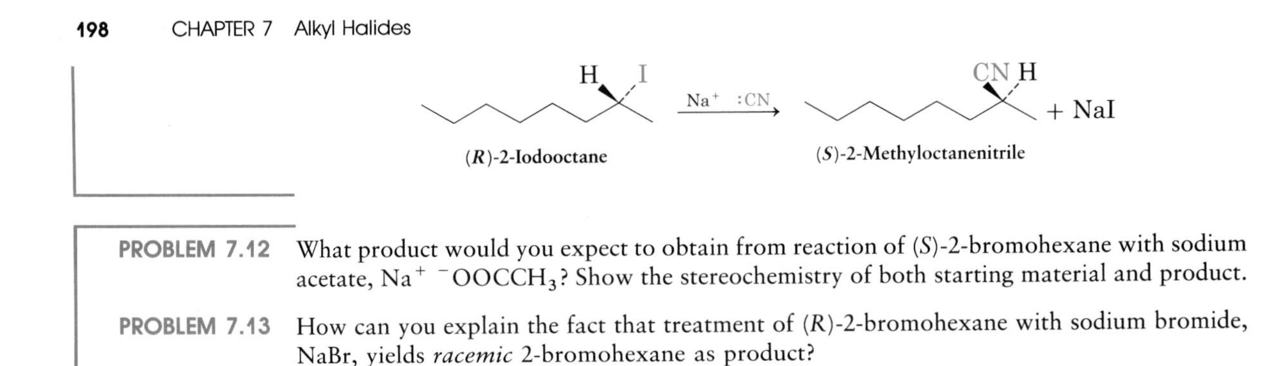

PROBLEM 7.12 What product would you expect to obtain from reaction of (*S*)-2-bromohexane with sodium acetate, $Na^{+}\ ^{-}OOCCH_3$? Show the stereochemistry of both starting material and product.

PROBLEM 7.13 How can you explain the fact that treatment of (*R*)-2-bromohexane with sodium bromide, NaBr, yields *racemic* 2-bromohexane as product?

Steric Effects in S_N2 Reactions

Since an attacking nucleophile must approach the substrate closely to expel the leaving group in an S_N2 reaction, we might expect that the ease of approach depends on the steric accessibility of the substrate. Bulky substrates in which the halide-bearing carbon atom is shielded from attack by the rest of the molecule should react more slowly than substrates in which the carbon is more accessible (Figure 7.5).

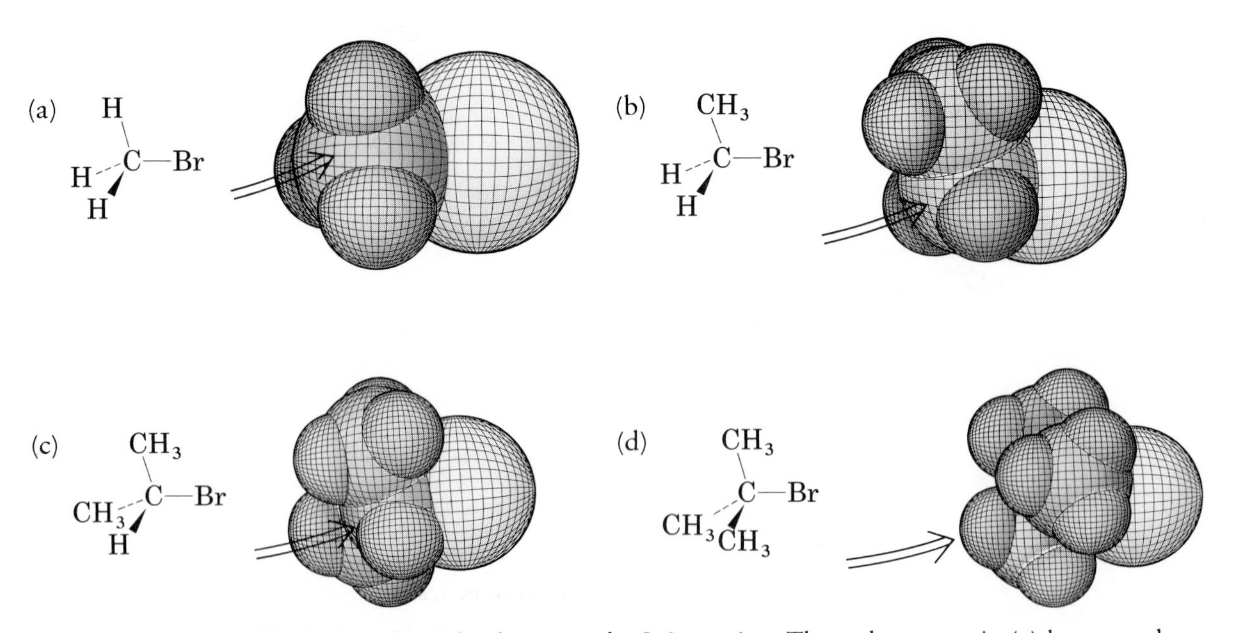

FIGURE 7.5 Steric hindrance to the S_N2 reaction. The carbon atom in (a) bromomethane is readily accessible, resulting in a fast S_N2 reaction, but the carbon atoms in (b) bromoethane, (c) 2-bromopropane, and (d) 2-bromo-2-methylpropane are successively less accessible, resulting in successively slower S_N2 reactions.

Studies of numerous S_N2 reactions have shown that the rates do indeed depend on the steric nature of the substrates. The relative reactivities are as follows:

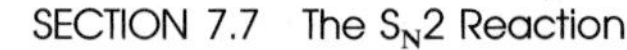

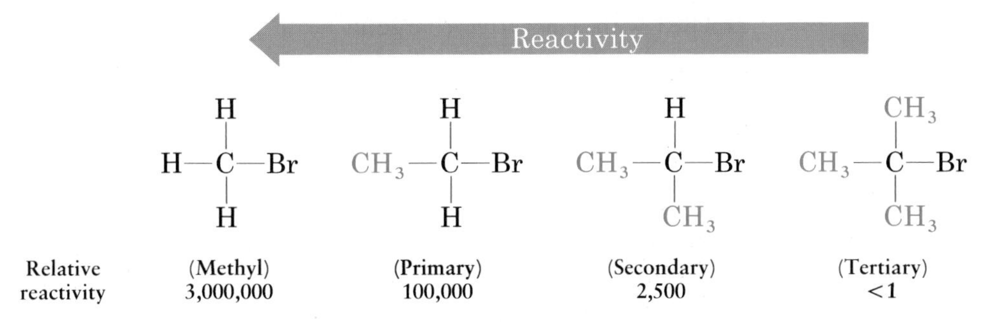

Methyl halides are by far the most reactive substrates in S_N2 reactions, followed by primary alkyl halides such as ethyl and propyl. Alkyl branching next to the leaving group, as in secondary halides, slows the reaction greatly; further branching, as in tertiary halides, effectively halts the reaction.

Although not shown in the reactivity list, *vinylic* halides ($R_2C{=}CRX$) and *aryl* halides (Ar–X) are completely unreactive toward S_N2 displacements. This lack of reactivity is probably due to steric hindrance because the incoming nucleophile would have to approach in the plane of the carbon–carbon double bond or ring to carry out a back-side displacement.

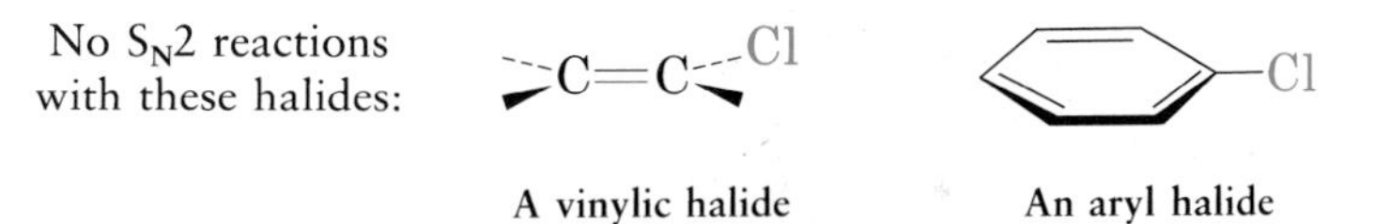

A vinylic halide An aryl halide

PRACTICE PROBLEM 7.7 Which would you expect to be faster, reaction of OH^- ion with 1-bromopentane or with 2-bromopentane?

SOLUTION Since 1-bromopentane is a 1° halide and 2-bromopentane is a 2° halide, reaction with the less hindered 1-bromopentane should be faster.

primary → $CH_3CH_2CH_2CH_2CH_2Br$ (1-Bromopentane)

Br, secondary → $CH_3CH_2CH_2CHBrCH_3$ (2-Bromopentane)

PROBLEM 7.14 Which of these reactions would you expect to be faster?

(a) Reaction of CN^- (cyanide ion) with $CH_3CHBrCH_3$ or with $CH_3CH_2CH_2Br$?
(b) Reaction of I^- with $(CH_3)_2CHCH_2Cl$ or with $H_2C{=}CHCl$?

The Leaving Group in S_N2 Reactions

Another variable that can affect the S_N2 reaction is the nature of the group displaced by the attacking nucleophile—the leaving group. Because the leaving group is expelled with a negative charge in most S_N2 reactions, we might expect the best leaving groups to be those that best stabilize the negative charge. Furthermore, because the stability of an anion is related to its basicity, we can also say that the best leaving groups should be the weakest bases.

The reason that stable anions (weak bases) make good leaving groups is that the charge is distributed over both the attacking nucleophile and the leaving group in the transition state for an S_N2 reaction. The greater the extent of charge stabilization by the leaving group, the more stable the transition state, and the more rapid the reaction.

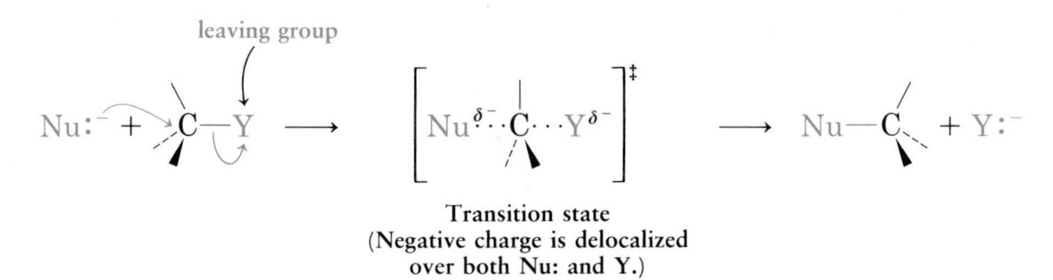

Transition state
(Negative charge is delocalized over both Nu: and Y.)

As the following reactivity list of leaving groups shows, weaker bases (anions derived from stronger acids) are indeed the best leaving groups. Of course, it's just as important to know which are poor leaving groups as to know which are good, and the reactivity list clearly indicates that fluoride ion (F^-), acetate ion (CH_3COO^-, AcO^-), hydroxide ion (HO^-), alkoxide ion (RO^-), and amide ion (H_2N^-) are not displaced by nucleophiles. In other words, alkyl fluorides, esters, alcohols, ethers, and amines don't undergo S_N2 reactions under normal circumstances.

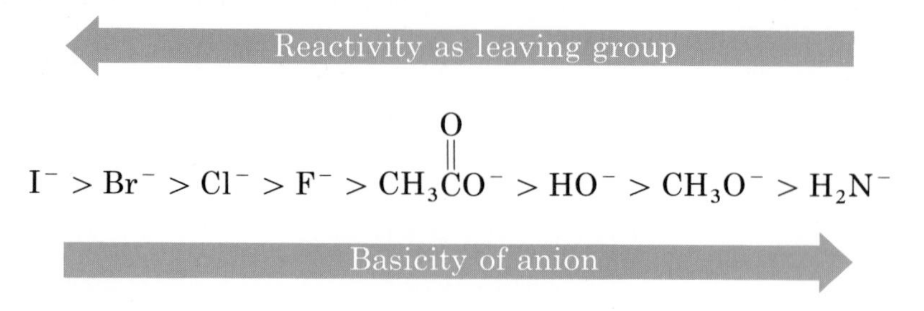

PROBLEM 7.15 Rank the following compounds in order of their expected reactivity toward S_N2 reaction:

$$CH_3I,\ CH_3OH,\ CH_3Br,\ CH_3COOCH_3$$

7.8 THE S_N1 REACTION

Although most nucleophilic substitution reactions take place by the S_N2 mechanism, an alternative called the S_N1 mechanism can also occur. In general, S_N1 reactions take place when a tertiary substrate is treated under neutral or acidic conditions in a hydroxylic solvent like water or alcohol. We saw in Section 7.3, for example, that alkyl halides can be prepared from alcohols by treatment with HCl or HBr. Tertiary alcohols react rapidly, but primary and secondary alcohols are much slower.

$$\underset{\textbf{An alcohol}}{R{-}OH} + HBr \longrightarrow \underset{\textbf{An alkyl halide}}{R{-}Br} + H_2O$$

$$\text{Reactivity order} \quad 3^\circ > 2^\circ > 1^\circ > \text{methyl}$$

What's going on here? Clearly, a nucleophilic substitution reaction is taking place—a −Br group is replacing an −OH—yet the reactivity order is backward from the normal S_N2 order because tertiary alcohols react fastest. The mechanism of the reaction is shown in Figure 7.6.

The −OH group is first protonated by HBr.

Spontaneous dissociation of the protonated alcohol occurs in a slow, rate-limiting step to yield a carbocation intermediate plus water.

The carbocation intermediate reacts with bromide ion in a fast step to yield the neutral substitution product.

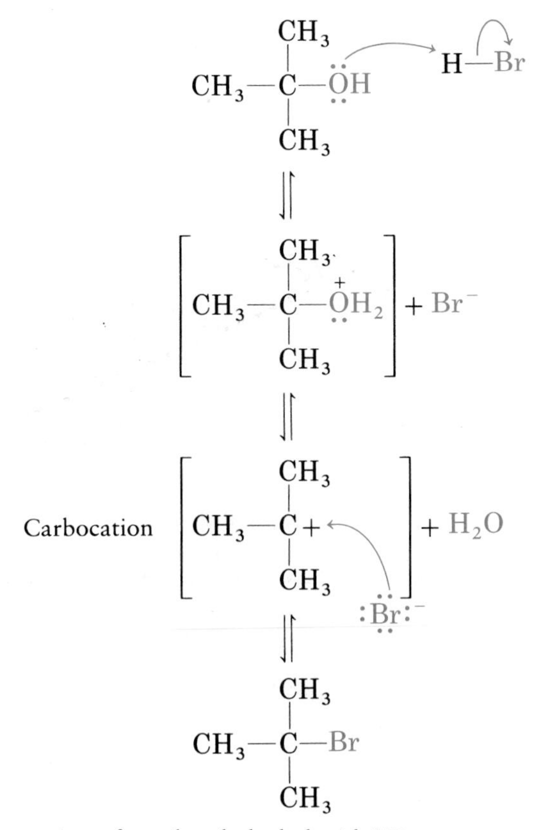

FIGURE 7.6 The mechanism of the S_N1 reaction of *tert*-butyl alcohol with HBr

Unlike an S_N2 reaction, in which the leaving group is displaced *at the same time* that the incoming nucleophile is approaching, an S_N1 reaction takes place by the spontaneous loss of the leaving group *before* the incoming nucleophile approaches. Loss of the leaving group generates a carbocation intermediate that then reacts with nucleophile in a second step to yield the substitution product.

This mechanism explains perfectly why tertiary alcohols react with HBr so much more rapidly than primary or secondary ones. S_N1 reactions occur only when stable carbocation intermediates are formed, and the more stable the carbocation intermediate, the faster the S_N1 reaction. Thus, the reactivity order of alcohols with HBr is exactly the same as the stability order of carbocations (Section 4.3).

Rates of S_N1 Reactions

Unlike an S_N2 reaction, whose rate depends on the concentrations of two reactants, the rate of an S_N1 reaction depends only on the alkyl halide concentration and is *independent of nucleophile concentration.* Thus, the derivation of the "1" in S_N1. S_N1 reactions are said to be **unimolecular** because only one molecule is involved in the step whose rate is measured.

unimolecular describing a step that involves only one molecule

The observation that S_N1 reactions are unimolecular means that the alkyl halide must undergo a spontaneous reaction without assistance from the nucleophile, exactly what the mechanism shown in Figure 7.6 accounts for.

PROBLEM 7.16 What effect would the following changes have on the rate of the reaction of *tert*-butyl alcohol with HBr?

(a) The HBr concentration is tripled.
(b) The HBr concentration is halved, and the *tert*-butyl alcohol concentration is doubled.

Stereochemistry of S_N1 Reactions

If S_N1 reactions occur through carbocation intermediates, as shown in Figure 7.6, the stereochemical consequences should differ from those for S_N2 reactions. Since carbocations are planar and sp^2-hybridized, they are achiral. Thus, if we carry out an S_N1 reaction on a single enantiomer of a chiral starting material and go through an achiral carbocation intermediate, the product must be optically inactive. The symmetrical intermediate carbocation can be attacked by a nucleophile equally well from either side, leading to a racemic mixture of enantiomers (Figure 7.7).

The prediction that S_N1 reactions on chiral substrates should lead to racemic products is exactly what's observed. For example, reaction of optically active (*R*)-1-phenyl-1-butanol with HCl gives a racemic alkyl chloride product:

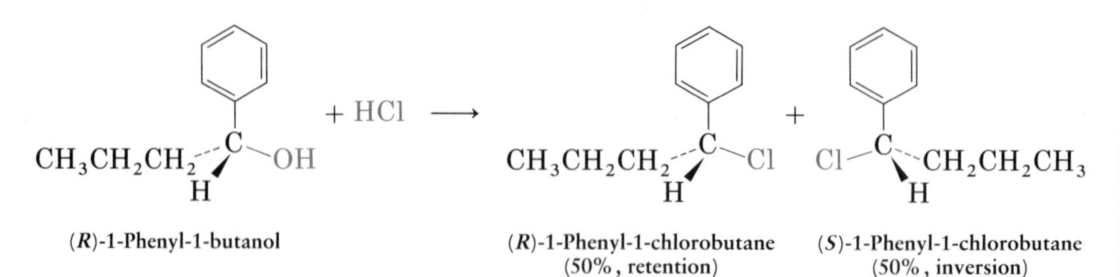

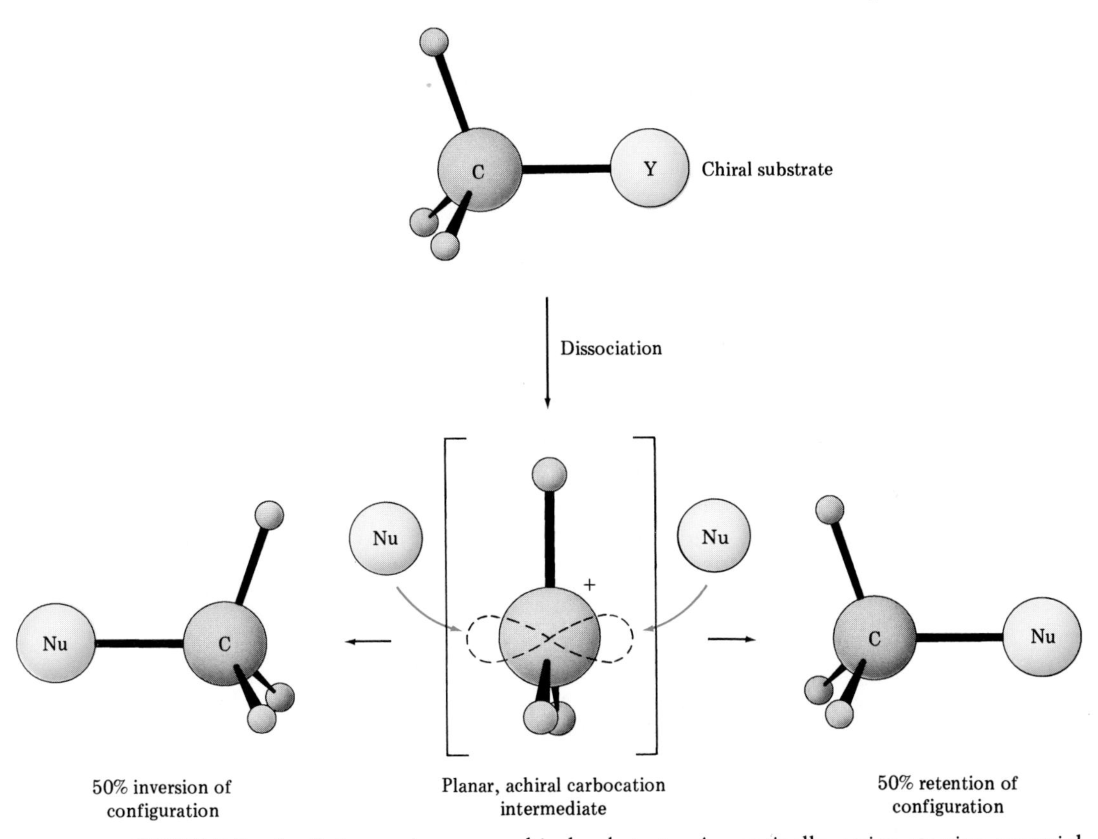

FIGURE 7.7 An S_N1 reaction on a chiral substrate. An optically active starting material must give a racemic product.

PRACTICE PROBLEM 7.8 What stereochemistry would you expect from the S_N1 reaction of (*R*)-3-bromo-3-methylhexane with methanol to yield 3-methoxy-3-methylhexane?

SOLUTION First draw the starting alkyl halide showing its correct stereochemistry. Then replace the bromine with a methoxy group ($-OCH_3$) to give racemic product.

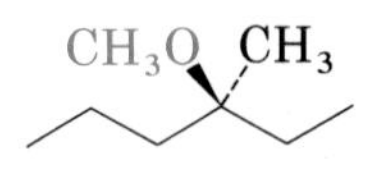

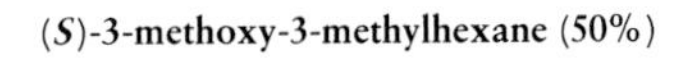

H_3C Br

(*R*)-3-Bromo-3-methylhexane

$+ CH_3OH \longrightarrow$

H_3C OCH_3

(*R*)-3-methoxy-3-methylhexane (50%)

PROBLEM 7.17 What product would you expect to obtain from S_N1 reaction of (*S*)-3-methyl-3-octanol [(*S*)-3-hydroxy-3-methyloctane] with HBr? Show the stereochemistry of both starting material and product.

The Leaving Group in S_N1 Reactions

During the discussion of S_N2 reactions in the previous section, we reasoned that the best leaving groups should be those that are most stable. That is, better leaving groups are weaker bases (anions of stronger acids). The same reactivity order is found for S_N1 reactions because the leaving group is intimately involved in the step whose rate is measured. Thus, we find the S_N1 reactivity order of leaving groups to be

$$I^- > Br^- > H_2O \approx Cl^- > F^- > CH_3\overset{\overset{\displaystyle O}{\|}}{C}O^-$$

← Reactivity as leaving group

Note that in the S_N1 reaction, which is often carried out under acidic conditions, neutral water can act as a leaving group. This is exactly what happens when a tertiary alcohol reacts with HX to yield an alkyl halide, as discussed in Section 7.3. The alcohol is first protonated by HX, spontaneous dissociation into a carbocation plus water occurs, and the carbocation then reacts with halide ion.

$$(CH_3)_3C\text{—}\ddot{O}H + H\text{—}Cl \longrightarrow \left[(CH_3)_3C\text{—}\overset{+}{O}H_2 \longrightarrow H_2O + (CH_3)_3C^+\right] \xrightarrow{:\ddot{Cl}:^-} (CH_3)_3C\text{—}Cl$$

2-Methyl-2-propanol → 2-Methyl-2-chloropropane

7.9 ELIMINATION REACTIONS OF ALKYL HALIDES: THE E2 REACTION

When a nucleophile reacts with an alkyl halide, two kinds of reactions are possible. Often, as we've seen, the nucleophile substitutes for the halide ion in either an

S_N1 or an S_N2 reaction. Alternatively, though, an elimination reaction can occur, leading to formation of an alkene:

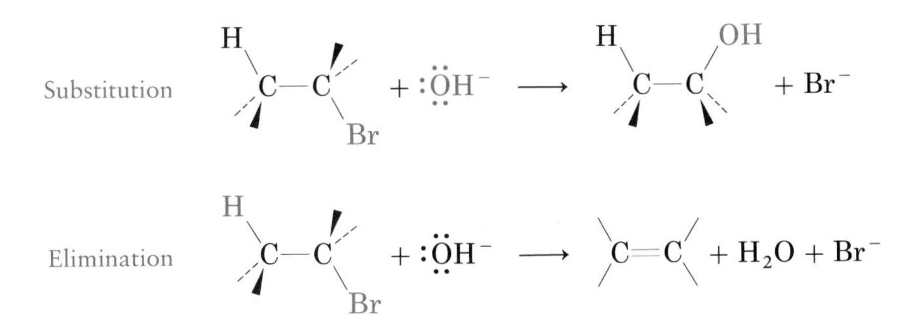

We saw in Section 4.9 that the elimination of HX from alkyl halides is an extremely useful method for preparing alkenes. The subject is complex, though, because eliminations can take place by several different mechanistic pathways just as substitutions can.

E2 reaction an elimination reaction that takes place in a single step through a bimolecular mechanism

The **E2 reaction** (for elimination, bimolecular) takes place when an alkyl halide is treated with a strong base such as hydroxide ion or alkoxide ion (RO^-). This is the most commonly occurring pathway for elimination, and the mechanism is shown in Figure 7.8.

Base (B:) attacks a neighboring C–H bond and begins to remove the H at the same time as the alkene double bond starts to form and the X group starts to leave.

Neutral alkene is produced when the C–H bond is fully broken and the X group has departed with the C–X bond electron pair.

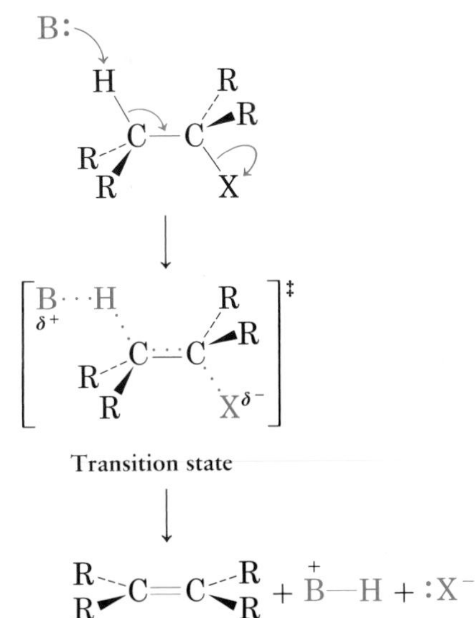

FIGURE 7.8 The mechanism of the E2 reaction. The reaction takes place in a single step, without intermediates. (Dotted lines indicate partial bonding in the transition state.)

Like the S_N2 reaction, the E2 reaction takes place in one step without intermediates. As the attacking base begins to abstract a proton from a carbon atom next to the leaving group, the C–H and C–X bonds begin to break at the same time that the C=C double bond begins to form. When the leaving group departs, it takes with it the two electrons from the former C–X bond.

One of the best pieces of evidence supporting this mechanism comes from measurements of reaction rates. Since both base and alkyl halide enter into the single step, E2 reactions show the same bimolecular behavior that S_N2 reactions do. A second piece of evidence involves the stereochemistry of E2 reactions. Eliminations almost always occur from an **anti periplanar geometry,** meaning that all reacting atoms lie in the same plane (*periplanar*) and that the H and X depart from opposite sides (*anti*) of the molecule:

anti periplanar geometry reaction geometry in which all reacting atoms lie in a plane, with one group on top and another group on the bottom of the molecule

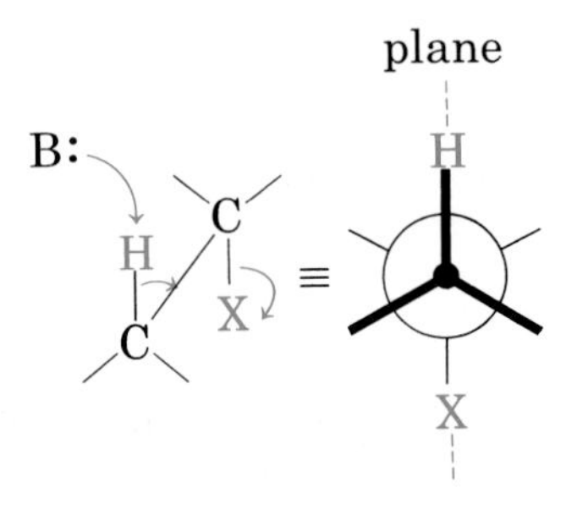

Anti periplanar geometry
(H, X, and both C's are in same plane)

What's so special about anti periplanar geometry? Because the original C–H and C–X sp^3 sigma orbitals in the starting material must overlap and become p orbitals in the alkene product, *there must also be partial overlap in the transition state.* This overlap in the transition state can only occur if the orbitals are in the same plane (are periplanar) to begin with (Figure 7.9).

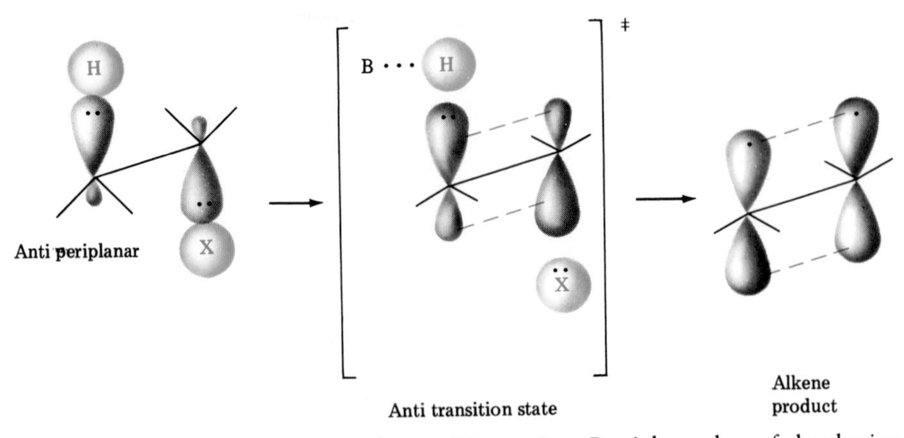

FIGURE 7.9 The transition state for an E2 reaction. Partial overlap of developing p orbitals in the transition state requires periplanar geometry.

Anti periplanar geometry for E2 reactions has stereochemical consequences that provide strong evidence for the proposed mechanism. To cite just one example, *meso*-1,2-dibromo-1,2-diphenylethane undergoes E2 elimination on treatment with base to give the pure *E* alkene, rather than a mixture of *E* and *Z* alkenes.

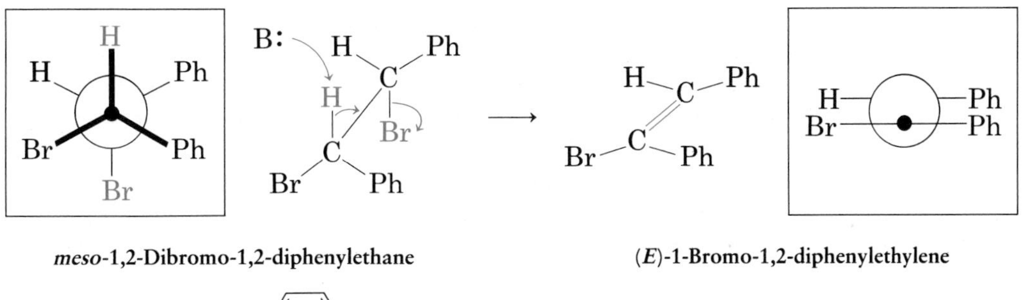

meso-1,2-Dibromo-1,2-diphenylethane

(*E*)-1-Bromo-1,2-diphenylethylene

where Ph = (phenyl ring)

PRACTICE PROBLEM 7.9 What stereochemistry would you expect for the alkene obtained by E2 elimination of (1*S*,2*S*)-1,2-dibromo-1,2-diphenylethane?

SOLUTION First we have to draw (1*S*,2*S*)-1,2-dibromo-1,2-diphenylethane so that we can see its stereochemistry and so that the −H and −Br groups to be eliminated are anti periplanar (molecular models are extremely helpful here). Keeping all substituents in approximately their same positions, we then eliminate HBr and see what alkene results. The product is (*Z*)-1-bromo-1,2-diphenylethylene.

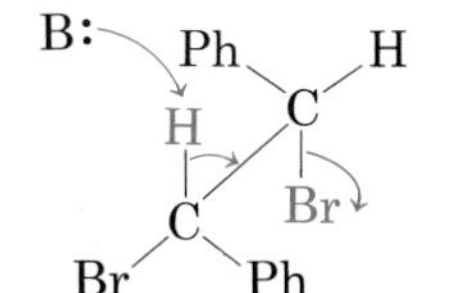

$\longrightarrow$

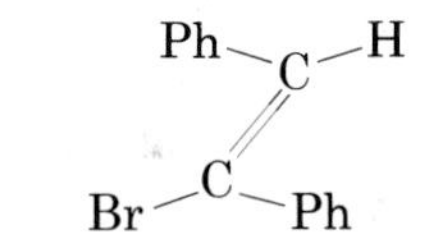

(1*S*,2*S*)-1,2-Dibromo-1,2-diphenylethane

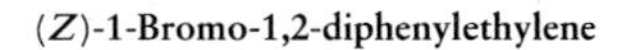

(*Z*)-1-Bromo-1,2-diphenylethylene

PROBLEM 7.18 Ignoring double-bond stereochemistry, what elimination products would you expect from these reactions? (Hint: Remember Zaitsev's rule; Section 4.9.)

(a) $CH_3CH_2CH(Br)CH(CH_3)CH_3$

(b) $CH_3CH(CH_3)CH_2{-}C(Cl)(CH_3){-}CH(CH_3)CH_3$

(c) Cyclohexyl−$CH(Br)CH_3$

PROBLEM 7.19 What stereochemistry would you expect for the alkene obtained by E2 elimination of (1*R*,2*R*)-1,2-dibromo-1,2-diphenylethane? Draw a Newman projection of the reacting conformation.

7.10 ELIMINATION REACTIONS OF ALKYL HALIDES: THE E1 REACTION

E1 reaction an elimination reaction that takes place in two steps through a unimolecular mechanism

Just as the S_N2 reaction is analogous to the E2 reaction, the S_N1 reaction also has a close analog: the **E1 reaction** (elimination, unimolecular). The mechanism of an E1 reaction is shown in Figure 7.10 for the elimination of HCl from 2-chloro-2-methylpropane.

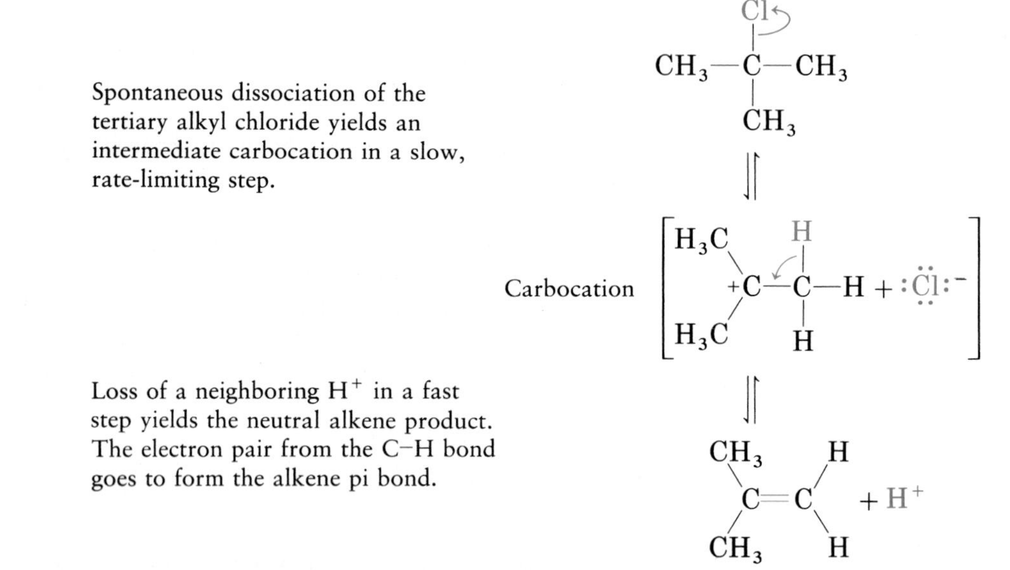

Spontaneous dissociation of the tertiary alkyl chloride yields an intermediate carbocation in a slow, rate-limiting step.
Carbocation
Loss of a neighboring H^+ in a fast step yields the neutral alkene product. The electron pair from the C–H bond goes to form the alkene pi bond.

FIGURE 7.10 Mechanism of the E1 reaction. Two steps are involved, and a carbocation intermediate is formed.

E1 reactions are unimolecular processes that occur by spontaneous dissociation of an alkyl halide and subsequent loss of a proton from the intermediate carbocation. The E1 mechanism normally takes place when an alkyl halide substrate is heated in a hydroxylic solvent such as water or methanol without any added base. In practice, we often find that E1 reactions occur in competition with S_N1 reactions; mixtures of substitution and elimination products are almost always obtained. For example, warming 2-chloro-2-methylpropane to 65°C in 80% aqueous ethanol yields a 64:36 mixture of 2-methyl-2-propanol (by S_N1 reaction) and 2-methylpropene (by E1 reaction).

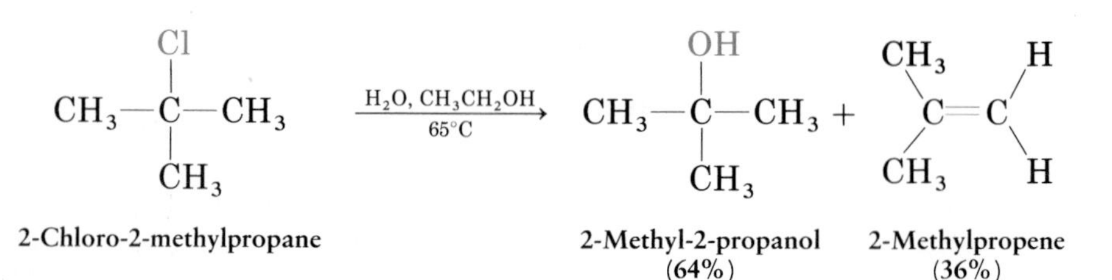

PROBLEM 7.20 What effect on the rate of an E1 reaction of 2-bromo-2-methylpropane would you expect if the concentration of the alkyl halide were tripled?

7.11 A SUMMARY OF REACTIVITY: S_N1, S_N2, E1, E2

Having seen four different modes of reaction between an alkyl halide and a nucleophile, you may well wonder how to predict what will take place in any given case. Will substitution or elimination occur? Will the reaction be unimolecular or bimolecular? There are no rigid answers to these questions, but we can make some broad generalizations:

1. *Primary alkyl halides* react by either S_N2 or E2 mechanisms because they are relatively unhindered and because their dissociation would give unstable primary carbocations. If a good nucleophile such as I^-, Br^-, RS^-, NH_3, or CN^- is used, only S_N2 substitution occurs. If a strong base such as hydroxide ion or an alkoxide ion (RO^-) is used, a small amount of competitive E2 elimination also occurs.
2. *Secondary alkyl halides* can react by any of the four possible mechanisms although S_N2 and E2 reactions are favored, just as they are for primary alkyl halides. If a good nucleophile ion is used, S_N2 substitution occurs. If a strong base like hydroxide ion is used, E2 elimination occurs.
3. *Tertiary alkyl halides* can react by any of three pathways: S_N1, E1, and E2. Strong bases like hydroxide ion always cause E2 elimination, but heating in hydroxylic solvents without any added base leads to a mixture of S_N1 and E1 products.

These generalizations are summarized in Table 7.2.

TABLE 7.2 Correlation of structure and reactivity for substitution and elimination reactions

Halide type	S_N1	S_N2	E1	E2
RCH_2X (primary)	Does not occur	Highly favored	Does not occur	Occurs when strong bases are used
R_2CHX (secondary)	Can occur with benzylic and allylic halides	Favored when good nucleophiles are used	Can occur with benzylic and allylic halides	Favored when strong bases are used
R_3CX (tertiary)	Favored in hydroxylic solvents	Does not occur	Occurs in competition with S_N1 reaction	Favored when bases are used

PRACTICE PROBLEM 7.10 Tell what kind of reaction this is:

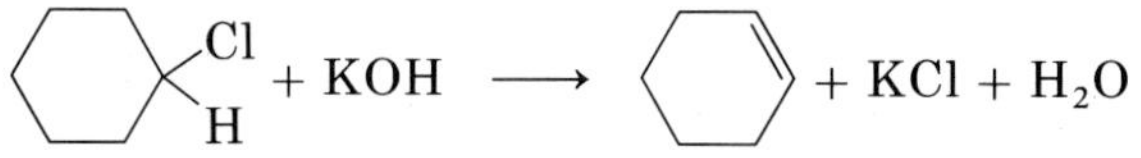

$$\text{(cyclohexyl–CH(Cl)(H))} + KOH \longrightarrow \text{cyclohexene} + KCl + H_2O$$

SOLUTION Because a secondary alkyl halide is undergoing loss of HCl on treatment with a strong base, this is an E2 reaction.

PROBLEM 7.21 Tell whether these reactions are S_N1, S_N2, E1, or E2:

(a) 1-bromobutane + NaN_3 $\longrightarrow$ 1-azidobutane

(b)
$$CH_3CH_2\underset{}{\overset{Cl}{\overset{|}{C}}}HCH_2CH_3 + KOH \longrightarrow CH_3CH_2CH{=}CHCH_3$$

(c)
$$\text{(1-chloro-1-methylcyclohexane)} + CH_3COOH \longrightarrow \text{(1-acetoxy-1-methylcyclohexane, OOCCH}_3\text{, CH}_3\text{)} + HCl$$

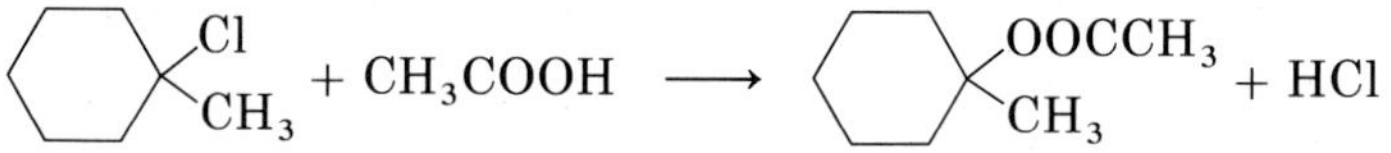

7.12 BIOLOGICAL SUBSTITUTION REACTIONS

Many biological processes occur by reaction pathways analogous to those carried out in the laboratory. Thus, a number of reactions occurring in living organisms take place by nucleophilic substitution reactions. Perhaps the most common of all biological substitution reactions is the *methylation*—the transfer of a methyl group from an electrophilic donor to a nucleophile.

$$\sim\!Y{-}CH_3 + :Nu^- \longrightarrow \sim\!Y:^- + CH_3{-}Nu$$

Methyl donor **Methylated nucleophile**

Although a laboratory chemist would probably use iodomethane for such a reaction, living organisms operate more subtly. The large and complex molecule *S*-adenosylmethionine is the biological methyl-group donor. Since the sulfur atom of *S*-adenosylmethionine has a positive charge (a *sulfonium ion*), it's an excellent leaving group for S_N2 displacements on the methyl carbon.

One example of the action of *S*-adenosylmethionine in biological methylations takes place in the adrenal medulla during the formation of adrenaline from norepinephrine (Figure 7.11).

After becoming used to simple alkyl halides like iodomethane, it's something of a shock to encounter molecules as complex as *S*-adenosylmethionine. From the chemical standpoint though, iodomethane and *S*-adenosylmethionine are doing exactly the same thing: transferring a methyl group in an S_N2 reaction. The same chemical principles apply to both.

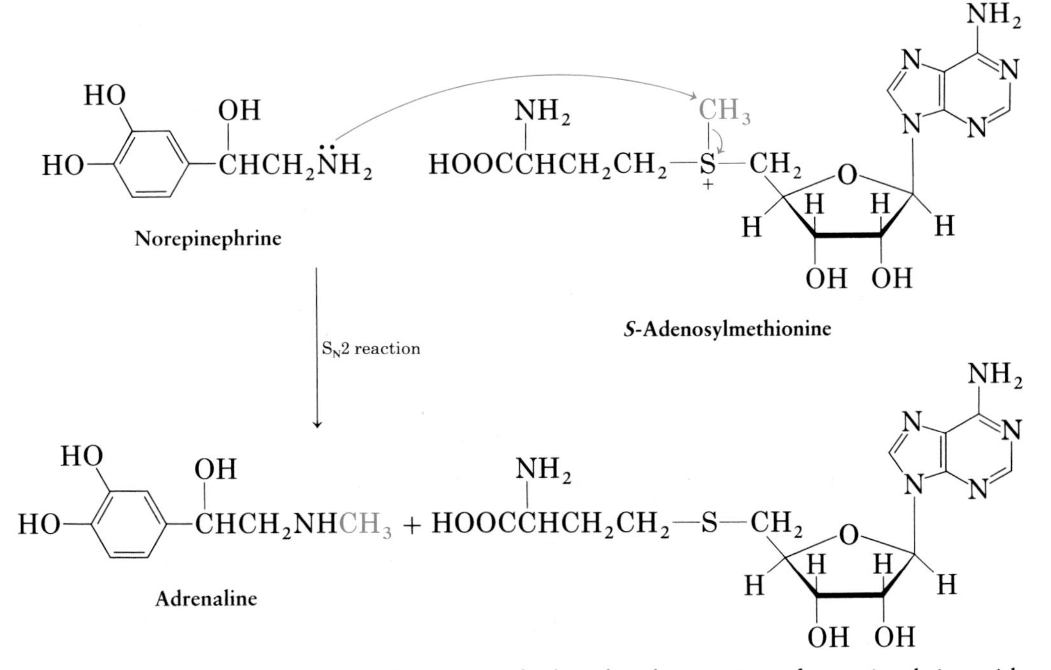

FIGURE 7.11 The biological formation of adrenaline by reaction of norepinephrine with *S*-adenosylmethionine

INTERLUDE

Alkyl Halides and the Ozone Hole

The aerosol can is a fixture of modern life, something we take for granted to spray our deodorants, paints, and insect repellents. In the early 1970s, though, it became apparent that the proliferation of aerosol sprays was leading to a serious environmental problem.

The volatile propellents used in aerosols at the time were various alkyl halides called *Freons*. The Freons are *chlorofluorocarbons,* simple alkanes in which all of the hydrogens have been replaced by either chlorine or fluorine. Fluorotrichloromethane (CCl_3F; Freon 11) and dichlorodifluoromethane (CCl_2F_2; Freon 12) are two of the most common Freons. The advantage of using Freons as aerosol propellents is that they're chemically inert and nonflammable. Thus, they don't react with the contents of the can, they leave no residue, they have no odor, and they're safe. They do, however, escape into the atmosphere where ultimately they find their way into the stratosphere.

The *ozone layer* is an atmospheric band extending from about 20 to 40 km above the earth's surface. Although ozone (O_3) is toxic in high concentrations, it is critically important in the upper atmosphere because it acts to shield the surface of the earth from intense solar ultraviolet radiation. If the ozone layer were depleted or destroyed, more ultraviolet radiation would reach the earth, causing an increased incidence of skin cancers and eye cataracts. Unfortunately, destruction of ozone is exactly what chlorofluorocarbons do. Beginning around 1976, a disturbing amount of ozone depletion, the so-called ozone hole, began showing up over the South Pole (Figure 7.12). Estimates of the extent of ozone destruction differ, but a recent report predicted a 5–9% depletion over the next 50 years.

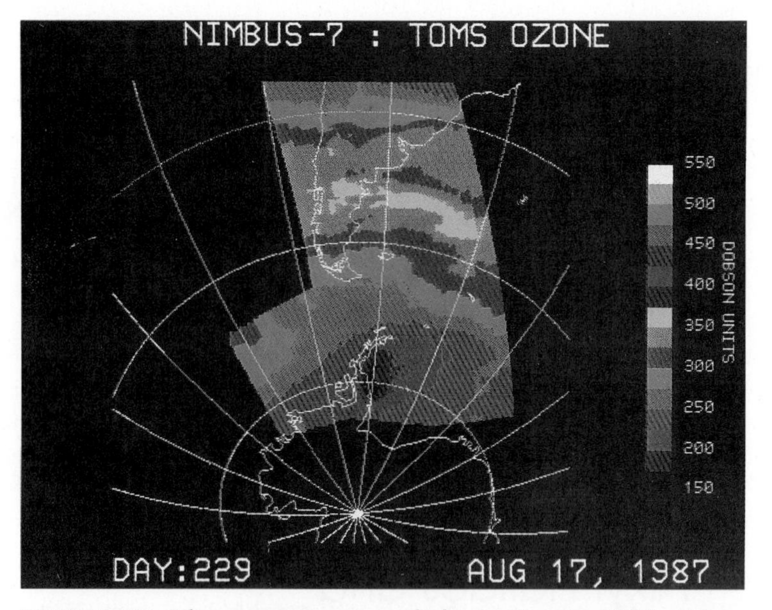

FIGURE 7.12 The antarctic ozone hole in August 1987. Ozone values in the hole indicated by the black portion in the center of the shaded area are up to 50% lower than normal values.

The mechanism of ozone destruction by chlorofluorocarbons involves radical reactions of the same sort we saw in the radical chlorination of methane (Section 7.2). Ultraviolet light ($h\nu$) striking a Freon molecule causes breakage of a carbon–chlorine bond, producing a chlorine radical. This radical then reacts with ozone to yield oxygen and ClO:

$$CCl_2F_2 \xrightarrow{h\nu} \cdot CClF_2 + \cdot Cl$$
$$\cdot Cl + O_3 \longrightarrow O_2 + ClO$$

Recognition of the problem led the U.S. government in 1980 to ban the use of Freons for aerosol propellents, though they are still widely used as refrigerants in automobile air conditioners. Worldwide action to reduce chlorofluorocarbon use finally began in September 1987, when an international agreement was reached by the European Community and 24 other nations.

SUMMARY AND KEY WORDS

Alkyl halides can be prepared by radical chlorination or bromination of alkanes, but product mixtures always result. Alkyl halides are best prepared from alcohols by treatment either with HX (for tertiary alcohols) or with $SOCl_2$ or PBr_3 (for primary and secondary alcohols).

Alkyl halides react with magnesium metal to form organomagnesium halides, or **Grignard reagents.** These organometallic compounds react with acids to yield the corresponding alkanes.

Treatment of an alkyl halide with a nucleophile/base results either in substitution or in elimination. Both kinds of reaction are of great importance and generality in organic chemistry. **Nucleophilic substitution reactions** occur by two mechanisms, S_N2 and S_N1. In the **S_N2 reaction,** the entering nucleophile attacks the alkyl halide from a direction 180° away from the **leaving group,** resulting in an inversion of configuration at the carbon atom. S_N2 reactions are strongly inhibited by increasing steric bulk of the reagents and are favored only for primary and secondary substrates. In the **S_N1 reaction,** the substrate spontaneously dissociates to a carbocation followed by rapid attack of nucleophile. In consequence, S_N1 reactions take place with racemization of configuration at the carbon atom and are favored only for tertiary substrates.

Elimination reactions also occur by two mechanisms, E2 and E1. In the **E2 reaction,** a base abstracts a proton at the same time that the leaving group departs. The E2 reaction takes place with **anti periplanar** geometry and occurs when a substrate is treated with a strong base. In the **E1 reaction,** the substrate spontaneously dissociates to form a carbocation that can subsequently lose a neighboring proton. The reaction occurs on tertiary substrates in neutral or acidic hydroxylic solvents.

SUMMARY OF REACTIONS

1. Synthesis of alkyl halides

(a) Radical chlorination of alkanes (Section 7.2)

$$-\overset{|}{\underset{|}{C}}-H + Cl_2 \xrightarrow{h\nu} -\overset{|}{\underset{|}{C}}-Cl + HCl$$

Reaction is very unselective

(b) Alkyl halides from alcohols (Section 7.3)

(1) Reaction of tertiary alcohol with HX

$$R_3C{-}OH + HX \longrightarrow R_3C{-}X + H_2O \qquad X = Cl, Br$$

(2) Reaction of primary and secondary alcohols with PBr_3 and $SOCl_2$

$$ROH + PBr_3 \longrightarrow RBr$$
$$ROH + SOCl_2 \longrightarrow RCl$$

2. Reactions of alkyl halides

(a) Formation and protonation of Grignard reagents (Section 7.4)

$$RX + Mg \longrightarrow RMgX$$
$$RMgX \longrightarrow RH$$

(b) S_N2 reaction: back-side attack of nucleophile on alkyl halide (Section 7.6 and 7.7)

$$Nu{:}^- + {-}C{-}X \longrightarrow Nu{-}C{-} + X{:}^-$$

Substrate must be primary or secondary

(c) S_N1 reaction: carbocation intermediate is involved (Section 7.8)

$$H{-}Nu + R_3C{-}X \longrightarrow [R_3C^+] \longrightarrow R_3C{-}Nu + HX$$

Substrate must be tertiary or (occasionally) secondary

(d) E2 reaction (Section 7.9)

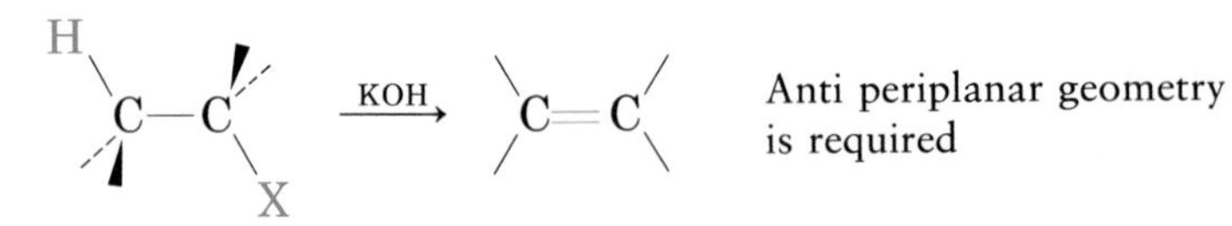

(e) E1 reaction (Section 7.10)

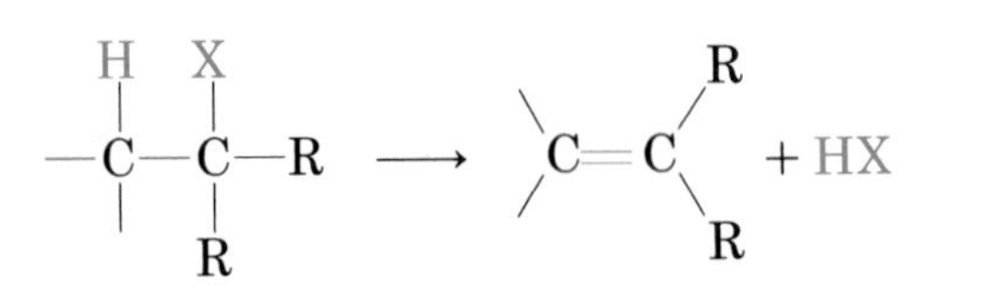

Best for tertiary substrates in neutral or acidic solvents. Carbocation intermediate is involved.

ADDITIONAL PROBLEMS

7.22 Name these alkyl halides according to IUPAC rules.

(a) $(CH_3)_2CHCHBrCHBrCH_2CH(CH_3)_2$ (b) $CH_3CH{=}CHCH_2CHICH_3$
(c) $(CH_3)_2CBrCH_2CH_2CHClCH(CH_3)_2$ (d) $CH_3CH_2CH(CH_2Br)CH_2CH_2CH_3$

7.23 Draw structures corresponding to these IUPAC names.

(a) 2,3-Dichloro-4-methylhexane (b) 4-Bromo-4-ethyl-2-methylhexane
(c) 3-Iodo-2,2,4,4-tetramethylpentane

7.24 Although radical chlorination of alkanes is usually unselective, chlorination of propene, $CH_3CH{=}CH_2$, occurs almost exclusively on the methyl group rather than on the double bond. Draw resonance structures of the allyl radical, $CH_2{=}CHCH_2\cdot$, to account for this result.

7.25 Draw resonance structures of the benzyl radical $C_6H_5CH_2\cdot$, to account for the fact that radical chlorination of toluene occurs exclusively on the methyl group rather than on the ring.

7.26 How would you prepare these compounds, starting with cyclopentene and any other reagents needed?

(a) Chlorocyclopentane (b) Cyclopentanol
(c) Cyclopentylmagnesium chloride (d) Cyclopentane

7.27 Predict the product(s) of these reactions.

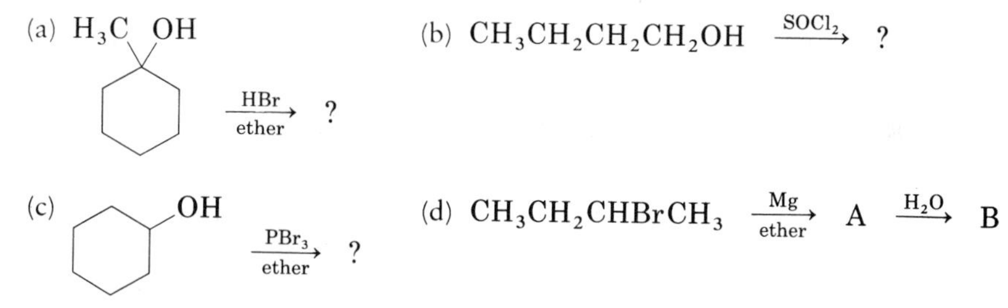

7.28 Which alkyl halide in each pair will react faster in an S_N2 reaction with hydroxide ion?

(a) Bromobenzene or benzyl bromide, $C_6H_5CH_2Br$
(b) CH_3Cl or $(CH_3)_3CCl$
(c) $CH_3CH{=}CHBr$ or $H_2C{=}CHCH_2Br$

7.29 How might you prepare these molecules using a nucleophilic substitution reaction at some step?

(a) CH_3CH_2Br (b) $CH_3CH_2CH_2CH_2CN$

(c) $CH_3O\overset{CH_3}{\underset{CH_3}{C}}CH_3$ (d) $CH_2CH_2CH_2N{=}\overset{+}{N}{=}N^-$

(e) CH_3CH_2SH (f) $CH_3\overset{O}{\overset{\|}{C}}OCH_3$

7.30 What products do you expect from reaction of 1-bromopropane with these reagents?

(a) NaI (b) NaCN (c) NaOH (d) Mg, then H_2O (e) $NaOCH_3$

7.31 Order these compounds with respect to both S_N1 and S_N2 reactivity.

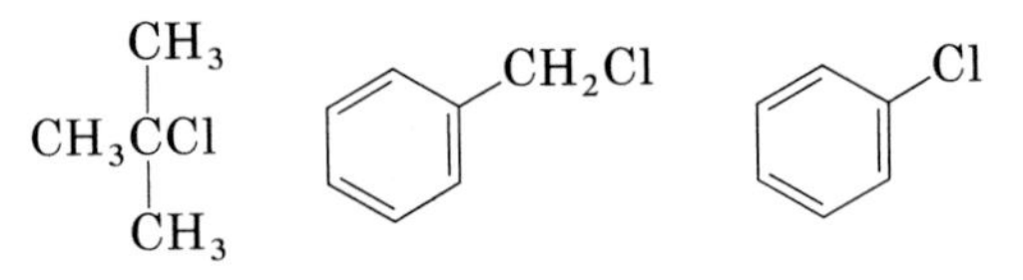

7.32 Order each set of compounds with respect to S_N2 reactivity.

(a) $(CH_3)_3CCl$, $CH_3CH_2CH_2Cl$, $CH_3CH_2CHClCH_3$
(b) $(CH_3)_2CHCHBrCH_3$, $(CH_3)_2CHCH_2Br$, CH_3Br

7.33 Predict the product and give the stereochemistry of reactions of these nucleophiles with (*R*)-2-bromooctane.

(a) $^-$:CN (b) $CH_3CO\ddot{O}:^-$ (c) $:\ddot{Br}:^-$

7.34 Ethers can be prepared by S_N2 reaction of alkoxide ions with alkyl halides: $R{-}O^- + R'{-}Br \rightarrow R{-}O{-}R' + Br^-$. Suppose you want to prepare cyclohexyl methyl ether. Which route would be better, reaction of methoxide ion, CH_3O^-, with bromocyclohexane or reaction of cyclohexoxide with bromomethane? Explain.

7.35 How could you prepare diethyl ether, $CH_3CH_2OCH_2CH_3$, starting from ethyl alcohol and any inorganic reagents needed? [See Problem 7.34.]

7.36 How could you prepare cyclohexane starting from 3-bromocyclohexene?

7.37 The S_N2 reaction can occur *intramolecularly* (within the same molecule). What product would you expect from treatment of 4-bromo-1-butanol with base?

$$BrCH_2CH_2CH_2CH_2OH \xrightarrow{\text{base}} [BrCH_2CH_2CH_2CH_2O^- Na^+] \longrightarrow ?$$

7.38 In light of your answer to Problem 7.37, propose a synthesis of 1,4-dioxane starting from 1,2-dibromoethane?

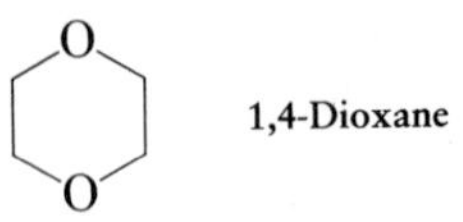

7.39 Propose a structure for an alkyl halide that can give a mixture of three alkenes on E2 reaction.

7.40 Heating either *tert*-butyl chloride or *tert*-butyl bromide with ethanol yields the same reaction mixture of about 80% *tert*-butyl ethyl ether [$(CH_3)_3COCH_2CH_3$] and 20% 2-methylpropene. Explain.

7.41 What effect would you expect these changes to have on the rate of the reaction of 1-iodo-2-methylbutane with cyanide ion?

(a) CN^- concentration is halved and 1-iodo-2-methylbutane concentration is doubled.
(b) Both CN^- and 1-iodo-2-methylbutane concentrations are tripled.

7.42 What effect would you expect on the rate of reaction of ethyl alcohol with 2-iodo-2-methylbutane if the concentration of 1-iodo-2-methylbutane were tripled?

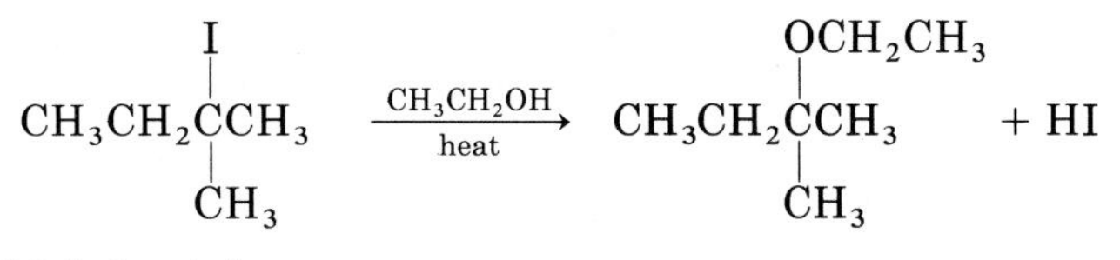

2-Iodo-2-methylbutane

7.43 Identify these reactions as S_N1, S_N2, E1, or E2.

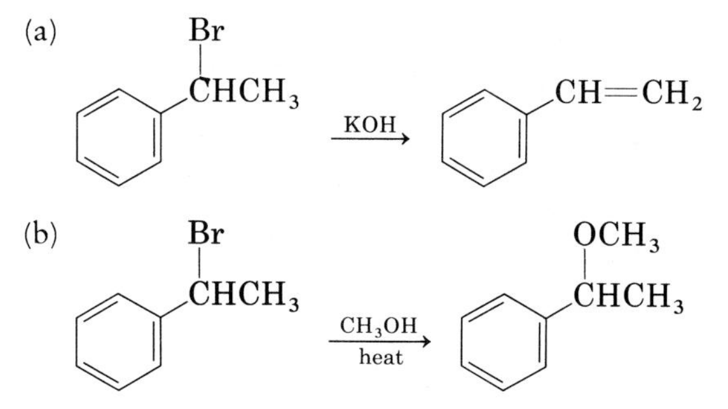

7.44 How can you explain the fact that *trans*-1-bromo-2-methylcyclohexane yields the non-Zaitsev elimination product 3-methylcyclohexene on treatment with base?

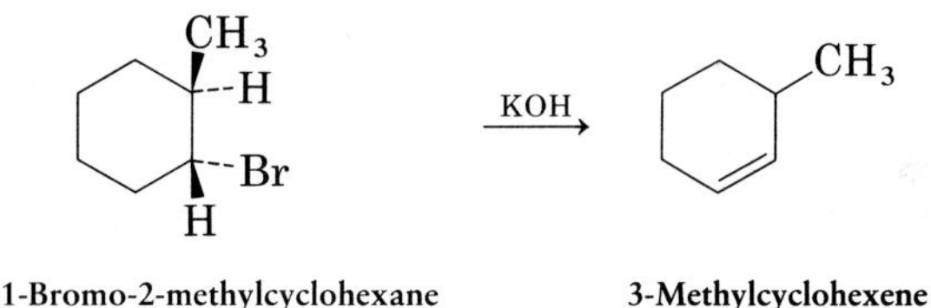

***trans*-1-Bromo-2-methylcyclohexane** **3-Methylcyclohexene**

7.45 Propose a structure for an alkyl halide that gives (Z)-2,3-diphenyl-2-butene on E2 reaction.

7.46 Describe the effects of the substrate structure on both S_N2 and S_N1 reactions.

7.47 Predict the major alkene product from these eliminations.

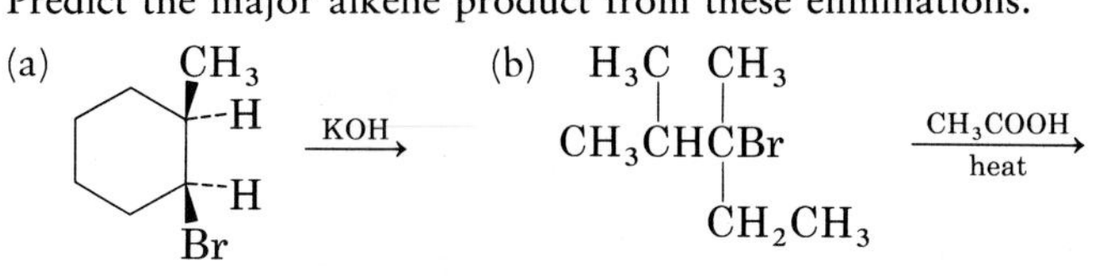

7.48 (2*R*,3*S*)-2-bromo-3-phenylbutane undergoes E2 reaction on treatment with sodium ethoxide to yield (*Z*)-2-phenyl-2-butene.

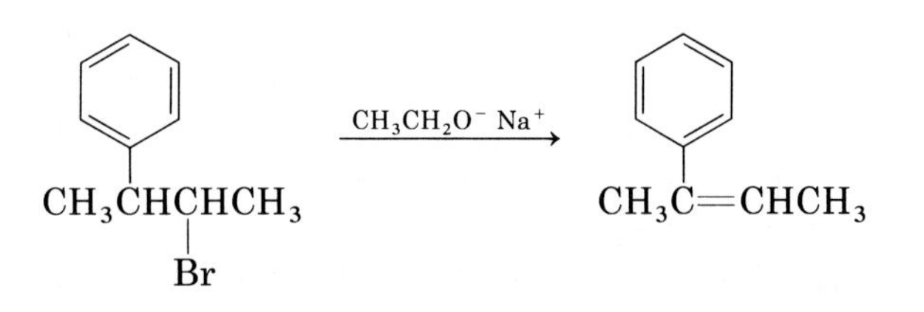

Formulate the reaction, showing the proper stereochemistry. Explain the observed result using Newman projections.

7.49 In light of your answer to Problem 7.48, which alkene, *E* or *Z*, would you expect from the E2 reaction of (2*R*,3*R*)-2-bromo-3-phenylbutane?

7.50 Optically active 2-butanol slowly becomes racemic upon standing in dilute sulfuric acid. Propose a mechanism to account for this racemization.

7.51 Draw the eight diastereomers of 1,2,3,4,5,6-hexachlorocyclohexane. One isomer loses HCl in an E2 reaction nearly 1000 times more slowly than the others. Which isomer reacts so slowly, and why?

7.52 Compound **A** is optically inactive and has the formula $C_{16}H_{16}Br_2$. On treatment with strong base, **A** gives hydrocarbon **B**, $C_{16}H_{14}$, which absorbs 2 equiv of hydrogen when reduced over a palladium catalyst and which reacts with ozone to give two carbonyl-containing products. One product, **C**, is an aldehyde with the formula C_7H_6O. The other product is glyoxal, OHCCHO. Formulate the reactions involved and suggest structures for **A**, **B**, and **C**. What is the stereochemistry of **A**?

7.53 Consider the following cleavage reaction of a methyl ester:

$$CH_3CH_2CH_2\overset{\overset{\displaystyle O}{\|}}{C}{-}O{-}CH_3 \xrightarrow{LiI} CH_3CH_2CH_2\overset{\overset{\displaystyle O}{\|}}{C}{-}O^-\ Li^+ + CH_3I$$

Use the following evidence to propose a mechanism for the reaction: (1) The rate of this reaction depends both on ester concentration and on iodide-ion concentration. (2) The corresponding ethyl ester cleaves approximately 10 times more slowly than the methyl ester.

CHAPTER 8

Alcohols, Ethers, and Phenols

alcohol
a compound with an $-OH$ group bonded to a saturated, sp^3-hybridized carbon atom

Alcohols are compounds that have hydroxyl groups bonded to saturated, sp^3-hybridized carbon atoms; **phenols** have hydroxyl groups bonded to an aromatic ring; and **ethers** have an oxygen atom bonded to two organic groups. All three classes of compounds can be thought of as organic derivatives of water in which one or both of the water hydrogens are replaced by an organic substituent ($H-O-H$ becomes $R-O-H$, $Ar-O-H$, or $R-O-R'$).

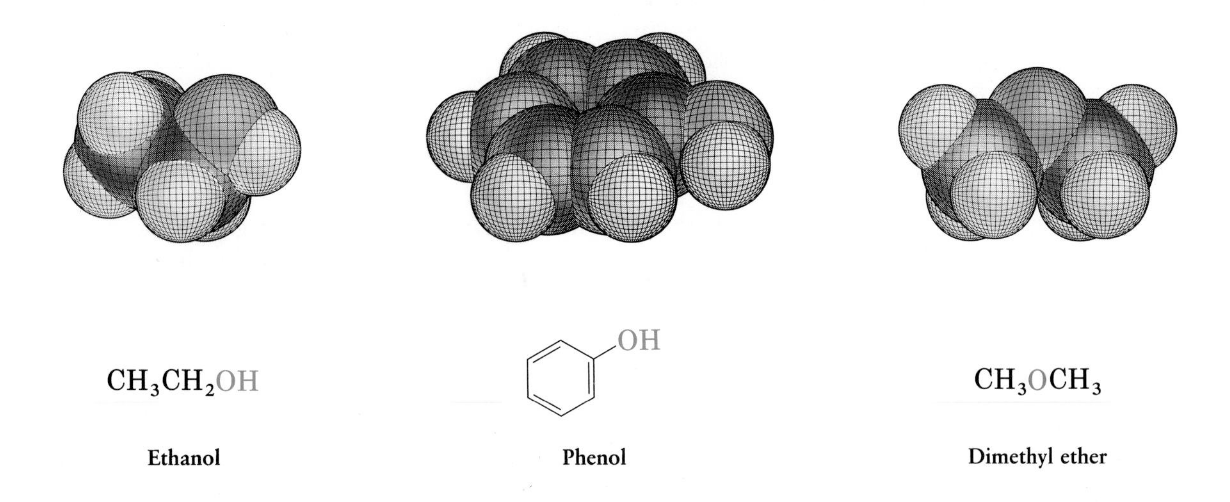

Ethanol

Phenol

Dimethyl ether

phenol
a compound with an $-OH$ group bonded to an aromatic ring

ether
a compound with two organic groups bonded to the same oxygen atom

Alcohols, phenols, and ethers occur widely in nature and have many industrial and pharmaceutical applications. Ethanol, for instance, is a fuel additive, an industrial solvent, and a beverage; menthol, an alcohol isolated from peppermint oil, is a flavoring agent; BHT (butylated hydroxytoluene) is a food additive that prolongs shelf life and protects against oxidation; and diethyl ether, the familiar "ether" of medical use, was once popular as an anesthetic agent but is now mainly used as an industrial solvent.

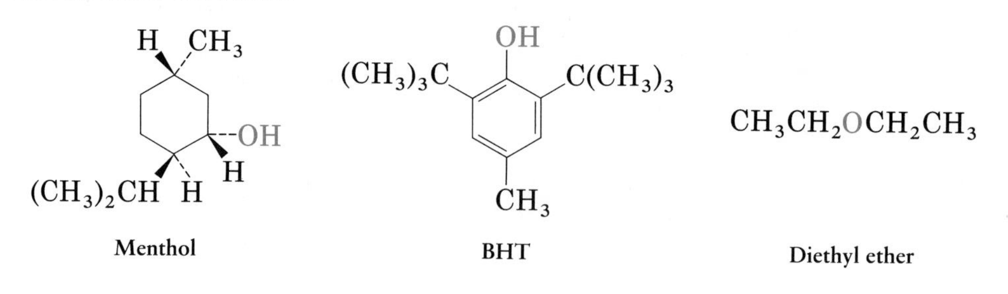

8.1 NAMING ALCOHOLS, PHENOLS, AND ETHERS

Alcohols

Alcohols are classified as either primary (1°), secondary (2°), or tertiary (3°), depending on the number of carbon substituents bonded to the hydroxyl-bearing carbon.

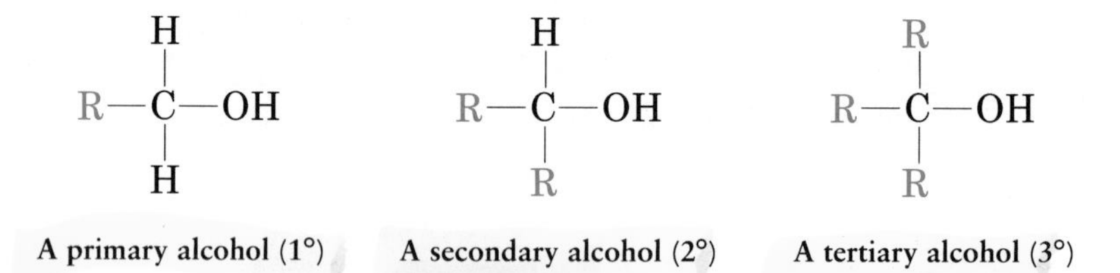

Simple alcohols are named in the IUPAC system as derivatives of the parent alkane:

1. Select the longest carbon chain that *contains the hydroxyl group* and derive the parent name by replacing the *-e* ending of the corresponding alkane with *-ol*.
2. Number the carbons of the chain beginning at the end nearer the hydroxyl group.
3. Number all substituents according to their position on the chain and write the name, listing the substituents in alphabetical order.

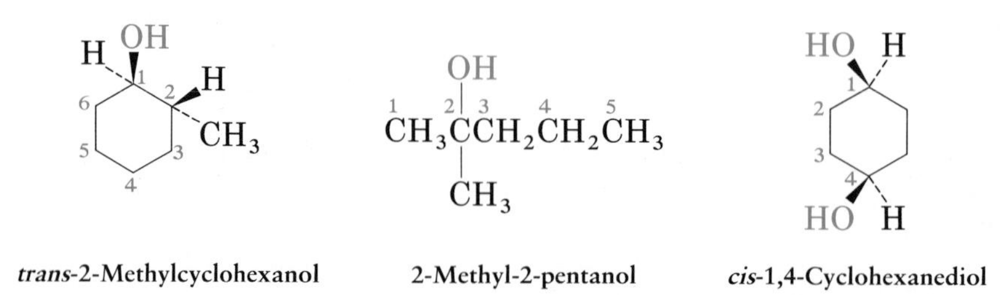

Certain well-known alcohols also have common names. For example:

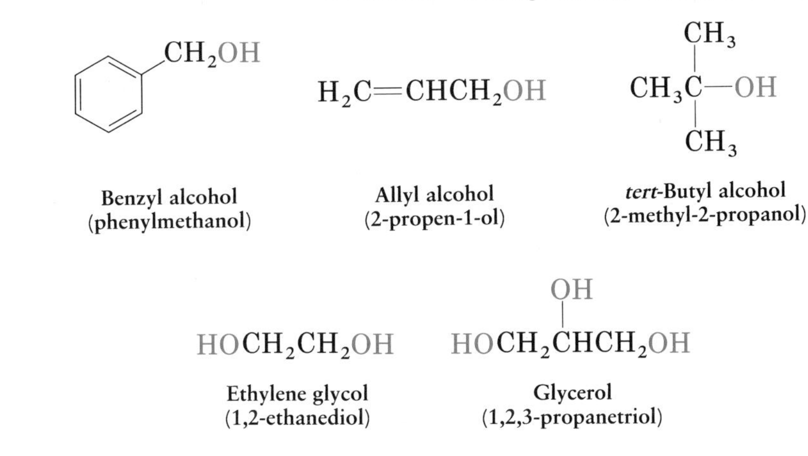

Benzyl alcohol (phenylmethanol) Allyl alcohol (2-propen-1-ol) *tert*-Butyl alcohol (2-methyl-2-propanol)

Ethylene glycol (1,2-ethanediol) Glycerol (1,2,3-propanetriol)

Phenols

The word *phenol* is used both as the name of a specific substance (hydroxybenzene) and as the family name for all hydroxy-substituted aromatic compounds. Phenols are named as substituted aromatic compounds according to the rules discussed in Section 5.1. Note, however, that -phenol is used as the parent name rather than -benzene. For example:

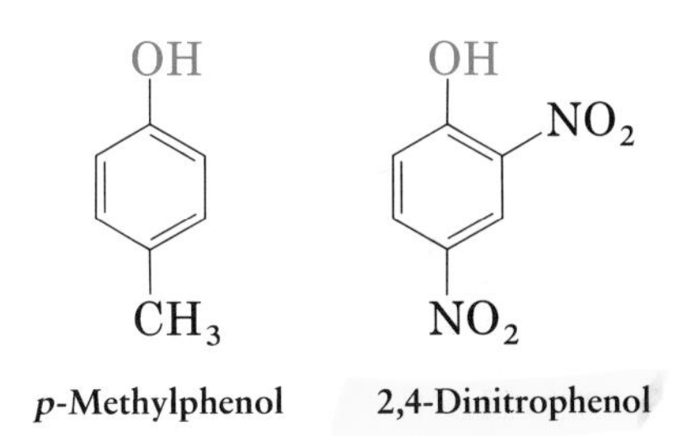

p-Methylphenol 2,4-Dinitrophenol

Ethers

Two systems of ether nomenclature are allowed by IUPAC rules. Simple ethers that contain no other functional groups are named by identifying the two organic residues and adding the word *ether*. For example:

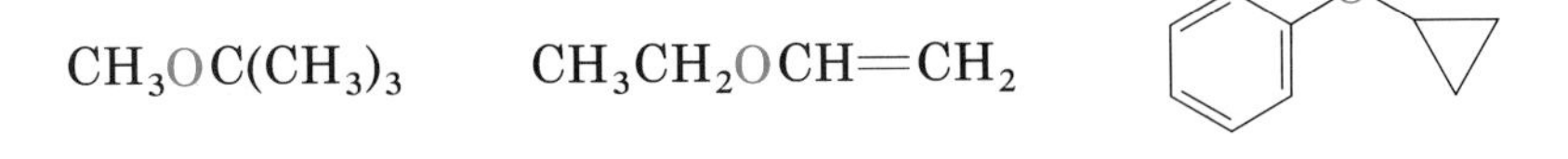

tert-Butyl methyl ether Ethyl vinyl ether Cyclopropyl phenyl ether

If more than one ether linkage is present in the molecule or if other functional groups are present, the ether group is considered as an *alkoxy* substituent on the parent compound. For example:

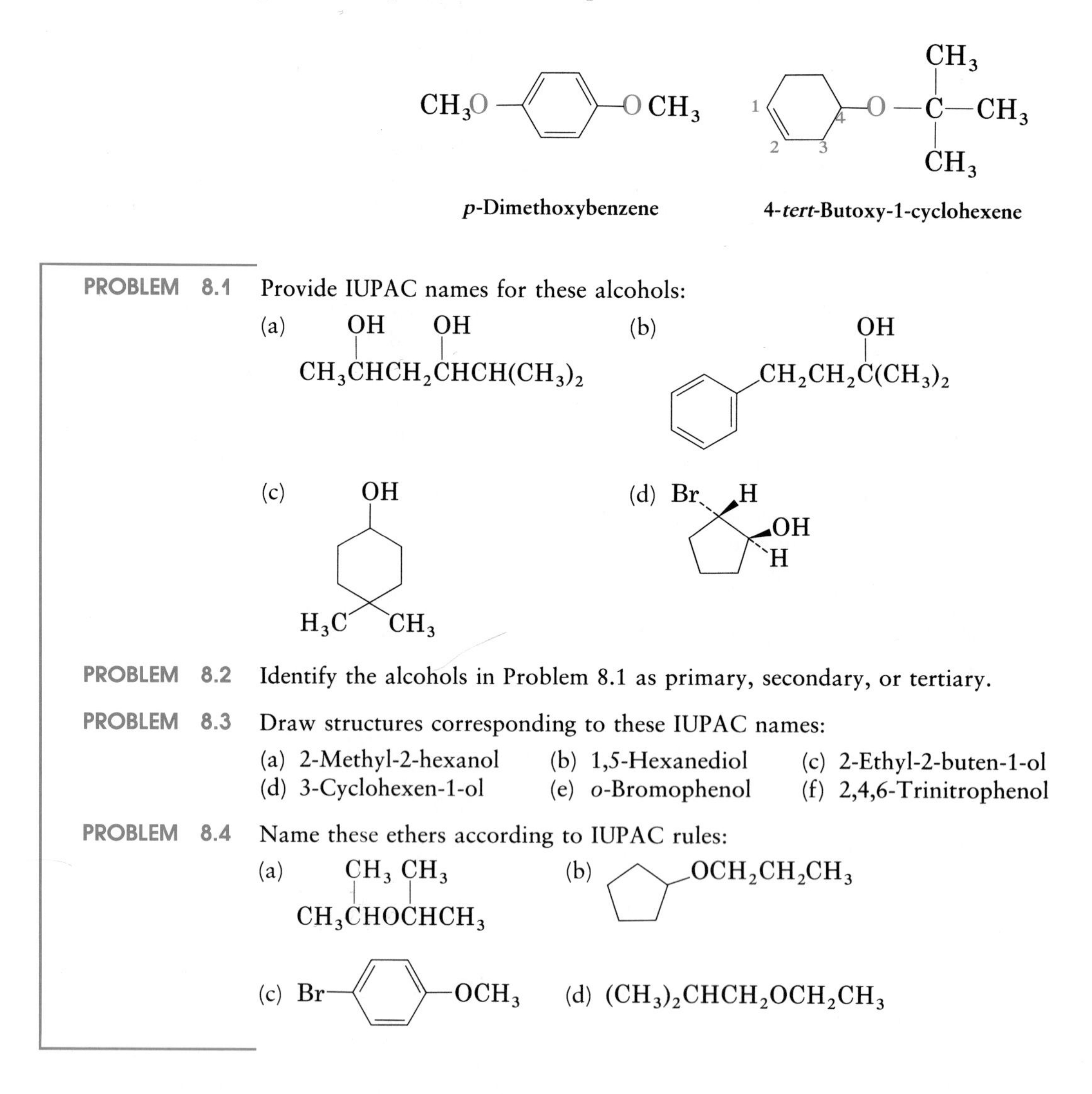

p-Dimethoxybenzene

4-*tert*-Butoxy-1-cyclohexene

PROBLEM 8.1 Provide IUPAC names for these alcohols:

(a) $CH_3CH(OH)CH_2CH(OH)CH(CH_3)_2$

(b) $C_6H_5CH_2CH_2C(OH)(CH_3)_2$

(c) 4,4-dimethylcyclohexanol structure (OH; H_3C, CH_3)

(d) Br, H, OH, H on cyclopentane

PROBLEM 8.2 Identify the alcohols in Problem 8.1 as primary, secondary, or tertiary.

PROBLEM 8.3 Draw structures corresponding to these IUPAC names:

(a) 2-Methyl-2-hexanol (b) 1,5-Hexanediol (c) 2-Ethyl-2-buten-1-ol
(d) 3-Cyclohexen-1-ol (e) *o*-Bromophenol (f) 2,4,6-Trinitrophenol

PROBLEM 8.4 Name these ethers according to IUPAC rules:

(a) $CH_3CH(CH_3)OCH(CH_3)CH_3$

(b) cyclopentyl–$OCH_2CH_2CH_3$

(c) Br–C_6H_4–OCH_3

(d) $(CH_3)_2CHCH_2OCH_2CH_3$

8.2 PROPERTIES OF ALCOHOLS, PHENOLS, AND ETHERS: HYDROGEN BONDING

As mentioned earlier, alcohols, phenols, and ethers can be thought of as organic derivatives of water in which one or both of the hydrogens have been replaced by organic residues. Thus, all three classes of compounds have nearly the same geometry as water. The R–O–H or R–O–R′ bonds have an approximately tetrahedral bond angle (112° in dimethyl ether, for example), and the oxygen atom is sp^3 hybridized.

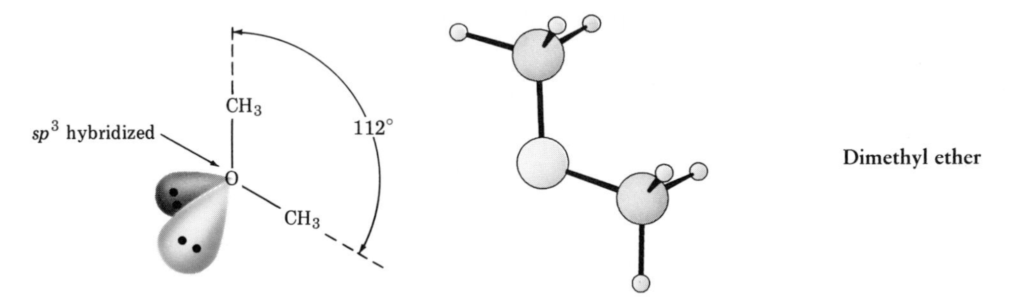

Alcohols and phenols are quite different from the hydrocarbons and alkyl halides we've studied thus far. As shown in Table 8.1, alcohols have higher boiling points than alkanes or haloalkanes of similar molecular weight. For example, the molecular weights of 1-propanol (mol wt = 60), butane (mol wt = 58), and chloroethane (mol wt = 65) are similar, but 1-propanol boils at 97°C, compared with −0.5°C for the alkane and 12.5°C for the chloroalkane. Similarly, phenols have higher boiling points than aromatic hydrocarbons. Phenol itself, for example, boils at 182°C whereas toluene boils at 110.6°C.

TABLE 8.1 Boiling points of some alkanes, chloroalkanes, and alcohols (°C)

Alkyl group, R	*Alkane,* R—H	*Chloroalkane,* R—Cl	*Alcohol,* R—OH
CH_3—	−162	−24	64.5
CH_3CH_2—	−88.5	12.5	78.3
$CH_3CH_2CH_2$—	−42	46.6	97
$(CH_3)_2CH$—	−42	36.5	82.5
$CH_3CH_2CH_2CH_2$—	−0.5	83.5	117
$(CH_3)_3C$—	−12	51	83

The reason for their unusually high boiling points is that alcohols and phenols, like water, are highly associated in solution because of the formation of hydrogen bonds. The positively polarized hydroxyl hydrogen atom of one molecule forms a weak hydrogen bond to the negatively polarized oxygen atom of another molecule (Figure 8.1). Although hydrogen bonds have a strength of only about 5 kcal/mol (versus 100 kcal/mol for a typical O–H bond), the presence of a great

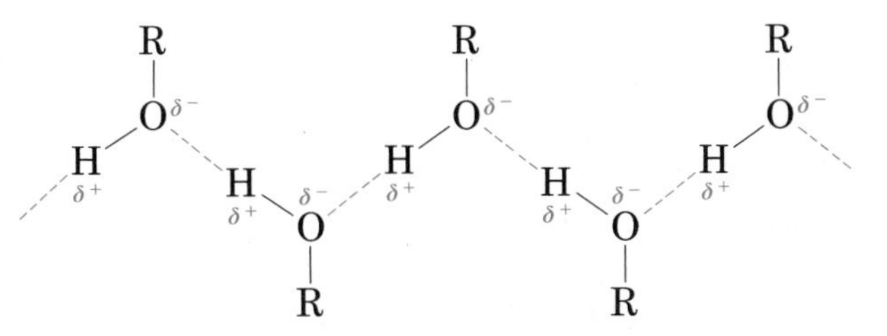

FIGURE 8.1 Hydrogen bonding in alcohols and phenols

many hydrogen bonds in solution means that extra energy is required to break them during the boiling process. Ethers, because they lack hydroxyl groups, can't form hydrogen bonds and therefore have relatively low boiling points.

8.3 ACIDS AND BASES: A REVIEW

We'll see in the next section that acidity is one of the most important characteristics of alcohols and phenols, but before doing so, let's review some fundamental ideas about acids and bases. There are two main ways of defining acidity: the Brønsted–Lowry definition, and the Lewis definition.

Brønsted–Lowry Acids and Bases

Brønsted–Lowry acid a substance that donates a hydrogen ion, H^+

Brønsted–Lowry base a substance that accepts a hydrogen ion, H^+

conjugate base the product that results when an acid loses H^+

conjugate acid the product that results when a base accepts H^+

According to the **Brønsted–Lowry definition,** an acid is a substance that donates a proton (hydrogen ion, H^+), and a base is a substance that accepts a proton. When hydrogen chloride dissolves in water, for example, HCl donates a proton and water accepts the proton. The products of the reaction are H_3O^+ and Cl^-. Chloride ion (Cl^-), the product that results when the acid donates a proton, is called the **conjugate base** of the acid. H_3O^+, the product that results when the base accepts a proton, is called the **conjugate acid** of the base.

For example:

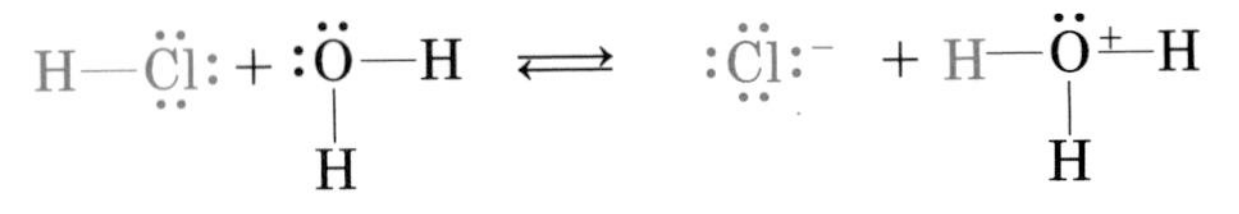

Acid — Base — Conjugate base — Conjugate acid

In a general sense:

$$H{-}A + :B \rightleftharpoons A:^- + H{-}B^+$$

An acid — A base — Conjugate base — Conjugate acid

acidity constant, K_a a value that expresses the strength of an acid in water solution

Acids differ in their proton-donating ability. Stronger acids such as HCl react almost completely with water whereas weaker acids such as acetic acid (CH_3COOH) react only slightly. The exact strength of a given acid in water solution is expressed by its **acidity constant, K_a.**[1]

For the reaction

$$HA + H_2O \rightleftharpoons H_3O^+ + A^-$$

$$K_{eq} = \frac{[H_3O^+][A^-]}{[HA][H_2O]} \quad \text{and} \quad K_a = \frac{[H_3O^+][A^-]}{[HA]}$$

[1] Remember that brackets [ ] refer to the molar concentration of the species indicated. Note also that the concentration of water $[H_2O]$ is left out of the expression for K_a, since it remains effectively constant.

Stronger acids have their equilibria toward the right and thus have larger acidity constants whereas weaker acids have their equilibria toward the left and have smaller acidity constants. We usually express acid strengths as pK_a values, where the pK_a is equal to the negative logarithm of the acidity constant:

$$pK_a = -\log K_a$$

A stronger acid (larger acidity constant K_a) has a *lower* pK_a; a weaker acid (smaller K_a) has a *higher* pK_a. Table 8.2 lists the pK_a of some common acids in order of their strength.

TABLE 8.2 Relative strength of some common acids

	Acid	*Name*	pK_a	*Conjugate base*	*Name*	
Weaker acid	CH_3CH_2OH	Ethanol	16.00	$CH_3CH_2O^-$	Ethoxide ion	Stronger base
↓	H_2O	Water	15.74	HO^-	Hydroxide ion	↑
	HCN	Hydrocyanic acid	9.2	CN^-	Cyanide ion	
	CH_3COOH	Acetic acid	4.72	CH_3COO^-	Acetate ion	
	HF	Hydrofluoric acid	3.2	F^-	Fluoride ion	
	HNO_3	Nitric acid	−1.3	NO_3^-	Nitrate ion	
Stronger acid	HCl	Hydrochloric acid	−7.0	Cl^-	Chloride ion	Weaker base

It's important to realize that there is an inverse relationship between the acid strength of an acid and the base strength of its conjugate base. Thus, the conjugate base of a stronger acid must be a *weaker* base since it has little affinity for a proton, and the conjugate base of a weaker acid must be a *stronger* base since it has a greater affinity for a proton. For example, chloride ion, the conjugate base of the stronger acid HCl, is a weak base because it has little affinity for a proton. Hydroxide ion, however, the conjugate base of the weak acid H_2O, is a stronger base with a greater affinity for a proton.

In general, an acid will donate a proton to the conjugate base of any acid with a higher pK_a. Conversely, the conjugate base of an acid will abstract a proton from any acid with a lower pK_a. For example, the data in Table 8.2 indicate that hydroxide ion will react with acetic acid, CH_3COOH, to yield acetate ion, CH_3COO^-, and water. Since water (pK_a = 15.74) is a weaker acid than acetic acid (pK_a = 4.72), hydroxide ion has a greater affinity for a proton than acetate ion has.

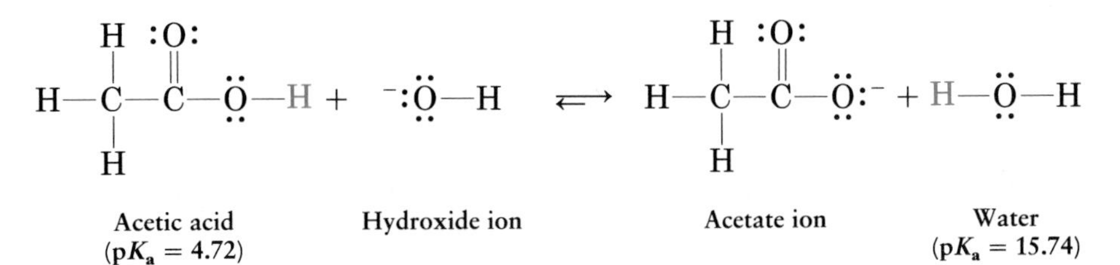

PRACTICE PROBLEM 8.1 Water has $pK_a = 15.74$, and acetylene has $pK_a = 25$. Which of the two is more acidic? Would you expect hydroxide ion to react with acetylene?

$$H{-}C{\equiv}C{-}H + H{-}O^- \longrightarrow H{-}C{\equiv}C{:}^- + H{-}O{-}H \quad ??$$

SOLUTION In comparing two acids, the one with the lower pK_a is stronger. Thus, water is a stronger acid than acetylene. Since water gives up a proton more easily than acetylene, the $H{-}O^-$ ion has less affinity for a proton than the $H{-}C{\equiv}C{:}^-$ ion. In other words, the anion of acetylene is a stronger base than hydroxide ion, and the reaction will not proceed as written.

PROBLEM 8.5 Formic acid, HCOOH, has $pK_a = 3.7$, and picric acid, $C_6H_3N_3O_7$, has $pK_a = 0.3$. Which is the stronger acid?

PROBLEM 8.6 Amide ion, H_2N^-, is a much stronger base than hydroxide ion, HO^-. Which would you expect to be a stronger acid, $H_2N{-}H$ (ammonia) or $HO{-}H$ (water)? Explain.

PROBLEM 8.7 Is either of these reactions likely to take place, according to the pK_a data in Table 8.2?

(a) $H{-}CN + CH_3COO^-Na^+ \longrightarrow Na^+\ {}^-CN + CH_3COO{-}H$

(b) $CH_3CH_2O{-}H + Na^+\ {}^-CN \longrightarrow CH_3CH_2O^-Na^+ + H{-}CN$

Lewis Acids and Bases

Lewis acid a substance that accepts an electron pair

Lewis base a substance that donates an electron pair

The Lewis definition of acids and bases differs from the Brønsted–Lowry definition in that it isn't limited to proton donors and acceptors. A **Lewis acid** is any substance that accepts an electron pair, and a **Lewis base** is any substance that donates an electron pair. Lewis acids include not only proton donors but also many other species. Thus, a proton (H^+) is a Lewis acid because it needs a pair of electrons to fill its vacant 1*s* orbital. Compounds such as BF_3 and $AlCl_3$ are Lewis acids because they too can accept electron pairs from Lewis bases to fill vacant valence orbitals.

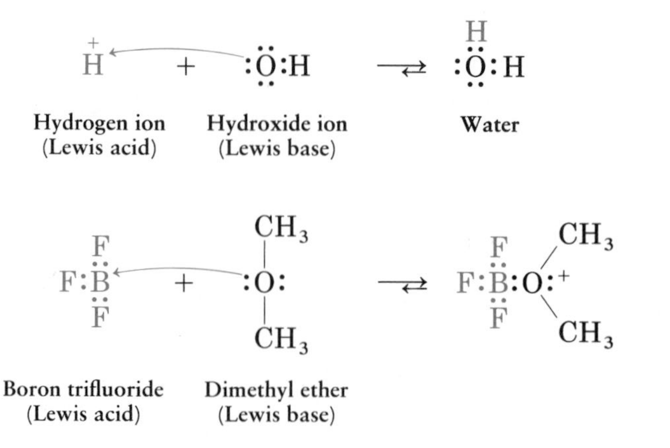

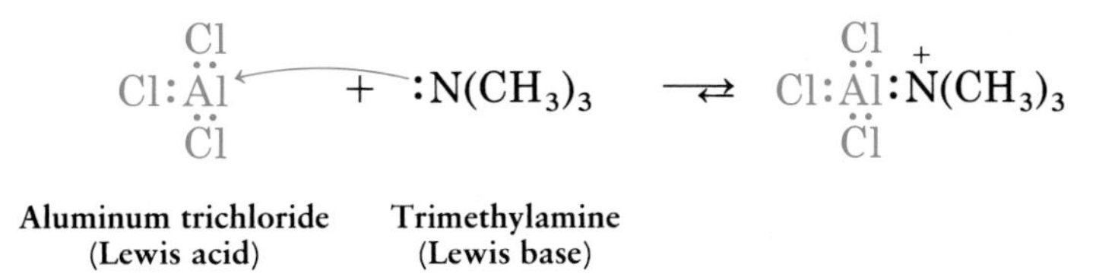

A Lewis base has a lone pair of electrons that it can donate to a Lewis acid for use in forming a bond. Thus, H_2O, with its two lone pairs of electrons on oxygen, acts as a Lewis base by donating an electron pair to a proton in forming the hydronium ion, H_3O^+:

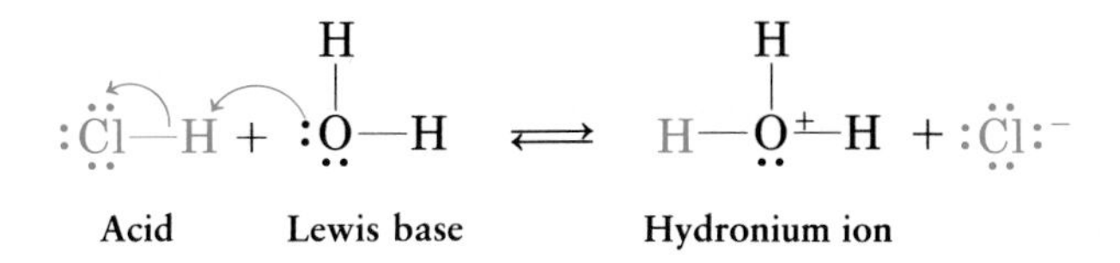

In a more general sense, most oxygen- and nitrogen-containing organic compounds such as dimethyl ether and ethyl alcohol are Lewis bases because each of their oxygen atoms has two lone pairs. Trimethylamine is a Lewis base because its nitrogen atom has a lone pair.

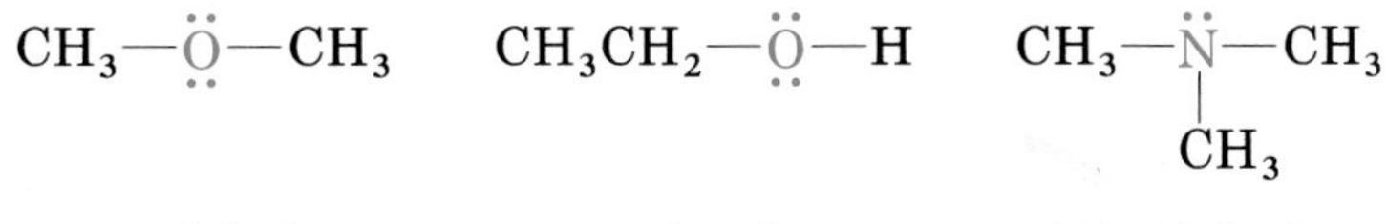

PRACTICE PROBLEM 8.2 Show how acetone can act as a Lewis base.

$$CH_3-\overset{\overset{\Large O}{\|}}{C}-CH_3 \qquad \textbf{Acetone}$$

SOLUTION The oxygen atom of acetone has two lone pairs of electrons that it can donate to a Lewis acid like H^+.

$$CH_3-\overset{\overset{:\ddot{O}:}{\|}}{C}-CH_3 + H-A \rightleftharpoons CH_3-\overset{\overset{:\overset{+}{O}-H}{\|}}{C}-CH_3 + A^-$$

PROBLEM 8.8 Which of the following are Lewis acids and which are Lewis bases?

(a) $CH_3CH_2-\ddot{\underset{\cdot\cdot}{O}}-H$ (b) $CH_3-\ddot{N}H-CH_3$ (c) $MgBr_2$

(d) $CH_3-\underset{\underset{\Large CH_3}{|}}{B}-CH_3$ (e) $H-\underset{\underset{\Large H}{|}}{\overset{+}{C}}-H$ (f) $CH_3-\underset{\underset{\Large CH_3}{|}}{\ddot{P}}-CH_3$

8.4 PROPERTIES OF ALCOHOLS AND PHENOLS: ACIDITY

Like water, alcohols and phenols are weakly acidic. In dilute aqueous solution, alcohols and phenols dissociate to a slight extent by donating a proton to water.

$$\mathrm{R\ddot{\underset{\cdot\cdot}{O}}{-}H + H_2\ddot{O}\!: \rightleftharpoons R\ddot{\underset{\cdot\cdot}{O}}\!:^- + H_3O\!:^+}$$

Table 8.3 gives the pK_a values of some common alcohols and phenols in comparison with water and HCl.

TABLE 8.3 Acidity constants of some alcohols and phenols

Alcohol or phenol	pK_a	
$(CH_3)_3COH$	18.00	Weaker acid
CH_3CH_2OH	16.00	↓
[HOH, water][a]	[15.74]	↓
CH_3OH	15.54	↓
p-Methylphenol	10.17	↓
Phenol	9.89	↓
p-Bromophenol	9.25	↓
p-Nitrophenol	7.15	↓
[HCl, hydrochloric acid][a]	[−7.00]	Stronger acid

[a] Values for water and hydrochloric acid shown for reference.

The data in Table 8.3 show that alcohols are about as acidic as water. They are generally much weaker than carboxylic acids or mineral acids, and they don't react with weak bases like bicarbonate ion. Alcohols do, however, react with alkali metals such as sodium and potassium to yield alkoxide salts that are themselves strong bases.

$$2\,CH_3OH + 2\,Na \longrightarrow 2\,CH_3O^-Na^+ + H_2$$

Methanol → Sodium methoxide

$$2\,(CH_3)_3COH + 2\,K \longrightarrow 2\,(CH_3)_3CO^-K^+ + H_2$$

tert-Butyl alcohol → Potassium *tert*-butoxide

Phenols are much more acidic than alcohols; indeed, some nitro-substituted phenols approach or surpass the acidity of carboxylic acids. One practical consequence of this acidity is that phenols are soluble in dilute aqueous sodium hydroxide.

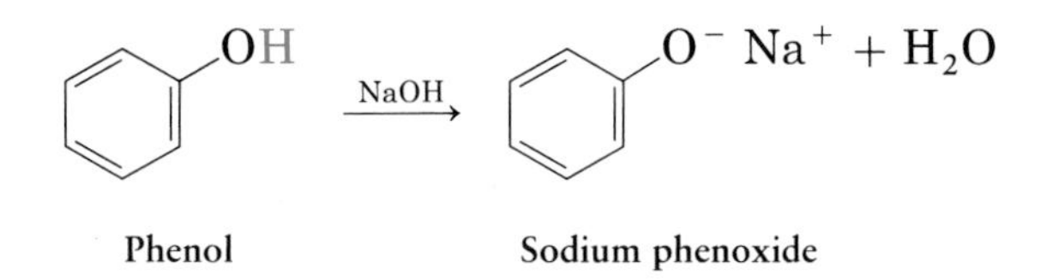

Phenol Sodium phenoxide

Phenols are more acidic than alcohols because the phenoxide anion is resonance-stabilized by the aromatic ring. Sharing the negative charge increases the stability of the phenoxide anion and the acidity of the corresponding phenol.

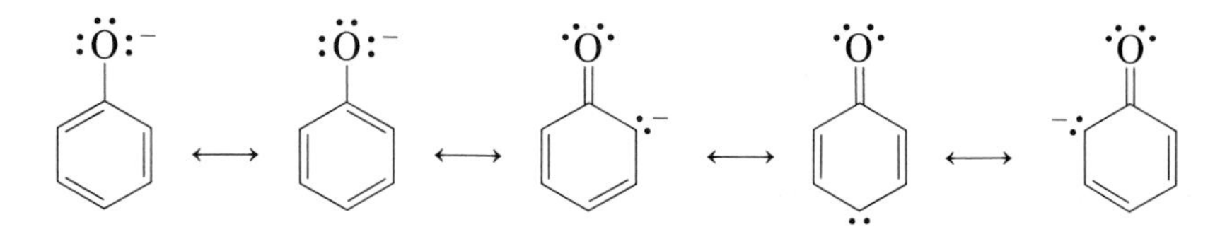

Substituted phenols can be either more or less acidic than phenol itself, depending on their structure. Phenols with an electron-withdrawing substituent are more acidic because the substituent stabilizes the corresponding phenoxide anion. Phenols with an electron-donating substituent, however, are less acidic because the substituent destabilizes the phenoxide anion.

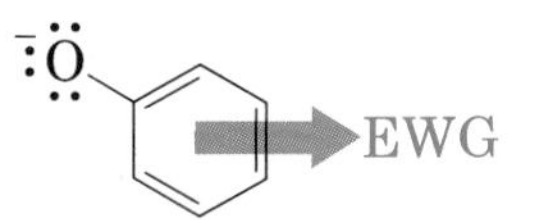

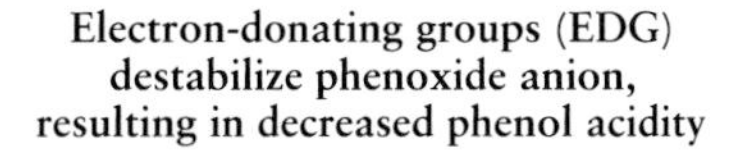

Electron-withdrawing groups (EWG) stabilize phenoxide anion, resulting in increased phenol acidity

Electron-donating groups (EDG) destabilize phenoxide anion, resulting in decreased phenol acidity

PRACTICE PROBLEM 8.3 Which would you expect to be more acidic, *p*-methylphenol or *p*-cyanophenol?

SOLUTION We know from their effects on aromatic substitution (Section 5.9) that methyl is an activating group (electron donor) whereas cyano is a deactivating group (electron acceptor). Thus *p*-cyanophenol is more acidic.

PROBLEM 8.9 Rank the compounds in each group in order of increasing acidity.

(a) Methanol, phenol, *p*-nitrophenol, *p*-methylphenol
(b) Benzyl alcohol, *p*-bromophenol, 2,4-dibromophenol, *p*-methoxyphenol

PROBLEM 8.10 Draw as many resonance structures as you can for the anion of *p*-cyanophenol.

8.5 SYNTHESIS OF ALCOHOLS

Alcohols occupy a central position in organic chemistry. They can be prepared from a variety of functional-group families (alkenes, alkyl halides, ketones, aldehydes, and esters, among others), and they can be transformed into an equally wide assortment of other families. Let's review briefly some of the methods of alcohol preparation we've already seen.

Alcohols can be prepared by hydration of alkenes. Treatment of the alkene with sulfuric acid and water leads to the Markovnikov (more highly substituted) product (Section 4.4).

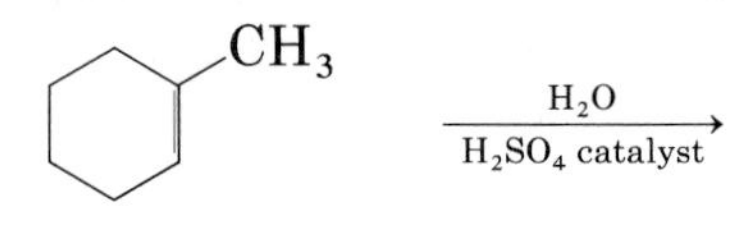

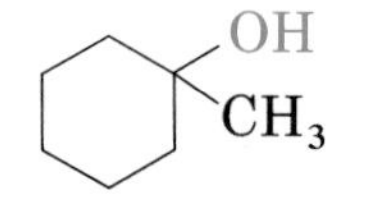

1-Methylcyclohexene 1-Methylcyclohexanol

1,2-Diols can be prepared by direct hydroxylation of an alkene with basic potassium permanganate (Section 4.7). The reaction takes place with syn stereochemistry.

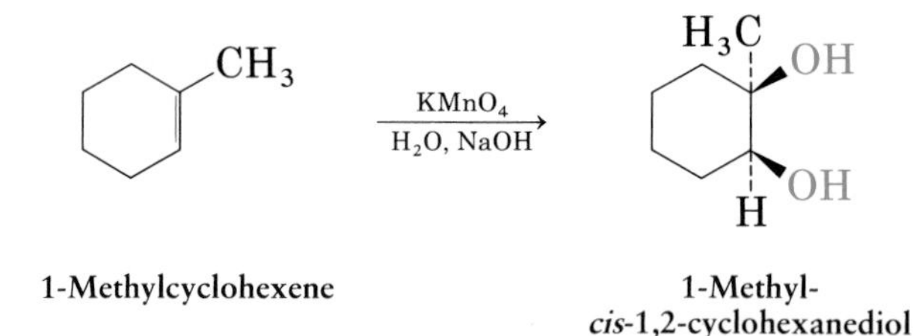

1-Methylcyclohexene 1-Methyl-*cis*-1,2-cyclohexanediol

In addition to being prepared from alkenes, we'll see in the next section that alcohols can also be prepared from all kinds of carbonyl-containing compounds.

8.6 ALCOHOLS FROM CARBONYL COMPOUNDS

The most valuable method for preparing alcohols is by reduction of carbonyl compounds:

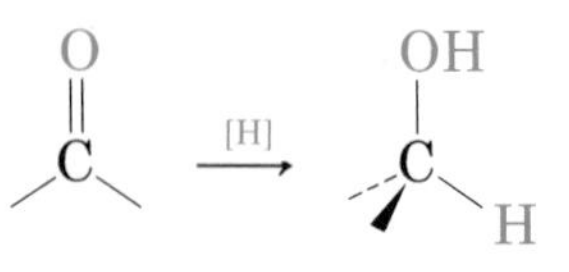

where [H] is a reducing agent.

Reduction of Aldehydes and Ketones

Aldehydes and ketones are easily reduced to yield alcohols.

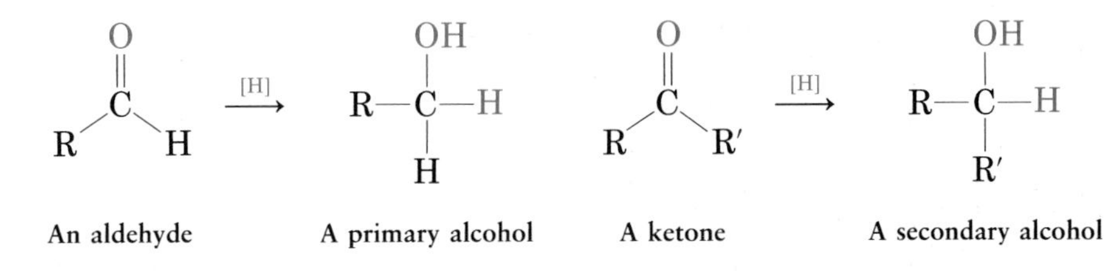

Although many reducing reagents are available, sodium borohydride, $NaBH_4$, is usually chosen for ketone and aldehyde reductions because of its safety and ease of handling. Aldehydes are reduced by $NaBH_4$ to give primary alcohols, and ketones are reduced to give secondary alcohols. High yields are usually obtained, as the following examples indicate.

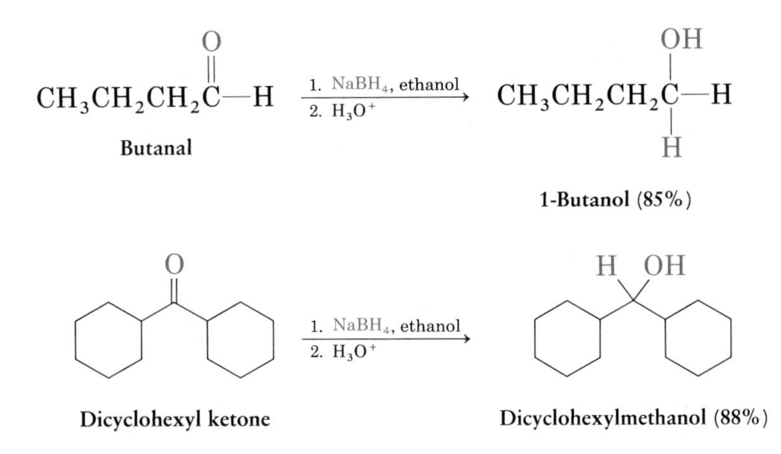

Lithium aluminum hydride, $LiAlH_4$, is another reducing agent that is sometimes used for ketone and aldehyde reductions. Far more powerful and reactive than $NaBH_4$, $LiAlH_4$ is also far more dangerous: It reacts violently with water and ethanol, decomposes explosively when heated above 120°C, and should be handled only by skilled persons.

Reduction of Esters and Carboxylic Acids

Esters and carboxylic acids are reduced to primary alcohols:

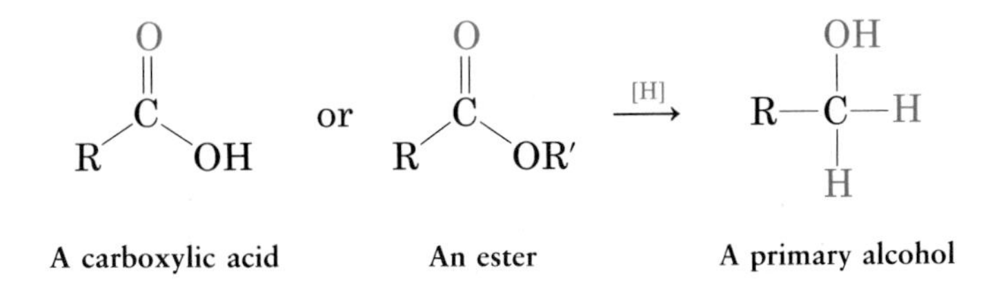

Since these reactions are more difficult than the corresponding reductions of ketones and aldehydes, $LiAlH_4$ is used rather than $NaBH_4$. Note that only one hydrogen is added to the carbonyl carbon atom during reductions of ketones and aldehydes but that two hydrogens become bonded to the carbonyl carbon during ester and carboxylic acid reductions.

Ester reduction:

$$CH_3CH_2CH{=}CH\overset{\overset{\displaystyle O}{\|}}{C}OCH_3 \xrightarrow[\text{2. } H_3O^+]{\text{1. } LiAlH_4\text{, ether}} CH_3CH_2CH{=}CHCH_2OH$$

Methyl 2-pentenoate — 2-Penten-1-ol (91%)

Carboxylic acid reduction:

$$CH_3(CH_2)_7CH{=}CH(CH_2)_7\overset{\overset{\displaystyle O}{\|}}{C}OH \xrightarrow[\text{2. } H_3O^+]{\text{1. } LiAlH_4\text{, ether}} CH_3(CH_2)_7CH{=}CH(CH_2)_7CH_2OH$$

Oleic acid — 9-Octadecen-1-ol (87%)

PRACTICE PROBLEM 8.4 Predict the product of this reaction:

$$CH_3CH_2CH_2\overset{\overset{\displaystyle O}{\|}}{C}CH_2CH_3 \xrightarrow{NaBH_4} ?$$

SOLUTION We know that ketones are reduced by treatment with $NaBH_4$ to yield secondary alcohols. Thus, reduction of 3-hexanone yields 3-hexanol.

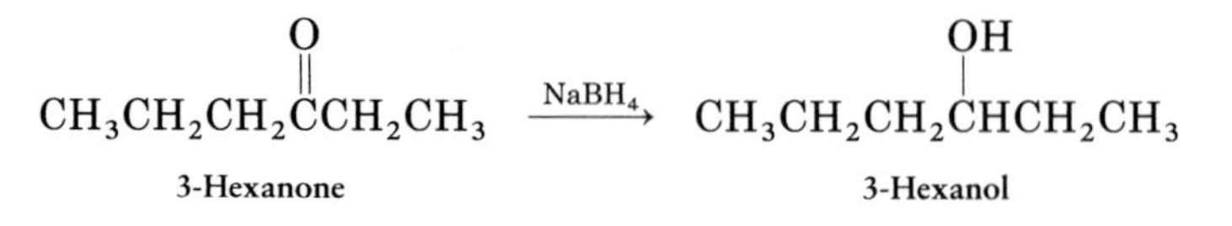

PROBLEM 8.11 How would you carry out these reactions?

(a) $$CH_3\overset{\overset{\displaystyle O}{\|}}{C}CH_2CH_2\overset{\overset{\displaystyle O}{\|}}{C}OCH_3 \xrightarrow{?} CH_3\overset{\overset{\displaystyle OH}{|}}{C}HCH_2CH_2\overset{\overset{\displaystyle O}{\|}}{C}OCH_3$$

(b) $$CH_3\overset{\overset{\displaystyle O}{\|}}{C}CH_2CH_2\overset{\overset{\displaystyle O}{\|}}{C}OCH_3 \xrightarrow{?} CH_3\overset{\overset{\displaystyle OH}{|}}{C}HCH_2CH_2CH_2OH$$

PROBLEM 8.12 What carbonyl compounds give these alcohols on reduction with $LiAlH_4$? Show all possibilities.

(a) $C_6H_5CH_2OH$ (benzene ring–CH_2OH) (b) $C_6H_5\overset{\overset{\displaystyle OH}{|}}{C}HCH_3$ (c) cyclohexane ring bearing OH and H

8.7 ETHERS FROM ALCOHOLS: THE WILLIAMSON ETHER SYNTHESIS

Williamson ether synthesis the reaction of an alkoxide ion with an alkyl halide to yield an ether

Metal alkoxides react with alkyl halides to yield ethers, a reaction known as the **Williamson ether synthesis.** Although it was discovered more than 100 years ago, the Williamson synthesis is still the best method for the preparation of both symmetrical and unsymmetrical ethers.

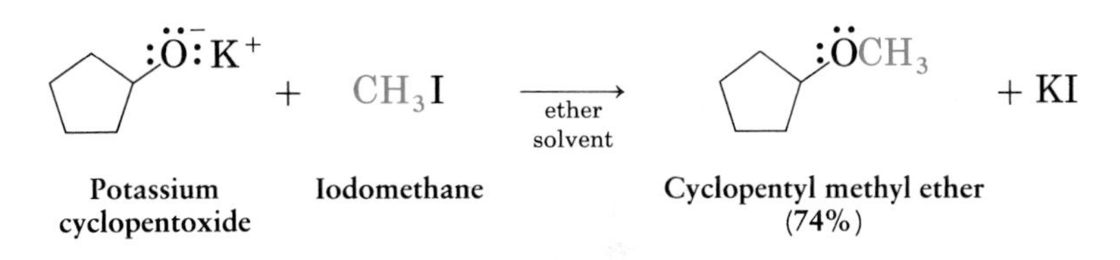

The alkoxide ion needed in the reaction is usually prepared by reaction of an alcohol with sodium or potassium (Section 8.4):

$$2\ ROH + 2\ Na \longrightarrow 2\ RO^-\ Na^+ + H_2$$

The Williamson synthesis is an S_N2 reaction (Section 7.7) that occurs by nucleophilic displacement of halide ion by the alkoxide ion nucleophile. Thus, the reaction is subject to all of the normal S_N2 limitations. Primary alkyl halides work best because competitive E2 elimination of HX can occur with more hindered substrates. For this reason, unsymmetrical ethers are best prepared by reaction of the more hindered alkoxide partner with the less hindered halide partner, rather than vice versa. For example, *tert*-butyl methyl ether is best synthesized by reaction of *tert*-butoxide ion with iodomethane, rather than by reaction of methoxide ion with 2-chloro-2-methylpropane.

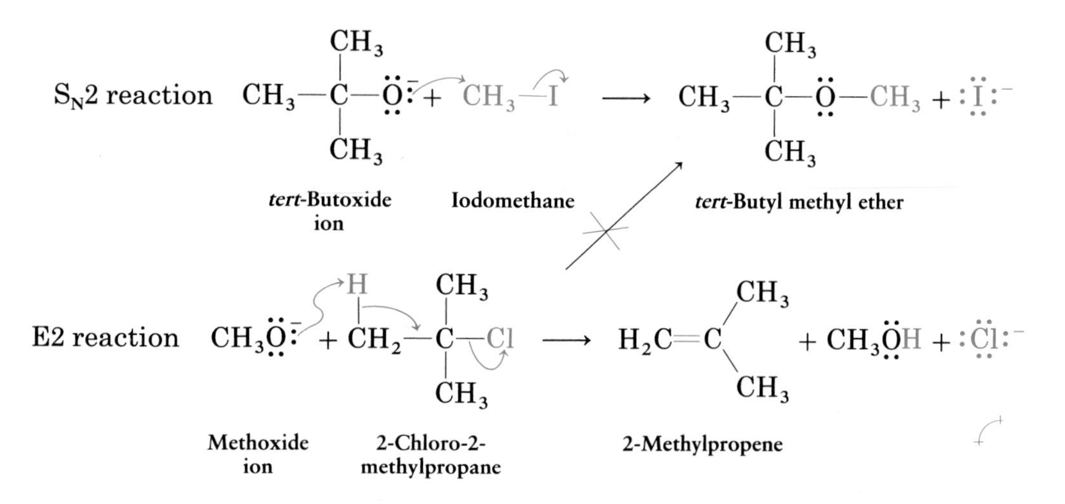

PROBLEM 8.13 Treatment of cyclohexanol with Na gives an alkoxide ion that undergoes reaction with iodoethane to yield cyclohexyl ethyl ether. Write the reaction, showing all the steps.

PROBLEM 8.14 How would you prepare these ethers?

(a) Methyl propyl ether (b) Anisole (methyl phenyl ether)

(c) 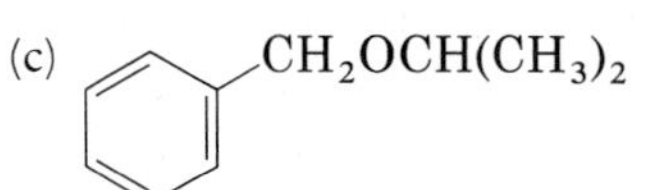

PROBLEM 8.15 Rank these compounds in order of their expected reactivity toward alkoxide ion nucleophiles in the Williamson ether synthesis: bromoethane, 2-bromopropane, chloroethane, 2-chloro-2-methylpropane.

8.8 REACTIONS OF ALCOHOLS

Dehydration

Alcohols can be dehydrated to give alkenes (Section 4.9). Tertiary alcohols lose water when treated with mineral acid under fairly mild conditions, but primary and secondary alcohols require higher temperature. The mechanism of this dehydration is simply an E1 reaction (Section 7.10), as shown in Figure 8.2. Strong acid protonates the alcohol oxygen, and the protonated intermediate spontaneously

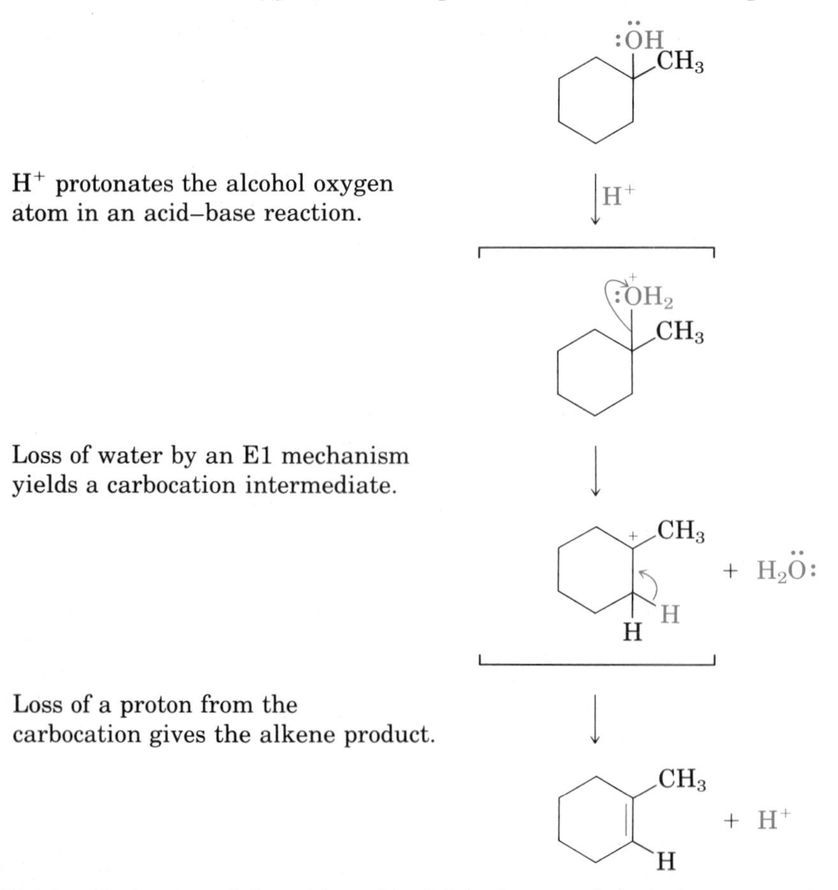

:ÖH
CH_3
H$^+$ protonates the alcohol oxygen atom in an acid–base reaction.
H^+
$:\ddot{O}H_2$
CH_3
Loss of water by an E1 mechanism yields a carbocation intermediate.
CH_3
+ $H_2\ddot{O}$:
H
H
Loss of a proton from the carbocation gives the alkene product.
CH_3
+ H^+
H

FIGURE 8.2 Mechanism of the acid-catalyzed dehydration of alcohols

loses water by an E1 mechanism to generate a carbocation. Loss of a proton from a neighboring carbon atom then yields the alkene product.

Once the acid-catalyzed dehydration is recognized to be an E1 reaction, the reason why tertiary alcohols react fastest becomes clear: Tertiary substrates always react fastest in E1 reactions because they lead to stable tertiary carbocation intermediates.

Conversion into Alkyl Halides and Ethers

Alcohols can be converted into both alkyl halides (Section 7.3) and ethers (Section 8.7). Tertiary alcohols are readily transformed into alkyl halides by an S_N1 mechanism on treatment with either HCl or HBr at 0°C. Primary and secondary alcohols are much more resistant to reaction with halogen acids, however, and are best converted into halides by treatment with either $SOCl_2$ or PBr_3.

Oxidation of Alcohols

The most important reaction of alcohols is their oxidation to carbonyl compounds. Primary alcohols yield aldehydes or carboxylic acids, and secondary alcohols yield ketones, but tertiary alcohols don't normally react with oxidizing agents.

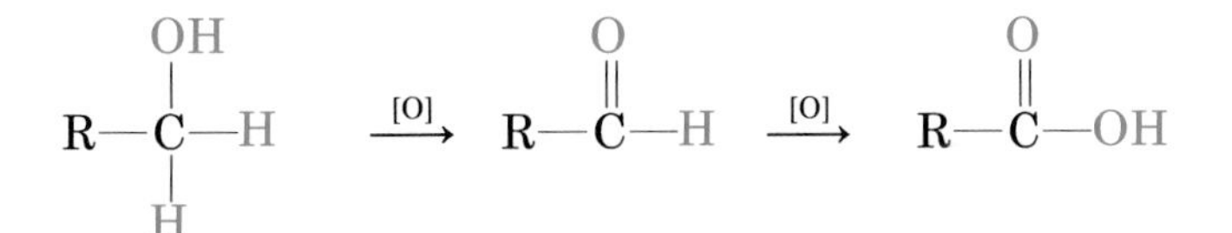

A primary alcohol **An aldehyde** **A carboxylic acid**

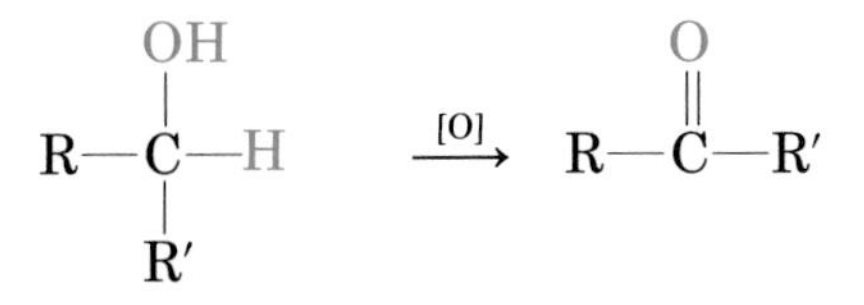

A secondary alcohol **A ketone**

[O] = **An oxidizing reagent**

Primary alcohols are oxidized either to aldehydes or to carboxylic acids, depending on the reagents chosen and on the conditions used. The best method for preparing aldehydes from primary alcohols on a laboratory scale (as opposed to an industrial scale) is by use of pyridinium chlorochromate (PCC), $C_5H_6NCrO_3Cl$, in dichloromethane solvent. This reagent is too expensive for large-scale use in industry, though.

$$CH_3(CH_2)_5CH_2OH \xrightarrow[CH_2Cl_2]{PCC} CH_3(CH_2)_5\overset{\overset{\displaystyle O}{\|}}{C}H$$

1-Heptanol **Heptanal** (78%)

Many oxidizing agents, such as chromium trioxide (CrO_3) in aqueous sulfuric acid (Jones' reagent), oxidize primary alcohols to carboxylic acids. Although aldehydes are intermediates in these oxidations, they can't usually be isolated because they are further oxidized too rapidly.

$$CH_3(CH_2)_8CH_2OH \xrightarrow[\text{acetone}]{\text{Jones' reagent } (CrO_3,\ H_2SO_4,\ H_2O)} CH_3(CH_2)_8\overset{O}{\overset{\|}{C}}OH$$

1-Decanol — Decanoic acid (93%)

Secondary alcohols are oxidized easily to produce ketones. For large-scale oxidations, an inexpensive reagent such as sodium dichromate in aqueous acetic acid might be used, though pyridinium chlorochromate and Jones' reagent also work:

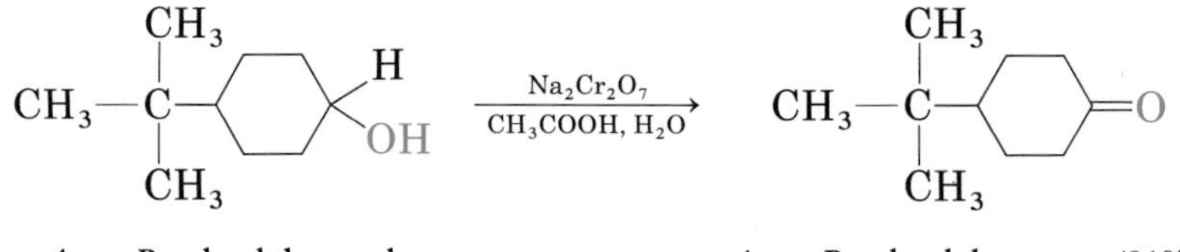

4-*tert*-Butylcyclohexanol — 4-*tert*-Butylcyclohexanone (91%)

PRACTICE PROBLEM 8.5 What product would you expect from reaction of benzyl alcohol with Jones' reagent?

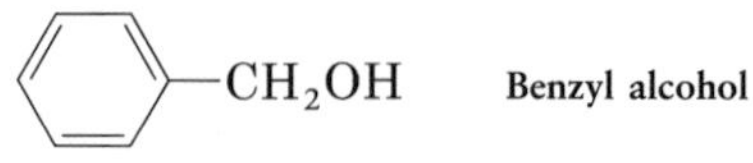

SOLUTION We know that treatment of primary alcohols with Jones' reagent yields carboxylic acids. Thus, oxidation of benzyl alcohol should yield benzoic acid.

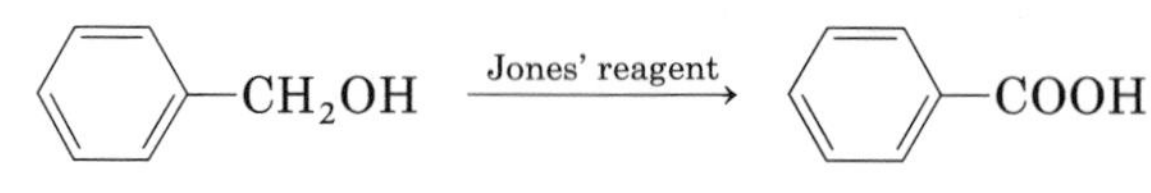

Benzyl alcohol — Benzoic acid

PROBLEM 8.16 What alcohols would give these products on oxidation?

(a) [phenyl methyl ketone structure: $C_6H_5COCH_3$] (b) $CH_3\overset{CH_3}{\overset{|}{C}}HCHO$ (c) [cyclopentanone structure]

PROBLEM 8.17 What products would you expect to obtain from oxidation of these alcohols with Jones' reagent?

(a) Cyclohexanol (b) 1-Hexanol (c) 2-Hexanol

PROBLEM 8.18 What products would you expect to obtain from oxidation of the alcohols in Problem 8.17 with pyridinium chlorochromate (PCC)?

8.9 SYNTHESIS AND REACTIONS OF PHENOLS

Phenols are synthesized from aromatic starting materials by a two-step sequence. The starting compound is first sulfonated by treatment with SO_3/H_2SO_4, and the arenesulfonic acid product is then converted into a phenol by high-temperature reaction with NaOH or KOH.

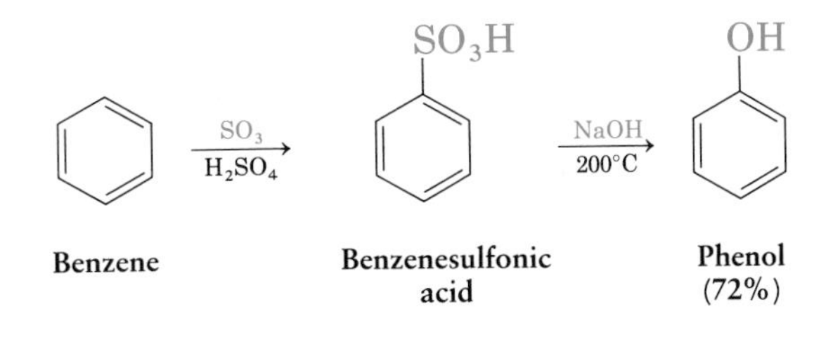

PROBLEM 8.19 *p*-Cresol (*p*-methylphenol) is used industrially both as an antiseptic and as a starting material to prepare the food additive BHT. How could you synthesize *p*-cresol from benzene?

Alcohol-Like Reactions

Phenols and alcohols are very different. Phenols can't be dehydrated by treatment with acid and can't be converted into alkyl halides by treatment with HX. Phenols can, however, be converted into ethers by reaction with alkyl halides in the presence of base. Williamson ether synthesis with phenols occurs easily because phenols are more acidic than alcohols and are therefore more easily converted into their anions.

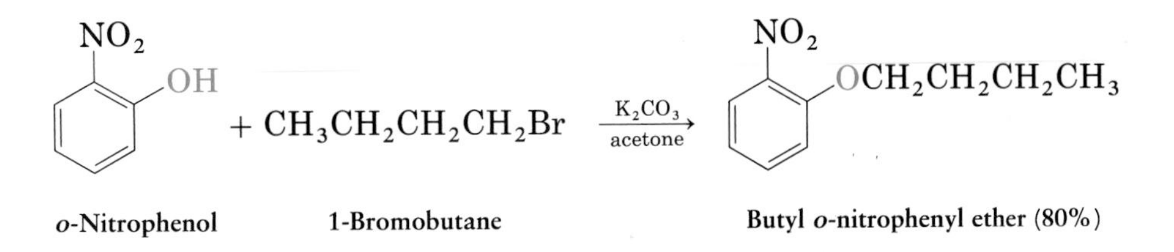

Electrophilic Aromatic Substitution Reactions

The hydroxyl group is a strongly activating, ortho-para-directing substituent in electrophilic aromatic substitution reactions (Sections 5.9 and 5.10). As a result, phenols are reactive substrates for electrophilic halogenation, nitration, and sulfonation.

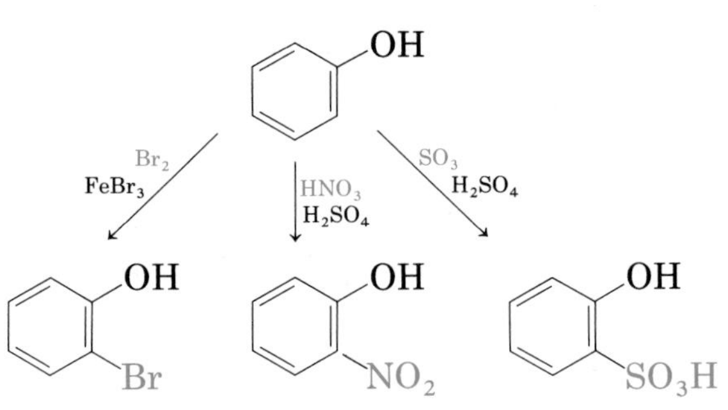

Oxidation of Phenols: Quinones

quinone a compound that contains a cyclohexadienedione functional group

Treatment of a phenol with a strong oxidizing agent yields a **quinone,** a cyclohexadienedione. Older procedures employed sodium dichromate as oxidant, but potassium nitrosodisulfonate [Fremy's salt, $(KSO_3)_2NO$] is now preferred. The reaction takes place under mild conditions through a radical mechanism to give a good yield of the quinone product.

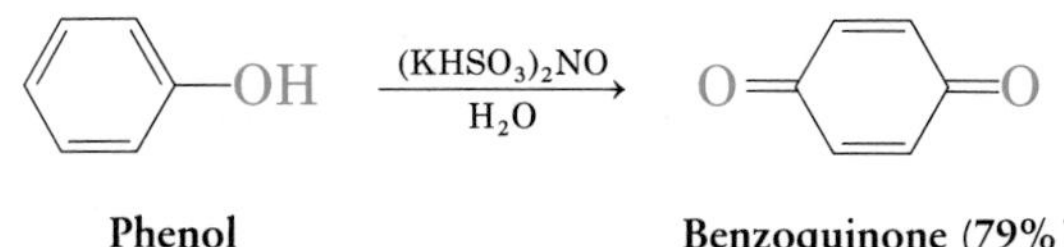

Phenol **Benzoquinone** (79%)

hydroquinone a compound that contains a *p*-dihydroxybenzene unit

Quinones are an interesting and valuable class of compounds because of their oxidation–reduction properties. They can easily be reduced to **hydroquinones** (*p*-dihydroxybenzenes) by $NaBH_4$ or $SnCl_2$, and hydroquinones can easily be oxidized back to quinones by Fremy's salt.

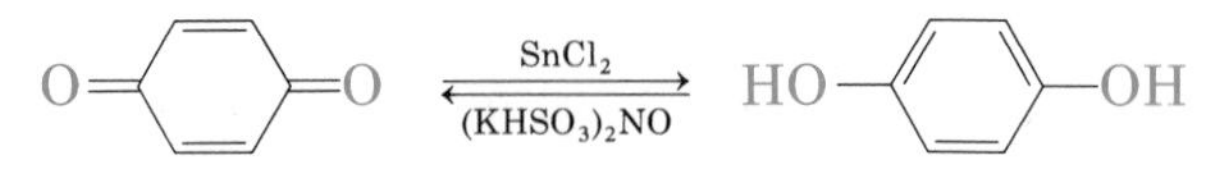

Benzoquinone **Hydroquinone**

8.10 REACTIONS OF ETHERS: ACIDIC CLEAVAGE

Ethers are unusually stable toward most reagents, a property that accounts for their wide use as inert reaction solvents. Halogens, mild acids, bases, and nucleophiles have no effect on most ethers. In fact, ethers undergo only one reaction of general use—cleavage by strong acids. Aqueous HI is the usual reagent for cleaving ethers, though aqueous HBr can also be used.

$$CH_3CH_2OCH(CH_3)_2 \xrightarrow{HI,\ H_2O} CH_3CH_2I + (CH_3)_2CHOH$$

Ethyl isopropyl ether **Iodoethane** **Isopropyl alcohol**

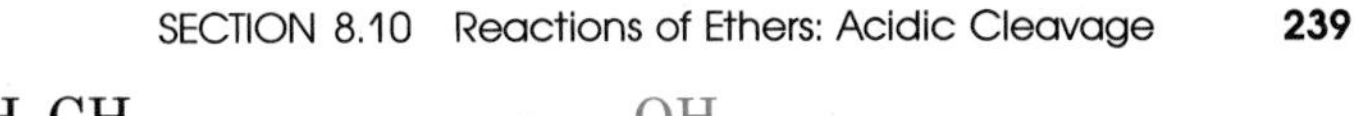

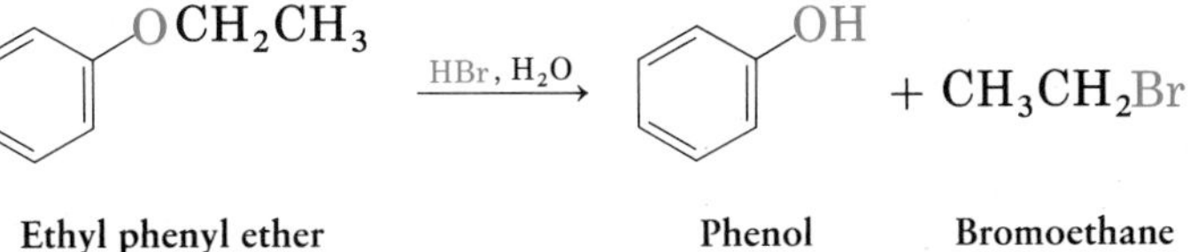

Ethyl phenyl ether — Phenol — Bromoethane

Acidic ether cleavages are typical nucleophilic substitution reactions. They take place through either S_N1 or S_N2 pathways, depending on the structure of the ether. Primary and secondary alkyl ethers react by an S_N2 pathway, in which nucleophilic iodide ion or bromide ion attacks the protonated ether at the less highly substituted site. The ether oxygen atom stays with the more hindered alkyl group, and the halide joins to the less hindered group. For example, ethyl isopropyl ether yields isopropyl alcohol and iodoethane on cleavage by HI.

$$\underset{\text{Ethyl isopropyl ether}}{CH_3\overset{CH_3}{\overset{|}{C}}H{-}\ddot{\underset{\cdot\cdot}{O}}{-}CH_2CH_3} \xrightarrow{H-I} \left[CH_3\overset{CH_3}{\overset{|}{C}}H{-}\overset{H}{\overset{|}{\underset{\cdot\cdot}{O}^{+}}}{-}CH_2CH_3\right] \xrightarrow{:\ddot{I}:^-} \underset{\text{Isopropyl alcohol}}{CH_3\overset{CH_3}{\overset{|}{C}}HOH} + \underset{\text{Iodoethane}}{CH_3CH_2I}$$

Tertiary ethers cleave by an S_N1 or E1 mechanism since they can produce stable intermediate carbocations. These reactions are often fast and take place at room temperature or below.

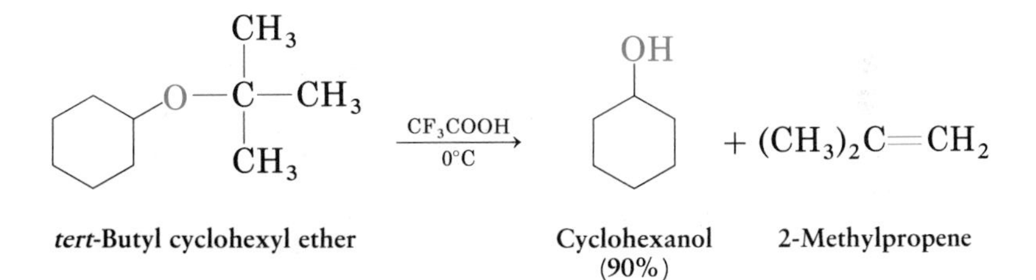

tert-Butyl cyclohexyl ether — Cyclohexanol (90%) — 2-Methylpropene

PRACTICE PROBLEM 8.6 What products would you expect from the reaction of methyl cyclopentyl ether with HI?

SOLUTION Iodide ion attacks the methyl group rather than the secondary cyclopentyl group in an S_N2 reaction, giving iodomethane and cyclopentanol:

$$\text{(cyclopentyl)}{-}OCH_3 \xrightarrow{HI} \text{(cyclopentyl)}{-}OH + CH_3I$$

PROBLEM 8.20 What products do you expect from reaction of these ethers with HI?
(a) $CH_3CH_2OCH_2CH_3$ (b) Cyclohexyl ethyl ether (c) $(CH_3)_3COCH_2CH_3$

PROBLEM 8.21 Write a detailed mechanism for the acid-catalyzed cleavage of *tert*-butyl cyclohexyl ether to yield cyclohexanol and 2-methylpropene. What kind of reaction is occurring?

8.11 CYCLIC ETHERS: EPOXIDES

For the most part, cyclic ethers behave like acyclic ethers. The chemistry of the ether functional group is the same whether it's in an open chain or in a ring. Thus, common cyclic ethers such as tetrahydrofuran (THF) and dioxane are often used as solvents because of their inertness.

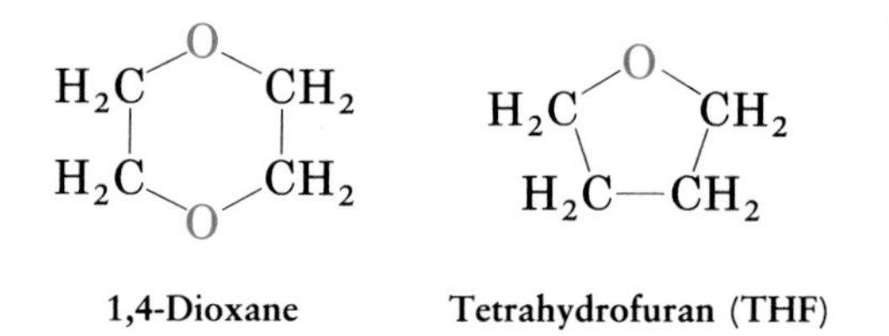

1,4-Dioxane Tetrahydrofuran (THF)

epoxide a three-membered, oxygen-containing ring

oxirane an alternative name for an epoxide

The one group of cyclic ethers that behaves differently from open-chain ethers is the group of three-membered-ring compounds called **epoxides,** or **oxiranes.** Epoxides are usually prepared by treating an alkene with a peroxyacid, RCO_3H. Magnesium monoperoxyphthalate (MMPP) is the preferred reagent, because it is more stable and easily handled than most other peroxyacids.

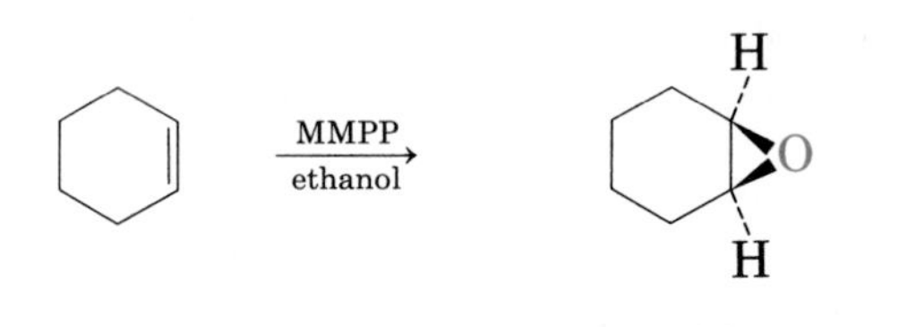

Cyclohexene 1,2-Epoxycyclohexane (85%) (cyclohexene oxide)

where magnesium monoperoxyphthalate (MMPP) is

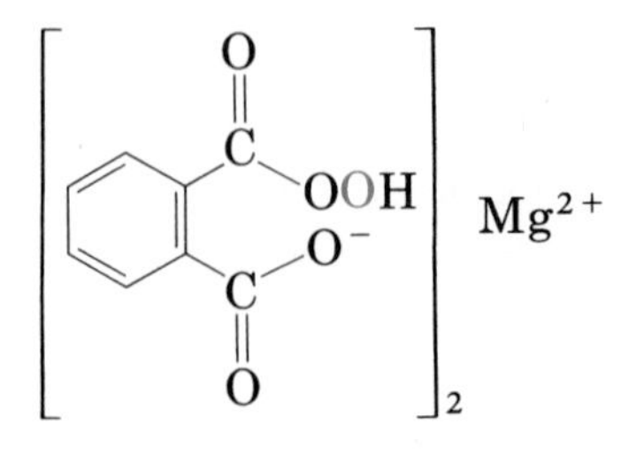

PROBLEM 8.22 What product do you expect from reaction of *cis*-2-butene with magnesium monoperoxyphthalate, assuming syn stereochemistry?

PROBLEM 8.23 Reaction of *trans*-2-butene with magnesium monoperoxyphthalate yields a different epoxide from that obtained by reaction of the cis isomer (Problem 8.22). How can you account for this?

8.12 RING-OPENING REACTIONS OF EPOXIDES

Epoxide rings are opened by treatment with acid in much the same way that other ethers are cleaved. The major difference is that epoxides react under much milder conditions because of ring strain. Dilute aqueous mineral acid at room temperature is sufficient to hydrolyze epoxides to 1,2-diols (also called *vicinal glycols*). Two million tons of ethylene glycol, most of it used as automobile antifreeze, are produced every year by acid-catalyzed hydration of ethylene oxide.

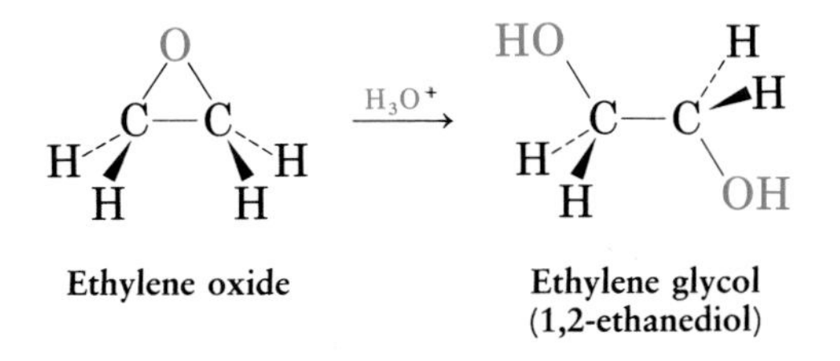

Acid-induced epoxide ring-opening takes place by S_N2 attack of a nucleophile on the protonated epoxide, in a manner analogous to the final step of alkene bromination in which a three-membered-ring bromonium ion is opened by nucleophilic attack (Section 4.5).

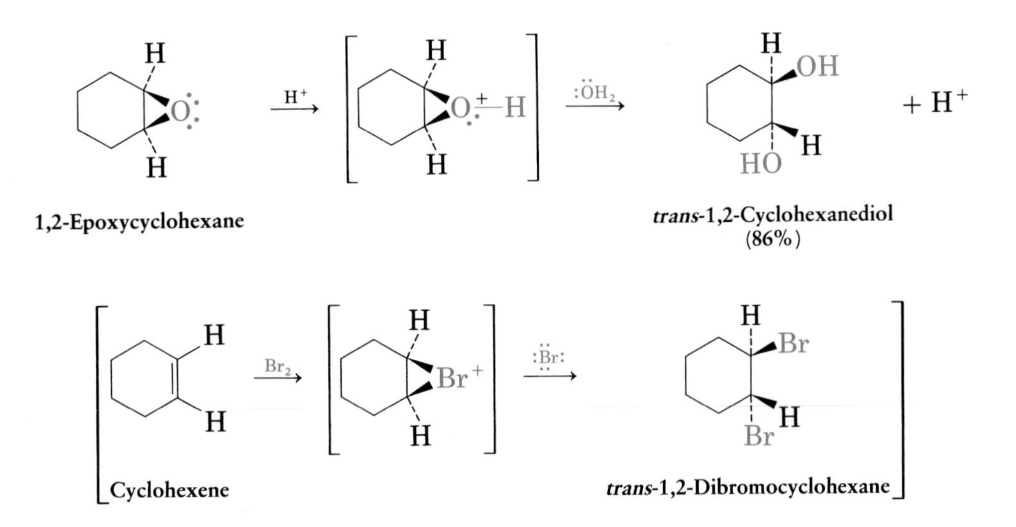

Epoxide opening is also involved in the mechanism by which the polycyclic aromatic hydrocarbons (PAHs) in chimney soot and cigarette smoke cause cancer. Benzo[a]pyrene, one of the best studied PAHs, is converted by metabolic oxidation into a diol epoxide. In the body, the epoxide ring reacts with an amino group in cellular DNA to give an altered DNA that is covalently bound to the PAH. With its DNA thus altered, the cell is unable to reproduce normally.

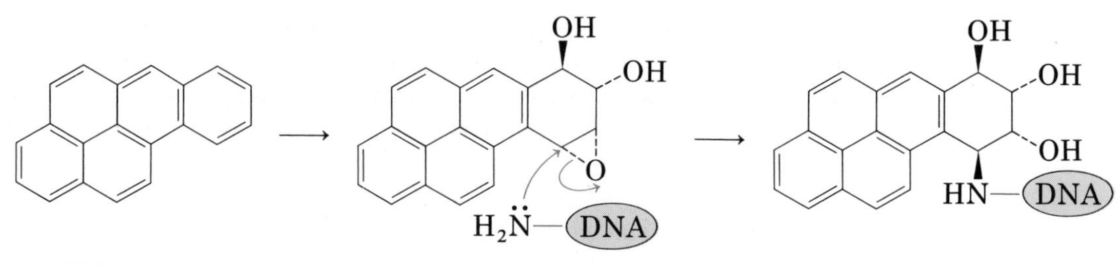

1,2-Benzpyrene A diol epoxide

PROBLEM 8.24 Show the steps involved in the acidic hydrolysis of *cis*-2,3-epoxybutane to yield 2,3-butanediol. What is the stereochemistry of the product if the ring-opening takes place by normal backside S_N2 attack?

PROBLEM 8.25 Answer Problem 8.24 for the acidic hydrolysis of *trans*-2,3-epoxybutane. Is the same product formed?

8.13 THIOLS AND SULFIDES

thiol
a compound with the –SH functional group

sulfide
a compound that has two organic groups bonded to a sulfur atom

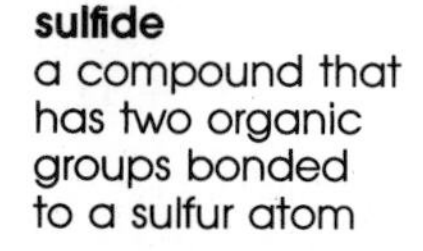

mercapto group
an alternative name for the thiol group, –SH

Sulfur is the element just below oxygen in the periodic table, and many oxygen-containing organic compounds have sulfur analogs. For instance, **thiols,** R–SH, are sulfur analogs of alcohols, and **sulfides,** R–S–R′, are sulfur analogs of ethers. Thiols are named in the same way as alcohols, with the suffix *-thiol* used in place of *-ol*. The –SH group itself is referred to as a **mercapto group.**

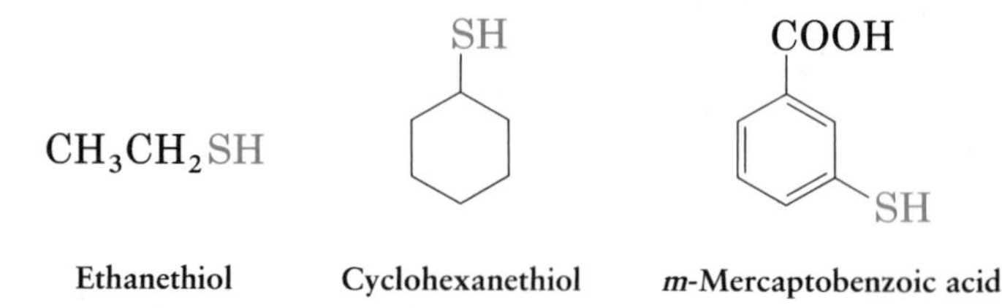

Ethanethiol Cyclohexanethiol *m*-Mercaptobenzoic acid

Sulfides are named in the same way as ethers, with *sulfide* used in place of *ether* for simple compounds and with *alkylthio* used in place of *alkoxy* for more complex substances.

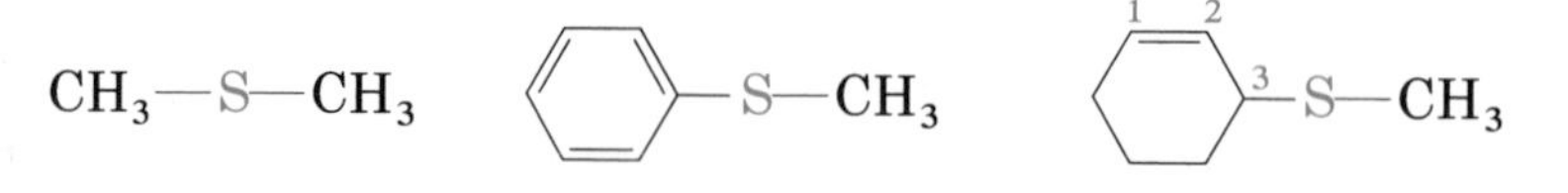

Dimethyl sulfide Methyl phenyl sulfide 3-(Methylthio)cyclohexene

Thiols are usually prepared from the corresponding alkyl halide by S_N2 displacement with a sulfur nucleophile such as hydrosulfide anion, HS^-.

$$CH_3(CH_2)_6CH_2Br + Na^+ :\ddot{S}H^- \longrightarrow CH_3(CH_2)_6CH_2\ddot{S}H + NaBr$$

1-Bromooctane Sodium hydrosulfide 1-Octanethiol

Sulfides are best prepared by treating a primary or secondary alkyl halide with a thiolate ion, RS^-. Reaction occurs by an S_N2 mechanism that is analogous to the Williamson ether synthesis (Section 8.7). Thiolate anions are among the best nucleophiles known, and product yields are usually high in these sulfide-forming reactions.

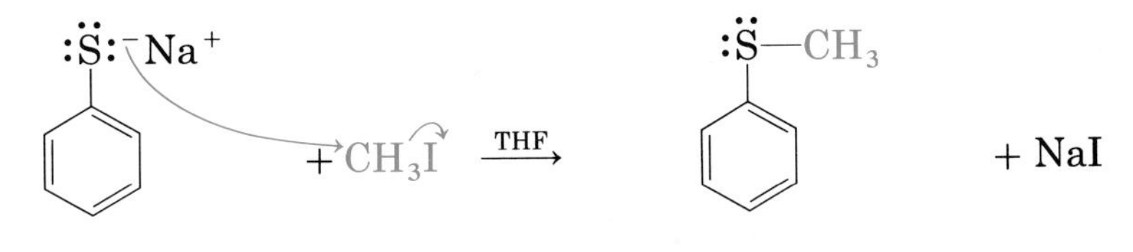

Sodium benzenethiolate **Methyl phenyl sulfide (96%)**

The outstanding characteristic of thiols is their appalling odor: Skunk scent is due primarily to the simple thiols 3-methyl-1-butanethiol and 2-butene-1-thiol. Thiols can be oxidized by mild reagents such as bromine to yield disulfides, R–S–S–R. The reaction is easily reversed, and disulfides can be reduced back to thiols by treatment with zinc metal and acetic acid.

$$2\,R{-}SH \underset{Zn,\ H^+}{\overset{Br_2}{\rightleftharpoons}} R{-}S{-}S{-}R + 2\,HBr$$

A thiol **A disulfide**

We'll see in Section 15.6 that the thiol–disulfide interconversion is extremely important in biochemistry because disulfide "bridges" form the cross-links between protein chains that help stabilize the three-dimensional conformations of proteins.

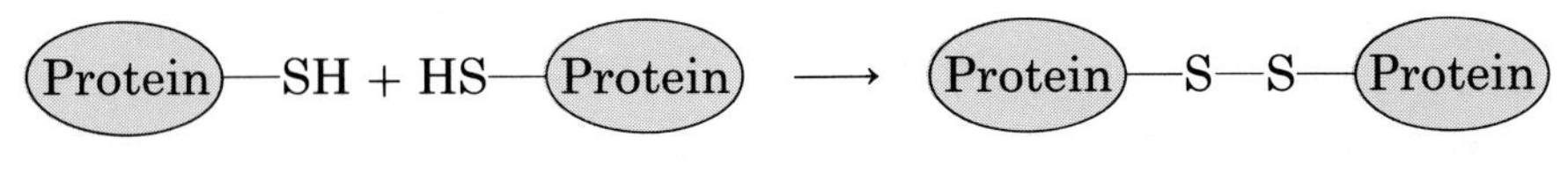

A cross-linked protein

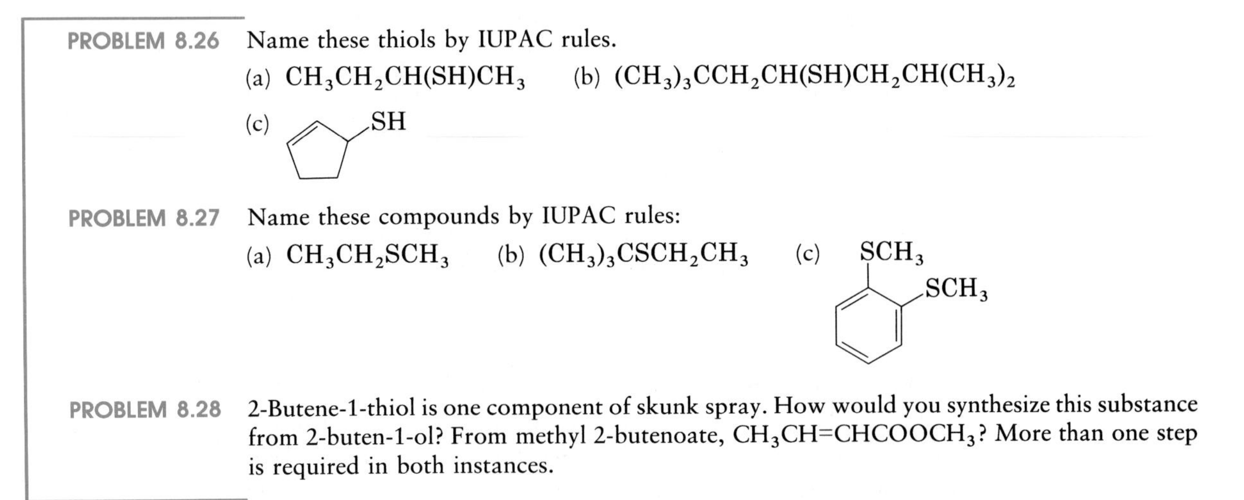

PROBLEM 8.26 Name these thiols by IUPAC rules.

(a) $CH_3CH_2CH(SH)CH_3$ (b) $(CH_3)_3CCH_2CH(SH)CH_2CH(CH_3)_2$

(c) [cyclopentene ring with SH]

PROBLEM 8.27 Name these compounds by IUPAC rules:

(a) $CH_3CH_2SCH_3$ (b) $(CH_3)_3CSCH_2CH_3$ (c) [benzene ring with two adjacent SCH_3 groups]

PROBLEM 8.28 2-Butene-1-thiol is one component of skunk spray. How would you synthesize this substance from 2-buten-1-ol? From methyl 2-butenoate, $CH_3CH{=}CHCOOCH_3$? More than one step is required in both instances.

INTERLUDE

Industrial Uses Of Simple Alcohols and Phenols

Methanol and ethanol are two of the most important of all industrial organic chemicals; both are manufactured on a vast scale for a variety of uses. Prior to the development of the modern chemical industry, methanol was prepared by heating wood in the absence of air and thus came to be called *wood alcohol*. Now, however, approximately 1.2 billion gallons of methanol are manufactured each year in the United States by catalytic reduction of carbon monoxide with hydrogen gas.

$$CO + 2\,H_2 \xrightarrow[\text{zinc oxide/chromia}]{400^\circ C} CH_3OH$$

Methanol is toxic to humans, causing blindness in low doses and death in larger amounts. Industrially, it is used both as a solvent and as a starting material for making formaldehyde, CH_2O, and acetic acid, CH_3COOH.

The production of ethanol by fermentation of grains and sugars is one of the oldest organic chemistry processes known, going back at least to the ancient Greeks. Fermentation is carried out by adding yeast to an aqueous sugar solution, where enzymes break down carbohydrates into ethanol and CO_2:

$$\underset{\text{A carbohydrate}}{C_6H_{12}O_6} \xrightarrow{\text{yeast}} 2\,CH_3CH_2OH + 2\,CO_2$$

Nearly 300 million gallons of ethanol are produced every year in the United States for use as a solvent or starting material. Only about 5% of this industrial ethanol now comes from fermentation, though. Most is obtained by high-temperature, acid-catalyzed hydration of ethylene.

$$H_2C{=}CH_2 + H_2O \xrightarrow[\text{catalyst}]{\text{acid}} CH_3CH_2OH$$

Phenol and several of its substituted analogs are also used industrially. Historically, the outbreak of World War I provided the stimulus for industrial preparation of large amounts of phenol, which was needed as starting material for synthesizing the explosive, picric acid (2,4,6-trinitrophenol). Today, approximately 1.5 million tons of phenol per year are manufactured for use in such products as Bakelite resin and adhesives for binding plywood.

In addition to its use in resins and adhesives, phenol is a starting material for the synthesis of chlorinated phenols and of the food preservative butylated hydroxytoluene (BHT). Pentachlorophenol ("Penta"), a widely used wood preservative, is prepared by reaction of phenol with excess chlorine; 2,4-

dichlorophenol is used to prepare the herbicide 2,4-dichlorophenoxyacetic acid ("2,4-D"); and 2,4,5-trichlorophenol is used to prepare the antiseptic agent hexachlorophene.

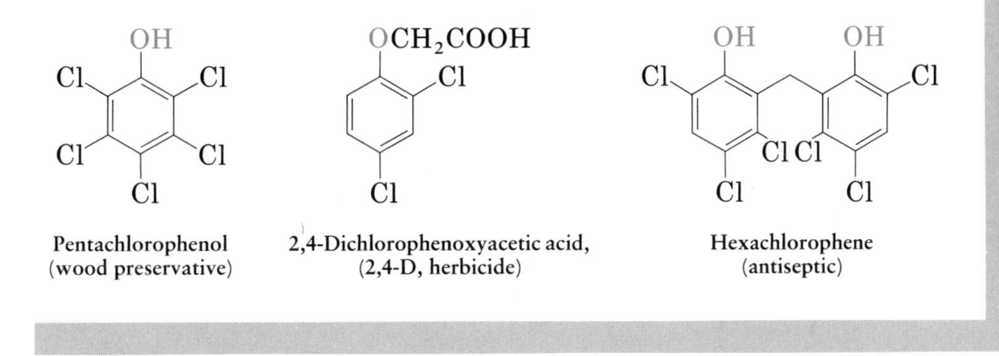

Pentachlorophenol (wood preservative)

2,4-Dichlorophenoxyacetic acid, (2,4-D, herbicide)

Hexachlorophene (antiseptic)

SUMMARY AND KEY WORDS

Ethers, alcohols, and phenols are organic derivatives of water in which one or both of the water hydrogens have been replaced by organic groups.

Ethers have two organic groups bonded to an oxygen. The groups may be alkyl, alkenyl, or aryl, and the oxygen atom may be in a ring or in an open chain. Ethers are prepared by S_N2 reaction of an alkoxide ion with a primary alkyl halide—the **Williamson synthesis.** Ethers are inert to most reagents but are attacked by strong acids to give cleavage products. **Epoxides** differ from other ethers in their ease of cleavage. The high reactivity of the strained three-membered ether ring allows epoxide rings to be opened easily.

Alcohols are compounds that have a hydroxyl group bonded to an alkyl residue. They can be prepared in many ways, including hydroxylation of alkenes and hydration of alkenes. The most important method of alcohol synthesis involves **reduction** of carbonyl compounds. Aldehydes, esters, and carboxylic acids yield primary alcohols on reduction; ketones yield secondary alcohols.

Alcohols and phenols are weak acids. A **Brønsted–Lowry acid** is a compound that can donate a proton (hydrogen ion, H^+); a **Brønsted–Lowry base** is a compound that can accept a proton. The exact strength of an acid or base is expressed by the acidity constant K_a. A **Lewis acid** is a compound that has a low-energy unfilled orbital and can accept an electron pair. A **Lewis base** is a compound that donates an unshared electron pair. Many organic molecules that contain oxygen and nitrogen are weak Lewis bases.

Alcohols can be converted into their **alkoxide anions** on treatment with strong base or with alkali metals. Alcohols can also be dehydrated to yield alkenes, can be transformed into alkyl halides by treatment with PBr_3 or $SOCl_2$, and can be converted into ethers by reaction of their anions with alkyl halides. The most important reaction of alcohols is their **oxidation** to yield carbonyl compounds. Primary alcohols give either aldehydes or carboxylic acids, secondary alcohols yield ketones, and tertiary alcohols are not oxidized.

Phenols are aromatic counterparts of alcohols and are prepared by reaction of arenesulfonic acids with NaOH at high temperature. Although similar to alcohols in some respects, phenols are more acidic because phenoxide anions are stabilized by resonance. Phenols undergo electrophilic aromatic substitution and can be oxidized to **quinones.**

Sulfides (R–S–R) and **thiols** (R–S–H) are sulfur analogs of ethers and alcohols. Thiols are prepared by S_N2 reaction of an alkyl halide with HS^-, and sulfides are prepared by further alkylation of the thiol with a second molecule of alkyl halide.

SUMMARY OF REACTIONS

1. Synthesis of alcohols (Section 8.6)

   (a) Reduction of aldehydes to yield primary alcohols

   $$\mathrm{R\overset{O}{\overset{\|}{C}}H} \xrightarrow{\mathrm{NaBH_4}} \mathrm{RCH_2OH}$$

   (b) Reduction of ketones to yield secondary alcohols

   $$\mathrm{R\overset{O}{\overset{\|}{C}}R'} \xrightarrow{\mathrm{NaBH_4}} \mathrm{R\overset{OH}{\overset{|}{C}}HR'}$$

   (c) Reduction of esters to yield primary alcohols

   $$\mathrm{R\overset{O}{\overset{\|}{C}}OR'} \xrightarrow{\mathrm{LiAlH_4}} \mathrm{RCH_2OH}$$

   (d) Reduction of carboxylic acids to yield primary alcohols

   $$\mathrm{R\overset{O}{\overset{\|}{C}}OH} \xrightarrow{\mathrm{LiAlH_4}} \mathrm{RCH_2OH}$$

2. Synthesis of ethers (Section 8.7)

   $$\mathrm{RO^-Na^+ + R'Br} \xrightarrow[\text{reaction}]{S_N2} \mathrm{ROR'}$$

3. Synthesis of phenols (Section 8.9)

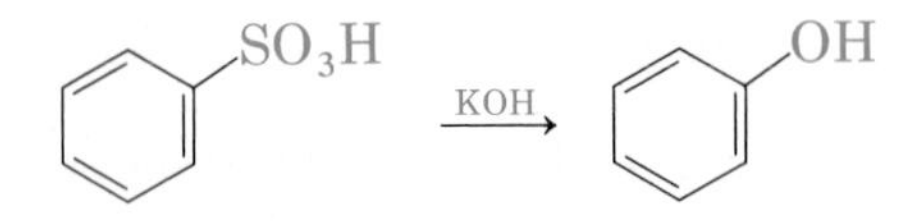

4. Synthesis of epoxides (Section 8.11)

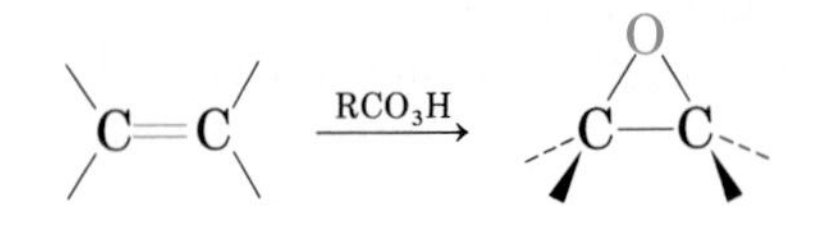

5. Synthesis of thiols (Section 8.13)

$$Na^+ \ ^-SH + RBr \xrightarrow[\text{reaction}]{S_N2} RSH$$

6. Synthesis of sulfides (Section 8.13)

$$RS^-Na^+ + R'Br \xrightarrow[\text{reaction}]{S_N2} RSR'$$

7. Reactions of alcohols

(a) Conversion into ethers (Section 8.7)

$$2\,ROH + 2\,Na \longrightarrow 2\,RO^-Na^+ + H_2$$

$$RO^-Na^+ + R'Br \xrightarrow[\text{reaction}]{S_N2} ROR'$$

(b) Dehydration to yield alkenes (Section 8.8)

$$-\overset{H}{\underset{|}{\overset{|}{C}}}-\overset{OH}{\underset{|}{\overset{|}{C}}}- \xrightarrow{H_2SO_4} \;>C{=}C<\; + H_2O$$

(c) Oxidation to yield carbonyl compounds (Section 8.8)

$$RCH_2OH \xrightarrow[\text{chlorochromate}]{\text{pyridinium}} R\overset{O}{\overset{\|}{C}}H \quad \text{(aldehyde)}$$

$$RCH_2OH \xrightarrow[\text{reagent}]{\text{Jones'}} R\overset{O}{\overset{\|}{C}}OH \quad \text{(carboxylic acid)}$$

$$R\overset{OH}{\overset{|}{C}}HR' \xrightarrow[\text{chlorochromate}]{\text{pyridinium}} R\overset{O}{\overset{\|}{C}}R' \quad \text{(ketone)}$$

8. Reactions of ethers; acidic cleavage (Section 8.10)

$$ROR' + HI \longrightarrow ROH + R'I$$

ADDITIONAL PROBLEMS

8.29 Draw structures corresponding to these IUPAC names:

(a) Ethyl isopropyl ether (b) 3,4-Dimethoxybenzoic acid
(c) 2-Methyl-2,5-heptanediol (d) *trans*-3-Ethylcyclohexanol
(e) 4-Allyl-2-methoxyphenol (eugenol, from oil of cloves)

8.30 Name these compounds according to IUPAC rules:

(a) $HOCH_2CH_2\overset{CH_3}{\overset{|}{C}}HCH_2OH$ (b) $CH_3\underset{HO}{\underset{|}{C}}H\underset{CH_2CH_2CH_3}{\underset{|}{C}}HCH_2CH_3$

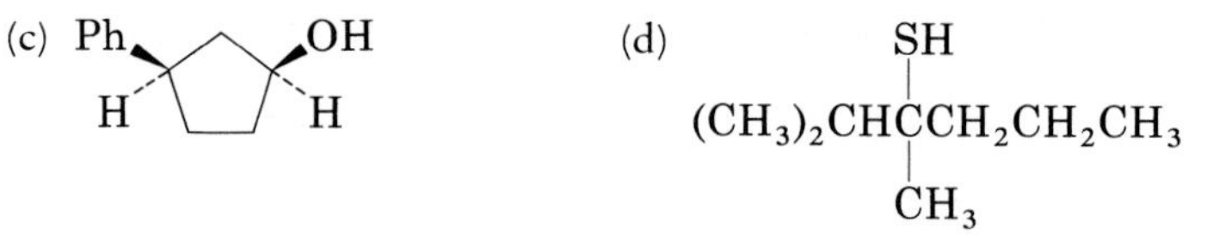

8.31 Draw and name the eight isomeric alcohols that have the formula $C_5H_{12}O$.

8.32 Which of the eight alcohols you identified in Problem 8.31 would react with Jones' reagent? Show the products you would expect from each reaction.

8.33 Predict the likely products of these cleavage reactions:

(a) $CH_3CH_2OCH(CH_3)CH_3 \xrightarrow{HI,\ H_2O}$ (b) $(CH_3)_3CCH_2OCH_3 \xrightarrow{HI,\ H_2O}$

8.34 What reagents would you use to carry out these transformations?

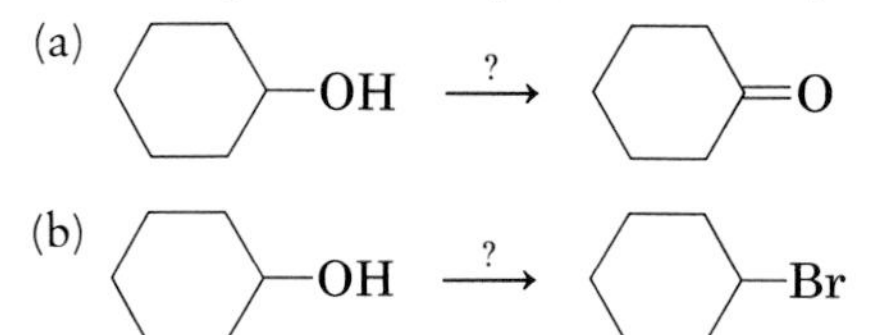

(c) $CH_3CH_2CH_2OH \xrightarrow{?} CH_3CH_2CHO$

(d) $CH_3CH_2CH_2OH \xrightarrow{?} CH_3CH_2COOH$

(e) $CH_3CH_2CH_2OH \xrightarrow{?} CH_3CH_2CH_2O^-Na^+$

(f) $CH_3CH_2CH_2OH \xrightarrow{?} CH_3CH_2CH_2Cl$

8.35 How would you prepare these compounds from 2-phenylethanol?

(a) Benzoic acid (b) Ethylbenzene
(c) 2-Bromo-1-phenylethane (d) Phenylacetic acid ($C_6H_5CH_2COOH$)
(e) Phenylacetaldehyde ($C_6H_5CH_2CHO$)

8.36 Give the structures of the major products you would obtain from reaction of phenol with these reagents:

(a) Br_2 (1 mol) (b) Br_2 (3 mol) (c) NaOH, then CH_3I (d) $(KSO_3)_2NO$

8.37 What products would you obtain from reaction of 1-butanol with these reagents?

(a) PBr_3 (b) CrO_3, H_2O, H_2SO_4 (c) Na (d) Pyridinium chlorochromate

8.38 What products would you obtain from reaction of 1-methylcyclohexanol with these reagents?

(a) HBr (b) H_2SO_4 (c) CrO_3 (d) Na (e) Product of (d), then CH_3I

8.39 What alcohols would you oxidize to obtain these products?

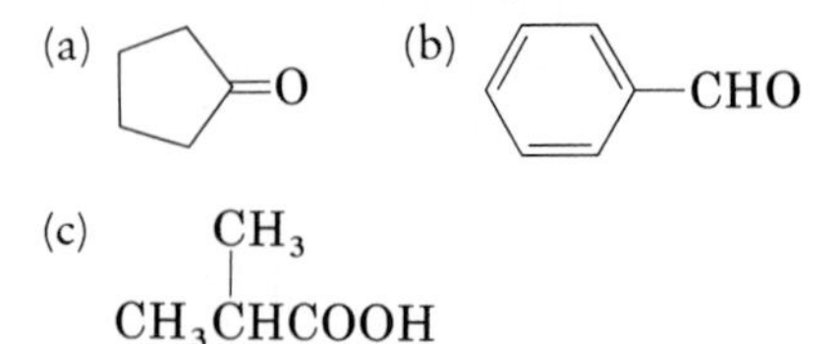

(c) $CH_3CH(CH_3)COOH$

8.40 Show the alcohols you would obtain by reduction of these carbonyl compounds.

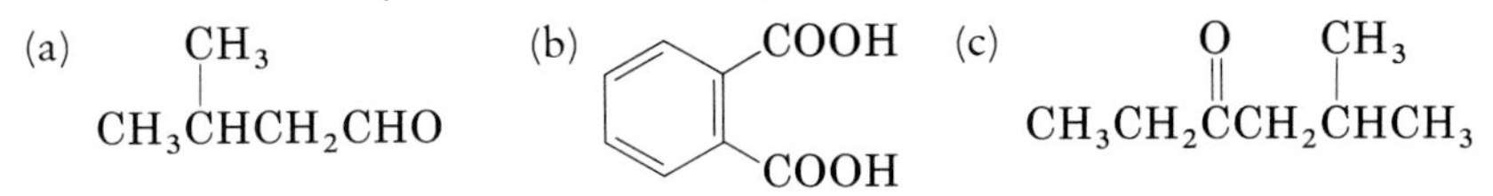

8.41 When 4-chloro-1-butanol is treated with a strong base such as sodium hydride, NaH, tetrahydrofuran is produced. Suggest a mechanism for this reaction.

$$ClCH_2CH_2CH_2CH_2OH \xrightarrow[\text{ether}]{\text{NaH}} \text{(tetrahydrofuran)} + H_2 + NaCl$$

8.42 Ammonia, $H_2N{-}H$, has $pK_a \approx 36$, and acetone has $pK_a \approx 20$. Will the following reaction take place? Explain.

$$H_3C{-}\overset{O}{\overset{\|}{C}}{-}CH_3 + Na^+ \ {}^-{:}NH_2 \xrightarrow{?} H_3C{-}\overset{O}{\overset{\|}{C}}{-}CH_2{:}^- \ Na^+ + NH_3$$

Acetone

8.43 Which is more acidic, *p*-methylphenol or *p*-(trifluoromethyl)phenol? Explain.

8.44 Classify these reagents as either Lewis acids or Lewis bases:

(a) $AlBr_3$ (b) $CH_3CH_2NH_2$ (c) BH_3
(d) HF (e) $CH_3{-}S{-}CH_3$ (f) $TiCl_4$

8.45 Rank these substances in order of increasing acidity.

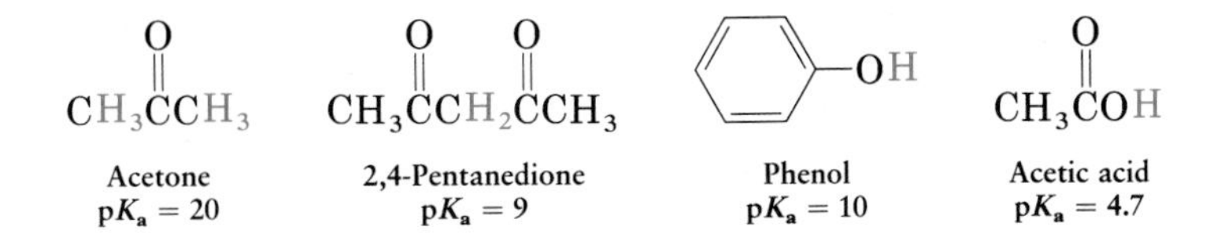

8.46 Which, if any, of the substances in Problem 8.45 are strong enough acids to react completely with NaOH? (The pK_a of H_2O is 15.7.)

8.47 Is *tert*-butoxide anion a strong enough base to react with water? In other words, does the following reaction take place as written? (The pK_a of *tert*-butyl alcohol is 18.)

$$(CH_3)_3CO^-Na^+ + H_2O \xrightarrow{?} (CH_3)_3COH + NaOH$$

8.48 Sodium bicarbonate, $NaHCO_3$, is the sodium salt of carbonic acid (H_2CO_3), $pK_a \approx 6.4$. Which of the substances shown in Problem 8.45 will react with sodium bicarbonate?

8.49 Assume that you have two unlabeled bottles, one that contains phenol ($pK_a \approx 10$) and one that contains acetic acid ($pK_a \approx 4.7$). In light of your answer to Problem 8.48, propose a simple way to tell what is in each bottle.

8.50 Identify the acids and bases in these reactions:

(a) $CH_3OH + H^+ \longrightarrow CH_3\overset{+}{O}H_2$

(b) $CH_3OH + {}^-NH_2 \longrightarrow CH_3O^- + NH_3$

(c) $CH_3\overset{O}{\overset{\|}{C}}CH_3 + TiCl_4 \longrightarrow CH_3{-}\overset{\overset{+}{O}{-}\overset{-}{Ti}Cl_4}{\overset{\|}{C}}{-}CH_3$

(d) 2,2,6,6-H cyclohexanone (H, H, H, H) + NaH ⟶ enolate anion (H, H, H) Na^+ + H_2

8.51 Starting from benzene, how would you prepare benzyl phenyl ether, $C_6H_5OCH_2C_6H_5$? More than one step is required.

8.52 Since all hamsters look pretty much alike, pairing and mating is governed by chemical means of communication. Investigations have shown that dimethyl disulfide, CH_3SSCH_3, is secreted by female hamsters as a sex attractant for males. How would you synthesize dimethyl disulfide in the laboratory if you wanted to trick your hamster?

8.53 *p*-Nitrophenol ($pK_a = 7.15$) is much more acidic than phenol ($pK_a = 9.89$). Draw as many resonance structures as you can for the *p*-nitrophenoxide anion.

8.54 The herbicide 2,4,5-T (2,4,5-trichlorophenoxyacetic acid) can be prepared by heating a mixture of 2,4,5-trichlorophenol and $ClCH_2COOH$ with NaOH. Show the structure of 2,4,5-T and explain how it is formed.

8.55 Starting from benzene, how would you prepare the 2,4,5-trichlorophenol needed for manufacture of 2,4,5-T (Problem 8.54)?

8.56 *tert*-Butyl ethers can be prepared by the reaction of an alcohol with 2-methylpropene in the presence of an acid catalyst. Propose a mechanism for this reaction.

8.57 How would you prepare these ethers?

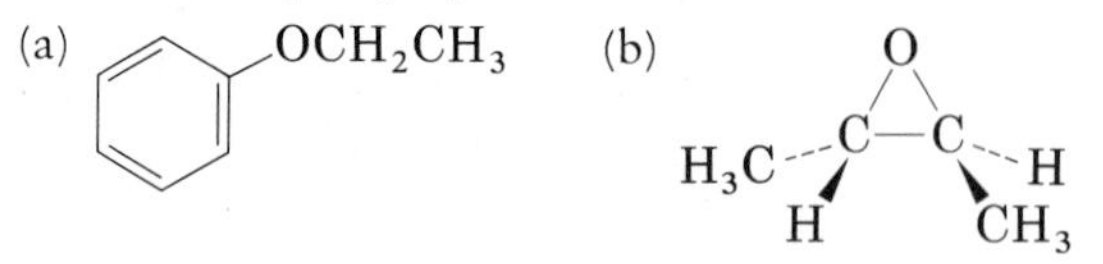

8.58 What product would you expect from reaction of tetrahydrofuran with hot aqueous HI?

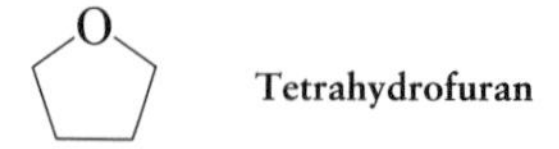

8.59 Methyl phenyl ether can be cleaved to yield iodomethane and lithium phenoxide upon heating with LiI. Propose a mechanism for this reaction.

8.60 The Zeisel method, a procedure for determining the number of methoxyl groups ($CH_3O{-}$) in a compound, involves heating a weighed amount of compound with HI. Ether cleavage occurs, and the iodomethane formed is distilled off and passed into a solution of $AgNO_3$. The silver iodide that precipitates is then weighed, and the percentage of methoxy groups in the sample is thereby determined. For example, 1.06 g of vanillin, the material responsible for the characteristic odor of vanilla, yields 1.60 g AgI. If vanillin has a molecular weight of 152, how many methoxyls does it contain?

8.61 When 2-methyl-2,5-pentanediol is treated with sulfuric acid, dehydration occurs to yield 2,2-dimethyltetrahydrofuran. Suggest a mechanism for this reaction.

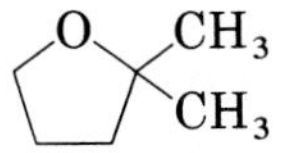

2,2-Dimethyltetrahydrofuran

CHAPTER 9

Aldehydes and Ketones: Nucleophilic Addition Reactions

carbonyl group the carbon–oxygen double bond, C=O

In this and the next two chapters, we'll discuss the most important functional group in organic chemistry: the **carbonyl group,** C=O (car-bo-*neel*). Carbonyl compounds are everywhere in nature. Most biologically important molecules contain carbonyl groups, as do many pharmaceutical agents and many of the synthetic chemicals that touch our everyday lives. Acetic acid (the chief component of vinegar), acetaminophen (an over-the-counter headache remedy), and Dacron (the polyester material used in clothing) all contain different kinds of carbonyl groups.

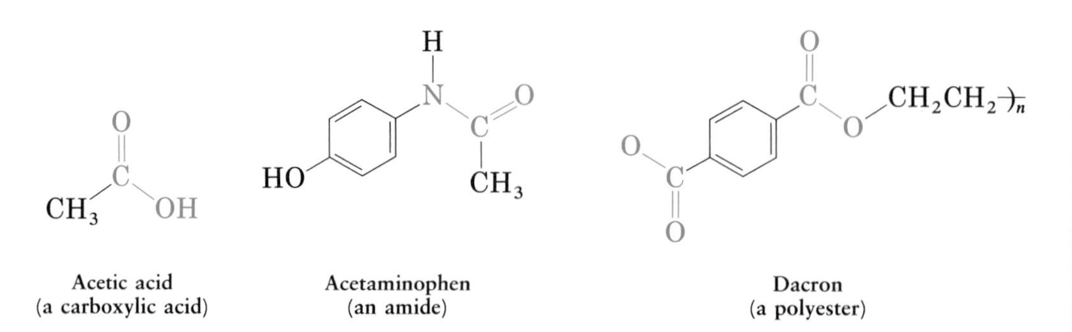

Acetic acid (a carboxylic acid)

Acetaminophen (an amide)

Dacron (a polyester)

9.1 KINDS OF CARBONYL COMPOUNDS

Table 9.1 shows some of the many kinds of carbonyl compounds. All contain an acyl fragment, $R-\overset{\displaystyle O}{\overset{\|}{C}}-$, bonded to another residue that may be a carbon, hydrogen, oxygen, halogen, sulfur, or other substituent.

It turns out to be very useful to classify carbonyl compounds into two general categories, based on the kinds of chemistry they undergo:

TABLE 9.1 Some types of carbonyl compounds

Name	*General formula*	*Name ending*
Aldehyde	$R-C(=O)-H$	*-al*
Ketone	$R-C(=O)-R'$	*-one*
Carboxylic acid	$R-C(=O)-O-H$	*-oic acid*
Acid chloride	$R-C(=O)-Cl$	*-yl* or *-oyl chloride*
Acid anhydride	$R-C(=O)-O-C(=O)-R'$	*-oic anhydride*
Ester	$R-C(=O)-O-R'$	*-oate*
Lactone (cyclic ester)	$>C-C(=O)-O$ (ring)	None
Amide	$R-C(=O)-N<$	*-amide*

Aldehydes (RCHO)

Ketones ($R_2C=O$)

The acyl groups in these two families are bonded to substituents (−H and −R, respectively) that can't stabilize a negative charge and therefore can't serve as leaving groups. Aldehydes and ketones behave similarly and undergo many of the same reactions.

Carboxylic acids (RCOOH)

Esters (RCOOR′)

Acid chlorides (RCOCl)

Acid anhydrides (RCOOCOR′)

Amides ($RCONH_2$)

The acyl groups in carboxylic acids and their derivatives are bonded to substituents (oxygen, halogen, nitrogen) that can stabilize a negative charge and so can serve as leaving groups in substitution reactions. The chemistry of these compounds is therefore similar.

PROBLEM 9.1 Propose structures for molecules that meet these descriptions.

(a) A ketone, $C_5H_{10}O$ (b) An aldehyde, $C_6H_{10}O$
(c) A keto aldehyde, $C_6H_{10}O_2$ (d) A cyclic ketone, C_5H_8O

9.2 STRUCTURE AND PROPERTIES OF CARBONYL GROUPS

The carbon–oxygen double bond of carbonyl groups is similar in many respects to the carbon–carbon double bond of alkenes (Figure 9.1). The carbonyl carbon atom is sp^2 hybridized and forms three sigma bonds. The fourth valence electron remains in a carbon p orbital and forms a pi bond to oxygen by overlap with an oxygen p orbital. The oxygen also has two nonbonding pairs of electrons, which occupy its remaining two orbitals. Like alkenes, carbonyl compounds are planar about the double bond and have bond angles of approximately 120°.

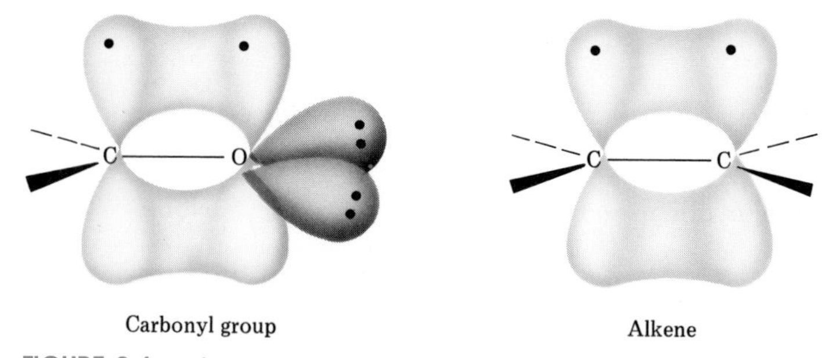

FIGURE 9.1 Electronic structure of the carbonyl group

Carbon–oxygen double bonds are polarized because of the high electronegativity of oxygen relative to carbon. Since the carbonyl carbon is positively polarized, it is an electrophilic (Lewis acidic) site and is attacked by nucleophiles. Conversely, the carbonyl oxygen is negatively polarized and is a nucleophilic (Lewis basic) site. We'll see in this and the next two chapters that most carbonyl-group reactions can be understood in terms of bond polarities.

δ^- :O: ‖ δ^+ C

Nucleophilic oxygen reacts with acids and electrophiles.

Electrophilic carbon reacts with bases and nucleophiles.

9.3 NAMING ALDEHYDES AND KETONES

Systematic names for aldehydes are derived by replacing the terminal *-e* of the alkane name with *-al*. The longest chain selected for the base name must contain the –CHO group, and the CHO carbon is always numbered as carbon 1. For example:

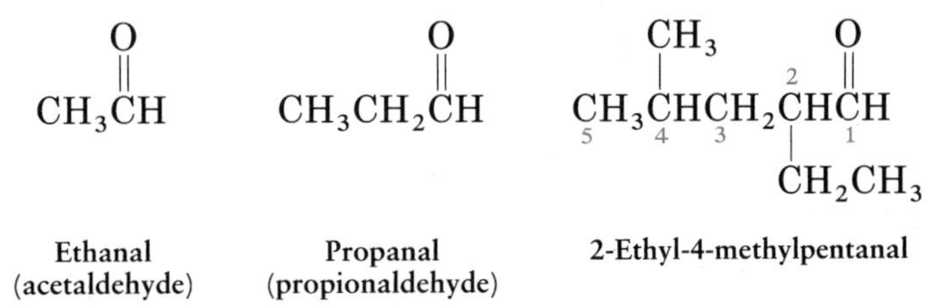

Ethanal (acetaldehyde) **Propanal** (propionaldehyde) **2-Ethyl-4-methylpentanal**

Note that the longest chain in 2-ethyl-4-methylpentanal is a hexane, but that this chain doesn't include the CHO group.

For more complex aldehydes, in which the –CHO group is attached to a ring, the suffix *-carbaldehyde* is used:

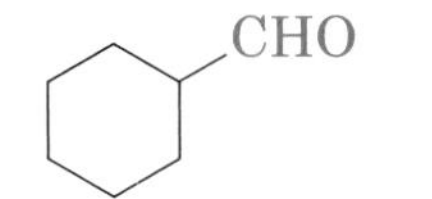

Cyclohexanecarbaldehyde

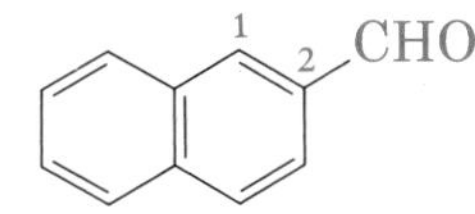

2-Naphthalenecarbaldehyde

Certain simple and well-known aldehydes also have common names, as indicated in Table 9.2.

TABLE 9.2 Common names of some simple aldehydes

Formula	*Trivial name*	*Systematic name*
HCHO	Formaldehyde	Methanal
CH_3CHO	Acetaldehyde	Ethanal
CH_3CH_2CHO	Propionaldehyde	Propanal
$CH_3CH_2CH_2CHO$	Butyraldehyde	Butanal
$CH_3CH_2CH_2CH_2CHO$	Valeraldehyde	Pentanal
$H_2C{=}CHCHO$	Acrolein	2-Propenal
C_6H_5CHO (benzene ring–CHO)	Benzaldehyde	Benzenecarbaldehyde

Ketones are named by replacing the terminal *-e* of the corresponding alkane name with *-one* (pronounced *own*). The chain selected for the base name is the longest one that contains the ketone group, and the numbering begins at the end nearer the carbonyl carbon. For example:

$$H_3C{-}\overset{O}{\overset{\|}{C}}{-}CH_3$$

Propanone (acetone)

$$\underset{1}{C}H_3\underset{2}{C}H_2\underset{3}{\overset{O}{\overset{\|}{C}}}\underset{4}{C}H_2\underset{5}{C}H_2\underset{6}{C}H_3$$

3-Hexanone

$$\underset{6}{C}H_3\underset{5}{C}H{=}\underset{4}{C}H\underset{3}{C}H_2\underset{2}{\overset{O}{\overset{\|}{C}}}\underset{1}{C}H_3$$

4-Hexen-2-one

A few ketones also have common names:

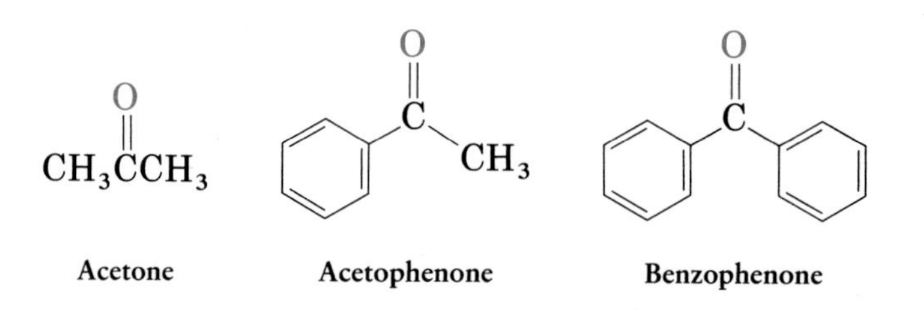

Acetone Acetophenone Benzophenone

When it becomes necessary to refer to the $-COR$ group as a substituent, the term *acyl* (a-sil) is used. Similarly, $-CHO$ is called a *formyl* group, $-COCH_3$ is an *acetyl* group, and $-COAr$ is an *aroyl* group.

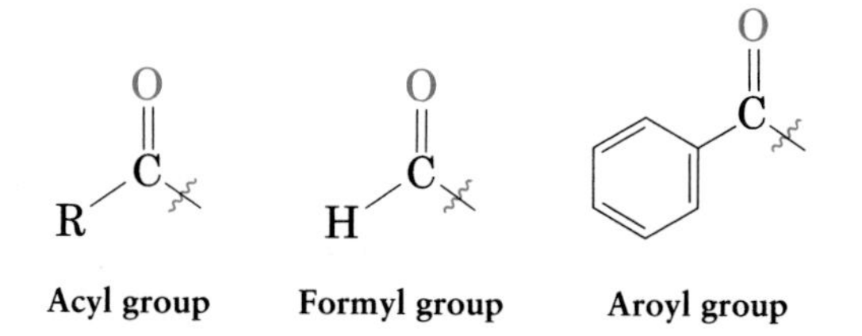

Acyl group Formyl group Aroyl group

Occasionally, the doubly bonded oxygen must be considered a substituent, and the prefix *oxo-* is used. For example:

$$\underset{6}{CH_3}\underset{5}{CH_2}\underset{4}{CH_2}\underset{3}{\overset{O}{\overset{\|}{C}}}\underset{2}{CH_2}\underset{1}{\overset{O}{\overset{\|}{C}}}OCH_3$$

Methyl 3-oxohexanoate

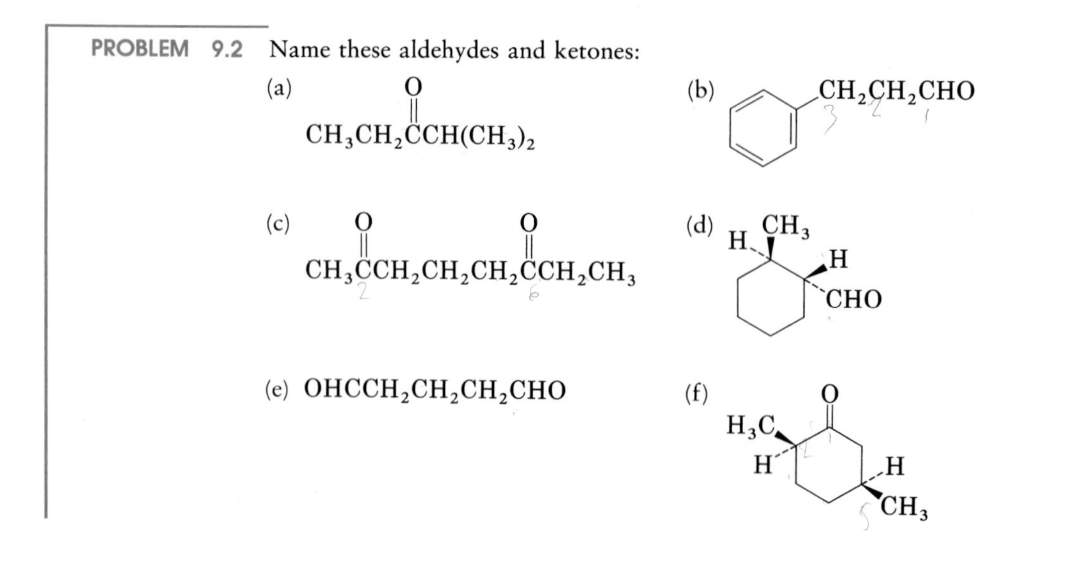

PROBLEM 9.3 Draw structures corresponding to these IUPAC names:

(a) 3-Methylbutanal
(b) 3-Methyl-3-butenal
(c) 4-Chloro-2-pentanone
(d) Phenylacetaldehyde
(e) 2,2-Dimethylcyclohexanecarbaldehyde
(f) 1,3-Cyclohexanedione

9.4 SYNTHESIS OF ALDEHYDES

We've already discussed two good methods of aldehyde synthesis: oxidation of primary alcohols and cleavage of alkenes. Let's review briefly:

1. Primary alcohols can be oxidized to give aldehydes (Section 8.8). The reaction is often carried out using pyridinium chlorochromate (PCC) in dichloromethane solution.

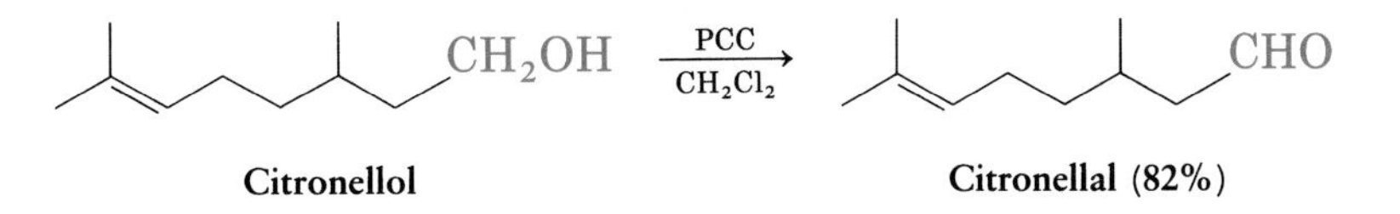

Citronellol **Citronellal** (82%)

2. Alkenes with at least one vinylic proton (remember: *vinylic* means on the double-bond carbon) undergo oxidative cleavage when treated with ozone to yield aldehydes (Section 4.7).

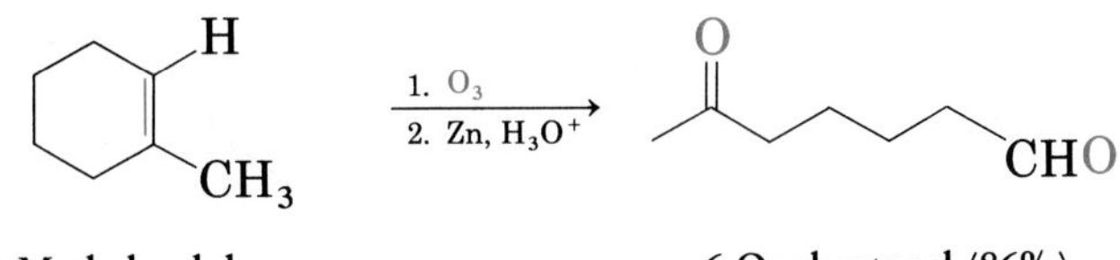

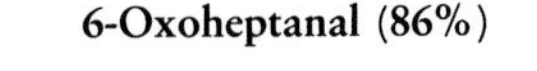

1-Methylcyclohexene **6-Oxoheptanal** (86%)

Yet another method of aldehyde synthesis is the reduction of certain carboxylic acid derivatives to yield aldehydes. For example, carboxylic acid chlorides are converted into aldehydes by catalytic hydrogenation.

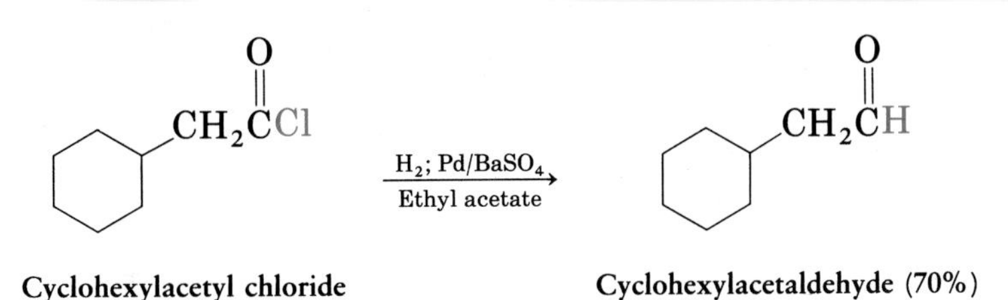

Cyclohexylacetyl chloride **Cyclohexylacetaldehyde** (70%)

PROBLEM 9.4 Show how you might prepare pentanal from these starting materials:

(a) 1-Pentanol (b) 1-Hexene (c) 5-Decene (d) $CH_3CH_2CH_2CH_2COCl$

9.5 SYNTHESIS OF KETONES

For the most part, methods of ketone synthesis are similar to those for aldehydes:

1. Secondary alcohols are oxidized to give ketones (Section 8.8). Pyridinium chlorochromate (PCC), CrO_3, and $Na_2Cr_2O_7$ are all effective.

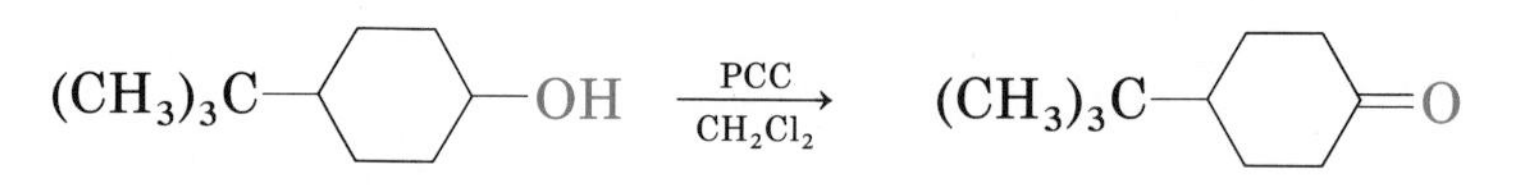

4-*tert*-Butylcyclohexanol   4-*tert*-Butylcyclohexanone (90%)

2. Ozonolysis of alkenes yields ketones if one of the double-bond carbon atoms is disubstituted (Section 4.7).

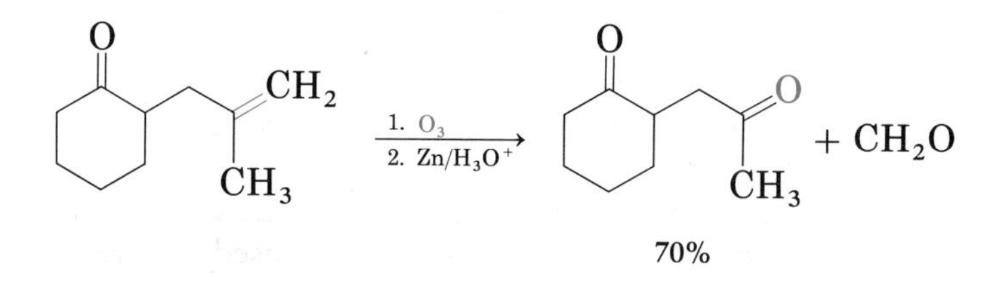

3. Terminal alkynes undergo hydration to yield methyl ketones. The reaction is catalyzed by mercuric sulfate (Section 4.15).

$$CH_3(CH_2)_3C{\equiv}CH \xrightarrow[Hg(OAc)_2]{H_3O^+} CH_3(CH_2)_3\overset{\overset{\large O}{\|}}{C}{-}CH_3$$

1-Hexyne   2-Hexanone (78%)

4. Aromatic rings undergo Friedel–Crafts acylation with an acid chloride to yield alkyl aryl ketones (Section 5.8).

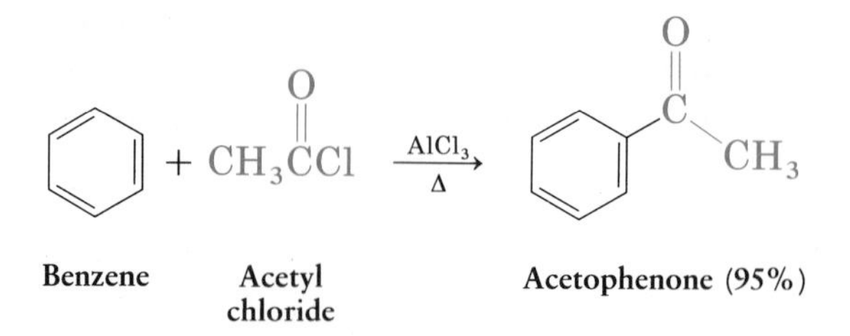

PROBLEM 9.5 Show how you could prepare 2-hexanone from these starting materials:

(a) 2-Hexanol (b) 1-Hexyne (c) 2-Methyl-1-hexene

PROBLEM 9.6 How would you carry out these reactions? More than one step may be required.

(a) 3-hexene $\longrightarrow$ 3-hexanone (b) benzene $\longrightarrow$ 1-phenylethanol

9.6 OXIDATION OF ALDEHYDES

Aldehydes are readily oxidized to yield carboxylic acids, RCHO → RCOOH, but ketones are usually unreactive toward oxidation. This reactivity difference is a consequence of the structural difference between the two functional groups: Aldehydes have a −CHO proton that can be removed during oxidation, but ketones don't.

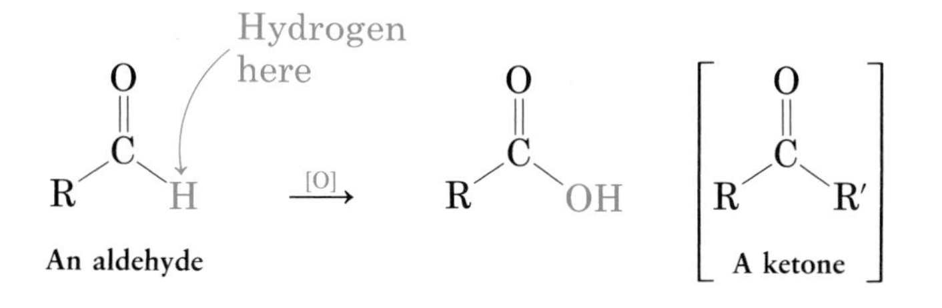

Tollens' reagent a solution of Ag^+ in aqueous NH_3; useful for oxidizing aldehydes to carboxylic acids

One of the simplest methods for oxidizing an aldehyde is to use silver oxide, Ag_2O, in dilute aqueous ammonia (the **Tollens reagent**). As the oxidation proceeds, silver metal is deposited on the walls of the reaction flask as a shiny mirror. In fact, the Tollens reagent can be used to detect the presence of an aldehyde functional group in a sample of unknown structure: A small amount of the unknown is dissolved in ethanol in a test tube, and a few drops of Tollens' reagent are added. If the test tube becomes silvery, the unknown is an aldehyde.

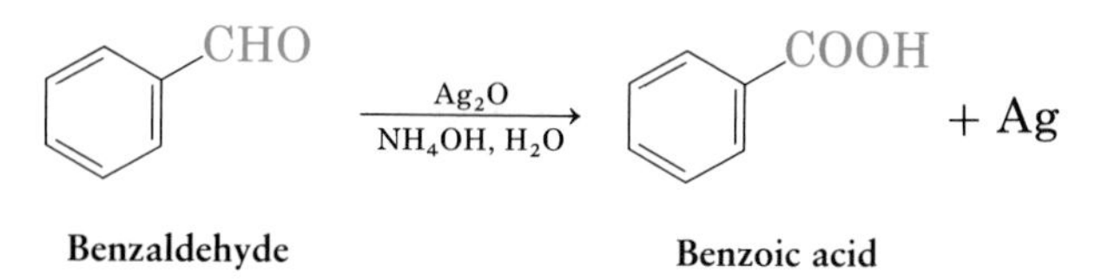

PRACTICE PROBLEM 9.1 What product would you obtain from reaction of 3-methylbutanal with Tollens' reagent?

SOLUTION Write the structure of the aldehyde starting material, and then replace the hydrogen bonded to the carbonyl group by −OH.

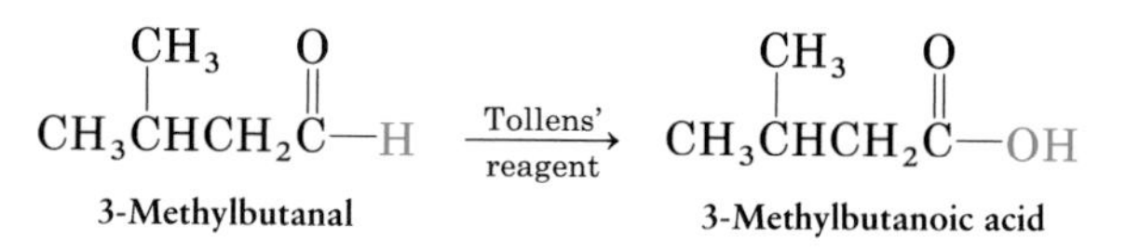

PROBLEM 9.7 Predict the products of the reaction of these substances with the Tollens' reagent:

(a) Pentanal (b) 2,2-Dimethylhexanal (c) Cyclohexanone

9.7 REACTIONS OF ALDEHYDES AND KETONES: NUCLEOPHILIC ADDITIONS

nucleophilic addition reaction a reaction that involves the addition of a nucleophile to a carbonyl group

carbanion a substance that has a trivalent, negatively charged carbon atom

The most important reaction of ketones and aldehydes is the **nucleophilic addition reaction.** As the name implies, a nucleophilic addition reaction involves the addition of a nucleophile (:Nu) to the carbonyl group. Hydroxide ion (HO^-), hydride ion (H^-), carbon anions (**carbanions,** R_3C^-), water (H_2O), ammonia (H_3N), and alcohols (ROH) are several of many possibilities.

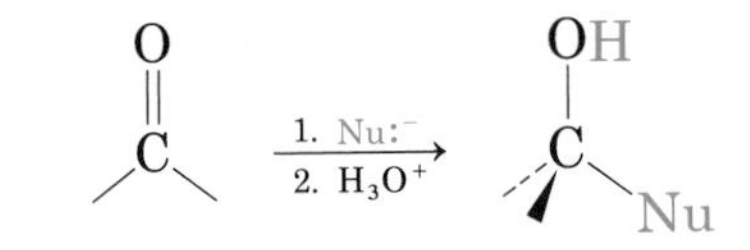

The general mechanism of nucleophilic addition is shown in Figure 9.2. As indicated, the nucleophile forms a new bond to the carbonyl-group carbon, the carbon–oxygen double bond breaks, and a proton bonds to the oxygen.

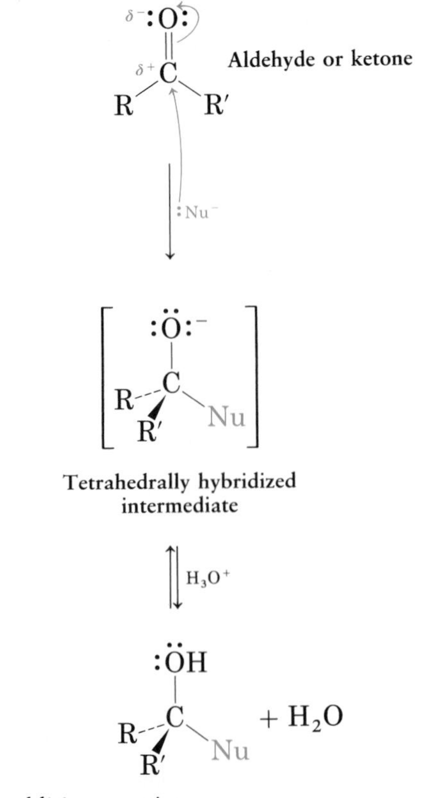

An electron pair from the nucleophile attacks the electrophilic carbonyl carbon, pushing an electron pair from the C=0 bond out onto oxygen. The carbonyl carbon rehybridizes from sp^2 to sp^3.

Protonation of the anion resulting from nucleophilic attack yields the neutral alcohol addition product.

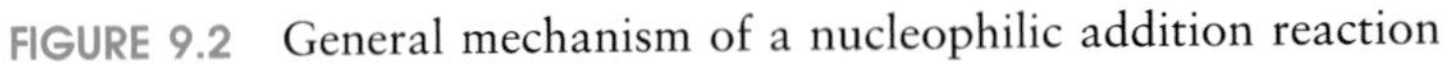
FIGURE 9.2 General mechanism of a nucleophilic addition reaction

Aldehydes are generally more reactive than ketones in nucleophilic additions because the presence of two relatively large substituents in ketones versus only one large substituent in aldehydes means that attacking nucleophiles are able to approach aldehydes more readily (Figure 9.3).

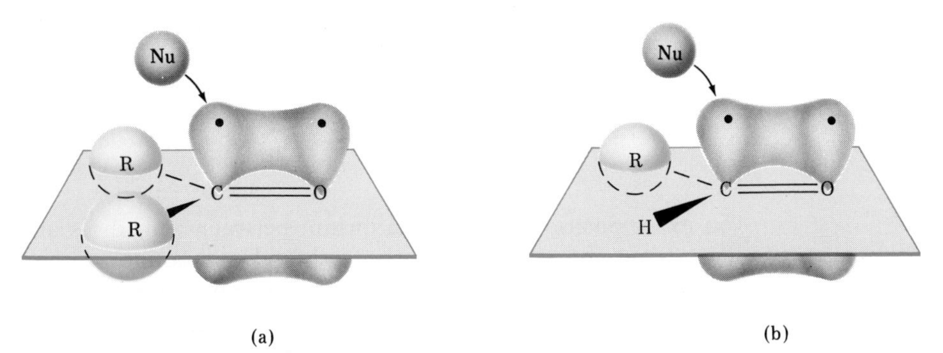

FIGURE 9.3 Nucleophilic attack on a ketone (a) is sterically hindered because of the two relatively large substituents. An aldehyde (b) has only one large substituent and is less hindered.

PRACTICE PROBLEM 9.2 What product would you expect from addition of the nucleophile hydroxide ion, HO^-, to acetaldehyde?

SOLUTION Hydroxide ion adds to the carbonyl carbon atom, giving an alkoxide–ion intermediate that is protonated to yield a 1,1-dialcohol.

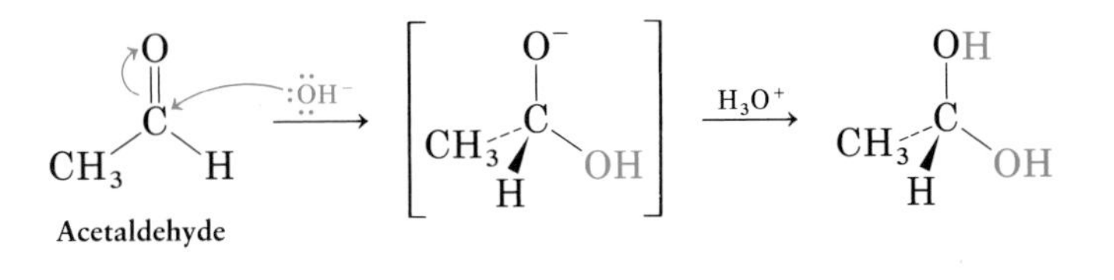

PROBLEM 9.8 What product would you expect if the nucleophile cyanide ion, CN^-, were to add to acetone, and the intermediate were to be protonated?

PROBLEM 9.9 The reduction of a ketone to a secondary alcohol on treatment with $NaBH_4$ (Section 8.6) is a nucleophilic addition reaction in which the nucleophile hydride ion (H^-) adds to the carbonyl group. Show the mechanism of this reduction.

PROBLEM 9.10 Which would you expect to be more reactive toward nucleophilic additions, propanal or 2,2-dimethylpropanal? Explain.

9.8 NUCLEOPHILIC ADDITION OF H_2O: HYDRATION

geminal diols referring to two groups attached to the same carbon atom

Aldehydes and ketones undergo a reversible reaction with water to yield 1,1-diols, or **geminal** (gem) **diols.**

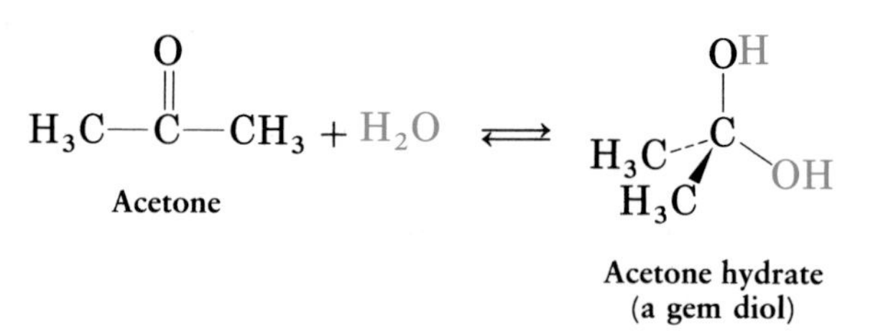

The exact position of the equilibrium between gem diols and ketones/aldehydes depends on the structure of the carbonyl compound. Although the equilibrium strongly favors the carbonyl compound in most cases, the gem diol is favored for a few simple aldehydes. For example, an aqueous solution of acetone consists of about 0.1% gem diol and 99.9% ketone whereas an aqueous solution of formaldehyde consists of 99.9% gem diol and 0.1% aldehyde.

The nucleophilic addition of water to ketones and aldehydes is slow in pure water but is catalyzed by both acid and base. Although these catalysts don't change the *position* of the equilibrium, they strongly affect the speed with which the hydration reaction occurs.

The base-catalyzed reaction takes place in several steps, as shown in Figure 9.4. The attacking nucleophile is the negatively charged hydroxide ion.

Hydroxide ion nucleophile adds to the ketone or aldehyde carbonyl group to yield an alkoxide ion intermediate.

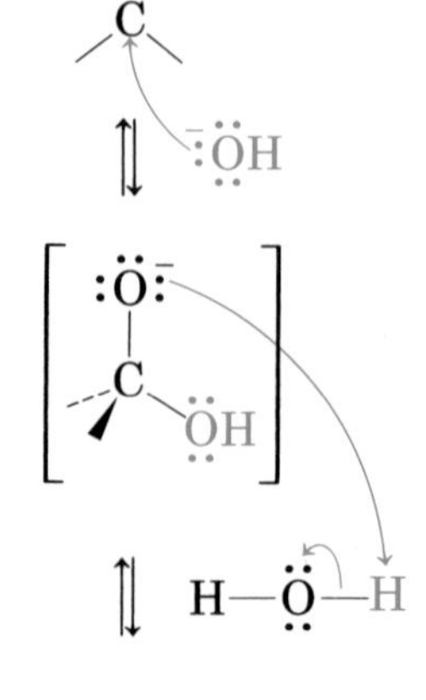

The basic alkoxide ion intermediate abstracts a proton (H^+) from water to yield gem diol product and regenerate hydroxide ion catalyst.

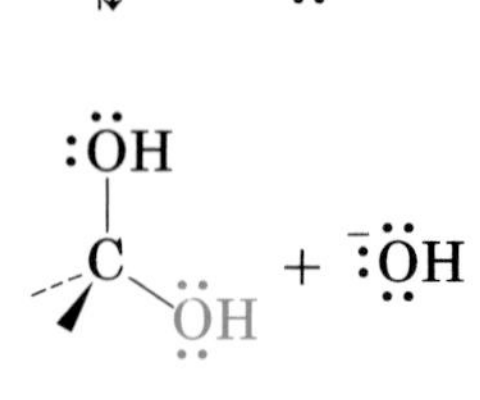

FIGURE 9.4 Mechanism of the base-catalyzed hydration reaction

The acid-catalyzed reaction also takes place in several steps. The acid catalyst first protonates the Lewis-basic oxygen atom of the carbonyl group, and subsequent nucleophilic addition of water yields a protonated gem diol. Loss of a proton then gives the neutral gem diol product (Figure 9.5).

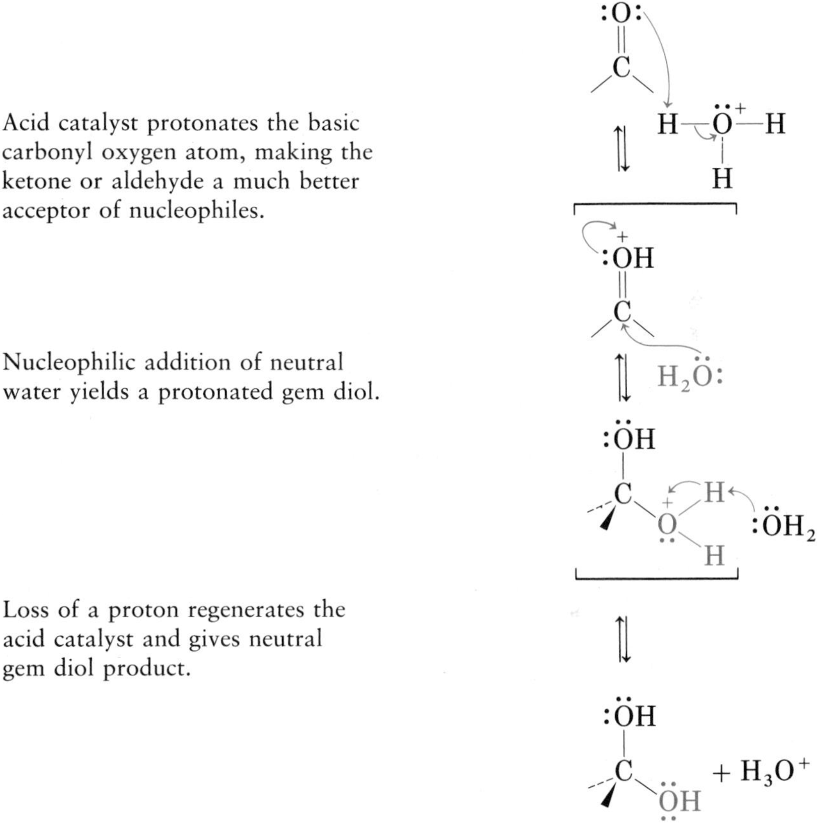

Acid catalyst protonates the basic carbonyl oxygen atom, making the ketone or aldehyde a much better acceptor of nucleophiles.
Nucleophilic addition of neutral water yields a protonated gem diol.
Loss of a proton regenerates the acid catalyst and gives neutral gem diol product.
:O:
C
H—O—H
H
:OH
H_2O:
:OH
H
:OH_2
+ H_3O^+
OH

FIGURE 9.5 Mechanism of the acid-catalyzed hydration reaction

Note the differences between the acid-catalyzed and base-catalyzed processes. The *base*-catalyzed reaction takes place rapidly because hydroxide ion is a much better nucleophilic *donor* than neutral water. The *acid*-catalyzed reaction takes place rapidly because the carbonyl compound is converted by protonation into a much better electrophilic *acceptor*.

PROBLEM 9.11 When dissolved in water, trichloroacetaldehyde (chloral, CCl_3CHO) exists primarily as the gem diol, chloral hydrate, $CCl_3CH(OH)_2$ (better known by the non-IUPAC name of "knockout drops"). Show the structure of chloral hydrate.

PROBLEM 9.12 The oxygen in water is primarily (99.8%) ^{16}O, but water enriched with the heavy isotope ^{18}O is also available. When a ketone or aldehyde is dissolved in $H_2{}^{18}O$, the isotopic label becomes incorporated into the carbonyl group: $R_2C{=}O + H_2O^* \rightarrow R_2C{=}O^* + H_2O$; (where $O^* = {}^{18}O$.) Explain.

9.9 NUCLEOPHILIC ADDITION OF GRIGNARD REAGENTS: ALCOHOL FORMATION

Grignard reagents, RMgX, react with ketones and aldehydes by a nucleophilic addition pathway to yield alcohols. As we saw in Section 7.4, Grignard reagents are prepared by reaction of alkyl, aryl, or vinylic halides with magnesium in ether.

$$\underset{\text{An organohalide}}{\text{R—X}} \xrightarrow[\text{ether}]{\text{Mg}} \underset{\text{A Grignard reagent}}{\overset{\delta^-}{\text{R}}\text{—Mg}\overset{\delta^+}{\text{X}}}$$

The carbon–magnesium bond of Grignard reagents is polarized in such a way that the carbon atom is both nucleophilic and basic. Grignard reagents therefore react as if they were carbanions, R^-, and they undergo nucleophilic addition to ketones and aldehydes just as water does. Nucleophilic addition first produces a tetrahedrally hybridized magnesium-alkoxide intermediate, which is then protonated to yield the neutral alcohol upon treatment with aqueous acid.

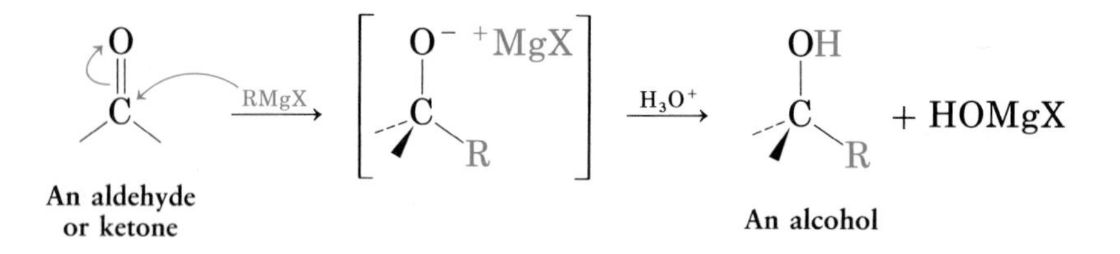

An aldehyde or ketone

An alcohol

A large number of alcohol products can be obtained from Grignard reactions, depending on the reagents used. For example, formaldehyde, CH_2O, reacts with Grignard reagents to give primary alcohols, RCH_2OH.

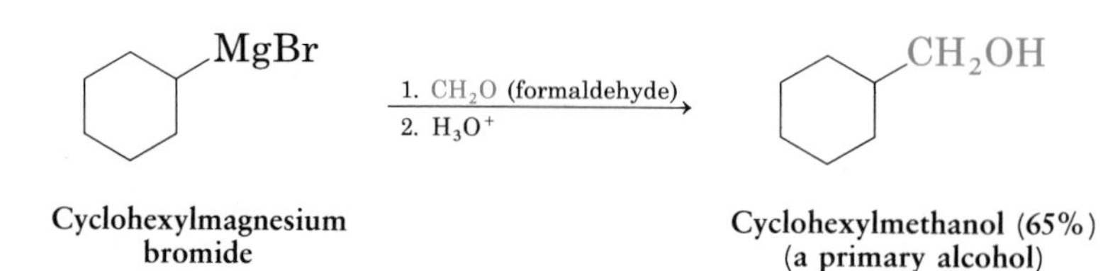

Cyclohexylmagnesium bromide

Cyclohexylmethanol (65%) (a primary alcohol)

Aldehydes react with Grignard reagents to give secondary alcohols, and ketones react similarly to yield tertiary alcohols:

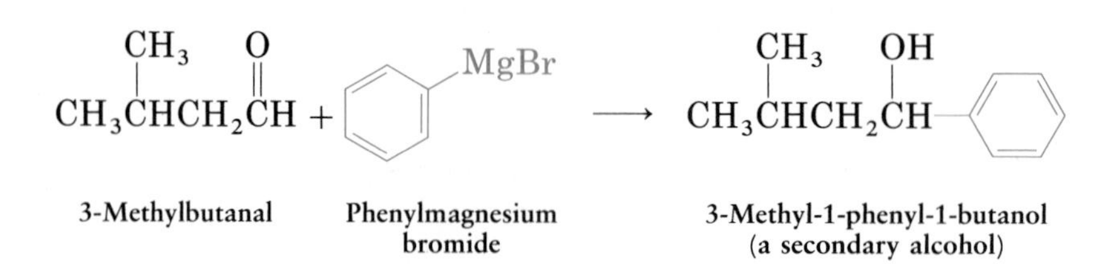

3-Methylbutanal

Phenylmagnesium bromide

3-Methyl-1-phenyl-1-butanol (a secondary alcohol)

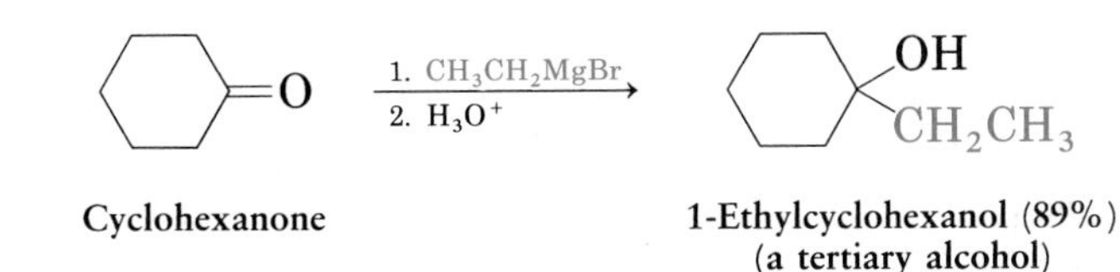

Although broad in scope, the Grignard reaction also has limitations. For example, Grignard reagents can't be prepared from organohalides if there are other reactive functional groups in the same molecule. A compound that is both an alkyl halide and a ketone won't form a Grignard reagent—it reacts with itself instead. Similarly, a compound that is both an alkyl halide and a carboxylic acid, alcohol, or amine can't form a Grignard reagent because the acidic RCOOH, ROH, or RNH_2 protons in the molecule simply react with the basic Grignard reagent as it's formed.

In general, Grignard reagents can't be prepared from compounds that have these functional groups in the molecule:

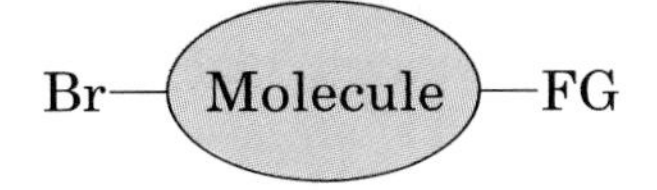

where FG = $-OH$, $-NH_2$, $-SH$, $-COOH$, $-NO_2$, $-CHO$, $-COR$, $-CN$, or $-CONH_2$.

PRACTICE PROBLEM 9.3 How can you use the addition of a Grignard reagent to a ketone to synthesize 2-phenyl-2-propanol?

SOLUTION First draw the structure of the product and identify the groups bonded to the alcohol carbon atom. In this instance, there are two methyl groups ($-CH_3$) and one phenyl ($-C_6H_5$). One of the three will have come from a Grignard reagent, and the remaining two will have come from a ketone. Thus, the possibilities are addition of methylmagnesium bromide to acetophenone and addition of phenylmagnesium bromide to acetone:

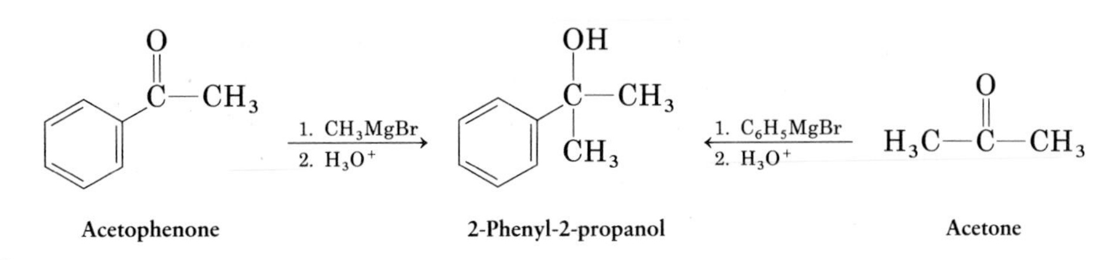

PROBLEM 9.13 Show the products obtained from addition of methylmagnesium bromide to these compounds:

(a) Cyclopentanone (b) Benzophenone (diphenyl ketone) (c) 3-Hexanone

PROBLEM 9.14 How can you use a Grignard addition reaction to prepare these alcohols?

(a) 2-Methyl-2-propanol (b) 1-Methylcyclohexanol (c) 3-Methyl-3-pentanol

9.10 NUCLEOPHILIC ADDITION OF AMINES: IMINE FORMATION

imine
a compound with the C=N functional group

Ammonia and primary amines, RNH_2, add to aldehydes and ketones to yield **imines,** $R_2C{=}NR'$. Imines are formed by nucleophilic addition to the carbonyl group by the nucleophilic amine, followed by loss of water from the amino alcohol addition product.

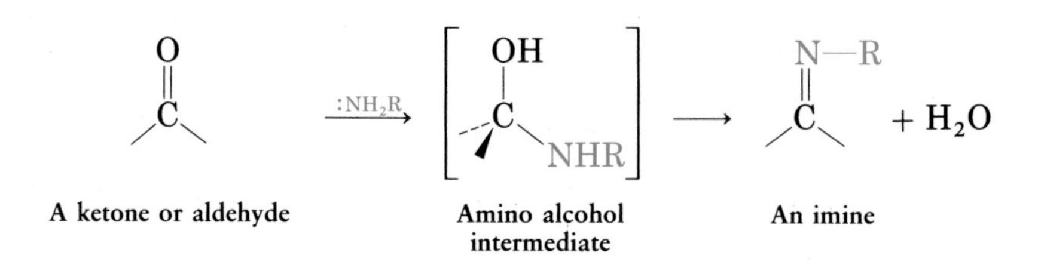

A ketone or aldehyde — Amino alcohol intermediate — An imine

Imine derivatives, such as oximes, and 2,4-dinitrophenylhydrazones (abbreviated as 2,4-DNPs) are also easily prepared by reaction of a ketone or aldehyde with the appropriate $H_2N{-}Y$ compound. These imines, which are usually crystalline, easy-to-handle materials, are often prepared as a means of converting liquid ketones or aldehydes into solid derivatives.

$$\text{Cyclohexanone}{=}O + :NH_2OH \longrightarrow \text{Cyclohexane}{=}N{-}OH + H_2O$$

Cyclohexanone — Hydroxylamine — Cyclohexanone oxime (mp 90°C)

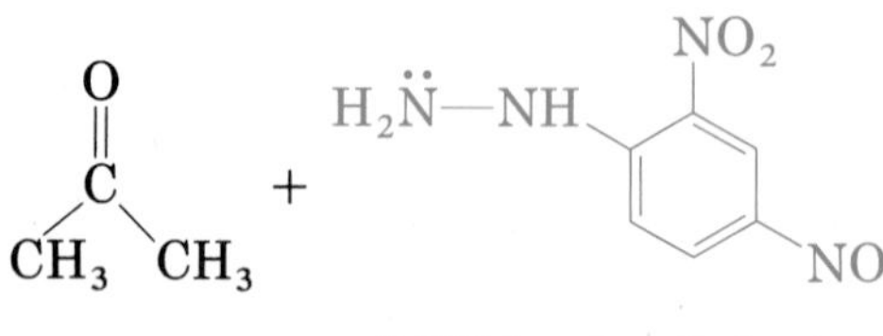

2,4-Dinitrophenylhydrazine

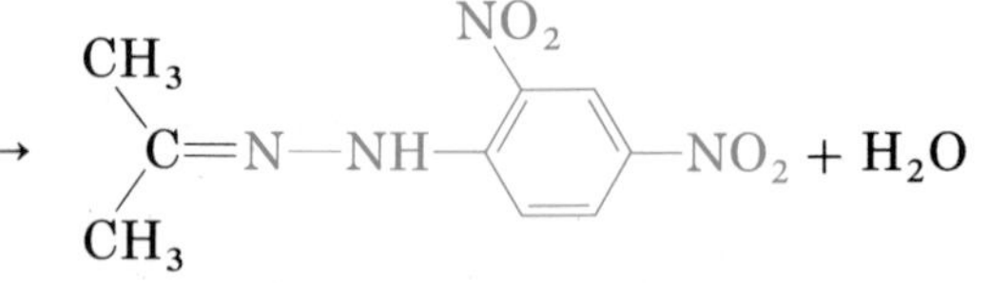

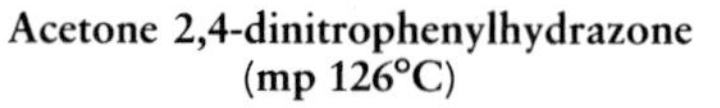
Acetone 2,4-dinitrophenylhydrazone (mp 126°C)

Wolff–Kishner reaction
a reaction for reducing a ketone or aldehyde to an alkane by reaction with KOH and hydrazine

An important variant of imine formation involves the treatment of a ketone or aldehyde with hydrazine, $H_2N{-}NH_2$, in the presence of strong base. This process, called the **Wolff–Kishner** reaction after its codiscoverers, is an extremely useful method for converting ketones or aldehydes into alkanes, $R_2C{=}O \rightarrow R_2CH_2$.

The Wolff–Kishner reaction involves initial formation of an imine intermediate called a *hydrazone*, followed by loss of nitrogen and formation of the alkane product. The reaction is very general and can be used for most aldehydes and ketones.

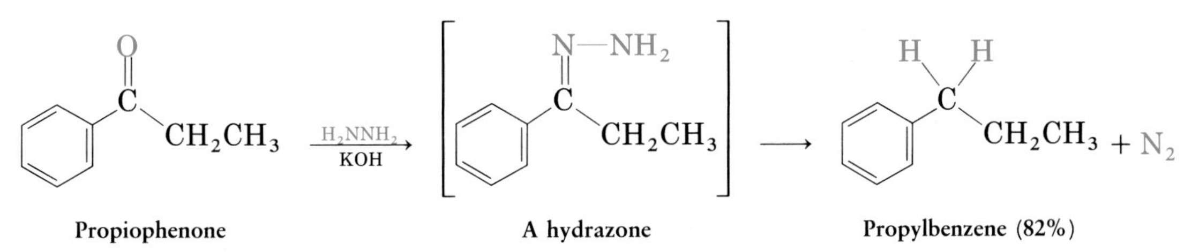

PRACTICE PROBLEM 9.4 What product would you expect from the reaction of 2-butanone with hydroxylamine, NH_2OH?

SOLUTION Take oxygen from the ketone and two hydrogens from the amine to form water, and then join the fragments that remain:

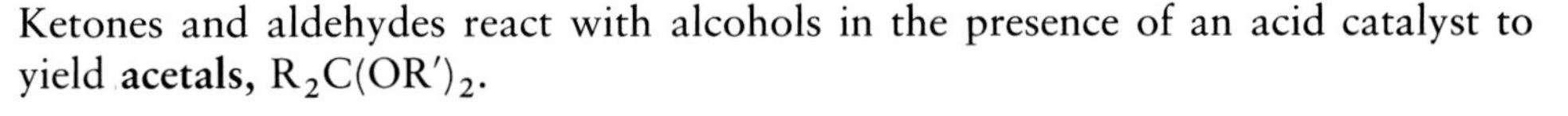

$$CH_3CH_2\overset{O}{\overset{\|}{C}}CH_3 + H_2NOH \longrightarrow CH_3CH_2\overset{NOH}{\overset{\|}{C}}CH_3 + H_2O$$

PROBLEM 9.15 Write the products you would expect to obtain from treatment of cyclohexanone with these reagents:

(a) H_2NOH (b) 2,4-Dinitrophenylhydrazine (c) N_2H_4, KOH (d) $NaBH_4$

PROBLEM 9.16 Show how you could prepare butylbenzene from benzene by carrying out a Friedel–Crafts acylation reaction (Section 5.8) followed by a Wolff–Kishner reduction.

9.11 NUCLEOPHILIC ADDITION OF ALCOHOLS: ACETAL FORMATION

acetal a compound that has two –OR groups bonded to the same carbon atom

Ketones and aldehydes react with alcohols in the presence of an acid catalyst to yield **acetals,** $R_2C(OR')_2$.

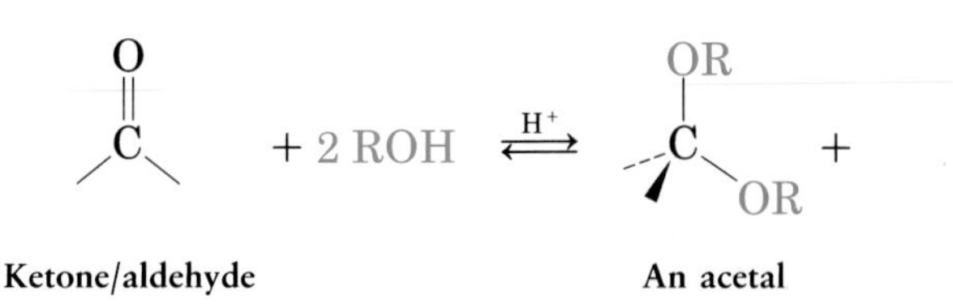

Ketone/aldehyde An acetal

hemiacetal a compound that has one –OR group and one –OH group attached to the same carbon atom

Acetal formation involves the acid-catalyzed nucleophilic addition of an alcohol to the carbonyl group in a manner similar to that of acid-catalyzed hydration (Section 9.8). The initial nucleophilic addition step yields a hydroxy ether called a **hemiacetal** that reacts further with a second equivalent of alcohol to yield the acetal plus water. For example, reaction of cyclohexanone with methanol yields the dimethyl acetal.

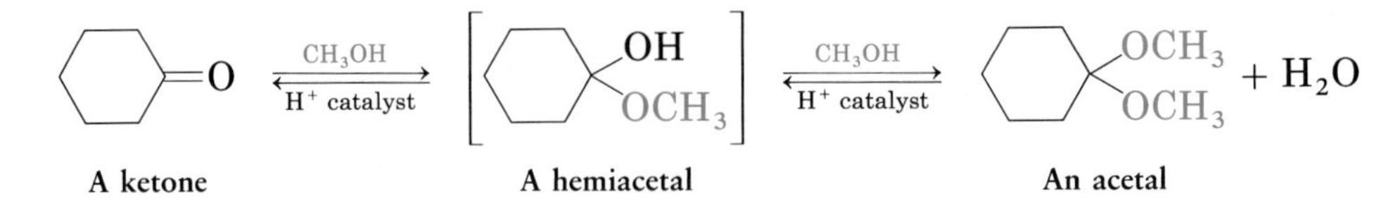

All the steps during acetal formation are reversible. Thus, the reaction can be made to go either forward (from carbonyl compound to acetal) or backward (from acetal to carbonyl compound) by changing the reaction conditions. The forward reaction is accomplished under conditions that remove water from the medium and thus drive the reaction to the right. The backward reaction is accomplished by treating the acetal with mineral acid in the presence of a large excess of water.

protecting group a group that is temporarily introduced into a molecule to protect a functional group from reaction elsewhere in the molecule

Acetals are valuable to organic chemists because they can serve as **protecting groups** for ketones and aldehydes. To see what this means, imagine that you are faced with having to reduce the keto ester ethyl 4-oxopentanoate to obtain the keto alcohol 5-hydroxy-2-pentanone. This reaction can't be done in a single step because of the presence of the ketone carbonyl group in the molecule. If you were to treat ethyl 4-oxopentanoate with $LiAlH_4$, both ketone and ester groups would be reduced.

$$CH_3\overset{O}{\overset{\|}{C}}CH_2CH_2\overset{O}{\overset{\|}{C}}OCH_2CH_3 \xrightarrow{?} CH_3\overset{O}{\overset{\|}{C}}CH_2CH_2CH_2OH$$

Ethyl 4-oxopentanoate **5-Hydroxy-2-pentanone**

This situation isn't unusual: It often happens that one functional group in a complex molecule interferes with intended chemistry on another functional group elsewhere in the molecule. In such situations, you can often circumvent the problem by protecting the interfering functional group to render it unreactive, carrying out the desired reaction, and then removing the protecting group.

Ketones and aldehydes can be protected by converting them into acetals. Acetals, like other ethers, are stable to bases, reducing agents, and Grignard reagents, but they are acid sensitive. Thus, you can selectively reduce the ester group in ethyl-4-oxopentanoate by converting the keto group into an acetal, treating the compound with $LiAlH_4$ in ether, and then removing the acetal protecting group by treatment with aqueous acid.

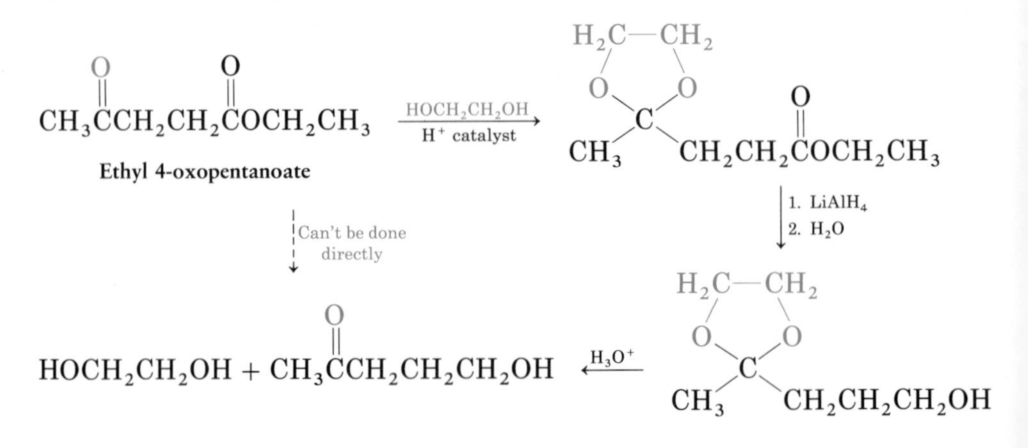

PRACTICE PROBLEM 9.5 What product would you obtain from acid-catalyzed reaction of 2-methylcyclopentanone with methanol?

SOLUTION Replace the oxygen of the ketone with two $-OCH_3$ groups from the alcohol:

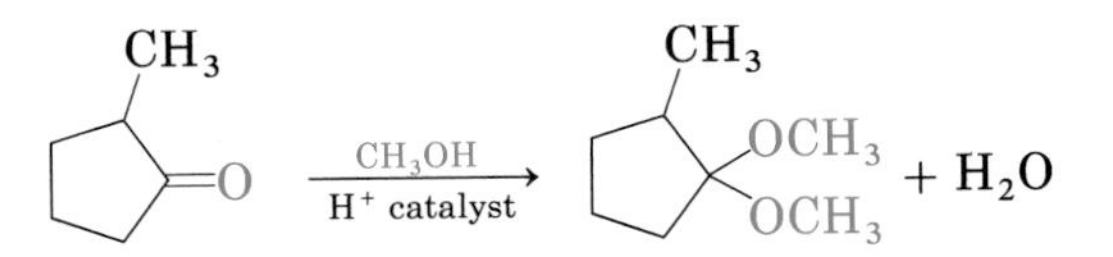

PROBLEM 9.17 What product would you expect from the acid-catalyzed reaction of cyclohexanone and ethanol?

PROBLEM 9.18 Show the mechanism of the acid-catalyzed formation of a *cyclic* acetal from ethylene glycol and acetone.

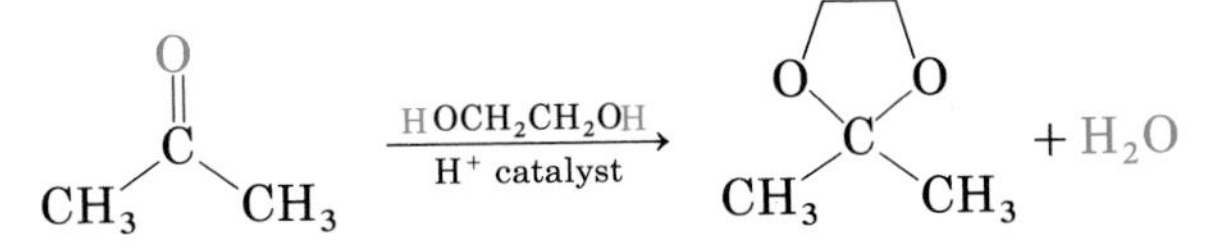

PROBLEM 9.19 Show how you might carry out the following transformation. A protection step is needed.

$$\text{HCCH}_2\text{CH}_2\text{COCH}_3 \ (\text{with C=O at both carbonyl carbons}) \longrightarrow \text{HCCH}_2\text{CH}_2\text{CH}_2\text{OH} \ (\text{with C=O at the aldehyde carbon})$$

9.12 NUCLEOPHILIC ADDITION OF PHOSPHORUS YLIDES: THE WITTIG REACTION

Wittig reaction the reaction of a ketone or aldehyde with a phosphorus ylide to yield an alkene

ylide a compound that has adjacent plus and minus charges

Ketones and aldehydes are converted into alkenes by means of the **Wittig reaction.** In this process, a phosphorus **ylide,** $R_2\overset{-}{C}-\overset{+}{P}(C_6H_5)_3$, adds to the ketone or aldehyde to yield an intermediate that reacts further to yield an alkene. (An ylide, pronounced **ill-id**, is a dipolar compound that has adjacent plus and minus charges.) The overall result is replacement of the carbonyl oxygen atom by the organic fragment originally bonded to phosphorus.

$$\text{>C=O} + R_2\overset{-}{\ddot{C}}-\overset{+}{P}(C_6H_5)_3 \longrightarrow \left[\begin{array}{cc} :\ddot{O}: & \overset{+}{P}(C_6H_5)_3 \\ | & | \\ -C- & C(R)_2 \end{array}\right] \longrightarrow \text{>C=CR}_2 + (C_6H_5)_3P{=}O$$

Ketone/aldehyde An ylide Alkene Triphenylphosphine oxide

The phosphorus ylides necessary for the Wittig reaction are easily prepared by S_N2 reaction of primary alkyl halides with triphenylphosphine, $(C_6H_5)_3P\colon$, followed by treatment with base. The proton on the carbon atom next to the positively charged phosphorus is acidic and can be removed by a strong base such as sodium hydride (NaH) or butyllithium (BuLi) to generate the ylide. For example:

$$(C_6H_5)_3P\colon + CH_3{-}Br \xrightarrow{S_N2 \text{ reaction}}$$

Triphenylphosphine **Bromomethane**

$$(C_6H_5)_3\overset{+}{P}{-}CH_3 \quad \ddot{Br}^{-} \xrightarrow[\text{THF}]{\text{BuLi}} (C_6H_5)_3\overset{+}{P}{-}\overset{-}{\ddot{C}}H_2$$

Methyltriphenylphosphonium bromide (99%) **Methylenetriphenylphosphorane**

The great value of the Wittig reaction is that pure alkenes of known structure are prepared. The alkene double bond is always exactly where the carbonyl group was in the precursor, and no product mixtures (other than *E,Z* isomers) are formed. For example, reaction of cyclohexanone with methylenetriphenylphosphorane yields the single pure alkene product, methylenecyclohexane, whereas the alternative synthesis by addition of methylmagnesium bromide to cyclohexanone, followed by acid-catalyzed dehydration, leads to a mixture of two alkenes.

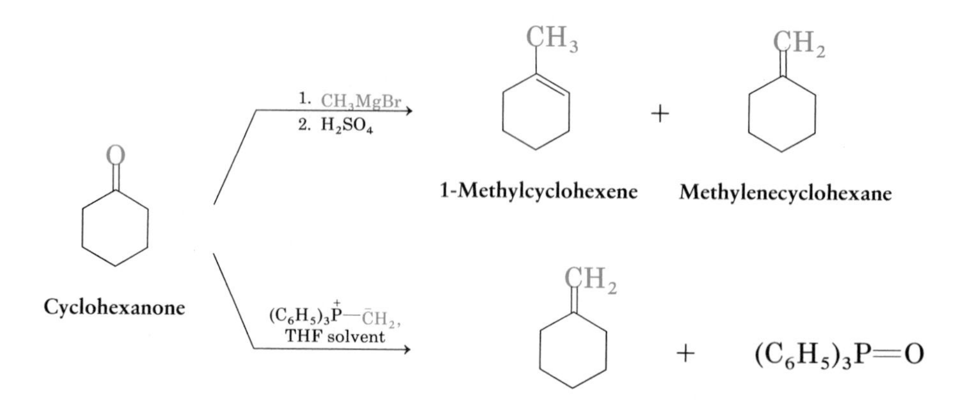

PRACTICE PROBLEM 9.6 Show the product resulting from reaction of acetophenone with the Wittig reagent prepared from bromoethane.

SOLUTION The Wittig reagent from bromoethane is ethylidenetriphenylphosphorane:

$$(C_6H_5)_3P\colon + CH_3CH_2Br \longrightarrow$$

$$(C_6H_5)_3\overset{+}{P}{-}CH_2CH_3 \quad Br^{-} \xrightarrow{\text{BuLi}} (C_6H_5)_3\overset{+}{P}{-}\overset{=}{C}HCH_3$$

Ethylidenetriphenylphosphorane

Reaction of a Wittig reagent with a ketone yields an alkene in which the ketone oxygen atom has been replaced by the group bonded to phosphorus.

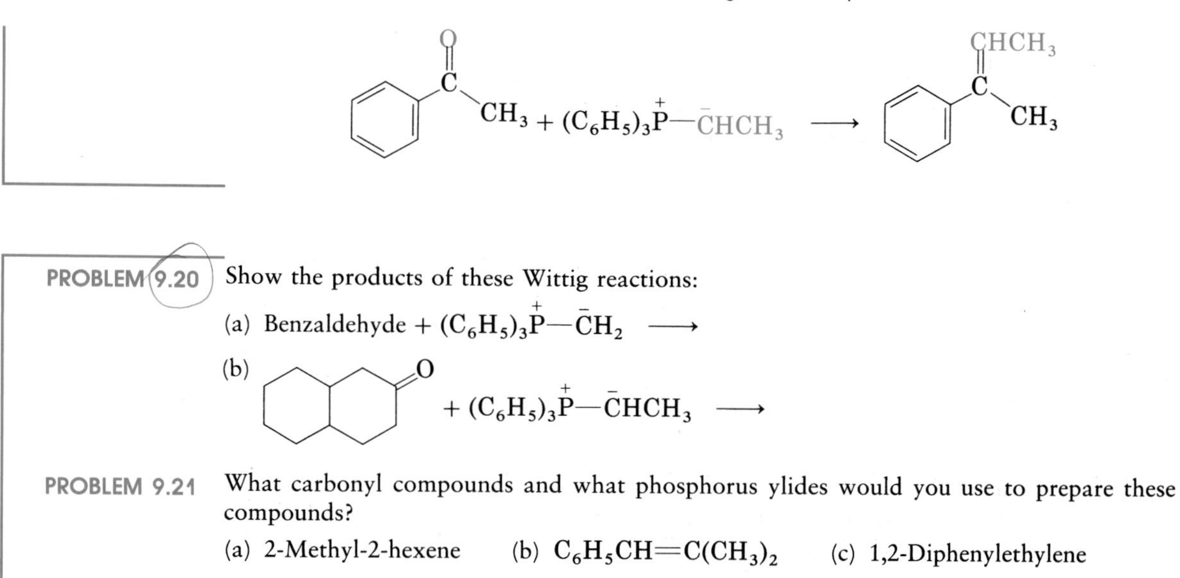

PROBLEM 9.20 Show the products of these Wittig reactions:

(a) Benzaldehyde + $(C_6H_5)_3\overset{+}{P}-\overset{-}{C}H_2 \longrightarrow$

(b) [bicyclic ketone] + $(C_6H_5)_3\overset{+}{P}-\overset{-}{C}HCH_3 \longrightarrow$

PROBLEM 9.21 What carbonyl compounds and what phosphorus ylides would you use to prepare these compounds?

(a) 2-Methyl-2-hexene (b) $C_6H_5CH{=}C(CH_3)_2$ (c) 1,2-Diphenylethylene

9.13 SOME BIOLOGICAL NUCLEOPHILIC ADDITION REACTIONS

Nature synthesizes the molecules of life using many of the reactions that chemists use in the laboratory. This is particularly true of carbonyl-group reactions, where nucleophilic addition steps are an important part of the biosynthesis of many vital molecules.

One of the pathways by which amino acids are made involves a nucleophilic addition reaction of α-keto acids. To choose a specific case, the bacterium *Bacillus subtilis* synthesizes alanine from pyruvic acid and ammonia. The key step in this biological transformation is the nucleophilic addition of ammonia to the ketone carbonyl group of pyruvic acid to give an imine that is further reduced by enzymes.

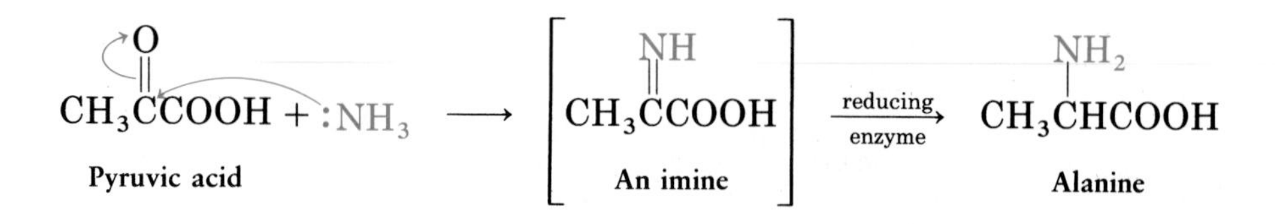

Other examples of nucleophilic carbonyl addition occur frequently in carbohydrate chemistry. For example, the six-carbon sugar, glucose, acts in some respects as if it were an aldehyde. Thus, glucose can be oxidized to yield a carboxylic acid. Spectroscopic examination of glucose, however, shows that no aldehyde group is present; instead, glucose exists as a *cyclic hemiacetal*. The hydroxyl group at carbon 5 adds to the aldehyde at carbon 1 in an internal nucleophilic addition step.

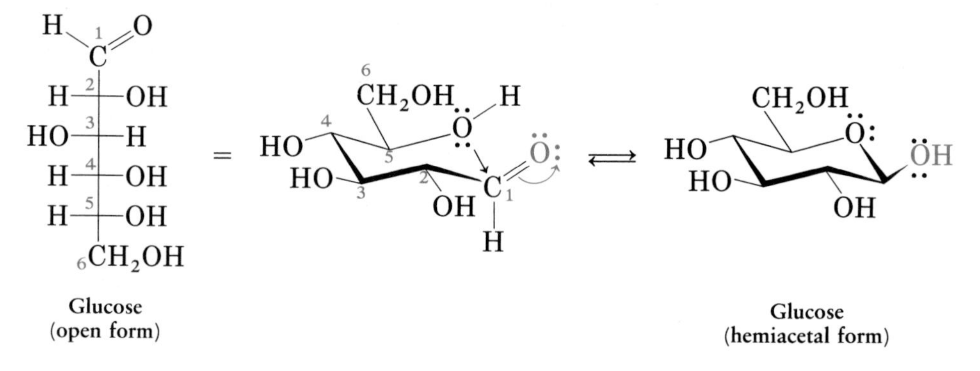

Glucose
(open form)

Glucose
(hemiacetal form)

Further reaction between molecules of glucose leads to the carbohydrate polymer cellulose. Cellulose, which constitutes the major building block of plant cell walls, consists simply of glucose units joined by acetal linkages between carbon 1 of one glucose with the hydroxyl group at carbon 4 of another glucose.

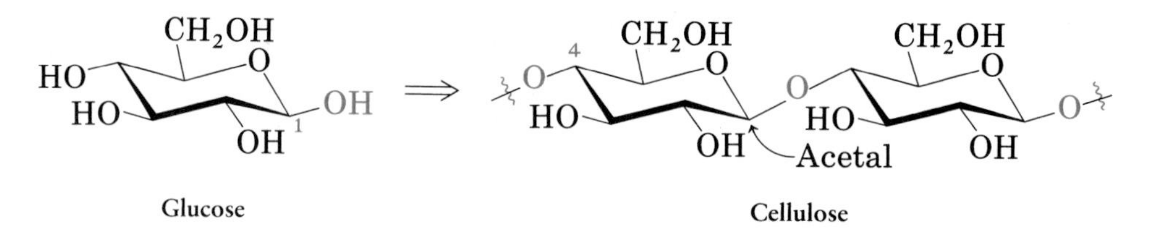

Glucose

Cellulose

We'll study this and other reactions of carbohydrates in more detail in Chapter 14.

INTERLUDE

Chemical Warfare in Nature

Among many known nucleophilic additions is the reaction of ketones and aldehydes with HCN (hydrogen cyanide) to yield cyano alcohols, or **cyanohydrins** [RCH(OH)CN].

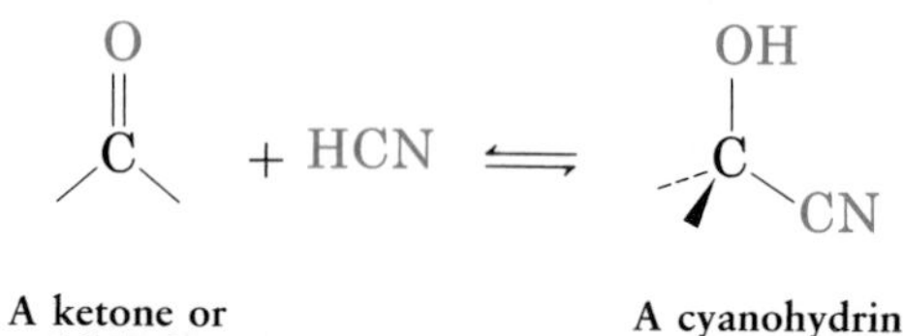

A ketone or aldehyde

A cyanohydrin

The reaction of HCN with ketones and aldehydes to yield cyanohydrins is of more than just chemical interest: Cyanohydrins also play an interesting role in the chemical defense mechanisms of certain plants and insects against predators. For example, when the millipede *Apheloria corrugata* is attacked by ants, it secretes mandelonitrile and an enzyme that catalyzes the decomposition of mandelonitrile into benzaldehyde and HCN. The millipede actually protects itself by discharging poisonous HCN at attackers.

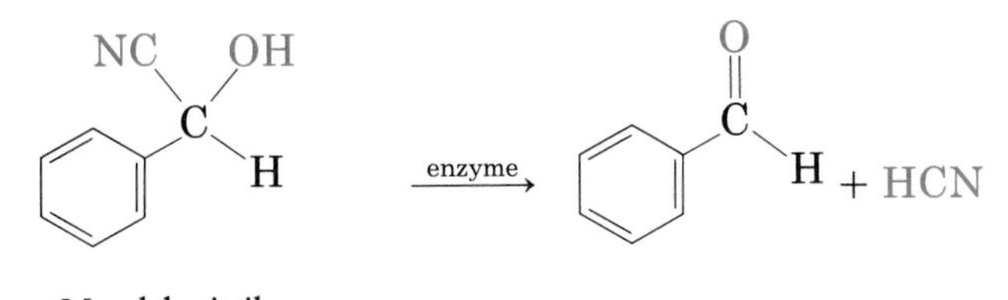

Mandelonitrile
(from ***Apheloria corrugata***)

In a similar vein, the pits of apricots and peaches contain a group of substances called *cyanogenic glycosides*. These compounds, of which amygdalin (Laetrile) is notorious because of its claimed anticancer activity, consist of benzaldehyde cyanohydrin bonded to simple sugars such as glucose. When eaten, the sugar unit is cleaved off, and HCN is released. Predators soon learn to avoid these seeds.

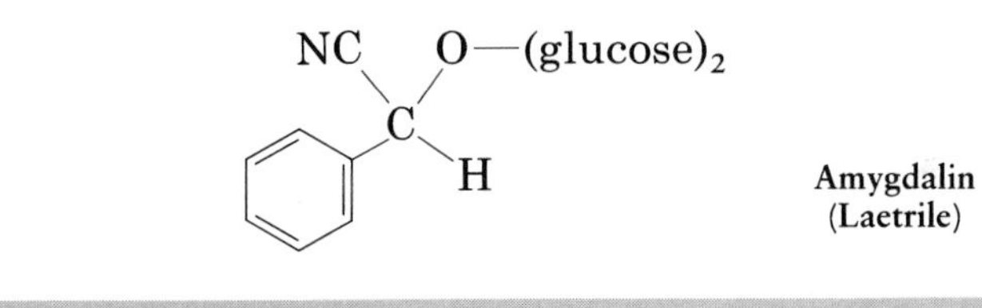

Amygdalin
(Laetrile)

SUMMARY AND KEY WORDS

Carbonyl compounds can be classified into two general categories:

R_2CO RCHO	**Ketones** and **aldehydes** are similar in their reactivity and are distinguished by the fact that the substituents on the acyl carbon can't serve as leaving groups.
RCOOH RCOOR′ $RCONH_2$ RCOCl RCOOCOR	**Carboxylic acids** and their derivatives—**esters, amides, anhydrides,** and **acid chlorides**—are distinguished by the fact that the substituents on the acyl carbon *can* serve as leaving groups.

Structurally, a carbon–oxygen double bond is similar to a carbon–carbon double bond. The carbonyl carbon atom is sp^2 hybridized and forms both an sp^2 sigma bond and a *p* pi bond to oxygen. Carbonyl groups are strongly polarized because of the electronegativity of oxygen. Thus, carbonyl carbons are strongly electrophilic.

Aldehydes are usually prepared by oxidative cleavage of alkenes or by oxidation of primary alcohols. Ketones are similarly prepared by oxidative cleavage of alkenes or by oxidation of secondary alcohols.

Ketones and aldehydes behave similarly in much of their chemistry though aldehydes are generally more reactive than ketones. Both undergo **nucleophilic addition reactions,** and a variety of product types can be prepared. For example, ketones and aldehydes are reduced by $NaBH_4$ or $LiAlH_4$ to yield secondary and primary alcohols, respectively. Addition of Grignard reagents also leads to alcohols. Primary amines add to ketones and aldehydes to give **imines, $R_2C{=}NR'$.** If hydrazine, H_2NNH_2, is used as the amine nucleophile, the initially formed imine intermediate undergoes further reaction with base to yield an alkane (the **Wolff–Kishner reaction**). Alcohols add to ketones and aldehydes to yield **acetals, $R_2C(OR'')_2$,** which are valuable as carbonyl **protecting groups.** Phosphorus ylides (**Wittig reagents**) add to aldehydes and ketones to give alkenes.

SUMMARY OF REACTIONS

1. Reaction of ketones and aldehydes with Grignard reagents to yield alcohols (Section 9.9)

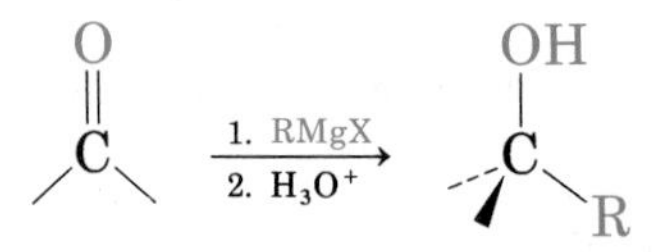

2. Reaction of ketones and aldehydes with amines to yield imines (Section 9.10)

$$>C{=}O + RNH_2 \longrightarrow >C{=}NR + H_2O$$

3. Wolff–Kishner reaction to yield alkanes (Section 9.10)

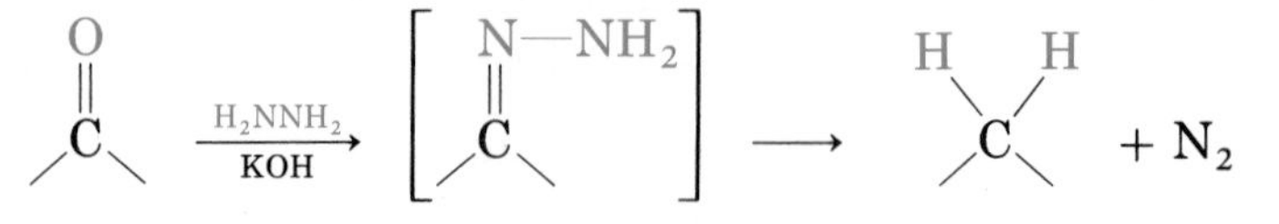

4. Reaction of ketones and aldehydes with alcohols to yield acetals (Section 9.11)

$$>C{=}O + ROH \underset{\text{catalyst}}{\overset{H^+}{\rightleftharpoons}} \left[>C(OH)(OR) \right] \rightleftharpoons >C(OR)(OR) + H_2O$$

5. Wittig reaction to yield alkenes (Section 9.12)

$$>C{=}O + (C_6H_5)_3\overset{+}{P}{-}\overset{-}{C}R_2 \longrightarrow >C{=}CR_2 + (C_6H_5)_3PO$$

ADDITIONAL PROBLEMS

9.22 Identify the different kinds of carbonyl groups in these molecules:

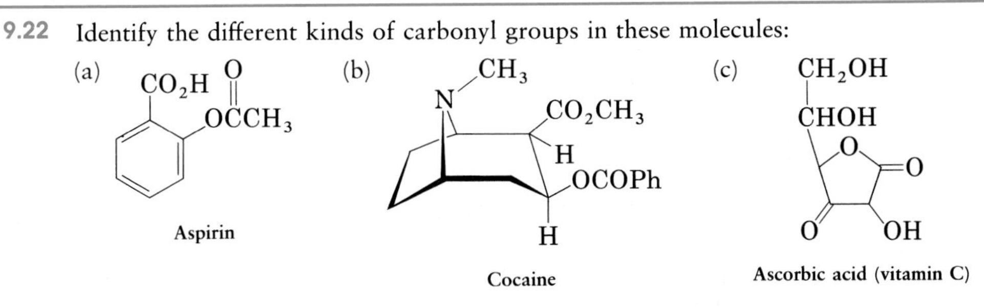

9.23 What is the structural difference between aldehydes and ketones?

9.24 Draw structures corresponding to these names:

(a) Bromoacetone
(b) 3-Methyl-2-butanone
(c) 3,5-Dinitrobenzaldehyde
(d) 3,5-Dimethylcyclohexanone
(e) 2,2,4,4-Tetramethyl-3-pentanone
(f) Butanedial
(g) (*S*)-2-Hydroxypropanal
(h) 3-Phenyl-2-propenal

9.25 Draw and name the seven ketones and aldehydes with the formula $C_5H_{10}O$.

9.26 Draw structures of molecules that meet these descriptions:

(a) A cyclic ketone, C_6H_8O
(b) A diketone, $C_6H_{10}O_2$
(c) An aryl ketone, C_9H_{10}
(d) A 2-bromoaldehyde, C_5H_9BrO

9.27 Give IUPAC names for these structures:

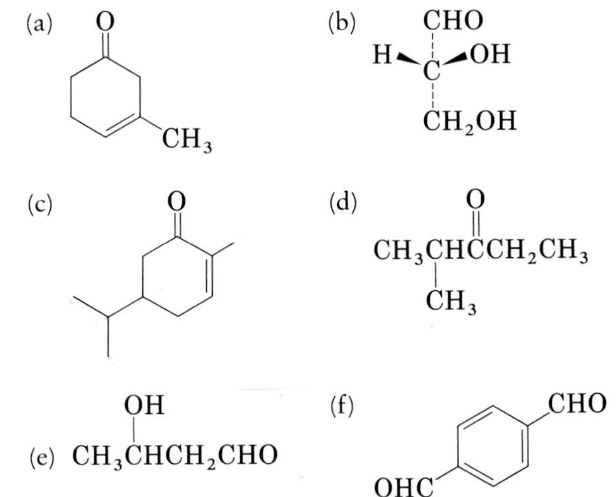

9.28 How can you explain the observation that S_N2 reaction of (dibromomethyl)benzene, $C_6H_5CHBr_2$, with NaOH yields benzaldehyde rather than (dihydroxymethyl)benzene, $C_6H_5CH(OH)_2$?

9.29 Predict the products of the reaction of phenylacetaldehyde, $C_6H_5CH_2CHO$, with these reagents:

(a) $NaBH_4$, then H_3O^+
(b) Tollens' reagent
(c) NH_2OH
(d) CH_3MgBr, then H_3O^+

(e) CH_3OH, H^+ catalyst (f) H_2NNH_2, KOH

(g) $(C_6H_5)_3\overset{+}{P}—\overset{-}{C}H_2$

9.30 Answer Problem 9.29 for reaction of acetophenone, $C_6H_5COCH_3$.

9.31 Reaction of 2-butanone with $NaBH_4$ yields an alcohol product having a new chiral center. What stereochemistry would you expect the product to have? (Hint: Review Section 6.13.)

9.32 In light of your answer to Problem 9.31, what stereochemistry would you expect the product from reaction of phenylmagnesium bromide with 2-butanone to have?

9.33 Starting from 2-cyclohexenone and any other reagents needed, how would you prepare these substances? More than one step may be required.

(a) Cyclohexene (b) 1-Methylcyclohexanol
(c) Cyclohexanol (d) 1-Phenyl-2-cyclohexen-1-ol

9.34 Show how the Wittig reaction can be used to prepare these alkenes. Identify the alkyl halide and the carbonyl components that would be used.

(a) $C_6H_5CH{=}CH—CH{=}CHC_6H_5$

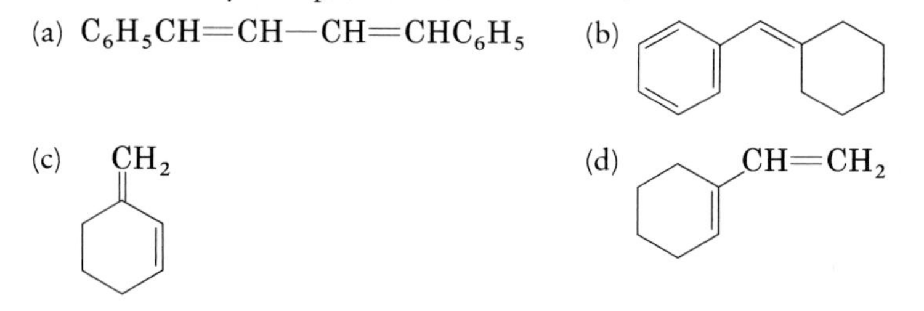

9.35 Why do you suppose tri*phenyl*phosphine is used in the Wittig reaction rather than, say, tri*methyl*phosphine? What problems might you run into if trimethylphosphine were used?

9.36 β-Carotene, the orange pigment of carrots and a dietary source of vitamin A, can be prepared by a double Wittig reaction between 2 equiv of β-ionylideneacetaldehyde and a diylide. What is the structure of the diylide?

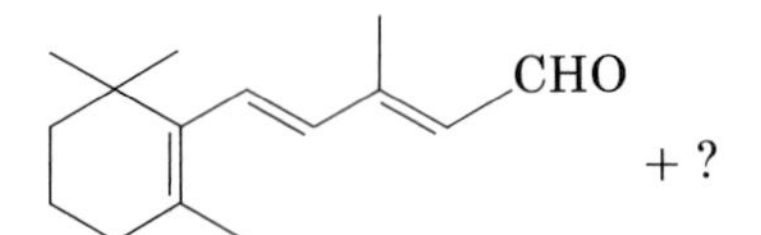

β-Ionylideneacetaldehyde

↓

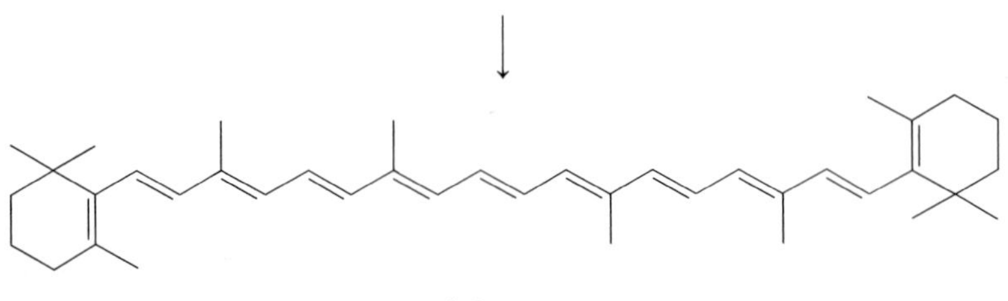

β-Carotene

9.37 Use a Grignard reaction on a ketone or aldehyde to synthesize these compounds:

(a) 2-Pentanol (b) 2-Phenyl-2-butanol
(c) 1-Ethylcyclohexanol (d) Diphenylmethanol

9.38 Show the structures of the starting alcohols and ketones or aldehydes you would use to make these acetals:

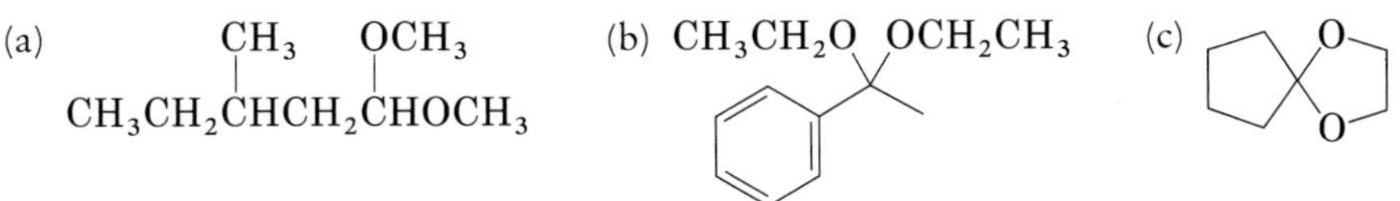

9.39 How would you synthesize these compounds from cyclohexanone?

(a) 1-Methylcyclohexene (b) *cis*-1,2-Cyclohexanediol
(c) 1-Bromo-1-methylcyclohexane (d) 1-Cyclohexylcyclohexanol

9.40 How can you explain the observation that treatment of 4-hydroxycyclohexanone with one equivalent of methylmagnesium bromide yields none of the expected addition product whereas treatment with an excess of Grignard reagent leads to a good yield of 1-methyl-1,4-cyclohexanediol?

9.41 Carvone is the major constituent of spearmint oil. What products would you expect from reaction of carvone with these reagents?

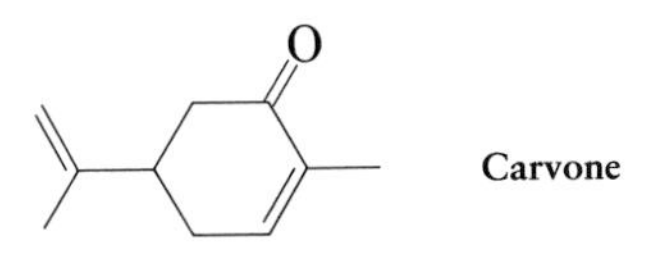

(a) $LiAlH_4$, then H_3O^+ (b) C_6H_5MgBr, then H_3O^+ (c) H_2, Pd catalyst
(d) CH_3OH, H^+ (e) $(C_6H_5)_3\overset{+}{P}—\overset{-}{C}H_2$

9.42 When 4-hydroxybutanal is treated with methanol in the presence of an acid catalyst, 2-methoxytetrahydrofuran is obtained. Propose a mechanism to account for this result.

$$HOCH_2CH_2CH_2CHO \xrightarrow[H^+]{CH_3OH}$$ [tetrahydrofuran ring with OCH_3 at C-2]

9.43 Using your knowledge of the reactivity differences between aldehydes and ketones, show how the following two selective reductions might be carried out. One of the schemes requires a protection step.

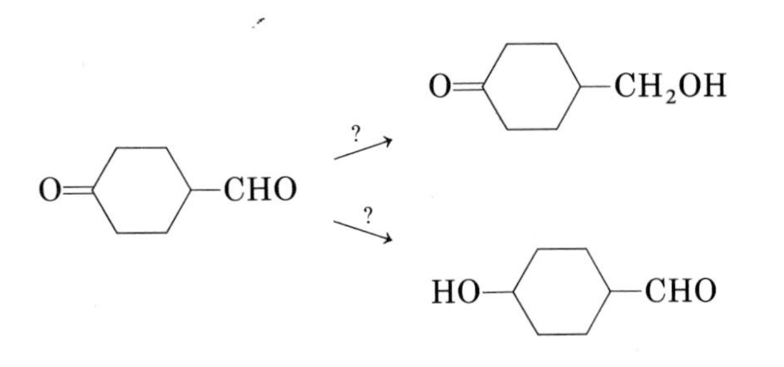

9.44 Treatment of a ketone or aldehyde with a thiol in the presence of an acid catalyst yields a thioacetal, $R_2C(SR')_2$. To what other reaction is this thioacetal formation analogous?

9.45 When crystals of pure α-glucose are dissolved in water, isomerization slowly occurs to produce β-glucose. How does this isomerization occur?

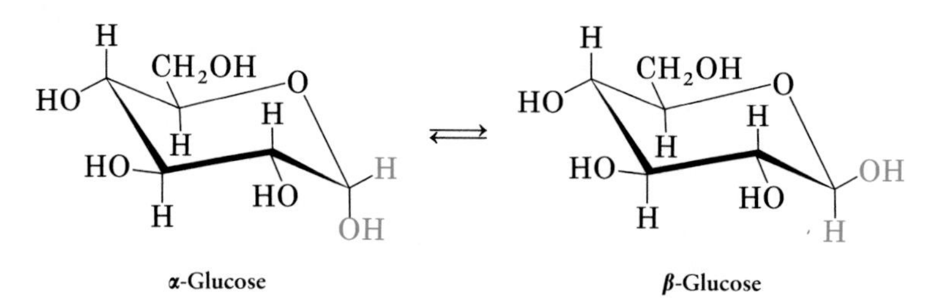

α-Glucose β-Glucose

9.46 The Wittig reaction can be used to prepare aldehydes by using (methoxymethylene)triphenylphosphorane as the Wittig reagent and treating the product with aqueous acid:

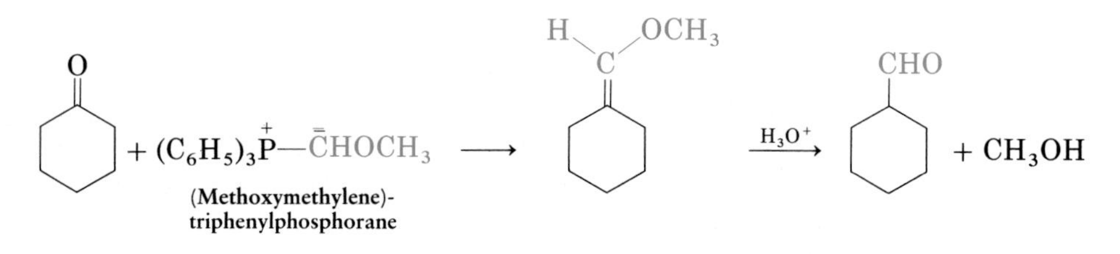

How would you prepare the necessary ylide?

9.47 Ketones react with dimethylsulfonium methylide to yield epoxides by a mechanism that involves an initial nucleophilic addition followed by an intramolecular S_N2 substitution. Formulate the mechanism.

$$\text{Cyclohexanone} + :\bar{C}H_2—\overset{+}{S}(CH_3)_2 \longrightarrow \text{(epoxide, } CH_2\text{)} + CH_3—S—CH_3$$

Dimethylsulfonium methylide

CHAPTER 10 Carboxylic Acids and Derivatives

Carboxylic acids and their derivatives are carbonyl compounds in which the acyl group is bonded to an electronegative atom such as oxygen, halogen, nitrogen, or sulfur. Although there are many kinds of carboxylic acid derivatives, we'll be concerned only with four of the more common ones in addition to the acids themselves: acid halides, acid anhydrides, esters, and amides. In contrast to aldehydes and ketones, these compounds contain an acyl group, RCO–, bonded to a substituent that can serve as a leaving group in substitution reactions.

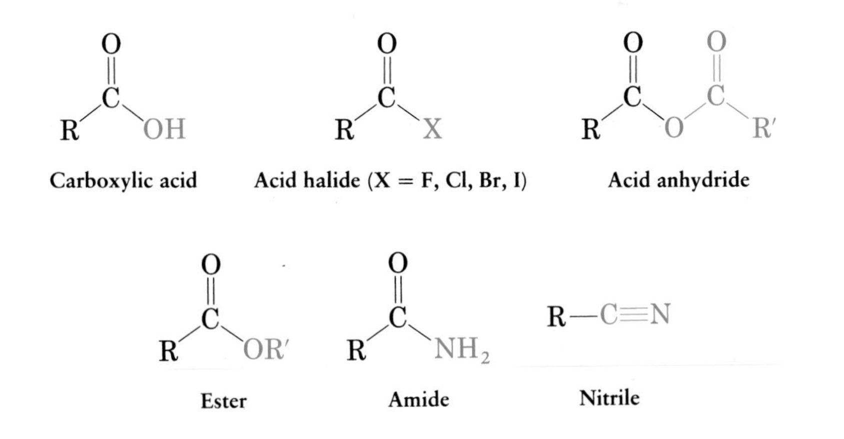

10.1 NAMING CARBOXYLIC ACIDS AND DERIVATIVES

Carboxylic Acids: RCOOH

IUPAC rules allow for two systems of acid nomenclature. Simple open-chain carboxylic acids are named by replacing the terminal *-e* of the alkane name with *-oic acid*. The carboxyl carbon atom is always numbered C1.

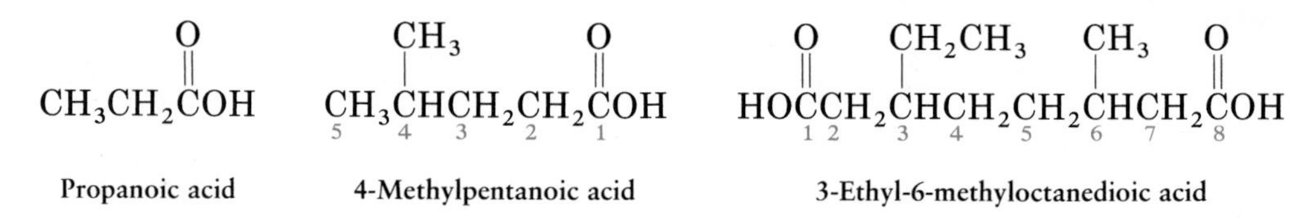

Propanoic acid **4-Methylpentanoic acid** **3-Ethyl-6-methyloctanedioic acid**

Alternatively, compounds that have a $-COOH$ group bonded to a ring are named by using the suffix *-carboxylic acid*. In this alternate system, the carboxylic acid carbon is attached to C1 on the ring but is not itself numbered.

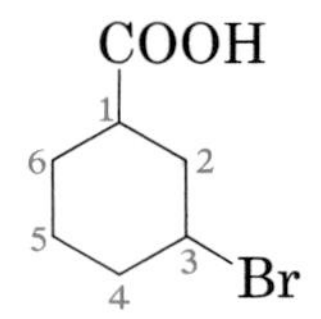

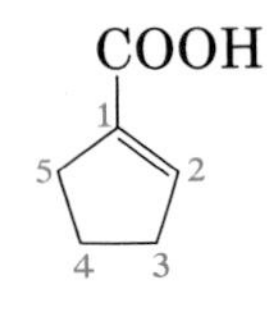

3-Bromocyclohexanecarboxylic acid **1-Cyclopentenecarboxylic acid**

Because many carboxylic acids were among the first organic compounds to be isolated and purified, there are a large number of common names for acids (Table 10.1). We'll use systematic names in this book, with the exception of formic (methanoic) acid, HCOOH, and acetic (ethanoic) acid, CH_3COOH, whose names are so well known that it makes little sense to refer to them in any other way.

TABLE 10.1 Some common names of carboxylic acids and acyl groups

Carboxylic acid		*Acyl group*	
Structure	*Name*	*Name*	*Structure*
HCOOH	Formic	Formyl	HCO—
CH_3COOH	Acetic	Acetyl	CH_3CO—
CH_3CH_2COOH	Propionic	Propionyl	CH_3CH_2CO—
$CH_3CH_2CH_2COOH$	Butyric	Butyryl	$CH_3(CH_2)_2CO$—
HOOCCOOH	Oxalic	Oxalyl	—OCCO—
$HOOCCH_2COOH$	Malonic	Malonyl	—$OCCH_2CO$—
$HOOCCH_2CH_2COOH$	Succinic	Succinyl	—$OC(CH_2)_2CO$—
$H_2C{=}CHCOOH$	Acrylic	Acryloyl	$H_2C{=}CHCO$—
COOH	Benzoic	Benzoyl	O ‖ C—

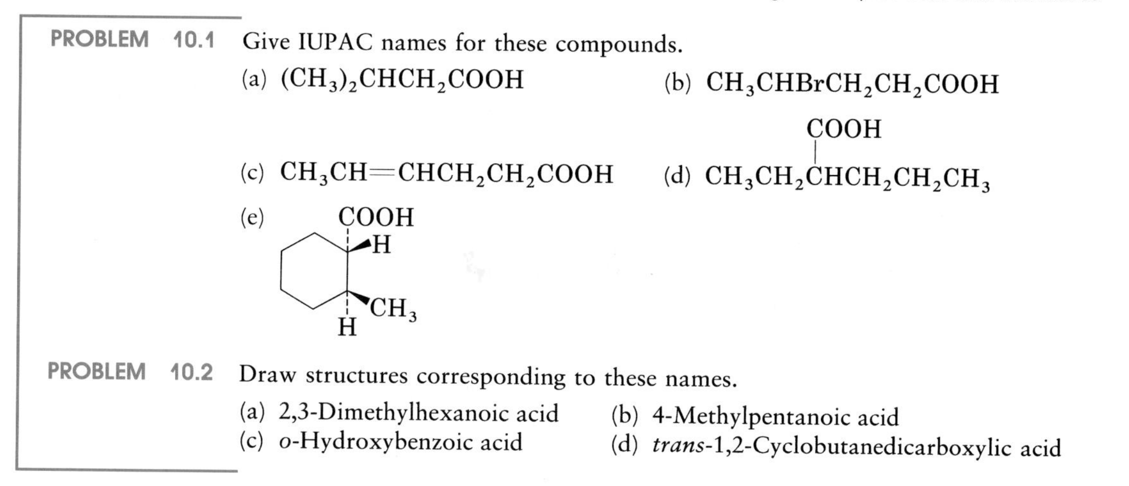

PROBLEM 10.1 Give IUPAC names for these compounds.

(a) $(CH_3)_2CHCH_2COOH$

(b) $CH_3CHBrCH_2CH_2COOH$

(c) $CH_3CH{=}CHCH_2CH_2COOH$

(d) $CH_3CH_2CH(COOH)CH_2CH_2CH_3$

(e)

PROBLEM 10.2 Draw structures corresponding to these names.

(a) 2,3-Dimethylhexanoic acid
(b) 4-Methylpentanoic acid
(c) *o*-Hydroxybenzoic acid
(d) *trans*-1,2-Cyclobutanedicarboxylic acid

Acid Halides: RCOCl

Carboxylic acid halides are named by identifying the acyl group and then the halide. The acyl group name is derived from the acid name by replacing the *-ic acid* ending with *-yl,* or the *-carboxylic acid* ending with *-carbonyl.* For example:

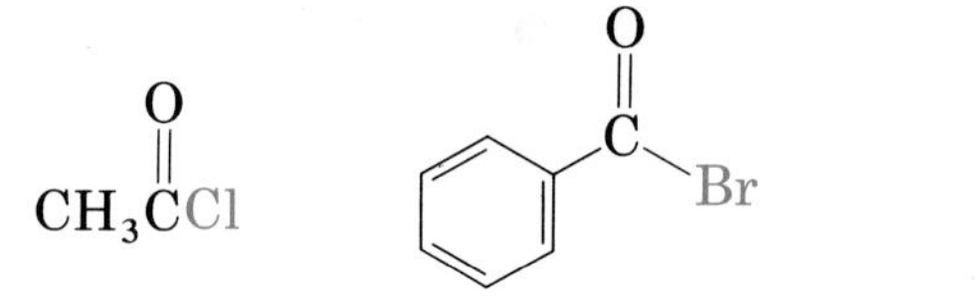

Acetyl chloride
(from acetic acid)

Benzoyl bromide
(from benzoic acid)

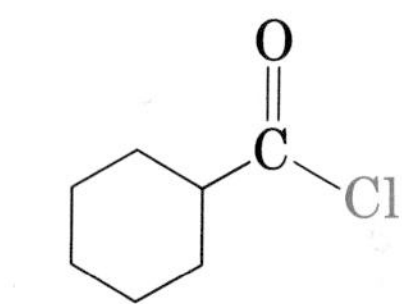

Cyclohexanecarbonyl chloride
(from cyclohexanecarboxylic acid)

Acid Anhydrides: RCO_2COR

Symmetrical anhydrides of simple carboxylic acids and cyclic anhydrides of dicarboxylic acids are named by replacing the word *acid* with *anhydride:*

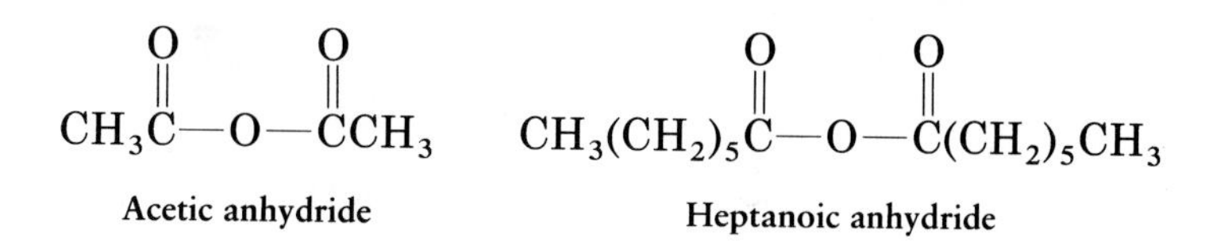

Acetic anhydride

Heptanoic anhydride

Amides: $RCONH_2$

Amides with an unsubstituted $-NH_2$ group are named by replacing the *-oic acid* or *-ic acid* ending with *-amide,* or by replacing the *-carboxylic acid* ending with *-carboxamide:*

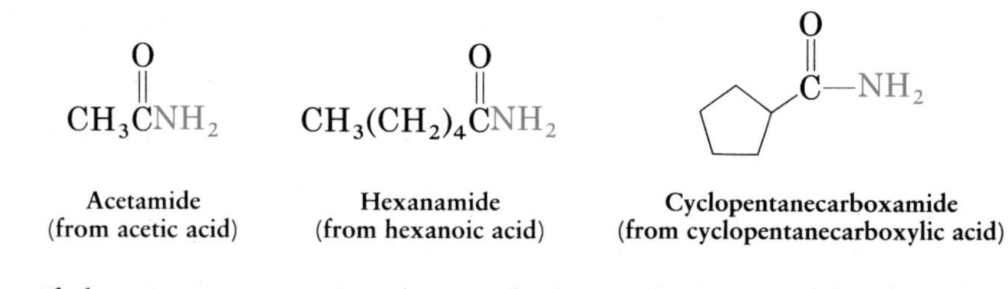

Acetamide (from acetic acid) Hexanamide (from hexanoic acid) Cyclopentanecarboxamide (from cyclopentanecarboxylic acid)

If the nitrogen atom is substituted, the amide is named by first identifying the substituent group and then citing the parent name. The substituents are preceded by the letter *N* to identify them as being directly attached to nitrogen.

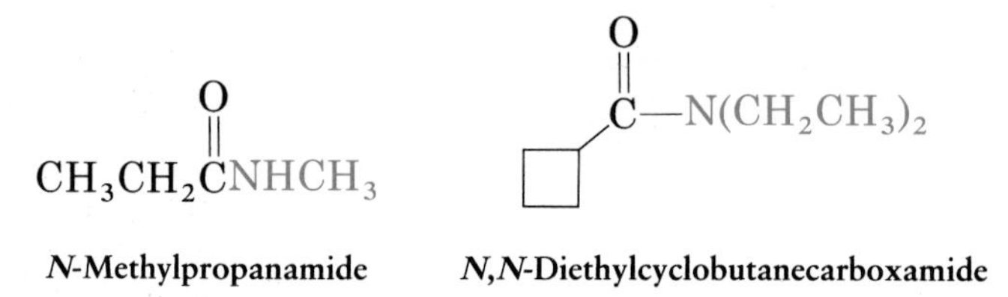

N-Methylpropanamide *N,N*-Diethylcyclobutanecarboxamide

Esters: RCO_2R'

Systematic names for esters are derived by first giving the name of the alkyl group attached to oxygen and then identifying the carboxylic acid. In so doing, the *-ic acid* ending is replaced by *-ate:*

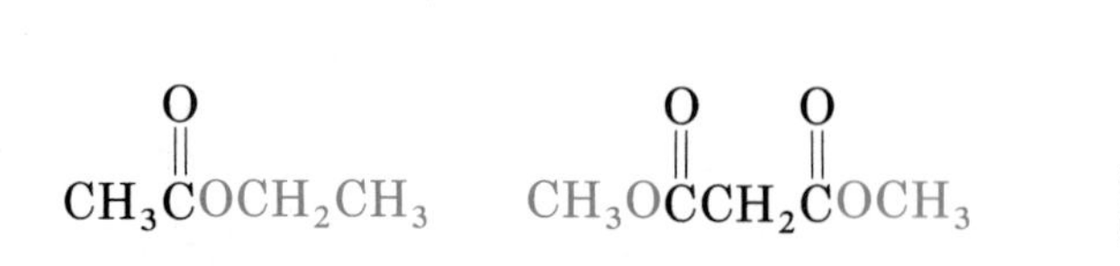

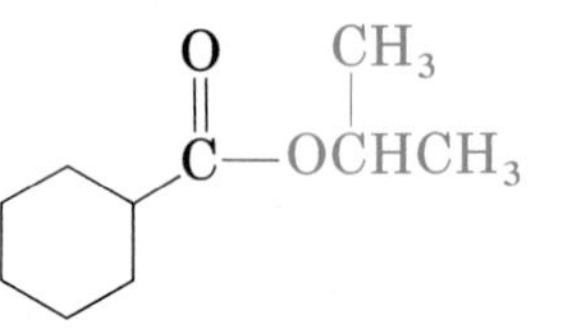

Ethyl acetate (the ethyl ester of acetic acid) Dimethyl malonate (the dimethyl ester of malonic acid) Isopropyl cyclohexanecarboxylate (the isopropyl ester of cyclohexanecarboxylic acid)

Nitriles: R—C≡N

nitrile a compound with a carbon–nitrogen triple bond

Compounds containing the —C≡N functional group are known as **nitriles.** Simple acyclic nitriles are named by adding *-nitrile* as a suffix to the alkane name, with the nitrile carbon itself numbered C1.

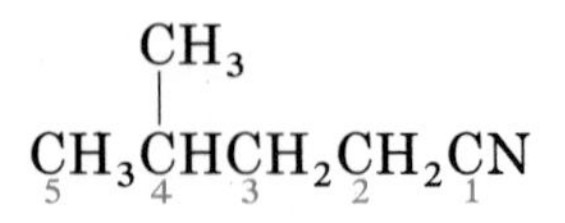

4-Methylpentanenitrile

More complex nitriles are named as derivatives of carboxylic acids by replacing the *-ic acid* or *-oic acid* ending with *-onitrile,* or by replacing the *-carboxylic* *acid* ending with *-carbonitrile*. In this system, the nitrile carbon atom is attached to C1 but is not itself numbered:

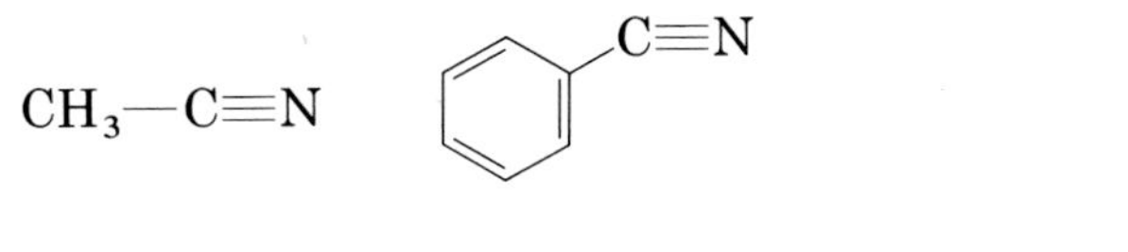

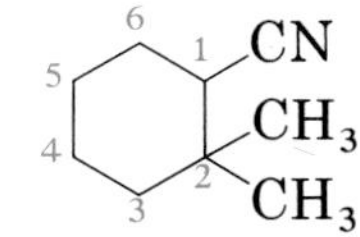

Acetonitrile (from acetic acid) | Benzonitrile (from benzoic acid) | 2,2-Dimethylcyclohexanecarbonitrile (from 2,2-dimethylcyclohexanecarboxylic acid)

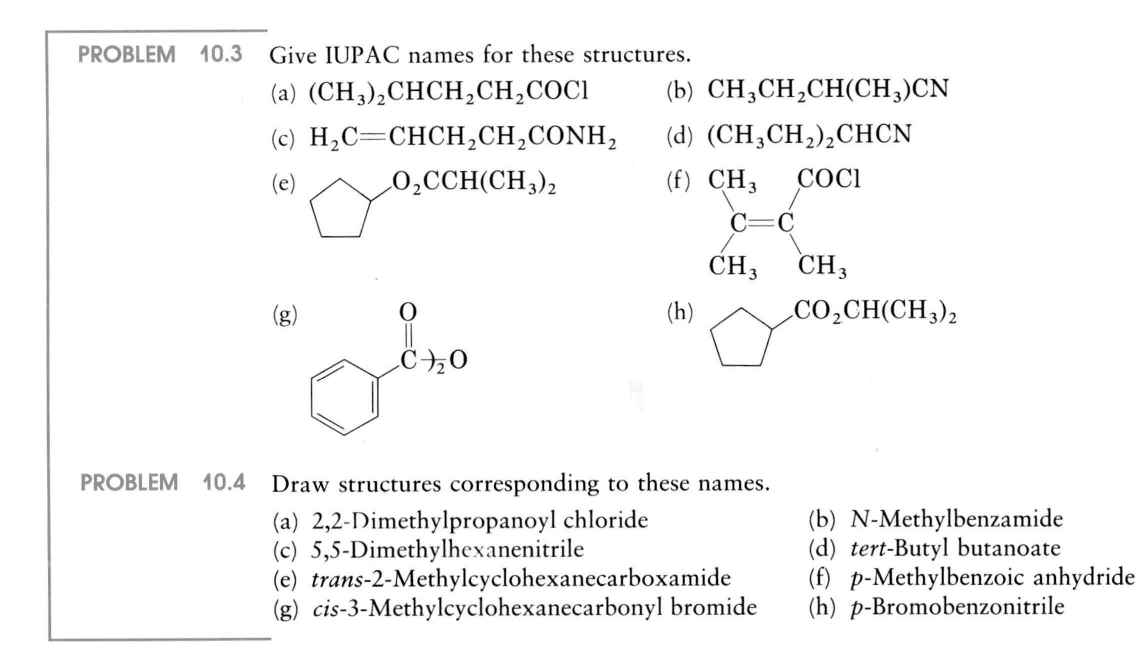

PROBLEM 10.3 Give IUPAC names for these structures.

(a) $(CH_3)_2CHCH_2CH_2COCl$ (b) $CH_3CH_2CH(CH_3)CN$

(c) $H_2C{=}CHCH_2CH_2CONH_2$ (d) $(CH_3CH_2)_2CHCN$

(e) [cyclopentyl]$O_2CCH(CH_3)_2$ (f) $(CH_3)_2C{=}C(CH_3)COCl$

(g) $(C_6H_5C(=O)\text{-})_2O$ (h) [cyclopentyl]$CO_2CH(CH_3)_2$

PROBLEM 10.4 Draw structures corresponding to these names.

(a) 2,2-Dimethylpropanoyl chloride
(b) *N*-Methylbenzamide
(c) 5,5-Dimethylhexanenitrile
(d) *tert*-Butyl butanoate
(e) *trans*-2-Methylcyclohexanecarboxamide
(f) *p*-Methylbenzoic anhydride
(g) *cis*-3-Methylcyclohexanecarbonyl bromide
(h) *p*-Bromobenzonitrile

10.2 OCCURRENCE, STRUCTURE, AND PROPERTIES OF CARBOXYLIC ACIDS

Carboxylic acids occupy a central place among acyl derivatives, both in nature and in the laboratory. For example, vinegar, produced by spoilage of wine, is simply a dilute solution of acetic acid, CH_3COOH; butanoic acid, $CH_3CH_2CH_2COOH$, is responsible for the rancid odor of sour butter; and hexanoic acid (caproic acid), $CH_3(CH_2)_4COOH$, is partially responsible for the unmistakable aroma of goats (Latin *caper* "goat").

Since the carboxylic acid functional group, $-COOH$, is structurally similar to both ketones and alcohols, we might expect to see similar properties. As in ketones, the carboxyl carbon is sp^2 hybridized. Carboxylic acid groups are therefore planar, with C–C–O and O–C–O bond angles of approximately 120°. The structure of acetic acid is shown in Figure 10.1.

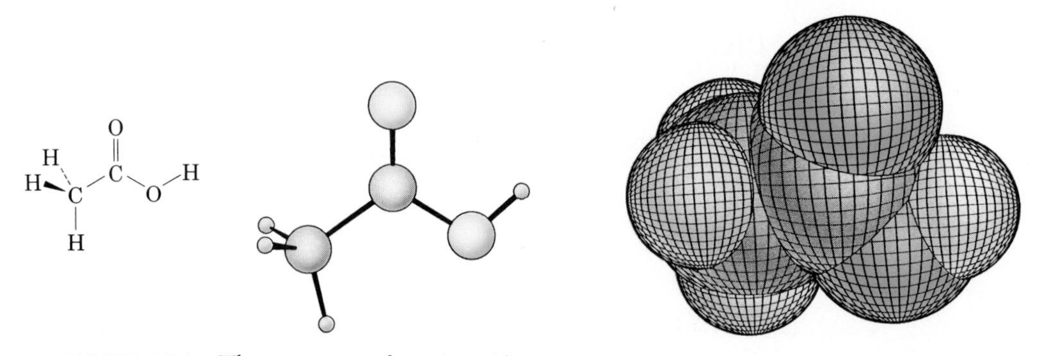

FIGURE 10.1 The structure of acetic acid

Like alcohols, carboxylic acids are strongly associated because of intermolecular hydrogen bonding. Most carboxylic acids exist as dimers held together by two hydrogen bonds:

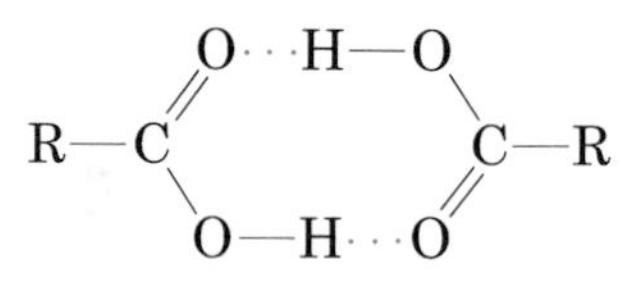

A carboxylic acid dimer

This strong hydrogen bonding has a noticeable effect on boiling points: Carboxylic acids normally boil at much higher temperatures than alkanes or alkyl halides of similar molecular weight. Acetic acid, for example, boils at 118°C, whereas chloropropane boils at 46.6°C.

10.3 ACIDITY OF CARBOXYLIC ACIDS

Since they are acidic, carboxylic acids react with bases such as sodium hydroxide to give metal carboxylate salts. Although carboxylic acids with more than six carbon atoms are only slightly soluble in water, alkali metal salts of carboxylic acids are generally quite water-soluble because of their ionic nature.

$$\mathrm{R{-}\overset{\overset{\displaystyle O}{\|}}{C}{-}O{-}H + NaOH \xrightarrow{H_2O} R{-}\overset{\overset{\displaystyle O}{\|}}{C}{-}O^-\ Na^+ + H_2O}$$

For most carboxylic acids, the acidity constant K_a is on the order of 10^{-5}. Acetic acid, for example, has $K_a = 1.8 \times 10^{-5}$, which corresponds to a pK_a of 4.72. In practical terms, a K_a value near 10^{-5} means that only about 1% of the molecules in a 0.1 M aqueous solution are dissociated, as opposed to the 100% dissociation observed for strong mineral acids like HCl and H_2SO_4. As indicated by the list of K_a values in Table 10.2, though, there is a considerable range in the strengths of various carboxylic acids. Trichloroacetic acid ($K_a = 0.23$), for example, is more than 12,000 times as strong as acetic acid ($K_a = 1.8 \times 10^{-5}$). How can we account for such differences?

TABLE 10.2 Acid strengths of some carboxylic acids

Name	K_a	pK_a	
HCl (hydrochloric acid)[a]	(10^7)	(−7)	Stronger acid
CCl_3COOH	0.23	0.64	↑
$CHCl_2COOH$	5.5×10^{-2}	1.26	
$CH_2ClCOOH$	1.4×10^{-3}	2.85	
HCOOH	1.77×10^{-4}	3.75	
C_6H_5COOH	6.46×10^{-5}	4.19	
$H_2C{=}CHCOOH$	5.6×10^{-5}	4.25	
CH_3COOH	1.8×10^{-5}	4.72	
CH_3CH_2OH (ethanol)[a]	(10^{-16})	(16)	Weaker acid

[a] Values for HCl and ethanol are shown for reference.

Since the dissociation of a carboxylic acid is an equilibrium process, any substituent that stabilizes the carboxylate anion by withdrawing electrons favors increased dissociation and increases acidity. Thus, introducing an electron-withdrawing chlorine atom makes chloroacetic acid stronger than acetic acid by a factor of 75. Introducing two electronegative substituents makes dichloroacetic acid some 3000 times as strong as acetic acid, and introducing three substituents makes trichloroacetic acid more than 12,000 times as strong.

Although weaker than mineral acids, carboxylic acids are nevertheless much stronger acids than alcohols. For example, K_a for ethanol is approximately 10^{-16}, making ethanol a weaker acid than acetic acid by a factor of 10^{11}.

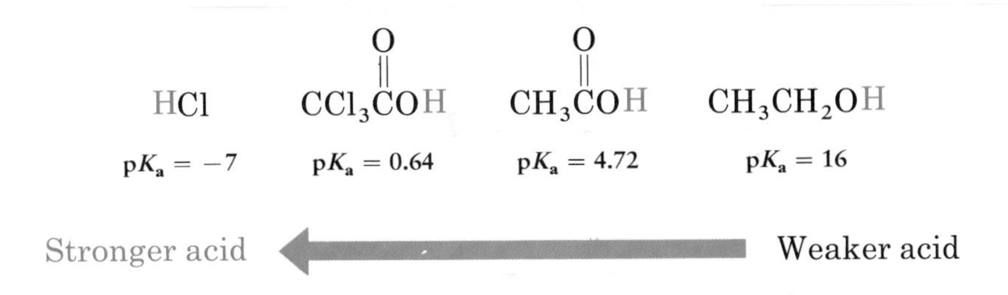

Why are carboxylic acids so much more acidic than alcohols even though both contain O–H groups? The easiest way to answer this question is to look at the relative stability of carboxylate anions versus alkoxide anions. In alkoxides, the negative charge is *localized* on the one oxygen atom. In carboxylate anions, however, the negative charge is *delocalized,* or spread out over both oxygen atoms, resulting in greater stability of the ion. In resonance terms (Section 4.12), a carboxylate anion is a stabilized resonance hybrid of two equivalent line-bond structures.

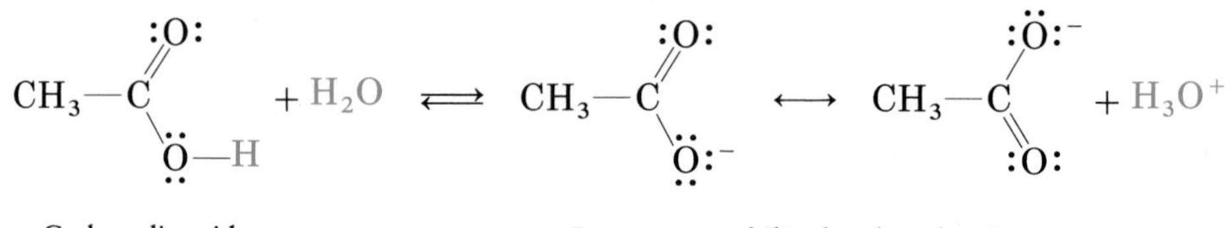

Carboxylic acid

Resonance-stabilized carboxylate ion
(two equivalent resonance forms)

PRACTICE PROBLEM 10.1 Which would you expect to be the stronger acid, benzoic acid or *p*-nitrobenzoic acid?

SOLUTION We know from its effect on aromatic substitution (Section 5.9) that nitro is an electron-withdrawing group that can stabilize a negative charge. Thus, *p*-nitrobenzoic acid is stronger than benzoic acid.

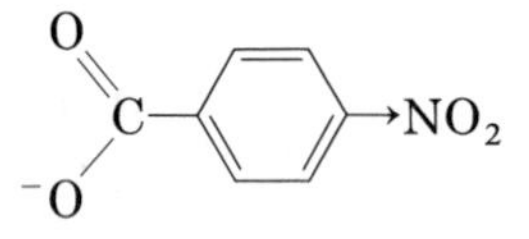

Nitro group withdraws electrons from ring and stabilizes negative charge.

PROBLEM 10.5 Draw structures for the products of these reactions:

(a) Benzoic acid + $NaOCH_3$ $\longrightarrow$
(b) $(CH_3)_3CCOOH$ + KOH $\longrightarrow$

PROBLEM 10.6 Rank these compounds in order of increasing acidity: sulfuric acid, methanol, phenol, *p*-nitrophenol, acetic acid

PROBLEM 10.7 Rank these compounds in order of increasing acidity:

(a) CH_3CH_2COOH, $BrCH_2COOH$, $BrCH_2CH_2COOH$
(b) Benzoic acid, ethanol, *p*-cyanobenzoic acid

10.4 SYNTHESIS OF CARBOXYLIC ACIDS

We've already seen most of the common methods for preparing carboxylic acids, but let's review them briefly:

1. Oxidation of substituted alkylbenzenes with potassium permanganate or sodium dichromate gives substituted benzoic acids (Section 5.12).

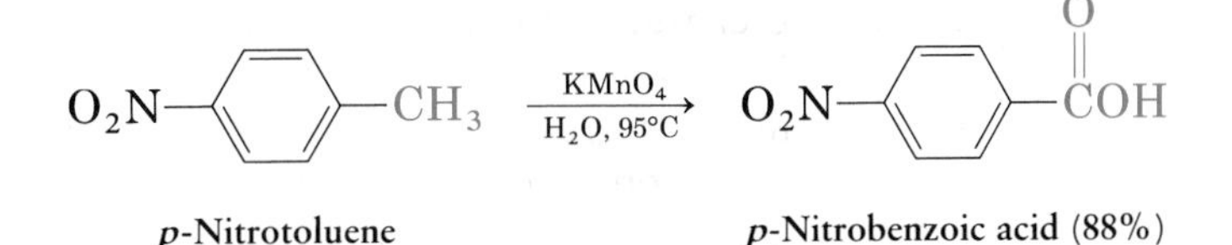

2. Oxidation of primary alcohols and aldehydes yields carboxylic acids (Sections 8.6 and 9.6). Primary alcohols are often oxidized with chromium trioxide or sodium dichromate; aldehydes are oxidized with Tollens' reagent ($AgNO_3$ in NH_4OH).

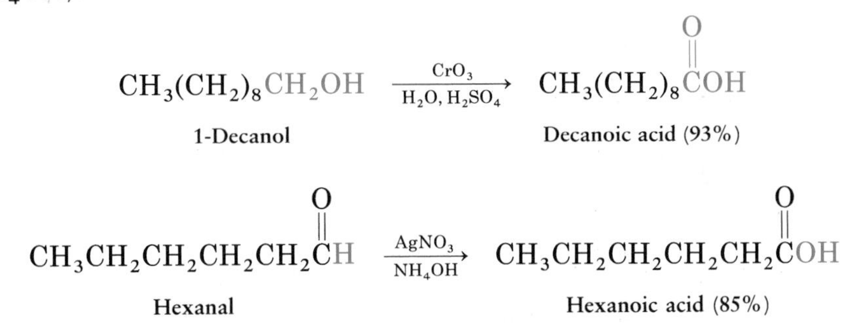

Hydrolysis of Nitriles

Carboxylic acids can be prepared from nitriles, R−C≡N, by reaction with aqueous acid or base (*hydrolysis*). Since nitriles themselves are usually prepared by S_N2 reaction between an alkyl halide and cyanide ion, CN^-, the two-step sequence of cyanide-ion displacement followed by nitrile hydrolysis is an excellent method for converting an alkyl halide into a carboxylic acid (RBr → RC≡N → RCOOH). A good example of the reaction occurs in the commercial synthesis of the antiarthritis drug, Fenoprofen.

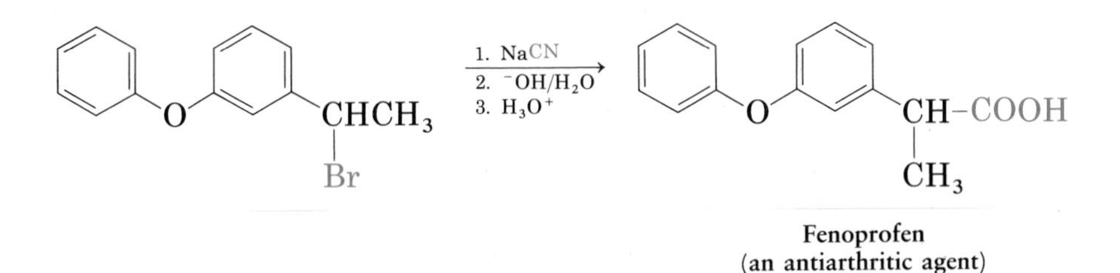

Fenoprofen
(an antiarthritic agent)

The method works best with primary alkyl halides because an E2 elimination reaction can occur when a secondary or tertiary alkyl halide is used (Section 7.9).

Carboxylation of Grignard Reagents

carboxylation the addition of CO_2 to a molecule

Yet another method of preparing carboxylic acids is by reaction of Grignard reagents, RMgX, with carbon dioxide. This **carboxylation** reaction is carried out either by pouring a solution of the Grignard reagent over dry ice (solid CO_2) or by bubbling a stream of dry CO_2 gas through the Grignard reagent solution.

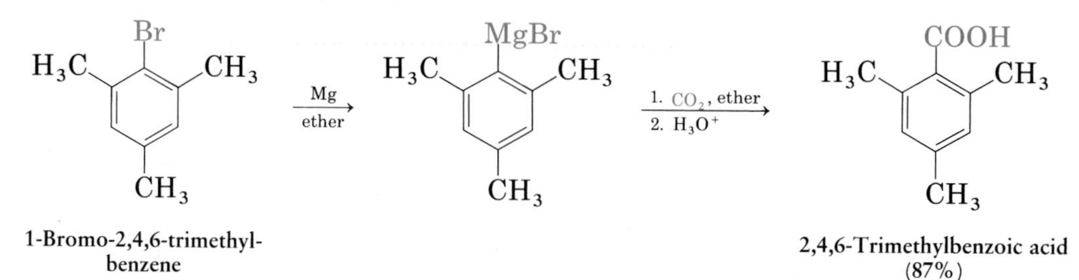

PRACTICE PROBLEM 10.2 How would you convert 2-chloro-2-methylpropane into 2,2-dimethylpropanoic acid?

SOLUTION Since 2-chloro-2-methylpropane is a tertiary alkyl halide, it won't undergo S_N2 substitution with cyanide ion. Thus, the only way you could carry out the desired reaction is to convert the alkyl halide into a Grignard reagent and then add CO_2.

$$(CH_3)_3CCl + Mg \longrightarrow (CH_3)_3CMgCl \xrightarrow[2.\ H_3O^+]{1.\ CO_2} (CH_3)_3CCOOH$$

PROBLEM 10.8 Predict the products of these reactions:

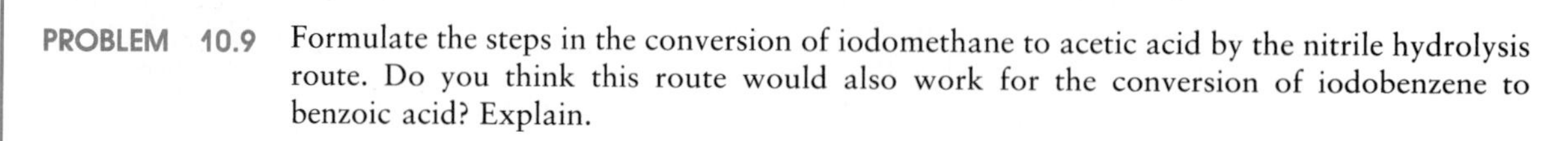

PROBLEM 10.9 Formulate the steps in the conversion of iodomethane to acetic acid by the nitrile hydrolysis route. Do you think this route would also work for the conversion of iodobenzene to benzoic acid? Explain.

PROBLEM 10.10 Formulate all the steps in the conversion of iodobenzene to benzoic acid by the Grignard carboxylation route. Do you think this route would also work for the conversion of iodomethane to acetic acid?

10.5 NUCLEOPHILIC ACYL SUBSTITUTION REACTIONS

nucleophilic acyl substitution reaction a substitution reaction that replaces one nucleophile bonded to a carbonyl group by another

We saw in Chapter 9 that nucleophilic addition to the polar C=O bond is a general feature of ketone and aldehyde chemistry. Carboxylic acids and their derivatives also react with nucleophiles, but the initially formed intermediate expels the substituent originally bonded to the carbonyl carbon, leading to a net **nucleophilic acyl substitution reaction** and the formation of a new carbonyl compound (Figure 10.2).

What is the reason for the different behaviors of ketones/aldehydes and carboxylic acid derivatives? The differences are simply a consequence of structure. Carboxylic acid derivatives have an acyl function bonded to a group −Y that can leave, as a stable anion. As soon as addition of a nucleophile occurs, the group Y

Ketone or aldehyde: nucleophilic addition

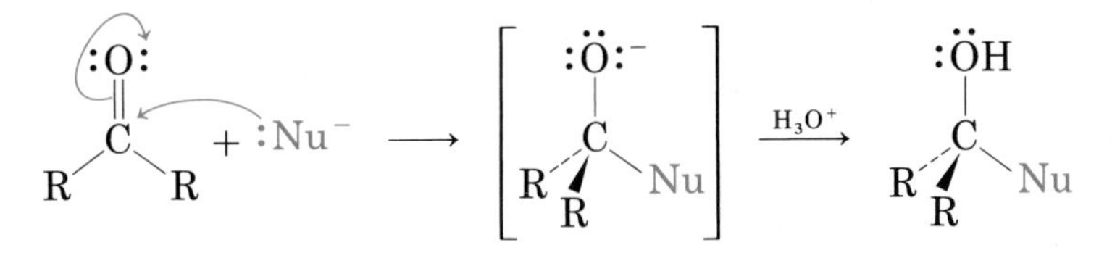

Carboxylic acid: nucleophilic substitution

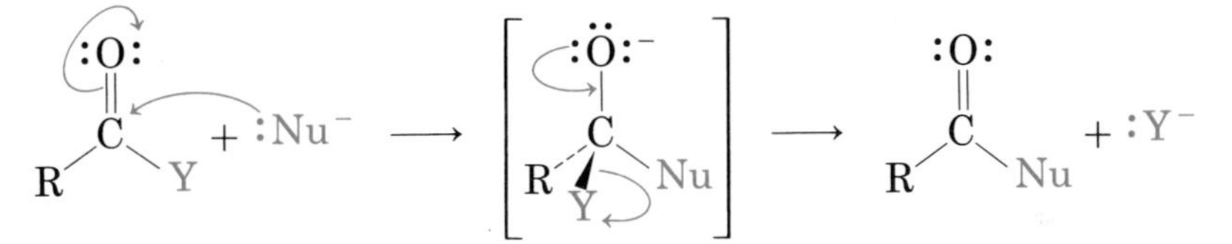

FIGURE 10.2 The general mechanisms of nucleophilic addition and nucleophilic acyl substitution reactions

leaves, and a new carbonyl compound forms. Ketones and aldehydes have no such leaving group, however, and therefore don't undergo substitution.

Note that the overall nucleophilic substitution reaction of acyl derivatives is superficially similar to what occurs in S_N2 reactions of alkyl halides (Section 7.7) in that a leaving group is replaced by an incoming nucleophile. The *mechanisms* of the two reactions are very different, however: S_N2 reactions occur in a single step by backside displacement of a leaving group whereas nucleophilic acyl substitutions take place in two steps through a tetrahedrally hybridized intermediate.

In comparing the reactivity of different acyl derivatives, the more highly polar a compound is, the more reactive it is. Thus, acid chlorides are the most reactive compounds because the electronegative chlorine atom strongly polarizes the carbonyl group whereas amides are the least reactive compounds:

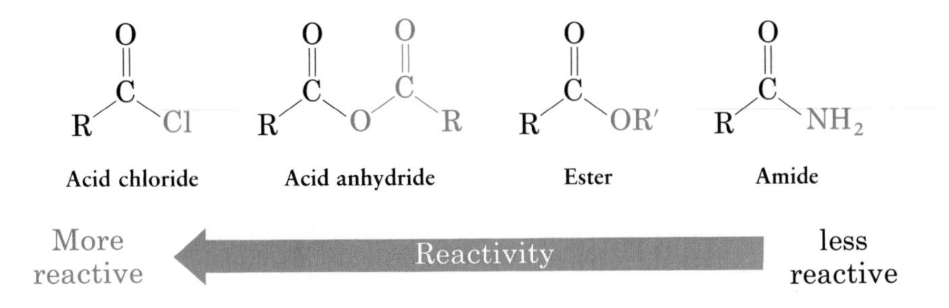

An important consequence of these reactivity differences is that *it's usually possible to convert a more reactive acid derivative into a less reactive one.* Acid chlorides, for example, can be converted into esters and amides, but amides and esters can't be converted into acid chlorides. Remembering the reactivity order is therefore a useful way to keep track of a large number of reactions (Figure 10.3).

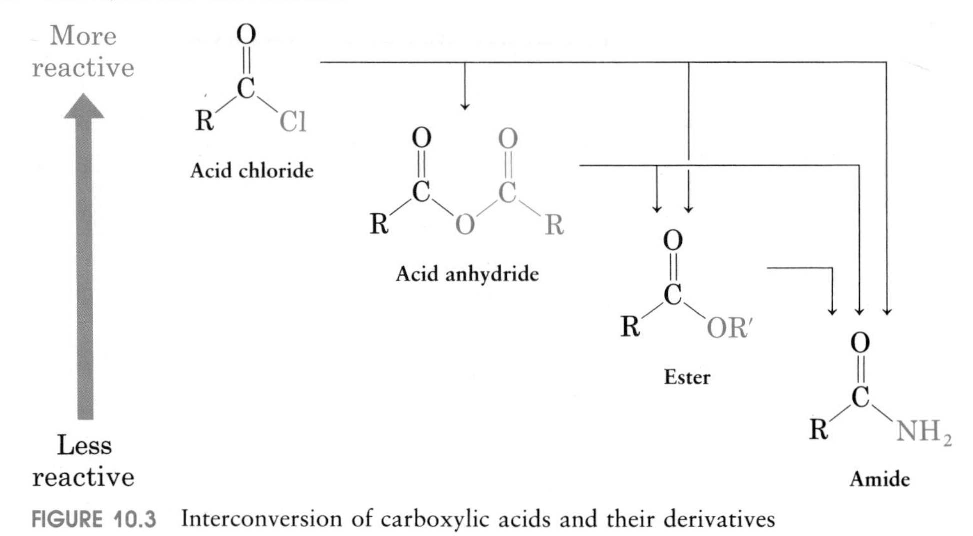

FIGURE 10.3 Interconversion of carboxylic acids and their derivatives

PRACTICE PROBLEM 10.3 Which is more reactive in a nucleophilic acyl substitution reaction with hydroxide ion, CH_3CONH_2 or CH_3COCl?

SOLUTION Since Cl is more electronegative than N, the carbonyl group of an acid chloride is more polar than the carbonyl group of an amide, and acid chlorides are therefore more reactive than amides.

PROBLEM 10.11 Which of the following compounds are more reactive in nucleophilic acyl substitution reactions:

(a) CH_3COCl or CH_3COOCH_3
(b) $(CH_3)_2CHCONH_2$ or $CH_3CH_2COOCH_3$
(c) CH_3COOCH_3 or $CH_3COOCOCH_3$
(d) CH_3COOCH_3 or CH_3CHO

PROBLEM 10.12 How can you account for the fact that methyl trifluoroacetate, CF_3COOCH_3, is more reactive than methyl acetate, CH_3COOCH_3, in nucleophilic acyl substitution reactions?

10.6 REACTIONS OF CARBOXYLIC ACIDS

Reduction: Conversion of Acids into Alcohols

We saw in Section 8.6 that carboxylic acids are reduced by lithium aluminum hydride ($LiAlH_4$) to yield primary alcohols. The reaction is sometimes difficult and often requires heating to go to completion.

$$\underset{\text{Oleic acid}}{CH_3(CH_2)_7CH{=}CH(CH_2)_7\overset{\overset{O}{\parallel}}{C}OH} \xrightarrow[2.\ H_3O^+]{1.\ LiAlH_4} \underset{\textit{cis}\text{-9-Octadecen-1-ol (87\%)}}{CH_3(CH_2)_7CH{=}CH(CH_2)_7CH_2OH}$$

Conversion of Acids into Acid Chlorides

The most important reactions of carboxylic acids are those that convert the carboxyl group into other acid derivatives by nucleophilic acyl substitution. Acid chlorides, anhydrides, esters, and amides can all be prepared from carboxylic acids.

Acid chlorides are prepared by treatment of carboxylic acids with either thionyl chloride, $SOCl_2$, or phosphorus trichloride, PCl_3. The net effect is substitution of the acid –OH group by –Cl. For example:

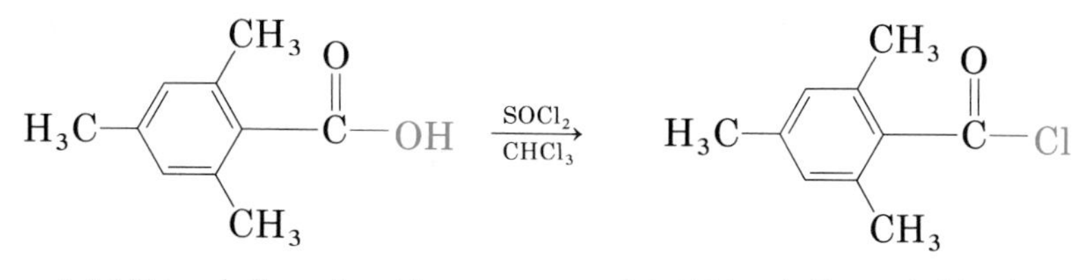

2,4,6-Trimethylbenzoic acid   2,4,6-Trimethylbenzoyl chloride (90%)

Conversion of Acids into Acid Anhydrides

Acid anhydrides, which have the general structure R–C(=O)–O–C(=O)–R, are formally derived from two molecules of carboxylic acid by removing one molecule of water. Anhydrides are difficult to prepare directly from the corresponding acids, and few of them are commercially available.

Conversion of Acids into Esters

One of the most important reactions of carboxylic acids is their conversion into esters. Among the many excellent methods for accomplishing this transformation is the S_N2 reaction between a carboxylate anion nucleophile and a primary alkyl halide, which we saw in Section 7.8:

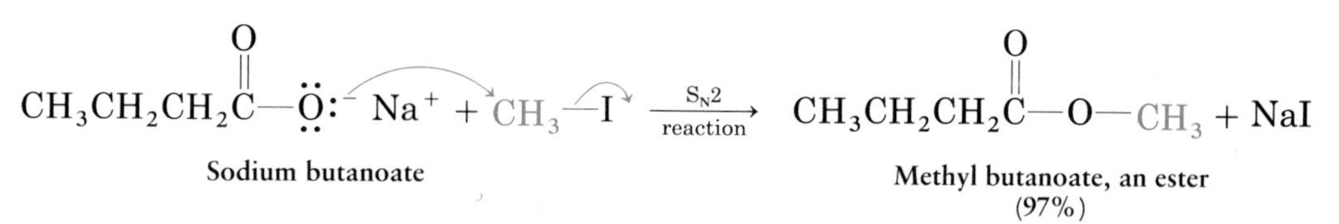

Sodium butanoate   Methyl butanoate, an ester (97%)

Fischer esterification reaction the conversion of a carboxylic acid into an ester by reaction with alcohol and an acid catalyst

Alternatively, esters can be synthesized by a nucleophilic acyl substitution reaction of a carboxylic acid with an alcohol. Called the **Fischer esterification reaction,** this method involves heating the carboxylic acid with a small amount of mineral acid catalyst in an alcohol solvent.

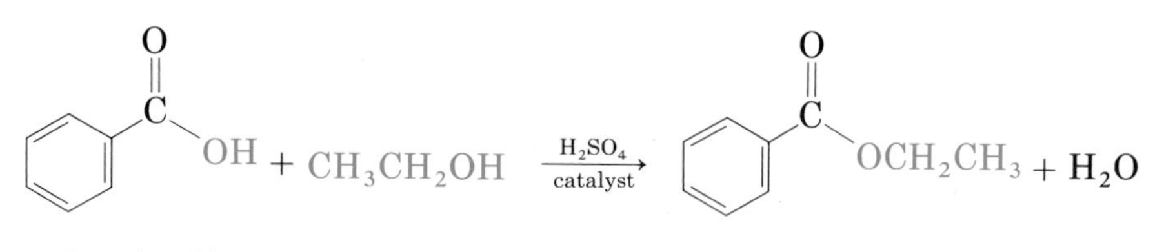

Benzoic acid   Ethyl benzoate (91%)

The Fischer esterification reaction, whose mechanism is shown in Figure 10.4, is a nucleophilic acyl substitution process. The catalyst first protonates an oxygen atom of the $-COOH$ group, which makes it much more reactive toward nucleophiles. An alcohol molecule then adds to the protonated carboxylic acid, and subsequent loss of water yields the ester product.

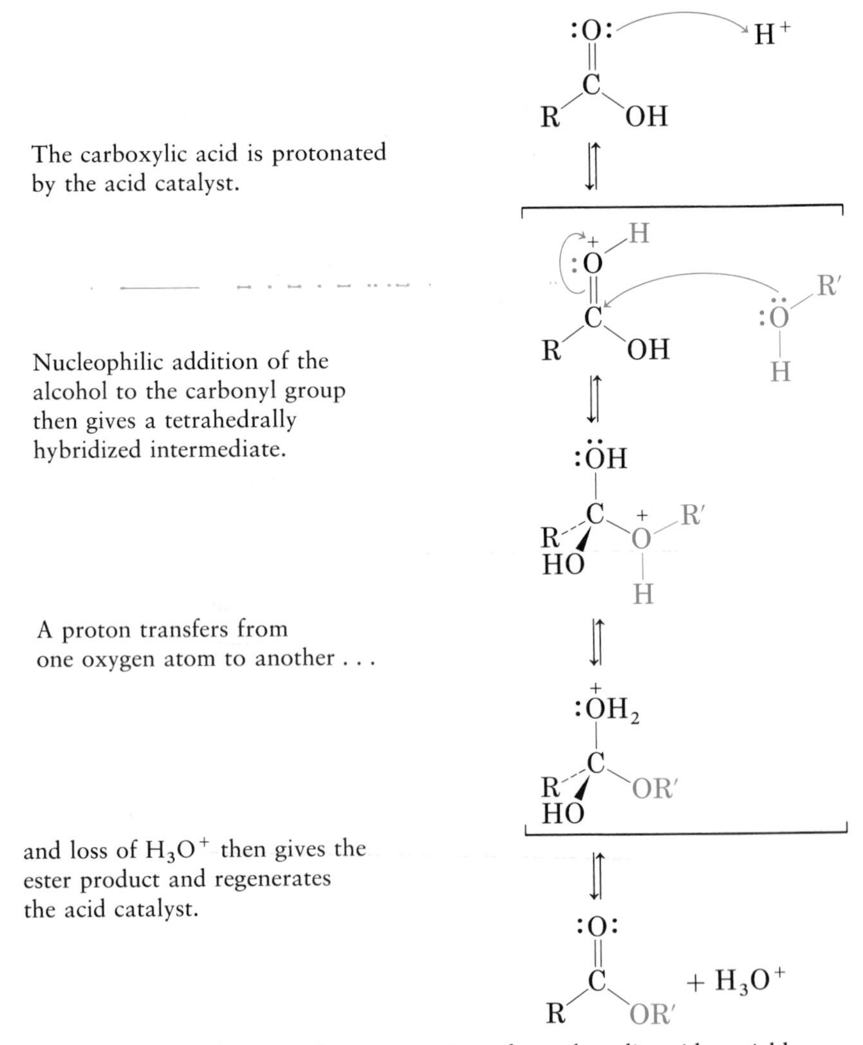

The carboxylic acid is protonated by the acid catalyst.
Nucleophilic addition of the alcohol to the carbonyl group then gives a tetrahedrally hybridized intermediate.
A proton transfers from one oxygen atom to another . . .
and loss of H_3O^+ then gives the ester product and regenerates the acid catalyst.
H^+
R
OH
R'
H
OR'
HO
$+ H_3O^+$

FIGURE 10.4 The Fischer esterification reaction of a carboxylic acid to yield an ester

All steps in the Fischer esterification reaction are reversible, and the position of the equilibrium can be driven to either side depending on the reaction conditions. Ester formation is favored when alcohol is used as solvent, but carboxylic acid is favored when water is used as solvent.

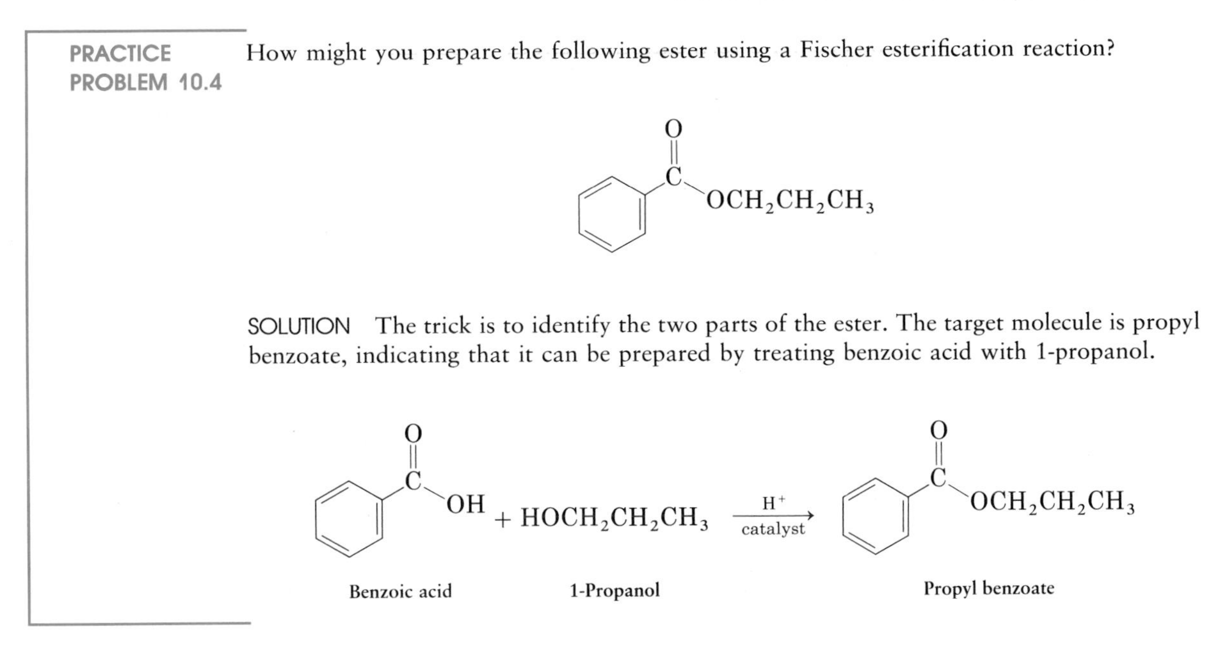

PRACTICE PROBLEM 10.4 How might you prepare the following ester using a Fischer esterification reaction?

SOLUTION The trick is to identify the two parts of the ester. The target molecule is propyl benzoate, indicating that it can be prepared by treating benzoic acid with 1-propanol.

PROBLEM 10.13 What products would you obtain by treating benzoic acid with these reagents? Formulate the reactions.

(a) $SOCl_2$ (b) CH_3OH, HCl (c) $LiAlH_4$ (d) NaOH

PROBLEM 10.14 Show how you might prepare these esters using Fischer esterification reactions:

(a) Butyl acetate (b) Methyl butanoate

PROBLEM 10.15 If 5-hydroxypentanoic acid is treated with an acid catalyst, an intramolecular esterification reaction occurs. What is the structure of the product? (*Intramolecular* means within the same molecule.)

Conversion of Acids into Amides

Amides are carboxylic acid derivatives in which the acid hydroxyl group has been replaced by a nitrogen substituent, $-NH_2$ (or $-NHR$, $-NR_2$). Amides are difficult to prepare directly from acids because amines are bases (Section 12.4), which convert acidic carboxyl groups into their carboxylate anions. Since the carboxylate anion has a negative charge, it no longer undergoes attack by nucleophiles except at high temperatures.

$$R-\overset{O}{\overset{\|}{C}}-OH + :NH_3 \longrightarrow R-\overset{O}{\overset{\|}{C}}-O^-\ NH_4^+$$

10.7 CHEMISTRY OF ACID HALIDES

Synthesis of Acid Halides

Acid chlorides are prepared from carboxylic acids by reaction with thionyl chloride ($SOCl_2$) or phosphorus trichloride (PCl_3), as we saw in the previous section.

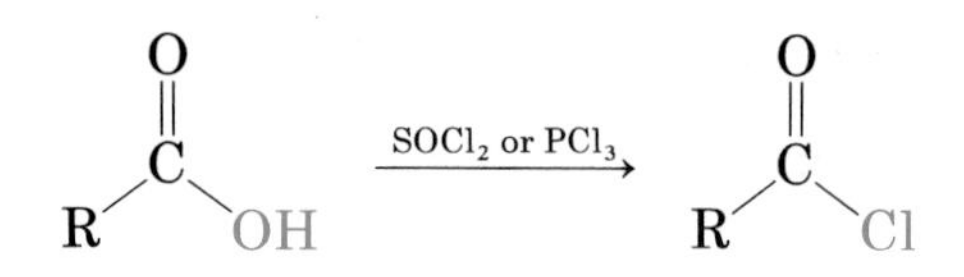

Reactions of Acid Halides

Acid halides are among the most reactive carboxylic acid derivatives and can be converted into many other kinds of substances. Most reactions of acid halides occur by nucleophilic acyl substitution mechanisms. As illustrated in Figure 10.5, the halide ion can be replaced by −OH to yield an acid, by −OR to yield an ester, by −OOCR to yield an anhydride, or by $-NH_2$ to yield an amide. Although Figure 10.5 illustrates these reactions only for acid chlorides, they also take place with other acid halides.

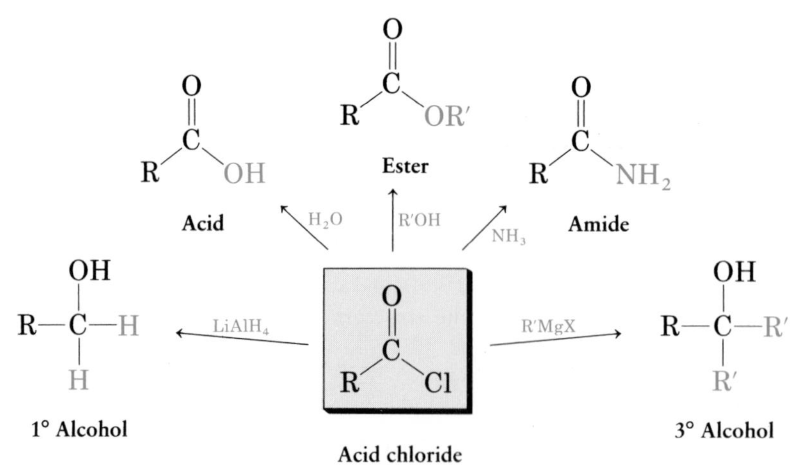

FIGURE 10.5 Some nucleophilic acyl substitution reactions of acid chlorides

Hydrolysis: Conversion of acid chlorides into acids Acid chlorides react with water to yield carboxylic acids. This hydrolysis reaction is a typical nucleophilic acyl substitution process that is initiated by attack of the nucleophile, water, on the acid chloride carbonyl group. The initially formed tetrahedral intermediate undergoes loss of HCl to yield the product.

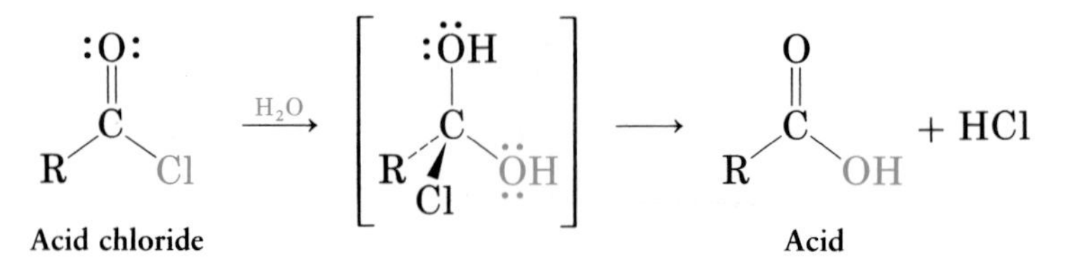

Alcoholysis: Conversion of acid halides into esters Acid chlorides react with alcohols to yield esters in a reaction analogous to their reaction with water to yield acids.

$$\underset{\text{Acetyl chloride}}{CH_3\overset{O}{\overset{\|}{C}}Cl} + \underset{\text{1-Butanol}}{CH_3CH_2CH_2CH_2OH} \xrightarrow[\text{solvent}]{\text{pyridine}} \underset{\text{Butyl acetate (90\%)}}{CH_3\overset{O}{\overset{\|}{C}}OCH_2CH_2CH_2CH_3} + HCl$$

Since HCl is generated as a byproduct of alcoholysis, the reaction is usually carried out in the presence of an amine base such as pyridine to react with the HCl as it's formed and prevent it from causing side reactions. If this weren't done, the HCl might react with the alcohol to form an alkyl chloride.

PRACTICE PROBLEM 10.5 Show how you could prepare ethyl benzoate by reaction of an acid chloride with an alcohol.

SOLUTION As its name implies, ethyl benzoate can be made by reaction of ethyl alcohol with the acid chloride of benzoic acid:

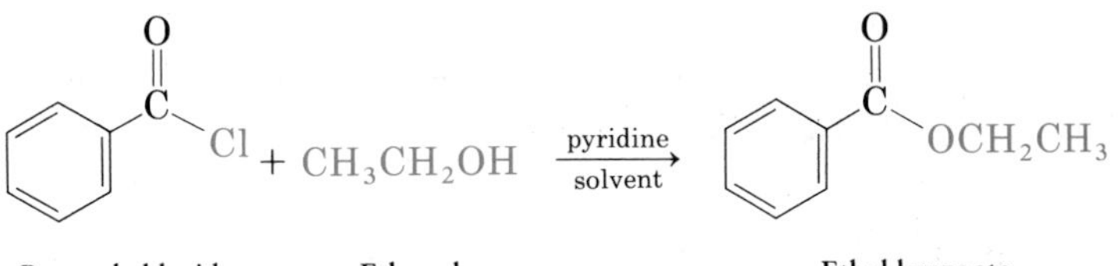

Benzoyl chloride Ethanol Ethyl benzoate

PROBLEM 10.16 How could you prepare these esters using the reaction of an acid chloride with an alcohol?
(a) $CH_3CH_2COOCH_3$ (b) $CH_3COOCH_2CH_3$ (c) Cyclohexyl acetate

Aminolysis: Conversion of acid chlorides into amides Acid chlorides react rapidly with ammonia and with amines to give amides. Both mono- and disubstituted amines can be used. For example, the sedative trimetozine is prepared commercially by reaction of 3,4,5-trimethoxybenzoyl chloride with the amine, morpholine. Note that one equivalent of NaOH is added to react with the HCl generated.

$$\underset{\text{2-Methylpropanoyl chloride}}{(CH_3)_2CH\overset{O}{\overset{\|}{C}}Cl} + 2\,{:}NH_3 \xrightarrow{H_2O} \underset{\substack{\text{2-Methylpropanamide}\\ \text{(83\%)}}}{(CH_3)_2CH\overset{O}{\overset{\|}{C}}NH_2} + \overset{+}{N}H_4\overset{-}{Cl}$$

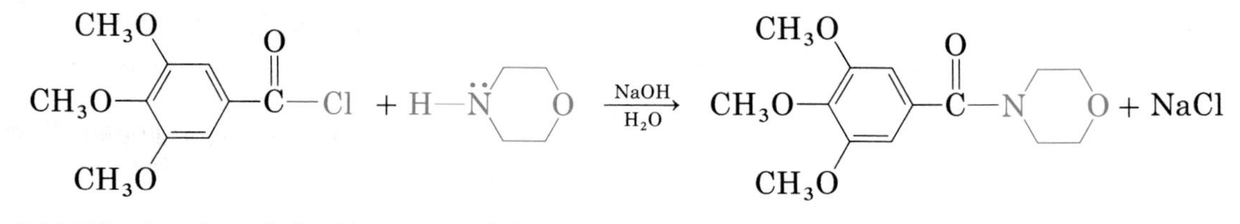

3,4,5-Trimethoxybenzoyl chloride Morpholine Trimetozine (an amide)

PRACTICE PROBLEM 10.6 Show how would you prepare *N*-methylpropanamide by reaction of an acid chloride with an amine.

SOLUTION Reaction of methylamine with propanoyl chloride gives *N*-methylpropanamide.

$$CH_3CH_2\overset{O}{\overset{\|}{C}}Cl + 2\ CH_3NH_2 \longrightarrow CH_3CH_2\overset{O}{\overset{\|}{C}}NHCH_3 + CH_3NH_3^+\,Cl^-$$

Propanoyl chloride Methylamine *N*-Methylpropanamide

PROBLEM 10.17 Write the steps in the mechanism of the reaction between morpholine and 3,4,5-trimethoxybenzoyl chloride to yield trimetozine.

PROBLEM 10.18 What amines would react with what acid chlorides to give these amide products?

(a) $CH_3CH_2CONH_2$ (b) $(CH_3)_2CHCH_2CONHCH_3$
(c) *N,N*-Dimethylpropanamide (d) *N,N*-Diethylbenzamide

10.8 CHEMISTRY OF ACID ANHYDRIDES

Synthesis of Acid Anhydrides

The best method of preparing acid anhydrides is by a nucleophilic acyl substitution reaction of an acid chloride with a carboxylic acid anion. Both symmetrical and unsymmetrical acid anhydrides can be prepared in this way.

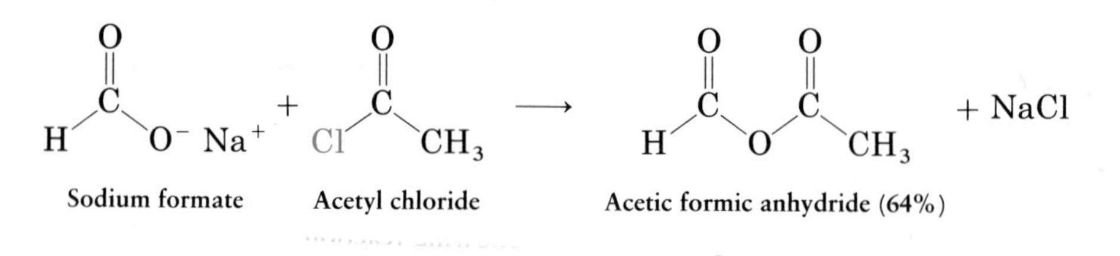

Sodium formate Acetyl chloride Acetic formic anhydride (64%)

Reactions of Acid Anhydrides

The chemistry of acid anhydrides is similar to that of acid chlorides. Although anhydrides react more slowly than acid chlorides, the kinds of reactions the two functional groups undergo are the same. Thus, acid anhydrides react with water to form acids, with alcohols to form esters, and with amines to form amides (Figure 10.6).

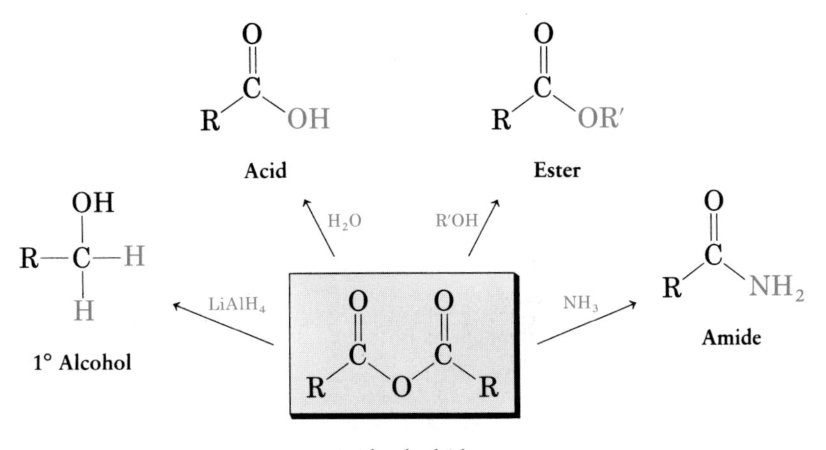

FIGURE 10.6 Some reactions of acid anhydrides

Acetic anhydride is often used to prepare acetate esters of complex alcohols and to prepare substituted acetamides from amines. For example, aspirin (an ester) is prepared by reaction of acetic anhydride with *o*-hydroxybenzoic acid. Similarly, acetaminophen (an amide) is prepared by reaction of acetic anhydride with *p*-hydroxyaniline.

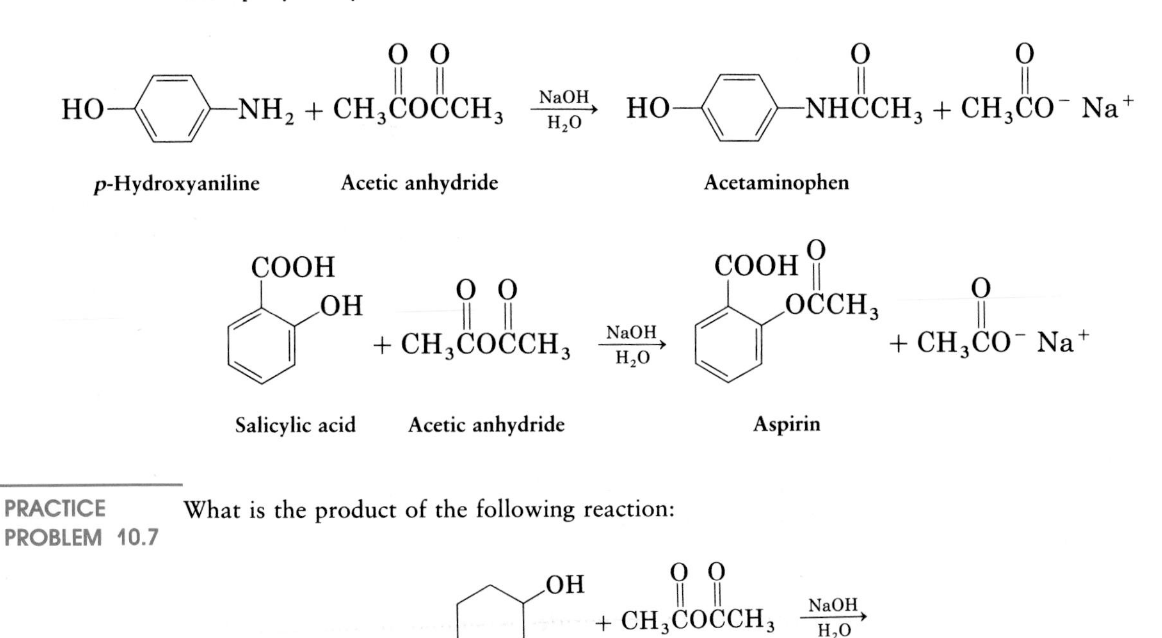

PRACTICE PROBLEM 10.7 What is the product of the following reaction:

SOLUTION Reaction of cyclohexanol with acetic anhydride yields cyclohexyl acetate by nucleophilic acyl substitution of the alcohol group for the acetate group of the anhydride.

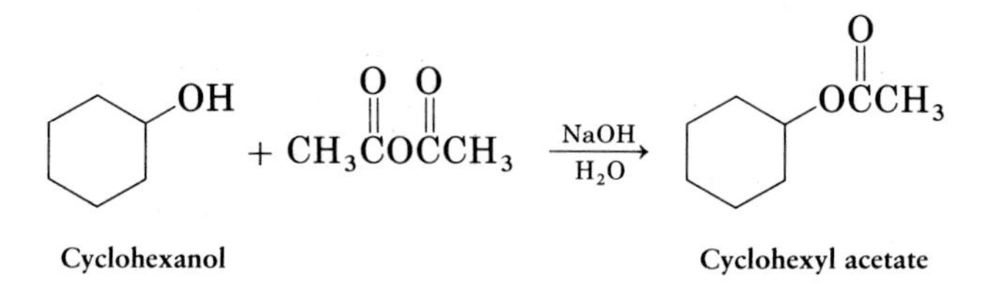

PROBLEM 10.19 Write the steps in the mechanism of the reaction between *p*-ethoxyaniline and acetic anhydride to prepare phenacetin.

PROBLEM 10.20 What product would you expect to obtain from reaction of one equivalent of methanol with a cyclic anhydride such as phthalic anhydride?

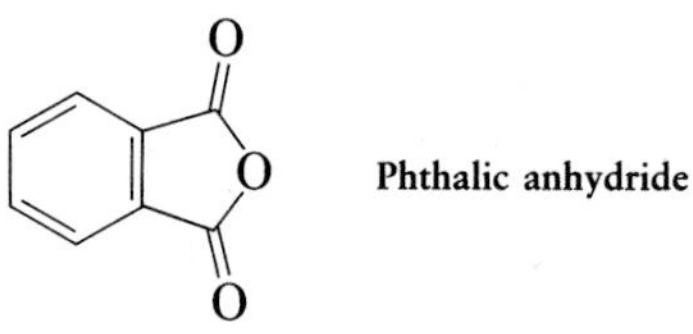

10.9 CHEMISTRY OF ESTERS

Esters are among the most important and widespread of naturally occurring compounds. Many simple esters are pleasant-smelling liquids that are responsible for the fragrant odors of fruits and flowers. For example, methyl butanoate has been isolated from pineapple oil, and isopentyl acetate has been found in banana oil. The ester linkage is also present in animal fats and other biologically important molecules.

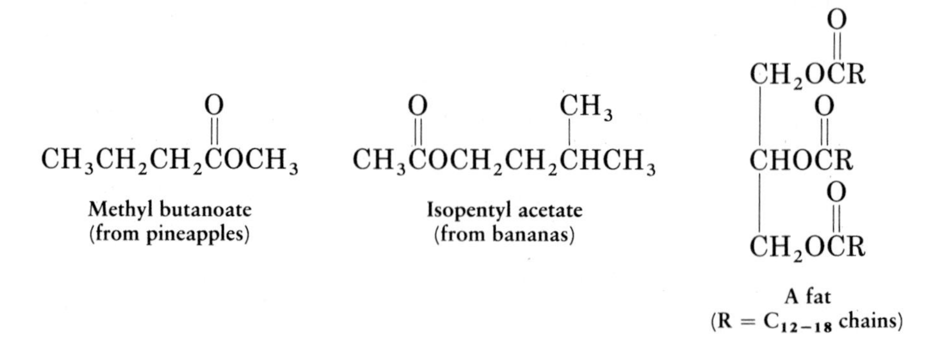

Synthesis of Esters

Esters are usually prepared either from acids or from acid chlorides by the methods already discussed. Thus, carboxylic acids are converted directly into esters either by S_N2 reaction of a carboxylate salt with a primary alkyl halide or by reaction of the acid with an alcohol (Section 10.6). Acid chlorides are converted into esters by reaction with an alcohol in the presence of base (Section 10.7).

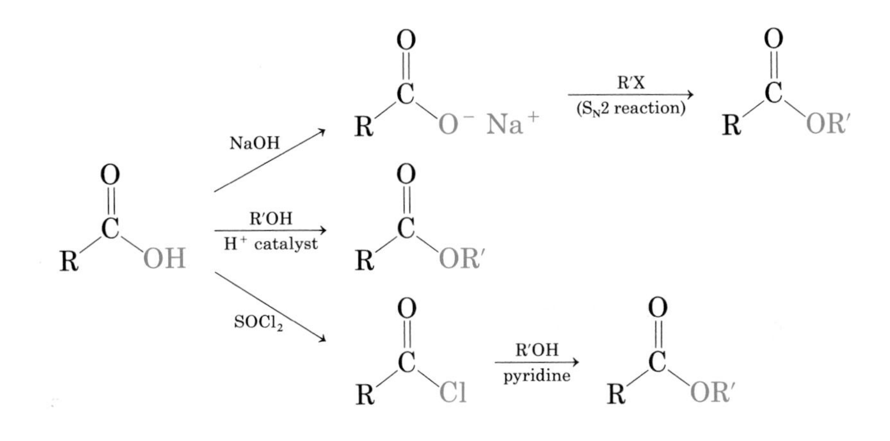

Reactions of Esters

Esters show the same kinds of chemistry we've seen for other acyl derivatives, but they're less reactive toward nucleophiles than acid chlorides or anhydrides. Figure 10.7 shows some general reactions of esters.

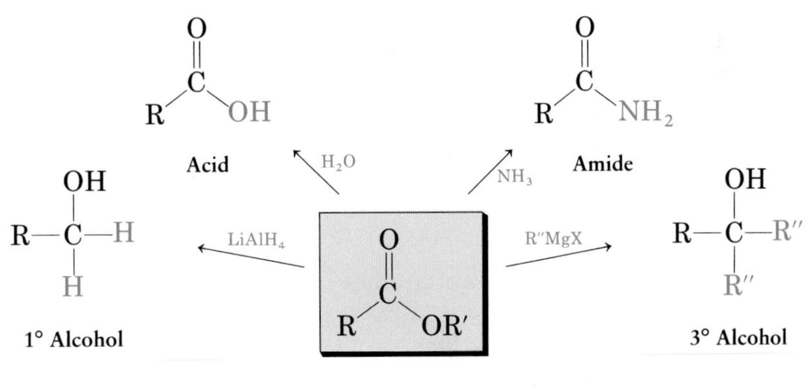

FIGURE 10.7 Some general reactions of esters

Hydrolysis: Conversion of esters into acids Esters are hydrolyzed either by aqueous base or by aqueous acid to yield carboxylic acid plus alcohol.

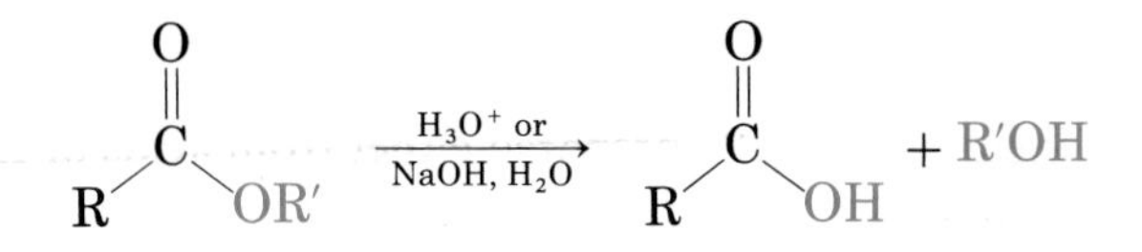

saponification a term for the base-induced hydrolysis of an ester to yield a carboxylate anion

Hydrolysis in basic solution is called **saponification** (Latin *sapo*, "soap"). (As we'll see in Section 16.3, the boiling of animal fat with alkali to make soap is a saponification since fats have ester linkages.) Ester hydrolysis occurs by a typical nucleophilic acyl substitution pathway in which hydroxide-ion nucleophile adds to the ester carbonyl group yielding a tetrahedral intermediate. Loss of alkoxide ion then gives a carboxylic acid, which is deprotonated to give the acid salt:

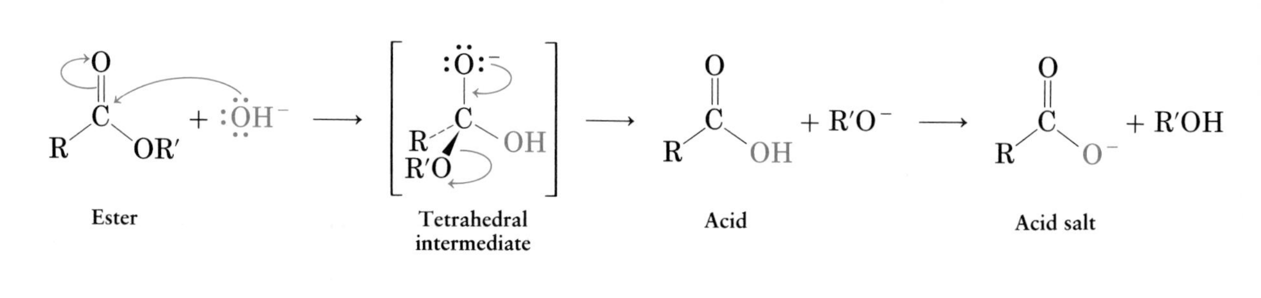

PRACTICE PROBLEM 10.8

Write the products of the following saponification reaction:

$$\mathrm{CH_3\overset{\overset{CH_3}{|}}{C}HCH_2\overset{\overset{O}{||}}{C}OCH_2CH_3 \xrightarrow[2.\ H_3O^+]{1.\ NaOH,\ H_2O}}$$

Ethyl 3-methylbutanoate

SOLUTION Esters are cleaved by aqueous base into their acid and alcohol components by breaking the bond between the carbonyl carbon and the alcohol oxygen:

$$\mathrm{CH_3\overset{\overset{CH_3}{|}}{C}HCH_2\overset{\overset{O}{||}}{C}OCH_2CH_3 \xrightarrow[2.\ H_3O^+]{1.\ NaOH,\ H_2O} CH_3\overset{\overset{CH_3}{|}}{C}HCH_2\overset{\overset{O}{||}}{C}OH + CH_3CH_2OH}$$

Ethyl 3-methylbutanoate → 3-Methylbutanoic acid + Ethanol

PROBLEM 10.21 Show the products of hydrolysis of these esters:

(a) Isopropy acetate (b) Methyl cyclohexanecarboxylate

(c) cyclopentyl–$\mathrm{O\overset{\overset{O}{||}}{C}C(CH_3)_3}$

PROBLEM 10.22 Why do you suppose saponification of esters is not reversible? In other words, why doesn't treatment of a carboxylic acid with alkoxide ion give an ester?

Aminolysis: Conversion of esters into amides Esters react with ammonia and amines to yield amides. The reaction is not often used, however, because higher yields are usually obtained starting from the acid chloride rather than from the ester.

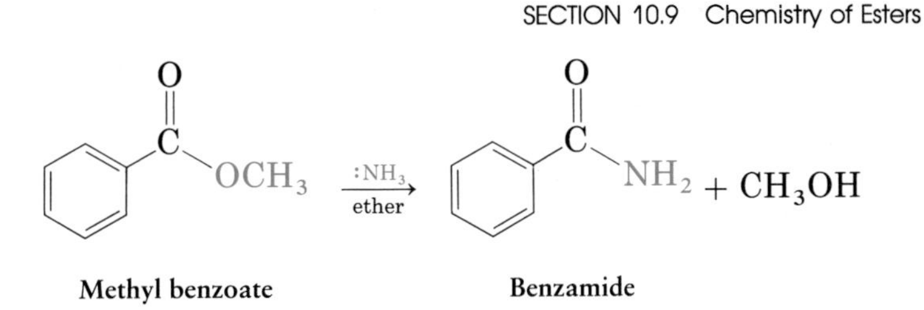

Reduction: Conversion of esters into alcohols Esters are easily reduced by treatment with $LiAlH_4$ to yield primary alcohols (Section 8.6).

$$\underset{\text{Ethyl-2-pentenoate}}{CH_3CH_2CH{=}CH\overset{\overset{O}{\|}}{C}OCH_2CH_3} \xrightarrow[\text{2. } H_3O^+]{\text{1. } LiAlH_4\text{, ether}} \underset{\text{2-Penten-1-ol (91\%)}}{CH_3CH_2CH{=}CHCH_2OH} + CH_3CH_2OH$$

Hydride ion first adds to the carbonyl group, followed by elimination of alkoxide ion to yield an aldehyde intermediate. Further reduction of the aldehyde gives the primary alcohol.

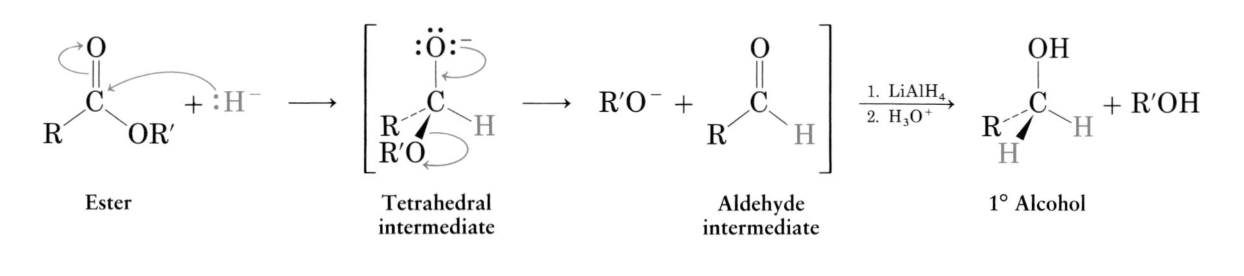

PRACTICE PROBLEM 10.9 What products would you obtain by reduction of propyl benzoate with $LiAlH_4$?

SOLUTION Reduction of esters with $LiAlH_4$ yields two molecules of alcohol product, one from the acyl portion of the ester and one from the alkoxy portion. Thus, reduction of propyl benzoate yields benzyl alcohol (from the acyl group) and 1-propanol (from the alkoxyl group).

$$\underset{\text{Propyl benzoate}}{C_6H_5\overset{\overset{O}{\|}}{C}OCH_2CH_2CH_3} \xrightarrow[\text{2. } H_3O^+]{\text{1. } LiAlH_4} \underset{\text{Benzyl alcohol}}{C_6H_5CH_2OH} + \underset{\text{1-Propanol}}{HOCH_2CH_2CH_3}$$

PROBLEM 10.23 Show the products you would obtain by reduction of these esters with $LiAlH_4$.

(a) $CH_3CH_2CH_2CH(CH_3)COOCH_3$ (b) Phenyl benzoate

PROBLEM 10.24 What product would you expect from the reaction of a cyclic ester such as butyrolactone with $LiAlH_4$?

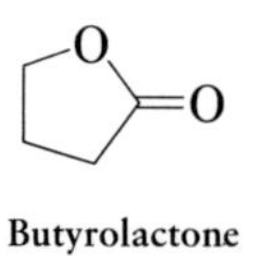

Butyrolactone

Reaction of esters with Grignard reagents Grignard reagents react with esters to yield tertiary alcohols in which two of the substituents are identical. For example:

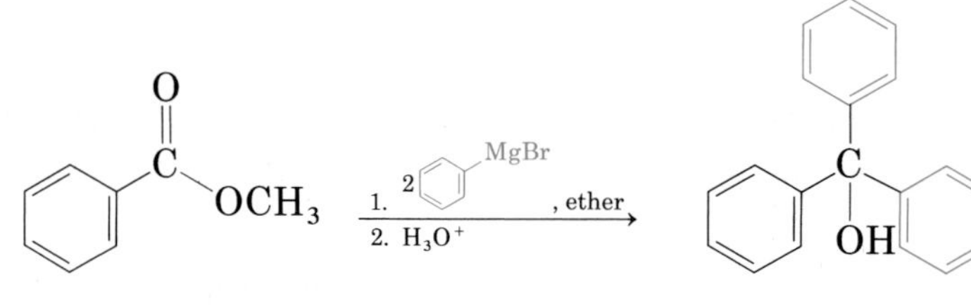

Methyl benzoate

Triphenylmethanol (96%)

PRACTICE PROBLEM 10.10 How could you use the reaction of a Grignard reagent with an ester to prepare 1,1-diphenyl-1-propanol?

SOLUTION The product of reaction between a Grignard reagent and an ester is a tertiary alcohol in which the alcohol carbon and one of the attached groups have come from the ester, and in which the remaining two groups bonded to the alcohol carbon have come from the Grignard reagent. Since 1,1-diphenyl-1-propanol has two phenyl groups and one ethyl group bonded to the alcohol carbon, it must have been prepared from reaction of a phenylmagnesium halide on an ester of propanoic acid.

$$C_6H_5MgX + CH_3CH_2COOR \longrightarrow C_6H_5-\underset{C_6H_5}{\overset{OH}{C}}-CH_2CH_3$$

1,1-Diphenyl-1-propanol

PROBLEM 10.25 What ester and what Grignard reagent might you use to prepare these alcohols?

(a) 2-Phenyl-2-propanol (b) 1,1-Diphenylethanol (c) 3-Ethyl-3-heptanol

10.10 CHEMISTRY OF AMIDES

Synthesis of Amides

Amides are usually prepared by reaction of an acid chloride with an amine, as we saw in Section 10.7. Ammonia, monosubstituted amines, and disubstituted amines all undergo this reaction.

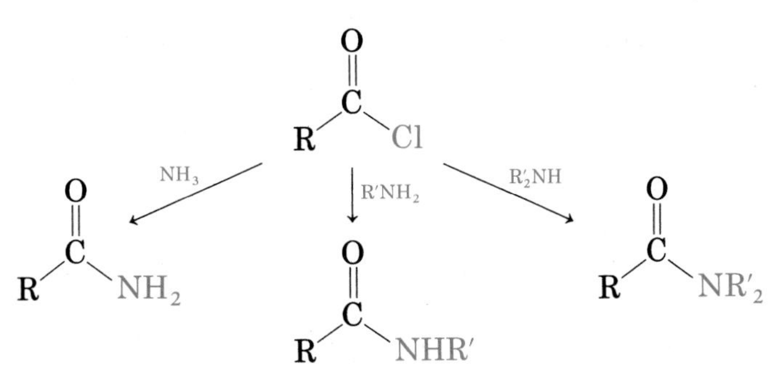

Reactions of Amides

Amides are much less reactive than acid chlorides, acid anhydrides, and esters. Thus, the amide linkage is stable enough to serve as the basic unit from which proteins are made (Section 15.5).

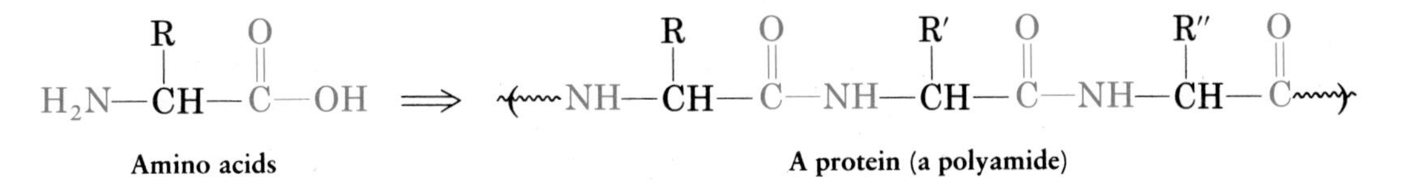

Amides undergo hydrolysis to yield carboxylic acids plus amine upon heating in either aqueous acid or base. Although the reaction is difficult and requires prolonged heating, the overall transformation is a typical nucleophilic acyl substitution of $-OH$ for $-NH_2$.

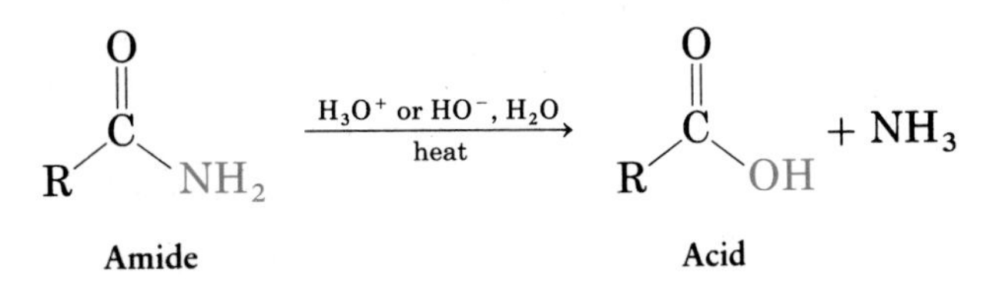

Like other carboxylic acid derivatives, amides are reduced by lithium aluminum hydride. The product of this reduction, however, is an *amine* rather than an alcohol.

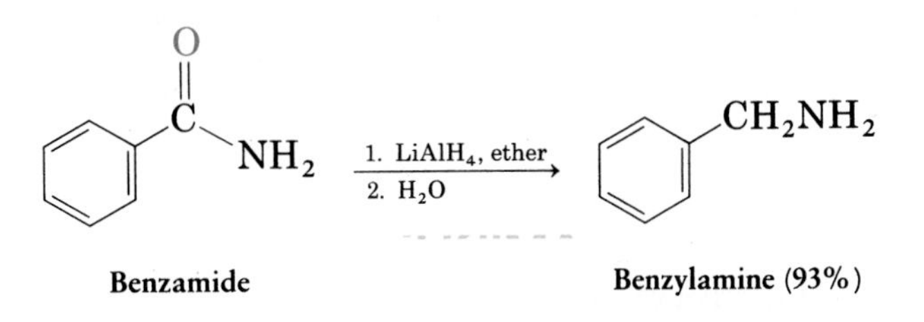

The effect of amide reduction is to convert the amide carbonyl group into a methylene group ($C{=}O \rightarrow CH_2$). This kind of reaction is specific for amides and doesn't occur with other carboxylic acid derivatives.

PRACTICE PROBLEM 10.11 How could you prepare *N*-ethylaniline by reduction of an amide with $LiAlH_4$?

SOLUTION Since reduction of an amide with $LiAlH_4$ yields an amine, the starting material for synthesis of *N*-ethylaniline must have been *N*-phenylacetamide.

$$C_6H_5NHCOCH_3 \xrightarrow[2.\ H_2O]{1.\ LiAlH_4,\ ether} C_6H_5NHCH_2CH_3$$

N-Phenylacetamide → *N*-Ethylaniline

PROBLEM 10.26 How would you convert *N*-ethylbenzamide into these products?

(a) Benzoic acid (b) Benzyl alcohol
(c) *N*-Ethylbenzylamine, $C_6H_5CH_2NHCH_2CH_3$

PROBLEM 10.27 The lithium aluminum hydride reduction of amides to yield amines is equally effective with both acyclic and cyclic amides (*lactams*). What product would you obtain from reduction of 5,5-dimethyl-2-pyrrolidone with $LiAlH_4$?

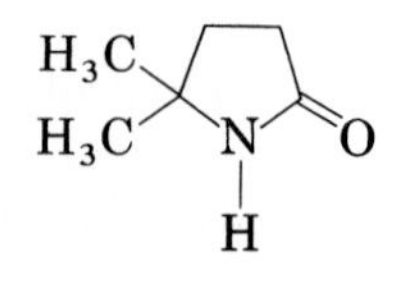

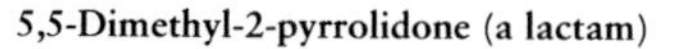
5,5-Dimethyl-2-pyrrolidone (a lactam)

10.11 CHEMISTRY OF NITRILES

Nitriles, R−C≡N, aren't related to carboxylic acids in the same sense that acyl derivatives are, but the chemistries of nitriles and carboxylic acids are so similar that the two classes of compounds should be considered together.

Synthesis of Nitriles

The simplest and best method of preparing nitriles is by the S_N2 displacement reaction of cyanide ion on a primary alkyl halide (Section 7.7).

$$RCH_2Br + Na^+CN^- \xrightarrow[\text{reaction}]{S_N2} RCH_2CN + NaBr$$

Reactions of Nitriles

The chemistry of nitriles is similar in many respects to the chemistry of carbonyl compounds. Like carbonyl groups, nitriles are strongly polarized. Thus, the nitrile-group carbon atom is electrophilic and undergoes attack by nucleophiles to yield an sp^2-hybridized intermediate imine anion that is analogous to the sp^3-hybridized intermediate alkoxide anion formed by addition of a nucleophile to a carbonyl group (Figure 10.8). Once formed, the intermediate imine anion can then go on to yield further products.

Carbonyl compound

$$\overset{\delta^-}{O}=\overset{\delta^+}{C}< \xrightarrow{:Nu^-} \left[-\overset{O^-}{\underset{|}{C}}-Nu \right] \longrightarrow \text{Products}$$

Nitrile

$$R-\overset{\delta^+}{C}\equiv\overset{\delta^-}{N} \xrightarrow{:Nu^-} \left[R-\overset{N^-}{\overset{\|}{C}}-Nu \right] \longrightarrow \text{Products}$$

FIGURE 10.8 The similar reactions of carbonyl compounds and nitriles with nucleophiles

Hydrolysis: Conversion of nitriles into carboxylic acids Nitriles are hydrolyzed in either acidic or basic solution to yield carboxylic acids and ammonia (or an amine).

$$R-C\equiv N \xrightarrow[H_2O,\ NaOH]{H_3O^+ \text{ or}} R-\overset{O}{\overset{\|}{C}}-OH + NH_3$$

Reduction: Conversion of nitriles into amines Reduction of nitriles with $LiAlH_4$ gives primary amines, just as reduction of esters gives primary alcohols.

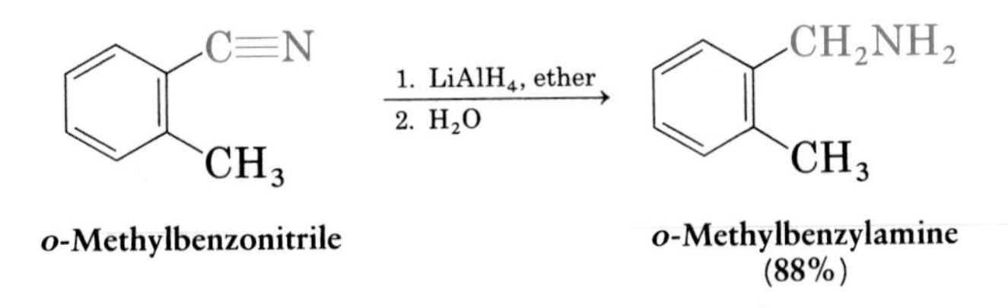

Reaction of nitriles with Grignard reagents Grignard reagents, RMgX, add to nitriles to give intermediate imine anions that can be hydrolyzed to yield ketones.

$$R-C\equiv N: + \ddot{R}'^{-}\overset{+}{M}gX \longrightarrow \left[R-\overset{:\ddot{N}^{-}\overset{+}{M}gX}{\overset{\|}{C}}-R' \right] \xrightarrow{H_3O^+} R-\overset{O}{\overset{\|}{C}}-R' + :NH_3$$

Nitrile — Imine anion — Ketone

For example, benzonitrile reacts with ethylmagnesium bromide to give propiophenone in high yield.

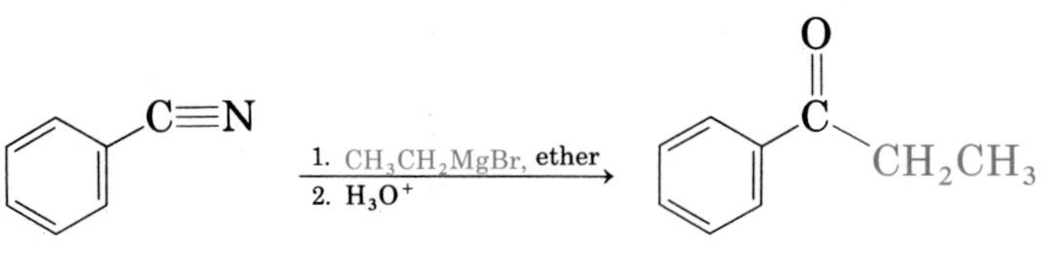

PRACTICE PROBLEM 10.12

Show how you could prepare 2-hexanone by reaction of a Grignard reagent on a nitrile.

SOLUTION There are two ways to prepare a ketone from a nitrile by Grignard addition:

$$\left.\begin{array}{l} CH_3C{\equiv}N + CH_3CH_2CH_2CH_2MgBr \\ CH_3MgBr + N{\equiv}CCH_2CH_2CH_2CH_3 \end{array}\right\} \longrightarrow CH_3\overset{O}{\overset{\|}{C}}CH_2CH_2CH_2CH_3$$

PROBLEM 10.28 What nitrile would you react with what Grignard reagent to prepare these ketones?

(a) $CH_3CH_2COCH_2CH_3$

(b) $CH_3CH_2COCH(CH_3)_2$

(c) Acetophenone (methyl phenyl ketone)

(d)

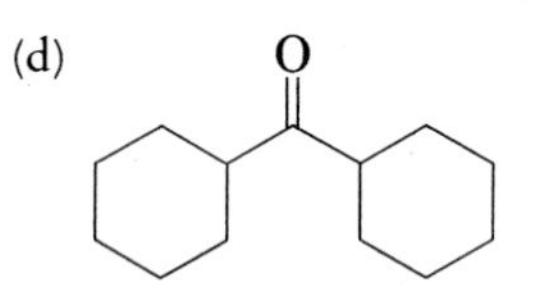

PROBLEM 10.29 By putting the proper reactions together in the appropriate sequence, you can prepare complex molecules from simple starting materials. How would you prepare the following molecules from the indicated starting materials? More than one step is required in each case.

(a) $(CH_3)_2CHCH_2CH_2NH_2$ from $(CH_3)_2CHCH_2I$

(b) 1-Phenyl-2-butanone from benzyl bromide, $C_6H_5CH_2Br$.

10.12 NYLON AND POLYESTER: STEP-GROWTH POLYMERS

chain-growth polymer a polymer produced by chain reaction of a monofunctional monomer

There are two main classes of synthetic polymers: chain-growth polymers and step-growth polymers. **Chain-growth polymers,** such as polyethylene and polystyrene, are prepared by chain-reaction processes in which an initiator first adds to the double bond of an alkene monomer to produce a reactive intermediate. This intermediate adds to a second alkene monomer unit, and the polymer chain lengthens as more monomer units add successively to the end of the growing chain.

step-growth polymer a polymer produced by stepwise reaction between two difunctional monomers

Step-growth polymers are produced by polymerization reactions between two difunctional molecules. Each new bond is formed in a discrete step, independent of all other bonds in the polymer; chain reactions aren't involved. The key bond-forming step is often a carbonyl nucleophilic acyl substitution.

A large number of step-growth polymers have been made; some of the commercially more important ones are shown in Table 10.3.

TABLE 10.3 **Some important step-growth polymers and their uses**

Monomer name	*Formula*	*Trade or common name of polymer*	*Uses*
Adipic acid Hexamethylenediamine	$HOOC(CH_2)_4COOH$ $H_2N(CH_2)_6NH_2$	Nylon 66	Fibers, clothing, tire cord, bearings
Ethylene glycol Dimethyl terephthalate	$HOCH_2CH_2OH$ $COOCH_3$ $COOCH_3$	Dacron, Terylene, Mylar	Fibers, clothing, tire cord, film
Caprolactam	O N—H	Nylon 6, Perlon	Fibers, large cast articles

Nylons

nylons polyamides prepared by reaction between a diacid and a diamide

The nylons, first synthesized by Wallace Carothers at the DuPont Company, are the best-known step-growth polymers. **Nylons** are polyamides prepared by reaction between a diamine and a diacid. For example, nylon 66 is prepared by heating the six-carbon adipic acid (hexanedioic acid) with the six-carbon hexamethylenediamine (1,6-hexanediamine) at 280°C.

$$\underset{\text{Hexamethylenediamine}}{H_2N(CH_2)_6NH_2} + \underset{\text{Adipic acid}}{HO\overset{O}{\overset{\|}{C}}(CH_2)_4\overset{O}{\overset{\|}{C}}OH} \longrightarrow \underset{\text{Nylon 66}}{\left(\overset{O}{\overset{\|}{C}}(CH_2)_4\overset{O}{\overset{\|}{C}}-NH(CH_2)_6NH\right)_n} + n\,H_2O$$

Nylons are used in engineering applications and in making fibers. A combination of high-impact strength and abrasion resistance makes nylon an excellent metal-substitute for bearings and gears. As fibers, nylon is used in a variety of applications, from clothing, to Aramid tire cord, to carpets, to Perlon mountaineering ropes.

Polyesters

polyesters polymers prepared by reaction between a diacid and a dialcohol

Just as polyamides (nylons) are made by reaction between diacids and diamines, **polyesters** are made by reaction between diacids and dialcohols. The most generally useful polyester is made by a nucleophilic acyl substitution reaction between dimethyl terephthalate (dimethyl 1,4-benzenedicarboxylate) and ethylene glycol. The product is used under the trade name Dacron to make clothing fiber and tire cord and under the name Mylar to make plastic film and recording tape. The tensile strength of polyester film is nearly equal to that of steel.

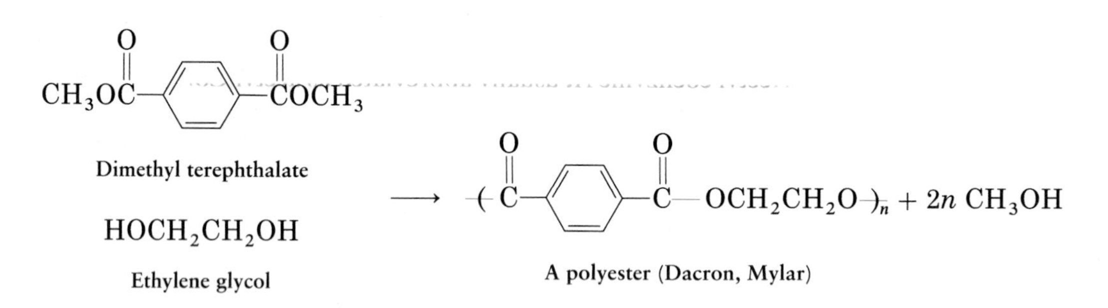

PRACTICE PROBLEM 10.13 Draw the structure of Qiana, a polyamide fiber made by reaction of hexanedioic acid with 1,4-cyclohexanediamine.

SOLUTION

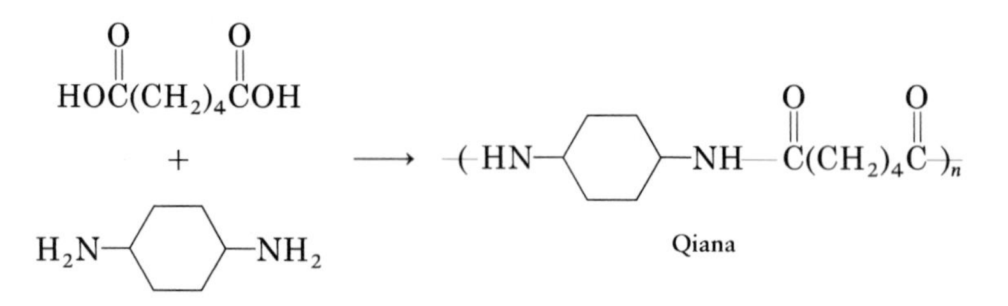

PROBLEM 10.30 Kevlar, a nylon polymer used to make bulletproof vests, is made by reaction of 1,4-benzenedicarboxylic acid with 1,4-benzenediamine. Show the structure of Kevlar.

INTERLUDE

Thiol Esters: Biological Carboxylic Acid Derivatives

Nucleophilic acyl substitution reactions take place in living organisms just as they take place in the chemical laboratory; the same principles apply in both cases. Nature, however, uses thiol esters, RCOSR′, rather than acid chlorides or acid anhydrides as its reagents. The pK_a of a typical alkanethiol (RSH) is about 10, placing thiols midway in acid strength between carboxylic acids (pK_a 5) and alcohols (pK_a 16). As a result, thiolate anions (RS^-) can act as leaving groups in nucleophilic acyl substitution reactions. Thiol esters aren't so reactive that they hydrolyze rapidly like anhydrides, yet they are more reactive than normal esters.

Acetyl coenzyme A, usually abbreviated as acetyl CoA, is the most common thiol ester in nature. Acetyl CoA is a much more complex molecule than acetyl chloride or acetic anhydride, yet it serves exactly the same purpose as either of these simpler reagents. Nature uses acetyl CoA as a reactive acetylating agent in nucleophilic acyl substitution reactions (an *acetylating* agent introduces an acetyl group, CH_3CO):

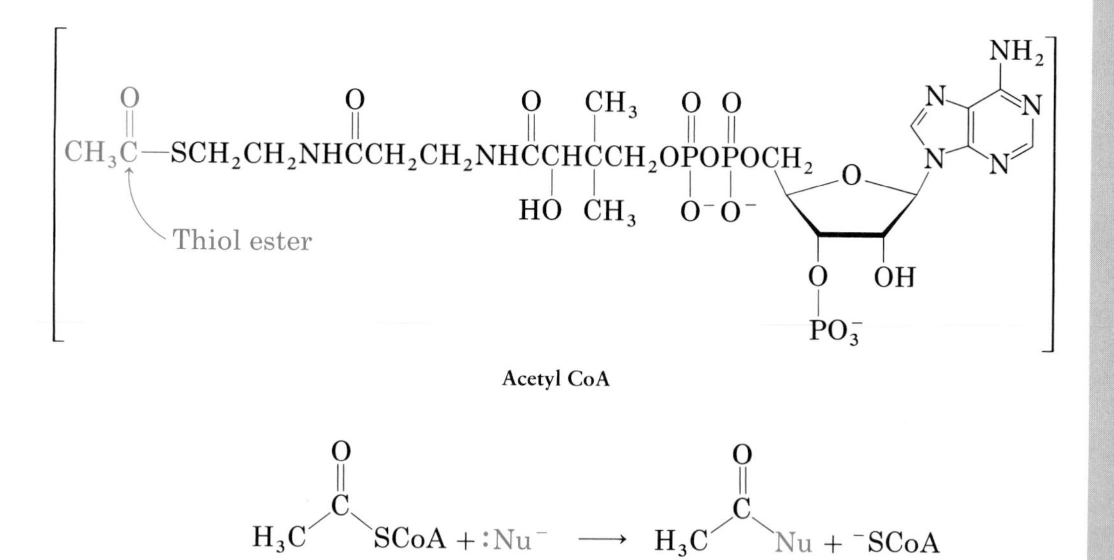

Acetyl CoA

As an example, *N*-acetylglucosamine, an important constituent of cell-surface membranes in mammals, is synthesized in nature by a reaction between glucosamine and acetyl CoA:

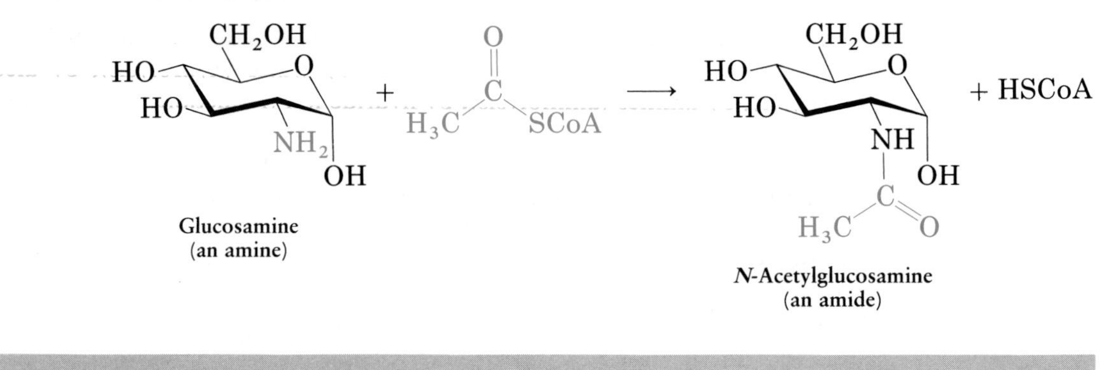

Glucosamine (an amine)

N-Acetylglucosamine (an amide)

SUMMARY AND KEY WORDS

Carboxylic acids are among the most important building blocks for synthesizing other molecules, both in nature and in the chemical laboratory. The distinguishing characteristic of carboxylic acids is their **acidity.** Although weaker than mineral acids like HCl, carboxylic acids nevertheless dissociate far more readily than alcohols. The reason for this difference lies in the stability of carboxylate ions: Carboxylate ions are stabilized by **resonance** between two equivalent forms:

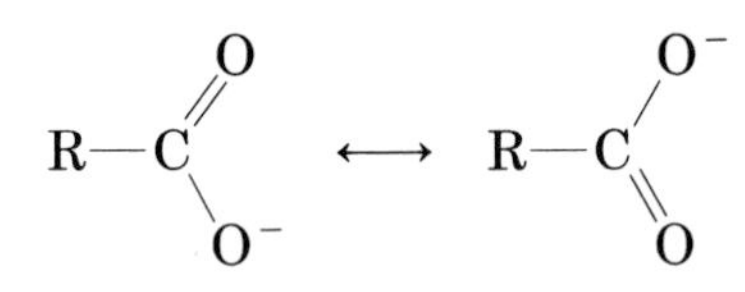

Most carboxylic acids have pK_a values near 5, but the exact acidity of an acid depends on its structure. Carboxylic acids substituted by an electron-withdrawing group are more acidic (have a lower pK_a) because their carboxylate ions are stabilized.

Carboxylic acids can be transformed into a variety of **acyl derivatives** in which the acid −OH group has been replaced by other substituents. **Acid chlorides, acid anhydrides, esters,** and **amides** are the most important acyl derivatives. The chemistries of these different acyl derivatives are similar and are dominated by a single general reaction type: the **nucleophilic acyl substitution reaction.** These substitutions take place by addition of a nucleophile to the polar carbonyl group of the acid derivative, followed by expulsion of the leaving group.

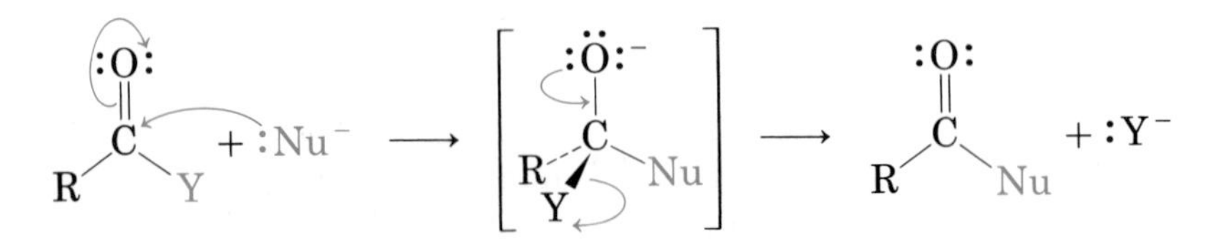

Carboxylic acid derivatives can undergo reaction with many different nucleophiles. Among the most important are substitution by water (hydrolysis), by alcohols (alcoholysis), by amines (aminolysis), by hydride ion (reduction), and by Grignard reagents.

Nitriles, R−C≡N, can also be considered as carboxylic acid derivatives because they undergo nucleophilic additions to the polar C≡N bond in the same way that carbonyl compounds do. The most important reactions of nitriles are their hydrolysis to yield carboxylic acids, their reduction to yield primary amines, and their reaction with Grignard reagents to yield ketones.

SUMMARY OF REACTIONS

1. Reactions of carboxylic acids (Section 10.6)

   (a) Conversion into acid chlorides

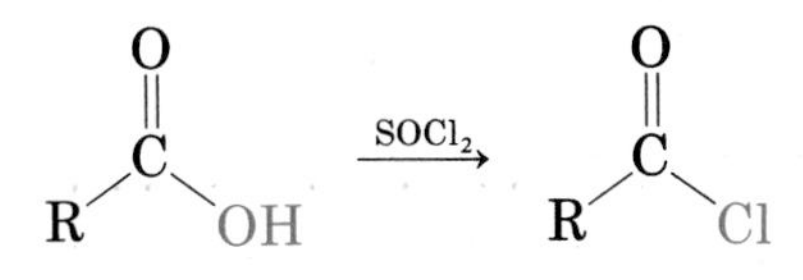

   (b) Conversion into esters

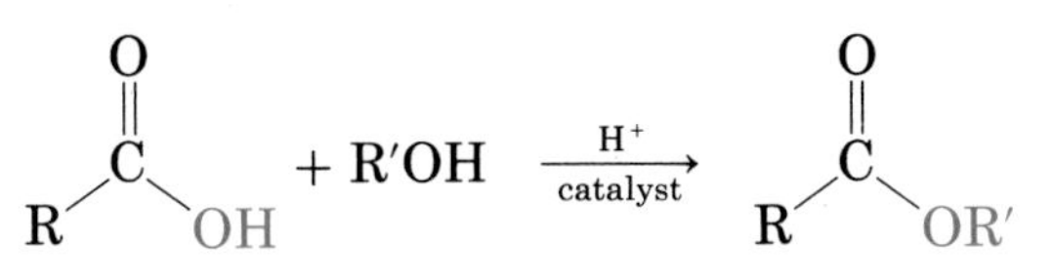

2. Reactions of acid halides (Section 10.7)

   (a) Conversion into carboxylic acids

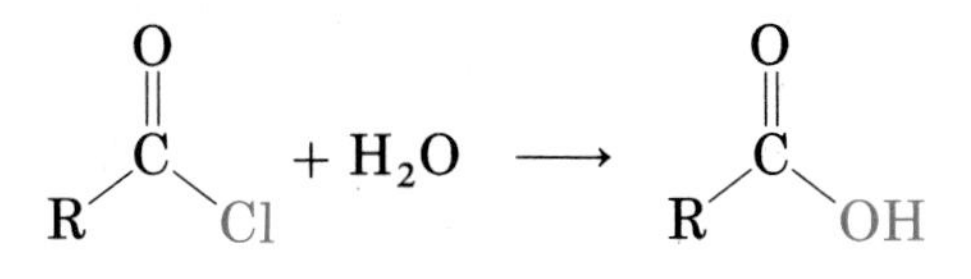

   (b) Conversion into esters

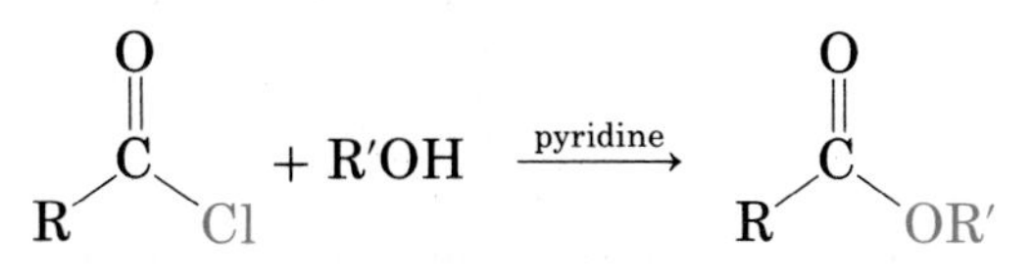

   (c) Conversion into amides

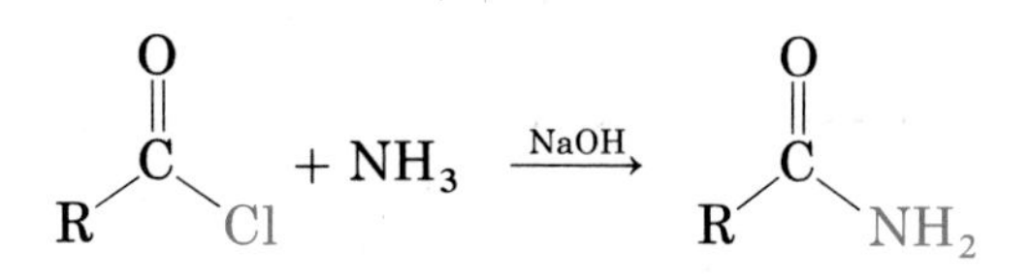

3. Reactions of acid anhydrides (Section 10.8)

(a) Conversion into esters

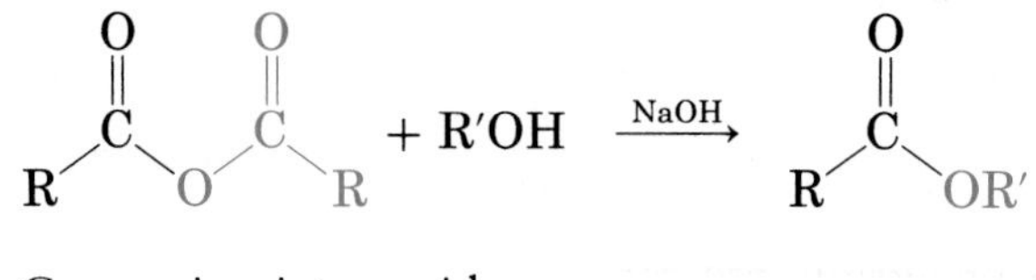

(b) Conversion into amides

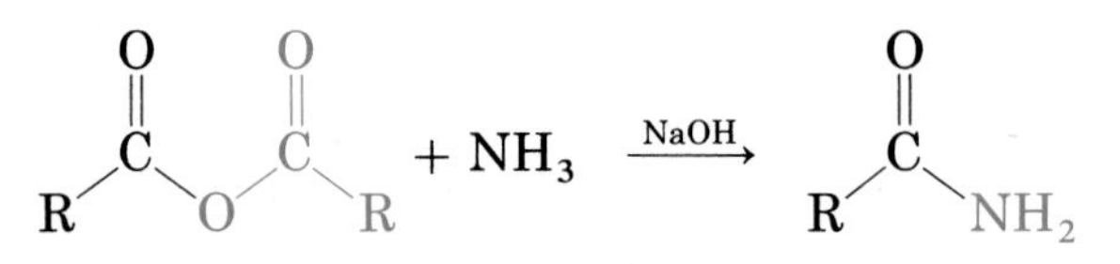

4. Reactions of esters (Section 10.9)

(a) Conversion into acids

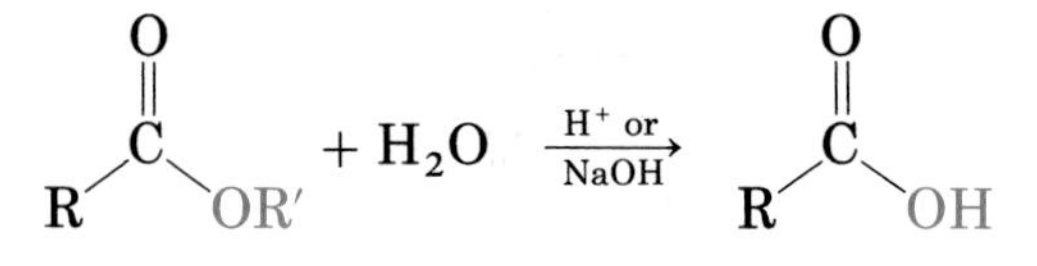

(b) Conversion into amides

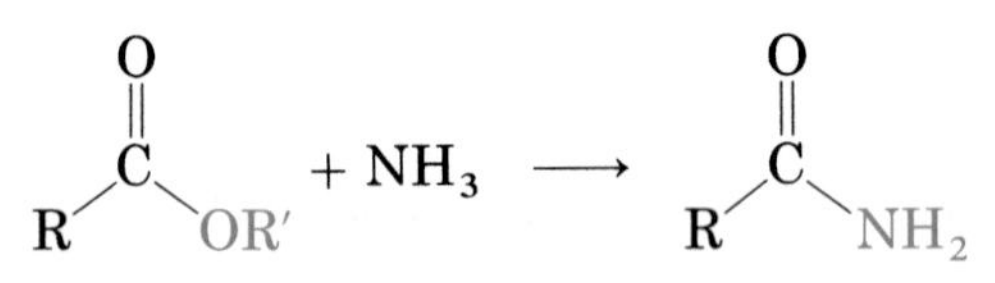

(c) Conversion into primary alcohols by reduction

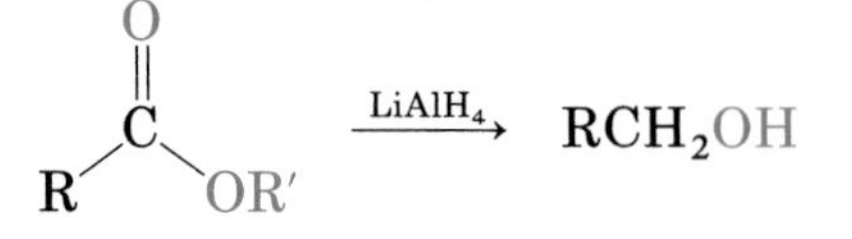

(d) Conversion into tertiary alcohols by Grignard reaction

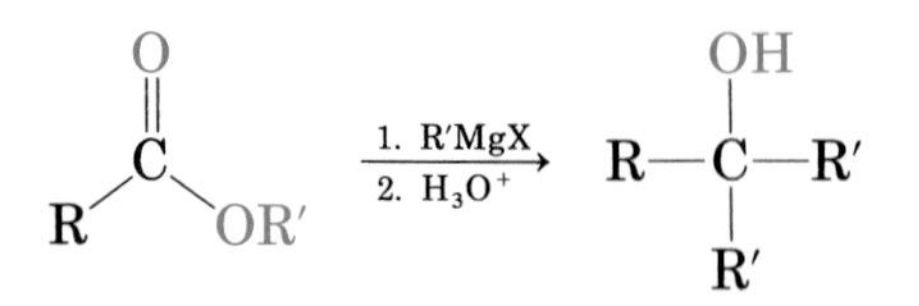

5. Reactions of amides (Section 10.10)

(a) Conversion into carboxylic acids

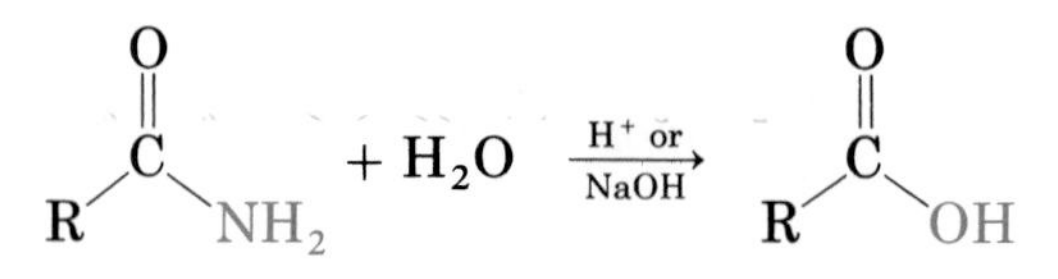

(b) Conversion into amines by reduction

$$\mathrm{RCONH_2} \xrightarrow[\text{2. } H_2O]{\text{1. } LiAlH_4} \mathrm{RCH_2NH_2}$$

6. Reactions of nitriles (Section 10.11)

(a) Conversion into carboxylic acids

$$\mathrm{R{-}C{\equiv}N + H_2O} \xrightarrow[\text{NaOH}]{H^+ \text{ or}} \mathrm{RCOOH}$$

(b) Conversion into amines by reduction

$$\mathrm{R{-}C{\equiv}N} \xrightarrow[\text{2. } H_2O]{\text{1. } LiAlH_4} \mathrm{RCH_2NH_2}$$

(c) Conversion into ketones by Grignard reaction

$$\mathrm{R{-}C{\equiv}N} \xrightarrow[\text{2. } H_2O]{\text{1. } R'MgX} \mathrm{RCOR'}$$

ADDITIONAL PROBLEMS

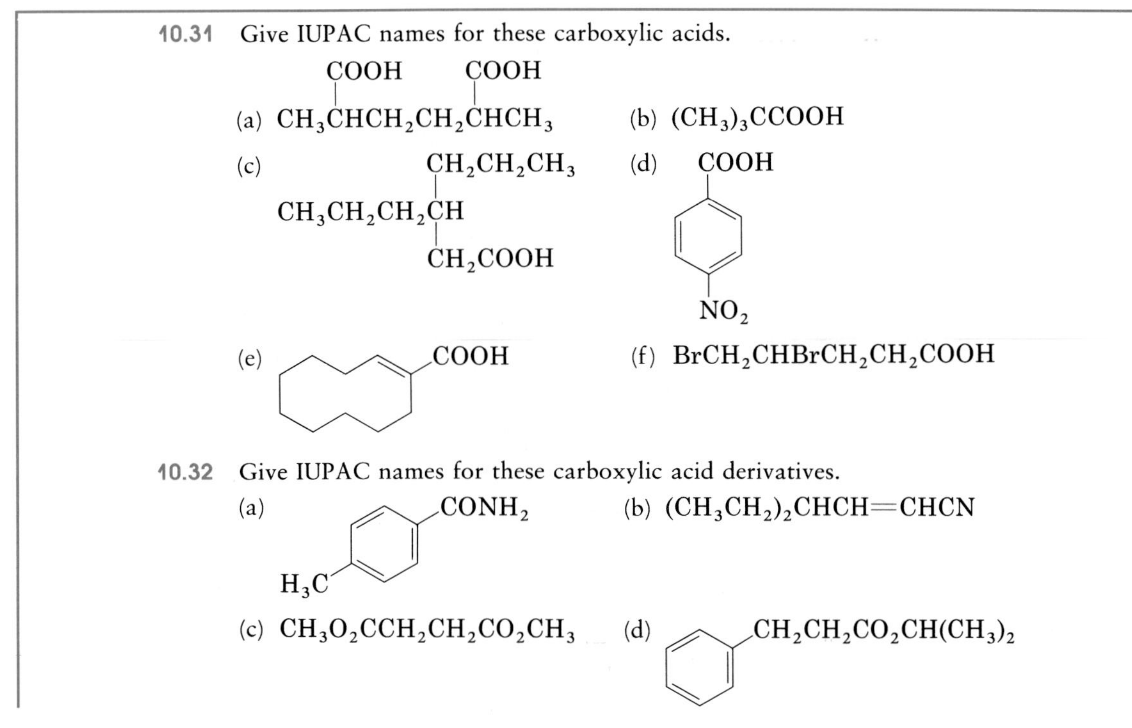

10.31 Give IUPAC names for these carboxylic acids.

(a) $CH_3CH(COOH)CH_2CH_2CH(COOH)CH_3$

(b) $(CH_3)_3CCOOH$

(c) $CH_3CH_2CH_2CH(CH_2CH_2CH_3)CH_2COOH$

(d) COOH, NO_2

(e) COOH

(f) $BrCH_2CHBrCH_2CH_2COOH$

10.32 Give IUPAC names for these carboxylic acid derivatives.

(a) $CONH_2$, H_3C

(b) $(CH_3CH_2)_2CHCH{=}CHCN$

(c) $CH_3O_2CCH_2CH_2CO_2CH_3$

(d) $CH_2CH_2CO_2CH(CH_3)_2$

(e) (g)

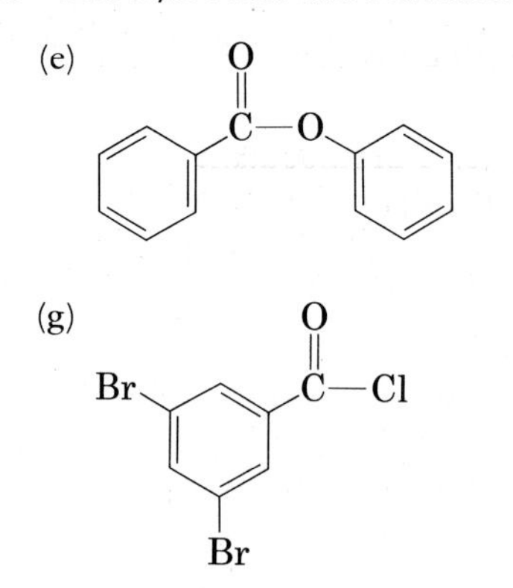

(f) $CH_3CHBrCH_2CONHCH_3$

(h)

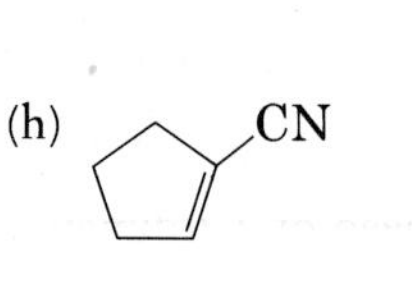

10.33 Draw structures corresponding to these IUPAC names:

(a) 4,5-Dimethylheptanoic acid
(b) *cis*-1,2-Cyclohexanedicarboxylic acid
(c) Heptanedioic acid
(d) Triphenylacetic acid
(e) 2,2-Dimethylhexanamide
(f) Phenylacetamide
(g) 2-Cyclobutenecarbonitrile
(h) Ethyl cyclohexanecarboxylate

10.34 Acetic acid boils at 118°C, but its ethyl ester boils at 77°C. Why is the boiling point of the acid so much higher, even though it has a lower molecular weight?

10.35 Draw and name the eight carboxylic acids with formula $C_6H_{12}O_2$.

10.36 Draw and name compounds that meet these descriptions:

(a) Three acid chlorides, C_6H_9ClO
(b) Three amides, $C_7H_{11}NO$
(c) Three nitriles, C_5H_7N
(d) Three esters, $C_5H_8O_2$

10.37 The following reactivity order has been found for the saponification of alkyl acetates by hydroxide ion:

$$CH_3COOCH_3 > CH_3COOCH_2CH_3 > CH_3COOCH(CH_3)_2 > CH_3COOC(CH_3)_3$$

How can you explain this reactivity order?

10.38 Citric acid has $pK_a = 3.14$, and tartaric acid has $pK_a = 2.98$. Which acid is stronger?

10.39 Order the compounds in each set with respect to increasing acidity:

(a) Acetic acid, chloroacetic acid, trifluoroacetic acid
(b) Benzoic acid, *p*-bromobenzoic acid, *p*-nitrobenzoic acid
(c) Acetic acid, phenol, cyclohexanol

10.40 How can you explain the fact that 2-chlorobutanoic acid has $pK_a = 2.86$, 3-chlorobutanoic acid has $pK_a = 4.05$, 4-chlorobutanoic acid has $pK_a = 4.82$, and butanoic acid itself has $pK_a = 4.82$?

10.41 Rank these compounds in order of their reactivity toward nucleophilic acyl substitution:

(a) CH_3COOCH_3 (b) CH_3COCl (c) CH_3CONH_2 (d) $CH_3COOCOCH_3$

10.42 How can you prepare acetophenone (methyl phenyl ketone) from these starting materials? More than one step may be required.

(a) Benzonitrile (b) Bromobenzene (c) Methyl benzoate (d) Benzene

10.43 How might you prepare these products starting with butanoic acid? More than one step may be required.

(a) 1-Butanol (b) Butanal (c) 1-Bromobutane
(d) Pentanenitrile (e) 1-Butene (f) Butylamine, $CH_3CH_2CH_2CH_2NH_2$

10.44 Predict the product of the reaction of *p*-methylbenzoic acid with each of these reagents:

(a) $LiAlH_4$ (b) CH_3OH, HCl (c) $SOCl_2$ (d) NaOH, then CH_3I

10.45 A chemist in need of 2,2-dimethylpentanoic acid decided to synthesize some by reaction of 2-chloro-2-methylpentane with NaCN, followed by hydrolysis of the product. After carrying out the reaction sequence, however, none of the desired product could be found. What do you suppose went wrong?

10.46 Which method of carboxylic acid synthesis, Grignard carboxylation or nitrile hydrolysis, would you use for each of the following reactions? Explain the reasons for each choice.

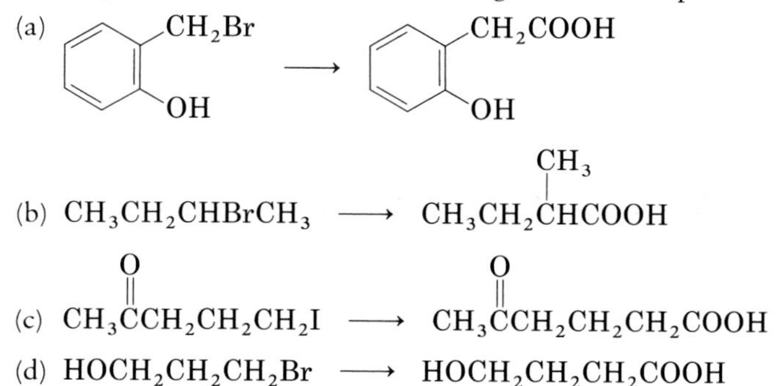

10.47 How can you explain the observation that an attempted Fischer esterification of 2,4,6-trimethylbenzoic acid with methanol/HCl is unsuccessful? No ester is obtained, and the starting acid is recovered unchanged.

10.48 Acid chlorides undergo reduction with $LiAlH_4$ in the same way that esters do to yield primary alcohols. What are the products of these reactions?

(a) $CH_3CH(CH_3)CH_2CH_2C(=O)Cl \xrightarrow[2.\ H_2O]{1.\ LiAlH_4}$

(b) 1-methylcyclopentanecarbonyl chloride (cyclopentane ring bearing COCl and CH_3) $\xrightarrow[2.\ H_2O]{1.\ LiAlH_4}$

10.49 The reaction of an acid chloride with $LiAlH_4$ to yield a primary alcohol (Problem 10.48) takes place in two steps. The first step is a nucleophilic acyl substitution of H^- for Cl^- to yield an aldehyde, and the second step is nucleophilic addition of H^- to the aldehyde to yield an alcohol. Write the mechanism of the reduction of CH_3COCl.

10.50 Acid chlorides undergo reaction with Grignard reagents at −78°C to yield ketones. Propose a mechanism for the reaction.

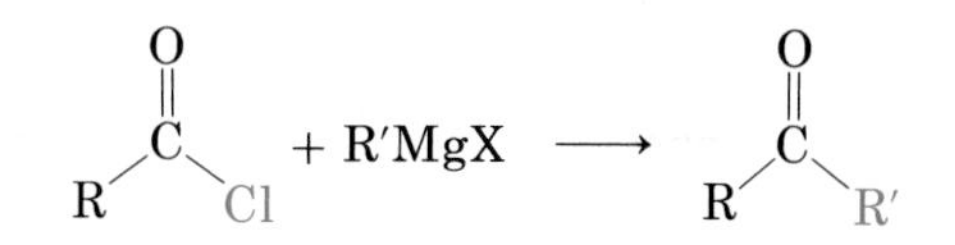

10.51 If the reaction of an acid chloride with a Grignard reagent (Problem 10.50) is carried out at room temperature, a tertiary alcohol is formed.

(a) Propose a mechanism for this reaction.

(b) What are the products of the reaction of methylmagnesium bromide, CH_3MgBr, with the acid chlorides listed in Problem 10.48?

10.52 When dimethyl carbonate, $CH_3OCOOCH_3$, is treated with phenylmagnesium bromide, triphenylmethanol is formed. Explain how this occurs.

10.53 Predict the product, if any, of reaction between propanoyl chloride and the following reagents. (See Problems 10.49 and 10.50.)

(a) Excess CH_3MgBr in ether (b) NaOH in H_2O (c) Methylamine
(d) $LiAlH_4$ (e) Cyclohexanol (f) Sodium acetate

10.54 Answer Problem 10.53 for reaction between methyl propanoate and the listed reagents.

10.55 What esters and what Grignard reagents would you use to make these alcohols.

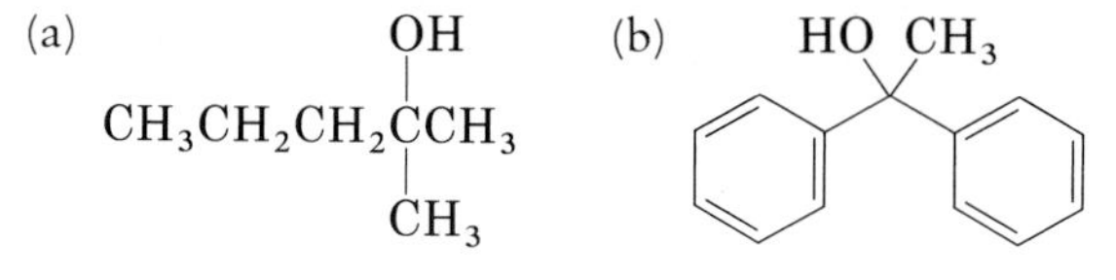

10.56 Show two ways to make these esters:

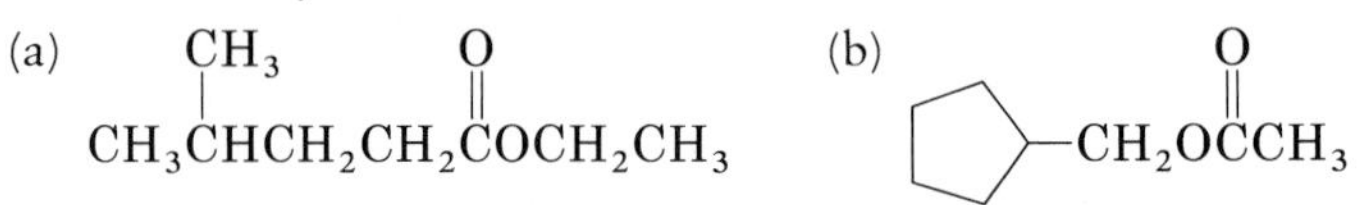

10.57 What products would you obtain on saponification of these esters.

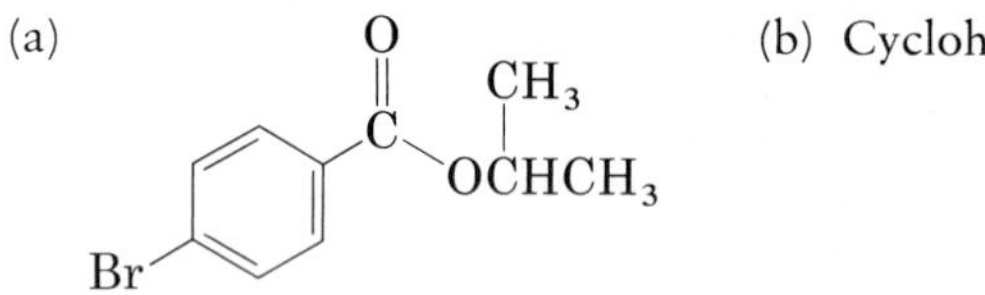

(b) Cyclohexyl propanoate

10.58 When *methyl* acetate is heated in pure ethanol containing a small amount of HCl catalyst, *ethyl* acetate results. Propose a mechanism for this reaction.

10.59 *tert*-Butoxycarbonyl azide, an important reagent used in protein synthesis, is prepared by treating *tert*-butoxycarbonyl chloride with sodium azide. Propose a mechanism for this reaction.

$$(CH_3)_3COCOCl + NaN_3 \longrightarrow (CH_3)_3COCON_3 + NaCl$$

10.60 Tranexamic acid, a drug useful for aiding blood clotting, is prepared commercially from *p*-methylbenzonitrile. Formulate the steps that are likely to be used in the synthesis. (Note: The cis and trans isomers of tranexamic acid are thermally interconvertible at 300°C, and the trans isomer is more stable.)

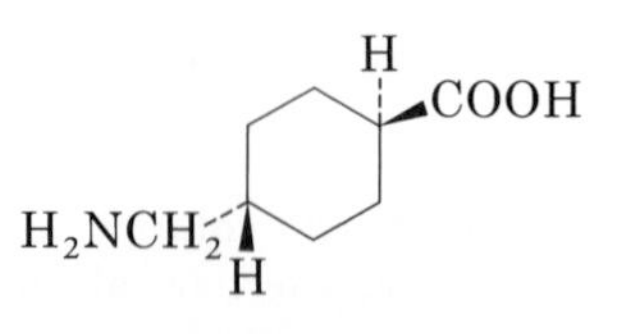

Tranexamic acid

10.61 What product would you expect to obtain upon treatment of the cyclic ester, butyrolactone, with excess phenylmagnesium bromide?

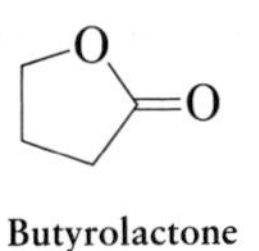

Butyrolactone

10.62 N,N-Diethyl-*m*-toluamide (DEET) is the active ingredient in many insect repellents. How might you synthesize this substance from *m*-bromotoluene?

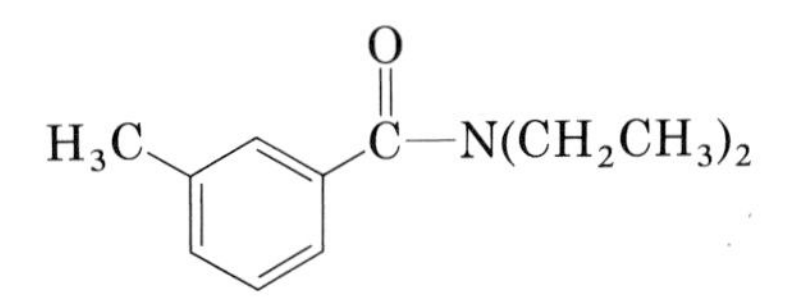

N,N-Diethyl-*m*-toluamide

10.63 In the iodoform reaction, a triiodomethyl ketone reacts with aqueous base to yield a carboxylate ion and iodoform (triiodomethane). Propose a mechanism for this reaction.

$$R-\overset{O}{\overset{\|}{C}}-CI_3 \xrightarrow{NaOH,\ H_2O} R-\overset{O}{\overset{\|}{C}}-O^- + CHI_3$$

10.64 The K_a for bromoacetic acid is approximately 1×10^{-3}. What percentage of the acid is dissociated in a 0.10 M aqueous solution?

CHAPTER 11

Carbonyl Alpha-Substitution Reactions and Condensation Reactions

Much of the chemistry of carbonyl compounds can be explained by just four fundamental reactions. We've already looked in detail at two of the four: the nucleophilic addition reaction (Chapter 9) and the nucleophilic acyl substitution reaction (Chapter 10). In this chapter, we'll look at the other two carbonyl-group reactions: the alpha-substitution reaction and the carbonyl condensation reaction.

alpha-substitution reaction a reaction that results in substitution of a hydrogen on the alpha carbon of a carbonyl compound

carbonyl condensation reaction a reaction between two carbonyl compounds such that the alpha carbon of one partner bonds to the carbonyl carbon of the other

Alpha-substitution reactions occur at the position *next to* the carbonyl group, the alpha (α) position. They involve the substitution of an α hydrogen atom by some other group:

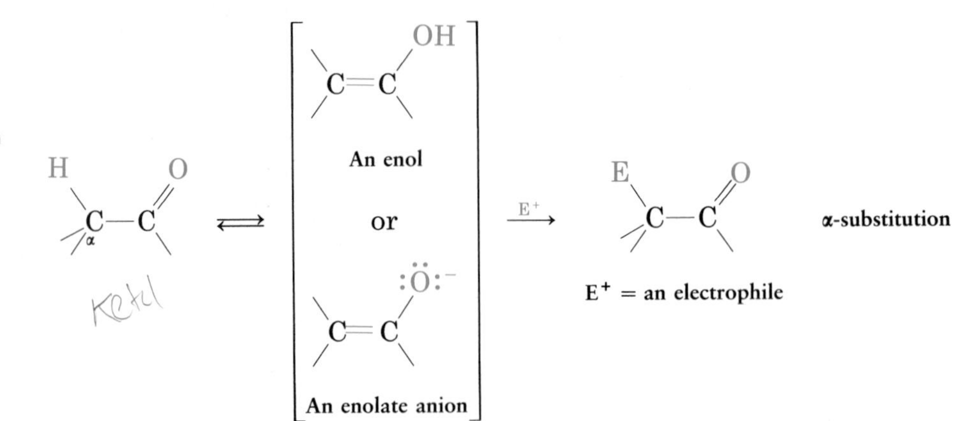

Carbonyl condensation reactions take place when *two* carbonyl compounds react with each other in such a way that the α carbon of one partner becomes bonded to the carbonyl carbon of the second partner:

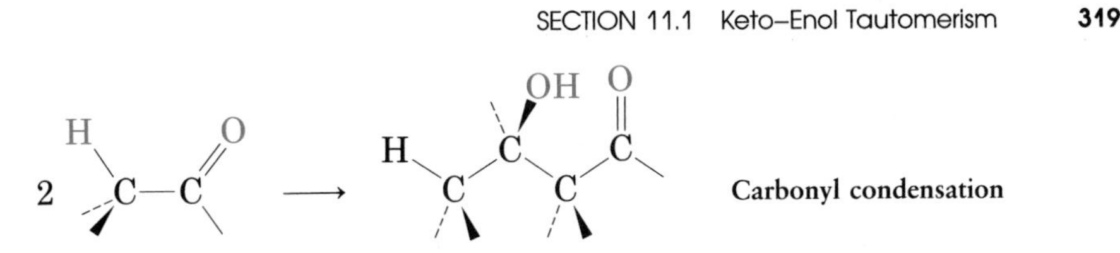

The key feature of both α-substitution reactions and carbonyl condensation reactions is that they take place by the formation of either *enol* or *enolate-ion* intermediates. Let's begin our study by learning more about these two species.

11.1 KETO–ENOL TAUTOMERISM

Carbonyl compounds that have hydrogen atoms on their alpha carbons are easily interconvertible with their corresponding enol (*ene* + *ol;* unsaturated alcohol) isomers. This interconversion between keto and enol forms is a special kind of isomerism called tautomerism (taw-*tom*-er-ism; Greek *tauto,* "the same," and *meros,* "part"). The individual isomers are called **tautomers** (*taw*-toe-mers).

tautomers
easily interconvertible, constitutional isomers

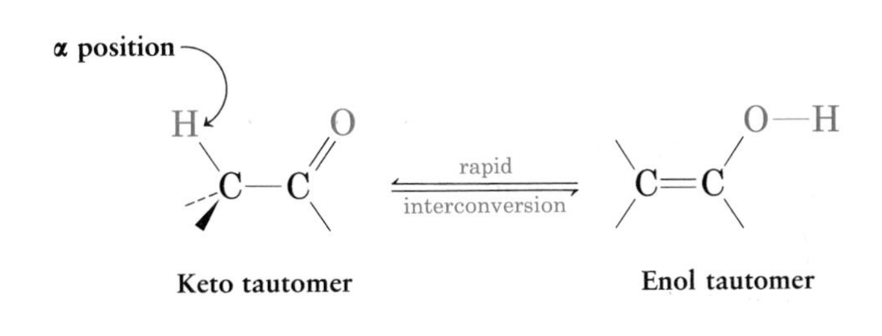

Keto tautomer

Enol tautomer

Note that two isomers must be *easily* interconvertible to be considered tautomers. Thus, keto and enol isomers of carbonyl compounds are tautomers, but two alkene isomers such as 1-butene and 2-butene aren't because they don't interconvert rapidly.

At equilibrium, most carbonyl compounds exist almost entirely in the keto form, and it's difficult to isolate the pure enol form. For example, cyclohexanone contains only about 0.00004% of its enol tautomer at room temperature, and acetone contains only about 0.0000005% enol. The percentage of enol tautomer is even less for carboxylic acids, esters, and amides. Even though enols are difficult to isolate and are present to only a small extent at equilibrium, they're nevertheless extremely important intermediates in much of the chemistry of carbonyl compounds.

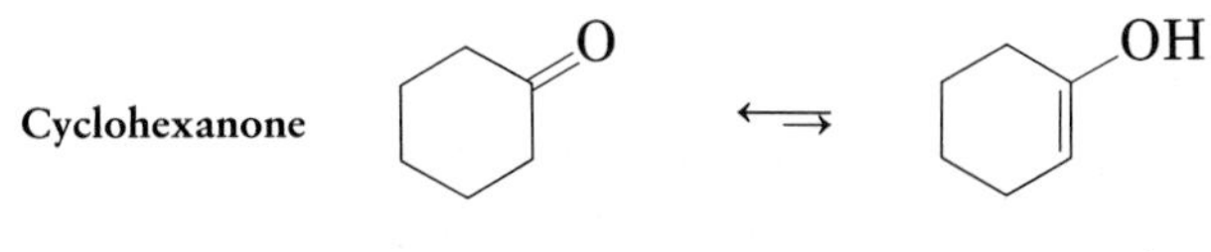

99.99996%

0.00004%

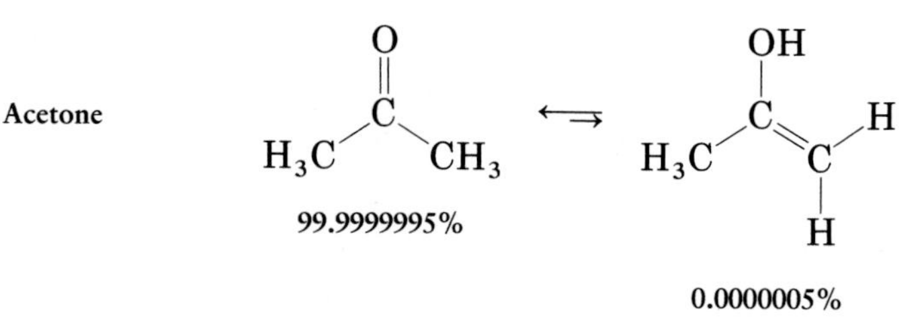

Keto–enol tautomerism of carbonyl compounds is catalyzed by both acids and bases. Acid catalysis involves protonation of the carbonyl oxygen atom (a Lewis base) to give an intermediate cation that can then lose a proton from the alpha carbon to yield the enol (Figure 11.1). This proton loss from the positively charged intermediate is analogous to what occurs during an E1 reaction when a carbocation loses a proton from the neighboring carbon to form an alkene (Section 7.10).

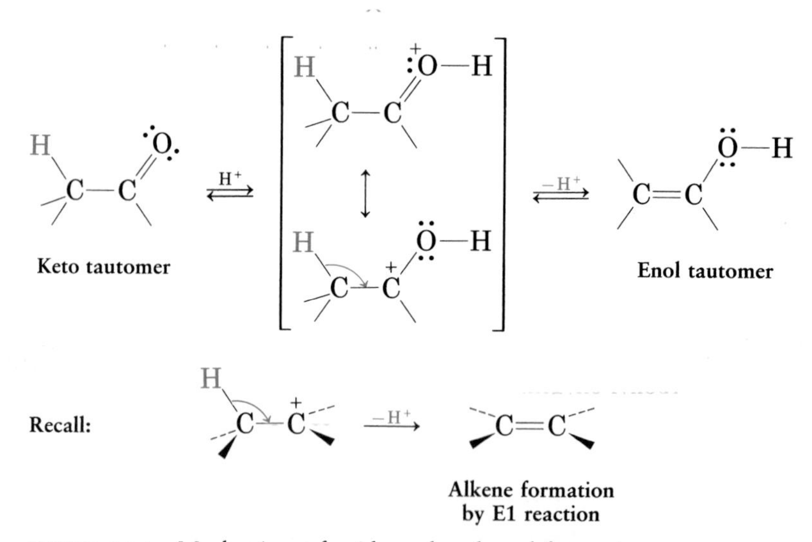

FIGURE 11.1 Mechanism of acid-catalyzed enol formation

Base-catalyzed enol formation occurs because the presence of a carbonyl group makes the hydrogens on the α carbon slightly acidic. Thus, a carbonyl compound can act as a weak acid and donate one of its α hydrogens to the base. The resultant resonance-stabilized anion, an **enolate ion,** is then reprotonated to yield a neutral compound. If protonation of the enolate ion takes place on the α carbon, the keto tautomer is regenerated, and no net change has taken place. If, however, protonation takes place on the oxygen atom, then an enol tautomer is formed (Figure 11.2).

enolate ion the anion of an enol; a resonance-stabilized alpha-keto carbanion

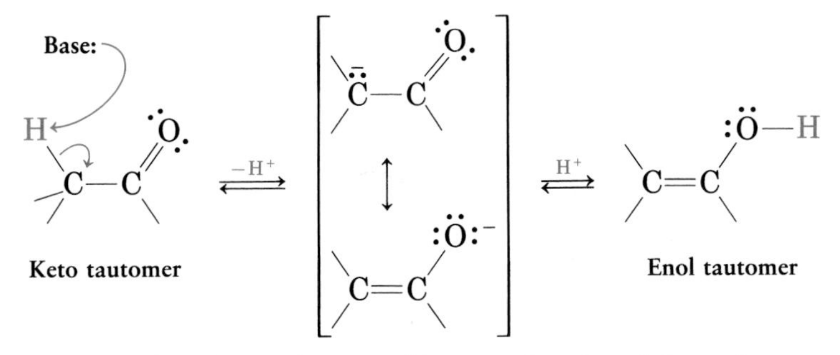

FIGURE 11.2 The mechanism of base-catalyzed enol formation

Note that only the protons on the α position of carbonyl compounds are acidic. The protons at beta, gamma, delta, and other positions aren't acidic because the resulting anions can't be resonance stabilized by the carbonyl group.

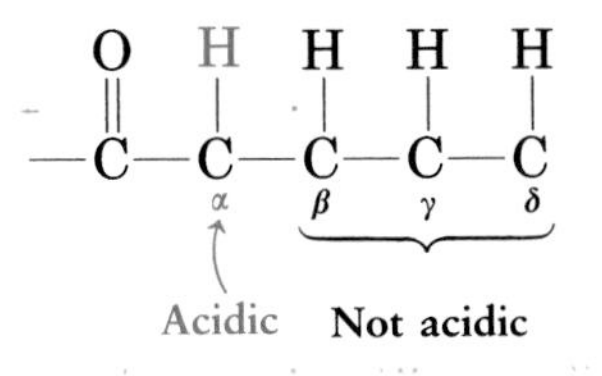

PRACTICE PROBLEM 11.1 Show the structure of the enol tautomer of propanal.

SOLUTION Enols are formed by removing a hydrogen from the carbon next to the carbonyl carbon, forming a double bond between the two carbons, and replacing the hydrogen on the carbonyl oxygen:

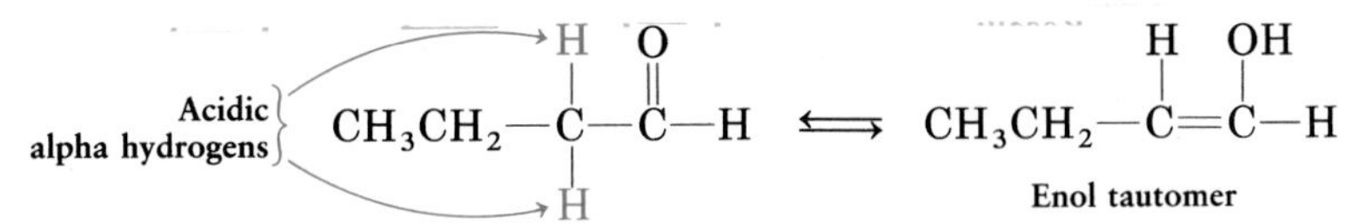

PROBLEM 11.1 Draw structures for the enol tautomers of these compounds.

(a) Cyclopentanone (b) Acetyl chloride (c) Ethyl acetate
(d) Acetic acid (e) Acetophenone (methyl phenyl ketone)

PROBLEM 11.2 How many acidic hydrogens does each of the molecules listed in Problem 11.1 have? Identify them.

PROBLEM 11.3 Account for the fact that 2-methylcyclohexanone can form two enol tautomers. Show the structures of both.

11.2 REACTIVITY OF ENOLS: THE MECHANISM OF α-SUBSTITUTION REACTIONS

What kind of chemistry should we expect of enols? Since their double bonds are electron-rich, enols behave as nucleophiles and react with electrophiles in much the same way as alkenes (Section 4.1). Because of electron donation from the oxygen lone-pair electrons, however, enols are even more reactive than alkenes.

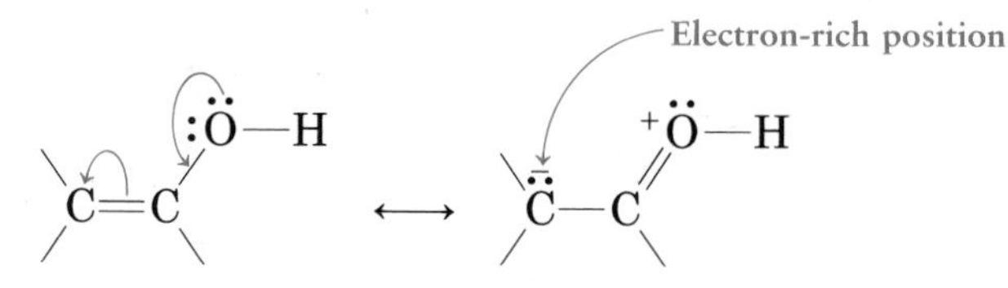

When an *alkene* reacts with an electrophile, the intermediate carbocation reacts with a nucleophile to give the addition product. When an *enol* reacts with an electrophile, however, the intermediate cation can lose the hydroxyl proton to regenerate a carbonyl compound. The net result of the reaction is α substitution (Figure 11.3).

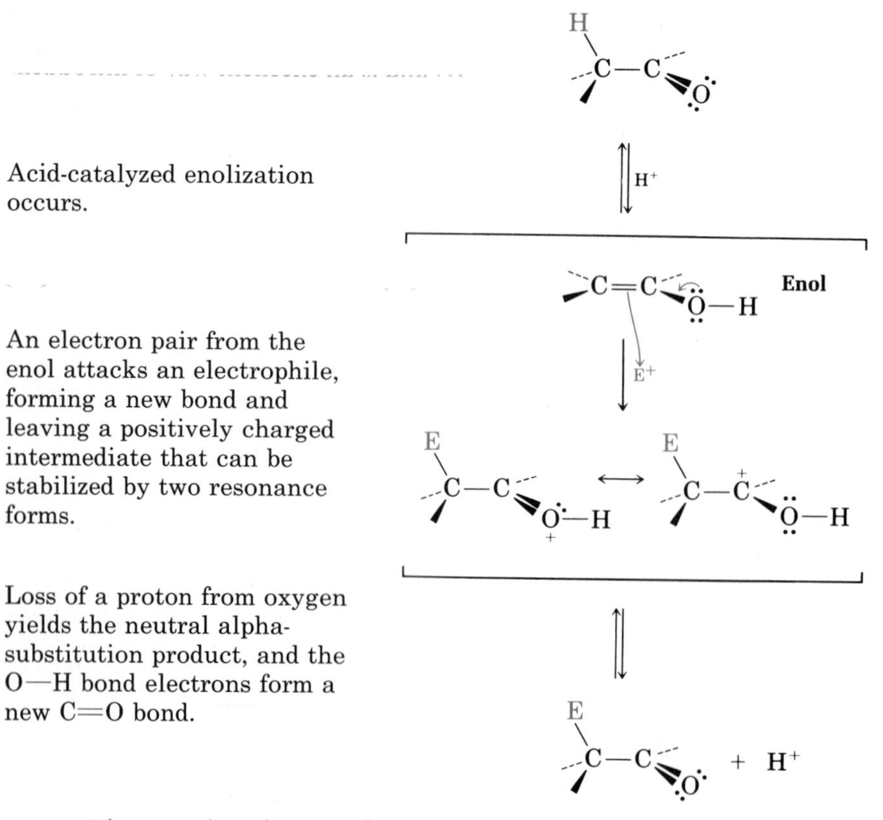

Acid-catalyzed enolization occurs.
An electron pair from the enol attacks an electrophile, forming a new bond and leaving a positively charged intermediate that can be stabilized by two resonance forms.
Loss of a proton from oxygen yields the neutral alpha-substitution product, and the O—H bond electrons form a new C=O bond.
H⁺
Enol
E⁺
+ H⁺

FIGURE 11.3 The general mechanism of a carbonyl α-substitution reaction

11.3 ALPHA HALOGENATION OF KETONES AND ALDEHYDES

Alpha halogenation provides one of the simplest examples of enol reactivity. Ketones and aldehydes are halogenated at their alpha positions by reaction with chlorine, bromine, or iodine in acidic solution. Bromine is most often used, and acetic acid is often employed as solvent. The reaction is a typical α-substitution process that proceeds through an enol intermediate.

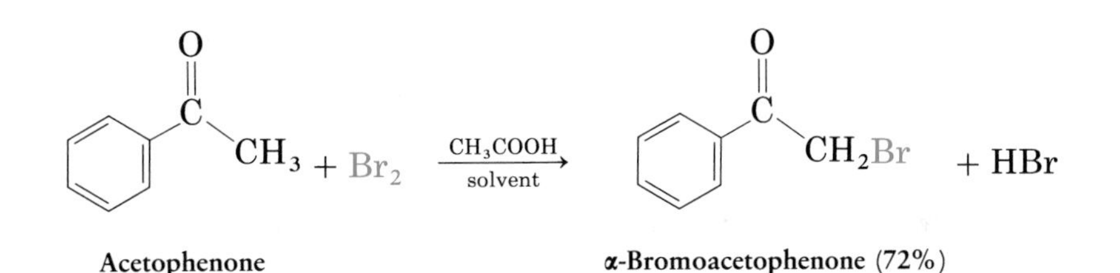

α-Bromo ketones are useful substances because they undergo elimination of HBr on treatment with base to yield α,β-unsaturated ketones. For example, 2-bromo-2-methylcyclohexanone gives 2-methyl-2-cyclohexenone in 62% yield when heated in pyridine. The reaction takes place by the normal E2 elimination pathway (Section 7.9) and is an excellent way of introducing carbon–carbon double bonds into molecules.

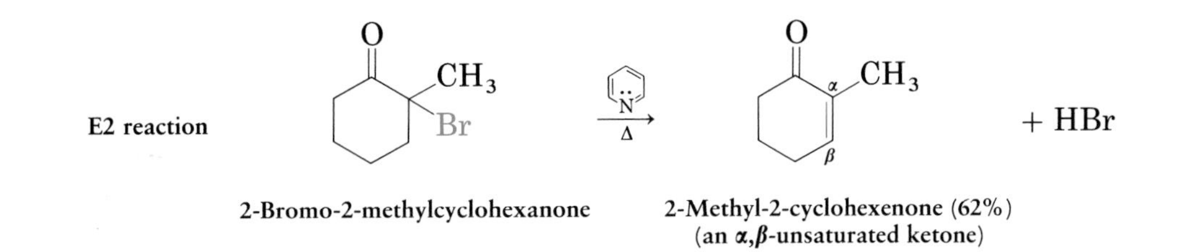

PRACTICE PROBLEM 11.2 What product would you obtain from reaction of cyclopentanone with Br_2 in acetic acid?

SOLUTION Locate the alpha hydrogens in the starting ketone and replace one of them by bromine to carry out an α-substitution reaction:

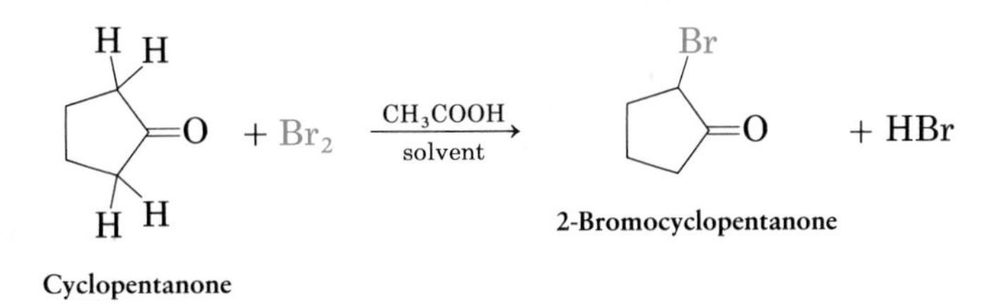

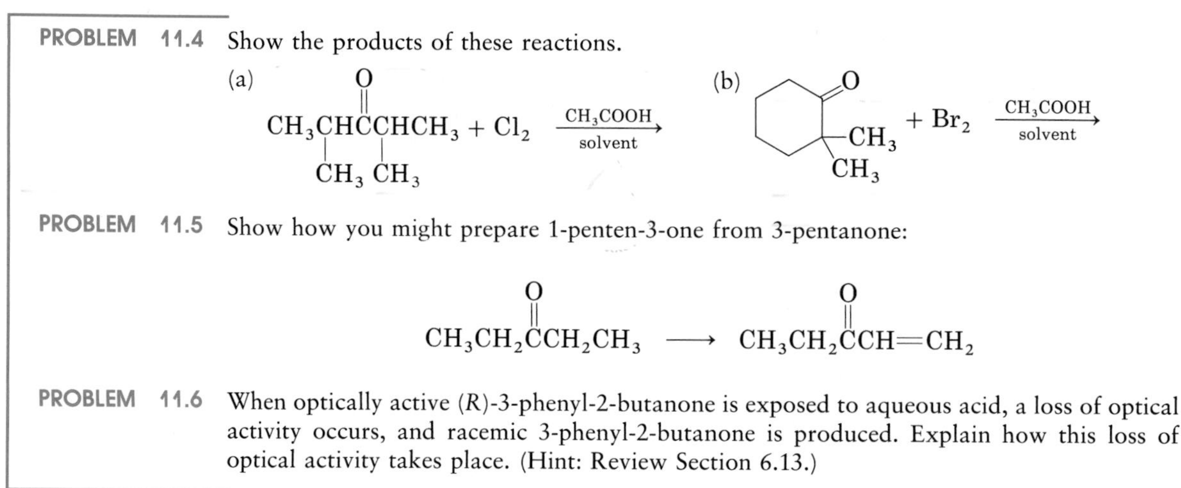

PROBLEM 11.4 Show the products of these reactions.

(a) $CH_3CH(CH_3)C(=O)CH(CH_3)CH_3 + Cl_2 \xrightarrow[\text{solvent}]{CH_3COOH}$

(b) 2,2-dimethylcyclohexanone $+ Br_2 \xrightarrow[\text{solvent}]{CH_3COOH}$

PROBLEM 11.5 Show how you might prepare 1-penten-3-one from 3-pentanone:

$$CH_3CH_2\overset{O}{\overset{\|}{C}}CH_2CH_3 \longrightarrow CH_3CH_2\overset{O}{\overset{\|}{C}}CH{=}CH_2$$

PROBLEM 11.6 When optically active (*R*)-3-phenyl-2-butanone is exposed to aqueous acid, a loss of optical activity occurs, and racemic 3-phenyl-2-butanone is produced. Explain how this loss of optical activity takes place. (Hint: Review Section 6.13.)

11.4 ACIDITY OF α-HYDROGEN ATOMS: ENOLATE ION FORMATION

During the discussion of base-catalyzed enol formation in Section 11.1, we said that carbonyl compounds act as weak acids. Strong bases can abstract acidic α protons from carbonyl compounds to form resonance-stabilized enolate ions.

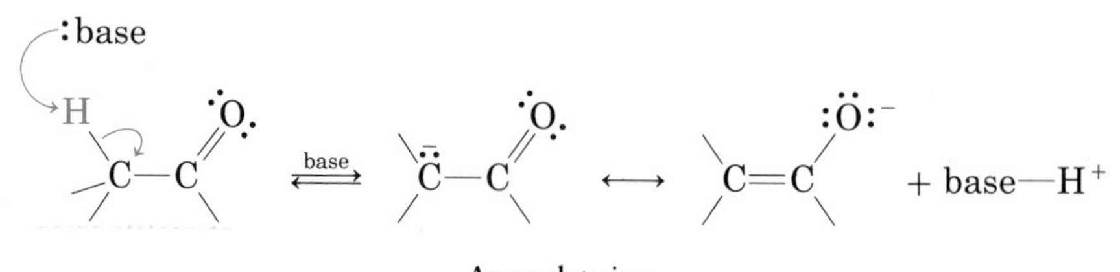

An enolate ion

Why are carbonyl compounds acidic? If we compare acetone, pK_a ~ 19, with ethane, pK_a ~ 50, we find that the presence of the neighboring carbonyl group increases the acidity of the neighboring C–H by a factor of 10^{30}.

$$\underset{\text{Ethane, p}K_a \approx 50}{CH_3CH_2\text{–}H} \quad \text{versus} \quad \underset{\text{Acetone, p}K_a \approx 19}{CH_3\text{—}\overset{O}{\overset{\|}{C}}\text{—}CH_2\text{–}H}$$

The best way to understand the acidity of carbonyl compounds is to look at an orbital picture of an enolate ion (Figure 11.4). Proton abstraction from a carbonyl compound occurs when the alpha C–H sigma bond is oriented parallel to

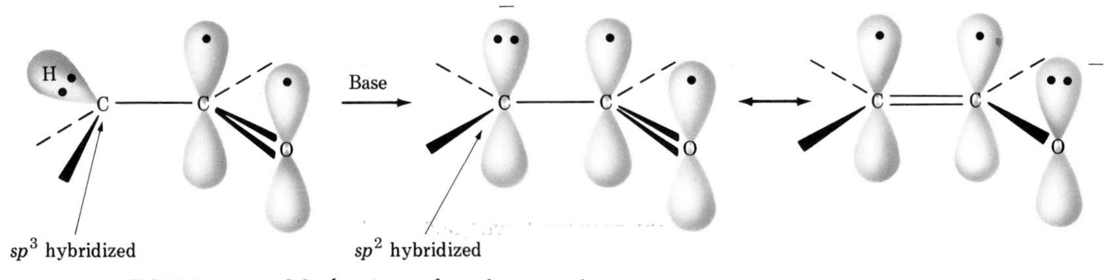

FIGURE 11.4 Mechanism of enolate-ion formation

the p orbitals of the carbonyl group. The α carbon of the product enolate ion is sp^2 hybridized and has a p orbital that overlaps the carbonyl p orbitals. Thus, the negative charge is shared by the electronegative oxygen atom, and the enolate ion is stabilized by resonance between two forms.

Carbonyl compounds are more acidic than alkanes for the same reason that carboxylic acids are more acidic than alcohols (Section 10.3). In both cases, the anions are stabilized by resonance. Enolate ions differ from carboxylate ions, though, because their two resonance forms aren't equivalent: The resonance form with the negative charge on the enolate oxygen atom is lower in energy than the form with the charge on carbon. The principle behind resonance stabilization is the same in both cases, however.

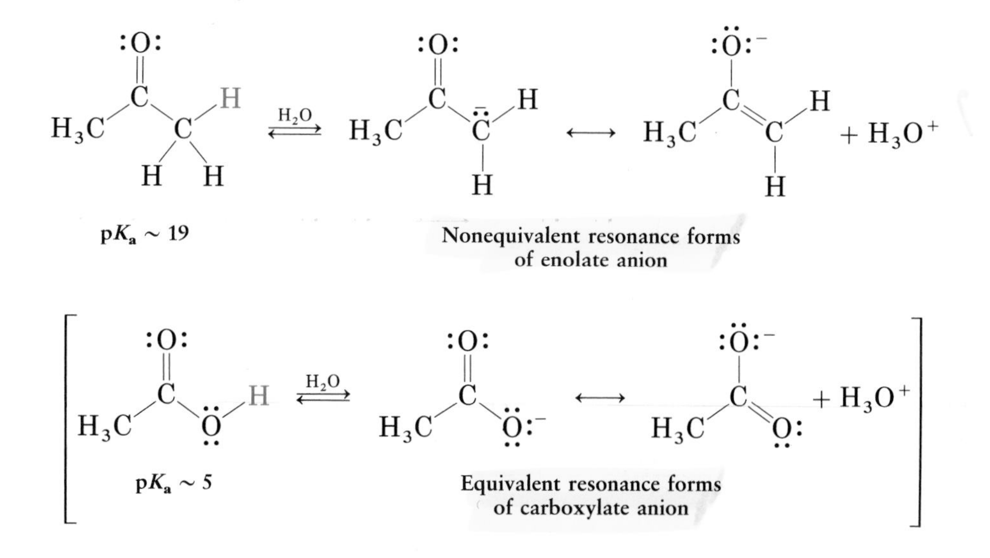

Since α-hydrogen atoms of carbonyl compounds are only weakly acidic compared with mineral acids or carboxylic acids, strong bases must be used to effect enolate-ion formation. If sodium ethoxide is used, ionization of acetone takes place only to the extent of about 0.1% since ethanol (pK_a = 16) is a stronger acid than acetone (pK_a = 19). If, however, a very powerful base such as sodium amide ($NaNH_2$, the sodium salt of ammonia) or sodium hydride (NaH, the sodium salt of H_2) is used, then a carbonyl compound is completely converted into its enolate ion.

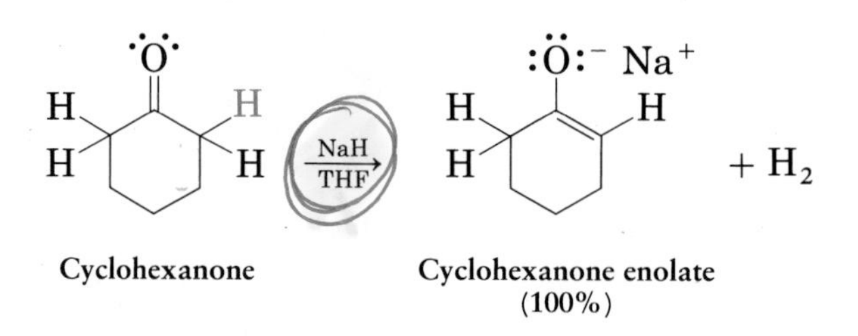

Cyclohexanone

Cyclohexanone enolate (100%)

All types of carbonyl compounds, including aldehydes, ketones, esters, acid chlorides, and amides, are much more acidic than alkanes. Table 11.1 lists the approximate pK_a values of various kinds of carbonyl compounds and shows how these values compare with other common acids.

TABLE 11.1 Acidity constants for some organic compounds

Compound type	*Compound*	pK_a	
Carboxylic acid	CH_3COOH	5	Stronger acid
1,3-Diketone	$CH_2(COCH_3)_2$	9	
1,3-Keto ester	$CH_3COCH_2CO_2C_2H_5$	11	
1,3-Diester	$CH_2(CO_2C_2H_5)_2$	13	
Water	HOH	16	
Primary alcohol	CH_3CH_2OH	16	
Acid chloride	CH_3COCl	16	
Aldehyde	CH_3CHO	17	
Ketone	CH_3COCH_3	19	
Ester	$CH_3CO_2C_2H_5$	25	
Nitrile	CH_3CN	25	
Dialkylamide	$CH_3CON(CH_3)_2$	30	
Ammonia	NH_3	35	Weaker acid

When a C–H bond is flanked by two carbonyl groups, acidity is enhanced even more. Thus, Table 11.1 shows that 1,3-diketones (called β-diketones), 1,3-keto esters (β-keto esters), and 1,3-diesters are much more acidic than water. The enolate ions derived from these β-dicarbonyl compounds are highly stabilized by delocalization of the negative charge onto both of the neighboring carbonyl oxygens. For example, there are three resonance forms for the enolate ion from 2,4-pentanedione:

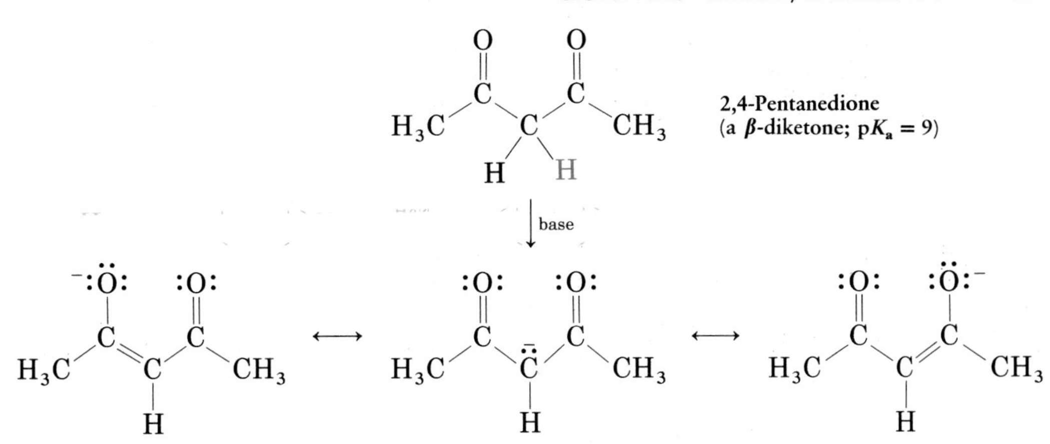

PRACTICE PROBLEM 11.3 Draw structures of the two enolate ions you could obtain by deprotonation of 3-methylcyclohexanone.

SOLUTION Locate the acidic hydrogens and then remove them one at a time to generate the possible enolate ions. In this case, 3-methylcyclohexanone can be deprotonated either at C2 or at C6.

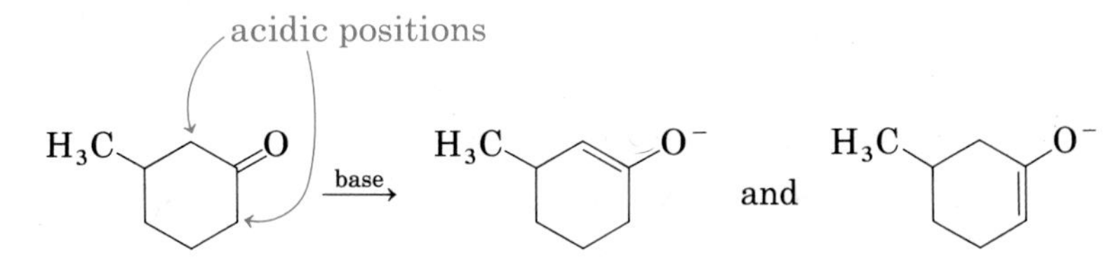

PROBLEM 11.7 Identify the acidic hydrogens in these molecules.

(a) CH_3CH_2CHO (b) $(CH_3)_3CCOCH_3$ (c) CH_3COOH
(d) $CH_3CH_2CH_2C{\equiv}N$ (e) 1,3-Cyclohexanedione

PROBLEM 11.8 Show the enolate ions you would obtain by deprotonation of these carbonyl compounds.

(a) Butanal (b) 2-Butanone (c) 2-Methylcyclohexanone

PROBLEM 11.9 Draw three resonance forms for the enolate ion you would obtain by deprotonation of methyl 3-oxopropanoate.

$$CH_3\overset{\overset{\displaystyle O}{\|}}{C}CH_2\overset{\overset{\displaystyle O}{\|}}{C}OCH_3$$ **Methyl 3-oxopropanoate**

11.5 REACTIVITY OF ENOLATE IONS

Enolate ions are more useful than enols for two reasons. First, pure enols can't normally be isolated: They're usually generated only as transient intermediates in low concentration. By contrast, stable solutions of pure enolate ions are easily prepared from most carbonyl compounds by treatment with a strong base. Second, enolate ions are much more reactive than enols. Whereas enols are neutral, enolate ions have a negative charge that makes them much better nucleophiles. Thus, the alpha position of enolate ions is highly reactive toward electrophiles.

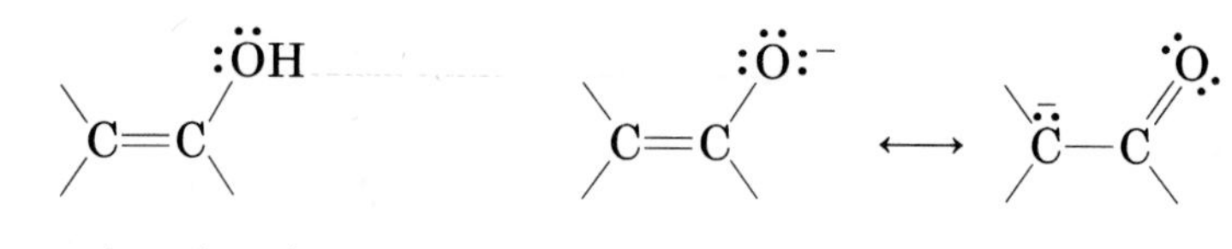

Enol: neutral, moderately reactive, very difficult to isolate

Enolate: negatively charged, very reactive, easily prepared

Since enolate ions are resonance hybrids of two nonequivalent forms, they can be thought of either as α-keto carbanions ($^{-}$C–C=O) or as vinylic alkoxides (C=C–O^{-}). Thus, enolate ions can react with electrophiles either on oxygen or on carbon. Reaction on oxygen yields an enol derivative whereas reaction on carbon yields an α-substituted carbonyl compound (Figure 11.5). Both kinds of reactivity are known, but reaction on carbon is the more commonly observed pathway.

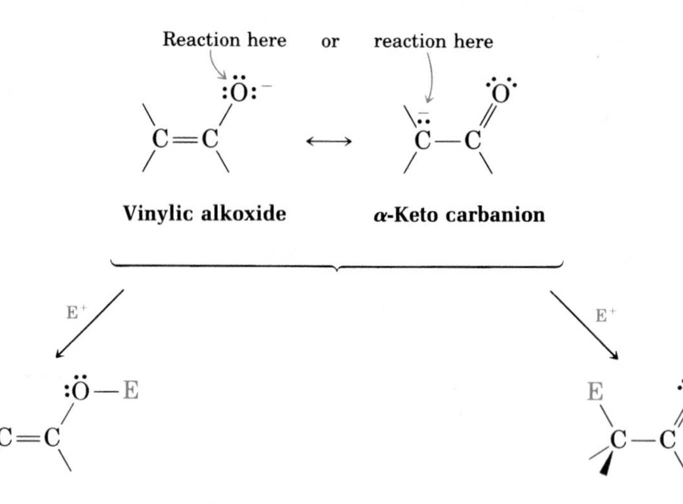

FIGURE 11.5 Two modes of enolate-ion reactivity. Reaction on carbon to yield an α-substituted carbonyl product is the more commonly followed path.

11.6 ALKYLATION OF ENOLATE IONS

alkylation the alpha substitution of a carbonyl compound by reaction of an enolate ion with an alkyl halide

One of the most important reactions of enolate ions is their **alkylation** by treatment with an alkyl halide. The alkylation reaction is extremely useful because it forms a new carbon–carbon bond, thereby joining two smaller pieces into one larger molecule. Alkylation occurs when the nucleophilic enolate ion reacts with the electrophilic alkyl halide in an S_N2 reaction, displacing the halide ion by back-side attack.

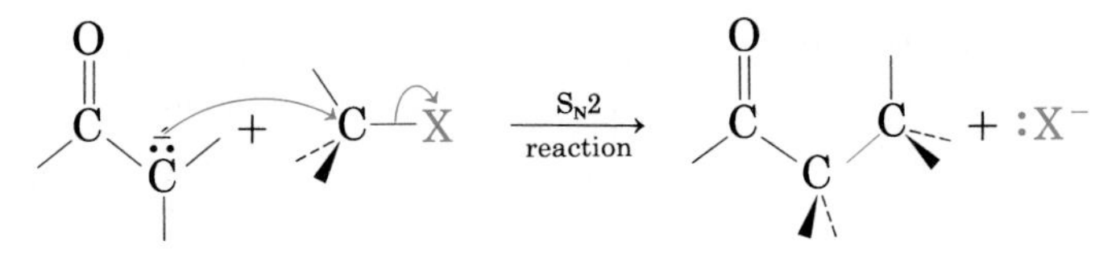

Like all S_N2 reactions (Section 7.7), alkylations with R–X are successful only when the alkyl group R is primary or methyl. Otherwise, a competing E2 elimination occurs if a secondary or tertiary halide is used. The leaving group X can be chloride, bromide, or iodide. Let's see some examples.

The Malonic Ester Synthesis

malonic ester synthesis a method for forming alpha substituted acetic acids by reaction of diethyl malonate with an alkyl halide, followed by decarboxylation

The **malonic ester synthesis,** one of the oldest and best known carbonyl alkylation reactions, is an excellent method for preparing substituted acetic acids from alkyl halides:

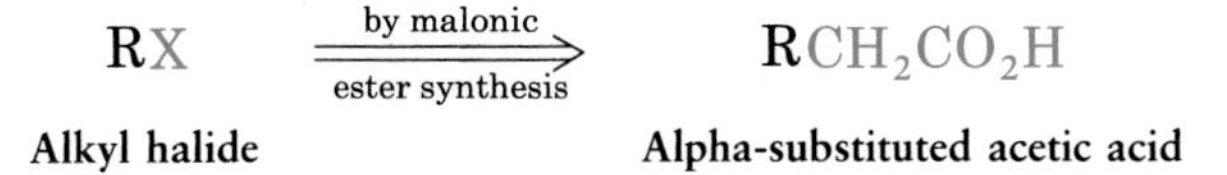

Diethyl propanedioate, commonly called diethyl malonate or malonic ester, is relatively acidic ($pK_a = 13$) because its α-hydrogen atoms are flanked by two carbonyl groups. Thus, malonic ester is easily converted into its enolate ion by reaction with sodium ethoxide in ethanol. The enolate ion, in turn, is readily alkylated by treatment with an alkyl halide, yielding an α-substituted malonic ester:

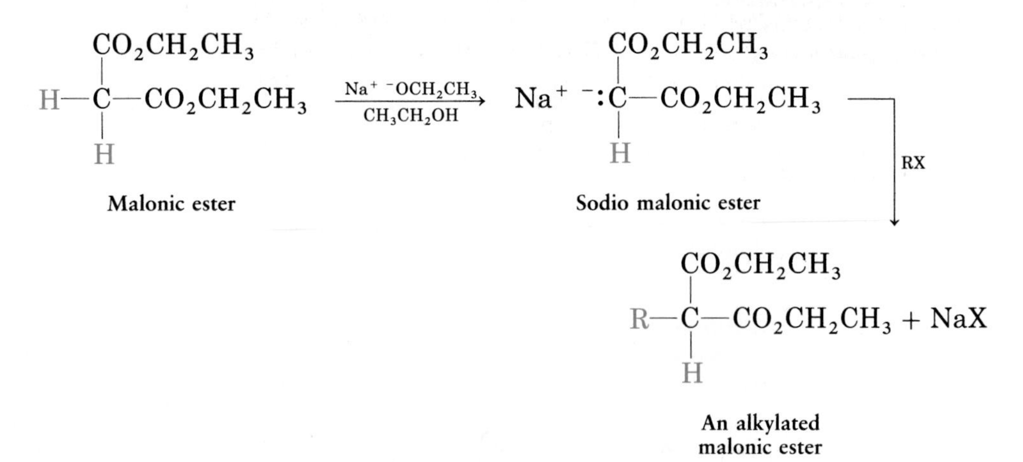

The product of a malonic ester alkylation has one acidic α-hydrogen left, and the alkylation process can therefore be repeated a second time to yield a dialkylated malonic ester:

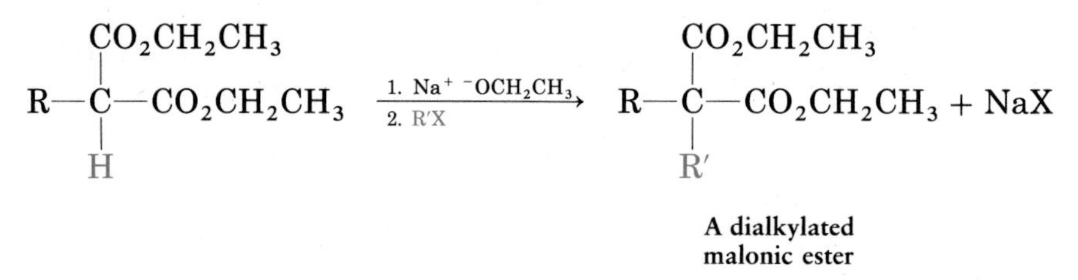

decarboxylation loss of carbon dioxide from a molecule

Once formed, alkylated malonic esters can be hydrolyzed and decarboxylated (**decarboxylation** means loss of carbon dioxide, CO_2) when heated with aqueous HCl. The product is a substituted carboxylic acid. Note that decarboxylation is not a general reaction of carboxylic acids but is a unique feature of compounds like malonic acids and related 3-oxo acids that have a second carbonyl group two atoms away:

$$\text{R—C(R')(CO}_2\text{CH}_2\text{CH}_3\text{)—CO}_2\text{CH}_2\text{CH}_3 \xrightarrow[\text{heat}]{\text{H}_3\text{O}^+} \text{R—CH(R')—COOH} + \text{CO}_2 + 2\ \text{CH}_3\text{CH}_2\text{OH}$$

The overall effect of the malonic ester synthesis is to convert an alkyl halide into a carboxylic acid and to lengthen the carbon chain by two atoms ($RX \rightarrow RCH_2COOH$). For example:

$$\underset{\text{1-Bromobutane}}{\text{CH}_3(\text{CH}_2)_2\text{CH}_2\text{Br}} + \text{Na}^+\ {}^-\text{:CH(COOCH}_3)_2 \longrightarrow \text{CH}_3(\text{CH}_2)_2\text{CH}_2\text{CH(COOCH}_3)_2$$

$$\downarrow \text{H}_3\text{O}^+\text{, heat}$$

$$\underset{\text{Hexanoic acid (75\%)}}{\text{CH}_3(\text{CH}_2)_2\text{CH}_2\text{CH}_2\text{COOH}} + \text{CO}_2 + 2\ \text{CH}_3\text{OH}$$

PRACTICE PROBLEM 11.4

How would you prepare heptanoic acid by a malonic ester synthesis?

SOLUTION The malonic ester synthesis converts an alkyl halide into a carboxylic acid with two more carbons in its chain. Thus, a seven-carbon acid chain must be derived from a five-carbon alkyl halide such as 1-bromopentane.

$$\text{CH}_3\text{CH}_2\text{CH}_2\text{CH}_2\text{CH}_2\text{Br} + \text{CH}_2(\text{COOCH}_2\text{CH}_3)_2 \xrightarrow[\text{2. H}_3\text{O}^+\text{, heat}]{\text{1. Na}^+\ {}^-\text{OCH}_2\text{CH}_3} \text{CH}_3\text{CH}_2\text{CH}_2\text{CH}_2\text{CH}_2\text{CH}_2\text{COOH}$$

PROBLEM 11.10 What alkyl halide would you use to prepare these compounds by a malonic ester synthesis?

(a) Butanoic acid (b) 3-Phenylpropanoic acid (c) 5-Methylhexanoic acid

PROBLEM 11.11 Show how you could use a malonic ester synthesis to prepare these compounds.

(a) 4-Methylpentanoic acid (b) 2-Methylpentanoic acid

The Acetoacetic Ester Synthesis

acetoacetic ester synthesis a method for forming alpha substituted acetones by reaction of ethyl acetoacetate with alkyl halides, followed by decarboxylation

The **acetoacetic ester synthesis** is a method for preparing α-substituted acetone derivatives from alkyl halides in the same way that the malonic ester synthesis prepares α-substituted acetic acids.

$$RX \xrightarrow[\text{ester synthesis}]{\text{Via acetoacetic}} RCH_2\overset{O}{\overset{\|}{C}}CH_3$$

α-substituted acetone

Ethyl 3-oxobutanoate, commonly called ethyl acetoacetate or acetoacetic ester, is similar to malonic ester in that its alpha hydrogens are flanked by two carbonyl groups. It is therefore easily converted into an enolate ion, which can be alkylated by reaction with an alkyl halide. A second alkylation can also be carried out, if desired, since acetoacetic ester has two acidic alpha protons that can be replaced.

$$H{-}\underset{CO_2C_2H_5}{\overset{H}{C}}{-}\overset{O}{\overset{\|}{C}}{-}CH_3 \xrightarrow[CH_3CH_2OH]{NaOCH_2CH_3} \left[H{-}\underset{CO_2C_2H_5}{\overset{..}{C}^{-}}{-}\overset{O}{\overset{\|}{C}}{-}CH_3\right] \xrightarrow{RX} R{-}\underset{CO_2C_2H_5}{\overset{H}{C}}{-}\overset{O}{\overset{\|}{C}}{-}CH_3$$

Acetoacetic ester — **A monoalkylated acetoacetic ester**

$$R{-}\underset{CO_2C_2H_5}{\overset{H}{C}}{-}\overset{O}{\overset{\|}{C}}{-}CH_3 \xrightarrow[CH_3CH_2OH]{NaOCH_2CH_3} \left[R{-}\underset{CO_2C_2H_5}{\overset{..}{C}^{-}}{-}\overset{O}{\overset{\|}{C}}{-}CH_3\right] \xrightarrow{R'X} R'{-}\underset{CO_2C_2H_5}{\overset{R'}{C}}{-}\overset{O}{\overset{\|}{C}}{-}CH_3$$

A monoalkylated acetoacetic ester — **A dialkylated acetoacetic ester**

Alkylated acetoacetic esters are hydrolyzed and decarboxylated by heating with aqueous HCl to yield α-substituted-acetone products. If a monoalkylated acetoacetic ester is heated with HCl, an α-monosubstituted acetone is formed; if a dialkylated acetoacetic ester is heated with HCl, an α,α-dialkylated acetone is formed.

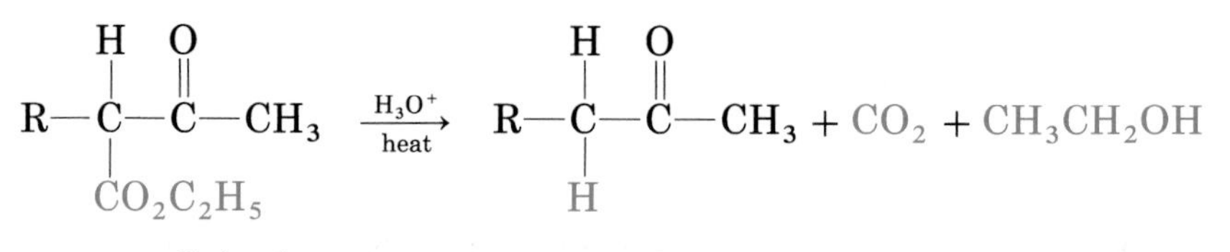

A monoalkylated acetoacetic ester **A monosubstituted acetone**

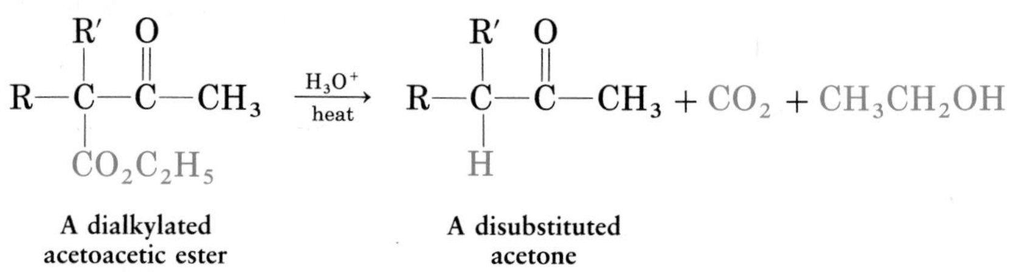

A dialkylated acetoacetic ester **A disubstituted acetone**

For example:

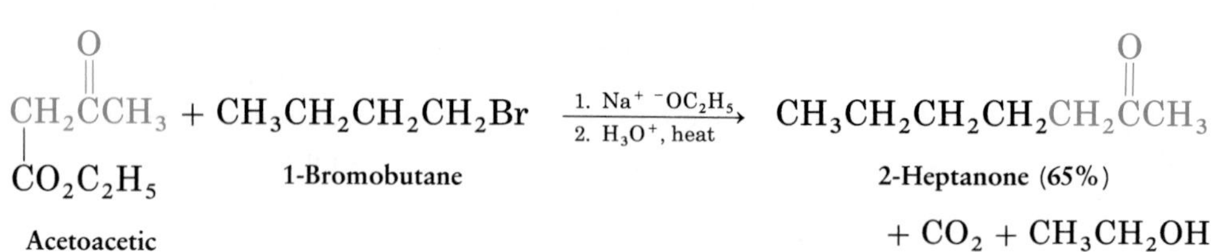

PRACTICE PROBLEM 11.5 How would you prepare 2-pentanone by an acetoacetic ester synthesis?

SOLUTION The acetoacetic ester synthesis yields a ketone product by adding three carbons to an alkyl halide:

$$\underbrace{CH_3\overset{O}{\overset{\|}{C}}CH_2}_{\text{These three carbons from acetoacetic ester}}\!-\!\underbrace{CH_2CH_3}_{\text{These carbons from alkyl halide}}$$

This bond formed

These three carbons from acetoacetic ester These carbons from alkyl halide

Thus, the acetoacetic ester synthesis of 2-pentanone would involve reaction of bromoethane:

$$CH_3\overset{\overset{O}{\|}}{C}CH_2COOCH_2CH_3 + CH_3CH_2Br \xrightarrow[2.\ H_3O^+]{1.\ Na^+\ ^-OCH_2CH_3} CH_3\overset{\overset{O}{\|}}{C}CH_2CH_2CH_3$$

PROBLEM 11.12 How would you prepare these compounds using an acetoacetic ester synthesis?

(a) 4-Phenyl-2-butanone (b) 5-Methyl-2-hexanone (c) 3-Methyl-2-hexanone

PROBLEM 11.13 Which of the following compounds can't be prepared by an acetoacetic ester synthesis? Explain.

(a) 2-Butanone (b) Phenylacetone
(c) Acetophenone (d) 3,3-Dimethyl-2-butanone

11.7 CARBONYL CONDENSATION REACTIONS

We've seen now that carbonyl compounds can behave either as electrophiles or as nucleophiles. In nucleophilic addition reactions and nucleophilic acyl substitution reactions, the carbonyl group behaves as an electrophile by accepting electrons from an attacking nucleophile. In α-substitution reactions, however, the carbonyl compound behaves as a nucleophile when it's converted into its enolate ion or enol tautomer.

Electrophilic carbonyl group is attacked by nucleophiles

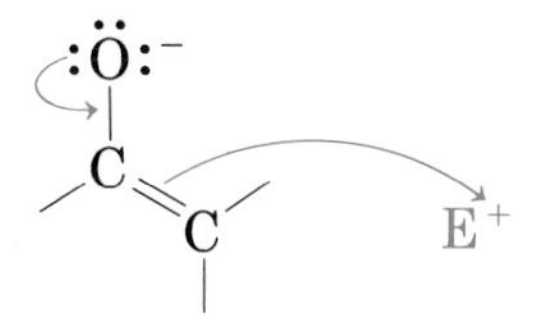

Nucleophilic enolate ion attacks electrophiles

Carbonyl condensation reactions, the fourth and last general category of carbonyl-group reactions involve both kinds of reactivity. These reactions take place between two carbonyl components and involve a combination of nucleophilic-addition and α-substitution steps. One component acts as an electron donor and undergoes an α-substitution process while the other component acts as an electron acceptor and undergoes a nucleophilic addition process. There are numerous variations of carbonyl condensation reactions depending on the exact structure of the two carbonyl components, but the general mechanism shown in Figure 11.6 remains the same.

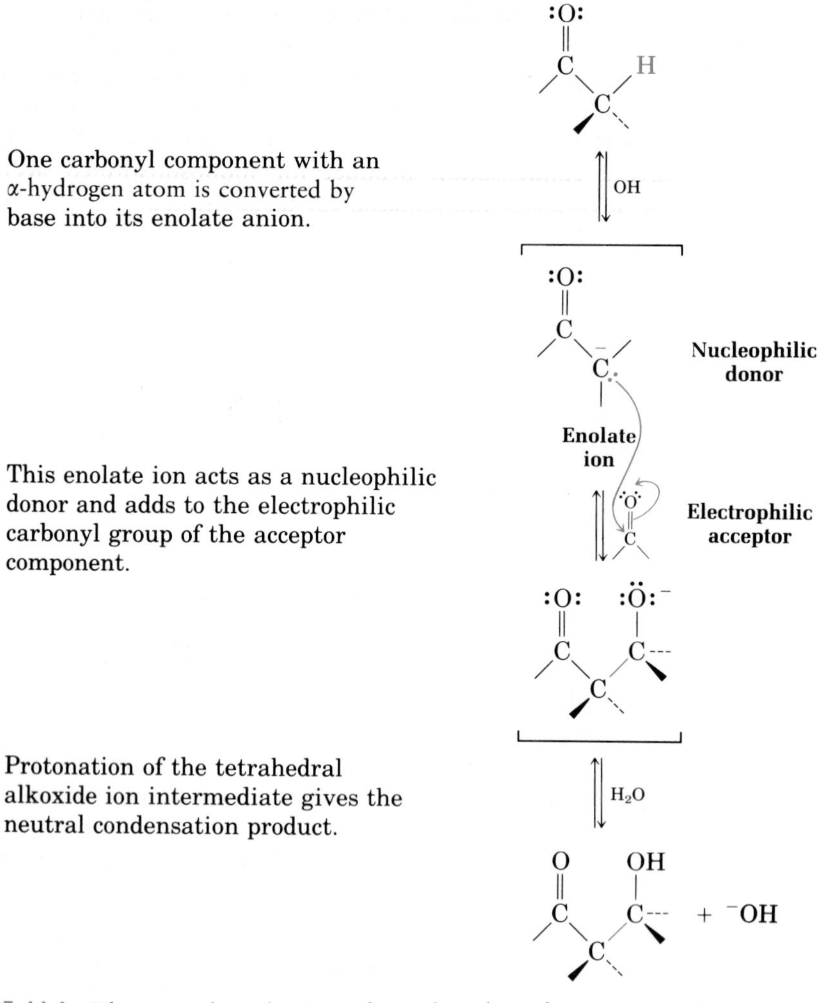

One carbonyl component with an α-hydrogen atom is converted by base into its enolate anion.
OH
Nucleophilic donor
Enolate ion
This enolate ion acts as a nucleophilic donor and adds to the electrophilic carbonyl group of the acceptor component.
Electrophilic acceptor
Protonation of the tetrahedral alkoxide ion intermediate gives the neutral condensation product.
H2O
+ ⁻OH

FIGURE 11.6 The general mechanism of a carbonyl condensation reaction

11.8 CONDENSATIONS OF ALDEHYDES AND KETONES: THE ALDOL REACTION

When acetaldehyde is treated in an alcoholic solvent with a basic catalyst such as sodium hydroxide or sodium ethoxide, a rapid and reversible condensation reaction occurs. The product is a β-hydroxy aldehyde product known commonly as *aldol* (*ald*ehyde + alcoh*ol*).

$$2\,CH_3\overset{O}{\overset{\|}{C}}H \underset{CH_3CH_2OH}{\overset{NaOCH_2CH_3}{\rightleftharpoons}} CH_3\underset{\beta}{\overset{OH}{\overset{|}{C}H}}-\underset{\alpha}{CH_2}\overset{O}{\overset{\|}{C}}H$$

Acetaldehyde — Aldol (a β-hydroxy aldehyde)

aldol reaction a carbonyl condensation between two ketones or aldehydes, leading to a beta-hydroxy ketone or aldehyde product

Called the **aldol reaction,** base-catalyzed dimerization is a general reaction of all ketones and aldehydes with α-hydrogen atoms. If the ketone or aldehyde doesn't have an α-hydrogen atom, though, aldol condensation can't occur. The exact position of the aldol equilibrium depends both on reaction conditions and on substrate structure. As the following examples indicate, the aldol equilibrium generally favors condensation product for monosubstituted acetaldehydes (RCH_2CHO) but favors starting material for disubstituted acetaldehydes (R_2CHCHO) and for ketones.

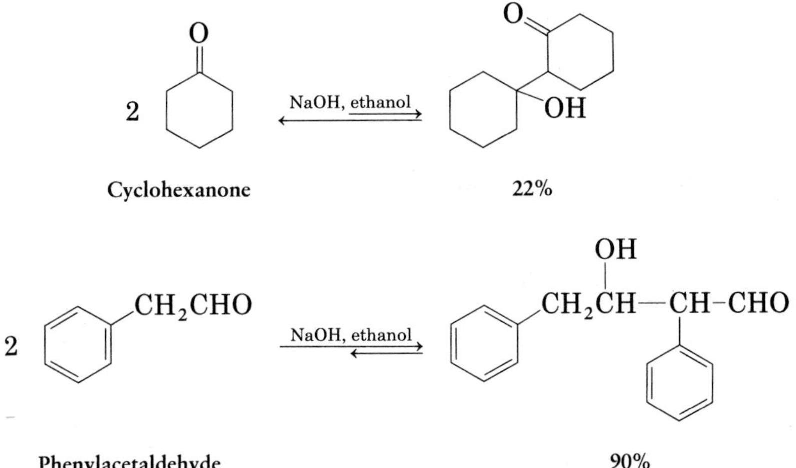

PRACTICE PROBLEM 11.6

What is the structure of the aldol product derived from propanal?

SOLUTION An aldol reaction combines two molecules of starting material, forming a bond between the α carbon of one partner and the carbonyl carbon of the second partner:

$$CH_3CH_2-\overset{\overset{\displaystyle O}{\|}}{\underset{\underset{\displaystyle H}{|}}{C}} + \underset{\underset{\displaystyle CH_3}{|}}{CH_2}-\overset{\overset{\displaystyle O}{\|}}{C}-H \xrightarrow{NaOH} CH_3CH_2-\overset{\overset{\displaystyle OH}{|}}{\underset{\underset{\displaystyle H}{|}}{C}}-\underset{\underset{\displaystyle CH_3}{|}}{CH}-\overset{\overset{\displaystyle O}{\|}}{C}-H$$

Bond formed here

PROBLEM 11.14 Which of these compounds can undergo the aldol reaction and which cannot? Explain.

(a) Cyclohexanone (b) Benzaldehyde
(c) 2,2,6,6-Tetramethylcyclohexanone (d) Formaldehyde

PROBLEM 11.15 Show the product of the aldol reaction of these compounds.

(a) Butanal (b) Cyclopentanone (c) Acetophenone

11.9 DEHYDRATION OF ALDOL PRODUCTS: SYNTHESIS OF ENONES

enone
a ketone that also contains a carbon–carbon double bond

The β-hydroxy ketones and β-hydroxy aldehydes formed in aldol reactions are easily dehydrated to yield conjugated **enones** (*ene* + *one*). In fact, it's this loss of water that gives the aldol condensation its name, since water condenses out of the reaction.

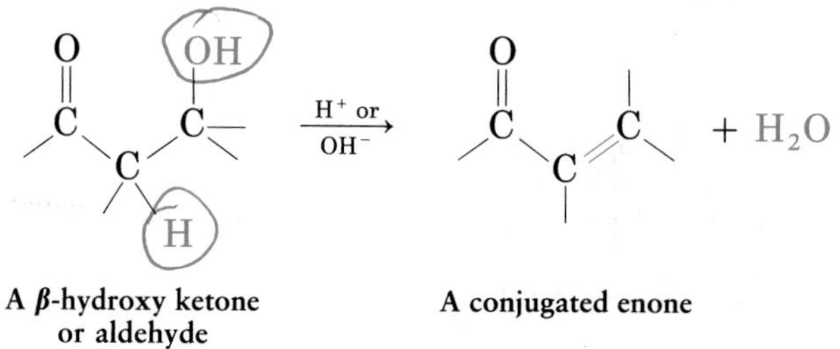

Although most alcohols are resistant to dehydration by dilute acid or base (Section 8.8), hydroxyl groups that are beta to a carbonyl group are special. Under basic conditions, an acidic α hydrogen is abstracted, and the resultant enolate ion expels hydroxide ion. Under acidic conditions, the hydroxyl group is protonated and then expelled by the neighboring enol.

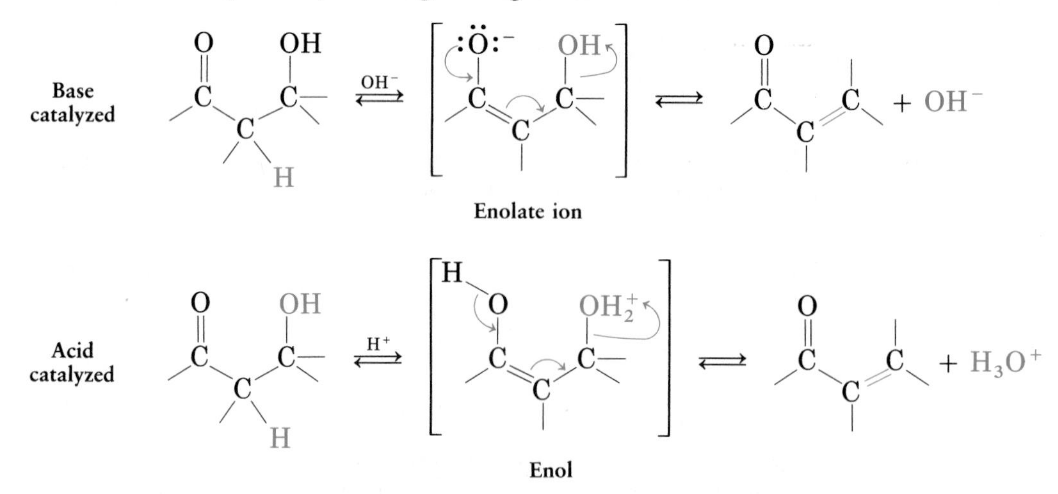

The conditions needed to cause aldol dehydration are often only a bit more vigorous (slightly higher temperature, for example) than the conditions needed for the aldol dimerization itself. As a result, conjugated enones are often obtained directly from aldol reactions; the intermediate β-hydroxy carbonyl compounds are usually not even isolated.

Conjugated enones are more stable than nonconjugated enones for the same reasons that conjugated dienes are more stable than nonconjugated dienes (Section 4.10). Interaction between the π electrons of the carbon–carbon double bond and the π electrons of the carbonyl group allows delocalization of the π electrons over all four atomic centers.

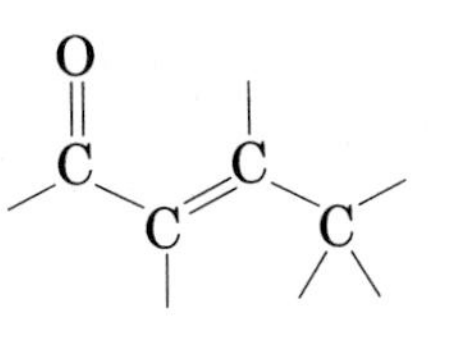

Conjugated enone
(more stable)

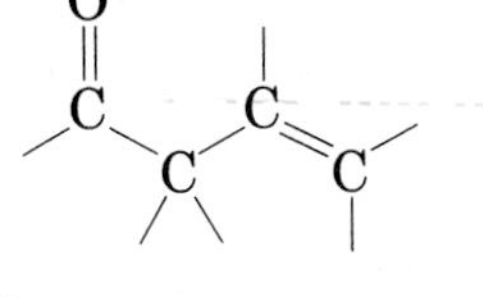

Nonconjugated enone
(less stable)

PRACTICE PROBLEM 11.7 What is the structure of the enone obtained from aldol condensation of acetaldehyde?

SOLUTION In the aldol reaction, H_2O is eliminated by removing two hydrogens from the acidic α position of one partner and the oxygen from the second partner:

$$H_3C{-}\overset{\displaystyle H}{\overset{|}{C}}{=}O + H_2\overset{\displaystyle H}{\overset{|}{C}}{-}CHO \xrightarrow{NaOH} H_3C{-}\overset{\displaystyle H}{\overset{|}{C}}{=}\overset{\displaystyle H}{\overset{|}{C}}{-}CHO + H_2O$$

2-Butenal

PROBLEM 11.16 Write the structures of the enone products you would obtain from aldol condensation of these compounds.

(a) Acetone (b) Cyclopentanone (c) Acetophenone (d) Propanal

PROBLEM 11.17 Aldol condensation of 2-butanone leads to a mixture of two enones (ignoring double-bond stereochemistry). Draw the two enones.

11.10 CONDENSATION OF ESTERS: THE CLAISEN CONDENSATION REACTION

Esters, like aldehydes and ketones, are weakly acidic. When an ester with an α hydrogen is treated with one equivalent of a base such as sodium methoxide, a reversible condensation reaction yields a β-keto ester product. For example, ethyl acetate yields ethyl acetoacetate on base treatment. This reaction between two ester components is known as the **Claisen condensation reaction.**

$$2\ CH_3\overset{\displaystyle O}{\overset{\|}{C}}OCH_2CH_3 \xrightarrow[2.\ H_3O^+]{1.\ Na^+\ {}^-OCH_2CH_3,\ \text{ethanol}} CH_3\underset{\beta}{\overset{\displaystyle O}{\overset{\|}{C}}}{-}\underset{\alpha}{CH_2}\overset{\displaystyle O}{\overset{\|}{C}}OCH_2CH_3 + CH_3CH_2OH$$

Ethyl acetate

Ethyl acetoacetate, a β-keto ester (75%)

Claisen condensation reaction a carbonyl condensation reaction between two esters, leading to formation of a beta-keto ester product

The mechanism of the Claisen reaction is similar to that of the aldol reaction, involving the nucleophilic addition of an ester enolate-ion donor to the carbonyl group of a second ester molecule (Figure 11.7). From the point of view of the donor component, the Claisen condensation is simply an α-substitution reaction. From the point of view of the acceptor component, the Claisen condensation is a nucleophilic acyl substitution reaction.

Methoxide ion base abstracts an acidic α-hydrogen atom from an ester molecule, yielding an ester enolate ion.

$$CH_3\overset{\overset{:O:}{\|}}{C}OCH_3 \quad \underset{}{\overset{^-OCH_3}{\rightleftharpoons}} \quad \left[^-:CH_2\overset{\overset{:O:}{\|}}{C}OCH_3 \right] + HOCH_3$$

The enolate ion adds to a second ester molecule by nucleophilic addition, yielding a tetrahedral intermediate.

$$CH_3\overset{\overset{:O:}{\|}}{C}OCH_3 \quad \rightleftharpoons \quad \left[CH_3\overset{\overset{:\ddot{O}:^-}{|}}{\underset{\underset{OCH_3}{|}}{C}}-CH_2\overset{\overset{:O:}{\|}}{C}OCH_3 \right]$$

Loss of methoxide ion from the tetrahedral intermediate yields methyl acetoacetate and regenerates the basic catalyst.

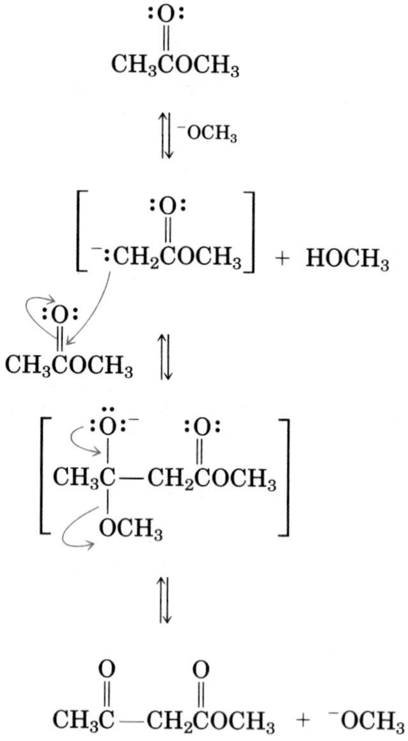

$$\rightleftharpoons \quad CH_3\overset{\overset{O}{\|}}{C}-CH_2\overset{\overset{O}{\|}}{C}OCH_3 + {}^-OCH_3$$

FIGURE 11.7 The mechanism of the Claisen condensation reaction

The only difference between an aldol condensation and a Claisen condensation involves the fate of the initially formed tetrahedral intermediate. The tetrahedral intermediate in the aldol reaction is protonated to give a stable alcohol product, exactly the behavior previously seen for ketones (Section 9.7). The tetrahedral intermediate in the Claisen reaction, however, expels an alkoxide leaving group to yield an acyl substitution product, exactly the behavior previously seen for esters (Section 10.5).

PRACTICE PROBLEM 11.8 What product would you obtain from Claisen condensation of methyl propanoate?

SOLUTION The Claisen condensation of an ester results in loss of one molecule of alcohol and formation of a product in which an acyl group of one reactant bonds to the α carbon of the second reactant:

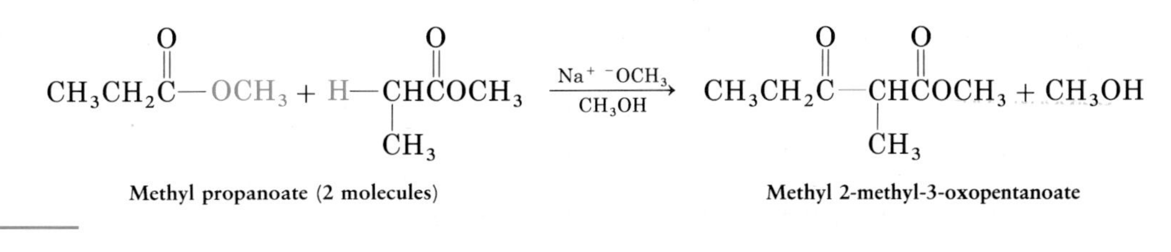

Methyl propanoate (2 molecules)

Methyl 2-methyl-3-oxopentanoate

PROBLEM 11.18 Which of these esters can't undergo a Claisen condensation? Explain.

(a) Methyl formate (b) Methyl propenoate (c) Methyl propanoate

PROBLEM 11.19 Show the products you would obtain by Claisen condensation of these esters.

(a) $(CH_3)_2CHCH_2COOCH_3$ (b) Methyl phenylacetate
(c) Methyl cyclohexylacetate

INTERLUDE

Biological Carbonyl Condensation Reactions

Carbonyl condensation reactions are used in nature for the biological synthesis of innumerable different molecules. Fats, amino acids, steroid hormones, and many other kinds of compounds are biosynthesized by plants and animals using carbonyl condensation reactions as the key step.

Nature uses the two-carbon acetate fragment of acetyl coenzyme A as the major building block for synthesis. We saw in the Chapter 10 Interlude that acetyl CoA can serve as an electrophilic acceptor for attack of nucleophiles at the acyl carbon. In addition, it can serve as a nucleophilic donor by loss of its acidic α proton to generate an enolate ion. The enolate ion of acetyl CoA can then add to another carbonyl group in a condensation reaction. For example, citric acid is biosynthesized by addition of acetyl CoA to the ketone carbonyl group of oxaloacetic acid (2-oxobutanedioic acid) in a type of aldol reaction.

$$CH_3—\overset{O}{\overset{\|}{C}}—S—CoA \longrightarrow :\overset{-}{C}H_2—\overset{O}{\overset{\|}{C}}—S—CoA$$

Acetyl CoA, a thiol ester

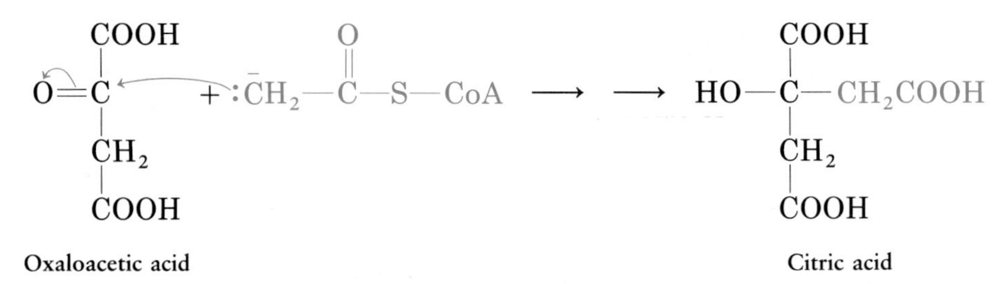

Acetyl CoA is also involved in the biosynthesis of steroids, fats, and other lipids. The key step in these biosyntheses is a Claisen-like condensation of acetyl CoA to yield acetoacetyl CoA.

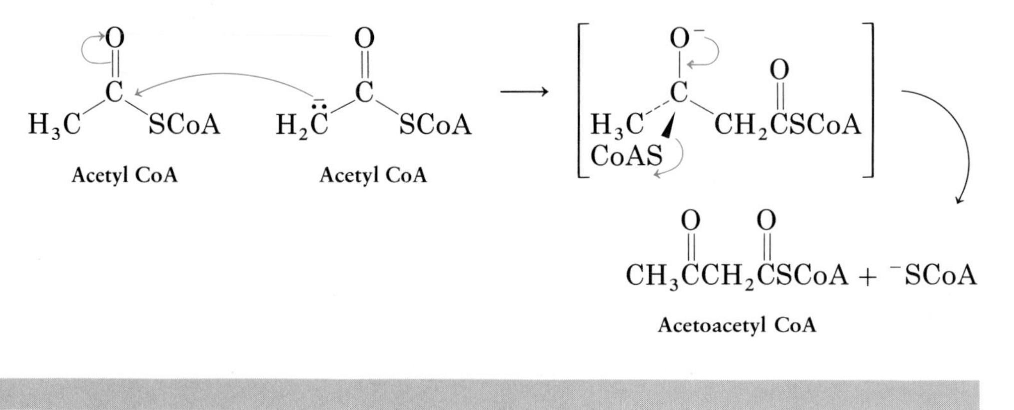

SUMMARY AND KEY WORDS

Alpha substitutions and **condensations** of carbonyl compounds represent two of the four fundamental reaction types in carbonyl-group chemistry. Alpha substitution reactions, which take place via **enol** or **enolate-ion** intermediates, result in the replacement of an α-hydrogen atom by some other substituent.

Carbonyl compounds are in equilibrium with their enols, a process known as **tautomerism.** Enol tautomers are normally present to only a small extent and can't usually be isolated in pure form. Nevertheless, enols react rapidly with a variety of electrophiles. For examples, ketones and aldehydes are rapidly halogenated at the α position by reaction with chlorine, bromine, or iodine in acetic acid solution.

Alpha-hydrogen atoms of carbonyl compounds are acidic and can be abstracted by strong bases to yield enolate ions. Ketones, aldehydes, esters, amides, and nitriles can all be deprotonated. The most important reaction of enolate ions is their **S_N2 alkylation** by alkyl halides. The nucleophilic enolate ion attacks an alkyl halide from the back side, displacing the leaving halide group and yielding an α-alkylated carbonyl product. The **malonic ester synthesis,** which involves alkylation of diethyl malonate with alkyl halides, provides a method for preparing monoalkylated or dialkylated acetic acids. The **acetoacetic ester synthesis** is a similar method for preparing monoalkylated or dialkylated acetone derivatives.

A **carbonyl condensation reaction** takes place between two carbonyl components and involves a combination of nucleophilic-addition and α-substitution steps. One carbonyl component (the donor) is converted into a nucleophilic enolate ion, which then adds to the electrophilic carbonyl group of the second component (the acceptor).

The **aldol reaction** is a carbonyl condensation that occurs between two ketone or aldehyde components. Aldol reactions are reversible; they lead first to β-hydroxy ketone products and then to α,β-unsaturated ketones. The **Claisen condensation reaction** is a carbonyl condensation that occurs between two ester components and that leads to a β-keto ester product.

SUMMARY OF REACTIONS

1. Halogenation of ketones and aldehydes (Section 11.3)

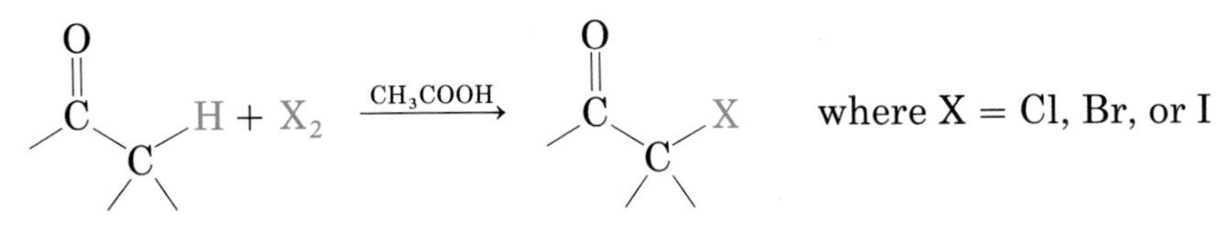

2. Malonic ester synthesis (Section 11.6)

$$H{-}\underset{CO_2R}{\overset{H}{C}}{-}CO_2R \xrightarrow[2.\ R'X]{1.\ base} R'{-}\underset{CO_2R}{\overset{H}{C}}{-}CO_2R \xrightarrow{H_3O^+} R'CH_2COOH$$

3. Acetoacetic ester synthesis (Section 11.6)

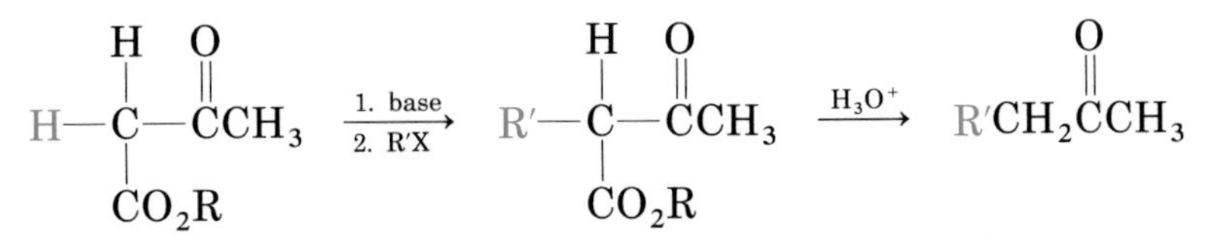

4. Aldol reaction of ketones and aldehydes (Section 11.8)

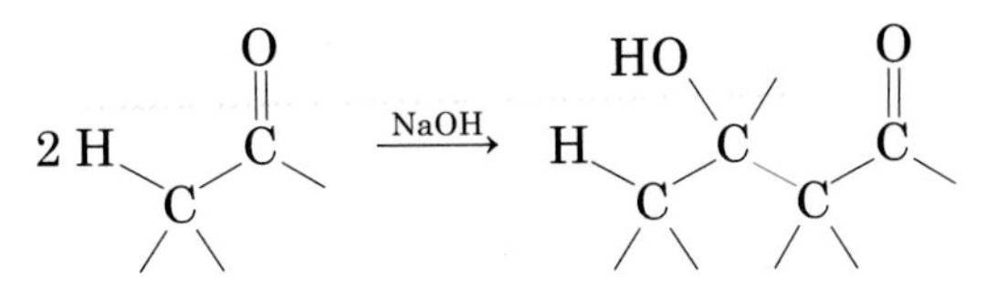

5. Claisen condensation reaction of esters (Section 11.10)

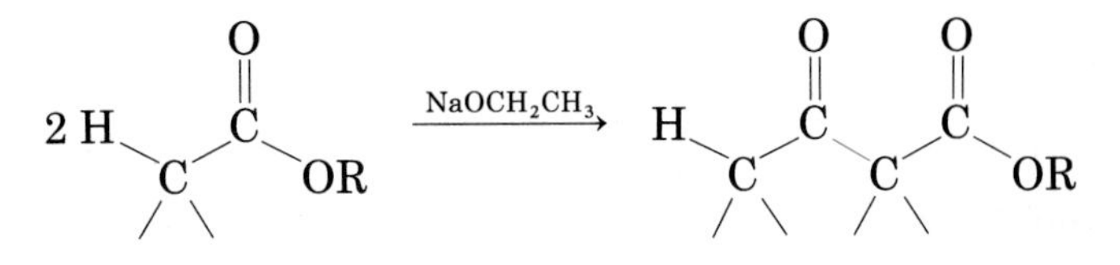

ADDITIONAL PROBLEMS

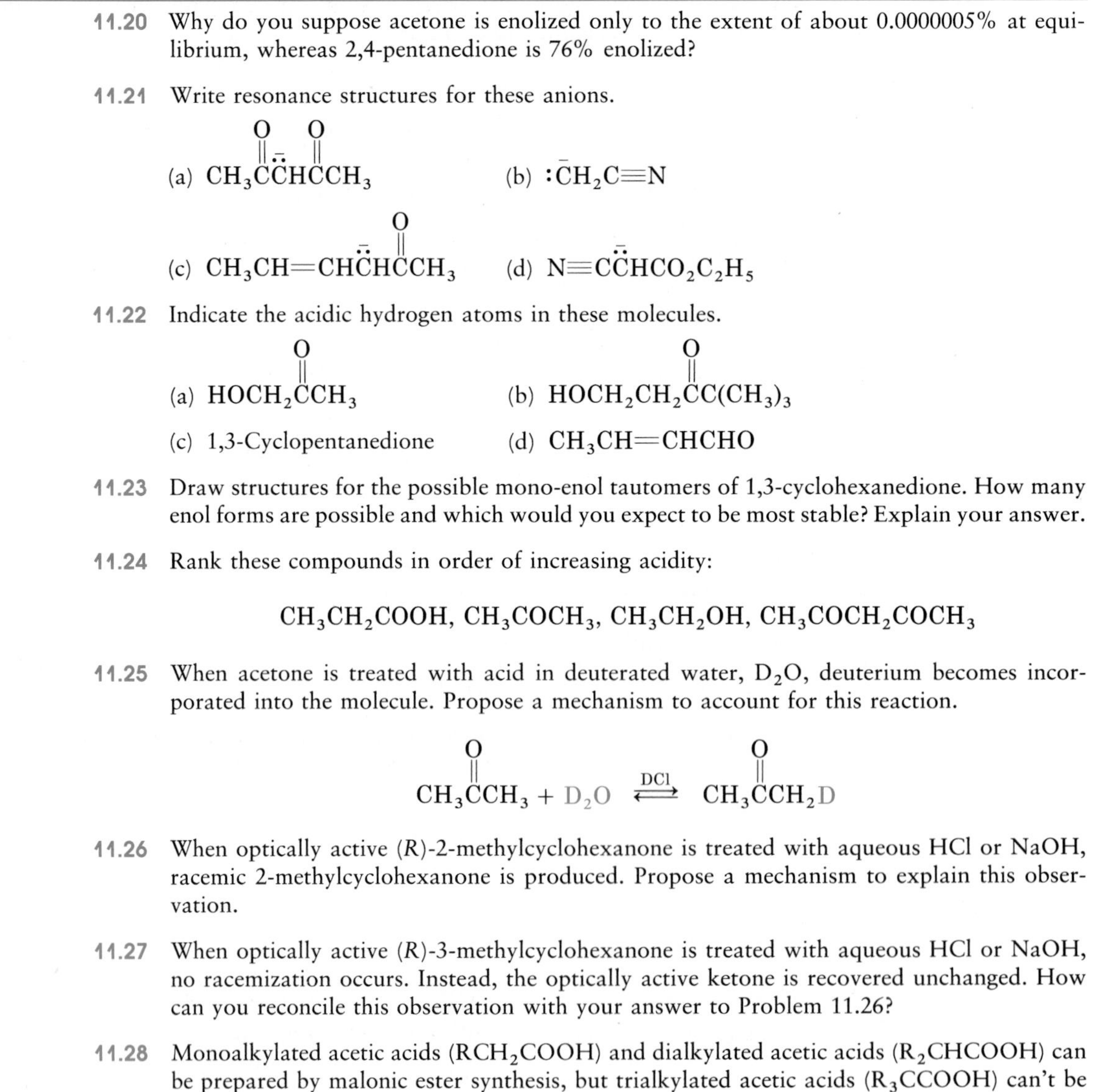

11.20 Why do you suppose acetone is enolized only to the extent of about 0.0000005% at equilibrium, whereas 2,4-pentanedione is 76% enolized?

11.21 Write resonance structures for these anions.

(a) $CH_3\overset{O}{\overset{\|}{C}}\overset{..}{\bar{C}}H\overset{O}{\overset{\|}{C}}CH_3$ (b) $:\bar{C}H_2C{\equiv}N$

(c) $CH_3CH{=}CH\overset{..}{\bar{C}}H\overset{O}{\overset{\|}{C}}CH_3$ (d) $N{\equiv}C\overset{..}{\bar{C}}HCO_2C_2H_5$

11.22 Indicate the acidic hydrogen atoms in these molecules.

(a) $HOCH_2\overset{O}{\overset{\|}{C}}CH_3$ (b) $HOCH_2CH_2\overset{O}{\overset{\|}{C}}C(CH_3)_3$

(c) 1,3-Cyclopentanedione (d) $CH_3CH{=}CHCHO$

11.23 Draw structures for the possible mono-enol tautomers of 1,3-cyclohexanedione. How many enol forms are possible and which would you expect to be most stable? Explain your answer.

11.24 Rank these compounds in order of increasing acidity:

$$CH_3CH_2COOH,\ CH_3COCH_3,\ CH_3CH_2OH,\ CH_3COCH_2COCH_3$$

11.25 When acetone is treated with acid in deuterated water, D_2O, deuterium becomes incorporated into the molecule. Propose a mechanism to account for this reaction.

$$CH_3\overset{O}{\overset{\|}{C}}CH_3 + D_2O \overset{DCl}{\rightleftharpoons} CH_3\overset{O}{\overset{\|}{C}}CH_2D$$

11.26 When optically active (*R*)-2-methylcyclohexanone is treated with aqueous HCl or NaOH, racemic 2-methylcyclohexanone is produced. Propose a mechanism to explain this observation.

11.27 When optically active (*R*)-3-methylcyclohexanone is treated with aqueous HCl or NaOH, no racemization occurs. Instead, the optically active ketone is recovered unchanged. How can you reconcile this observation with your answer to Problem 11.26?

11.28 Monoalkylated acetic acids (RCH_2COOH) and dialkylated acetic acids ($R_2CHCOOH$) can be prepared by malonic ester synthesis, but trialkylated acetic acids (R_3CCOOH) can't be prepared in this way. Explain.

11.29 Which of the following compounds would you expect to undergo aldol condensation? Draw the product in each case.

(a) 2,2-Dimethylpropanal (b) Cyclobutanone
(c) Benzophenone (diphenyl ketone) (d) Decanal

11.30 Which of the following esters can be prepared by a malonic ester synthesis? Show what reagents you would use:

(a) Ethyl pentanoate (b) Ethyl 3-methylbutanoate
(c) Ethyl 2-methylbutanoate (d) Ethyl 2,2-dimethylpropanoate

11.31 Nonconjugated β,γ-unsaturated ketones such as 3-cyclohexenone are in an acid-catalyzed equilibrium with their conjugated α,β-unsaturated isomers. Propose a mechanism for the acid-catalyzed interconversion of the two isomers.

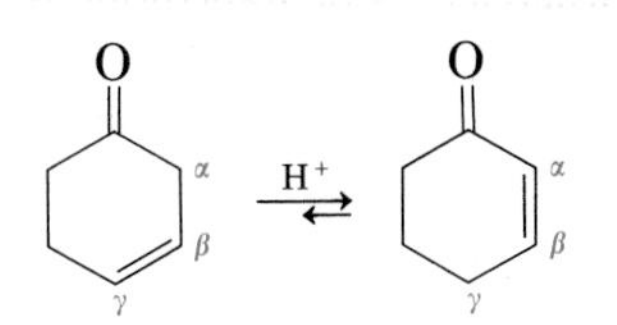

11.32 The α,β to β,γ interconversion of unsaturated ketones (Problem 11.31) is also catalyzed by base. Propose a mechanism for the reaction.

11.33 One consequence of the base-catalyzed α,β to β,γ isomerization of unsaturated ketones (Problem 11.32) is that C5-substituted 2-cyclopentenones can be interconverted with C2-substituted 2-cyclopentenones. Propose a mechanism to account for this isomerization.

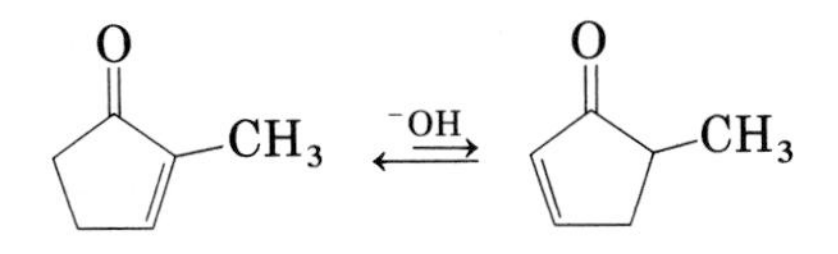

11.34 If a 1:1 mixture of ethyl acetate and ethyl propanoate is treated with base under Claisen condensation conditions, a mixture of four β-keto ester products is obtained. Show their structures.

11.35 If a mixture of ethyl acetate and ethyl benzoate is treated with base, a mixture of two Claisen condensation products is obtained. Explain.

11.36 Cinnamaldehyde, the aromatic constituent of cinnamon oil, can be synthesized by a mixed aldol reaction between benzaldehyde and acetaldehyde. Formulate the reaction. What other product would you expect to obtain?

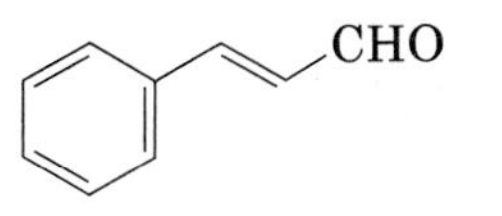

Cinnamaldehyde

11.37 How might you prepare these compounds using aldol condensation reactions?

(a) $C_6H_5C(CH_3){=}CHCOC_6H_5$ (b) 4-Methyl-3-penten-2-one

11.38 1-Butanol is synthesized commercially starting from acetaldehyde by a three-step route that involves an aldol reaction. How might you carry out this transformation?

$$CH_3CHO \xrightarrow{\text{3 steps}} CH_3CH_2CH_2CH_2OH$$

11.39 How would you prepare these compounds, using either an acetoacetic ester synthesis or a malonic ester synthesis?

(a) $(CH_3)_2C(COOCH_3)_2$ (b) $(CH_3)_2CHCOCH_3$

11.40 By starting with a dihalide, cyclic compounds can be prepared using the malonic ester synthesis. What product would you expect to obtain from reaction between diethyl malonate, 1,4-dibromobutane, and two equivalents of base?

11.41 In light of your answer to Problem 11.40, how might you use the acetoacetic ester synthesis to prepare cyclopentyl methyl ketone?

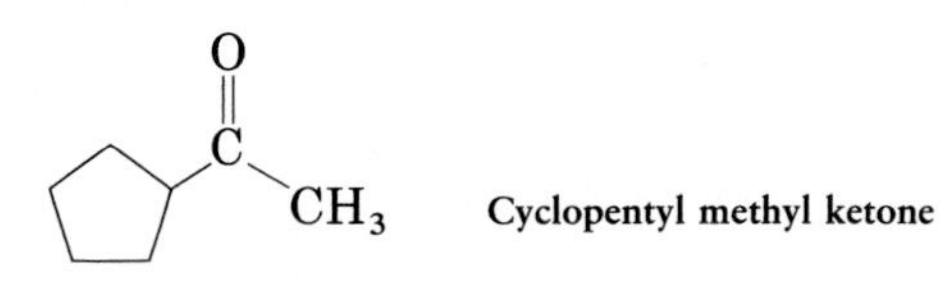

11.42 The aldol reaction can sometimes take place *internally* if a di-carbonyl compound is treated with base. What product would you expect to obtain from aldol cyclization of hexanedial, $OHCCH_2CH_2CH_2CH_2CHO$?

11.43 How can you account for the fact that *cis*- and *trans*-4-*tert*-butyl-2-methylcyclohexanone are interconverted by base treatment? Which of the two isomers do you think is more stable, and why? (Hint: See Section 2.11.)

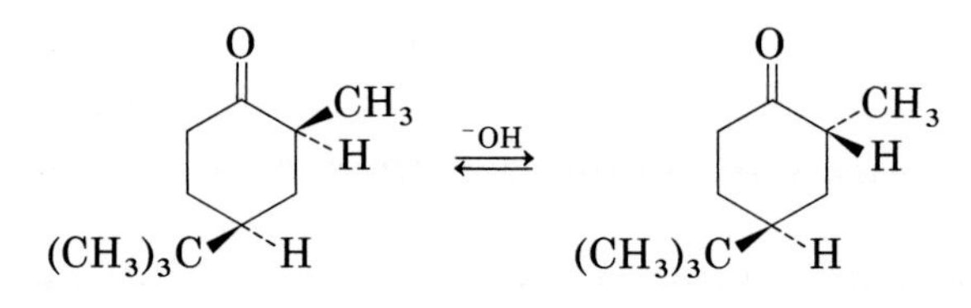

11.44 Show how you might convert geraniol, the chief constituent of rose oil, into either ethyl geranylacetate or geranylacetone.

$$CH_3\overset{CH_3}{\overset{|}{C}}=CHCH_2CH_2\overset{CH_3}{\overset{|}{C}}=CHCH_2OH$$

Geraniol

? ↙ ↘ ?

$$CH_3\overset{CH_3}{\overset{|}{C}}=CHCH_2CH_2\overset{CH_3}{\overset{|}{C}}=CHCH_2CH_2\overset{O}{\overset{\|}{C}}OCH_2CH_3$$

Ethyl geranylacetate

$$CH_3\overset{CH_3}{\overset{|}{C}}=CHCH_2CH_2\overset{CH_3}{\overset{|}{C}}=CHCH_2CH_2\overset{O}{\overset{\|}{C}}CH_3$$

Geranylacetone

11.45 The Claisen condensation is reversible. That is, a *β*-keto ester can be cleaved by base into two fragments. Show the mechanism by which the following cleavage occurs.

$$C_6H_5-\overset{O}{\overset{\|}{C}}CH_2\overset{O}{\overset{\|}{C}}OCH_2CH_3 \xrightarrow[H_2O]{NaOH} C_6H_5-\overset{O}{\overset{\|}{C}}O^-\ Na^+ + CH_3\overset{O}{\overset{\|}{C}}OCH_2CH_3$$

CHAPTER

12 Amines

amine
an organic derivative of ammonia

primary amine
an amine with one organic substituent on nitrogen, RNH_2

secondary amine
an amine with two organic substituents on nitrogen, R_2NH

tertiary amine
an amine with three organic substituents on nitrogen, R_3N

Amines are organic derivatives of ammonia in the same way that alcohols and ethers are organic derivatives of water. Amines are classified either as **primary** (RNH_2), **secondary** (R_2NH), or **tertiary** (R_3N), depending on the number of organic substituents attached to nitrogen. For example, methylamine (CH_3NH_2) is a primary amine and trimethylamine [$(CH_3)_3N$] is a tertiary amine. Note that this usage of the terms *primary, secondary,* and *tertiary* is different from our previous usage. When we speak of a tertiary alcohol or alkyl halide, we refer to the degree of substitution at the alkyl carbon atom, but when we speak of a tertiary amine, we refer to the degree of substitution at the nitrogen atom.

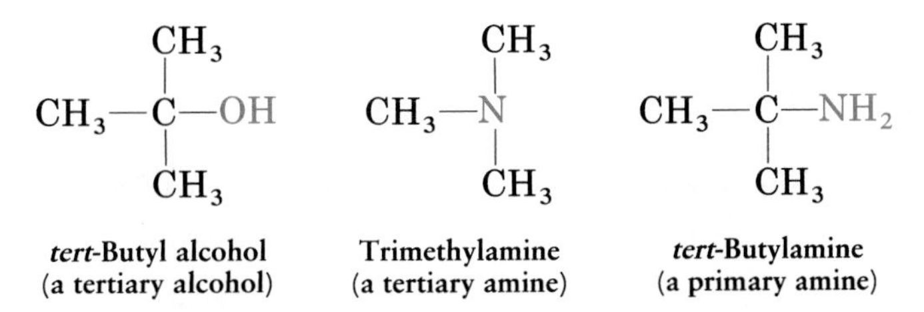

tert-Butyl alcohol (a tertiary alcohol) Trimethylamine (a tertiary amine) *tert*-Butylamine (a primary amine)

quaternary ammonium salt
a compound with four organic substituents attached to a positively charged nitrogen, $R_4N^+ X^-$

Compounds with four groups attached to nitrogen are also known, but the nitrogen atom must carry a positive charge. Such compounds are called **quaternary ammonium salts.**

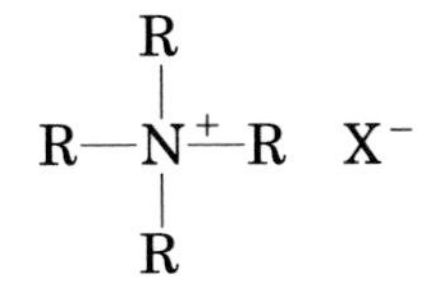

A quaternary ammonium salt

alkylamine an amine that has its nitrogen atom bonded to a saturated, alkyl-group carbon

arylamine an amine that has its nitrogen atom bonded to an aromatic ring

Amines can be either alkyl-substituted (**alkylamines**) or aryl-substituted (**arylamines**). Although much of the chemistry of the two classes is similar, we'll soon see that there are also important differences.

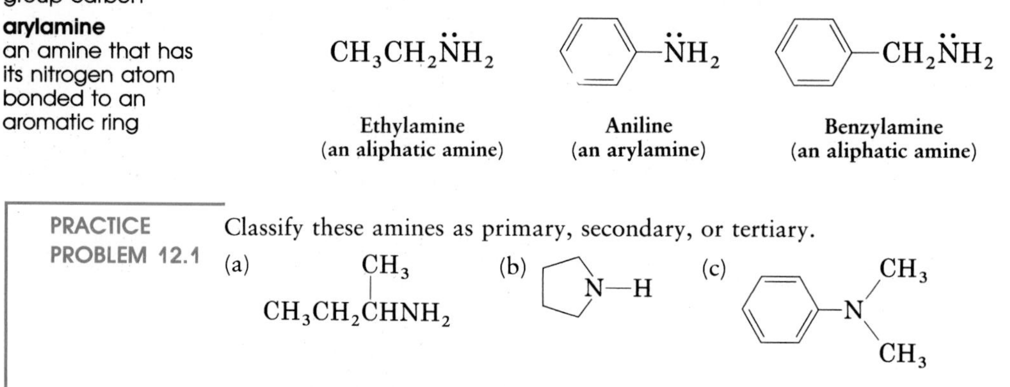

Ethylamine (an aliphatic amine) — **Aniline (an arylamine)** — **Benzylamine (an aliphatic amine)**

PRACTICE PROBLEM 12.1 Classify these amines as primary, secondary, or tertiary.

(a) $CH_3CH_2CH(CH_3)NH_2$ (b) pyrrolidine (N—H) (c) $C_6H_5N(CH_3)_2$

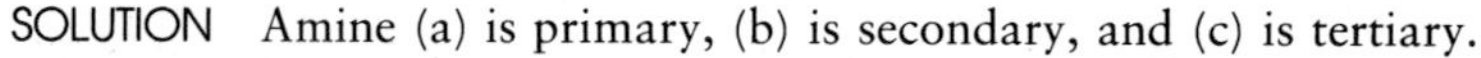

SOLUTION Amine (a) is primary, (b) is secondary, and (c) is tertiary.

PROBLEM 12.1 Classify the following compounds as either primary, secondary, or tertiary amines or as quaternary ammonium salts.

(a) $(CH_3)_2CHNH_2$ (b) $(CH_3CH_2)_2NH$

(c) $C_6H_{11}N(CH_3)_2$ (cyclohexyl) (d) $C_6H_5CH_2\overset{+}{N}(CH_3)_3\ I^-$

PROBLEM 12.2 Draw structures of compounds that meet these descriptions.

(a) A secondary amine with one isopropyl group
(b) A tertiary amine with one phenyl group and one ethyl group
(c) A quaternary ammonium salt with four different groups bonded to nitrogen

12.1 NAMING AMINES

Primary amines, RNH_2, are named in the IUPAC system in either of two ways. For simple amines, the suffix *-amine* is added to the name of the organic substituent.

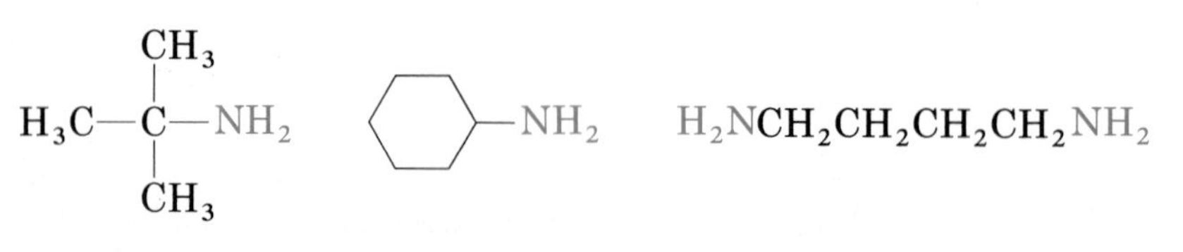

tert-**Butylamine** — **Cyclohexylamine** — **1,4-Butanediamine**

Amines that also have other functional groups are named by considering the $-NH_2$ as an amino substituent on the parent molecule.

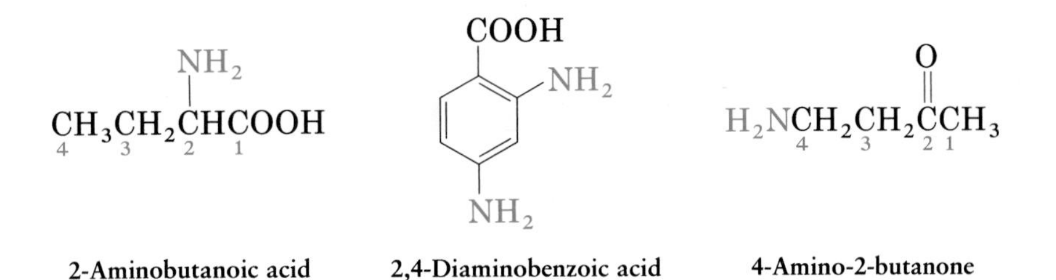

2-Aminobutanoic acid　　2,4-Diaminobenzoic acid　　4-Amino-2-butanone

Symmetrical secondary and tertiary amines are named by adding the prefix *di-* or *tri-* to the alkyl group.

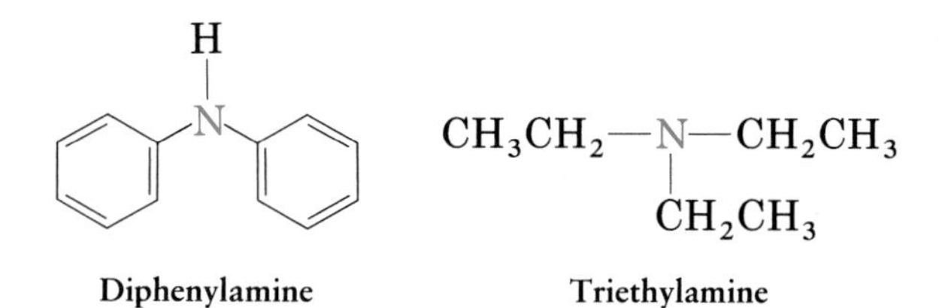

Diphenylamine　　Triethylamine

Unsymmetrically substituted secondary and tertiary amines are named as *N*-substituted primary amines. The largest organic group is chosen as the parent, and the other groups are considered as *N*-substituents on the parent (*N* since they're attached to nitrogen).

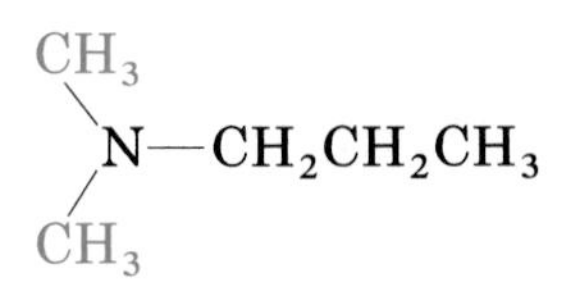

H₃C　CH₂CH₃ — N — cyclohexyl

***N,N*-Dimethylpropylamine**
(propylamine is the parent name; the two methyl groups are substituents on nitrogen)

***N*-Ethyl-*N*-methylcyclohexylamine**
(cyclohexylamine is the parent name; methyl and ethyl are *N*-substituents)

There are few common names for simple amines, but phenylamine is usually called *aniline*.

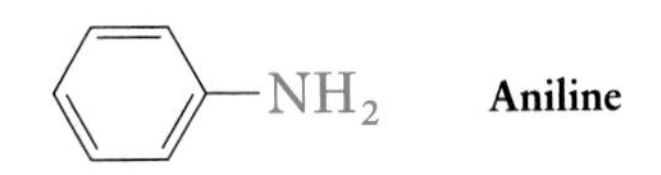

Aniline

heterocyclic amine an amine in which the nitrogen atom occurs in a ring

Heterocyclic amines, compounds in which the nitrogen atom is part of a ring, are also common, and each different heterocyclic ring system is given its own parent name. In all cases, the nitrogen atom is numbered as position 1.

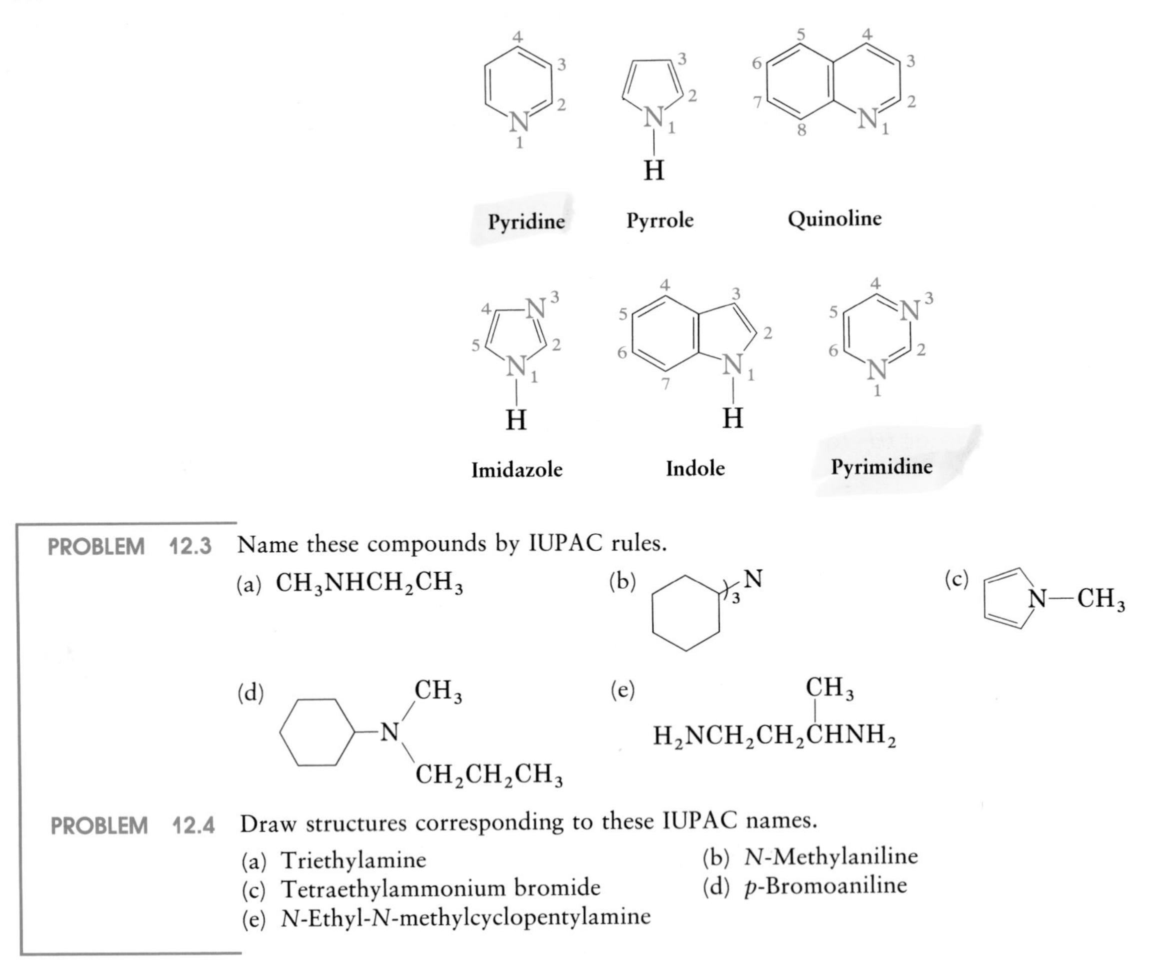

PROBLEM 12.3 Name these compounds by IUPAC rules.

(a) $CH_3NHCH_2CH_3$ (b) (c) (d) (e) $H_2NCH_2CH_2CH(CH_3)NH_2$

PROBLEM 12.4 Draw structures corresponding to these IUPAC names.

(a) Triethylamine
(b) *N*-Methylaniline
(c) Tetraethylammonium bromide
(d) *p*-Bromoaniline
(e) *N*-Ethyl-*N*-methylcyclopentylamine

12.2 STRUCTURE AND PROPERTIES OF AMINES

Bonding in amines is similar to bonding in ammonia. The nitrogen atom is sp^3 hybridized with the three substituents occupying three corners of a tetrahedron and the nitrogen's nonbonding lone pair of electrons occupying the fourth corner. As expected, the C–N–C bond angles are very close to the 109° tetrahedral value. For trimethylamine, the C–N–C angle is 108°, and the C–N bond length is 1.47 Å.

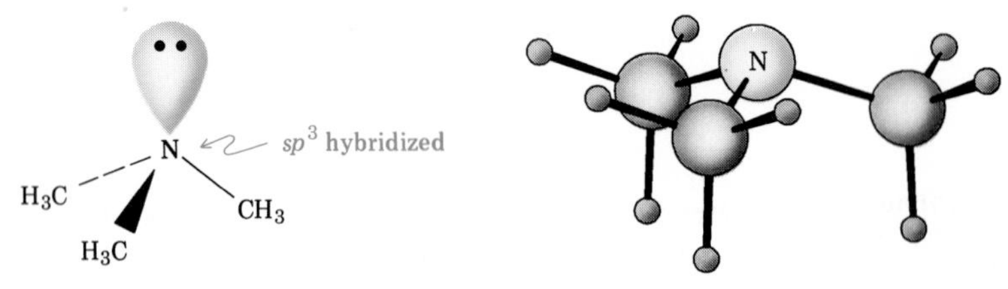

Trimethylamine

Amines are highly polar and therefore have higher boiling points than alkanes of similar molecular weights. Like alcohols, amines with fewer than five carbon atoms are generally water soluble. Also like alcohols, primary and secondary amines form hydrogen bonds and are highly associated in the liquid state.

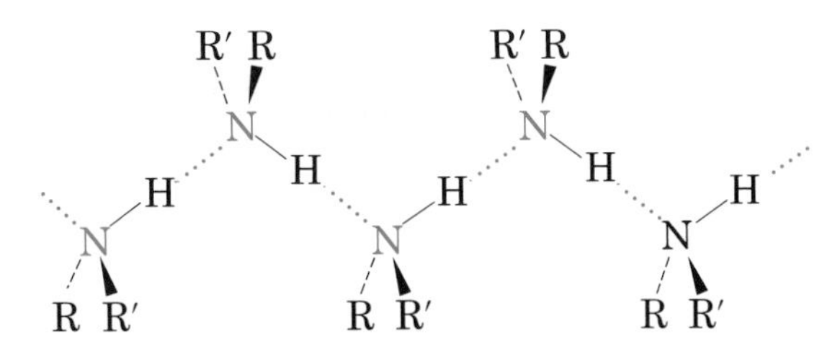

One other characteristic property of amines is their odor. Low-molecular-weight amines have a characteristic fishlike aroma, while diamines such as putrescine (1,4-butanediamine) have names that are self-explanatory.

12.3 AMINE BASICITY

The chemistry of amines is dominated by the presence of the nitrogen lone pair of electrons. Because of the lone pair, amines are both basic and nucleophilic. Amines react with Lewis acids to form acid–base salts, and they react with electrophiles in many of the polar reactions seen in past chapters.

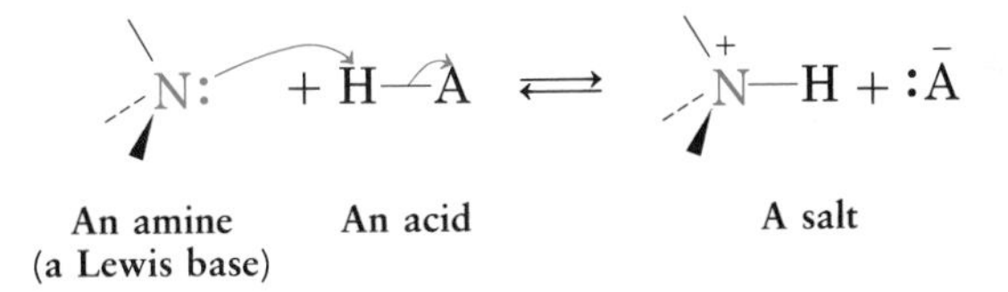

Amines are much more basic than alcohols, ethers, or water. When an amine is dissolved in water, an equilibrium is established in which water acts as a protic acid and donates a proton to the amine. Not all amines are equal in base strength though. The most convenient way to measure the basicity of an amine RNH_2 is to look at the acidity of the corresponding ammonium salt RNH_3^+. Since a more strongly basic amine holds a proton more tightly, its corresponding ammonium ion is less acidic (higher pK_a). A more weakly basic amine, however, holds a proton less tightly, and its corresponding ammonium ion is more acidic (lower pK_a).

If this ammonium salt has a lower pK_a (stronger acid), then this amine is a weaker base.

$$R{-}NH_3^+ + H_2O \rightleftharpoons R{-}\ddot{N}H_2 + H_3O^+$$

If this ammonium salt has a higher pK_a (weaker acid), then this amine is a stronger base.

Table 12.1 lists the pK_a of some common ammonium salts and indicates that substitution has relatively little effect on alkylamine basicity. The salts of most simple alkylamines have pK_a's in the narrow range 10–11, regardless of their exact structure.

TABLE 12.1 Base strengths of some common amines

Name	*Structure*	pK_a *of ammonium salt*
Ammonia	$:NH_3$	9.26
Primary amine		
Methylamine	$CH_3\ddot{N}H_2$	10.64
Ethylamine	$CH_3CH_2\ddot{N}H_2$	10.75
Aniline	C_6H_5—$\ddot{N}H_2$	4.63
Secondary amine		
Dimethylamine	$(CH_3)_2\ddot{N}H$	10.73
Diethylamine	$(CH_3CH_2)_2\ddot{N}H$	10.94
Tertiary amine		
Trimethylamine	$(CH_3)_3N:$	9.79
Triethylamine	$(CH_3CH_2)_3N:$	10.79

The most important conclusion from the data in Table 12.1 is that arylamines such as aniline are weaker bases than alkylamines. The nitrogen lone-pair electrons of arylamines are delocalized by orbital overlap with the pi orbitals of the aromatic ring and are therefore less available for bonding to an acid. In resonance terms, arylamines are more stable and less reactive than alkylamines because of their five resonance structures.

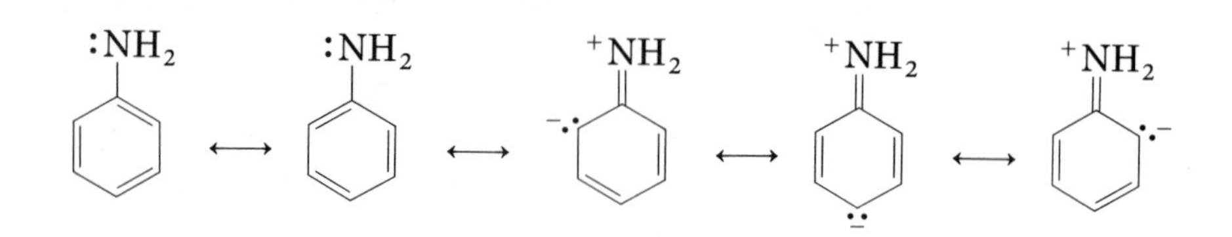

In contrast to amines, *amides* ($RCONH_2$) are completely nonbasic. Amides don't form salts when treated with acids, and their aqueous solutions are neutral. As with arylamines, the main reason for the decreased basicity of amides is that the nitrogen lone-pair electrons are delocalized by orbital overlap with the neighboring carbonyl-group pi orbital. The electrons are therefore much less available for bonding to an acid. In resonance terms, amides are more stable and less reactive than amines because they are hybrids of two resonance forms:

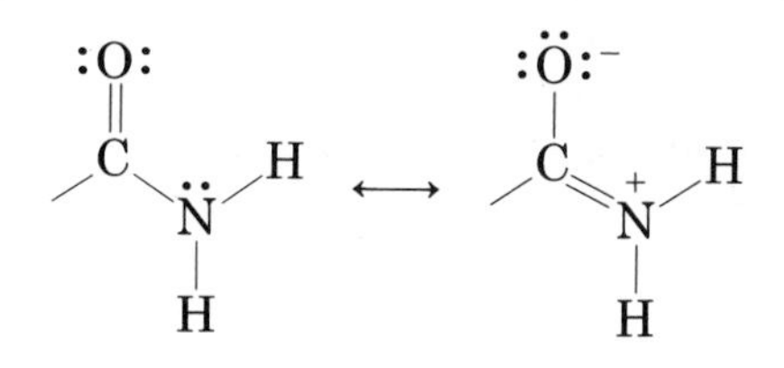

We can take advantage of the basicity of amines to purify them. For example, if we have a mixture of an amine and a neutral compound such as a ketone, we can dissolve the mixture in an organic solvent and treat it with aqueous acid. The basic amine dissolves in the acidic water, while the neutral ketone remains in the organic solvent. Separation and neutralization of the water then provides the pure amine (Figure 12.1).

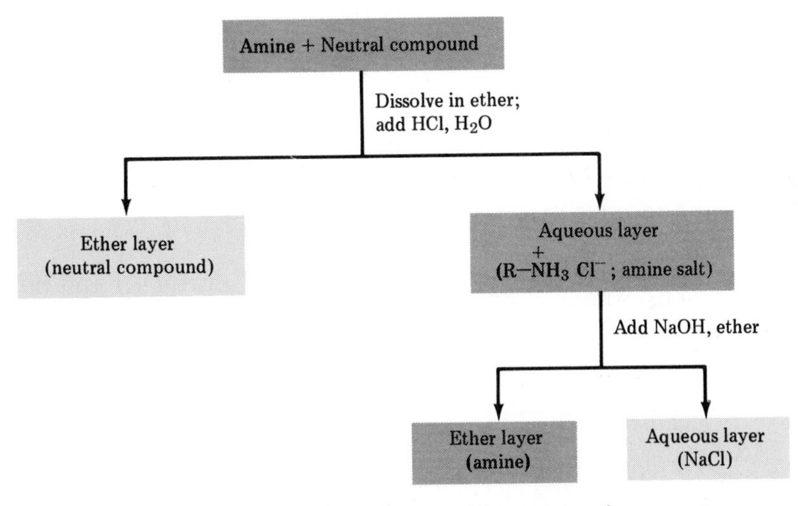

FIGURE 12.1 Separation and purification of an amine from a mixture

PRACTICE PROBLEM 12.2 Which would you expect to be the stronger base, aniline or *p*-nitroaniline?

SOLUTION Since a nitro group on a benzene ring is strongly electron withdrawing (Section 5.9), it pulls electrons from the $-NH_2$ group, making them less available for donation to acids and making *p*-nitroaniline a weaker base than aniline.

PRACTICE PROBLEM 12.3 Predict the product of this reaction.

$$CH_3CH_2NHCH_3 + HCl \longrightarrow ?$$

SOLUTION Amines are protonated by acids to yield ammonium salts.

$$CH_3CH_2NHCH_3 + HCl \longrightarrow CH_3CH_2\overset{+}{N}H_2CH_3\ Cl^-$$

PROBLEM 12.5 Which would you expect to be the stronger base, aniline or *p*-methylaniline? Explain.

PROBLEM 12.6 Predict the product of this reaction.

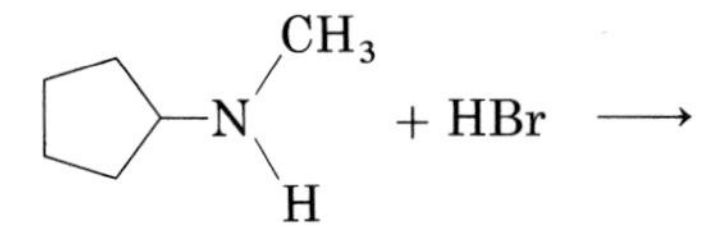

PROBLEM 12.7 Which compound in each of the following pairs is more basic?

(a) $CH_3CH_2NH_2$ or $CH_3CH_2CONH_2$ (b) NaOH or $C_6H_5NH_2$
(c) CH_3NHCH_3 or $CH_3NHC_6H_5$ (d) CH_3OCH_3 or $(CH_3)_3N$

12.4 RESOLUTION OF ENANTIOMERS BY USE OF AMINE SALTS

resolution the separation of a racemic mixture into two pure enantiomers

Amine basicity is often used to carry out the separation (**resolution**) of a racemic carboxylic acid mixture into its two pure enantiomers. [Recall from Section 6.10 that a racemic mixture is a 50:50 mixture of (+) and (−) enantiomers.]

$$50\%\ (+):50\%\ (-) \xrightarrow{\text{resolve}} \text{pure } (+) \text{ and pure } (-)$$

Racemic mixture of enantiomers | **Pure separate enantiomers**

Historically, Louis Pasteur was the first person to resolve a racemic mixture when he was able to crystallize a salt of (±)-tartaric acid and separate two kinds of crystals by hand (Section 6.5). Pasteur's method isn't generally applicable, though, since few racemic mixtures crystallize into separate mirror-image forms. The most commonly used method of resolution makes use of an acid–base reaction between a racemic mixture of carboxylic acids and a chiral amine.

To understand how this method of resolution works, let's see what happens when a racemic mixture of chiral acids such as (+)- and (−)-lactic acids reacts with an achiral amine base such as methylamine (Figure 12.2). Stereochemically, the situation is analogous to what happens when left and right hands (chiral) pick up a ball (achiral). Both left and right hands pick up the ball equally well, and the products, ball in right hand versus ball in left hand, are mirror images. In the same way, both (+)- and (−)-lactic acid react with methylamine equally well, and the product is a mixture of two salts: methylammonium (+)-lactate and methylammonium (−)-lactate. Just as with the chiral hands and achiral ball, the two salts are mirror images and we still have a racemic mixture.

Now let's see what happens when the racemic mixture of (+)- and (−)-lactic acids reacts with a single enantiomer of a chiral amine base such as (*R*)-1-phenylethylamine (Figure 12.3). Stereochemically, this situation is analogous to what happens when a hand (a chiral reagent) puts on a glove (also a chiral reagent).

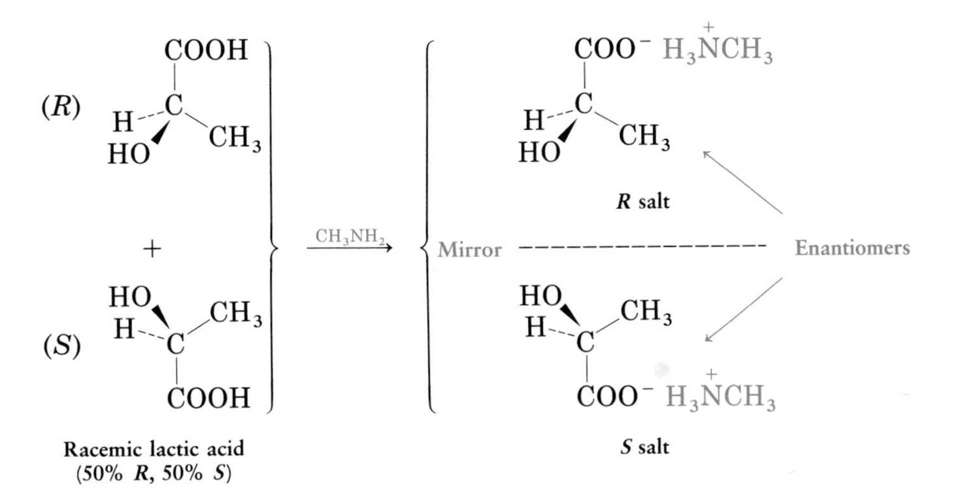

FIGURE 12.2 Reaction of racemic lactic acid with achiral methylamine leads to a racemic mixture of salts.

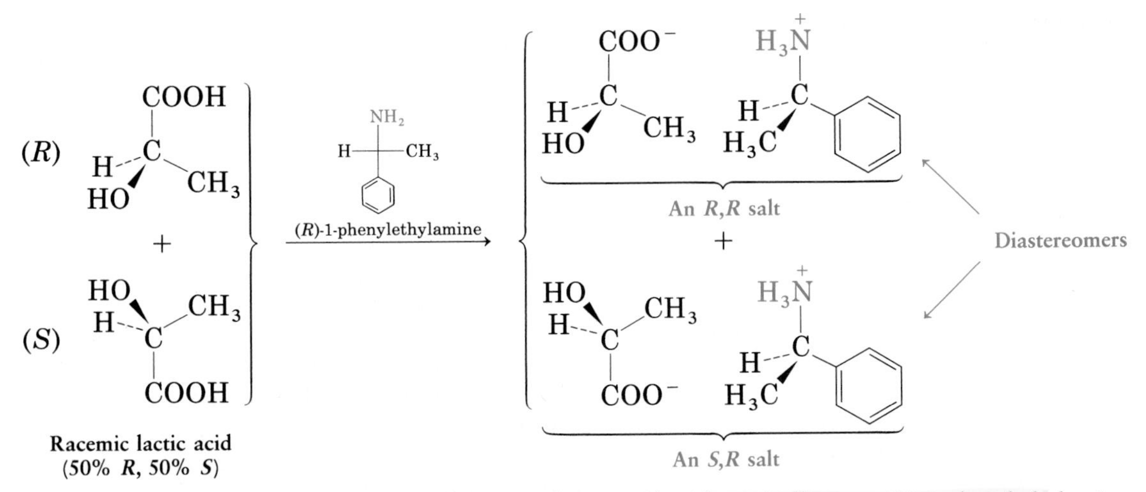

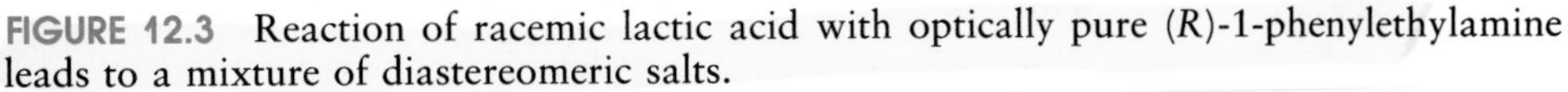
FIGURE 12.3 Reaction of racemic lactic acid with optically pure (*R*)-1-phenylethylamine leads to a mixture of diastereomeric salts.

Left and right hands don't put on the same glove in the same way. The products—right hand in right glove versus left hand in right glove—are not mirror images; they're altogether different.

In the same way, (+)- and (−)-lactic acid react with (*R*)-1-phenylethylamine to give different products. (*R*)-Lactic acid reacts with (*R*)-1-phenylethylamine to give the *R*,*R* salt, whereas (*S*)-lactic acid reacts with the same *R* amine to give the *S*,*R* salt. *These two salts are diastereomers* (Section 6.7). They are different compounds and have different chemical and physical properties. It may therefore be possible to separate them by fractional crystallization or by some other laboratory technique. Once separated, acidification of the two diastereomeric salts with mineral acid then allows us to recover and isolate the two pure enantiomers of lactic acid.

PRACTICE PROBLEM 12.4 Suppose that racemic lactic acid reacts with methanol to form the ester, methyl lactate. What stereochemistry would you expect the product(s) to have. What is the relationship of one product to another?

SOLUTION Reaction of a racemic acid with an achiral alcohol such as methanol yields a racemic mixture of mirror-image ester products:

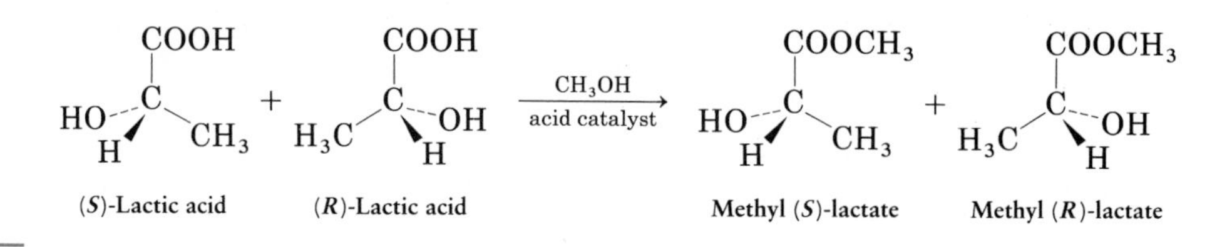

PROBLEM 12.8 Suppose that racemic lactic acid reacts with (*S*)-2-butanol to form an ester. What stereochemistry would you expect the product(s) to have? What is the relationship of one product to another.

PROBLEM 12.9 How might you use the reaction described in Problem 12.8 to resolve (±)-lactic acid?

12.5 SYNTHESIS OF AMINES

S_N2 Reactions of Alkyl Halides

Alkylamines are excellent nucleophiles in S_N2 reactions (Section 7.7). As a result, the simplest method of amine synthesis is by S_N2 reaction of ammonia or an alkylamine with an alkyl halide. If ammonia is used, a primary amine results; if a primary amine is used, a secondary amine results; and so on. Even tertiary amines react with alkyl halides to yield quaternary ammonium salts, $R_4N^+X^-$.

Ammonia	$\ddot{N}H_3 + R—X \longrightarrow$	$RNH_3^+X^-$	$\xrightarrow{NaOH}$	RNH_2	Primary
Primary	$R\ddot{N}H_2 + R—X \longrightarrow$	$R_2NH_2^+X^-$	$\xrightarrow{NaOH}$	R_2NH	Secondary
Secondary	$R_2\ddot{N}H + R—X \longrightarrow$	$R_3NH^+X^-$	$\xrightarrow{NaOH}$	R_3N	Tertiary
Tertiary	$R_3\ddot{N} + R—X \longrightarrow$	$R_4N^+X^-$		Quaternary ammonium salt	

S_N2 reaction

Unfortunately, these reactions don't stop cleanly after a single alkylation has occurred. Since primary, secondary, and tertiary amines all have similar reactivity, the initially formed monoalkylated product often undergoes further reaction to yield a mixture of products. For example, treatment of 1-bromooctane with a twofold excess of ammonia leads to a mixture containing only 45% yield of octylamine. A nearly equal amount of dioctylamine is produced by double alkylation, along with smaller amounts of trioctylamine and tetraoctylammonium bromide.

$$CH_3(CH_2)_6CH_2Br + :NH_3 \longrightarrow CH_3(CH_2)_6CH_2\ddot{N}H_2 + [CH_3(CH_2)_6CH_2]_2\ddot{N}H$$

1-Bromooctane — Octylamine (45%) — Dioctylamine (43%)

$$+ [CH_3(CH_2)_6CH_2]_3N: + [CH_3(CH_2)_6CH_2]_4\overset{+}{N}\overset{-}{Br}$$

Trace — Trace

PRACTICE PROBLEM 12.5 How could you prepare diethylamine starting from ammonia?

SOLUTION Look at the starting material (NH_3) and the product $(CH_3CH_2)_2NH$ and note the difference. Since two ethyl groups have become bonded to the nitrogen atom, the reaction must involve ammonia and two molecules of an ethyl halide:

$$2\ CH_3CH_2Br + NH_3 \longrightarrow (CH_3CH_2)_2NH$$

PROBLEM 12.10 Show how you could prepare these amines from ammonia.
(a) Triethylamine (b) Tetramethylammonium bromide

Reduction of Nitriles and Amides

We've already seen how amines can be prepared by reduction of nitriles (Section 10.11) and amides (Section 10.10) with $LiAlH_4$. The two-step sequence of reactions, initial S_N2 reaction of an alkyl halide with cyanide ion followed by reduction, is a good method for converting an alkyl halide into a primary amine having one more carbon atom. Amide reduction provides a method for converting carboxylic acids into amines with the same number of carbon atoms.

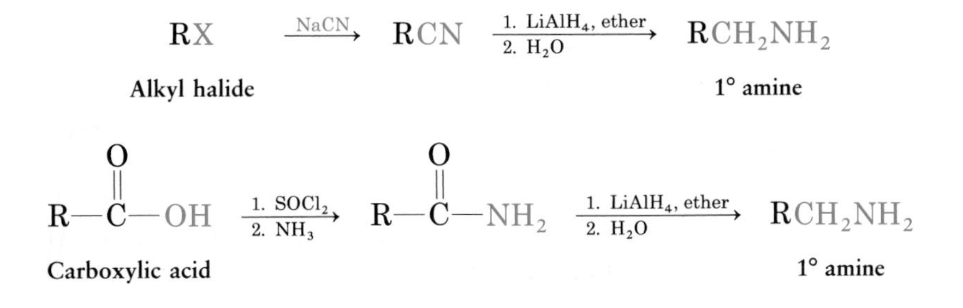

PRACTICE PROBLEM 12.6

What amide would you start with to prepare *N*-ethylcyclohexylamine?

SOLUTION Reduction of an amide with $LiAlH_4$ yields an amine in which the amide carbonyl group has been replaced by a methylene ($-CH_2-$) unit, $RCONR_2 \rightarrow RCH_2NR_2$. Since *N*-ethylcyclohexylamine has only one $-CH_2-$ carbon attached to its nitrogen (the ethyl group), the product must come from reduction of *N*-cyclohexylacetamide:

$$C_6H_{11}-NH-\overset{O}{\overset{\|}{C}}-CH_3 \xrightarrow[2.\ H_2O]{1.\ LiAlH_4} C_6H_{11}-NHCH_2CH_3$$

N-Cyclohexylacetamide *N*-Ethylcyclohexylamine

PRACTICE PROBLEM 12.7

What nitrile would yield butylamine on reaction with $LiAlH_4$?

SOLUTION Reduction of a nitrile with $LiAlH_4$ yields a primary amine whose $-CH_2NH_2$ part comes from the $-C{\equiv}N$. Thus, butylamine must have come from butanenitrile:

$$CH_3CH_2CH_2C{\equiv}N \xrightarrow[2.\ H_2O]{1.\ LiAlH_4} CH_3CH_2CH_2CH_2NH_2$$

Butanenitrile Butylamine

PROBLEM 12.11 Propose structures for amides that might be precursors of these amines.

(a) Propylamine (b) Dipropylamine (c) Benzylamine, $C_6H_5CH_2NH_2$

PROBLEM 12.12 Propose structures for nitriles that might be precursors of these amines.

(a) $CH_3\underset{}{\overset{CH_3}{\overset{|}{C}}}HCH_2CH_2NH_2$ (b) Benzylamine, $C_6H_5CH_2NH_2$

Reduction of Nitroarenes

Arylamines are prepared by nitration of an aromatic starting material, followed by reduction of the nitro group. The reduction step can be carried out in many different ways, depending on the circumstances. Catalytic hydrogenation over platinum works well but is sometimes incompatible with the presence elsewhere in the molecule of other reducible groups such as carbon–carbon double bonds. Iron, zinc, tin, and stannous chloride ($SnCl_2$) are also effective when used in aqueous acid.

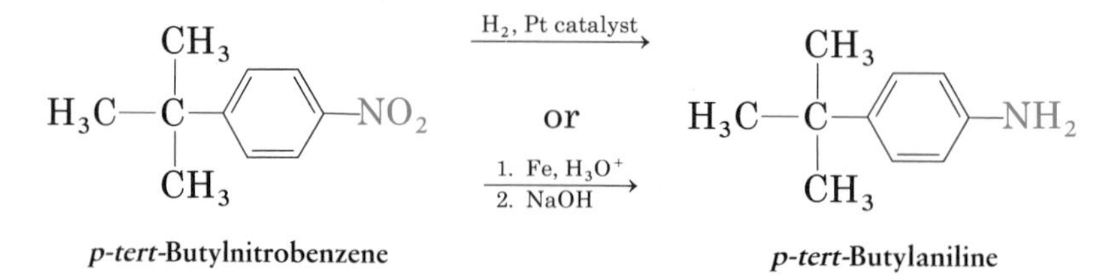

p-tert-Butylnitrobenzene *p-tert*-Butylaniline

PRACTICE PROBLEM 12.8

How could you synthesize *p*-methylaniline starting from benzene? More than one step is required.

SOLUTION A methyl group is introduced onto a benzene ring by a Friedel–Crafts reaction with $CH_3Cl/AlCl_3$ (Section 5.8), and an amino group is introduced onto a ring by nitration

and reduction. The overall sequence is

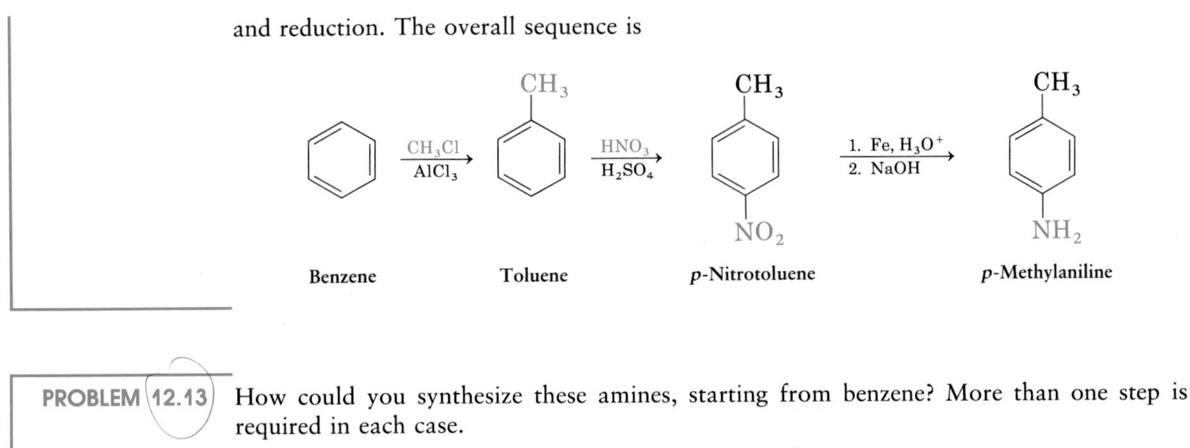

PROBLEM 12.13 How could you synthesize these amines, starting from benzene? More than one step is required in each case.

(a) *m*-Aminobenzoic acid (b) 2,4,6-Tribromoaniline

12.6 REACTIONS OF AMINES

We've already studied the two most important reactions of alkylamines: alkylation and acylation. As we saw in the previous section, primary, secondary, and tertiary amines can be alkylated by reaction with alkyl halides. Primary and secondary (but not tertiary) amines can also be acylated by nucleophilic acyl substitution reactions with acid chlorides or acid anhydrides (Sections 10.7 and 10.8). The products are amides.

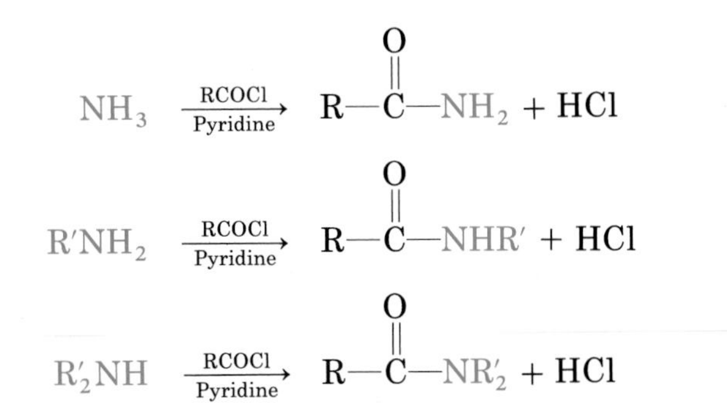

PROBLEM 12.14 Write an equation for the reaction of diethylamine with acetyl chloride to yield *N*,*N*-diethylacetamide.

Diazonium Salts: The Sandmeyer Reaction

Primary amines react with nitrous acid, HNO_2, to yield diazonium salts, $R-\overset{+}{N}\equiv N\ X^-$. Although alkyl diazonium salts are too reactive to be isolated, aryl diazonium salts, $Ar-\overset{+}{N}\equiv N\ X^-$, are more stable.

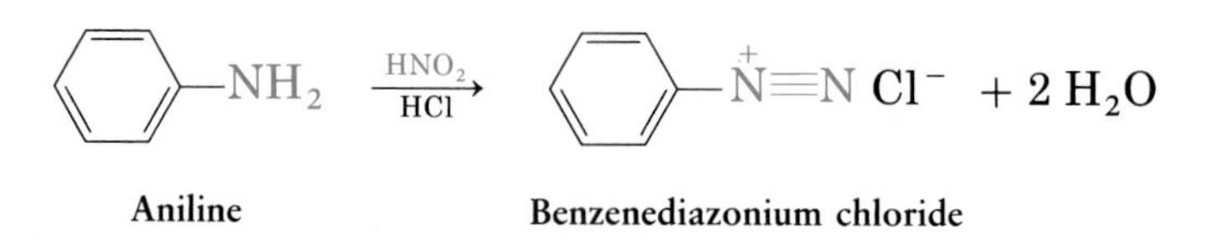

Aryl diazonium salts are extremely useful compounds because the diazonio group (N_2^+) can be replaced by nucleophiles in a substitution reaction ($Ar-N_2^+ + :Nu^- \rightarrow Ar-Nu + N_2$). Many different nucleophiles react with arenediazonium salts, and many different substituted benzenes can be prepared with the reaction. The overall sequence of (1) nitration, (2) reduction, (3) diazotization, and (4) nucleophilic replacement, is probably the single most versatile method for preparing substituted aromatic rings (Figure 12.4).

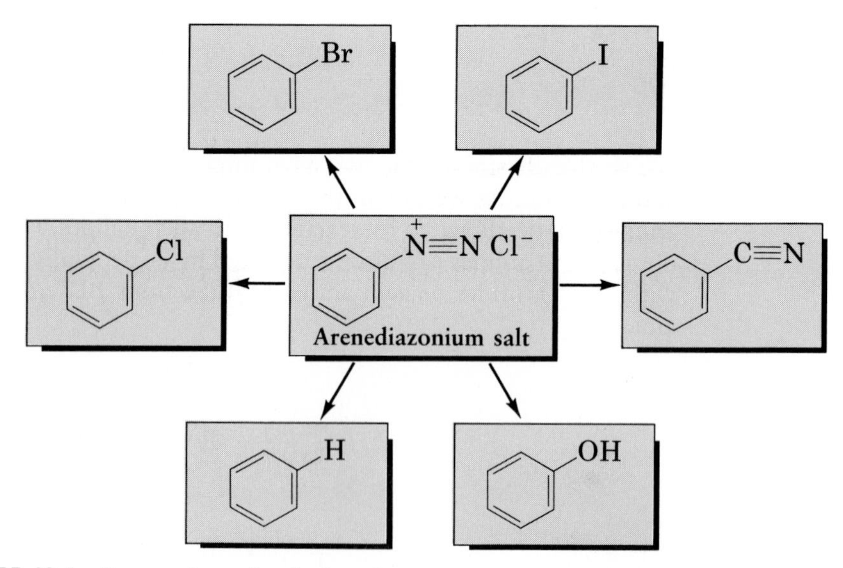

FIGURE 12.4 Preparation of substituted aromatic compounds by diazonio replacement reactions

Sandmeyer reaction the conversion of an arenediazonium salt into an aryl halide by reaction with a cuprous halide

Aryl chlorides and bromides are prepared by reaction of an aryl diazonium salt with HX in the presence of a small amount of cuprous halide (CuX) catalyst, a process called the **Sandmeyer reaction.** Aryl iodides are prepared by reaction with sodium iodide.

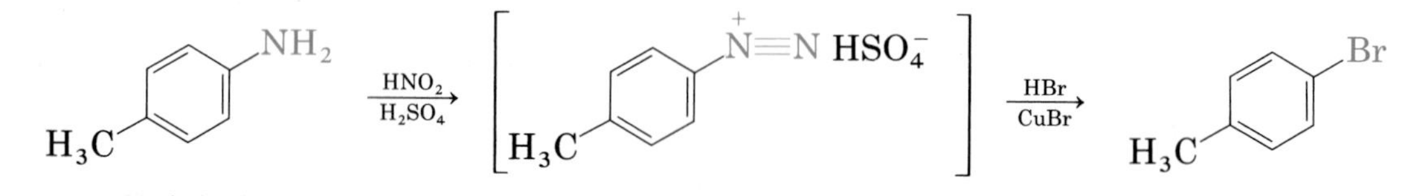

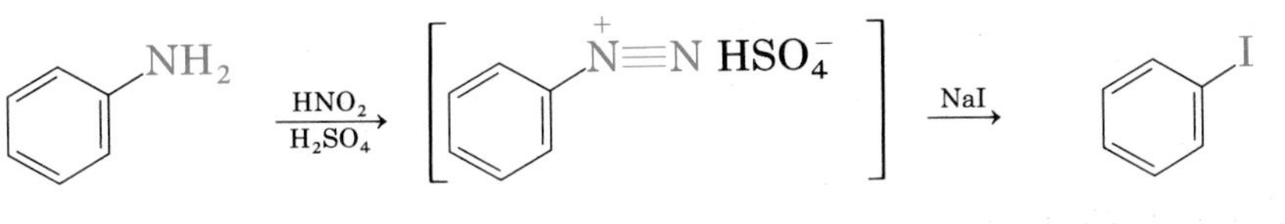

Aniline

Iodobenzene (67%)

Treatment of an arenediazonium salt with KCN and a small amount of cuprous cyanide, CuCN, yields a nitrile, ArCN. This reaction is particularly useful because it allows the replacement of a nitrogen substituent by a carbon substituent. The nitrile can then be elaborated into other functional groups such as carboxyl, −COOH. For example, hydrolysis of *o*-methylbenzonitrile, produced by Sandmeyer reaction of *o*-methylbenzenediazonium bisulfate with cuprous cyanide, yields *o*-methylbenzoic acid.

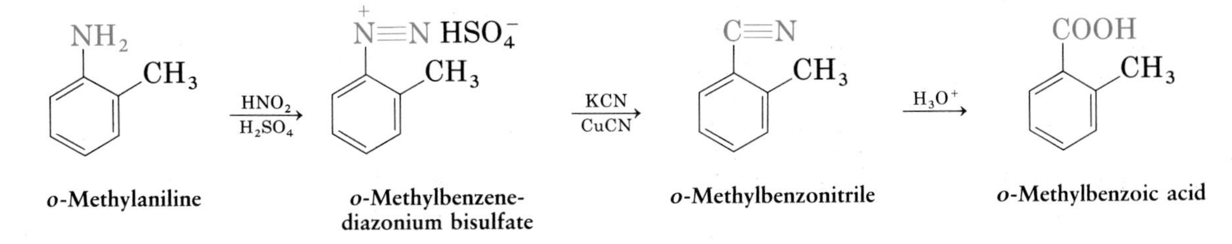

The diazonio group can also be replaced by −OH to yield phenols and by −H to yield arenes. Phenols are prepared by addition of the aryl diazonium salt to hot aqueous acid. For example:

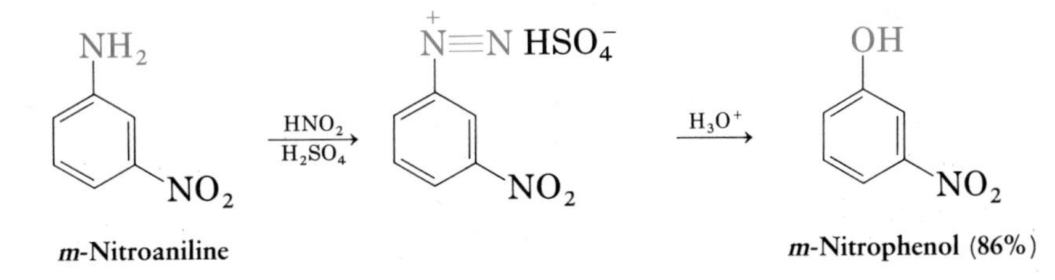

m-Nitroaniline

m-Nitrophenol (86%)

Arenes are produced by reaction of the diazonium salt with hypophosphorous acid, H_3PO_2. For example, *p*-methylaniline can be converted into 3,5-dibromotoluene by a sequence involving bromination, diazotization, and hypophosphorous acid treatment.

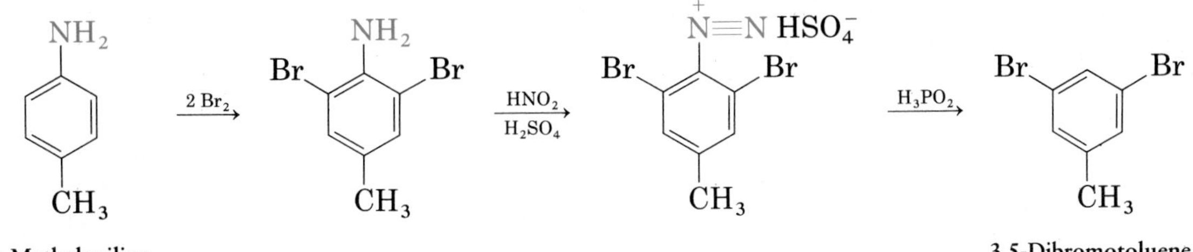

p-Methylaniline

3,5-Dibromotoluene

PRACTICE PROBLEM 12.9 How would you prepare *p*-methylphenol from benzene, using a diazonio replacement reaction?

SOLUTION Working backward, the immediate precursor of the target molecule might be *p*-methylbenzenediazonium ion, which could be prepared from *p*-nitrotoluene. *p*-Nitrotoluene, in turn, could be prepared by nitration of toluene, which could be prepared by Friedel–Crafts methylation of benzene.

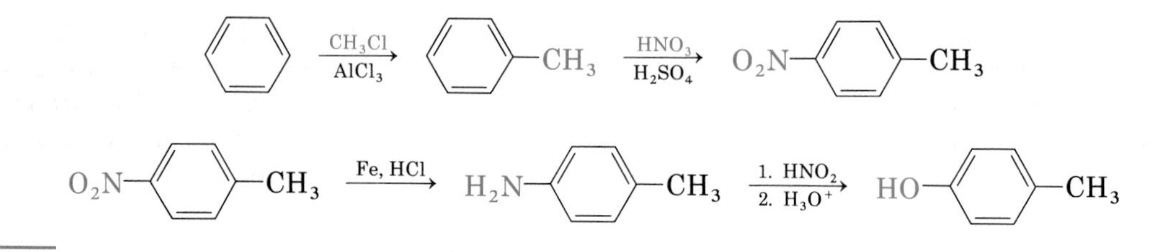

PROBLEM 12.15 How would you prepare *p*-bromobenzonitrile from bromobenzene using a diazonio replacement reaction?

PROBLEM 12.16 How would you prepare these compounds from benzene?
(a) *m*-Bromobenzoic acid (b) *m*-Bromochlorobenzene

12.7 HETEROCYCLIC AMINES

carbocycle a cyclic molecule that contains only carbon in its ring

heterocycle a cyclic molecule that contains one or more different atoms in addition to carbon in its ring

Cyclic organic compounds can be classed either as carbocycles or as heterocycles. **Carbocycles** contain only carbon atoms in their rings, but **heterocycles** contain one or more different atoms in addition to carbon. Heterocyclic amines are particularly common in organic chemistry, and many have important biological properties. For example, the antiulcer agent cimetidine and the sedative phenobarbital are heterocyclic amines.

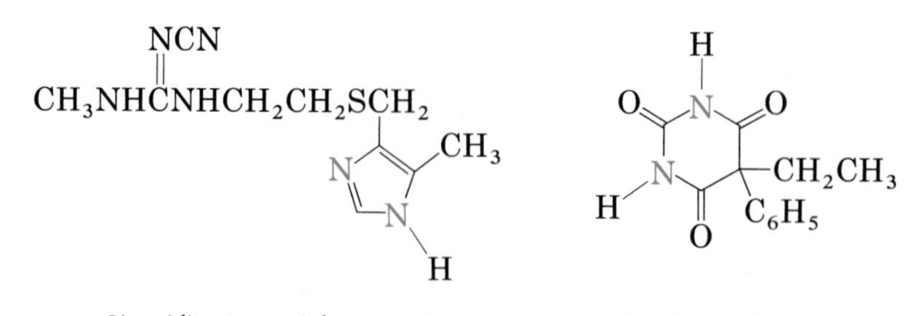

Cimetidine (an antiulcer agent) Phenobarbital (a sedative)

For the most part, heterocyclic amines have the same chemistry as their open-chain counterparts. In certain cases, though, particularly when the ring is unsaturated, heterocycles have unique and interesting properties. Let's look at several examples.

Pyrrole, A Five-Membered Aromatic Heterocycle

Pyrrole, a five-membered heterocyclic amine, has two double bonds and one nitrogen. Though pyrrole is both an amine and a conjugated diene, its chemistry isn't consistent with either of these structural features. Unlike most amines, pyrrole isn't basic; unlike most conjugated dienes, pyrrole doesn't undergo electrophilic addition reactions. How can we explain these observations?

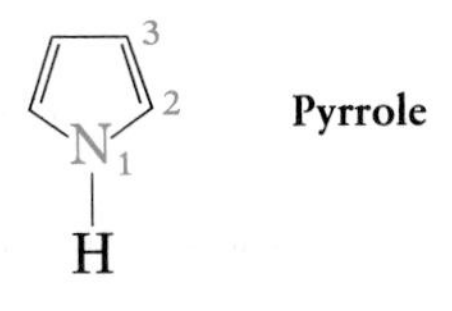

In fact, pyrrole is aromatic. Even though pyrrole has a five-membered ring, it has six π electrons in a cyclic conjugated π-orbital system just as benzene does (Section 5.4). Each of the four carbon atoms contributes one π electron, and the sp^2-hybridized nitrogen atom contributes two more (its lone pair). The six π electrons occupy *p*-orbitals with lobes above and below the plane of the flat ring, as shown in Figure 12.5.

N—H

Pyrrole

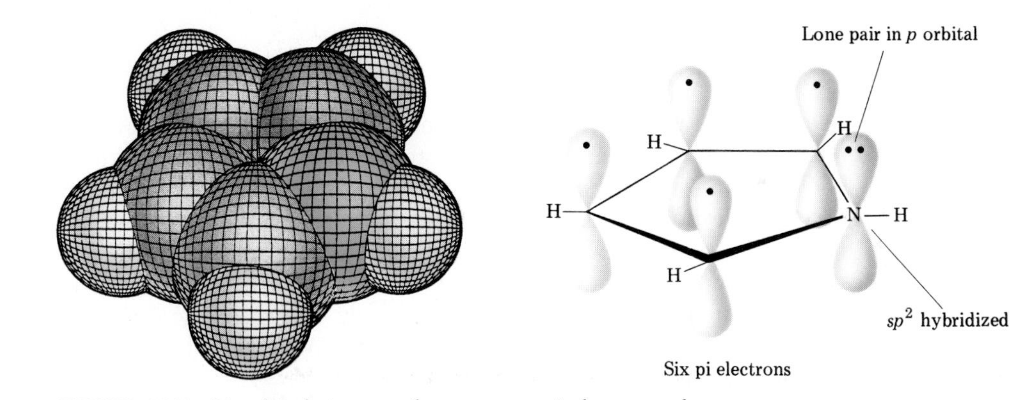

FIGURE 12.5 Pi orbitals in pyrrole, an aromatic heterocycle

Like benzene, pyrrole undergoes substitution of a ring hydrogen atom on reaction with electrophiles. Substitution normally occurs at the position next to nitrogen, as the following nitration shows.

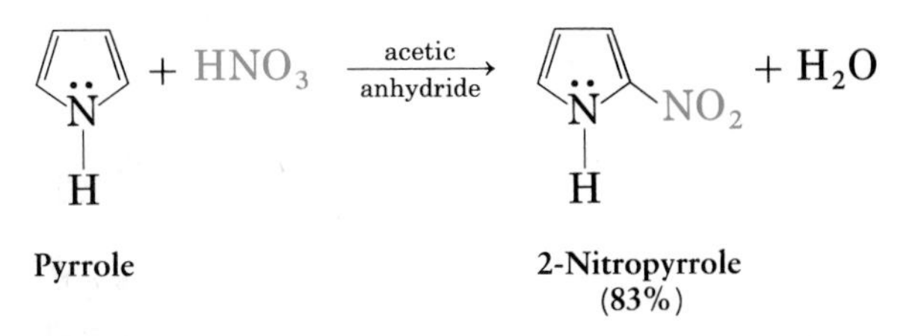

Substituted pyrrole rings form the basic building blocks from which a number of important plant and animal pigments are constructed. Among these is *heme*, an iron-containing tetrapyrrole in blood.

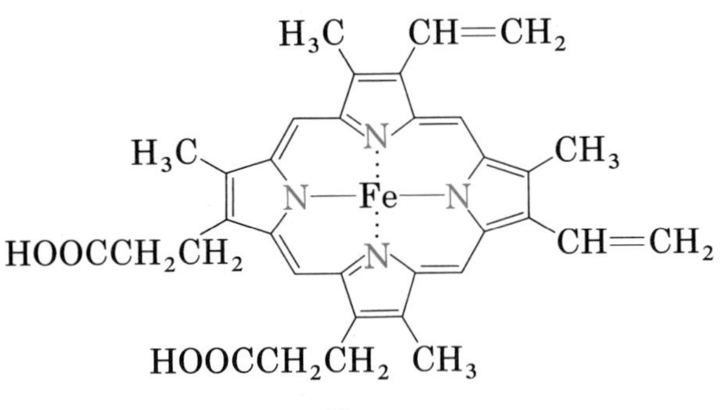

Heme

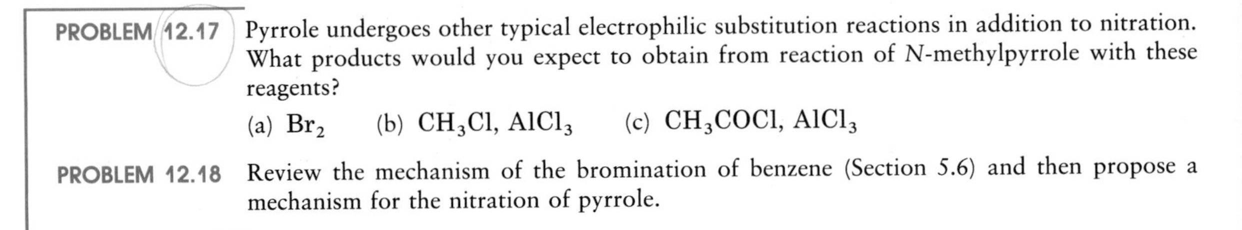

PROBLEM 12.17 Pyrrole undergoes other typical electrophilic substitution reactions in addition to nitration. What products would you expect to obtain from reaction of *N*-methylpyrrole with these reagents?

(a) Br_2 (b) CH_3Cl, $AlCl_3$ (c) CH_3COCl, $AlCl_3$

PROBLEM 12.18 Review the mechanism of the bromination of benzene (Section 5.6) and then propose a mechanism for the nitration of pyrrole.

Pyridine, A Six-Membered Aromatic Heterocycle

Pyridine is the nitrogen-containing heterocyclic analog of benzene. Like benzene, pyridine is a flat molecule with bond angles of approximately 120° and with carbon–carbon bond lengths of 1.39 Å, intermediate between normal single and double bonds. Also like benzene, pyridine is aromatic with six π electrons in a cyclic, conjugated π-orbital system. The sp^2-hybridized nitrogen atom and the five carbon atoms contribute one π electron each to the cyclic conjugated π-orbitals of the ring. Unlike the situation in pyrrole, however, the lone pair of electrons on the pyridine nitrogen atom isn't part of the π-orbital system but occupies an sp^2 orbital in the plane of the ring (Figure 12.6).

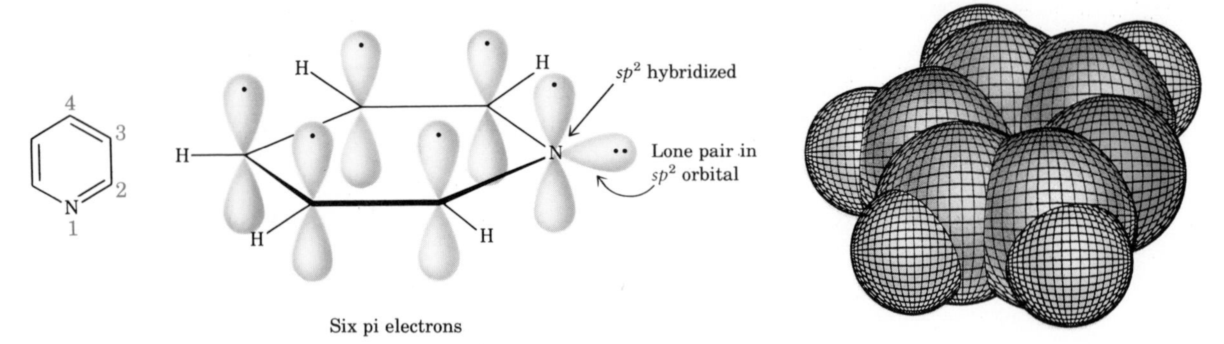

FIGURE 12.6 Electronic structure of pyridine

Substituted pyridines such as the B_6-complex vitamins pyridoxal and pyridoxine are important biologically. Present in yeast, cereal, and other foodstuffs, the B_6 vitamins play an important role in the synthesis of some amino acids.

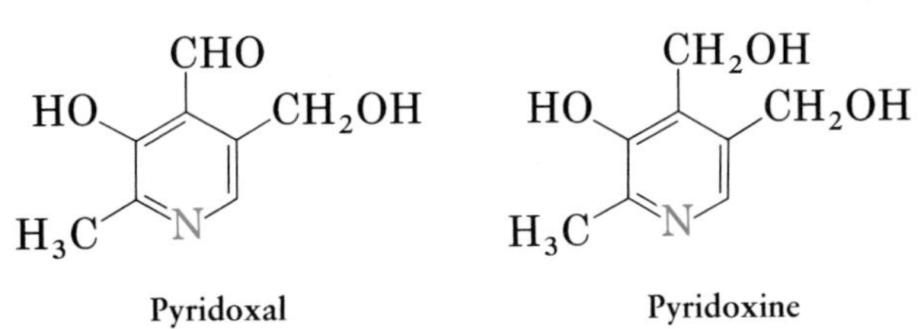

Pyridoxal Pyridoxine

PROBLEM 12.19 The five-membered heterocycle imidazole contains two nitrogen atoms, one "pyrrole-like" and one "pyridine-like." Draw an orbital picture of imidazole and indicate in which orbital each nitrogen has its lone-pair electrons.

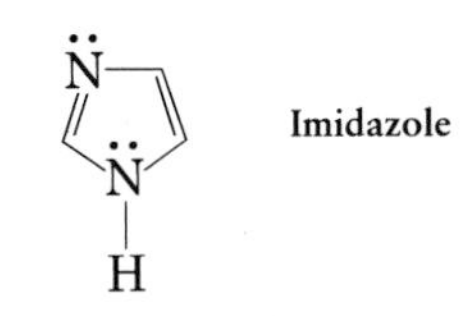

Imidazole

Fused-Ring Aromatic Heterocycles

fused-ring heterocycle a molecule that contains a heterocyclic ring joined to another ring

Quinoline, isoquinoline, and indole are **fused-ring heterocycles** that contain both a benzene ring and a heterocyclic aromatic ring. All three ring systems occur widely in nature, and many members of the class have useful biological activity. Thus, quinine, a quinoline derivative found in the bark of the South American cinchona tree, is an important antimalarial drug. Lysergic acid, an indole derivative found in the ergot fungus that grows on rotting grain, is the parent acid from which the psychoactive drug LSD is derived.

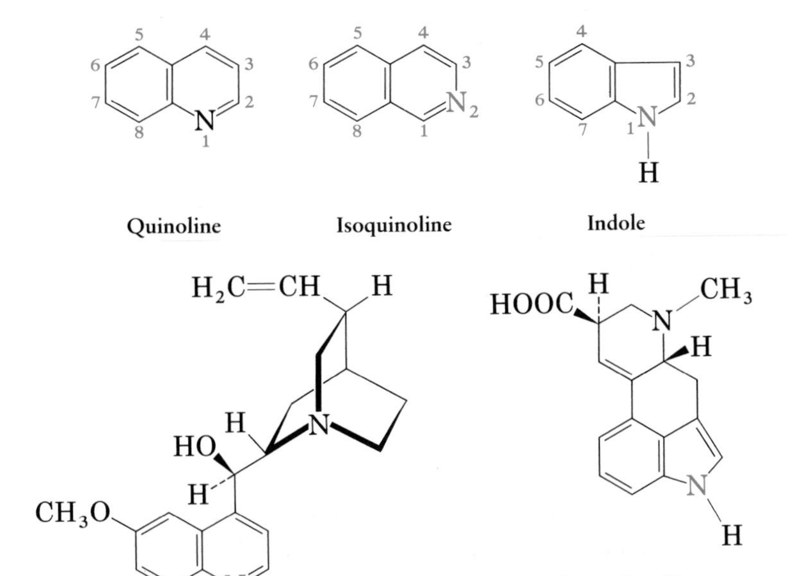

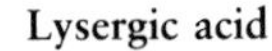
Lysergic acid

12.8 NATURALLY OCCURRING AMINES: MORPHINE ALKALOIDS

Amines were among the first organic compounds to be isolated in pure form, and an enormous variety of amines is found in both plants and animals. Morphine, for example, is a powerful analgesic agent (pain-killer) that is isolated from the opium poppy, *Papaver somniferum*. Once known as "vegetable alkali" because their water solutions are basic, naturally occurring amines such as morphine are now referred to as **alkaloids.**

alkaloid
a naturally occurring amine

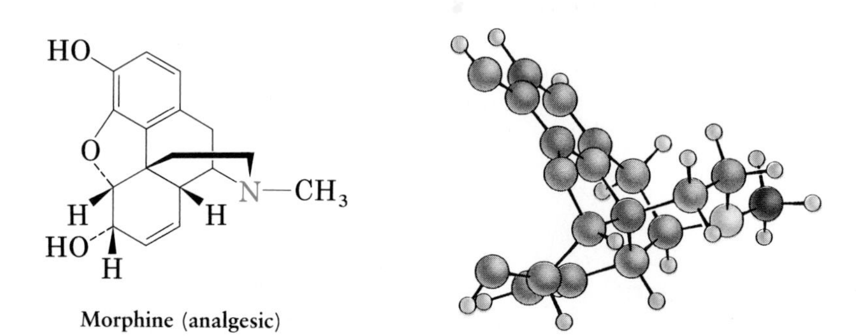

Morphine (analgesic)

The medical uses of the morphine family of alkaloids have been known at least since the seventeenth century, when crude extracts of the opium poppy were used for the relief of pain. Morphine was the first pure alkaloid to be isolated from the poppy, but its close relative, codeine, also occurs naturally. Codeine, which is simply the methyl ether of morphine, is used in prescription cough medicines. Heroin, another close relative of morphine, does not occur naturally but is synthesized by diacetylation of morphine.

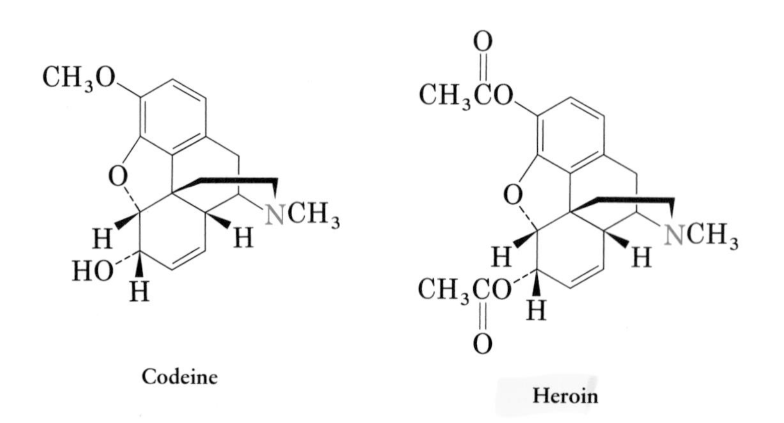

Codeine

Heroin

Morphine alkaloids comprise a class of extremely useful pharmaceutical agents, yet they also pose an enormous societal problem because of their addictive properties. Much effort has gone into understanding how morphine works and into developing modified morphine analogs that retain the desired pain-killing activity but don't cause addiction. Our present understanding is that morphine appears to bind to opiate receptor sites in the brain. It doesn't interfere with or lessen the transmission of a pain signal to the brain but changes the brain's reception of the signal. Hundreds of morphine-like molecules have been synthesized and tested for their analgesic properties.

Studies have shown that not all of the complex framework of morphine is necessary for biological activity. According to the "morphine rule," biological activity requires: (1) an aromatic ring attached to (2) a quaternary carbon atom, and (3) a tertiary amine situated (4) two carbon atoms farther away. For example, meperidine (Demerol) is widely used as a pain-killer, and methadone has been used as an antagonist in the treatment of heroin addiction to reverse the undesirable side effects of morphine.

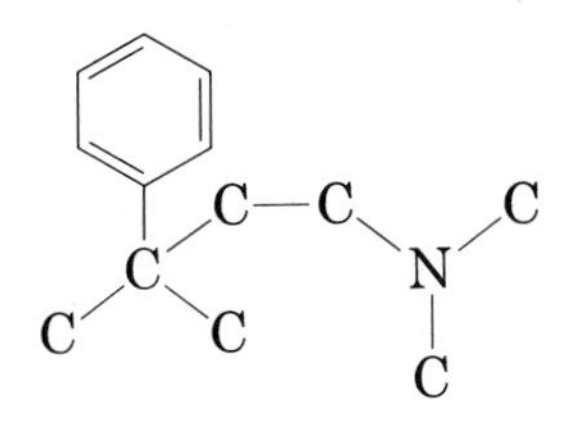

The morphine rule:
an aromatic ring, attached to a quaternary carbon, attached to two more carbons, attached to a tertiary amine.

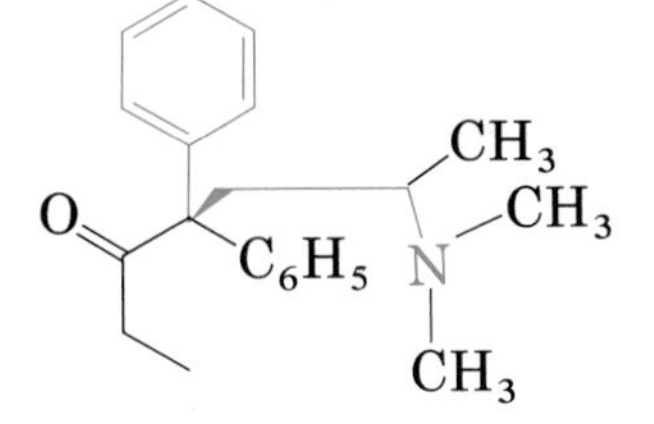

Methadone

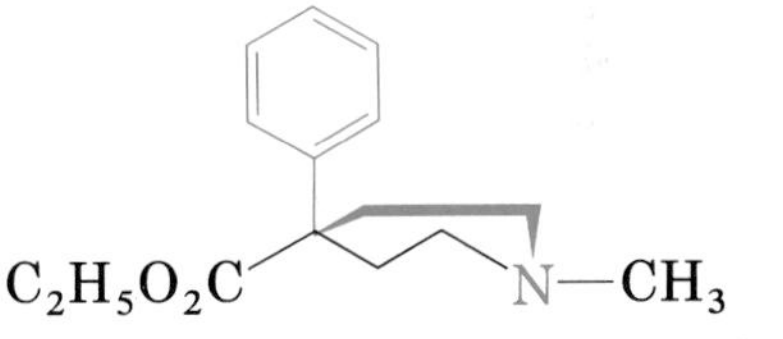

Meperidine

PROBLEM 12.20 Show how the morphine rule fits the structure of dextromethorphan, a common cough remedy.

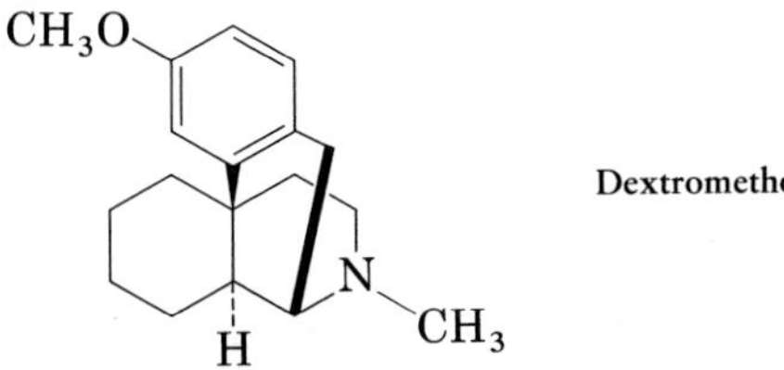

Dextromethorphan

INTERLUDE

Organic Dyes and the Chemical Industry

The founding of the modern organic chemical industry can be traced to the need for a single organic compound, aniline, and to the activities of one person, Sir William Henry Perkin. Perkin, a student at the Royal College of Chemistry in London, spent his free time working in an improvised home laboratory, and decided during Easter vacation in 1856 to examine the oxidation of aniline with potassium dichromate. Although the reaction appeared unpromising at first, yielding a tarry black product, Perkin was able by careful extraction with methanol to isolate a few percent yield of a beautiful purple pigment with the properties of a dye.

Since the only dyes known at the time were naturally occurring vegetable dyes like indigo, Perkin's synthetic purple dye, which he named *mauve,* created a sensation. Realizing the commercial possibilities, Perkin did what any young entrepreneur would do today: He resigned his post at the Royal College and, at the age of 18, formed a company to exploit his remarkable discovery. Since there had never before been a need for synthetic chemicals, no chemical industry existed at the time. Large-scale chemical manufacture was unknown, and Perkin therefore devised a procedure for preparing the needed quantities of aniline by nitration of benzene.

Subsequent work showed that Perkin's original mauve was in fact not derived from aniline at all but from a small amount of methylaniline impurity in his starting material. Pure aniline yields a similar dye, however, which came to be marketed under the name *pseudomauveine.*

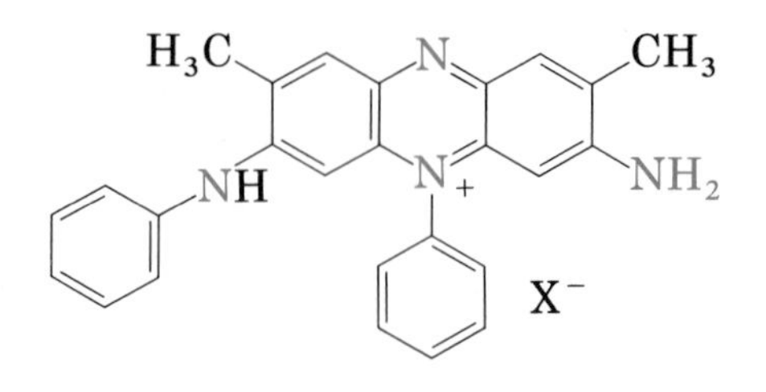

Perkin's mauve
(pseudomauveine has no methyl groups)

Today, dyestuff manufacture is a thriving and important part of the chemical industry, and many pigments such as *p*-(dimethylamino)azobenzene, used at one time as a yellow food-coloring agent under the name "butter yellow," are derived from aniline.

$$C_6H_5-N{=}N-C_6H_4-\ddot{N}(CH_3)_2$$

p-(Dimethylamino)azobenzene (yellow crystals, mp 127°C)

SUMMARY AND KEY WORDS

Amines are organo-substituted derivatives of ammonia. They are named in the IUPAC system either by adding the suffix *-amine* to the names of the alkyl substituents or by considering the amino group as a substituent on a more complex parent molecule.

Bonding in amines is similar to that in ammonia. The nitrogen atom is sp^3 hybridized; the three substituents are directed to three corners of a tetrahedron; and the lone pair of nonbonding electrons occupies the fourth corner of the tetrahedron.

The chemistry of amines is dominated by the presence of the lone-pair electrons on nitrogen. Thus, amines are both basic and nucleophilic. **Arylamines** are generally weaker bases than **alkylamines** because their lone-pair electrons are delocalized by orbital overlap with the aromatic pi-electron system.

The simplest method of amine synthesis involves S_N2 reaction of ammonia or an amine with an alkyl halide. Alkylation of ammonia yields a primary amine; alkylation of a primary amine yields a secondary amine; and so on. Amines can also be prepared from amides and nitriles by reduction with $LiAlH_4$. Arylamines are prepared by nitration of an aromatic ring, followed by reduction of the nitro group.

Many of the reactions that amines undergo are familiar from previous chapters. Thus, amines react with alkyl halides in S_N2 reactions and with acid chlorides in nucleophilic acyl substitution reactions. The most important reaction of arylamines is their conversion by nitrous acid into **aryl diazonium salts, $Ar{-}N_2^+\ X^-$**. The diazonio group can then be replaced by many other substituents (the **Sandmeyer reaction**) to give a variety of substituted aromatic compounds. Aryl chlorides, bromides, iodides, and nitriles can be prepared, as can phenols and arenes.

Heterocyclic amines, compounds in which the nitrogen atom is in a ring, have a great diversity in their structures and properties. For example, pyrrole, pyridine, indole, and quinoline all show aromatic properties.

SUMMARY OF REACTIONS

1. Synthesis of amines (Section 12.5)

   (a) Alkylamines by S_N2 reaction

$$NH_3 + RX \longrightarrow RNH_2$$
$$RNH_2 + RX \longrightarrow R_2NH$$

$$R_2NH + RX \longrightarrow R_3N$$
$$R_3N + RX \longrightarrow R_4N^+X^-$$

(b) Arylamines by reduction of nitroarenes

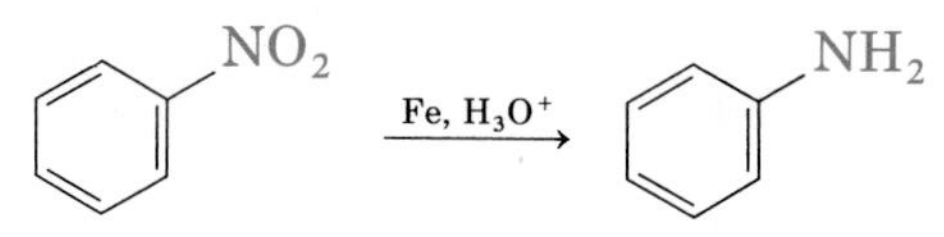

2. Reactions of amines (Section 12.6)
Formation and reactions of arenediazonium salts

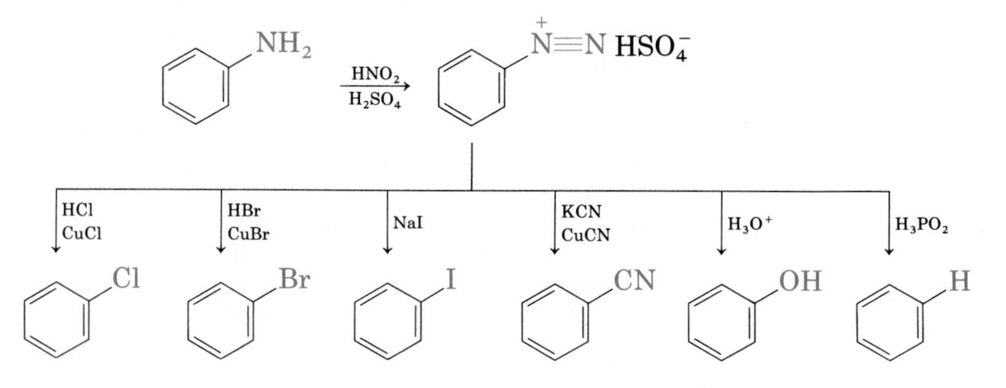

ADDITIONAL PROBLEMS

12.21 Classify each of the amine nitrogen atoms in the following substances as either primary, secondary, or tertiary.

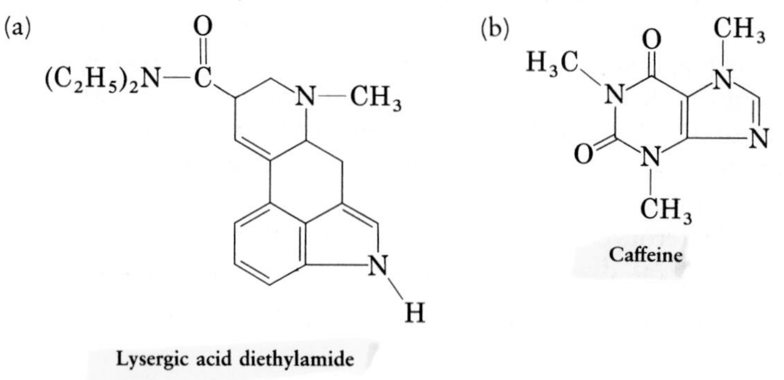

Lysergic acid diethylamide

Caffeine

12.22 Draw structures corresponding to these IUPAC names.

(a) *N*,*N*-Dimethylaniline
(b) *N*-Methylcyclohexylamine
(c) (Cyclohexylmethyl)amine
(d) (2-Methylcyclohexyl)amine
(e) 3-(*N*,*N*-Dimethylamino)propanoic acid

12.23 Name these compounds according to IUPAC rules.

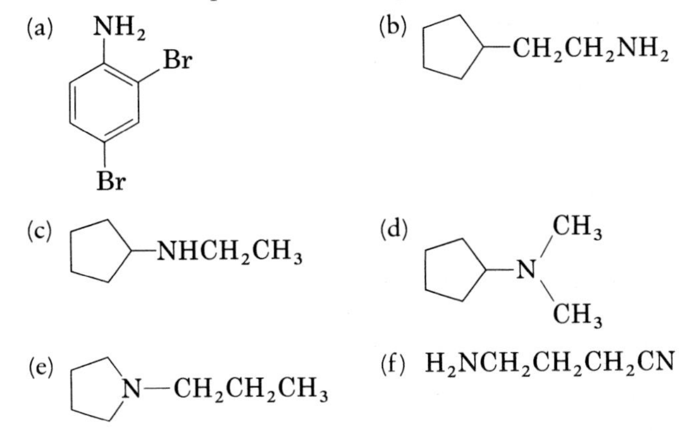

(f) $H_2NCH_2CH_2CH_2CN$

12.24 How can you explain the fact that trimethylamine (bp 3°C) has a lower boiling point than dimethylamine (bp 7°C)?

12.25 There are eight isomeric amines with the formula $C_4H_{11}N$. Draw them, name them, and classify each as primary, secondary, or tertiary.

12.26 Propose structures for amines that fit these descriptions.

(a) A secondary arylamine
(b) A 1,3,5-trisubstituted arylamine
(c) An achiral quaternary ammonium salt
(d) A five-membered heterocyclic amine

12.27 Show the products of these reactions.

(a) $CH_3CH_2CH_2NH_2 + CH_3Br \longrightarrow$
(b) Cyclohexylamine + HBr $\longrightarrow$
(c) $CH_3CH_2CONH_2 + LiAlH_4 \longrightarrow$
(d) Benzonitrile + $LiAlH_4 \longrightarrow$

12.28 How might you prepare these amines, starting from ammonia and any alkyl halides needed?

(a) Hexylamine
(b) Benzylamine
(c) Tetramethylammonium iodide
(d) *N*-Methylcyclohexylamine

12.29 How might you prepare each of these amines from 1-bromobutane?

(a) Butylamine (b) Dibutylamine (c) Pentylamine

12.30 How might you prepare each of the amines listed in Problem 12.29 from 1-butanol?

12.31 How would you prepare benzylamine, $C_6H_5CH_2NH_2$, from each of these starting materials?

(a) Benzamide (b) Benzoic acid (c) Nitrobenzene (d) Chlorobenzene

12.32 Write equations for the reaction of *p*-bromobenzenediazonium bisulfate with these reagents.

(a) H_3O^+ (b) HBr, CuBr (c) H_3PO_2 (d) KCN, CuCN

12.33 Show how you might prepare benzoic acid from aniline. A diazonio replacement reaction is needed.

12.34 How might you prepare pentylamine from these starting materials?

(a) Pentanamide (b) Pentanenitrile (c) Pentanoic acid

12.35 Which compound is more basic, $CH_3CH_2NH_2$ or $CF_3CH_2NH_2$? Explain.

12.36 Which compound is more basic, *p*-aminobenzaldehyde or aniline?

12.37 1,6-Hexanediamine, one of the starting materials used for the manufacture of nylon-6,6, can be synthesized by a route that begins with the addition of chlorine to 1,3-butadiene (Section 4.10). How would you carry out the complete synthesis?

12.38 Another method for making 1,6-hexanediamine (Problem 12.37) starts from adipic acid (hexanedioic acid). How would you carry out the synthesis?

12.39 Give the structures of the major organic products you would obtain from reaction of *m*-methylaniline with these reagents.

(a) Br_2 (1 mol) (b) CH_3I (excess) (c) CH_3COCl, pyridine

12.40 Draw structures for these amines.

(a) 2-Ethylpyrrole (b) 2,3-Dimethylaniline (c) 3-Methylindole

12.41 Furan, the ether analog of pyrrole, is aromatic in the same way that pyrrole is. Draw an orbital picture of furan and show how it has six electrons in its cyclic conjugated π orbitals.

12.42 By analogy with the chemistry of pyrrole, what product would you expect from reaction of furan (Problem 12.41) with Br_2?

12.43 How would you synthesize 1,3,5-tribromobenzene from benzene? A diazonio replacement reaction is needed.

12.44 We've seen that amines are basic and amides are neutral. *Imides,* compounds with two carbonyl groups flanking an N–H, are actually acidic. Show by drawing resonance structures of the anion why imides are acidic.

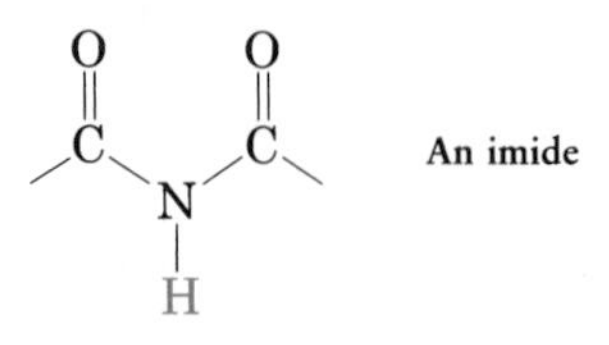

An imide

12.45 Tyramine is an alkaloid found, among other places, in mistletoe and in ripe cheese. How would you prepare tyramine from toluene?

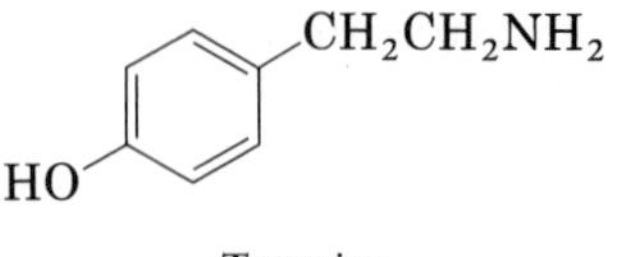

Tyramine

12.46 Atropine, $C_{17}H_{23}NO_3$, is a poisonous alkaloid isolated from the leaves and roots of *Atropa belladonna,* the deadly nightshade. In low doses, atropine acts as a muscle relaxant: 0.5 ng (nanogram, 10^{-9} g) is sufficient to cause pupil dilation. On reaction with aqueous NaOH, atropine yields tropic acid, $C_6H_5CH(CH_2OH)COOH$), and tropine, $C_8H_{15}NO$. Tropine, an optically inactive alcohol, yields tropidene on dehydration. Propose a suitable structure for atropine.

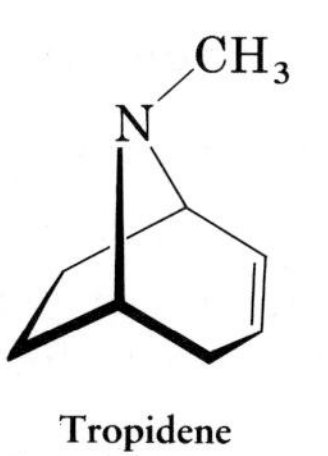

Tropidene

12.47 Choline, a component of the phospholipids in cell membranes, can be prepared by reaction of trimethylamine with ethylene oxide. Propose a mechanism for the reaction.

$$(CH_3)_3N + H_2C\overset{O}{\frown}CH_2 \xrightarrow{H_2O} (CH_3)_3\overset{+}{N}CH_2CH_2OH$$

Choline

CHAPTER 13

Structure Determination

Structure determination is central to organic chemistry. Every time a reaction is run, the products have to be isolated, purified, and identified. In the nineteenth and early twentieth centuries, determining the structure of an organic molecule was a time-consuming process requiring great skill and patience. In the past few decades, though, extraordinary advances have been made in chemical instrumentation. Sophisticated instruments are now available that greatly simplify structure determination.

What are the instruments for determining structures, and how are they used? We'll answer these questions by looking at three of the most useful methods of structure determination: infrared spectroscopy, ultraviolet spectroscopy, and nuclear magnetic resonance spectroscopy, each of which yields a different kind of structural information.

Infrared spectroscopy	What functional groups are present?
Ultraviolet spectroscopy	Is a conjugated pi-electron system present?
Nuclear magnetic resonance spectroscopy	What carbon–hydrogen framework is present?

13.1 INFRARED SPECTROSCOPY AND THE ELECTROMAGNETIC SPECTRUM

electromagnetic radiation different wavelengths of radiant energy that make up the electromagnetic spectrum

electromagnetic spectrum the total range of electromagnetic radiation

Infrared spectroscopy (IR) is a method of structure determination that involves the interaction of molecules with infrared radiant energy. Before beginning a study of infrared spectroscopy, though, we need to look into the nature of radiant energy and the electromagnetic spectrum.

Visible light, X-rays, microwaves, radio waves, and so forth are all different kinds of **electromagnetic radiation.** Collectively, they make up the **electromagnetic spectrum,** shown in Figure 13.1. As indicated, the electromagnetic spectrum is loosely divided into regions, with the familiar visible region accounting for only a small portion of the overall spectrum (from 3.8×10^{-5} to 7.8×10^{-5} cm in wavelength). The visible region is flanked by the infrared and the ultraviolet regions.

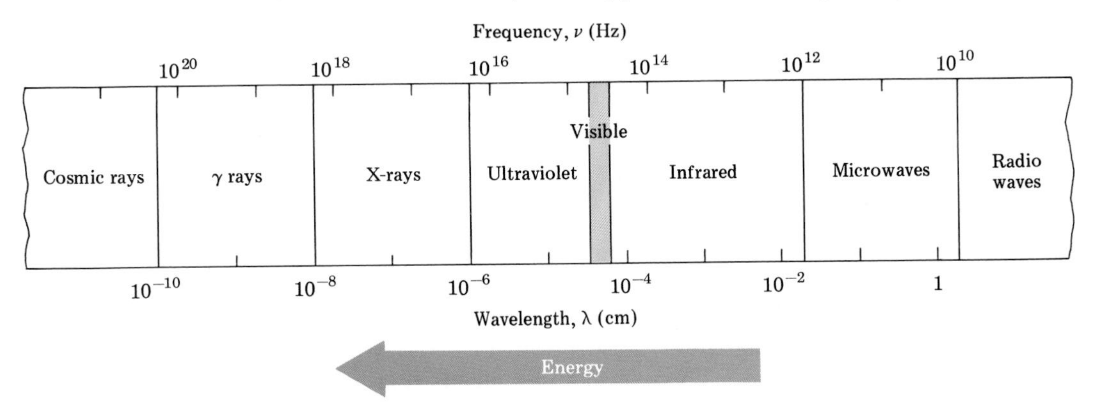

FIGURE 13.1 The electromagnetic spectrum

hertz (Hz)
a unit of wavelength measurement; cycles per second

Electromagnetic radiation can be thought of as having dual behavior. In some respects it has the properties of a particle (called a *photon*), yet in other respects it behaves as a wave traveling at the speed of light. Electromagnetic waves can be described by their wavelength (λ; Greek lambda) and frequency (ν; Greek nu). The wavelength is simply the length of one complete wave cycle from trough to trough; the frequency is the number of wave cycles that travel past a fixed point in a certain period of time (usually given in cycles per second, or **hertz,** Hz).

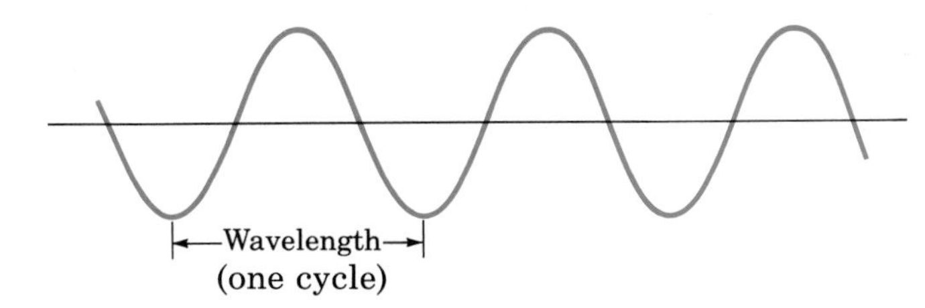

Electromagnetic energy is transmitted only in discrete energy bundles, called *quanta*. The amount of energy corresponding to 1 quantum of energy (or 1 photon) of a given frequency ν is expressed by the equation:

$$\varepsilon = h\nu = \frac{hc}{\lambda}$$

where

ε = energy of 1 photon (1 quantum)
h = Planck's constant (6.62×10^{-27} erg sec)
ν = frequency in hertz (cycles per second)
λ = wavelength in centimeters
c = speed of light (3×10^{10} cm/sec)

This equation says that the energy of a given photon varies *directly* with its frequency ν but *inversely* with its wavelength λ. High frequencies and short wavelengths correspond to high-energy radiation such as gamma rays; low frequencies and long wavelengths correspond to low-energy radiation such as radio waves.

When an organic compound is exposed to electromagnetic radiation, it absorbs energy of certain wavelengths and transmits energy of other wavelengths. Thus, if we irradiate an organic compound with energy of many wavelengths and determine which are absorbed and which are transmitted, we can determine the **absorption spectrum** of the compound. The results are displayed on a graph that plots the wavelength versus the amount of radiation transmitted through the sample.

absorption spectrum
a plot of wavelength versus the absorption that results when a compound is irradiated with electromagnetic radiation

The IR spectrum of ethanol is shown in Figure 13.2. The horizontal axis shows the wavelength in micrometers (μm), and the vertical axis shows the intensity of the corresponding energy absorption in % transmittance. The baseline corresponding to 0% absorption (or 100% transmittance) runs along the top of the chart, and a downward spike means that energy absorption has occurred at that wavelength.

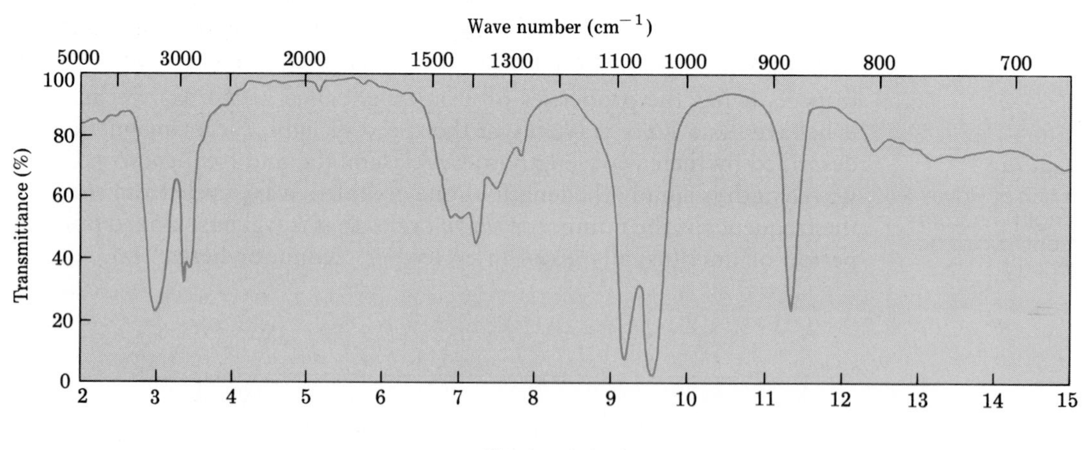

FIGURE 13.2 Infrared spectrum of ethanol. A transmittance of 100% means that all of the energy is passing through the sample. A lower transmittance means that some energy is being absorbed. Thus, each downward spike corresponds to an energy absorption.

PRACTICE PROBLEM 13.1

Which is higher in energy, FM radio waves with a frequency of 1.015×10^8 Hz (101.5 MHz) or visible light with a frequency of 5×10^{14} Hz?

SOLUTION The equation $\varepsilon = h\nu$ says that energy increases as frequency increases. Thus, visible light is higher in energy than radio waves.

PRACTICE PROBLEM 13.2

What is the wavelength of visible light with a frequency of 4.5×10^{14} Hz?

SOLUTION Frequency and wavelength are related by the equation $\lambda = c/\nu$, where c is the speed of light (3.0×10^{10} cm/sec).

$$\lambda = \frac{3.0 \times 10^{10} \text{ cm/sec}}{4.5 \times 10^{14} \text{ cycles/sec}} = 6.7 \times 10^{-5} \text{ cm/cycle}$$

PROBLEM 13.1 How does the energy of infrared radiation with $\lambda = 10^{-4}$ cm compare with that of an X-ray with $\lambda = 3 \times 10^{-7}$ cm?

PROBLEM 13.2 Which is higher in energy, radiation with $\nu = 4 \times 10^9$ Hz or radiation with $\lambda = 9 \times 10^{-4}$ cm?

13.2 INFRARED SPECTROSCOPY OF ORGANIC MOLECULES

wave number a unit of frequency measurement equal to the reciprocal of the wavelength measured in centimeters ($\tilde{\nu} = 1/\lambda$)

The infrared region of the electromagnetic spectrum covers the range from just above the visible (7.8×10^{-5} cm) to approximately 10^{-2} cm, but only the middle of the region is used by organic chemists (Figure 13.3). This midportion extends from 2.5×10^{-3} to 2.5×10^{-4} cm, and wavelengths are usually given in micrometers (1 μm = 10^{-6} m). Frequencies are usually given in **wave numbers** ($\tilde{\nu}$), rather than in hertz; the wave number is $1/\lambda$ and is expressed in units of cm^{-1}.

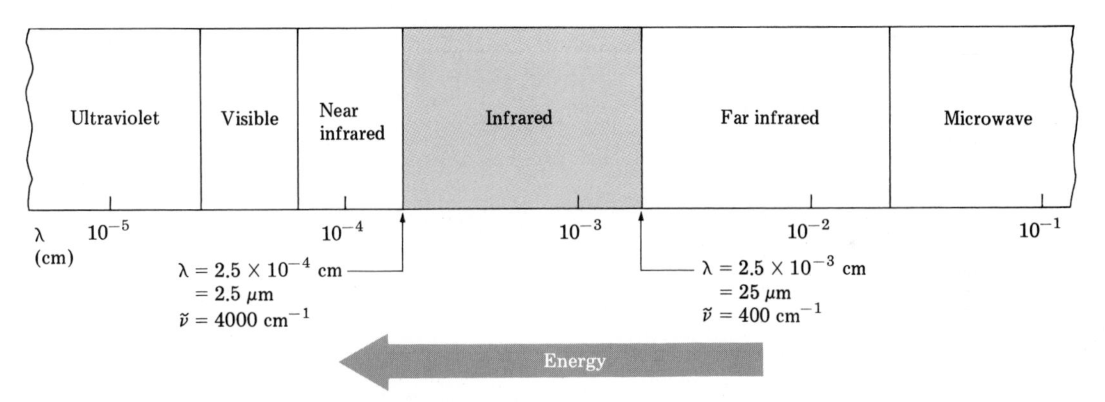

FIGURE 13.3 The infrared region of the electromagnetic spectrum

What causes a molecule to absorb certain wavelengths of infrared light but not others? All molecules contain energy—energy that causes bonds to stretch and bend, atoms to wag and rock, and other molecular vibrations to occur. The amount of energy a molecule contains, though, is not continuously variable but is *quantized*. That is, a molecule can stretch or bend only at specific frequencies that correspond to specific energy levels. When a molecule is irradiated with infrared light, the only wavelengths absorbed are those that correspond to the quantized stretching and bending energy levels in the molecule.

Since each infrared absorption corresponds to a specific molecular motion, we can see what kinds of motions a molecule has by observing its infrared spectrum. By then working backward and interpreting the molecule's motions, we can find out what kinds of functional groups are present in the molecule.

IR spectrum ⟶ What molecular motions? ⟶ What functional groups?

The full interpretation of an IR spectrum is difficult because most organic molecules have dozens of different bond stretchings, rotations, and bendings. Thus, an IR spectrum usually contains dozens of absorptions. Fortunately, we don't have to interpret an IR spectrum fully to get useful information because *most functional groups have characteristic infrared absorptions, which don't change from one compound to another*. The C=O absorption of ketones is almost always in the range 1690–1750 cm^{-1}; the O–H absorption of alcohols is almost always in the range 3200–3600 cm^{-1}; and the C=C absorptions of alkenes is almost always in the range 1640–1680 cm^{-1}. By learning to recognize these characteristic functional-group absorptions, it's possible to interpret infrared spectra.

Look at Figure 13.4 to see how IR spectra can be used. Although the IR spectra of 1-hexanol and 2-hexanone both contain many peaks, the characteristic absorptions of the various functional groups allow the compounds to be distinguished. 1-Hexanol shows a characteristic alcohol O–H absorption at 3300 cm^{-1} and a C–O absorption at 1060 cm^{-1}; 2-hexanone shows a characteristic ketone peak at 1710 cm^{-1}.

One other point about infrared spectroscopy: It's also possible to derive structural information from an IR spectrum by noticing which absorptions are *not*

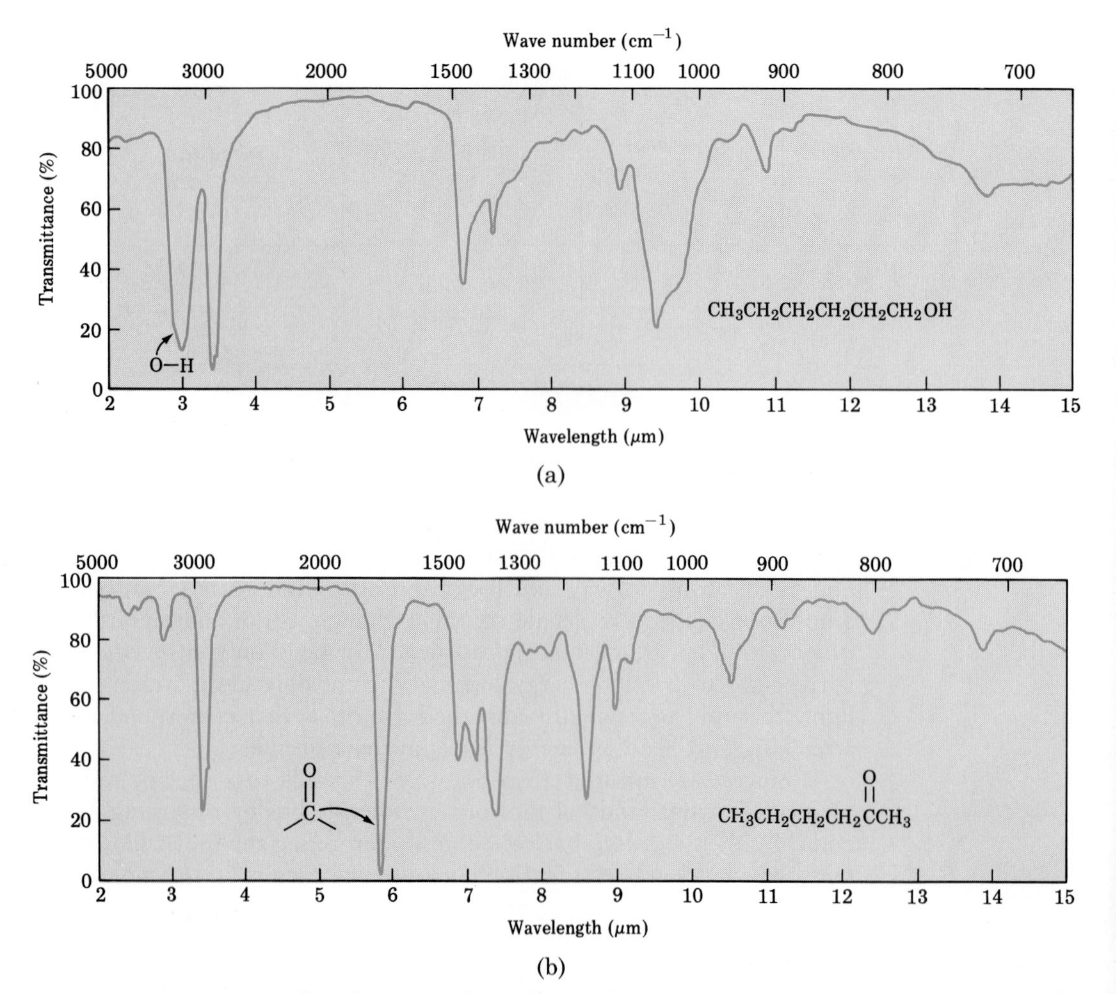

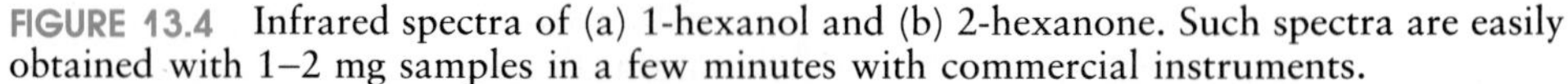
FIGURE 13.4 Infrared spectra of (a) 1-hexanol and (b) 2-hexanone. Such spectra are easily obtained with 1–2 mg samples in a few minutes with commercial instruments.

present. If the spectrum of an unknown has no absorption near 3400 cm^{-1}, the unknown isn't an alcohol; if the spectrum has no absorption near 1710 cm^{-1}, the unknown isn't a ketone; and so on. Table 13.1 lists the characteristic IR absorption frequencies of some of the most common functional groups.

TABLE 13.1 Characteristic infrared absorptions of some functional groups

Functional group class	*Band position (cm^{-1})*	*Intensity of absorption*
Alkanes, alkyl groups		
C—H	2850–2960	Medium to strong
Alkenes		
=C—H	3020–3100	Medium
C=C	1650–1670	Medium
Alkynes		
≡C—H	3300	Strong
—C≡C—	2100–2260	Medium
Alkyl halides		
C—Cl	600–800	Strong
C—Br	500–600	Strong
C—I	500	Strong
Alcohols		
O—H	3200–3600	Strong, broad
C—O	1050–1150	Strong
Aromatics		
(aromatic) C—H	3030	Medium
(benzene ring C=C)	1600, 1500	Strong
Amines		
N—H	3310–3500	Medium
C—N	1030, 1230	Medium
Carbonyl compounds[a]		
C=O	1670–1780	Strong
Carboxylic acids		
O—H	2500–3100	Strong, very broad
Nitriles		
C≡N	2210–2260	Medium
Nitro compounds		
NO_2	1540	Strong

[a] Acids, esters, aldehydes, and ketones.

To remember the positions of various IR absorptions it helps to divide the infrared range from 4000 to 200 cm^{-1} into four regions (Figure 13.5).

1. The region from 4000 to 2500 cm^{-1} corresponds to N–H, C–H, and O–H bond stretching and contracting motions. Both N–H and O–H bonds absorb in the 3300–3600 cm^{-1} range, whereas C–H bond stretching occurs near 3000 cm^{-1}. Since almost all organic compounds have C–H bonds, almost all IR spectra have an intense absorption in this region.
2. The region from 2500 to 2000 cm^{-1} is where triple-bond stretching occurs. Both nitriles (RC≡N) and alkynes (RC≡CR) show peaks here.
3. The region from 2000 to 1500 cm^{-1} contains C=O, C=N, and C=C double bond absorptions. Carbonyl groups generally absorb from 1670 to 1780 cm^{-1}, and alkene stretching normally occurs in the narrow range from 1640 to 1680 cm^{-1}. The exact position of a C=O absorption is often diagnostic of the exact kind of carbonyl group in the molecule. Esters usually absorb at 1735 cm^{-1}, aldehydes at 1725 cm^{-1}, and open-chain ketones at 1715 cm^{-1}.
4. The region below 1500 cm^{-1} is the so-called *fingerprint region*. A large number of absorptions due to various C–O, C–C, and C–N single bond vibrations occur here, forming a unique pattern that acts as an identifying "fingerprint" of each organic molecule.

Wave number (cm^{-1})

4000–2500	2500–2000	2000–1500	1500–400
C—H	C≡C	C=C	Fingerprint region
O—H	C≡N	C=O	
N—H		C=N	

FIGURE 13.5 Regions in the infrared spectrum

PRACTICE PROBLEM 13.3 Refer to Table 13.1 and make an educated guess about the functional groups that cause these IR absorptions:

(a) 1735 cm^{-1} (b) 3500 cm^{-1}

SOLUTION (a) An absorption at 1735 cm^{-1} is in the carbonyl-group region of the IR spectrum. (b) An absorption at 3500 cm^{-1} is in the –OH (alcohol) region.

PRACTICE PROBLEM 13.4 Acetone and 2-propen-1-ol (H_2C=$CHCH_2OH$) are isomers. How could you distinguish them by IR spectroscopy?

SOLUTION Acetone has a strong ketone carbonyl absorption at 1710 cm^{-1}. 2-Propen-1-ol has a hydroxyl absorption at 3500 cm^{-1} and an alkene absorption at 1660 cm^{-1}.

PROBLEM 13.3 What functional groups might molecules contain if they show IR absorptions at these frequencies?

(a) 1715 cm^{-1} (b) 1540 cm^{-1} (c) 2210 cm^{-1}
(d) 1720 and 2500–3100 cm^{-1} (e) 3500 and 1735 cm^{-1}

PROBLEM 13.4 How might you use IR spectroscopy to help you distinguish between these pairs of isomers?

(a) Ethanol and dimethyl ether (b) Cyclohexane and 1-hexene
(c) Propanoic acid and 3-hydroxypropanal

13.3 ULTRAVIOLET SPECTROSCOPY

The **ultraviolet (UV)** region of the electromagnetic spectrum extends from the low-wavelength end of the visible region (4×10^{-5} cm) to 10^{-6} cm. The portion of greatest interest to organic chemists, though, is the narrow range from 2×10^{-5} cm to 4×10^{-5} cm. Absorptions in this region are measured in nanometers (nm; 1 nm = 10^{-9} m = 10^{-7} cm). Thus, the ultraviolet range of interest is from 200 to 400 nm (Figure 13.6).

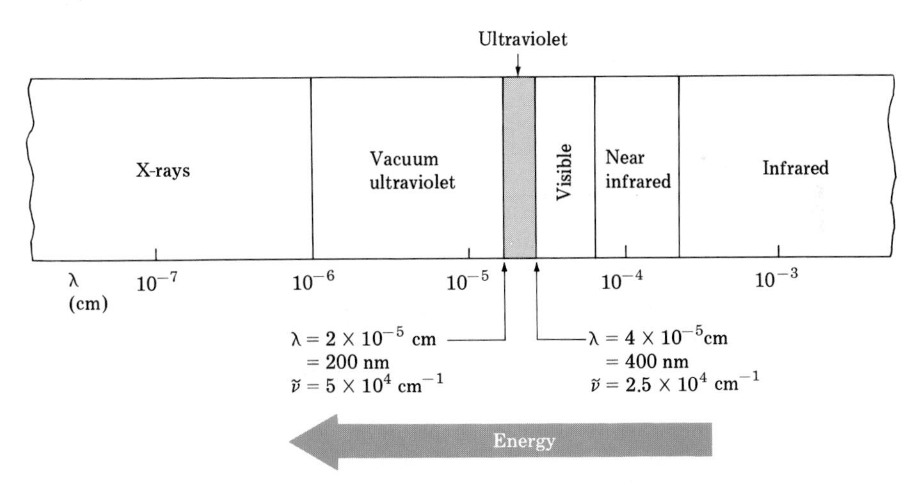

FIGURE 13.6 The ultraviolet (UV) region of the electromagnetic spectrum

We've seen that an organic molecule either absorbs or transmits electromagnetic radiation, depending on the radiation's energy level. With IR radiation, the energies absorbed correspond to the amounts necessary to increase molecular motions of functional groups. With ultraviolet irradiation, though, the energies absorbed correspond to the amounts necessary to raise the energy levels of π electrons.

Ultraviolet spectra are recorded by irradiating a sample with UV light of continuously changing wavelength. When the wavelength of light corresponds to the energy level required to excite an electron to a higher level, energy is absorbed. The absorption is detected and displayed on a chart that plots wavelength versus percentage radiation absorbed.

A typical UV spectrum, that of 1,3-butadiene, is shown in Figure 13.7. Unlike IR spectra, which often show many sharp lines, UV spectra are usually quite simple. Often, there's only a single broad peak, whose position is identified by noting the wavelength at the very top of the peak (λ_{max}). For 1,3-butadiene, $\lambda_{max} = 217$ nm.

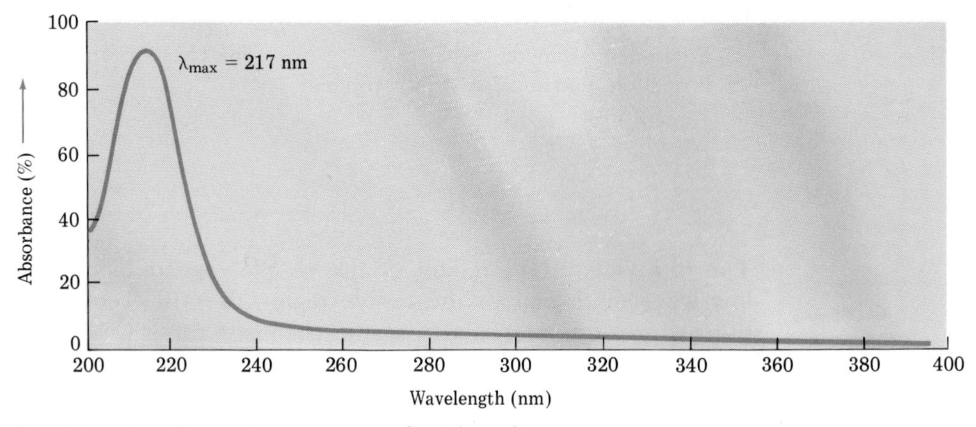

FIGURE 13.7 Ultraviolet spectrum of 1,3-butadiene

13.4 INTERPRETING ULTRAVIOLET SPECTRA: THE EFFECT OF CONJUGATION

The exact wavelength of radiation necessary to cause an electronic excitation in a conjugated molecule depends on the nature of the conjugated system. Working backward, we can obtain information about the nature of the conjugated pi-electron system in a molecule by measuring the molecule's UV spectrum.

One of the most important factors affecting the wavelength of UV absorption is the extent of conjugation. It turns out that the energy required for an electronic transition decreases as the extent of conjugation increases. Thus, 1,3-buta*di*ene shows an absorption at $\lambda_{max} = 217$ nm, 1,3,5-hexa*tri*ene absorbs at $\lambda_{max} = 258$ nm, and 1,3,5,7-octa*tetra*ene has $\lambda_{max} = 290$ nm. (Remember: Longer wavelength means lower energy.)

Other kinds of conjugated pi-electron systems besides dienes and polyenes also show ultraviolet absorptions. Conjugated enones such as 3-butene-2-one and aromatic molecules such as benzene have characteristic UV absorptions that aid in structure determination. The UV absorption maxima of some representative conjugated molecules are given in Table 13.2.

PRACTICE PROBLEM 13.5 1,5-Hexadiene and 1,3-hexadiene are isomers. How can you distinguish them by UV spectroscopy?

SOLUTION 1,3-Hexadiene is a conjugated diene, but 1,5-hexadiene is nonconjugated. Only the conjugated isomer shows a UV absorption above 200 nm.

TABLE 13.2 **Ultraviolet absorption maxima of some conjugated molecules**

Name	*Structure*	λ_{max} *(nm)*
Ethylene	$H_2C{=}CH_2$	171
2-Methyl-1,3-butadiene	$H_2C{=}C(CH_3){-}CH{=}CH_2$	220
1,3-Cyclohexadiene	(ring structure)	256
1,3,5-Hexatriene	$H_2C{=}CH{-}CH{=}CH{-}CH{=}CH_2$	258
3-Buten-2-one	$H_2C{=}CH{-}C(CH_3){=}O$	219
Benzene	(ring structure)	254

PROBLEM 13.5 Which of the following compounds show UV absorptions in the range 200–400 nm?

(a) 1,3-Cyclohexadiene (b) 1,4-Cyclohexadiene
(c) Methyl propenoate (d) *p*-Bromotoluene
(e) 2-Methylcyclohexanone (f) 2-Methyl-2-cyclohexenone

PROBLEM 13.6 How can you distinguish between 1,3-hexadiene and 1,3,5-hexatriene by UV spectroscopy?

13.5 NUCLEAR MAGNETIC RESONANCE SPECTROSCOPY

Of all techniques available for determining structure, **nuclear magnetic resonance (NMR) spectroscopy** is the most valuable. It's the method that organic chemists turn to first for information. We've seen that IR spectroscopy provides information about a molecule's functional groups and that UV spectroscopy provides information about a molecule's conjugated pi-electron system. NMR spectroscopy doesn't replace or duplicate either of these techniques; rather, it complements them by providing a "map" of the carbon–hydrogen framework of an organic molecule. Taken together, IR, UV, and NMR spectroscopies often make it possible to find the structures of extremely complex unknowns.

How does NMR spectroscopy work? Many kinds of nuclei, including ^{1}H and ^{13}C, behave as if they were a child's top spinning about an axis. Since they're positively charged, these spinning nuclei act like tiny magnets and interact with an external magnetic field (denoted H_0). In the absence of a strong external magnetic field, the nuclear spins of magnetic nuclei are oriented randomly. When a sample containing these nuclei is placed between the poles of a strong magnet, however, the nuclei adopt specific orientations, much as a compass needle orients itself in the earth's magnetic field.

A spinning ^1H or ^{13}C nucleus can orient so that its own tiny magnetic field is aligned either with (parallel to) or against (antiparallel to) the external field. The two orientations don't have the same energy and therefore aren't present in equal amounts in the sample. The parallel orientation is slightly lower in energy, making this spin state slightly favored over the antiparallel orientation (Figure 13.8).

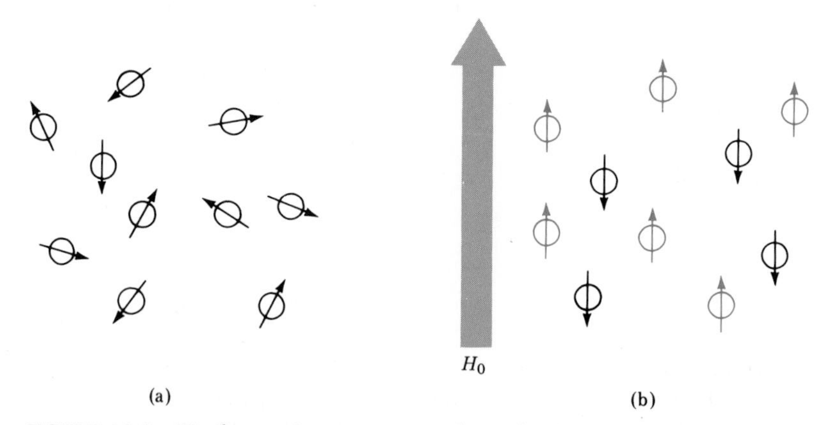

FIGURE 13.8 Nuclear spins are oriented randomly in the absence of a strong external magnetic field (a), but they have a specific orientation in the presence of an external magnetic field H_0 (b). Note that some of the spins (color) are aligned parallel to the external field, and others are antiparallel. The parallel spin state is lower in energy.

If such oriented nuclei are irradiated with electromagnetic radiation of the proper frequency, energy absorption occurs and the lower energy state "spin-flips" to the higher energy state. When this spin-flip occurs, the nuclei are said to be in resonance with the applied radiation, hence the name *nuclear magnetic resonance.*

The exact amount of radio-frequency (rf) energy necessary for resonance depends both on the strength of the external magnetic field and on the nuclei being irradiated. If a very strong magnetic field is applied, the energy difference between the two spin states is large, and higher-energy (higher-frequency) radiation is required. If a weaker magnetic field is applied, less energy is required to effect the transition between nuclear spin states.

In practice, superconducting magnets producing enormously powerful fields up to 140,000 gauss are sometimes used, but a field strength of 14,100 gauss is more common. At this magnetic-field strength, energy in the 60 MHz range (1 MHz = 1 megahertz = 1 million cycles per second) is required to bring a ^1H nucleus into resonance, and energy of 15 MHz is required to bring a ^{13}C nucleus into resonance.

PROBLEM 13.7 NMR spectroscopy uses electromagnetic radiation with a frequency of 6×10^7 Hz. Is this a greater or lesser amount of energy than that used by IR spectroscopy?

13.6 THE NATURE OF NMR ABSORPTIONS

From what's been said thus far, you might expect all protons in a molecule to absorb energy at the same frequency and all ^{13}C nuclei to absorb at the same frequency. If this were true, we would observe only a single NMR absorption peak in the ^{1}H or ^{13}C spectrum of a molecule, a situation that would be of little use for structure determination. In fact, the absorption frequency is not the same for all nuclei.

All nuclei in molecules are surrounded by electron clouds. When an external magnetic field is applied to a sample molecule, the electron clouds set up tiny local magnetic fields of their own. These local fields act in opposition to the applied field, so that the *effective* field experienced by the nucleus is a bit weaker than the applied field.

$$H_{\text{effective}} = H_{\text{applied}} - H_{\text{local}}$$

In describing this effect of local fields, we say that the carbon and hydrogen nuclei are *shielded* from the applied field by the electron clouds that surround them. All the carbons and all the hydrogens in the sample that have the same electronic environment are considered to be *equivalent*. They are said to be the same kind of carbon or hydrogen, and they respond to the applied field in the same way, giving rise to a single NMR signal. Nuclei that have different electronic environments are said to be *nonequivalent*. They are different kinds of carbon or hydrogen, and they respond to the applied field differently, giving rise to different NMR signals. Thus, the NMR spectrum of an organic compound provides a map of its carbon–hydrogen framework. With practice, it's possible to read the map and thereby derive structural information about an unknown molecule.

Figure 13.9 shows both the ^{1}H and the ^{13}C NMR spectra of methyl acetate, CH_3COOCH_3. The horizontal axis shows the difference in effective field strength felt by the nuclei, and the vertical axis indicates the intensity of absorption of energy. Each peak in the NMR spectrum corresponds to a different kind of hydrogen or carbon in a molecule. Note, though, that ^{1}H and ^{13}C spectra can't be observed at the same time on the same spectrometer because different amounts of energy are required to spin-flip the different kinds of nuclei; the two kinds of spectra must be recorded separately.

The ^{13}C spectrum of methyl acetate (Figure 13.9) shows three peaks, one for each of the three kinds of carbon atoms in the molecule. The ^{1}H spectrum shows only two peaks, however, even though methyl acetate has six protons. One peak is due to the CH_3CO protons and the other to the $COOCH_3$ protons. Since the three protons of each methyl group have the same chemical (and magnetic) environment, they are shielded to the same extent and absorb at the same place.

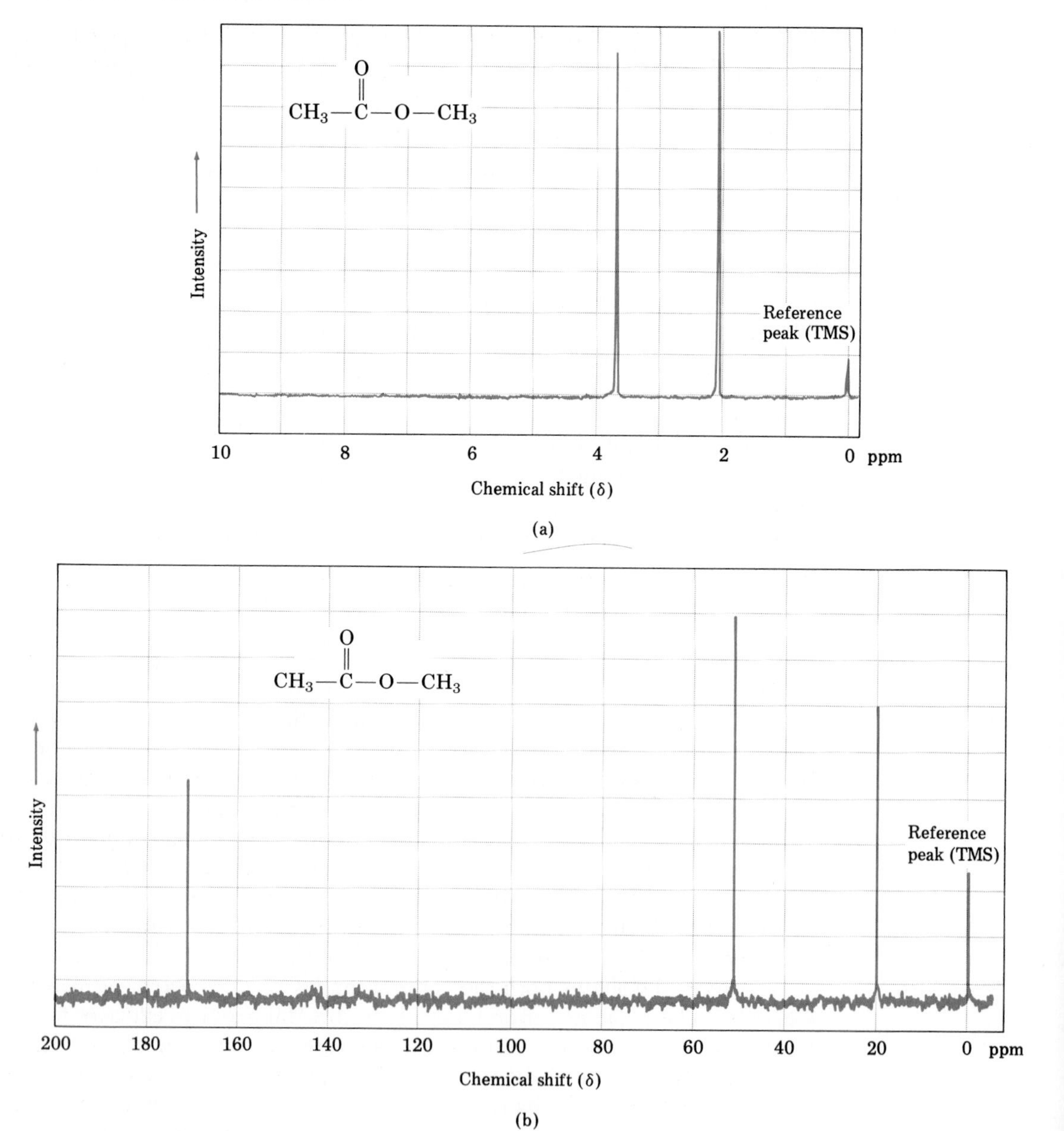

FIGURE 13.9 (a) The ^1H NMR spectrum and (b) the ^{13}C NMR spectrum of methyl acetate, CH_3COOCH_3

PRACTICE PROBLEM 13.6 How many signals would you expect *p*-dimethylbenzene to show in its ^1H and ^{13}C NMR spectra?

SOLUTION Because of the molecule's symmetry, the two methyl groups in *p*-dimethylbenzene are equivalent, and all four ring hydrogens are equivalent. Thus, there are only two absorptions in the ^1H NMR spectrum. Also because of symmetry, there are only three absorptions in the ^{13}C NMR spectrum: one for the two equivalent methyl-group carbons, one for the four equivalent =CH− ring carbons, and one for the two equivalent ring carbons bonded to the methyl groups.

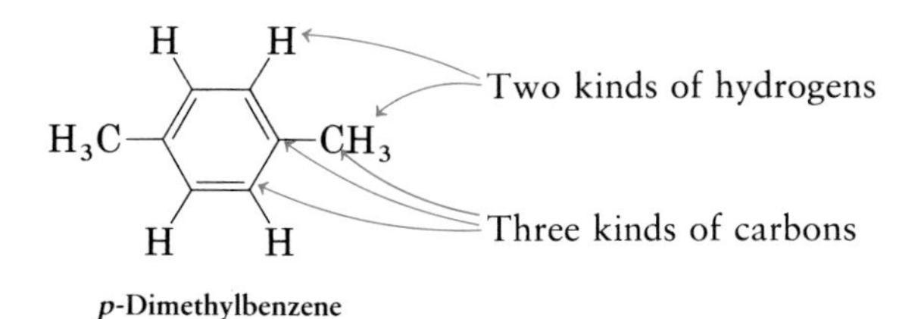

p-Dimethylbenzene

PROBLEM 13.8 How many signals would you expect each of these compounds to show in its ^{1}H and ^{13}C NMR spectra?

(a) Methane (b) Ethane (c) Propane
(d) Cyclohexane (e) Dimethyl ether (f) Benzene
(g) $(CH_3)_3COH$ (h) Chloroethane (i) $(CH_3)_2C{=}C(CH_3)_2$

PROBLEM 13.9 How can you explain the fact that 2-chloropropene shows signals for three kinds of protons in its ^{1}H NMR spectrum?

13.7 CHEMICAL SHIFTS

NMR spectra are displayed on charts that show the applied-field strength increasing from left to right (Figure 13.9). Thus, the left side of the chart is the low-field (or *downfield*) side, and the right side is the high-field (or *upfield*) side. To define the position of an absorption, the NMR chart is calibrated and a reference point is used. In practice, a small amount of tetramethylsilane [TMS, $(CH_3)_4Si$] is added to the sample so that a reference absorption line is produced when the spectrum is run. TMS is used as a reference for both ^{1}H and ^{13}C spectra because it gives rise to a single peak that occurs at a higher field (farther right on the chart) than other absorptions that are normally found in organic molecules.

chemical shift the position of an absorption on the NMR chart

delta scale the arbitrary scale used for defining the position of NMR absorptions. One delta unit is equal to one part per million of spectrometer frequency

The exact place on the chart at which a nucleus absorbs is called its **chemical shift.** By convention, the chemical shift of TMS is called zero, and other peaks normally occur at lower fields (to the left on the chart). For historical reasons, NMR charts are calibrated in units of frequency using the **delta scale.** One delta unit (δ) is equal to one part per million (ppm) of the spectrometer operating frequency. For example, if we were using a 60 MHz instrument to measure the ^{1}H NMR spectrum of a sample, 1 δ would be 1 ppm of 60,000,000 Hz, or 60 Hz. Similarly, if we were measuring the spectrum with a 100 MHz instrument, then 1 δ = 100 Hz.

Although this method of calibrating NMR charts may seem complex, there's a good reason for it: There are many different kinds of NMR spectrometers operating at many different frequencies and magnetic field strengths. By employing a system of measurement in which NMR absorptions are expressed in relative terms (parts per million relative to spectrometer frequency) rather than in absolute terms (Hz), comparisons of spectra obtained on different instruments are possible. *The chemical shift of an NMR absorption given in ppm or δ units is constant, regardless of the operating frequency of the instrument.* A ^{1}H nucleus that absorbs at 2.0 δ on a 60 MHz instrument (2 ppm × 60 MHz = 120 Hz to the left of TMS) also absorbs at 2.0 δ on a 300 MHz instrument (2.0 ppm × 300 MHz = 600 Hz to the left of TMS).

PRACTICE PROBLEM 13.7 Cyclohexane shows an absorption at 1.43 δ in its ^{1}H NMR spectrum. How many hertz away from TMS is this on a spectrometer operating at 60 MHz? on a spectrometer operating at 220 MHz?

SOLUTION On a 60 MHz spectrometer, 1 δ = 60 Hz. Thus, 1.43 δ = 86 Hz away from the TMS reference peak. On a 220 MHz spectrometer, 1 δ = 220 Hz and 1.43 δ = 315 Hz.

PROBLEM 13.10 When the ^{1}H NMR spectrum of acetone is recorded on a 60 MHz instrument, a single sharp resonance line at 2.1 δ is observed.

(a) How far away from TMS (in hertz) does the acetone absorption occur?
(b) What is the position of the acetone absorption in δ units on a 100 MHz instrument?
(c) How many hertz away from TMS does the absorption in the 100 MHz spectrum correspond to?

PROBLEM 13.11 The following ^{1}H NMR resonances were recorded on a spectrometer operating at 60 MHz. Convert each into δ units.

(a) $CHCl_3$, 436 Hz (b) CH_3Cl, 183 Hz
(c) CH_3OH, 208 Hz (d) CH_2Cl_2, 318 Hz

13.8 CHEMICAL SHIFTS IN ^{1}H NMR SPECTRA

Everything we've said thus far about NMR spectroscopy applies to both ^{1}H and ^{13}C spectra. Now let's focus only on ^{1}H NMR spectroscopy to see how it can be used in organic structure determination. Most ^{1}H NMR absorptions occur in the range from 0 to 8 δ, which can be divided into five regions that are characteristic of certain kinds of protons (Figure 13.10). Once the regions are memorized, it's possible to tell at a glance what kinds of protons a molecule contains.

Table 13.3 shows the correlation of ^{1}H chemical shift with environment in more detail. In general, protons bonded to saturated sp^3 carbons absorb upfield

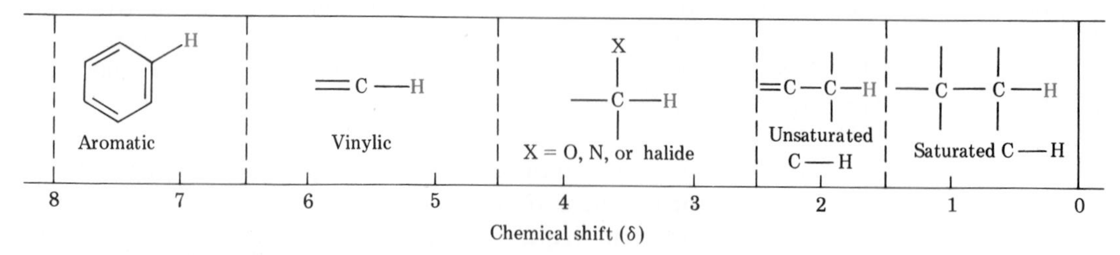

FIGURE 13.10 Chemical shifts of different kinds of protons

TABLE 13.3 Correlation of ^1H chemical shift with environment

Type of proton	*Formula*	*Chemical shift (δ)*
Reference peak	$(CH_3)_4Si$	0
Saturated primary	$—CH_3$	0.7–1.3
Saturated secondary	$—CH_2—$	1.2–1.4
Saturated tertiary	$—\overset{\vert}{\underset{\vert}{C}}—H$	1.4–1.7
Allylic primary	$>C{=}\underset{\vert}{C}—CH_3$	1.6–1.9
Methyl ketones	$—\overset{O}{\overset{\Vert}{C}}—CH_3$	2.1–2.4
Aromatic methyl	$Ar—CH_3$	2.5–2.7
Alkyl chloride	$Cl—\overset{\vert}{\underset{\vert}{C}}—H$	3.0–4.0
Alkyl bromide	$Br—\overset{\vert}{\underset{\vert}{C}}—H$	2.5–4.0
Alkyl iodide	$I—\overset{\vert}{\underset{\vert}{C}}—H$	2.0–4.0
Alcohol, ether	$—O—\overset{\vert}{\underset{\vert}{C}}—H$	3.3–4.0
Alkynyl	$—C{\equiv}C—H$	2.5–2.7
Vinylic	$>C{=}\underset{\vert}{C}—H$	5.0–6.5
Aromatic	$Ar—H$	6.5–8.0
Aldehyde	$—\overset{O}{\overset{\Vert}{C}}—H$	9.7–10.0
Carboxylic acid	$—\overset{O}{\overset{\Vert}{C}}—O—H$	11.0–12.0
Alcohol	$—\overset{\vert}{\underset{\vert}{C}}—O—H$	Extremely variable (2.5–5.0)

whereas protons bonded to sp^2 carbons absorb at lower fields. Protons on carbons that are bonded to electronegative atoms such as N, O, or halogen also absorb at lower fields.

PRACTICE PROBLEM 13.8 Methyl 2,2-dimethylpropanoate $(CH_3)_3COOCH_3$ has two peaks in its ^1H NMR spectrum. At what approximate chemical shifts do they come?

SOLUTION The CH_3O- protons absorb around 3.5–4.0 δ since they are on carbon bonded to oxygen. The $(CH_3)_3C-$ protons absorb around 1.0 δ since they are typical alkane-like protons. (See Figure 13.11)

PROBLEM 13.12 Each of the following compounds exhibits a single ^1H NMR peak. Approximately where would you expect each compound to absorb?

(a) Ethane (b) Acetone (c) Benzene (d) Trimethylamine

13.9 INTEGRATION OF ^1H NMR SPECTRA: PROTON COUNTING

Look at the ^1H NMR spectrum of methyl 2,2-dimethylpropanoate in Figure 13.11. There are two peaks, corresponding to the two kinds of protons present, but the peaks aren't the same size. The peak at 1.2 δ, due to the $(CH_3)_3C-$ protons, is larger than the peak at 3.7 δ due to the $-OCH_3$ protons.

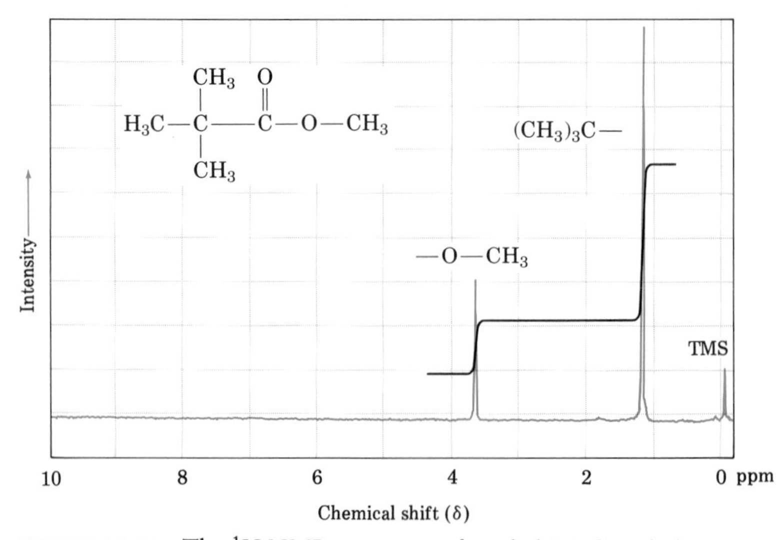

FIGURE 13.11 The ^1H NMR spectrum of methyl 2,2-dimethylpropanoate. Integrating the peaks in a "stair-step" manner shows that they have a 1:3 ratio, corresponding to the ratio of the numbers of protons responsible for each peak.

integration a means of electronically measuring the ratios of numbers of nuclei responsible for peaks in an NMR spectrum

The area under each peak is proportional to the number of protons causing that peak. By electronically measuring (**integrating**) the area under each peak, it's possible to measure the relative number of each kind of proton in a molecule. Integrated peak areas are presented on the chart in a "stair-step" manner, with the height of each step proportional to the number of protons under the peak. For example, the two peaks in methyl 2,2-dimethylpropanoate are found to have a 1:3 (or 3:9) ratio when integrated—exactly what we expect since the three $-OCH_3$ protons are equivalent and the nine $(CH_3)_3C-$ protons are equivalent.

PROBLEM 13.13 How many peaks would you expect to see in the ^1H NMR spectrum of *p*-dimethylbenzene (*p*-xylene)? What ratio of peak areas would you expect to find on integration of the spectrum? Refer to Table 13.3 for approximate chemical shift values, and sketch what the spectrum might look like.

13.10 SPIN–SPIN SPLITTING IN ^1H NMR SPECTRA

spin–spin splitting the splitting of an NMR absorption into multiple peaks because of the interaction of neighboring nuclear spins

coupling the interaction of neighboring nuclear spins that results in spin–spin splitting

In the ^1H NMR spectra we've seen thus far, each different kind of proton in a molecule has given rise to a single peak. It often happens, though, that the absorption of a proton splits into multiple peaks. For example, the ^1H NMR spectrum of chloroethane in Figure 13.12 indicates that the $-CH_2Cl$ protons appear as four peaks (a *quartet*) at 3.6 δ, and the $-CH_3$ protons appear as a *triplet* at 1.5 δ.

Known as **spin–spin splitting,** the phenomenon of multiple absorptions is due to the fact that the nuclear spin of one atom interacts, or **couples,** with the nuclear spin of another nearby atom. In other words, the tiny magnetic field of one nucleus affects the magnetic field felt by a neighboring nucleus.

To understand the reasons for spin–spin splitting, let's look at the $-CH_3$ protons in chloroethane. The three equivalent $-CH_3$ protons are neighbored by

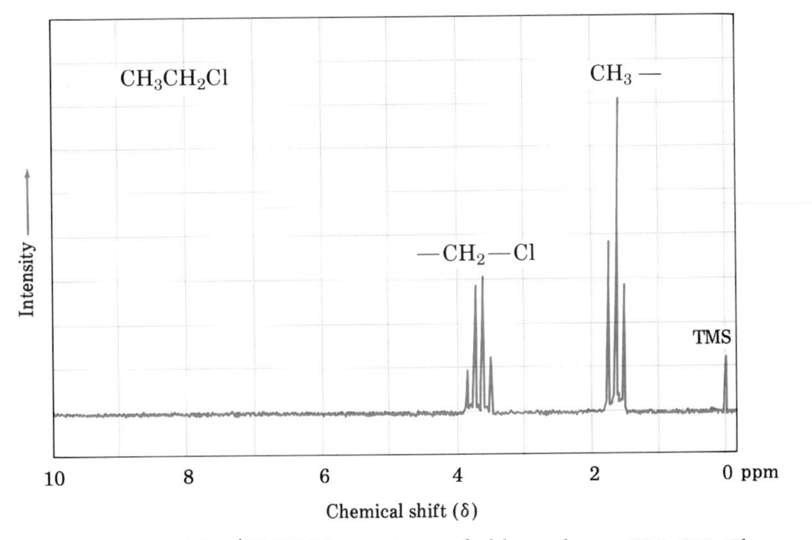

FIGURE 13.12 The ^1H NMR spectrum of chloroethane, CH_3CH_2Cl

two magnetic nuclei, the $-CH_2Cl$ protons. Each of the $-CH_2Cl$ protons has its own nuclear spin, which can align either with or against the applied magnetic field, producing a tiny effect that is felt by the neighboring $-CH_3$ protons.

There are three ways in which the $-CH_2Cl$ protons can align. If both protons align *with* the applied magnetic field, the total effective field felt by the neighboring $-CH_3$ protons is slightly larger than it would otherwise be. Consequently, the applied field necessary to cause resonance is slightly reduced. Alternatively, if one $-CH_2Cl$ proton aligns *with* and one aligns *against* the applied field (two possible ways), there is no effect on the neighboring CH_3- protons. Finally, if both $-CH_2Cl$ protons align *against* the applied field, the effective field felt by the $-CH_3$ protons is slightly smaller than it would otherwise be, and the applied field needed for resonance must be slightly increased. These three possible alignments of $-CH_2Cl$ spins cause the neighboring $-CH_3$ protons to appear as three peaks with a 1:2:1 ratio in the NMR spectrum. Figure 13.13 shows schematically how spin–spin splitting arises.

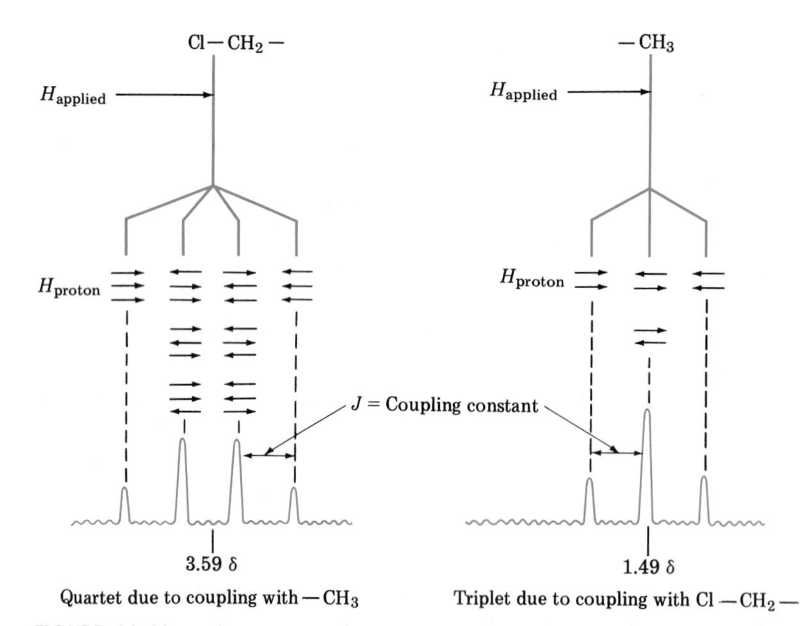

FIGURE 13.13 The origin of spin–spin splitting in chloroethane. The nuclear spins of neighboring protons (indicated by horizontal arrows) align either with or against the applied field, causing the splitting of absorptions into multiplets.

In the same way that the $-CH_3$ protons of chloroethane are split into a triplet in the NMR spectrum, the $-CH_2Cl$ protons are split into a *quartet*. The three spins of the neighboring $-CH_3$ protons align in four combinations: all three with the applied field, two with and one against (three possibilities), one with and two against (three possibilities), or all three against. Thus, four peaks are produced in a 1:3:3:1 ratio.

***n* + 1 rule**
the signal of a proton with *n* neighboring protons splits into *n* + 1 peaks in the NMR spectrum

As a general rule (the ***n* + 1 rule**), protons that have *n* neighboring protons show *n* + 1 peaks in their NMR spectrum. For example, the $-CHBr-$ proton in

2-bromopropane ($CH_3CHBrCH_3$) appears as a seven-line multiplet (a *septet*) in the NMR (Figure 13.14) because its signal is split by the six neighboring protons ($n + 1 = 7$ when $n = 6$). Similarly, the $-CH_3$ protons of 2-bromopropane appear as a doublet because their signal is split only by the single neighboring $-CHBr-$ proton.

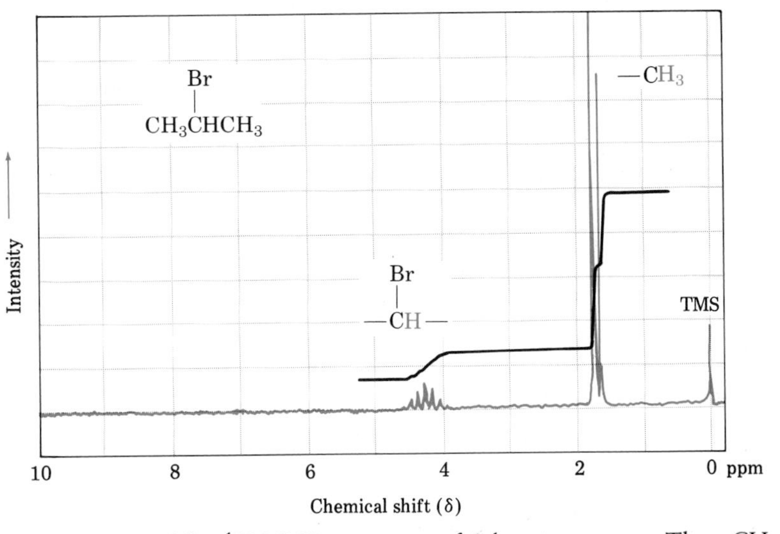

FIGURE 13.14 The ^{1}H NMR spectrum of 2-bromopropane. The $-CH_3$ proton signal is split into a doublet, and the $-CHBr-$ proton signal is split into a septet.

coupling constant (*J*) a measure of the amount of coupling between two neighboring nuclei

The distance between peaks in a multiplet is called the **coupling constant.** It is denoted J and is measured in hertz. Coupling constants generally fall in the range 0–18 Hz, though the exact value depends on the geometry of the molecule. A typical value for an open-chain alkane is 6–8 Hz. Note that the same coupling constant is shared by both groups of hydrogens whose spins are coupled. In chloroethane, for instance, the $-CH_2Cl$ protons are coupled to the $-CH_3$ protons with coupling constant $J = 7$ Hz. The $-CH_3$ protons are similarly coupled to the $-CH_2Cl$ protons with the same $J = 7$ Hz coupling constant.

Three important rules about spin–spin splitting are illustrated by the spectra of chloroethane in Figure 13.12 and 2-bromopropane in Figure 13.14:

1. Chemically equivalent protons don't show spin–spin splitting. The equivalent protons may be on the same carbon or on different carbons, but their signals still appear as singlets and don't split.

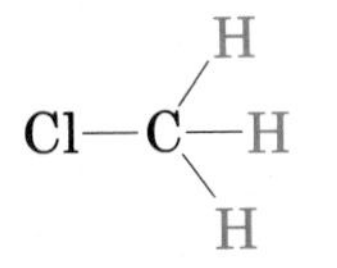

Three C—H protons are chemically equivalent; no splitting occurs.

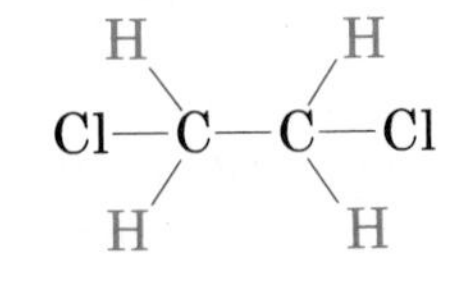

Four C—H protons are chemically equivalent; no splitting occurs.

2. A proton with n equivalent neighboring protons gives a signal that is split into a multiplet of $n + 1$ peaks with coupling constant J. Protons that are more than two carbon atoms apart usually don't split each other.

3. Two groups of protons coupled to each other have the same coupling constant J.

PRACTICE PROBLEM 13.9 Predict the splitting pattern for each kind of hydrogen in isopropyl propanoate, $CH_3CH_2COOCH(CH_3)_2$.

SOLUTION First find how many different kinds of protons are present (there are four). Then find out how many neighboring protons each of the kinds has and apply the $n + 1$ rule to determine the splitting patterns:

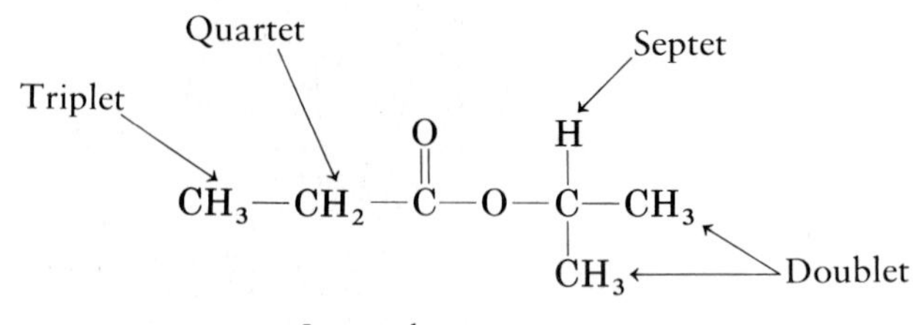

Isopropyl propanoate

PROBLEM 13.14 Predict the splitting patterns you would expect for each proton in these molecules:

(a) $(CH_3)_3CH$ (b) CH_3CHBr_2 (c) $CH_3OCH_2CH_2Br$
(d) $CH_3CH_2COOCH_3$ (e) $ClCH_2CH_2CH_2Cl$ (f) $(CH_3)_2CHCOOCH_3$

PROBLEM 13.15 Propose structures for compounds that show these ^1H NMR spectra:

(a) C_2H_6O; one singlet (b) $C_3H_6O_2$; two singlets
(c) C_3H_7Cl; one doublet and one septet

13.11 USE OF ^{1}H NMR SPECTRA

^{1}H NMR spectroscopy is used to help identify the product of nearly every reaction run in the laboratory. For example, we said in Section 4.2 that addition of HCl to alkenes occurs with Markovnikov orientation; that is, the more highly substituted chloroalkane is formed. With the help of ^{1}H NMR, we can now prove this statement.

Does addition of HCl to 1-methylcyclohexene yield 1-chloro-1-methylcyclohexane or 1-chloro-2-methylcyclohexane?

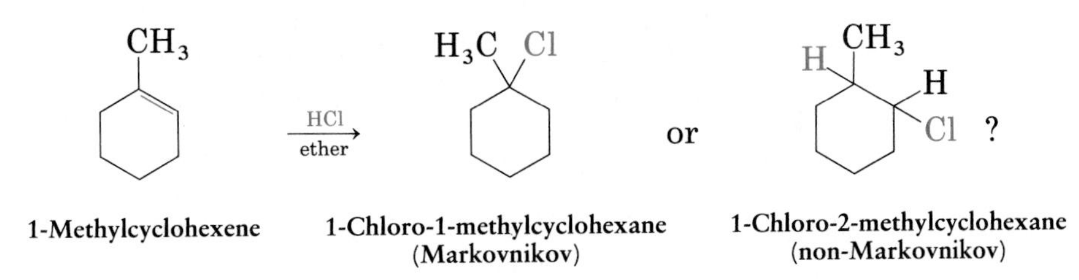

The ^1H NMR spectrum of the reaction product is shown in Figure 13.15. Although many of the ring protons overlap into a broad, poorly defined multiplet centered around 1.6 δ, the spectrum also shows a large singlet absorption in the saturated methyl region at 1.5 δ, indicating that the product has a methyl group bonded to a quaternary carbon (R_3C-CH_3). Furthermore, the spectrum shows no absorptions in the range 4–5 δ, where we would expect the signal of a R_2CHCl proton to occur. Thus, the reaction product must be 1-chloro-1-methylcyclohexane.

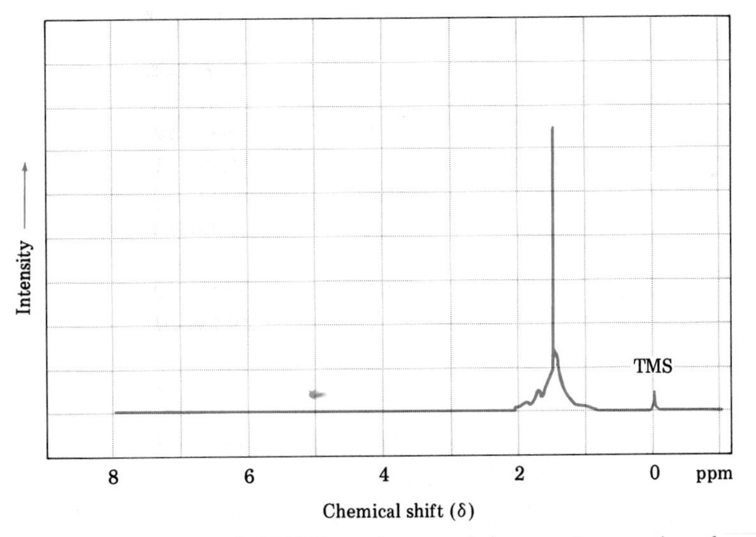

FIGURE 13.15 The ^1H NMR spectrum of the reaction product from HCl and 1-methylcyclohexene. The presence of the $-CH_3$ absorption at 1.6 δ and the absence of any absorptions near 4 δ identify the product as 1-chloro-1-methylcyclohexane.

13.12 ^{13}C NMR SPECTROSCOPY

In some ways, it's surprising that carbon NMR is even possible. After all, ^{12}C, the most abundant carbon isotope, has no nuclear spin and isn't observable by NMR. The only naturally occurring carbon isotope with a magnetic moment is ^{13}C, but its natural abundance is only 1.1%. Thus, only about 1 of every 100 carbon atoms in organic molecules is observable by NMR. Fortunately, the technical problems caused by this low abundance have been overcome by improved electronics and computer techniques, and ^{13}C NMR has now become a routine structural tool.

At its simplest, ^{13}C NMR makes it possible to count the number of carbon atoms in a molecule. In addition, it's possible to get information about the environment of each carbon by observing its chemical shift. As illustrated by the ^{13}C NMR spectrum of methyl acetate shown earlier (Figure 13.9), we normally observe a single, sharp resonance line for each kind of carbon atom in a molecule. Thus, methyl acetate has three nonequivalent carbon atoms and three peaks in its ^{13}C NMR spectrum. (Coupling between adjacent carbon atoms isn't seen because the low natural abundance of ^{13}C makes it unlikely that two such nuclei would be next to each other in a molecule.)

Most ^{13}C resonances are between 0 and 250 δ, with the exact chemical shift dependent on a carbon's environment in the molecule. Figure 13.16 shows how environment and chemical shift are correlated.

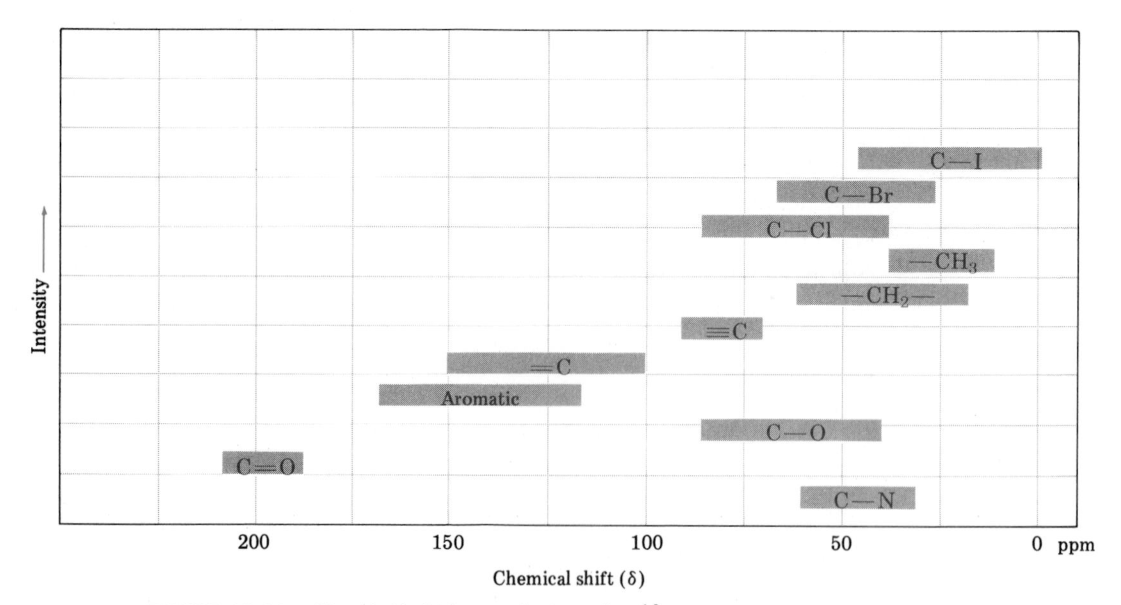

FIGURE 13.16 Chemical shift correlations for ^{13}C NMR

Although the factors that determine chemical shifts are complex, it's possible to make some generalizations. One rule is that carbons bonded to electronegative atoms like oxygen, nitrogen, and halogen absorb downfield (to the left) of normal alkane carbons. Another general rule is that sp^3-hybridized carbons absorb in the range 0–100 δ, and sp^2 carbons absorb in the range 100–200 δ. Carbonyl-group carbons are particularly distinct in the ^{13}C NMR spectrum and are easily observed at the extreme low-field side of the chart in the range 170–210 δ. For example, the ^{13}C NMR spectrum of *p*-bromoacetophenone in Figure 13.17 shows an absorption for the carbonyl carbon at 197 δ.

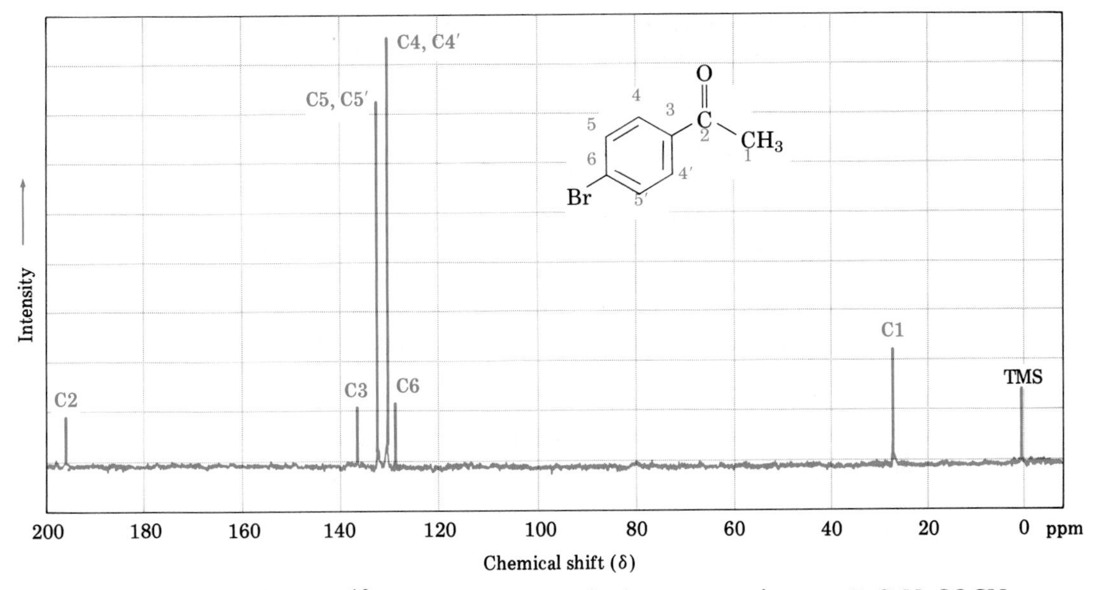

FIGURE 13.17 The ^{13}C NMR spectrum of *p*-bromoacetophenone, $BrC_6H_4COCH_3$

The ^{13}C NMR spectrum of *p*-bromoacetophenone is interesting for another reason as well. Note that only six absorptions are observed even though the molecule has eight carbons. *p*-Bromoacetophenone has a symmetry plane that makes carbons 4 and 4′, and carbons 5 and 5′, equivalent. Thus, the six ring carbons show only four absorptions in the range 128–137 δ. In addition, the $-CH_3$ carbon absorbs at 26 δ.

PRACTICE PROBLEM 13.10 How many resonance lines would you expect to see in the ^{13}C NMR spectrum of methylcyclopentane?

SOLUTION Methylcyclopentane has a symmetry plane. Thus, it has only four peaks in its ^{13}C NMR spectrum.

PROBLEM 13.16 How many resonance lines would you expect to observe in the ^{13}C NMR spectra of these compounds?

(a) Cyclopentane
(b) 1,3-Dimethylbenzene
(c) 1,2-Dimethylbenzene
(d) 1-Methylcyclohexene

PROBLEM 13.17 Propose structures for compounds whose ^{13}C NMR spectra fit these descriptions.

(a) A hydrocarbon with seven peaks in its spectrum
(b) A six-carbon compound with only five peaks in its spectrum
(c) A four-carbon compound with three peaks in its spectrum

INTERLUDE

Colored Organic Compounds

Why are some organic compounds colored but others aren't? Why is β-carotene (from carrots) orange but benzene is colorless? The answers have to do both with the structures of colored molecules and with the way we perceive light.

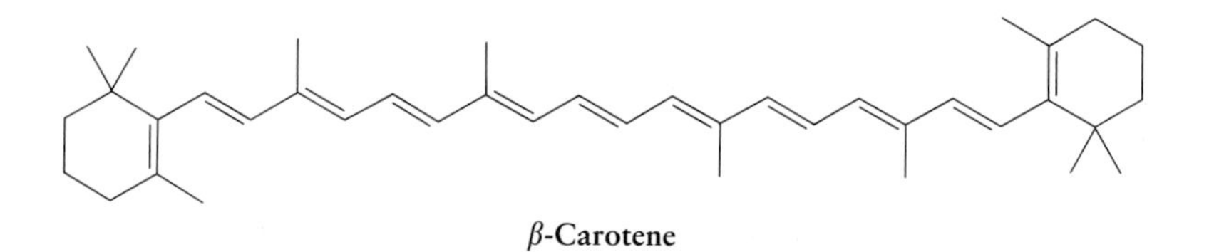

β-Carotene

The visible region of the electromagnetic spectrum extends from approximately 400 to 800 nm. Colored compounds like β-carotene have such extended systems of conjugation that their "UV" absorptions actually extend into the visible region. β-Carotene's absorption, for example, occurs at $\lambda_{max} = 455$ nm (Figure 13.18).

Ordinary "white" light from the sun or from a lamp consists of all wavelengths in the visible region. When white light strikes β-carotene, the wavelengths

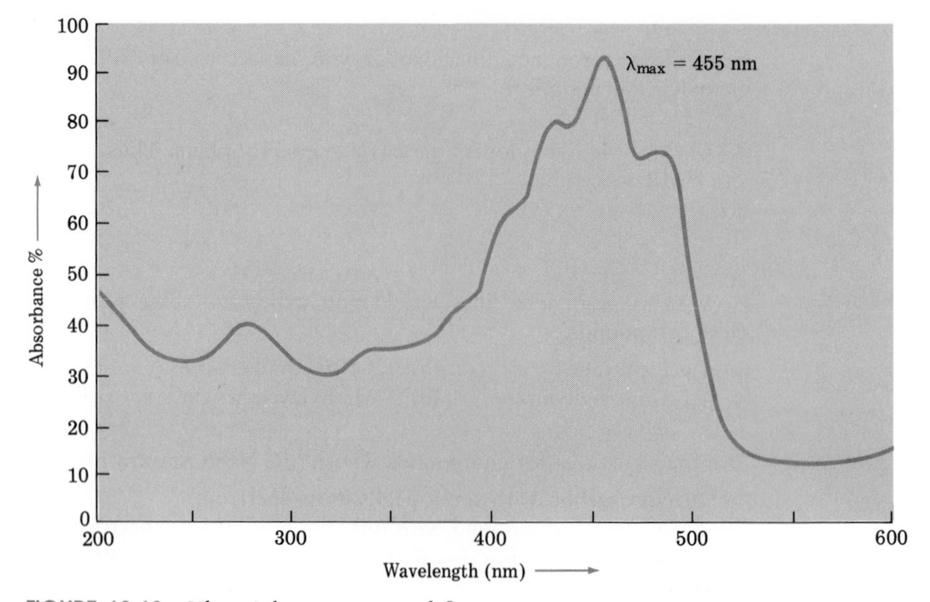

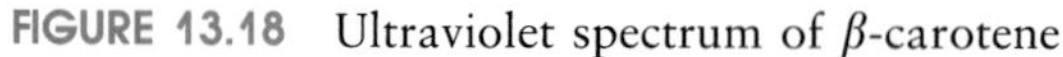

FIGURE 13.18 Ultraviolet spectrum of β-carotene

from 400 to 500 nm (blue) are absorbed while all other wavelengths reach our eyes. We therefore see the white light with the blue subtracted out and perceive a yellow-orange color. (The yellow-orange coloration accounts for the use of β-carotene as a coloring agent in margarine.) What's true for β-carotene is true for all other colored organic compounds. All have an extended system of pi-electron conjugation that gives rise to an absorption in the visible region of the electromagnetic spectrum.

SUMMARY AND KEY WORDS

Three main spectroscopic methods are used to determine the structures of organic molecules. Each of the three gives a different kind of information:

Infrared spectroscopy	What functional groups are present?
Ultraviolet spectroscopy	Is a conjugated pi-electron system present?
Nuclear magnetic resonance spectroscopy	What carbon–hydrogen framework is present?

When an organic molecule is irradiated with **infrared (IR)** radiation, frequencies of light corresponding to the energy levels of molecular bending and stretching motions are absorbed. Each kind of functional group has a characteristic set of infrared absorptions that allows it to be identified; for example, an alkene C=C bond absorbs in the range 1650–1670 cm^{-1}, a saturated ketone absorbs near 1715 cm^{-1}, and a nitrile absorbs near 2230 cm^{-1}. By observing which frequencies of IR radiation are absorbed by a molecule *and which are not,* the functional groups in a molecule can be identified.

Ultraviolet (UV) spectroscopy is applicable to conjugated pi-electron systems. When a conjugated molecule is irradiated with ultraviolet light, energy absorption occurs leading to excitation of pi electrons to higher energy levels. The greater the extent of conjugation, the longer the wavelength needed for excitation.

Nuclear magnetic resonance (NMR) spectroscopy is the most important of the common spectroscopic techniques. When ^{1}H and ^{13}C nuclei are placed in a magnetic field, their spins orient either with or against the field. On irradiation with radio frequency (rf) waves, energy is absorbed and the nuclear spins flip from the lower energy state to the higher energy state. This absorption of energy is detected, amplified, and displayed as an NMR spectrum. NMR spectra display four general features:

1. **Number of resonance lines.** Each nonequivalent kind of ^{1}H or ^{13}C nucleus in a molecule gives rise to a different resonance line.
2. **Chemical shift.** The exact position of each peak is called its chemical shift and is correlated to the chemical nature of each ^{1}H or ^{13}C nucleus. Most ^{1}H absorptions fall in the range 0–10 δ downfield from the TMS reference signal.

3. **Integration.** The area under each peak can be electronically integrated to determine the relative number of protons responsible for each peak in a spectrum.
4. **Spin–spin splitting.** Neighboring nuclear spins can **couple,** splitting NMR absorptions into multiplets. The NMR signal of a ^1H nucleus neighbored by n adjacent protons splits into $n + 1$ peaks with coupling constant J.

ADDITIONAL PROBLEMS

13.18 What kinds of functional groups might compounds contain if they show the following IR absorptions?

(a) 1670 cm^{-1} (b) 1735 cm^{-1}
(c) 1540 cm^{-1} (d) 1710 cm^{-1} and 2500–3100 cm^{-1} (broad)

13.19 At what approximate positions might the following compounds show IR absorptions?

(a) Benzoic acid (b) Methyl benzoate (c) *p*-Hydroxybenzonitrile
(d) 3-Cyclohexenone (e) Methyl 4-oxopentanoate

13.20 The following ^1H NMR absorptions, determined on a spectrometer operating at 60 MHz, are given in hertz downfield from the TMS standard. Convert the absorptions to δ units.

(a) 131 Hz (b) 287 Hz (c) 451 Hz

13.21 At what positions in hertz downfield from TMS standard would the NMR absorptions in Problem 13.20 appear on a spectrometer operating at 100 MHz?

13.22 These NMR absorptions, given in δ units, were obtained on a spectrometer operating at 80 MHz. Convert the chemical shifts from δ units into hertz downfield from TMS.

(a) 2.1 δ (b) 3.45 δ (c) 6.30 δ

13.23 If C–O single-bond stretching occurs at 1000 cm^{-1} and C=O double-bond stretching occurs at 1700 cm^{-1}, which of the two requires more energy? How does your answer correlate with the relative strengths of single and double bonds?

13.24 Tell what is meant by each of these terms:

(a) Chemical shift (b) Coupling constant (c) λ_{max}
(d) Spin–spin splitting (e) Wave number (f) Applied magnetic field

13.25 When measured on a spectrometer operating at 60 MHz, chloroform ($CHCl_3$) shows a single sharp absorption at 7.3 δ.

(a) How many parts per million downfield from TMS does chloroform absorb?
(b) How many hertz downfield from TMS does chloroform absorb if the measurement is carried out on a spectrometer operating at 360 MHz?
(c) What is the position of the chloroform absorption in δ units if it is measured on a 360 MHz spectrometer?

13.26 How many absorptions would you expect in the ^{13}C NMR spectra of these compounds?

(a) 1,1-Dimethylcyclohexane (b) Ethyl methyl ether (c) Cyclohexanone
(d) 2-Methyl-2-butene (e) *cis*-2-Pentene (f) *trans*-2-Pentene

13.27 How many types of nonequivalent protons are there in each of the molecules listed in Problem 13.26?

13.28 Describe the ^1H NMR spectra you would expect for these compounds:

(a) CH_3CHCl_2 (b) $CH_3COOCH_2CH_3$ (c) $(CH_3)_3CCH_2CH_3$

13.29 The following compounds all show a single line in their ^1H NMR spectra. List them in order of expected increasing chemical shift: CH_4, CH_2Cl_2, cyclohexane, CH_3COCH_3, $H_2C{=}CH_2$, benzene.

13.30 Propose structures for compounds that meet these descriptions:

(a) C_5H_8, with IR absorptions at 3300 and 2150 cm^{-1}
(b) C_4H_8O, with a strong IR absorption at 3400 cm^{-1}
(c) C_4H_8O, with a strong IR absorption at 1715 cm^{-1}
(d) C_8H_{10}, with IR absorptions at 1600 and 1500 cm^{-1}

13.31 How would you use infrared spectroscopy to distinguish between these pairs of isomers?

(a) $(CH_3)_3N$ and $CH_3CH_2NHCH_3$ (b) CH_3COCH_3 and $CH_2{=}CHCH_2OH$
(c) CH_3COCH_3 and CH_3CH_2CHO

13.32 How would you use ^1H NMR spectroscopy to distinguish between the isomer pairs shown in Problem 13.31?

13.33 How could you use ^{13}C NMR spectroscopy to distinguish between the isomers pairs shown in Problem 13.31?

13.34 Assume that you're carrying out the dehydration of 1-methylcyclohexanol to yield 1-methylcyclohexene. How could you use IR spectroscopy to determine when the reaction was complete? What characteristic absorptions would you expect for both starting material and product?

13.35 Dehydration of 1-methylcyclohexanol might lead to either of two isomeric alkenes, 1-methylcyclohexene or methylenecyclohexane. How could you use NMR spectroscopy (both ^1H and ^{13}C) to determine the structure of the product?

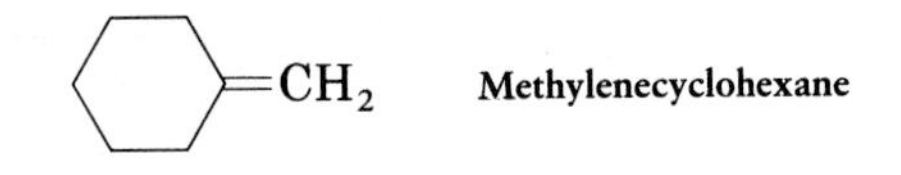

13.36 3,4-Dibromohexane can undergo base-induced double dehydrobromination to yield either 3-hexyne or 2,4-hexadiene. How could you use UV spectroscopy to help you identify the product? How could you use ^1H NMR spectroscopy?

13.37 Describe the ^1H and ^{13}C NMR spectra you expect for these compounds:

(a) $ClCH_2CH_2CH_2Cl$ (b) $CH_3COCH_2CH_2Cl$

13.38 Propose structures for compounds with the following formulas that show only one peak in their ^1H NMR spectra:

(a) C_5H_{12} (b) C_5H_{10} (c) $C_4H_8O_2$

13.39 Assume that you have a compound with formula C_3H_6O.

(a) Propose as many structures as you can that fit the molecular formula (there are seven).
(b) If your compound has an IR absorption at 1710 cm^{-1}, what can you conclude?
(c) If your compound has a single ^1H NMR absorption at 2.1 δ, what is its structure?

13.40 Propose structures for compounds that fit these ^1H NMR data:

(a) $C_5H_{10}O$
- 6 H doublet at 0.95 δ, $J = 7$ Hz
- 3 H singlet at 2.10 δ
- 1 H multiplet at 2.43 δ

(b) C_3H_5Br
- 3 H singlet at 2.32 δ
- 1 H singlet at 5.25 δ
- 1 H singlet at 5.54 δ

13.41 How can you use ^1H and ^{13}C NMR to help you distinguish among these four isomers?

$$\begin{array}{l} CH_2{-}CH_2 \\ |\qquad\ \ | \\ CH_2{-}CH_2 \end{array} \qquad H_2C{=}CHCH_2CH_3 \qquad CH_3CH{=}CHCH_3 \qquad \begin{array}{c} CH_3 \\ | \\ CH_3C{=}CH_2 \end{array}$$

13.42 How can you use ^1H NMR to help you distinguish between the following isomers?

3-Methyl-2-cyclohexenone 4-Cyclopentenyl methyl ketone

13.43 How can you use ^{13}C NMR to help you distinguish between the isomers in Problem 13.42?

13.44 How can you use UV spectroscopy to help you distinguish between the isomers in Problem 13.42?

13.45 The ^1H NMR spectrum of compound A, $C_3H_6Br_2$, is shown. Propose a plausible structure for A and explain how the peaks in the spectrum fit your structure.

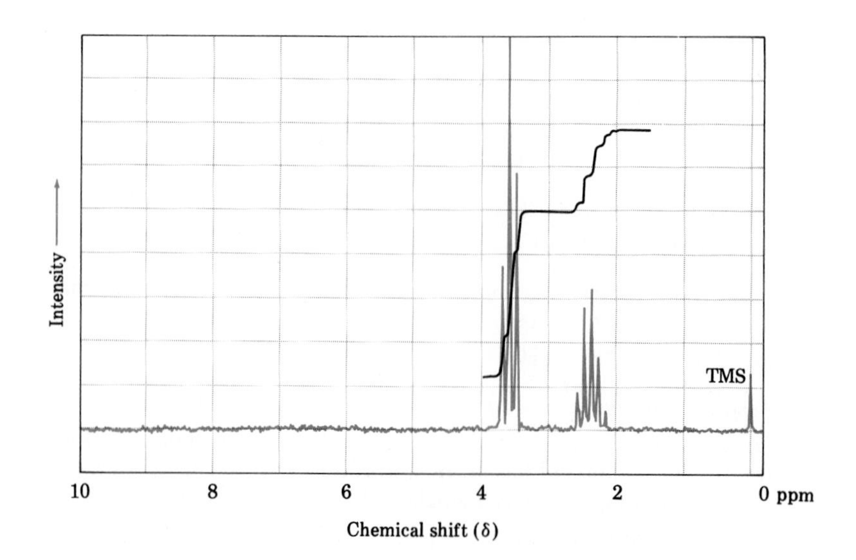

13.46 The compound whose ^1H NMR spectrum is shown has the formula $C_4H_7O_2Cl$ and has an IR absorption peak at 1740 cm^{-1}. Propose a plausible structure.

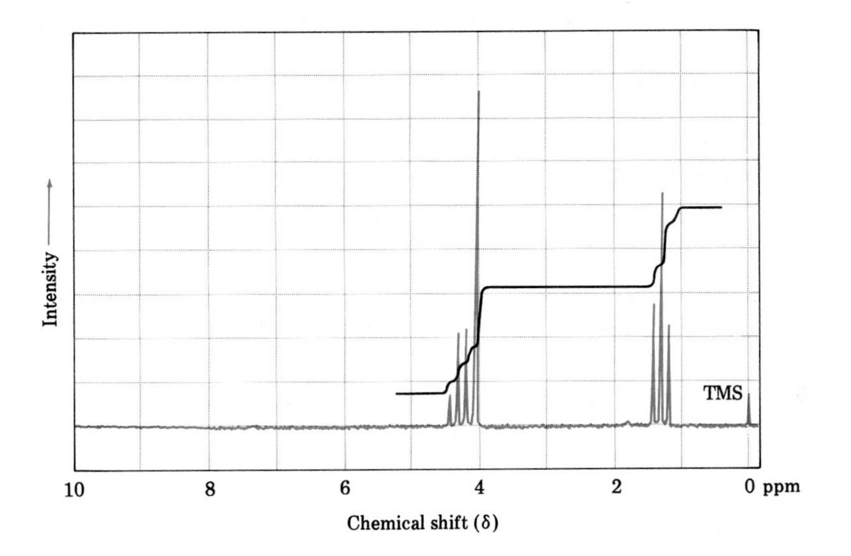

13.47 Propose a structure for a compound of formula C_4H_9Br that has the following ^1H NMR spectrum.

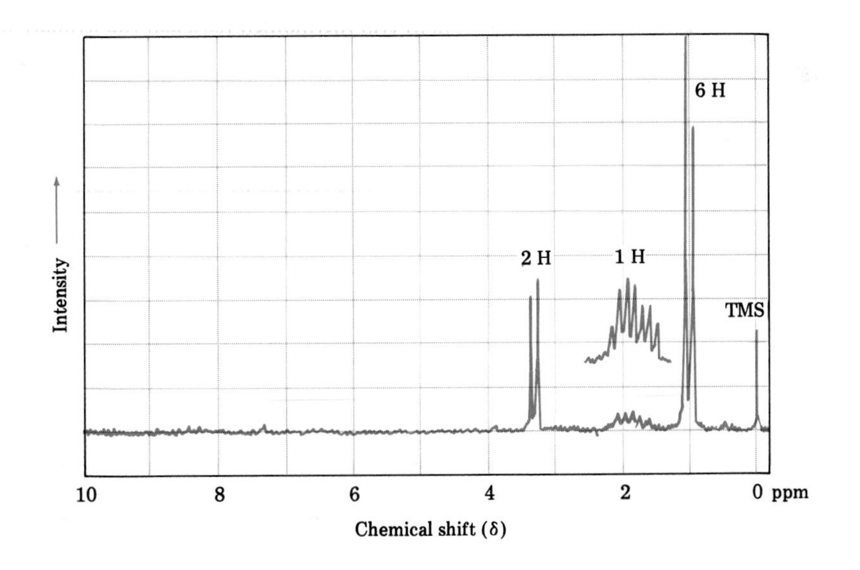

13.48 The energy of electromagnetic radiation expressed in units of kcal/mol can be determined by the formula

$$E = \frac{2.86 \times 10^{-3} \text{ kcal/mol}}{\lambda \text{ (in cm)}}$$

What is the energy of infrared radiation of wavelength 10^{-4} cm?

13.49 Using the formula given in Problem 13.48, calculate the energy required to effect the electronic excitation of 1,3-butadiene (λ_{max} = 217 nm).

13.50 Using the equation given in Problem 13.48, calculate the amount of energy required to spin-flip a proton in a spectrometer operating at 100 MHz. Does increasing the spectrometer frequency from 60 MHz to 100 MHz increase or decrease the amount of energy necessary for resonance?

CHAPTER 14

Biomolecules: Carbohydrates

Carbohydrates are everywhere in nature; they occur in every living organism and are essential to life. The sugar and starch in food and the cellulose in wood, paper, and cotton are nearly pure carbohydrate. Modified carbohydrates form part of the coating around living cells; other carbohydrates are found in DNA, which carries genetic information; and still others are invaluable as medicines.

carbohydrate a straight-chain polyhydroxy ketone or aldehyde such as glucose

The word **carbohydrate** derives historically from the fact that glucose, the first simple carbohydrate to be obtained pure, has the molecular formula $C_6H_{12}O_6$ and was originally thought to be a "hydrate of carbon," $C_6(H_2O)_6$. This view was soon abandoned, but the name persisted. Today, the term *carbohydrate* is used to refer loosely to the broad class of polyhydroxylated aldehydes and ketones commonly called sugars.

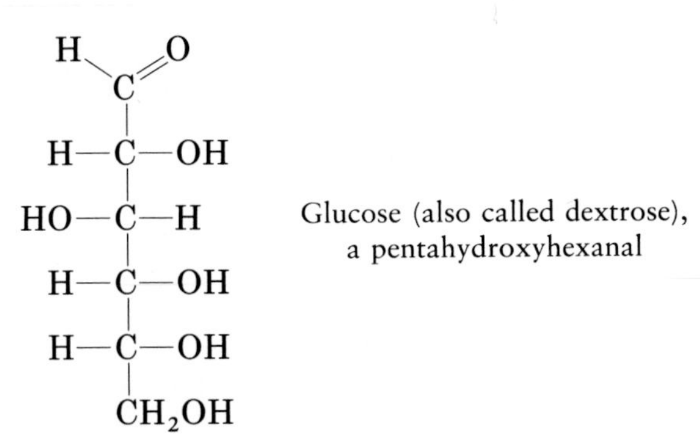

Glucose (also called dextrose), a pentahydroxyhexanal

Carbohydrates are synthesized by green plants during photosynthesis, a complex process during which carbon dioxide is converted into glucose. Many molecules of glucose are then chemically linked for storage by the plant in the form of either cellulose or starch. It has been estimated that more than 50% of the dry weight of the earth's biomass—all plants and animals—consists of glucose polymers. When eaten and then metabolized, carbohydrates provide the major source of energy required by organisms. Thus, carbohydrates act as the chemical intermediaries by which solar energy is stored and used to support life.

$$6\ CO_2 + 6\ H_2O \xrightarrow{\text{sunlight}} 6\ O_2 + \underset{\text{Glucose}}{C_6H_{12}O_6} \longrightarrow \text{cellulose, starch}$$

14.1 CLASSIFICATION OF CARBOHYDRATES

simple sugar a carbohydrate like glucose that can't be hydrolyzed to smaller molecules

monosaccharide a simple sugar

complex carbohydrate a carbohydrate composed of two or more simple sugars linked by an acetal bond

disaccharide a complex carbohydrate composed of two simple sugars bonded together

polysaccharide a complex carbohydrate composed of many simple sugars bonded together

aldose a simple sugar with an aldehyde carbonyl group

ketose a simple sugar with a ketone carbonyl group

Carbohydrates are generally classed into two groups, simple and complex. **Simple sugars,** or **monosaccharides,** are carbohydrates like glucose and fructose that can't be hydrolyzed into smaller molecules. **Complex carbohydrates** are made of two or more simple sugars linked together. For example, sucrose (table sugar) is a **disaccharide** (two sugars) made up of one glucose molecule linked to one fructose molecule. Similarly, cellulose is a **polysaccharide** (many sugars) made up of several thousand glucose molecules linked together. Hydrolysis of polysaccharides breaks them down into their constituent monosaccharide units.

$$1\ \text{Sucrose} \xrightarrow{H_3O^+} 1\ \text{Glucose} + 1\ \text{Fructose}$$

$$\text{Cellulose} \xrightarrow{H_3O^+} {\sim}3000\ \text{Glucose}$$

Monosaccharides are further classified as either **aldoses or ketoses.** The *-ose* suffix is used as the family-name ending for carbohydrates, and the *ald-* and *ket-* prefixes identify the nature of the carbonyl group (aldehyde or ketone). The number of carbon atoms in the monosaccharide is indicated by using *tri-*, *tetr-*, *pent-*, *hex-*, and so forth in the parent name. For example, glucose is an *aldohexose,* a six-carbon aldehydo sugar; fructose is a *ketohexose,* a six-carbon keto sugar; and ribose is an *aldopentose,* a five-carbon aldehydo sugar. Most of the commonly occurring simple sugars are either aldopentoses or aldohexoses.

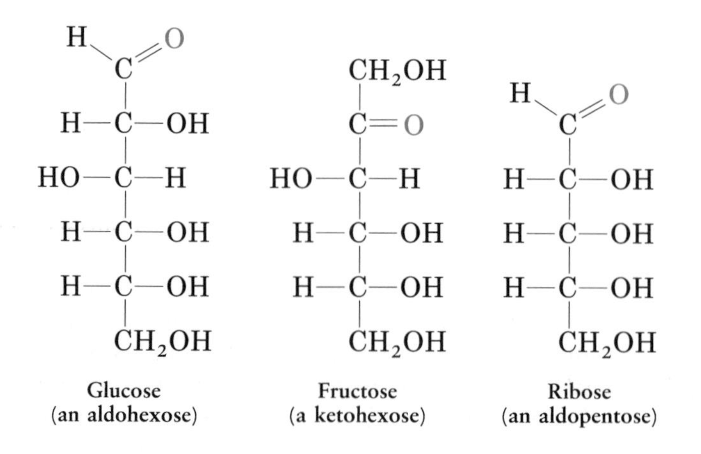

Glucose (an aldohexose) Fructose (a ketohexose) Ribose (an aldopentose)

PRACTICE PROBLEM 14.1 Classify this monosaccharide.

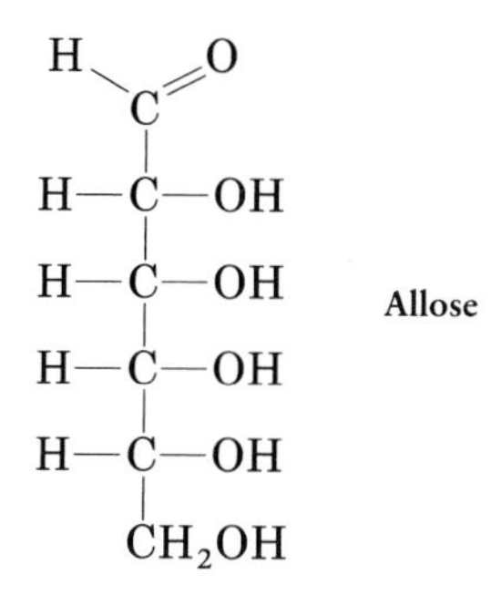

SOLUTION Since allose has six carbons and an aldehyde carbonyl group, it is an aldohexose.

PROBLEM 14.1 Classify the following monosaccharides.

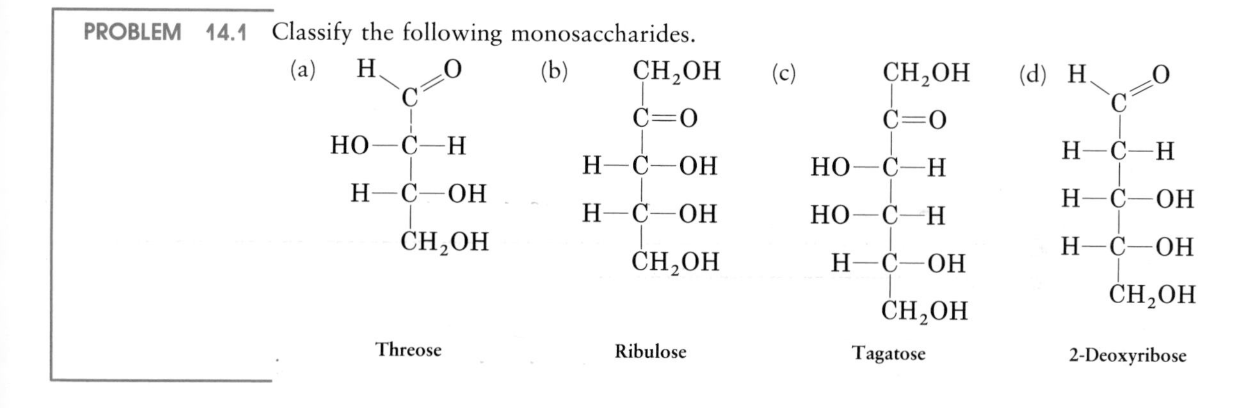

14.2 CONFIGURATIONS OF MONOSACCHARIDES: FISCHER PROJECTIONS

Fischer projection a method for depicting stereochemistry at a chiral carbon by showing a tetrahedral carbon as two crossed lines

Since all carbohydrates have chiral carbon atoms, it was recognized long ago that a standard method of representation is needed to designate carbohydrate stereochemistry. In 1891, Emil Fischer suggested a method based on the projection of a tetrahedral carbon atom onto a flat surface. These **Fischer projections** were soon adopted and are now a standard means of depicting stereochemistry at chiral centers.

A tetrahedral carbon atom in a Fischer projection is represented by two perpendicular lines. The horizontal lines represent bonds coming out of the page, and the vertical lines represent bonds going into the page. By convention, the carbonyl carbon is placed at or near the top in Fischer projections. Thus, (*R*)-

glyceraldehyde, the simplest monosaccharide, can be represented in the following way:

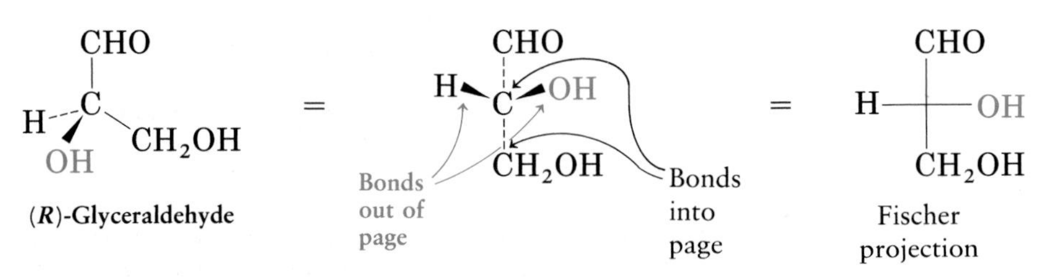

Carbohydrates with more than one chiral center are shown by "stacking" the atoms, one on top of the other. Once again, the carbonyl carbon is at or near the top of the Fischer projection. Molecular models are particularly helpful in visualizing these structures.

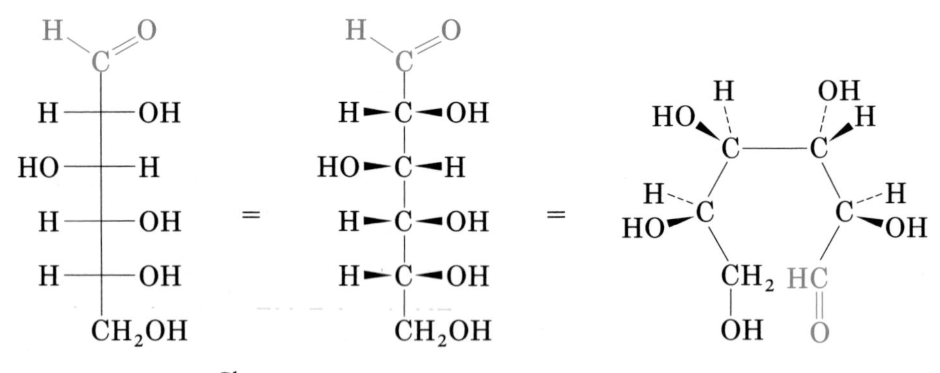

PRACTICE PROBLEM 14.2 Convert the following tetrahedral representation of (*R*)-2-butanol into a Fischer projection.

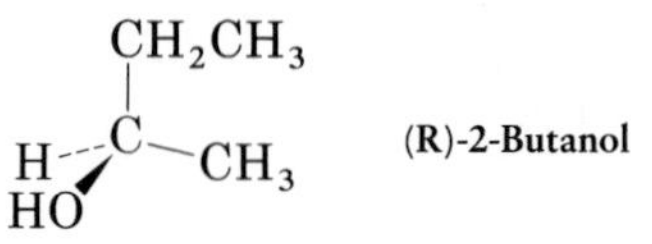

SOLUTION Orient the molecule so that two horizontal bonds are facing you and two vertical bonds are receding away from you. Then press the molecule flat into the paper, indicating the chiral carbon as the intersection of two crossed lines.

(*R*)-2-Butanol

PRACTICE PROBLEM 14.3 Convert the following Fischer projection of lactic acid into a standard tetrahedral representation, and indicate whether the molecule is (*R*) or (*S*).

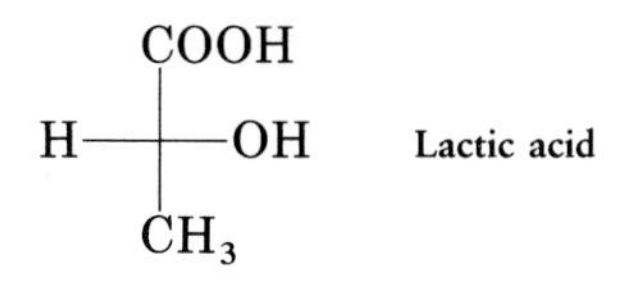

SOLUTION After placing a carbon atom at the intersection of the two crossed lines, imagine that the two horizontal bonds are coming toward you and the two vertical bonds are receding away from you. The projection represents (*R*)-lactic acid.

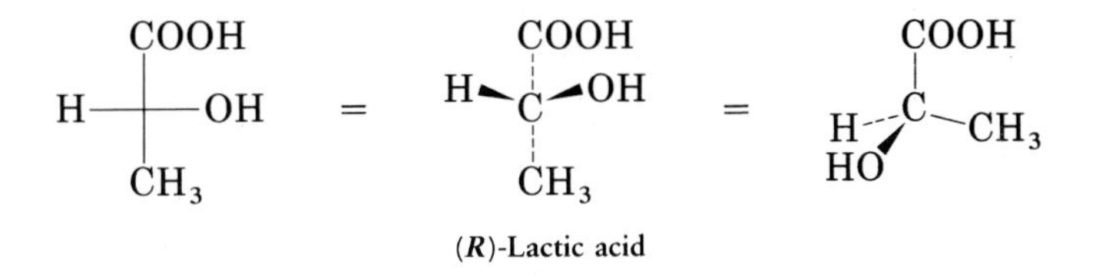

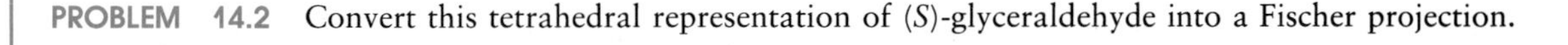

PROBLEM 14.2 Convert this tetrahedral representation of (*S*)-glyceraldehyde into a Fischer projection.

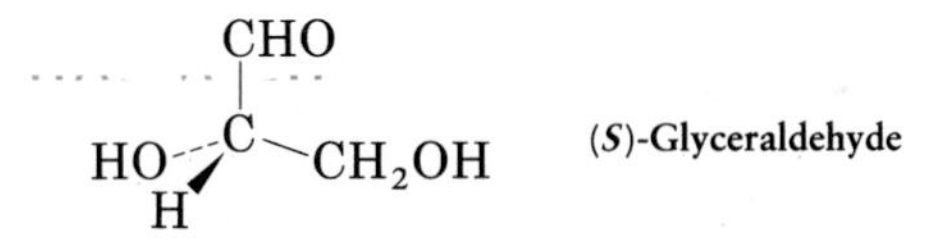

PROBLEM 14.3 Draw Fischer projections of both (*R*)-2-chlorobutane and (*S*)-2-chlorobutane.

PROBLEM 14.4 Convert these Fischer projections into tetrahedral representations and assign *R* or *S* stereochemistry to each.

(a) COOH (top), H_2N (left), H (right), CH_3 (bottom)

(b) CHO (top), H (left), OH (right), CH_3 (bottom)

(c) CH_3 (top), H (left), CHO (right), CH_2CH_3 (bottom)

14.3 D,L SUGARS

Glyceraldehyde has one chiral carbon atom and therefore has two enantiomeric (mirror-image) forms, but only the dextrorotatory enantiomer occurs naturally. That is, a sample of naturally occurring glyceraldehyde placed in a polarimeter rotates plane-polarized light in a clockwise direction, denoted (+). Since (+)-glyceraldehyde is known to have the *R* configuration at C2, it can be represented as in Figure 14.1. For historical reasons dating from long before the adoption of the *R*,*S* system, (*R*)-(+)-glyceraldehyde is also referred to as D-*glyceraldehyde* (D

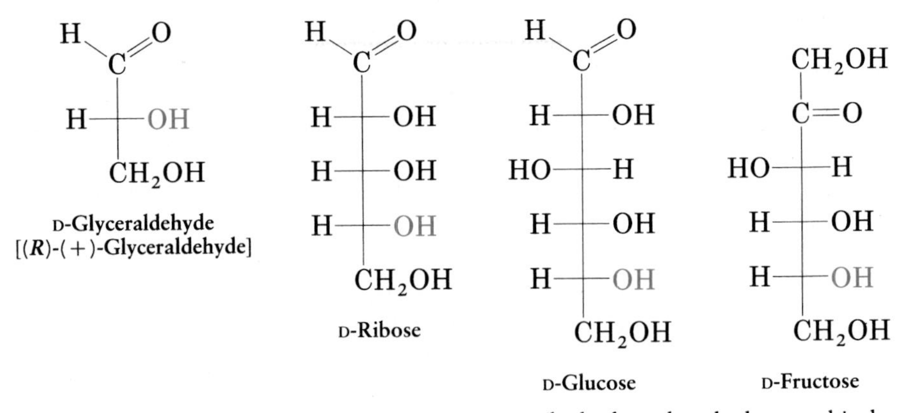

FIGURE 14.1 Some naturally occurring D sugars. The hydroxyl at the lowest chiral center is on the right in Fischer projections.

for dextrorotatory). The other enantiomer, (*S*)-(−)-glyceraldehyde, is known as L-*glyceraldehyde* (L for levorotatory).

Because of the way that monosaccharides are synthesized in nature, glucose, fructose, ribose, and most other naturally occurring monosaccharides have the same stereochemical configuration as D-glyceraldehyde at the chiral carbon atom farthest from the carbonyl group. In Fischer projections, therefore, most naturally occurring sugars have the hydroxyl group at the lowest chiral carbon atom pointing to the right (Figure 14.1). Such compounds are referred to as **D-sugars.**

D-sugar a sugar whose hydroxyl group at the chiral carbon farthest from the carbonyl group points to the right when the molecule is drawn in Fischer projection

L-sugar a sugar whose hydroxyl group at the chiral carbon farthest from the carbonyl group points to the left when the molecule is drawn in Fischer projection

In contrast to the D sugars, all **L sugars** have the hydroxyl group at the chiral center farthest from the carbonyl group on the *left* in Fischer projections. Thus, L sugars are mirror images (enantiomers) of D sugars. Note that the D and L notations have no relation to the direction in which a given sugar rotates plane-polarized light. A D sugar may be either dextrorotatory or levorotatory. The prefix D indicates only that the stereochemistry of the lowest chiral carbon atom is to the right in Fischer projection when the molecule is drawn in the standard way with the carbonyl group at or near the top.

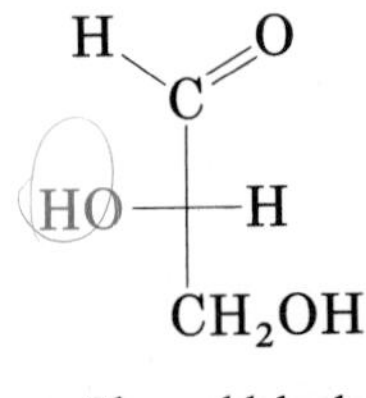

L-Glyceraldehyde
[(*S*)-(−)-Glyceraldehyde]

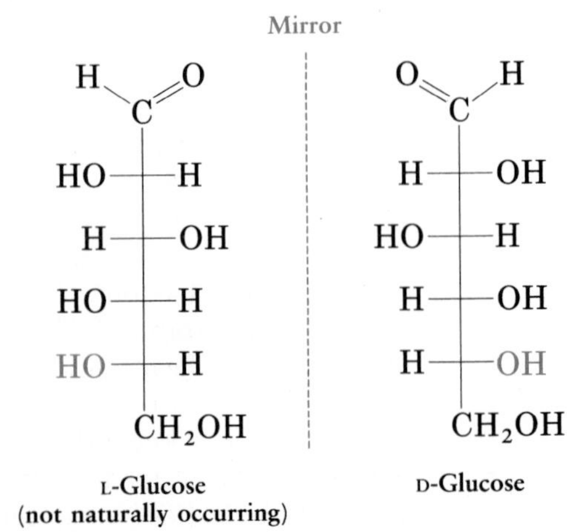

L-Glucose
(not naturally occurring)

D-Glucose

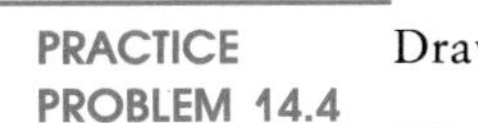

PRACTICE PROBLEM 14.4 Draw a Fischer projection of L-fructose.

SOLUTION Since L-fructose is the enantiomer (mirror image) of D-fructose, we simply look at the structure of D-fructose and then reverse the configuration at each chiral center.

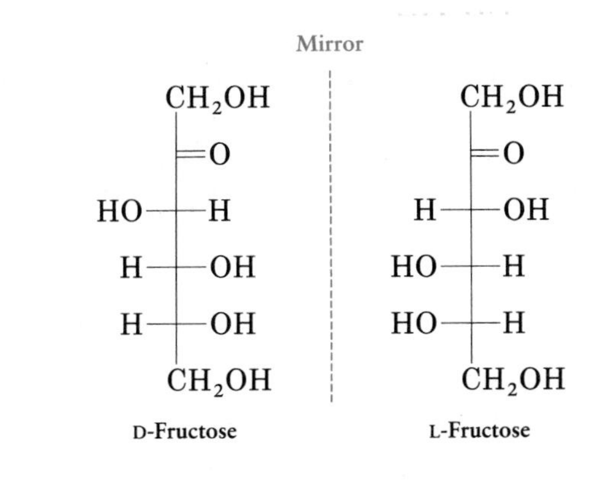

PROBLEM 14.5 Which of the following are L sugars and which are D sugars?

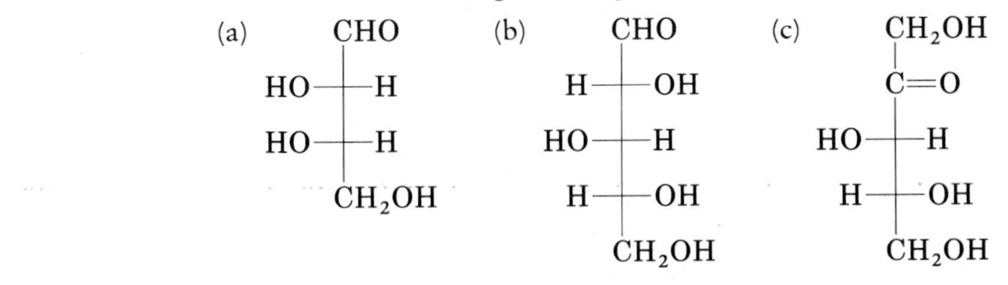

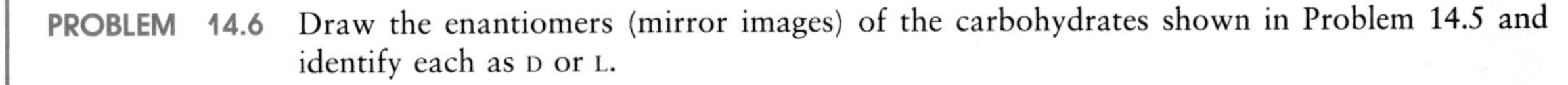

PROBLEM 14.6 Draw the enantiomers (mirror images) of the carbohydrates shown in Problem 14.5 and identify each as D or L.

14.4 CONFIGURATIONS OF ALDOSES

Aldotetroses are four-carbon sugars with two chiral centers. Thus, there are $2^2 = 4$ possible stereoisomeric aldotetroses, or two D,L pairs of enantiomers, called *erythrose* and *threose.*

Aldopentoses have three chiral centers and a total of $2^3 = 8$ possible stereoisomers, or four D,L pairs of enantiomers. These four pairs are called *ribose, arabinose, xylose,* and *lyxose.* All except lyxose occur widely in nature. D-Ribose is an important part of RNA (ribonucleic acid); L-arabinose is found in many plants; and D-xylose is found in wood.

Aldohexoses have four chiral centers, for a total of $2^4 = 16$ possible stereoisomers, or eight D,L pairs of enantiomers. The names of the eight are *allose, altrose, glucose, mannose, gulose, idose, galactose,* and *talose.* Of the eight, only D-glucose, from starch and cellulose, and D-galactose, from gums and fruit pectins, are widely distributed in nature. D-Mannose and D-talose also occur naturally, but in lesser abundance.

Fischer projections of the four-, five-, and six-carbon aldoses are shown in Figure 14.2 for the D-series. Starting from D-glyceraldehyde, we can construct the two D-aldotetroses by inserting a new chiral carbon atom just below the aldehyde carbon. Each of the two D-aldotetroses then leads to two D-aldopentoses (four total), and each of the four D-aldopentoses leads to two D-aldohexoses (eight total).

PROBLEM 14.7 Write Fischer projections for the following L sugars. Remember that an L sugar is the mirror image of the corresponding D sugar shown in Figure 14.2.

(a) L-Arabinose (b) L-Threose (c) L-Galactose

PROBLEM 14.8 How many aldoheptoses are possible? How many of them are D sugars and how many are L sugars?

PROBLEM 14.9 Draw Fischer projections for the two D-aldoheptoses (Problem 14.8) whose stereochemistry at C3, C4, C5, and C6 is the same as that of glucose at C2, C3, C4, and C5.

14.5 CYCLIC STRUCTURES OF MONOSACCHARIDES: HEMIACETAL FORMATION

We said during the discussion of carbonyl-group chemistry in Section 9.13 that alcohols undergo a rapid and reversible nucleophilic addition reaction with ketones and aldehydes to form hemiacetals:

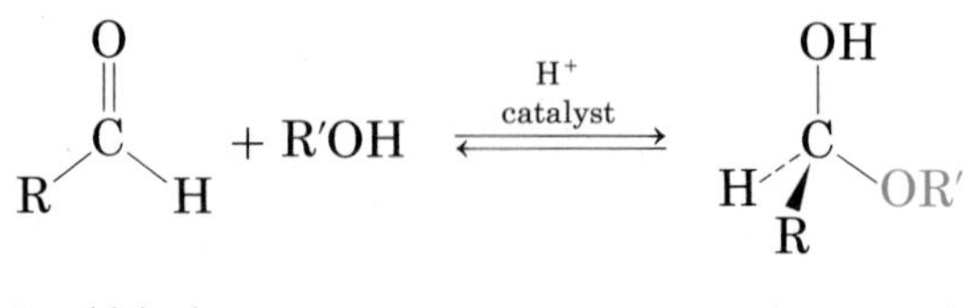

If both the hydroxyl and the carbonyl group are in the same molecule, an *intramolecular* nucleophilic addition can take place, leading to the formation of a *cyclic* hemiacetal. Five- and six-membered cyclic hemiacetals form particularly easily, and many carbohydrates therefore exist in an equilibrium between open-chain and cyclic forms. For example, glucose exists in aqueous solution primarily as the six-membered (**pyranose**) ring formed by intramolecular nucleophilic addition of the hydroxyl group at C5 to the C1 aldehyde group. Fructose, on the other hand, exists to the extent of about 20% as the five-membered (**furanose**) ring formed by addition of the hydroxyl group at C5 to the C2 ketone. The words *pyranose* for a six-membered ring and *furanose* for a five-membered ring are derived from the names of the simple cyclic ethers pyran and furan. The cyclic forms of glucose and fructose are shown in Figure 14.3.

pyranose the six-membered-ring structure of a simple sugar

furanose the five-membered-ring structure of a simple sugar

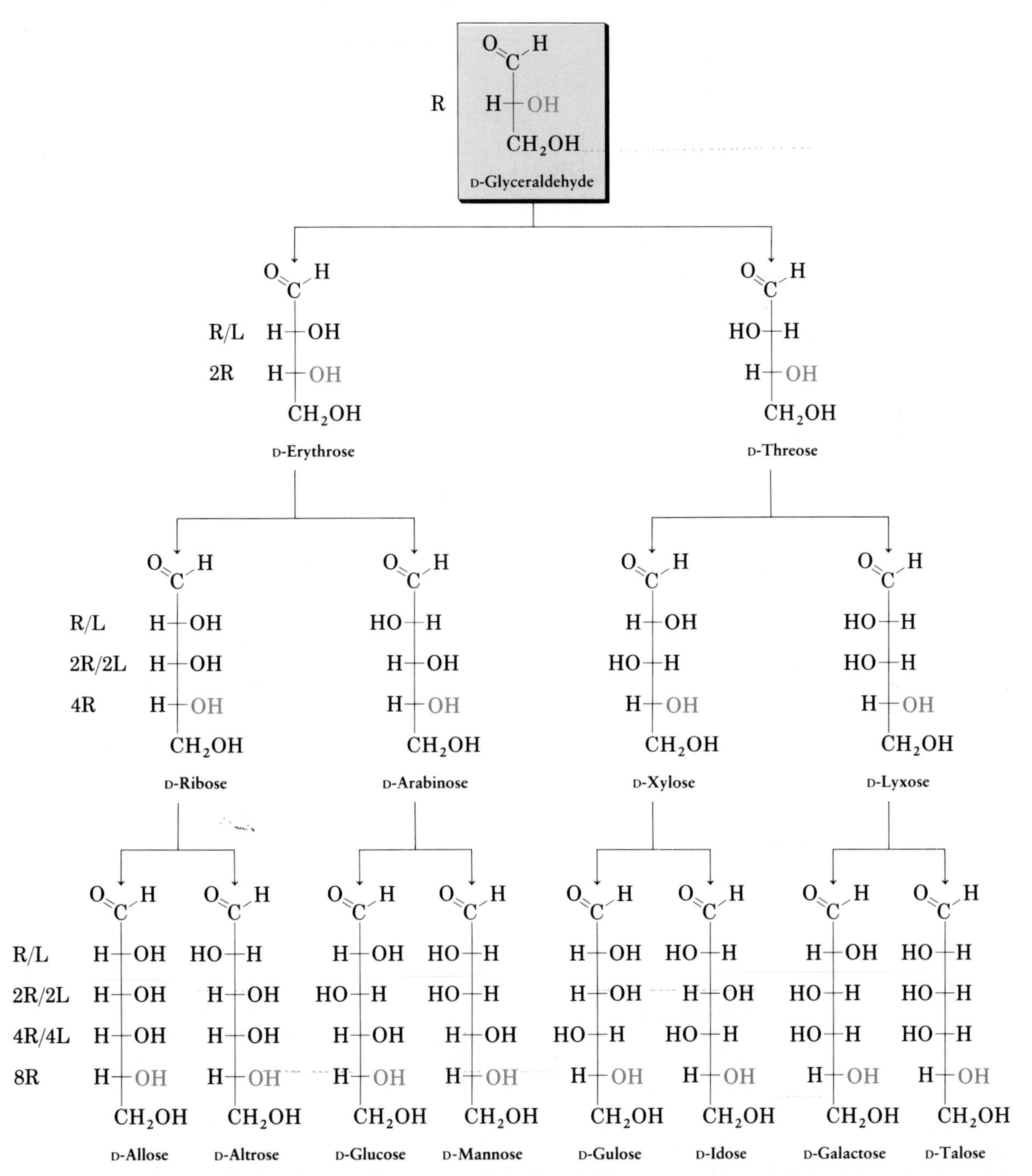

FIGURE 14.2 Configurations of D-aldoses: The structures are arranged in order from left to right so that the hydroxyl groups on C2 alternate right/left (R/L) in going across a series; the hydroxyl groups at C3 alternate two right/two left (2R/2L); the hydroxyl groups at C4 alternate four right/four left (4R/4L); and the hydroxyl groups at C5 are to the right in all eight (8R).

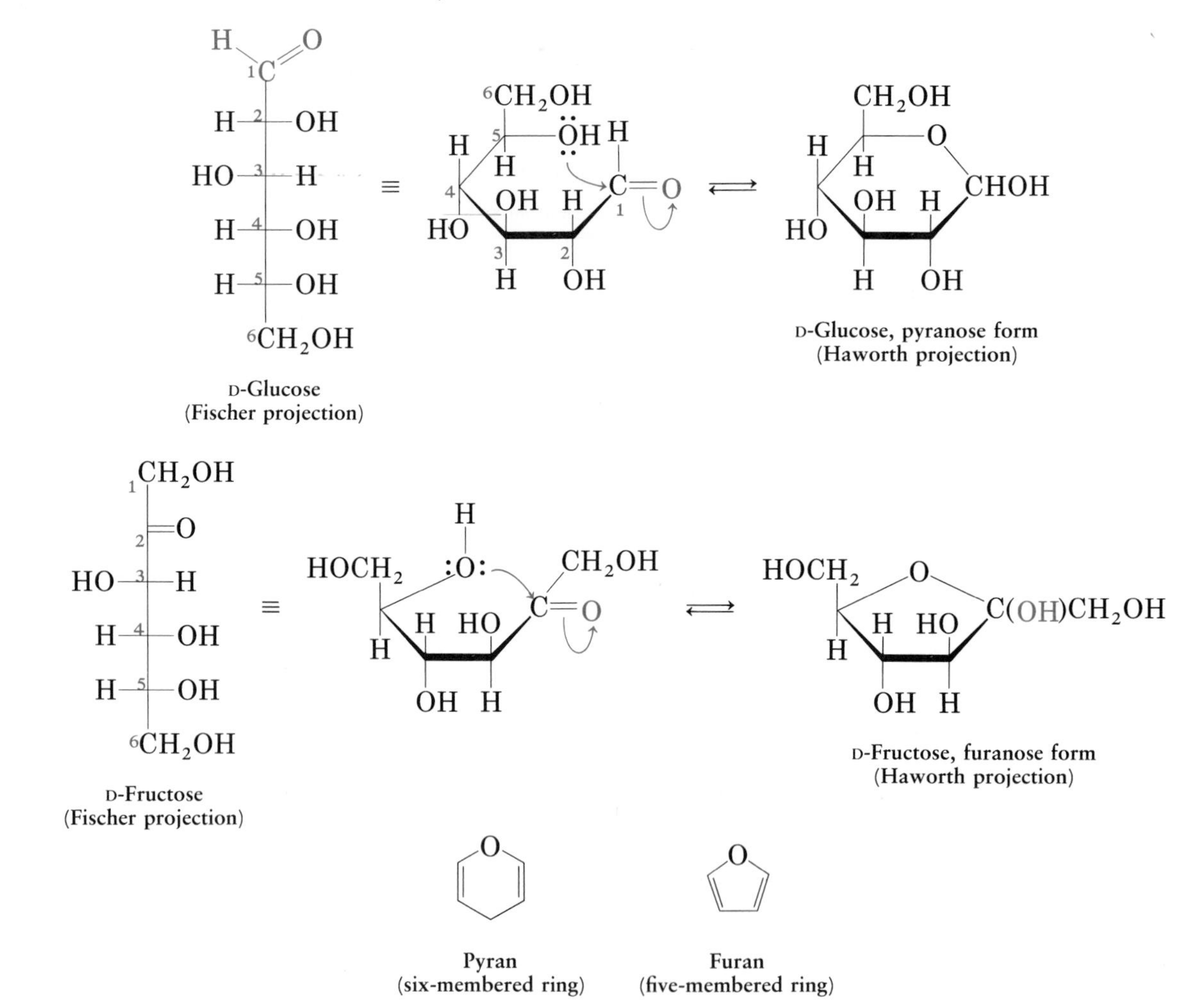

FIGURE 14.3 Glucose and fructose in their cyclic pyranose and furanose forms

Haworth projection a view of a furanose or pyranose sugar in which the ring is flat and viewed edge-on with the oxygen atom at the upper right

Pyranose and furanose rings are often represented using the **Haworth projections** shown in Figure 14.3, rather than Fischer projections. In a Haworth projection, the hemiacetal ring is drawn as if it were flat and is viewed edge-on, with the oxygen atom at the upper right. Although convenient, this view isn't really accurate because pyranose rings are actually chair-shaped like cyclohexane (Section 2.10), rather than flat. Nevertheless, Haworth projections are widely used because they make it possible to see at a glance the cis–trans relationships among hydroxyl groups on the ring.

When converting from one kind of projection to the other, remember that a hydroxyl on the *right* in a Fischer projection is *down* in a Haworth projection. Conversely, a hydroxyl on the *left* in a Fischer projection is *up* in a Haworth projection. For D sugars, the terminal $-CH_2OH$ group is always up in Haworth projections, whereas for L sugars, the $-CH_2OH$ group is down. Figure 14.4 illustrates the conversion for glucose.

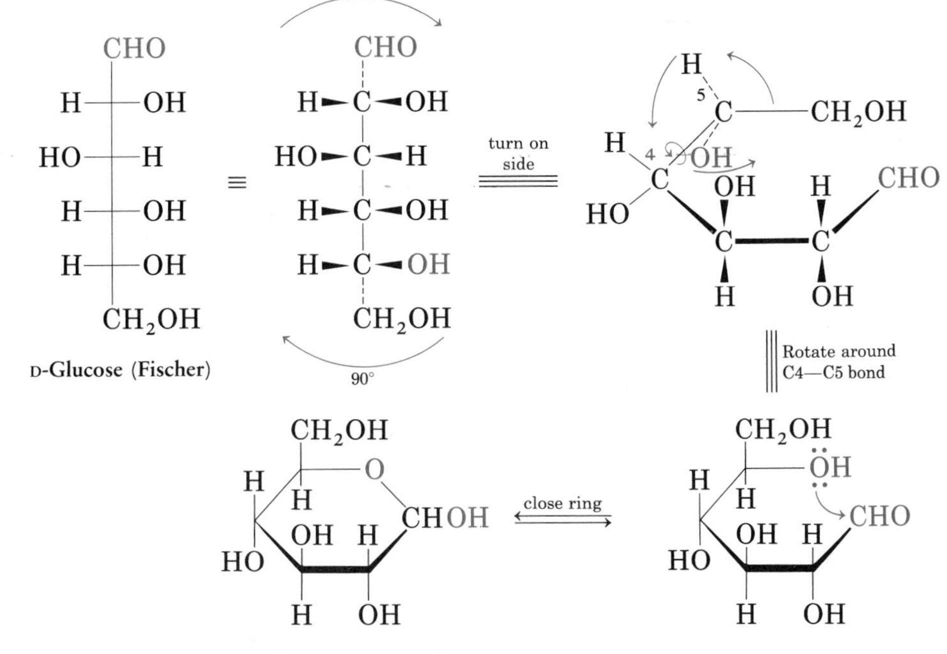

FIGURE 14.4 Interconversion of Fischer and Haworth projections of D-glucose

PRACTICE PROBLEM 14.5 D-Mannose differs from D-glucose in its stereochemistry at C2. Draw a Haworth projection of D-mannose in its pyranose form.

SOLUTION First draw a Fischer projection of D-mannose. Then lay it on its side and curl it around so that the aldehyde group (C1) is toward the front and the CH_2OH (C6) is toward the rear. Now connect the hydroxyl at C5 to the C1 carbonyl group to form a pyranose ring.

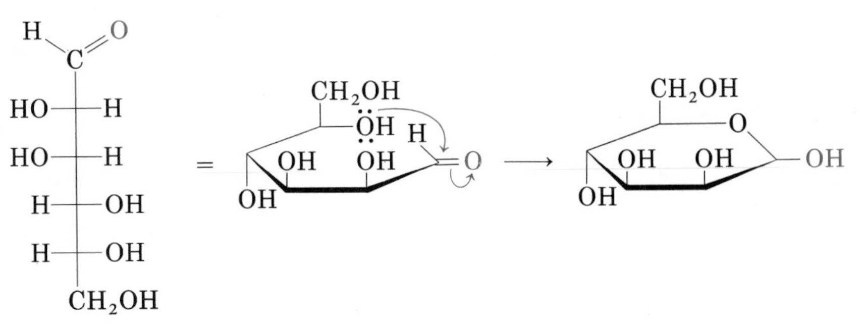

PROBLEM 14.10 D-Galactose differs from D-glucose in its stereochemistry at C4. Draw a Haworth projection of D-galactose in its pyranose form.

PROBLEM 14.11 Ribose exists largely in a furanose form produced by addition of the C4 hydroxyl group to the C1 aldehyde. Draw a Haworth projection of D-ribose in its furanose form.

14.6 MONOSACCHARIDE ANOMERS: MUTAROTATION

anomer a pyranose or furanose sugar whose hydroxyl group at C1 is either up (β) or down (α)

anomeric center the hemiacetal carbon in a pyranose or furanose sugar

When an open-chain monosaccharide cyclizes to a furanose or pyranose form, a new chiral center is formed at what used to be the carbonyl carbon. Two diastereomers called **anomers** (*an*-oh-mers) are produced, with the hemiacetal carbon referred to as the **anomeric center.** For example, glucose cyclizes reversibly in aqueous solution to yield a 36:64 mixture of two anomers. The minor anomer with the C1 –OH group trans to the $-CH_2OH$ substituent at C5 (and therefore down in a Haworth projection) is called the *alpha* (α) *anomer;* its complete name is α-D-glucopyranose. The major anomer with the C1 –OH group cis to the $-CH_2OH$ substituent at C5 (and therefore up in a Haworth projection), is called the *beta* (β) *anomer;* its complete name is β-D-glucopyranose.

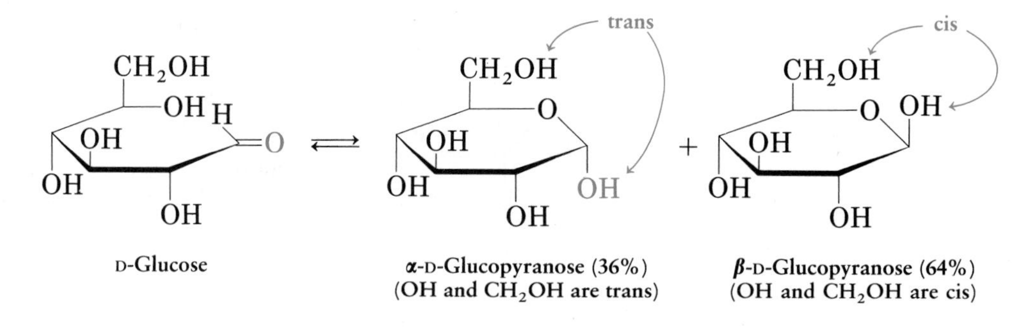

D-Glucose

α-D-Glucopyranose (36%) **(OH and CH_2OH are trans)**

β-D-Glucopyranose (64%) **(OH and CH_2OH are cis)**

Both anomers of D-glucose can be crystallized and purified. Pure α-D-glucose has a melting point of 146°C and a specific rotation $[\alpha]_D$ of +112.2°; pure β-D-glucose has a melting point of 148–155°C and a specific rotation of +18.7°. When a sample of either pure α-D-glucose or pure β-D-glucose is dissolved in water, however, both optical rotations slowly change and ultimately converge to a constant value of +52.6°. The specific rotation of the α-anomer solution decreases from +112.2° to +52.6°, and the specific rotation of the β-anomer solution increases from +18.7° to +52.6°. Known as **mutarotation,** this phenomenon is due to the slow conversion of the pure α and β enantiomers into the 36:64 equilibrium mixture.

mutarotation the change in optical rotation observed when a solution of a single sugar anomer equilibrates to a mixture of anomers

Mutarotation occurs by a reversible ring-opening of each anomer to the open-chain aldehyde form, followed by reclosure. Although equilibration is slow at neutral pH, it is catalyzed by both acid and base.

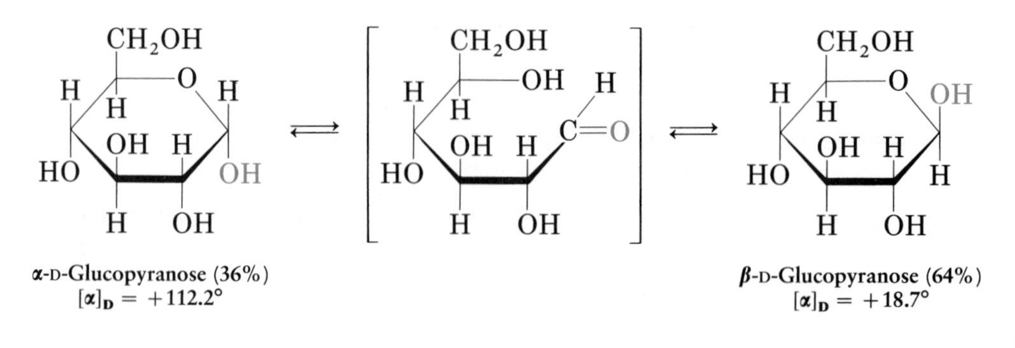

α-D-Glucopyranose (36%) $[\alpha]_D = +112.2°$

β-D-Glucopyranose (64%) $[\alpha]_D = +18.7°$

PRACTICE PROBLEM 14.6 Draw Haworth projections of the two pyranose anomers of D-galactose and identify each as α or β.

SOLUTION The alpha anomer has the $-OH$ group at C1 pointing down, and the beta anomer has the $-OH$ group at C1 pointing up.

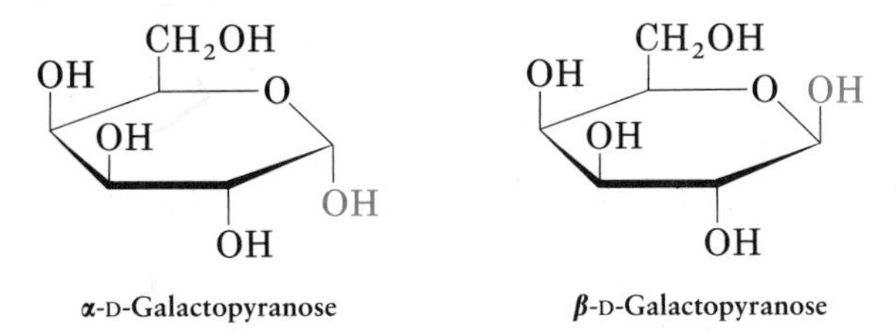

α-D-Galactopyranose β-D-Galactopyranose

PROBLEM 14.12 Draw the two anomers of D-fructose in their furanose forms.

PROBLEM 14.13 If the specific rotation of pure α-D-glucose is $+112.2°$ and the specific rotation of pure β-D-glucose is $+18.7°$, show how the equilibrium percentages of α and β anomers can be calculated from the equilibrium specific rotation of $+52.6°$.

PROBLEM 14.14 Many other sugars besides glucose exhibit mutarotation. For example, α-D-galactose has $[\alpha]_D = +150.7°$, and β-D-galactose has $[\alpha]_D = +52.8°$. If either anomer is dissolved in water and allowed to reach equilibrium, the specific rotation of the solution is $+80.2°$. What are the percentages of each anomer at equilibrium?

14.7 CONFORMATIONS OF MONOSACCHARIDES

Although Haworth projections are easy to draw, they don't give an accurate three-dimensional picture of molecular conformation. Pyranose rings, like cyclohexane rings (Section 2.10), have a chairlike geometry with axial and equatorial substituents. Any substituent that's up in a Haworth projection is also up in a chair conformation, and any substituent that's down in a Haworth projection is also down in the chair formulation. Haworth projections can be converted into chair representations by following three steps:

1. Draw the Haworth projection with the ring oxygen atom at the upper right.
2. Raise the leftmost carbon atom (C4) *above* the ring plane.
3. Lower the anomeric carbon atom (C1) *below* the ring plane.

Figure 14.5 shows how this is done for α-D-glucopyranose and β-D-glucopyranose.

Note that in β-D-glucopyranose, all the substituents on the ring are equatorial. Thus, β-D-glucopyranose is the least sterically crowded and most stable of the eight D-aldohexoses.

PROBLEM 14.15 Draw β-D-galactopyranose in its chair conformation. Label all the ring substituents as axial or equatorial.

PROBLEM 14.16 Draw β-D-mannopyranose in its chair conformation and label all substituents as axial or equatorial. Which would you expect to be more stable, mannose or galactose (Problem 14.15)?

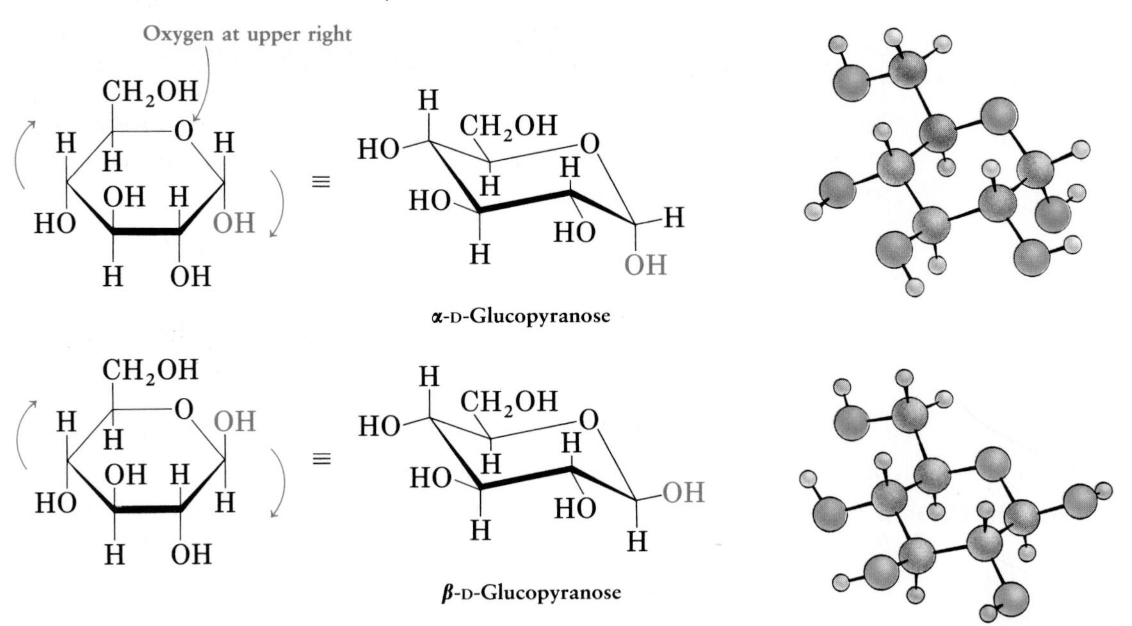

FIGURE 14.5 Chair representations of α-D-glucopyranose and β-D-glucopyranose

14.8 REACTIONS OF MONOSACCHARIDES

Since monosaccharides contain only two kinds of functional groups, carbonyls and hydroxyls, most of the chemistry of monosaccharides is the now-familiar chemistry of these two groups.

Ester and Ether Formation

Monosaccharides behave as simple alcohols in much of their chemistry. For example, carbohydrate hydroxyl groups can be converted into esters and ethers. Ester and ether derivatives of carbohydrates are often much easier to work with than the free sugars. Because of their many hydroxyl groups, monosaccharides are usually soluble in water but insoluble in organic solvents such as ether. Ester and ether derivatives, however, are soluble in organic solvents and are easily crystallized.

Esterification is carried out by treating the carbohydrate with an acid chloride or acid anhydride in the presence of a base. All the hydroxyl groups react, including the anomeric one. For example, β-D-glucopyranose is converted into its pentaacetate by treatment with acetic anhydride in pyridine solution.

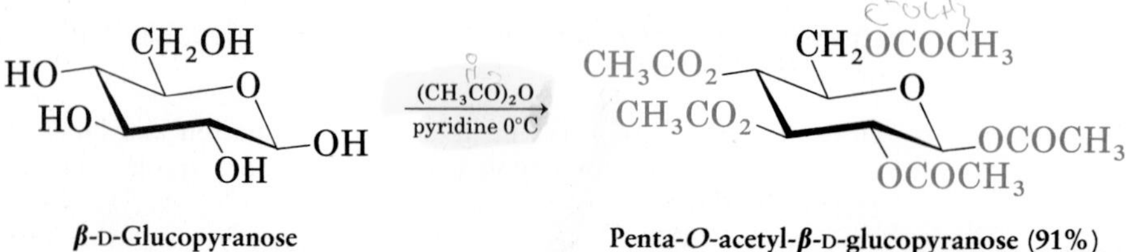

Carbohydrates are converted into ethers by treatment with an alkyl halide in the presence of base, the Williamson ether synthesis (Section 8.7). Silver oxide is a particularly mild and useful base for this reaction, since hydroxide and alkoxide bases tend to degrade the sensitive sugar molecules. For example, α-D-glucopyranose is converted into its pentamethyl ether in 85% yield on reaction with iodomethane and silver oxide.

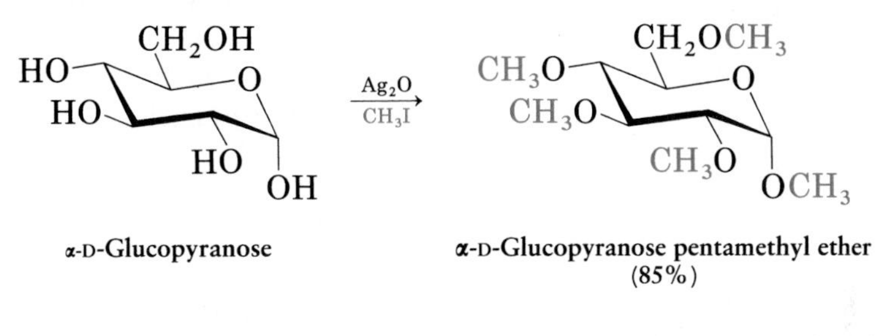

α-D-Glucopyranose α-D-Glucopyranose pentamethyl ether (85%)

PROBLEM 14.17 Draw the products you would obtain by reaction of β-D-ribofuranose with:
(a) CH_3I, Ag_2O (b) $(CH_3CO)_2O$, pyridine

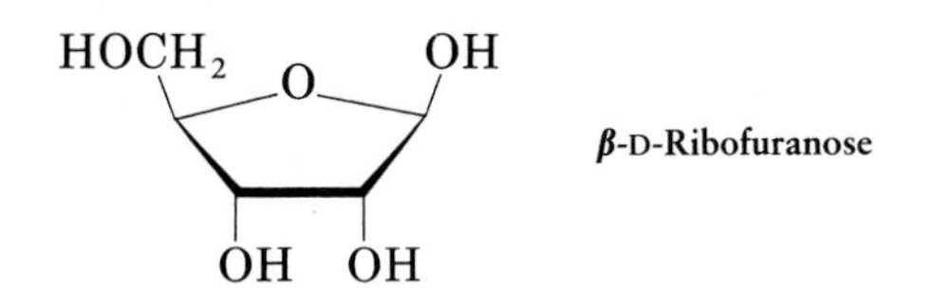

β-D-Ribofuranose

Glycoside Formation

We saw in Section 9.14 that treatment of a hemiacetal with an alcohol and an acid catalyst yields an acetal.

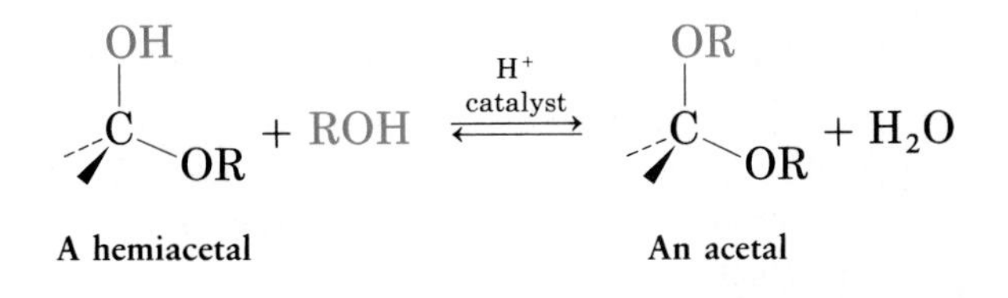

A hemiacetal An acetal

In the same way, treatment of a monosaccharide hemiacetal with an alcohol and an acid catalyst yields an acetal in which the anomeric hydroxyl group has been replaced by an alkoxy group. For example, reaction of glucose with methanol gives methyl β-D-glucopyranoside:

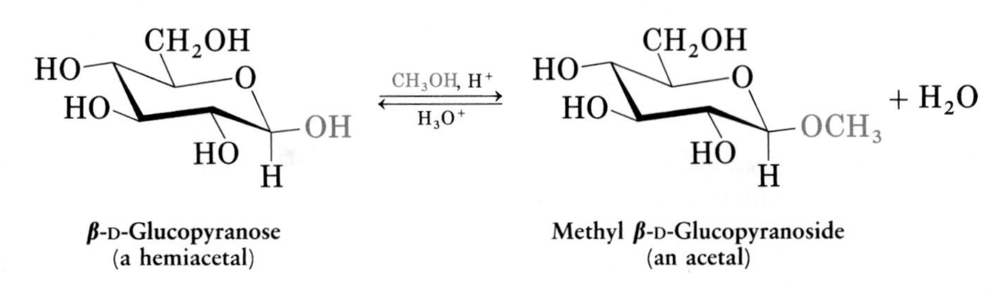

β-D-Glucopyranose (a hemiacetal) Methyl β-D-Glucopyranoside (an acetal)

glycoside a carbohydrate acetal formed by reaction of a carbohydrate with an alcohol

Carbohydrate acetals are called **glycosides.** They are named by first citing the alkyl group and then replacing the *-ose* ending of the sugar with *-oside*. Glycosides, like all acetals, are stable to water. They aren't in equilibrium with an open-chain form, and they don't show mutarotation. They can, however, be converted back to the free monosaccharide by hydrolysis with aqueous acid.

Glycosides are widespread in nature, and a great many biologically active molecules contain glycosidic linkages. For example, digitoxin, the active component of the digitalis preparations used for treatment of heart disease, is a glycoside consisting of a complex steroid alcohol linked to a trisaccharide (Figure 14.6). Note that the three sugars are also linked to each other by glycoside bonds.

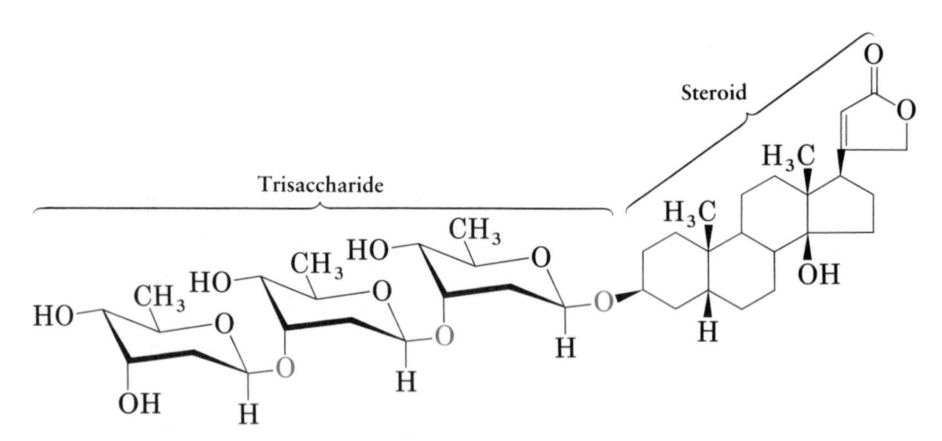

FIGURE 14.6 The structure of digitoxin, a complex glycoside

PRACTICE PROBLEM 14.7 What problem would you expect from acid-catalyzed reaction of β-D-ribofuranose with methanol?

SOLUTION Acid-catalyzed reaction of a monosaccharide with an alcohol yields a glycoside in which the anomeric $-OH$ group is replaced by the $-OR$ group of the alcohol:

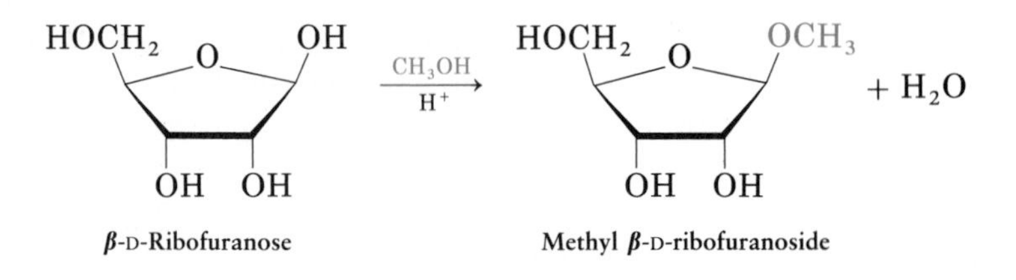

PROBLEM 14.18 Draw the product you would obtain from acid-catalyzed reaction of β-D-galactopyranose with ethanol.

Reduction of Monosaccharides

alditol a polyalcohol formed by reduction of a ketose or aldose

Treatment of an aldose or ketose with sodium borohydride reduces it to a polyalcohol called an **alditol.** The reaction occurs by interception of the open-chain form present in the aldehyde ⇄ hemiacetal equilibrium.

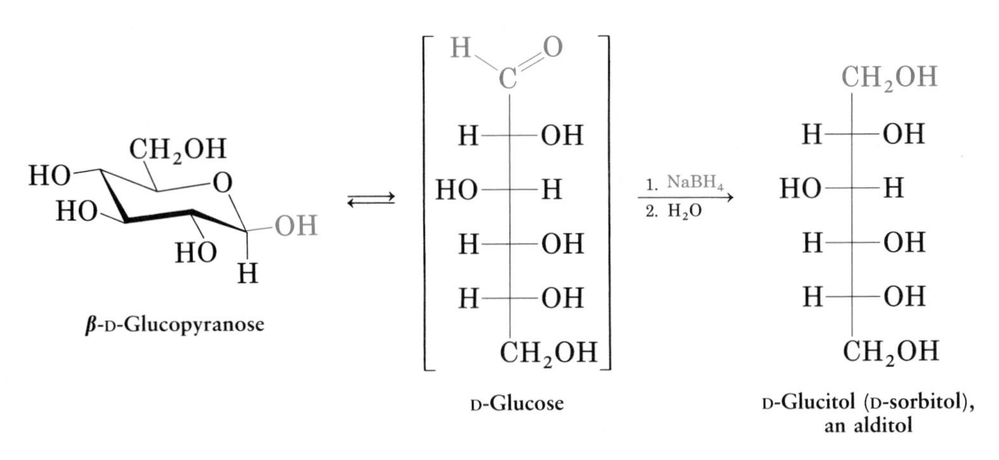

D-Glucitol, the alditol produced on reduction of D-glucose, is itself a naturally occurring substance that has been isolated from many fruits and berries. It is used under the name D-sorbitol as a sweetener and sugar substitute in many foods.

PRACTICE PROBLEM 14.8 Show the structure of the alditol you would obtain from reduction of D-galactose.

SOLUTION First draw D-galactose in its open-chain form and then convert the $-CHO$ group at C1 into a $-CH_2OH$ group.

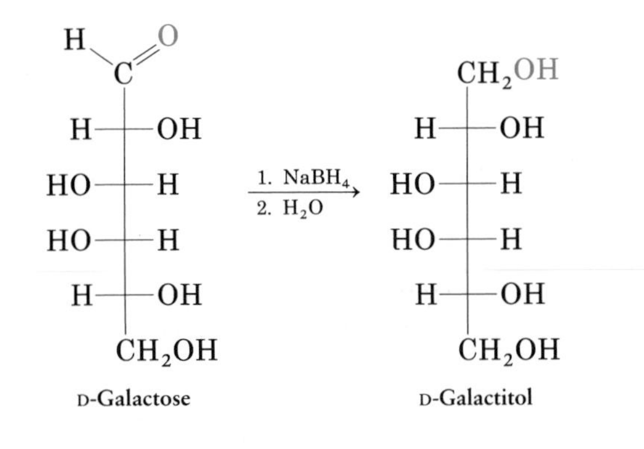

PROBLEM 14.19 How can you account for the fact that reduction of D-glucose leads to an optically active alditol (D-glucitol) whereas reduction of D-galactose leads to an optically inactive alditol?

PROBLEM 14.20 Reduction of L-gulose with $NaBH_4$ leads to the same alditol (D-glucitol) as reduction of D-glucose. Explain this result.

Oxidation of Monosaccharides

aldonic acid a polyhydroxy monocarboxylic acid formed by oxidation of an aldose

reducing sugar a sugar that reduces Tollens' and Fehling's reagents because it contains a hemiacetal group

Like other aldehydes, aldoses are easily oxidized to yield carboxylic acids called **aldonic acids.** Aldoses react with Tollens' reagent (Ag^+ in aqueous ammonia), Fehling's reagent (Cu^{2+} with aqueous sodium tartrate), and Benedict's reagent (Cu^{2+} with aqueous sodium citrate) to yield the oxidized sugar and a reduced metallic species. All three reactions serve as simple chemical tests for what are called **reducing sugars** (*reducing* because the sugar reduces the metallic oxidizing agent.)

If Tollens' reagent is used, metallic silver is produced as a shiny mirror on the walls of the reaction flask or test tube. If Fehling's or Benedict's reagent is used, a reddish precipitate of cuprous oxide signals a positive result. Some diabetes self-test kits sold in drugstores for home use employ Benedict's test. As little as 0.1% glucose in urine gives a positive test.

All aldoses are reducing sugars since they contain aldehyde carbonyl groups, but glycosides are nonreducing. Glycosides don't react with Tollens' or Fehling's reagents because the acetal group can't open to an aldehyde under basic conditions.

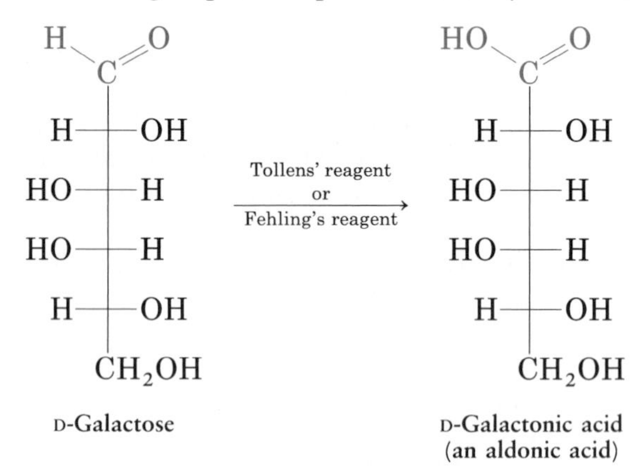

D-Galactose

D-Galactonic acid (an aldonic acid)

aldaric acid a polyhydroxy dicarboxylic acid formed by oxidation of an aldose

If warm dilute nitric acid is used as the oxidizing agent, aldoses are oxidized to dicarboxylic acids called **aldaric acids.** Both the aldehyde carbonyl and the terminal $-CH_2OH$ group are oxidized in this reaction.

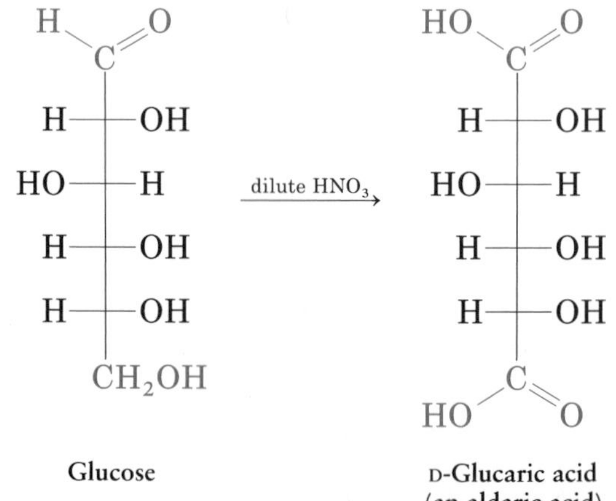

Glucose

D-Glucaric acid (an aldaric acid)

PROBLEM 14.21 D-Glucose yields an optically active aldaric acid on treatment with nitric acid, but D-allose yields an optically inactive aldaric acid. Explain.

PROBLEM 14.22 Which of the other six D-aldohexoses yield optically active aldaric acids, and which yield optically inactive aldaric acids? (See Problem 14.21.)

14.9 DISACCHARIDES

We saw in the previous section that reaction of a monosaccharide hemiacetal yields a glycoside in which the anomeric hydroxyl group is replaced by an alkoxyl substituent. If the alcohol is another sugar, the glycoside product is a disaccharide.

Cellobiose and Maltose

1,4′ link
a glycosidic link between the C1 carbonyl group of one sugar with the C4 hydroxyl group of another sugar

Disaccharides can contain a glycosidic acetal bond between the anomeric carbon of one sugar and a hydroxyl group at *any* position on the other sugar. A glycosidic link between C1 of the first sugar and C4 of the second sugar, called a **1,4′ link,** is particularly common. (The prime indicates that the 4′ position is on a sugar other than the nonprime 1 position.)

A glycosidic bond to the anomeric carbon can be either alpha or beta. *Cellobiose,* the disaccharide obtained by partial hydrolysis of cellulose, consists of two D-glucopyranoses joined by a 1,4′-β-glycoside bond. *Maltose,* the disaccharide obtained by partial hydrolysis of starch, consists of two D-glucopyranoses joined by a 1,4′-α-glycoside bond.

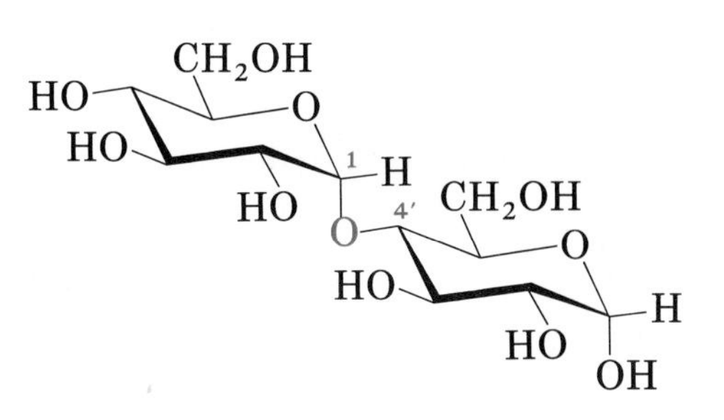

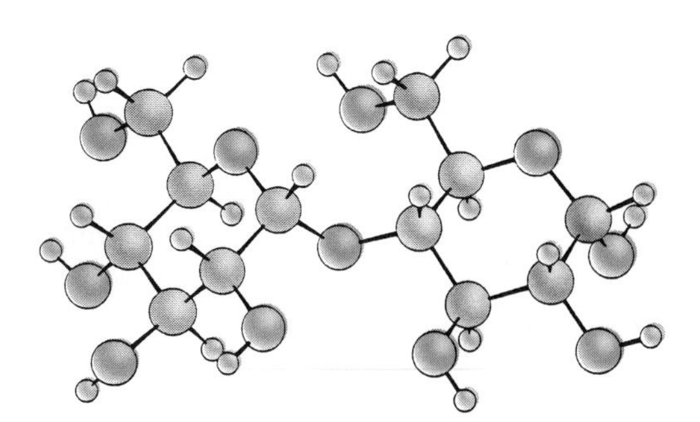

Maltose, a 1,4′-α-glycoside
[4-*O*-(α-D-glucopyranosyl)-α-D-glucopyranose]

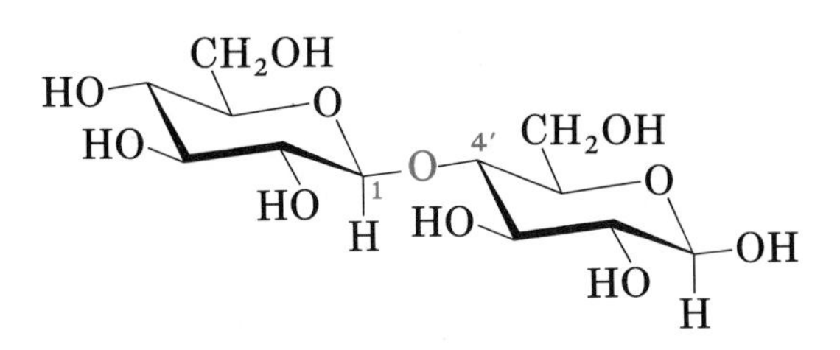

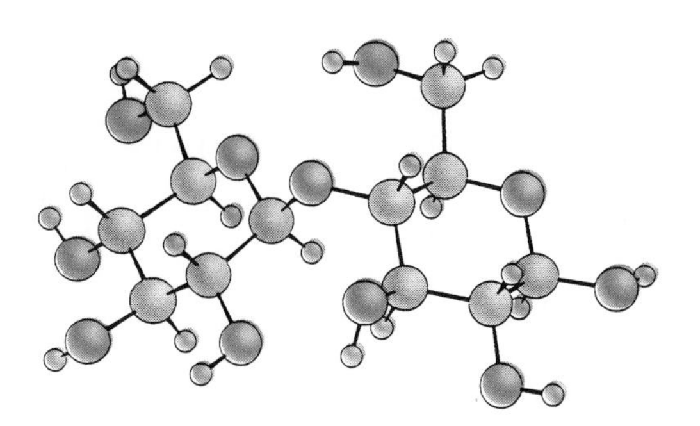

Cellobiose, a 1,4′-β-glycoside
[4-*O*-(β-D-glucopyranosyl)-β-D-glucopyranose]

Maltose and cellobiose are both reducing sugars because the right-hand saccharide units have hemiacetal groups. Both are in equilibrium with aldehyde forms, which can reduce Tollens' or Fehling's reagent. For a similar reason, both maltose and cellobiose exhibit mutarotation of the alpha and beta anomers of the glucopyranose unit on the right.

Despite the similarities of their structures, maltose and cellobiose are dramatically different biologically. Cellobiose can't be digested by humans and can't be fermented by yeast. Maltose, however, is digested without difficulty and is readily fermented.

PROBLEM 14.23 Draw the structures of the product obtained from reaction of cellobiose with these reagents.
(a) $NaBH_4$ (b) $AgNO_3$, H_2O, NH_3

Sucrose

Sucrose, ordinary table sugar, is probably the most abundant pure organic chemical in the world. Whether from sugar cane (20% by weight) or from sugar beets (15% by weight), and whether raw or refined, all table sugar is sucrose.

Sucrose is a disaccharide that yields one equivalent of glucose and one equivalent of fructose on hydrolysis of its glycoside link. This 1:1 mixture of glucose and fructose is often referred to as *invert sugar* because the sign of optical rotation changes (inverts) during the hydrolysis from sucrose, $[\alpha]_D = +66.5°$, to a glucose/fructose mixture, $[\alpha]_D = -22°$. Certain insects such as honeybees have enzymes called *invertases* that catalyze the hydrolysis of sucrose to glucose + fructose. Honey, in fact, is primarily a mixture of glucose, fructose, and sucrose.

Unlike most other disaccharides, sucrose isn't a reducing sugar and doesn't exhibit mutarotation. These observations imply that sucrose has no hemiacetal groups, suggesting that the glucose and fructose units must both be glycosides. This can happen only if the two sugars are joined by a glycoside link between anomeric carbons, C1 of glucose and C2 of fructose.

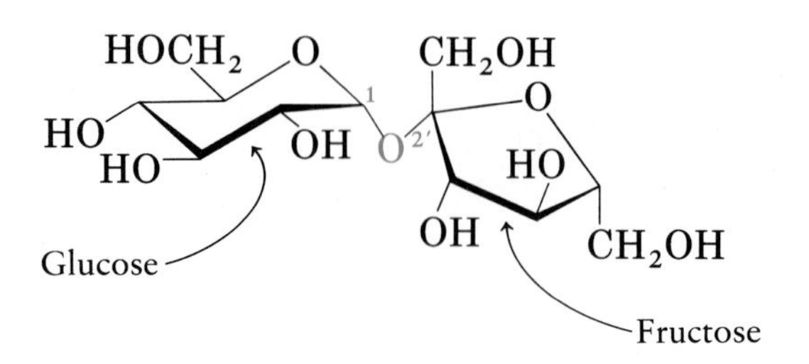

Sucrose, a 1,2′-glycoside

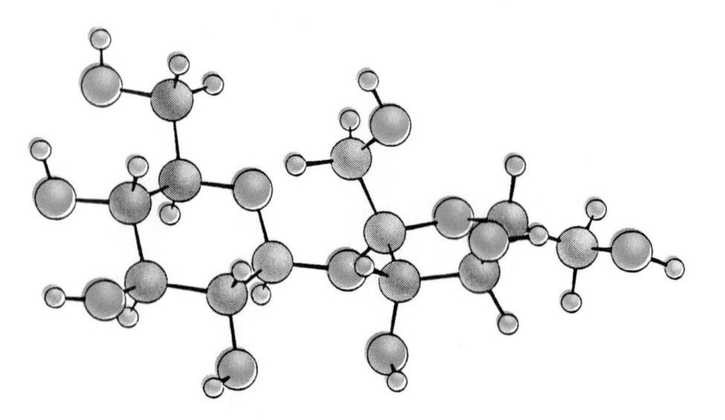

14.10 POLYSACCHARIDES

Polysaccharides are carbohydrates in which tens, hundreds, or even thousands of simple sugars are linked by glycoside bonds. Since these compounds have no free anomeric hydroxyls (except for one at the end of the chain), they aren't reducing sugars and don't show mutarotation. Cellulose and starch are the two most widely occurring polysaccharides.

Cellulose

Cellulose consists simply of D-glucose units linked by the 1,4′-β-glycoside bonds we saw in cellobiose. Several thousand glucose units are linked to form one large molecule, and different molecules can then interact to form a large aggregate structure held together by hydrogen bonds.

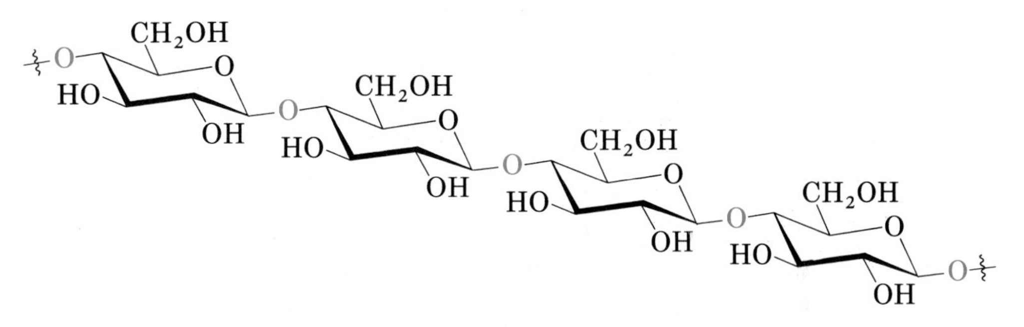

Cellulose, a 1,4′-*O*-(β-D-glucopyranoside) polymer

Nature uses cellulose primarily as a structural material to impart strength and rigidity to plants. Wood, leaves, grasses, and cotton are primarily cellulose. Cellulose also serves as raw material for the manufacture of cellulose acetate, known commercially as acetate rayon.

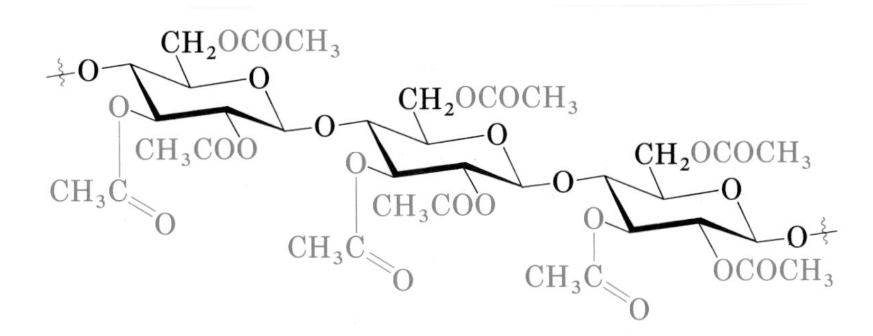

A segment of cellulose acetate (rayon)

Starch and Glycogen

Starch is a glucose polymer whose monosaccharide units are linked by the 1,4′-α-glycoside bonds we saw in maltose. Starch can be separated into two fractions called *amylopectin* and *amylose*. Amylose, which accounts for about 20% by weight of starch, consists of several hundred glucose molecules linked by 1,4′-α-glycoside bonds.

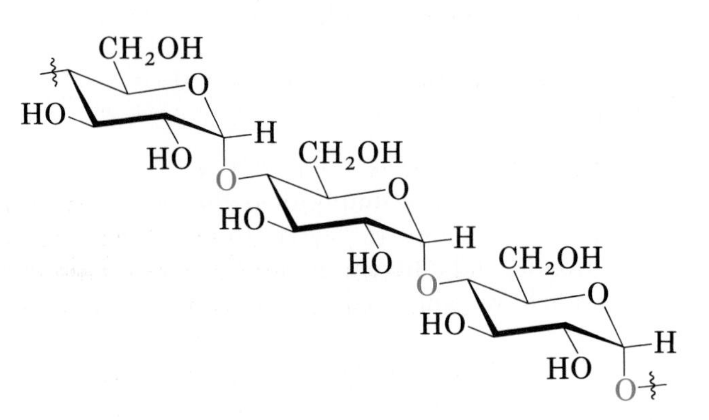

Amylose, a 1,4′-*O*-(α-D-glucopyranoside) polymer

Amylopectin, which accounts for the remaining 80% of starch, is more complex in structure than amylose. Unlike cellulose or amylose, which are linear polymers, amylopectin contains 1,6′-α-glycoside *branches* approximately every 25 glucose units. As a result, amylopectin has an exceedingly complex three-dimensional structure (Figure 14.7). Nature uses starch as the means by which plants store energy for later use. Potatoes, corn, and cereal grains contain large amounts of starch.

When eaten, starch is digested in the mouth and stomach by enzymes called *glycosidases,* which catalyze the hydrolysis of glycoside bonds and release individual

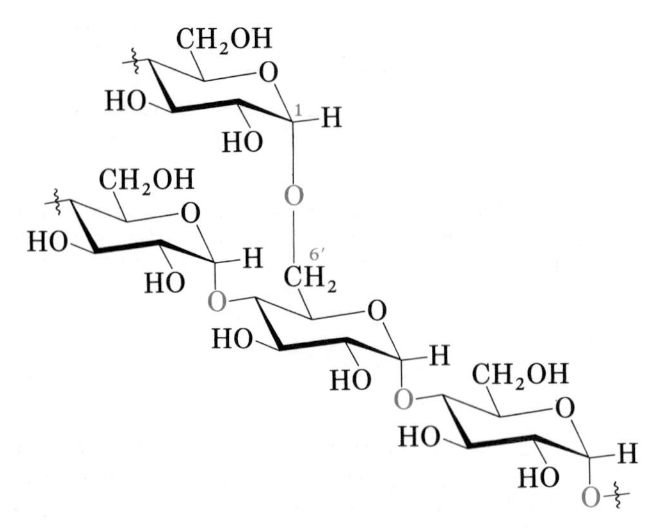

FIGURE 14.7 A 1,6′-α-glycoside branch in amylopectin

molecules of glucose. Like most enzymes, glycosidases are highly selective in their action. They hydrolyze only the α glycoside links in starch and leave the β glycoside links in cellulose untouched. Thus humans can eat potatoes and grains but not grass and wood.

Glycogen is a polysaccharide that serves the same energy-storage purpose in animals that starch serves in plants. Dietary carbohydrate that isn't needed for immediate energy is converted by the body to glycogen for long-term storage. Like the amylopectin found in starch, glycogen contains a complex three-dimensional structure with both 1,4′ and 1,6′ links (Figure 14.8). Glycogen molecules are larger than those of amylopectin—up to 100,000 glucose units—and contain even more branches.

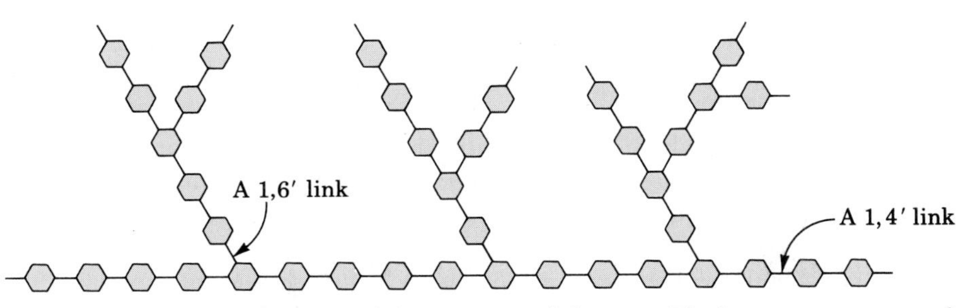

FIGURE 14.8 A representation of the structure of glycogen. The hexagons represent glucose units linked by 1,4′ and 1,6′ acetal bonds.

14.11 OTHER IMPORTANT CARBOHYDRATES

In addition to the common carbohydrates mentioned in previous sections, there are a variety of important carbohydrate-derived materials whose structures have been chemically modified. Their structural resemblance to sugars is clear, but they aren't simple aldoses or ketoses.

Deoxy Sugars

deoxy sugar a sugar with an −OH group missing from one carbon

Deoxy sugars differ from normal sugars by having one of their oxygen atoms "missing." In other words, an −OH group is replaced by an −H. 2-Deoxyribose, a sugar found in DNA (deoxyribonucleic acid), is the most important deoxy sugar. Note that 2-deoxyribose adopts a furanose (five-membered) form.

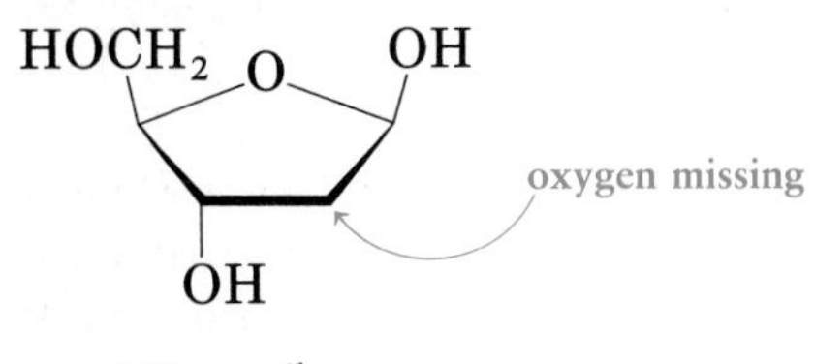

2-Deoxyribose

Amino Sugars

amino sugar a sugar with an $-OH$ group on one carbon replaced by a $-NH_2$ group

Amino sugars such as D-glucosamine have one of their $-OH$ groups replaced by an $-NH_2$. The *N*-acetyl amide derived from D-glucosamine is the monosaccharide unit from which *chitin,* the hard crust that protects insects and shellfish, is built. Still other amino sugars are found in antibiotics such as streptomycin and gentamicin.

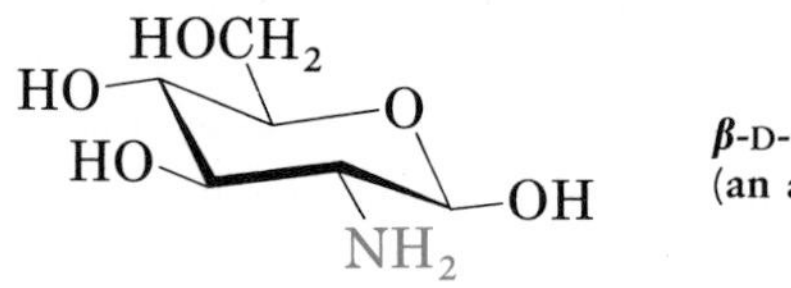

β-D-Glucosamine (an amino sugar)

14.12 CELL-SURFACE CARBOHYDRATES

For many years, carbohydrates were thought to be uninteresting compounds whose only biological purposes were as structural materials and energy sources. Although carbohydrates do indeed fill these roles, recent work has shown that they perform many other important biochemical functions as well. Polysaccharides are now known to be centrally involved in the critical process by which one cell type recognizes another. Small polysaccharide chains, covalently bound by glycoside links to hydroxyl groups on proteins (*glycoproteins*), act as biochemical labels on cell surfaces, as illustrated by the human blood-group antigens.

It has been known for over 80 years that human blood can be classified into four blood-group types—A, B, AB, and O—and that blood from a donor of one type can't be transfused into a recipient with another type unless the two types are compatible (Table 14.1). Should an incompatible mix be made, the red blood cells clump together, or *agglutinate.*

TABLE 14.1 Human-blood-group compatibilities

	Acceptor blood type			
Donor blood type	*A*	*B*	*AB*	*O*
A	o	x	o	x
B	x	o	o	x
AB	x	x	o	o
O	o	o	o	o

o = compatible; x = incompatible

The agglutination of incompatible red blood cells, which indicates that the recipient's immune system has recognized the presence of foreign cells in the body and has formed antibodies to them, results from the presence of polysaccharide markers on the surface of the cells. Type A, B, and O red blood cells each have characteristic markers called *antigenic determinants;* type AB cells have both type A and type B markers. The structures of all three blood-group determinants have been elucidated and are shown in Figure 14.9. All three contain *N*-acetylamino sugars as well as the unusual monosaccharide L-fucose.

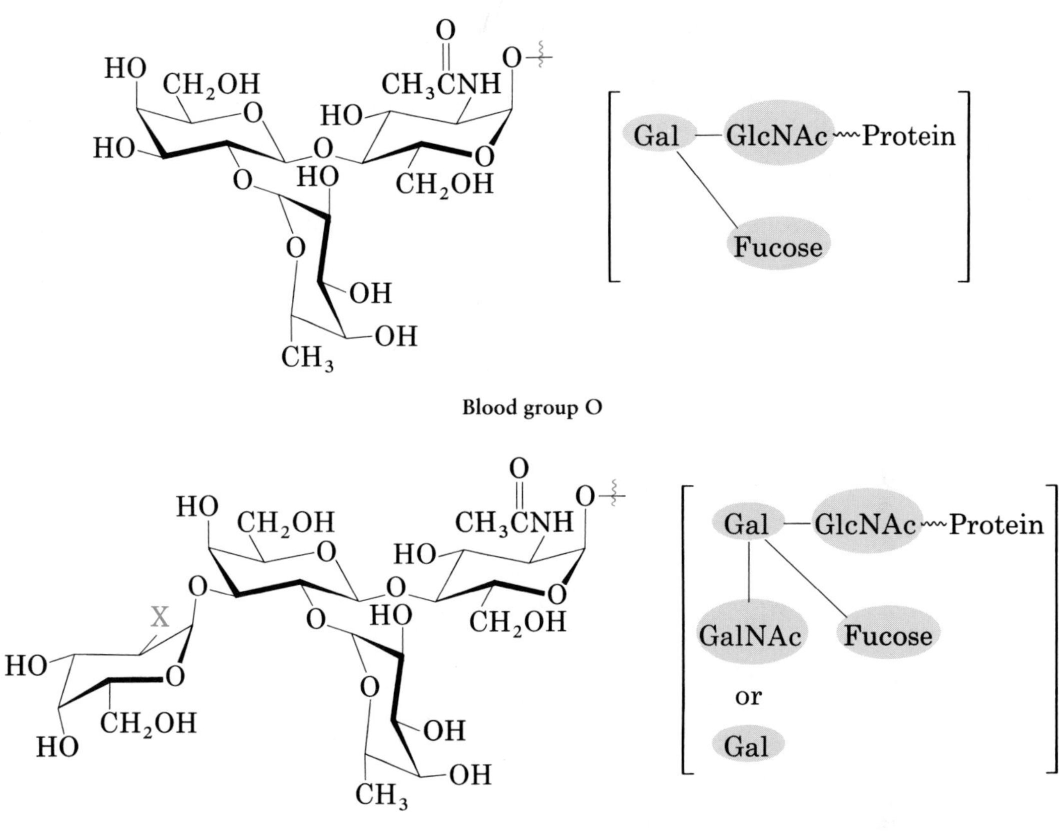

FIGURE 14.9 Structures of the A, B, and O blood-group antigenic determinants. (Gal is D-galactose; GlcNAc is *N*-acetylglucosamine; GalNAc is *N*-acetylgalactosamine.)

The antigenic determinant of blood-group O is a trisaccharide whereas the determinants of blood-groups A and B are tetrasaccharides. Type A and B determinants differ only in the substitution of an acetylamino group ($-NHCOCH_3$) for a hydroxyl in the terminal galactose residue.

INTERLUDE

Sweetness

Say the word *sugar* and people immediately think of sweet-tasting candies, donuts, and such. In fact most of the simple carbohydrates we've discussed in this chapter do taste sweet, but the degree of sweetness varies greatly from one sugar to another. With sucrose (table sugar) as a reference point, fructose is nearly twice as sweet, but lactose is only about one sixth as sweet. Comparisons are difficult, though, because sweetness is simply a matter of taste, and the ranking of sugars is a matter of personal opinion. Nevertheless, the ordering in the table is generally agreed on.

Sweetness of some sugars and sugar substitutes

Name	*Type*	*Sweetness*
Lactose	Disaccharide	0.16
Glucose	Monosaccharide	0.75
Sucrose	Disaccharide	**1.00**
Fructose	Monosaccharide	1.75
Cyclamate	Artificial	300
Aspartame	Artificial	1500
Saccharin	Artificial	3500

The desire of many people to cut their caloric intake has led to the development of the artificial sweeteners aspartame, saccharin, and cyclamate. All are far sweeter than natural sugars, but doubts have been raised as to their long-term safety. Cyclamates have been banned in the United States (but not in Canada), and saccharin has been banned in Canada (but not in the United States). None of the three has any structural resemblance to carbohydrates:

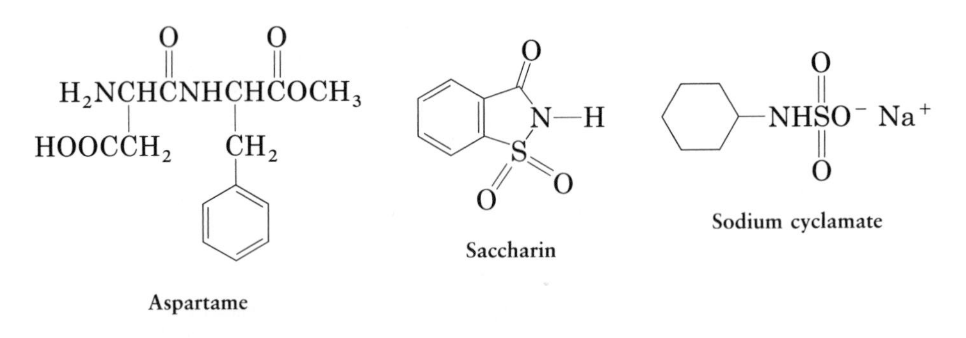

Aspartame

Saccharin

Sodium cyclamate

SUMMARY AND KEY WORDS

Carbohydrates are polyhydroxy aldehydes and ketones. They can be classified according to the number of carbon atoms and the kind of carbonyl group they contain. Thus, glucose is an **aldohexose,** a six-carbon aldehydo sugar. **Monosaccharides** are further classified as either D or L **sugars,** depending on the stereochemistry of the chiral carbon atom farthest from the carbonyl group. Most naturally occurring sugars are in the D series.

Monosaccharides normally exist as cyclic hemiacetals rather than as open-chain aldehydes or ketones. The hemiacetal linkage results from reaction of the carbonyl group with a hydroxyl group three or four carbon atoms away. A five-membered-ring hemiacetal is a **furanose;** a six-membered-ring hemiacetal is a **pyranose.** Cyclization leads to the formation of a new chiral center (the **anomeric center**) and to production of two diastereomeric hemiacetals called **alpha** and **beta anomers.**

Stereochemical relationships among monosaccharides are portrayed in several ways. **Fischer projections** display chiral carbon atoms as a pair of crossed lines. These projections are useful in allowing us to quickly relate one sugar to another, but cyclic **Haworth projections** provide a better view. Any group to the right in a Fischer projection is down in a Haworth projection.

Much of the chemistry of monosaccharides is the familiar chemistry of alcohol and carbonyl functional groups. Thus, the hydroxyl groups of carbohydrates form esters and ethers in the normal way. The carbonyl group of a monosaccharide can be reduced with sodium borohydride to yield an **alditol,** can be oxidized with Tollens' or Fehling's reagents to yield an **aldonic acid,** can be oxidized with warm nitric acid to yield an **aldaric acid,** and can be treated with an alcohol in the presence of acid catalyst to yield a **glycoside.**

Disaccharides are complex carbohydrates in which two simple sugars are linked by a glycoside bond between the anomeric carbon of one unit and a hydroxyl of the second unit. The two sugars can be the same, as in maltose and cellobiose, or different, as in sucrose. The glycoside bond can be either α (maltose) or β (cellobiose) and can involve any hydroxyl of the second sugar. A 1,4′ link is most common (cellobiose, maltose), but other links such as 1,6′ (amylopectin) and 1,2′ (sucrose) also occur. **Polysaccharides,** such as cellulose, starch, and glycogen, are used in nature both as structural materials and for long-term energy storage.

ADDITIONAL PROBLEMS

14.24 Classify the following sugars by type (for example, glucose is an aldohexose).

(a)
$$\begin{array}{c} CH_2OH \\ | \\ C{=}O \\ | \\ CH_2OH \end{array}$$

(b)
$$\begin{array}{rcl} & CH_2OH & \\ H- & + & -OH \\ & {=} & O \\ H- & + & -OH \\ & CH_2OH & \end{array}$$

(c)
$$\begin{array}{rcl} & CHO & \\ H- & + & -OH \\ HO- & + & -H \\ H- & + & -OH \\ HO- & + & -H \\ H- & + & -OH \\ & CH_2OH & \end{array}$$

14.25 Write open-chain structures for a ketotetrose and a ketopentose.

14.26 Write an open-chain structure for a deoxyaldohexose.

14.27 Write an open-chain structure for a five-carbon amino sugar.

14.28 The structure of ascorbic acid (vitamin C) is shown. Does ascorbic acid have a D or an L configuration?

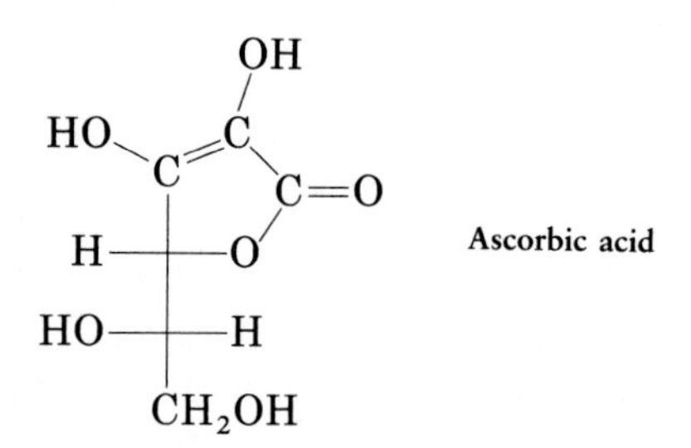
Ascorbic acid

14.29 Draw a Haworth projection of ascorbic acid (Problem 14.28)

14.30 Define the following terms, and give an example of each.

(a) Monosaccharide (b) Anomeric center (c) Haworth projection
(d) Fischer projection (e) Glycoside (f) Reducing sugar
(g) Pyranose form (h) 1,4′ Link (i) D-Sugar

14.31 The following cyclic structure is that of gulose. Is this a furanose or pyranose form? Is it an α or β anomer? Is it a D-sugar or an L-sugar?

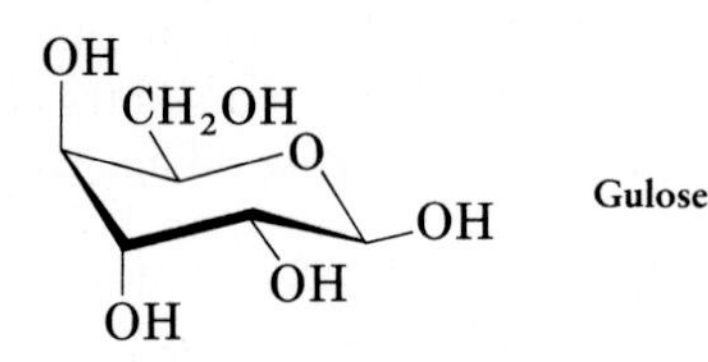
Gulose

14.32 Uncoil gulose (Problem 14.31) and write it in its open-chain form.

14.33 Draw D-ribulose in its five-membered cyclic β-hemiacetal form.

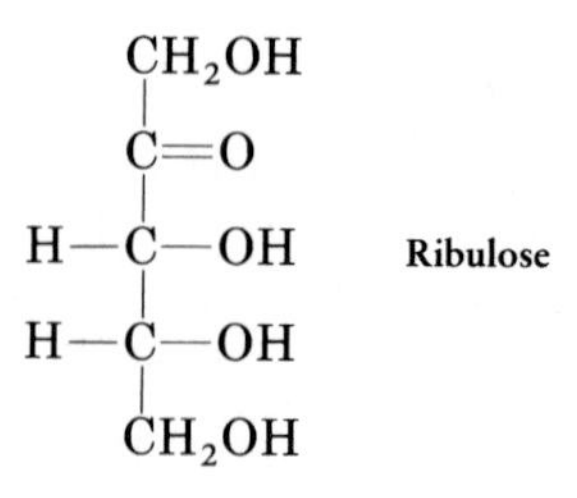
Ribulose

14.34 Look up the structure of D-talose in Table 14.2 and draw the β anomer in its pyranose form. Identify the ring substituents as axial or equatorial.

14.35 Draw structures for the products you would expect to obtain from reaction of β-D-talopyranose (Problem 14.34) with each of the following reagents.

(a) $NaBH_4$ (b) Warm dilute HNO_3 (c) $AgNO_3$, NH_3, H_2O
(d) CH_3CH_2OH, H^+ (e) CH_3I, Ag_2O (f) $(CH_3CO)_2O$, pyridine

14.36 What is the stereochemical relationship of D-allose to L-allose?

14.37 How many D-2-ketohexoses are there? Draw them.

14.38 One of the D-2-ketohexoses (Problem 14.37) is called *sorbose*. On treatment with $NaBH_4$, sorbose yields a mixture of gulitol and iditol. What is the structure of D-sorbose? (Gulitol and iditol are the alditols obtained by reduction of gulose and idose.)

14.39 Another D-2-ketohexose, *psicose*, yields a mixture of allitol and altritol when reduced with $NaBH_4$. What is the structure of psicose?

14.40 Fructose exists at equilibrium as an approximately 2:1 mixture of β-D-fructopyranose and β-D-fructofuranose. Draw both forms in Haworth projection.

14.41 Draw Fischer projections of these substances.

(a) (*R*)-2-Methylbutanoic acid (b) (*S*)-3-Methyl-2-pentanone

14.42 Convert these Fischer projections into tetrahedral representations.

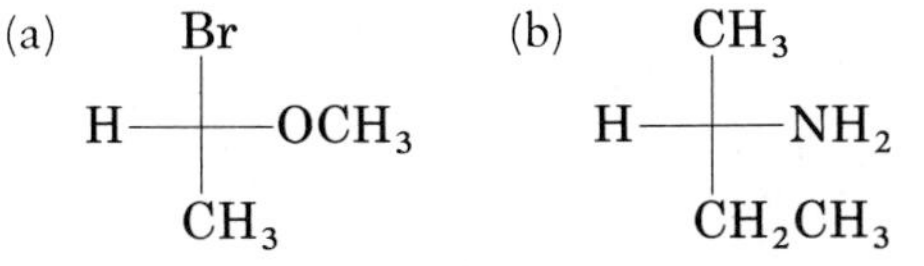

14.43 Which of the eight D-aldohexoses yield optically inactive (meso) alditols on reduction with $NaBH_4$?

14.44 What other D-aldohexose would give the same alditol as D-talose?

14.45 Which of the eight D-aldohexoses give the same aldaric acids as their L-enantiomers?

14.46 Which of the other three D-aldopentoses gives the same aldaric acid as D-lyxose?

14.47 The Ruff degradation is a method used to shorten an aldose chain by one carbon atom. The original C1 carbon atom is cleaved off, and the original C2 carbon atom becomes the aldehyde of the chain-shortened aldose. For example, D-glucose, an aldohexose, is converted by Ruff degradation into D-arabinose, an aldopentose. What other D-aldohexose would also yield D-arabinose on Ruff degradation?

14.48 D-Galactose and D-talose yield the same aldopentose on Ruff degradation (Problem 14.47). What does this tell you about the stereochemistries of galactose and talose? Which D-aldopentose is obtained?

14.49 The aldaric acid obtained by nitric-acid oxidation of D-erythrose, one of the D-aldotetroses, is optically inactive. The aldaric acid obtained from oxidation of the other D-aldotetrose, D-threose, is optically active, however. How does this information allow you to assign structures to the two D-aldotetroses?

14.50 Gentiobiose is a rare disaccharide found in saffron and gentian. It is a reducing sugar and forms only glucose on hydrolysis with aqueous acid. If gentiobiose contains a 1,6′-β-glycoside link, what is its structure?

14.51 The position of the glycosidic link between sugars in disaccharides can be determined by taking advantage of the known chemistry of acetal groups. For example, reaction of cellobiose with iodomethane and silver oxide yields an octamethyl ether derivative. Hydrolysis of the glycoside bonds in this derivative yields a tri-O-methylglucopyranose and a tetra-O-methylglucopyranose. Look up the structure of cellobiose and formulate the reactions by drawing structures for the octamethyl, tetramethyl, and trimethyl ethers. How can you use this information to determine the position of the glycoside link?

14.52 Trehalose is a nonreducing disaccharide found in the blood of insects. Reaction with iodomethane and silver oxide, followed by acidic hydrolysis, yields 2 equivalents of 2,3,4,6-tetra-O-methylglucose (see Problem 14.51). How many structures are possible for trehalose? If trehalose is cleaved by glycosidase enzymes that hydrolyze α-glycosides but not by enzymes that hydrolyze β-glycosides, which structure is correct?

14.53 Isotrehalose and neotrehalose are chemically similar to trehalose (Problem 14.52), except that neotrehalose is hydrolyzed only by β-glycosidases whereas isotrehalose is hydrolyzed by both α- and β-glycosidases. What are the structures of isotrehalose and neotrehalose?

14.54 The cyclitols are a group of carbocyclic sugar derivatives with the general formula 1,2,3,4,5,6-cyclohexanehexaol, that is, a cyclohexane ring with one hydroxyl on each carbon. Draw the structures of the nine stereoisomeric cyclitols in Haworth projection.

CHAPTER 15

Biomolecules: Amino Acids, Peptides, and Proteins

protein
a large biological polymer made of many amino acid units linked together by amide bonds

amino acid
a compound that contains both an amino group and a carboxylic acid group

Proteins are large biomolecules that occur in every living organism. They're of many types and have many biological functions. The keratin of skin and fingernails, the insulin that regulates glucose metabolism in the body, and the DNA polymerase that catalyzes the synthesis of DNA in cells are all proteins. Regardless of their appearance or function, all proteins are chemically similar: They're made up of many *amino acid* units linked together.

Amino acids, the building blocks from which proteins are made, are difunctional. They contain both a basic amino group and an acidic carboxyl group:

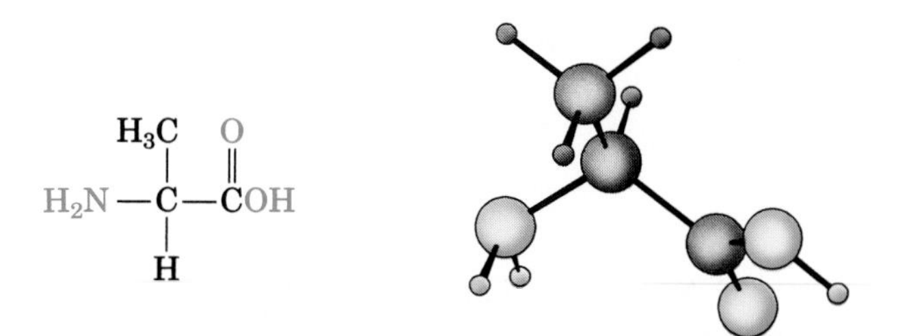

Alanine, an amino acid

peptide bond
the amide bond linking amino acids in proteins

dipeptide
a compound that has two amino acid units linked together

polypeptide
a compound that has up to 50 amino acid units linked together

Their value as building blocks for proteins derives from the fact that amino acids can link together into peptides by forming amide or **peptide bonds.** A **dipeptide** results when the $-NH_2$ of one amino acid is linked to the $-COOH$ of a second amino acid by an amide bond; a **tripeptide** results from linkage of three amino acids by two amide bonds; and so on. Any number of amino acids can link together to form a long chain. For classification purposes, chains with fewer than 50 amino acids are usually called **polypeptides,** and the term *protein* is used for longer chains.

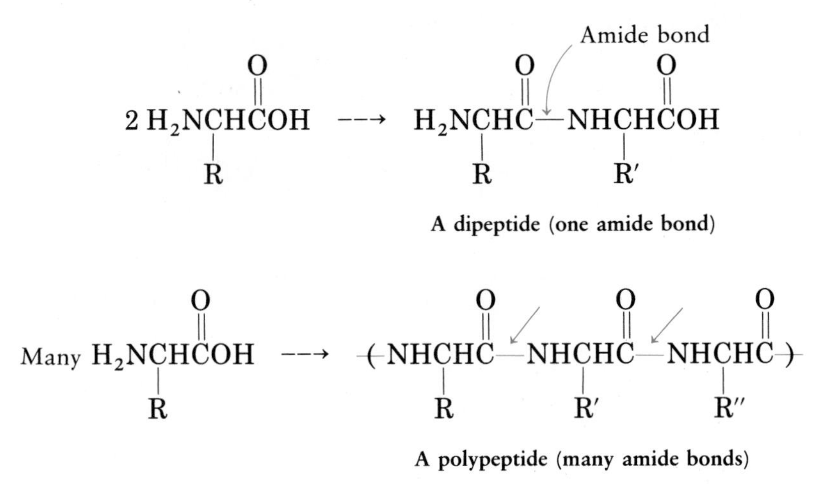

A dipeptide (one amide bond)

A polypeptide (many amide bonds)

15.1 STRUCTURES OF AMINO ACIDS

α-amino acid a compound in which the amino group is attached to the carbon atom next to the carboxyl group

The structures of the 20 amino acids commonly found in proteins are shown in Table 15.1. All 20 are **α-amino acids;** that is, the amino group in each is attached to the carbon atom alpha to (next to) the carbonyl group. Note that 19 of the 20 amino acids are primary amines, RNH_2, and differ only in the nature of their side chains. Proline, however, is a secondary amine whose nitrogen and α-carbon atom are part of a five-membered pyrrolidine ring. Proline can still form amide bonds like the other 19 α-amino acids, though.

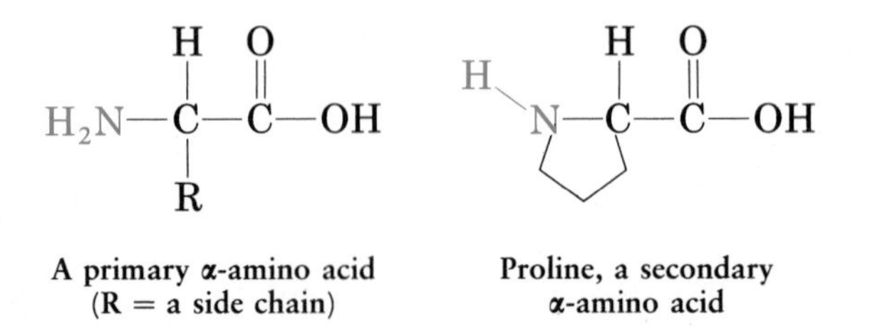

A primary α-amino acid (R = a side chain)

Proline, a secondary α-amino acid

Note also that each of the amino acids in Table 15.1 is referred to by a mnemonic three-letter shorthand code: Ala for alanine, Gly for glycine, and so on. In addition, a new one-letter code, shown in parentheses in the table, is currently gaining popularity.

With the exception of glycine, H_2NCH_2COOH, the α carbons of the amino acids are chiral. Two enantiomeric forms of each amino acid are therefore possible, but nature uses only a single enantiomer to construct proteins. In Fischer projections, naturally occurring amino acids are represented by placing the carboxyl group at the top as if drawing a carbohydrate (Section 14.2) and then placing the amino group on the left. Because of their stereochemical similarity to L-sugars (Section 14.3), the naturally occurring α-amino acids are often referred to as L-amino acids.

TABLE 15.1 **Structures of the 20 common amino acids found in proteins. Amino acids essential to human diets are shown in color.**

Name	*Abbreviations*	*Molecular weight*	*Structure*	*Isoelectric point*
Neutral amino acids				
Alanine	Ala (A)	89	$CH_3CHCOOH$ NH_2	6.0
Asparagine	Asn (N)	132	$H_2N\overset{O}{\overset{\Vert}{C}}CH_2CHCOOH$ NH_2	5.4
Cysteine	Cys (C)	121	$HSCH_2CHCOOH$ NH_2	5.0
Glutamine	Gln (Q)	146	$H_2N\overset{O}{\overset{\Vert}{C}}CH_2CH_2CHCOOH$ NH_2	5.7
Glycine	Gly (G)	75	CH_2COOH NH_2	6.0
Isoleucine	Ile (I)	131	$CH_3CH_2CHCHCOOH$ $H_3C\ \ NH_2$	6.0
Leucine	Leu (L)	131	$CH_3CHCH_2CHCOOH$ $CH_3\ \ NH_2$	6.0
Methionine	Met (M)	149	$CH_3SCH_2CH_2CHCOOH$ NH_2	5.7
Phenylalanine	Phe (F)	165	$C_6H_5{-}CH_2CHCOOH$ NH_2	5.5
Proline	Pro (P)	115	H_2 $C{-}CH{-}COOH$ $H_2C\quad\ \vert$ $C{-}N{-}H$ H_2	6.3
Serine	Ser (S)	105	$HOCH_2CHCOOH$ NH_2	5.7
Threonine	Thr (T)	119	$CH_3CHCHCOOH$ $HO\ \ NH_2$	5.6

(*Continued*)

TABLE 15.1 *(Continued)*

Name	*Abbreviations*		*Molecular weight*	*Structure*	*Isoelectric point*
Tryptophan	Trp	(W)	204	(indol-3-yl)$-CH_2CH(NH_2)COOH$	5.9
Tyrosine	Tyr	(Y)	181	$HO-C_6H_4-CH_2CH(NH_2)COOH$	5.7
Valine	Val	(V)	117	$(CH_3)_2CHCH(NH_2)COOH$	6.0
Acidic amino acids					
Aspartic acid	Asp	(D)	133	$HOOCCH_2CH(NH_2)COOH$	3.0
Glutamic acid	Glu	(E)	147	$HOOCCH_2CH_2CH(NH_2)COOH$	3.2
Basic amino acids					
Arginine	Arg	(R)	174	$H_2NC(=NH)NHCH_2CH_2CH_2CH(NH_2)COOH$	10.8
Histidine	His	(H)	155	(imidazol-4-yl, N–H)$-CH_2CH(NH_2)COOH$	7.6
Lysine	Lys	(K)	146	$H_2NCH_2CH_2CH_2CH_2CH(NH_2)COOH$	9.7

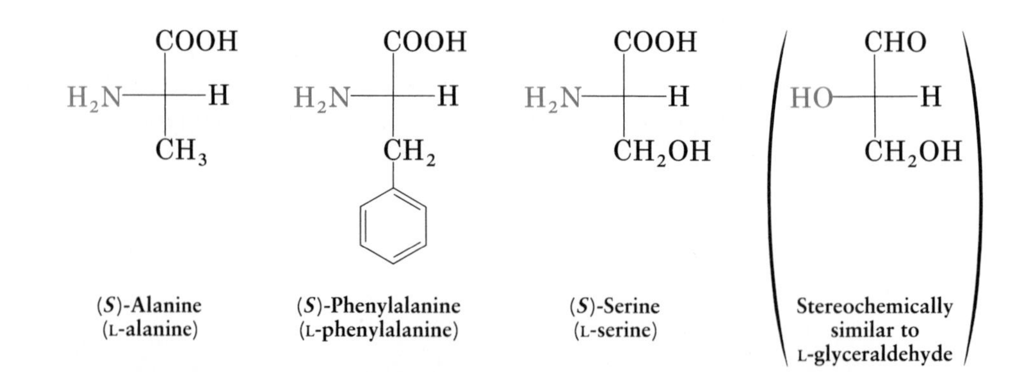

(*S*)-Alanine (L-alanine) (*S*)-Phenylalanine (L-phenylalanine) (*S*)-Serine (L-serine) (Stereochemically similar to L-glyceraldehyde)

The 20 common amino acids can be further classified as either neutral, basic, or acidic, depending on the nature of their side chains. Fifteen of the twenty have neutral side chains, but two (aspartic acid and glutamic acid) have an extra carboxylic acid function in their side chains, and three (lysine, arginine, and histidine) have basic amino groups in their side chains.

essential amino acid
an amino acid that must be obtained in the diet

All 20 of the amino acids are necessary for protein synthesis, but humans are thought to be able to synthesize only 10 (the exact number isn't known with certainty). The other 10 are called **essential amino acids** because they must be obtained from dietary sources. Failure to include an adequate dietary supply of any of these essential amino acids leads to poor growth and general failure to thrive.

PROBLEM 15.1 Look carefully at the 20 amino acids in Table 15.1. How many contain aromatic rings? How many contain sulfur? How many are alcohols? How many have hydrocarbon side chains?

PROBLEM 15.2 Eighteen of the nineteen L-amino acids have the *S*-configuration at the α-carbon. Cysteine is the only L-amino acid that has an *R* configuration. Explain.

PROBLEM 15.3 Draw L-alanine in the standard three-dimensional format using solid, wedged, and dashed lines.

15.2 DIPOLAR STRUCTURE OF AMINO ACIDS

zwitterion
a dipolar substance that has both plus and minus charges

Since amino acids contain both acidic and basic groups in the same molecule, they undergo an internal acid–base reaction and exist primarily as the dipolar ion or **zwitterion** (German *zwitter*, "hybrid"):

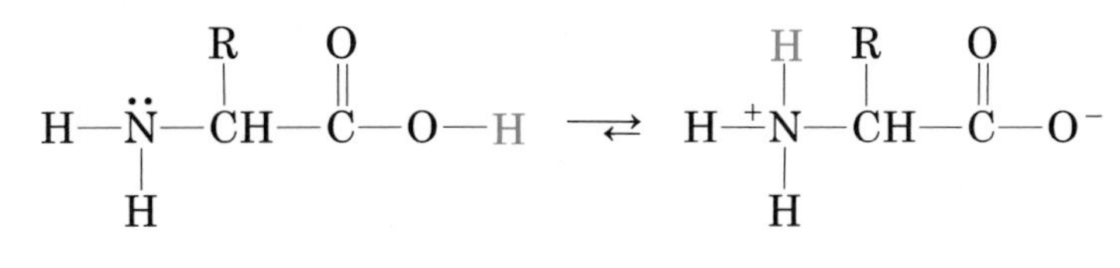

Amino acid zwitterions are a kind of internal salt and therefore have many of the physical properties associated with salts. Thus, amino acids are crystalline with high melting points and are soluble in water but insoluble in hydrocarbons. In addition, amino acids are *amphoteric;* they can react either as acids or as bases, depending on the circumstances. In aqueous acid solution, an amino acid zwitterion accepts a proton to yield a cation; in aqueous basic solution, the zwitterion loses a proton to form an anion.

In acid solution

$$H_3\overset{+}{N}{-}CHR{-}C(=O){-}O^- + H_3O^+ \rightleftharpoons H_3\overset{+}{N}{-}CHR{-}C(=O){-}OH + H_2O$$

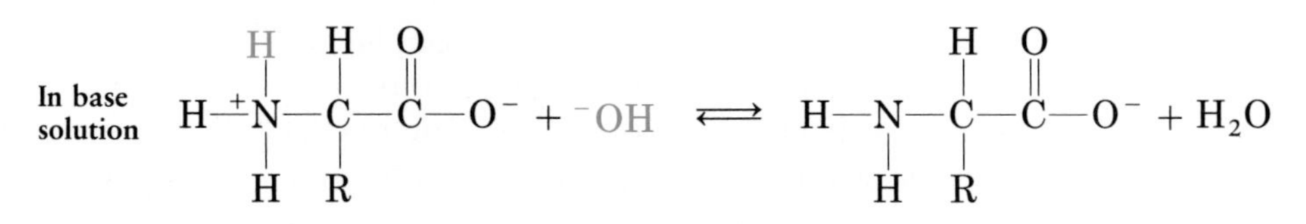

Note that it's the carboxylate anion, $-COO^-$, that acts as the basic site in the zwitterion and accepts the proton in acid solution. Similarly, it's the ammonium cation, $-NH_3^+$, that acts as the acidic site and donates a proton in basic solution.

PRACTICE PROBLEM 15.1 Write an equation for the reaction of glycine hydrochloride with (a) 1 equiv NaOH and (b) 2 equiv NaOH.

SOLUTION Glycine hydrochloride has the structure

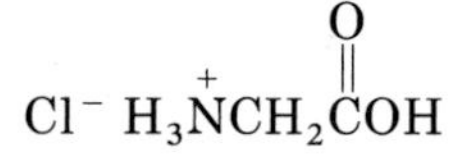

(a) Reaction with the first equivalent of NaOH removes the acidic $-COOH$ proton:

$$Cl^- \; H_3\overset{+}{N}CH_2COOH + NaOH \longrightarrow H_3\overset{+}{N}CH_2COO^- + H_2O + NaCl$$

(b) Reaction with a second equivalent of NaOH removes the $-NH_3^+$ proton:

$$H_3\overset{+}{N}CH_2COO^- + NaOH \longrightarrow H_2NCH_2COO^-Na^+ + H_2O$$

PROBLEM 15.4 Draw phenylalanine in its zwitterionic form.

PROBLEM 15.5 Write structural formulas for these equations.

(a) Phenylalanine + 1 equiv NaOH $\longrightarrow$?
(b) Product of (a) + 1 equiv HCl $\longrightarrow$?
(c) Product of (a) + 2 equiv HCl $\longrightarrow$?

15.3 ISOELECTRIC POINTS

isoelectric point the pH at which an amino acid exists primarily in its neutral zwitterionic form

In acid solution (low pH), an amino acid is protonated and exists primarily as a cation; in basic solution (high pH), an amino acid is deprotonated and exists primarily as an anion. Thus, at some intermediate pH, the amino acid must be exactly balanced between anionic and cationic forms and exist primarily as the neutral, dipolar zwitterion. This pH is called the amino acid's **isoelectric point.**

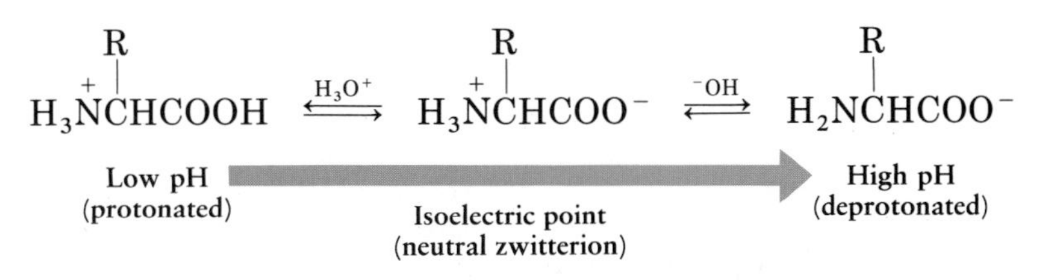

The isoelectric point of an amino acid depends on its structure; values for the 20 common amino acids are given in Table 15.1. The 15 amino acids with neutral side chains have isoelectric points near neutrality, in the pH range 5.0–6.5. (These values aren't exactly at neutral pH 7 because carboxyl groups are stronger acids in aqueous solution than amino groups are bases.) The two amino acids with acidic side chains have isoelectric points at lower (more acidic) pH, which suppresses dissociation of the extra $-COOH$ function, and the three amino acids with basic side chains have isoelectric points at higher (more basic) pH, which suppresses protonation of the extra amino function. For example, aspartic acid has its isoelectric point at pH 3.0, and lysine has its isoelectric point at pH 9.7.

We can take advantage of the differences in isoelectric points to separate a mixture of amino acids (or a mixture of proteins) into pure constituents. Using a technique known as **electrophoresis,** a solution of various amino acids is placed near the center of a strip of paper or gel. The paper or gel is moistened with an aqueous buffer of a particular pH, and electrodes are connected to the ends of the strip. When an electric potential is applied, those amino acids with negative charges (those that are deprotonated because their isoelectric points are below the pH of the buffer) migrate slowly towards the positive electrode. Similarly, those amino acids with positive charges (those that are protonated because their isoelectric points are above the pH of the buffer) migrate towards the negative electrode.

electrophoresis a technique for separating charged species by placing them in an electric field

Different amino acids migrate at different rates, depending on their isoelectric points and on the pH of the buffer. Thus, the different amino acids can be separated. Figure 15.1 illustrates this separation for a mixture of lysine (basic), glycine (neutral), and aspartic acid (acidic).

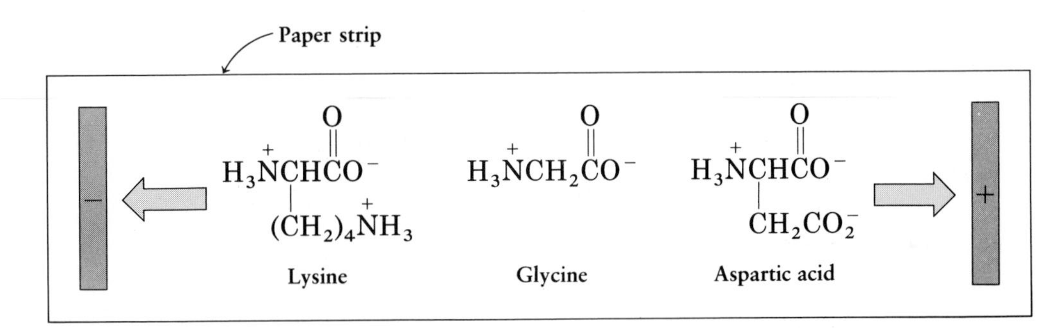

FIGURE 15.1 Separation of an amino acid mixture by electrophoresis. At pH 6.0, glycine molecules are primarily neutral and do not migrate; lysine molecules are largely protonated and migrate toward the negative electrode; aspartic acid molecules are largely deprotonated and migrate toward the positive electrode. (Lysine has its isoelectric point at pH 9.7, glycine at 6.0, and aspartic acid at 3.0.)

PRACTICE PROBLEM 15.2 Draw structures of the predominant forms of glycine at pH 3.0, pH 6.0, and pH 9.0.

SOLUTION According to Table 15.1, the isoelectric point of glycine is 6.0. At any pH lower than 6.0, glycine is protonated; at pH 6.0, glycine is zwitterionic; and at any pH higher than 6.0, glycine is deprotonated.

$$H_3\overset{+}{N}CH_2\overset{O}{\overset{\|}{C}}OH \qquad H_3\overset{+}{N}CH_2\overset{O}{\overset{\|}{C}}O^- \qquad H_2NCH_2\overset{O}{\overset{\|}{C}}O^-$$

At pH 3.0   At pH 6.0   At pH 9.0

PROBLEM 15.6 Draw the structures of the predominant forms of these amino acids.

(a) Lysine at pH 2.0 (b) Aspartic acid at pH 6.0
(c) Lysine at pH 11.0 (d) Alanine at pH 4.0

PROBLEM 15.7 For the mixtures of amino acids indicated, predict the direction of migration of each component (toward the positive or negative electrode) and the relative rate of migration.

(a) Valine, glutamic acid, and histidine at pH 7.6
(b) Glycine, phenylalanine, and serine at pH 5.7
(c) Glycine, phenylalanine, and serine at pH 6.0

15.4 PEPTIDES

residue
a common term for an amino acid unit

Peptides are amino acid polymers in which the amino acid units, also called **residues,** are linked together by amide bonds. The amino group of one residue forms an amide bond with the carboxyl of a second residue; the amino group of the second residue forms an amide bond with the carboxyl of a third; and so on. For example, alanylserine is the dipeptide formed when an amide bond is made between the alanine carboxyl and the serine amino group:

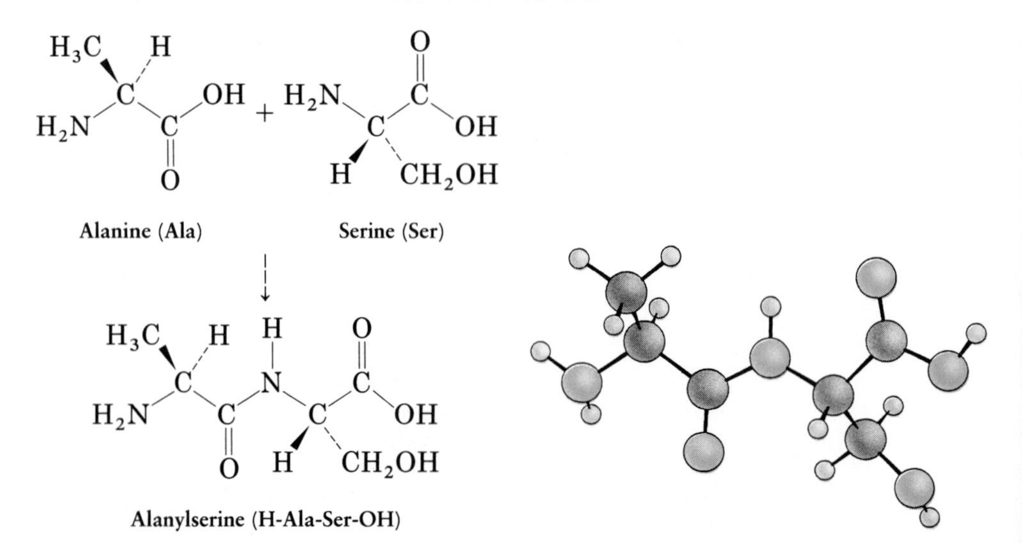

Alanine (Ala)   Serine (Ser)

Alanylserine (H-Ala-Ser-OH)

Note that two peptides can result from reaction between alanine and serine, depending on which carboxyl group reacts with which amino group. If the alanine amino group reacts with the serine carboxyl, serylalanine results:

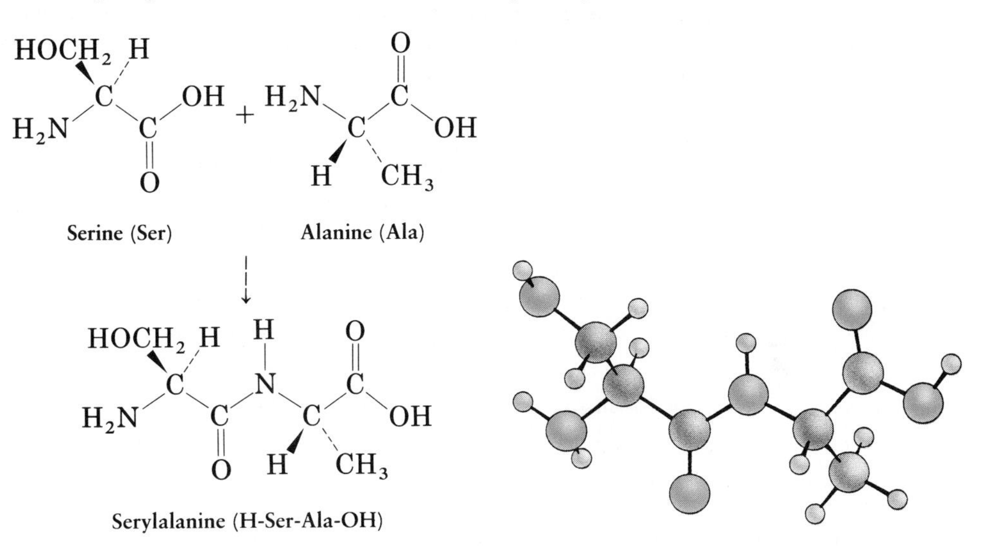

N-terminal amino acid the amino acid with a free $-NH_2$ group at one end of a protein chain

C-terminal amino acid the amino acid with a free $-COOH$ group at one end of a protein chain

By convention, peptides are always written with the **N-terminal amino acid** (the one with the free $-NH_2$ group) on the left and the **C-terminal amino acid** (the one with the free $-COOH$ group) on the right. The name of the peptide is usually indicated using the three-letter abbreviations listed in Table 15.1. An H- is often appended to the abbreviation of the leftmost amino acid to underscore its position as the N-terminal group, and an -OH is often appended to the abbreviation of the rightmost amino acid (C-terminal). Thus, serylalanine is abbreviated H-Ser-Ala-OH, and alanylserine is abbreviated H-Ala-Ser-OH.

The number of possible isomeric peptides increases rapidly as the number of amino acid units increases. There are six ways in which three amino acids can be joined, more than 40,000 ways in which the eight amino acids in the hormone angiotensin II can be joined (Figure 15.2), and an inconceivably vast number of ways in which the 1800 amino acids in myosin, the major protein of muscle fibers, can be arranged.

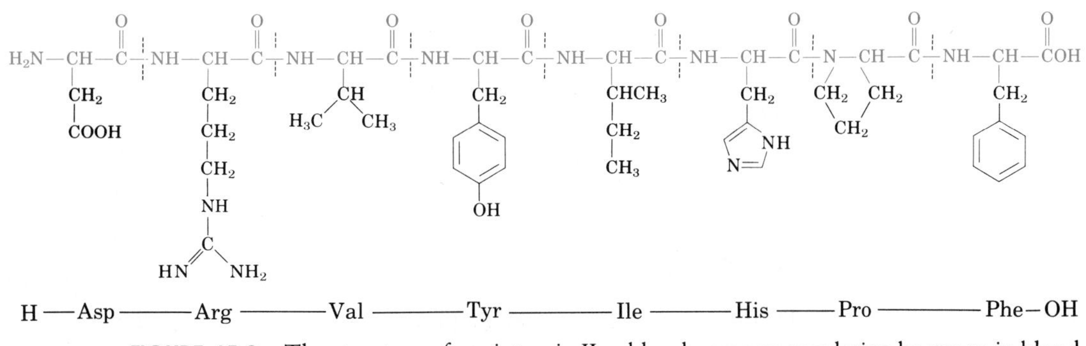

FIGURE 15.2 The structure of angiotensin II, a blood-pressure-regulating hormone in blood plasma

PRACTICE PROBLEM 15.3 Draw the full structure of H-Ala-Val-OH.

SOLUTION By convention, the N-terminal amino acid is written on the left and the C-terminal amino acid on the right. Thus, alanine is N-terminal, valine is C-terminal, and the amide bond is formed between the alanine $-COOH$ and the valine $-NH_2$.

$$\underset{\displaystyle CH_3}{H_2N-CH}-\overset{\displaystyle O}{\overset{\|}{C}}-NH-\underset{\displaystyle CH(CH_3)_2}{CH}-\overset{\displaystyle O}{\overset{\|}{C}}-OH \qquad \text{H-Ala-Val-OH}$$

PRACTICE PROBLEM 15.4 Name the six tripeptides that contain methionine, lysine, and isoleucine.

SOLUTION

H-Met-Lys-Ile-OH	H-Lys-Met-Ile-OH	H-Ile-Met-Lys-OH
H-Met-Ile-Lys-OH	H-Lys-Ile-Met-OH	H-Ile-Lys-Met-OH

PROBLEM 15.8 Draw full structures for the two peptides made from leucine and cysteine.

PROBLEM 15.9 Using the three-letter shorthand notations for each amino acid, name the six possible isomeric tripeptides that contain valine, tyrosine, and glycine.

PROBLEM 15.10 Draw the full structure of H-Met-Pro-Val-Gly-OH, and indicate where the amide bonds are.

15.5 COVALENT BONDING IN PEPTIDES

disulfide linkage a sulfur–sulfur bond between two cysteine residues in a protein

In addition to the amide bonds that link amino acid residues in peptides, a second kind of covalent bonding occurs when a **disulfide linkage,** RS–SR, is formed between two cysteine residues. The linkage is sometimes indicated by writing CyS, with a capital "S" (for sulfur), and then drawing a line from one CyS to the other: $\overset{\frown}{\text{CyS CyS}}$. As we saw in Section 8.14, disulfides are formed from thiols (RSH) by mild oxidation and are converted back to thiols by mild reduction.

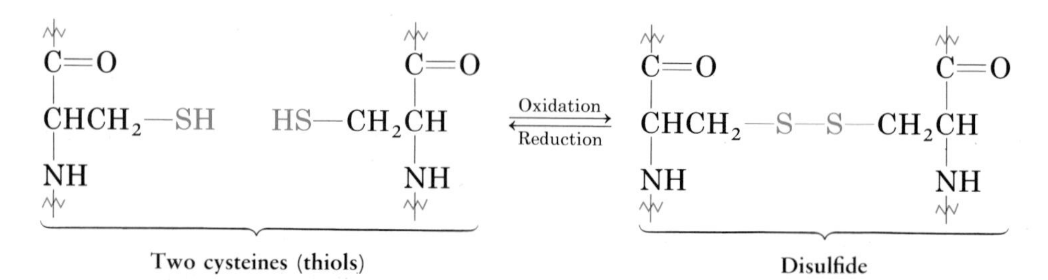

Disulfide bonds between cysteine residues in two separate peptide chains link the chains together. Alternatively, a disulfide bond between two cysteine residues

in the same chain creates a loop in the chain. Such is the case with the nonapeptide vasopressin, an antidiuretic hormone involved in controlling water balance in the body. Note that the C-terminal end of vasopressin occurs as the primary amide, $-CONH_2$, rather than as the free acid.

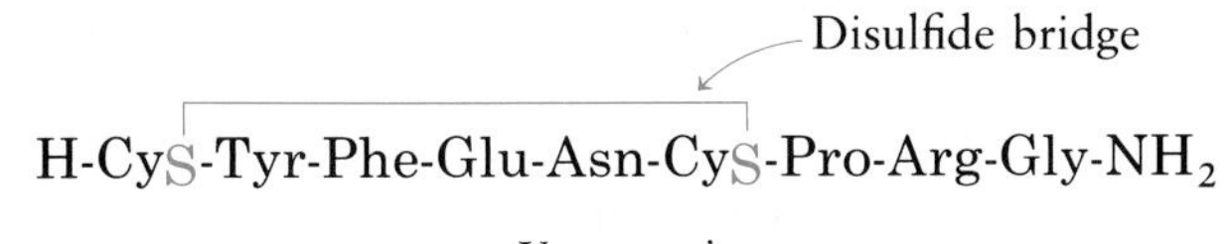

Vasopressin

15.6 PEPTIDE STRUCTURE DETERMINATION: AMINO ACID ANALYSIS

Determining the structure of a peptide requires answering three questions: What amino acids are present? How much of each is present? In what sequence do the amino acids occur in the peptide chain? The answers to the first two of these questions are provided by an instrument called an *amino acid analyzer.*

An amino acid analyzer is an automated instrument based on techniques worked out in the 1950s by W. Stein and S. Moore at the Rockefeller University. In preparation for analysis, the peptide is broken into its constituent amino acids by reducing all disulfide bonds and hydrolyzing all amide bonds with aqueous HCl. The resultant amino acid mixture is then analyzed by placing it at the top of a glass column filled with a special adsorbent material and pumping a series of aqueous buffers through the column. The various amino acids migrate down the column at different rates depending on their structures and are thus separated as they come out (*elute* from) the end of the column.

As each amino acid elutes from the end of the glass column, it is allowed to mix with a solution of *ninhydrin,* a reagent that forms a purple color when it reacts with α-amino acids. The purple color is detected by a spectrometer (Section 13.7), which measures its intensity and charts it as a function of time.

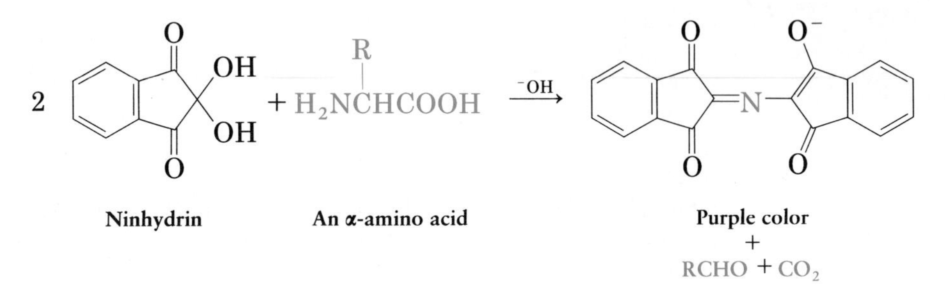

Ninhydrin **An α-amino acid** **Purple color** + $RCHO + CO_2$

Since the time required for any given amino acid to elute from a standard column is reproducible, the identity of all amino acids in a peptide is determined simply by noting the various elution times. The amount of each amino acid in the sample is determined by measuring the intensity of the purple color resulting from its reaction with ninhydrin. Thus, the identity and percentage composition of each amino acid in a peptide can be found. Figure 15.3 shows the results of amino acid analysis of a standard equimolar mixture of 17 α-amino acids.

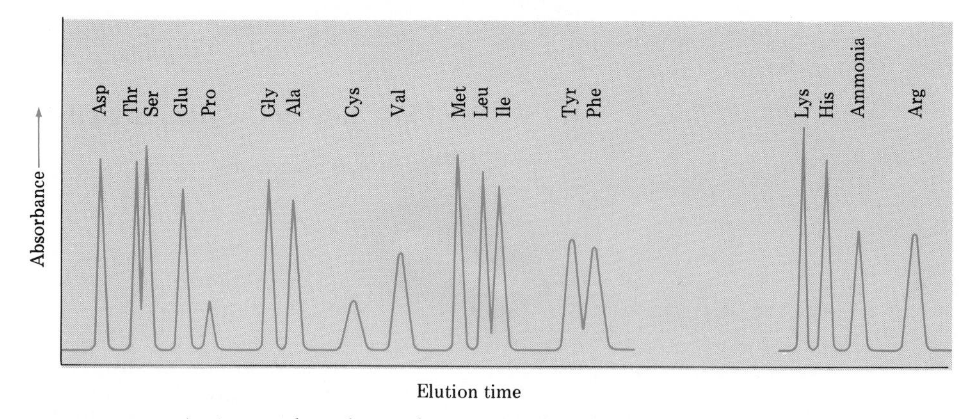

FIGURE 15.3 Amino acid analysis of an equimolar amino acid mixture

PROBLEM 15.11 Write an equation for the reaction of valine with ninhydrin.

15.7 PEPTIDE SEQUENCING: THE EDMAN DEGRADATION

Edman degradation a method for selectively cleaving the N-terminal amino acid from a peptide chain

After the identity and amount of each amino acid are known, the peptide is *sequenced* to find in what order the amino acids are linked. The general idea of peptide sequencing is to cleave selectively one amino acid residue at a time from the end of the peptide chain (either C terminus or N terminus). That terminal amino acid is then separated and identified, and the cleavage reaction is repeated on the chain-shortened peptide, until the entire peptide sequence is known. Most peptide sequencing is now done by **Edman degradation,** an efficient method of N-terminal analysis. Automated Edman *protein sequenators* are available that allow a series of 20 or more repetitive sequencing steps to be carried out automatically.

Edman degradation involves treatment of a peptide with phenyl isothiocyanate, $C_6H_5-N=C=S$, followed by mild acid hydrolysis, as shown in Figure 15.4. The first step attaches a marker to the $-NH_2$ group of the N-terminal amino acid, and the second step splits the N-terminal residue from the chain, yielding a *phenylthiohydantoin* derivative plus the chain-shortened peptide. The phenylthiohydantoin is then identified by comparison with known derivatives of the common amino acids, and the chain-shortened peptide is resubmitted to another round of Edman degradation.

Complete sequencing of large peptides and proteins by Edman degradation is impractical since the method is limited to about 20 cycles by buildup of unwanted byproducts. Instead, a large peptide chain is first cleaved by partial hydrolysis into a number of smaller fragments; then the sequence of each fragment is determined, and the individual pieces are fitted together.

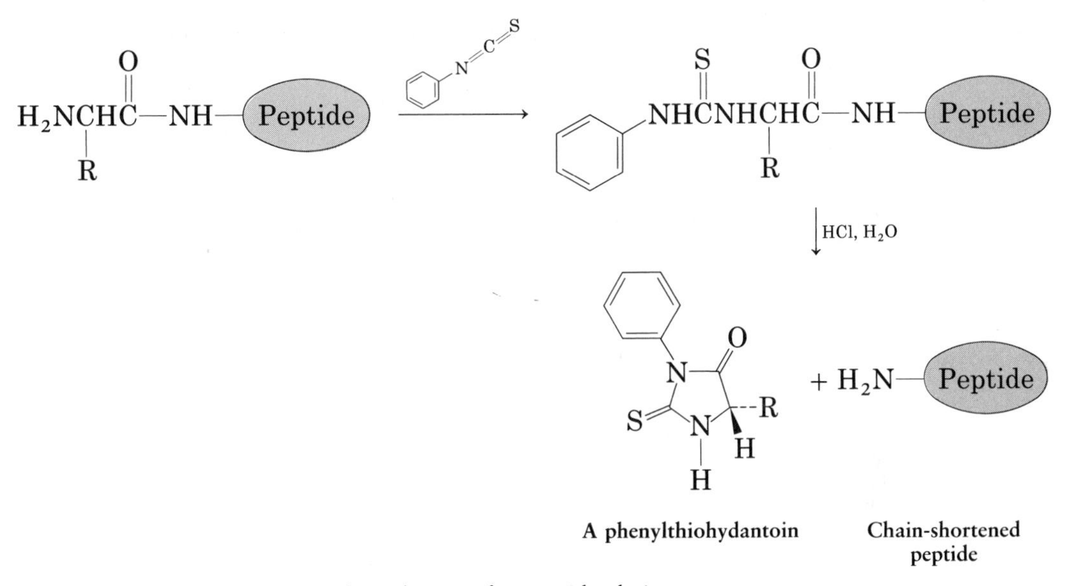

FIGURE 15.4 Edman degradation of a peptide chain

Partial hydrolysis of a peptide can be carried out either chemically with aqueous acid or enzymatically with enzymes such as trypsin and chymotrypsin. Acid hydrolysis is unselective and leads to a more-or-less random mixture of small fragments. Enzymic hydrolysis, however, is quite specific. Trypsin catalyzes hydrolysis only at the carboxyl side of the basic amino acids arginine and lysine; chymotrypsin cleaves only at the carboxyl side of the aryl-substituted amino acids phenylalanine, tyrosine, and tryptophan. For example:

H-Val-Phe-Leu-Met-Tyr-Pro-Gly-Trp-Cys-Glu-Asp-Ile-Lys-Ser-Arg-His-OH

Chymotrypsin cleaves these bonds. Trypsin cleaves these bonds.

As an example of peptide sequencing, lets look at a hypothetical structure determination of angiotensin II, a hormonal octapeptide involved in controlling hypertension by regulating the sodium–potassium salt balance in the body.

1. Amino acid analysis of angiotensin II shows the composition: Arg, Asp, His, Ile, Phe, Pro, Tyr, Val.
2. An N-terminal analysis by the Edman method shows that angiotensin II has an aspartic acid residue at the N terminus.
3. Partial hydrolysis of angiotensin II with dilute HCl might yield the following fragments, whose sequences could be determined by Edman degradation:
   a. H-Asp-Arg-Val-OH
   b. H-Ile-His-Pro-OH
   c. H-Arg-Val-Tyr-OH
   d. H-Pro-Phe-OH
   e. H-Val-Tyr-Ile-OH

4. Matching the overlapping regions of the various fragments provides the full sequence of angiotensin II:

```
a. H-Asp-Arg-Val-OH
c.       H-Arg-Val-Tyr-OH
e.           H-Val-Tyr-Ile-OH
b.                   H-Ile-His-Pro-OH
d.                          H-Pro-Phe-OH
   H-Asp-Arg-Val-Tyr-Ile-His-Pro-Phe-OH
```

Angiotensin II

The structure of angiotensin II is relatively simple—the entire sequence could easily be done by a protein sequenator—but the methods and logic used here are the same as those used to solve far more complex structures. Indeed, single protein chains with more than 400 amino acids have been sequenced by these methods.

PRACTICE PROBLEM 15.5 A hexapeptide with the composition Arg, Gly, Leu, Pro_3 has proline at both C-terminal and N-terminal positions. What is the structure of the hexapeptide if partial hydrolysis gives H-Gly-Pro-Arg-OH, H-Arg-Pro-OH, and H-Pro-Leu-Gly-OH?

SOLUTION Line up the overlapping fragments:

```
H-Pro-Leu-Gly-OH
        H-Gly-Pro-Arg-OH
                H-Arg-Pro-OH
```

The final sequence is H-Pro-Leu-Gly-Pro-Arg-Pro-OH.

PROBLEM 15.12 What fragments would result if angiotensin II were cleaved with trypsin? With chymotrypsin?

PROBLEM 15.13 Give the amino acid sequence of a hexapeptide containing Arg, Gly, Ile, Leu, Pro, and Val that produces these fragments on partial acid hydrolysis: H-Pro-Leu-Gly-OH, H-Arg-Pro-OH, H-Gly-Ile-Val-OH.

PROBLEM 15.14 Propose two structures for a tripeptide that gives Leu, Ala, and Phe on hydrolysis but doesn't react with phenyl isothiocyanate.

15.8 PEPTIDE SYNTHESIS

After a peptide's structure has been determined, synthesis is often the next goal. This might be done either as a final proof of structure or as a means of obtaining larger amounts of the peptide for biological evaluation. Although simple amides are usually formed by reaction between amines and acid chlorides (Section 10.7), peptide synthesis is much more complex because of the requirement for specificity.

Many different amide links must be formed, and they must be formed in a specific order rather than at random.

The solution to the specificity problem is *protection*. We can force a reaction to take only the desired course by protecting all of the amine and carboxylic acid functional groups except for those we want to react. For example, if we want to couple alanine with leucine to synthesize H-Ala-Leu-OH, we could protect the amino group of alanine and the carboxyl group of leucine to render them unreactive. With only the alanine carboxyl and the leucine amine available, we could form the desired amide bond and then remove the protecting groups.

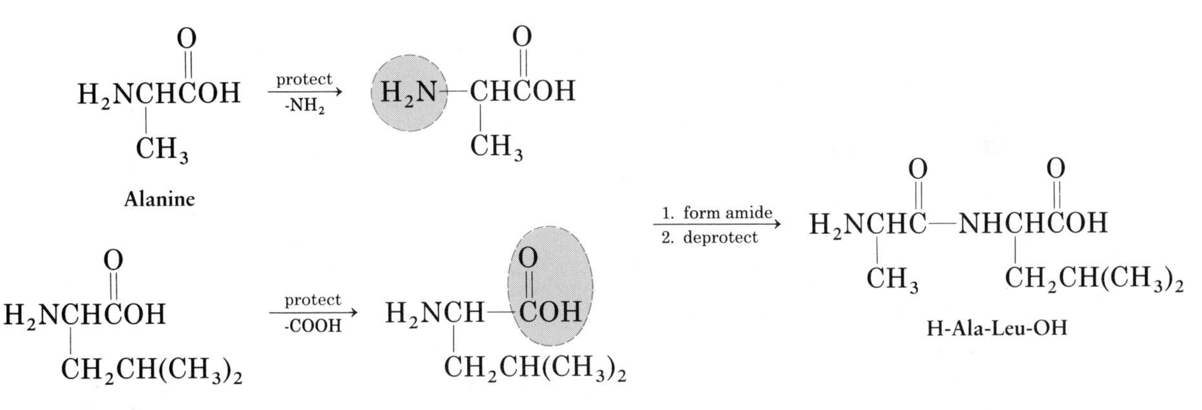

Carboxyl groups are often protected simply by converting them into methyl esters. Ester groups are easily made from carboxylic acids and are easily hydrolyzed by mild treatment with aqueous sodium hydroxide.

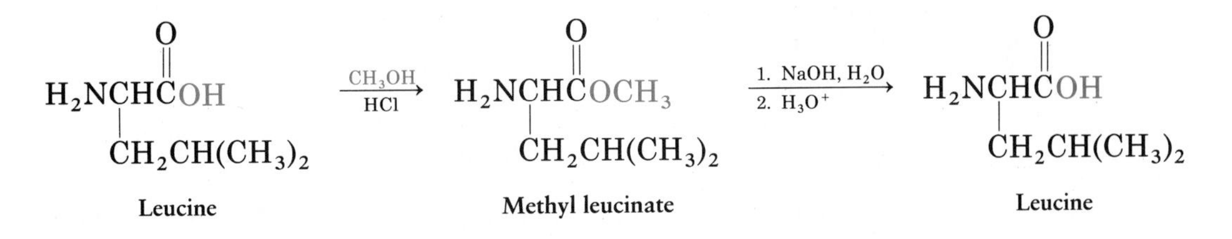

Amino groups are often protected as their *tert*-butoxycarbonyl amide (BOC) derivatives. The BOC protecting group is easily introduced by reaction of the amino acid with di-*tert*-butyl dicarbonate and is removed by brief treatment with a strong acid such as trifluoroacetic acid, CF_3COOH.

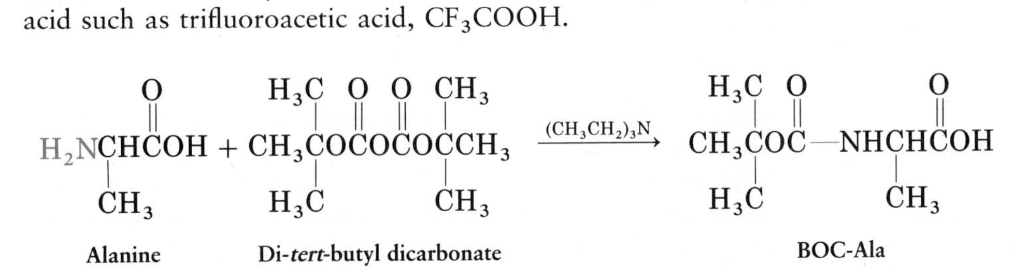

Formation of the peptide bond is accomplished by treating a mixture of the protected acid and amine components with dicyclohexylcarbodiimide (DCC). Although its mechanism of action is complex, DCC functions by first converting the acid into a reactive intermediate that then undergoes further nucleophilic acyl substitution reaction with the amine.

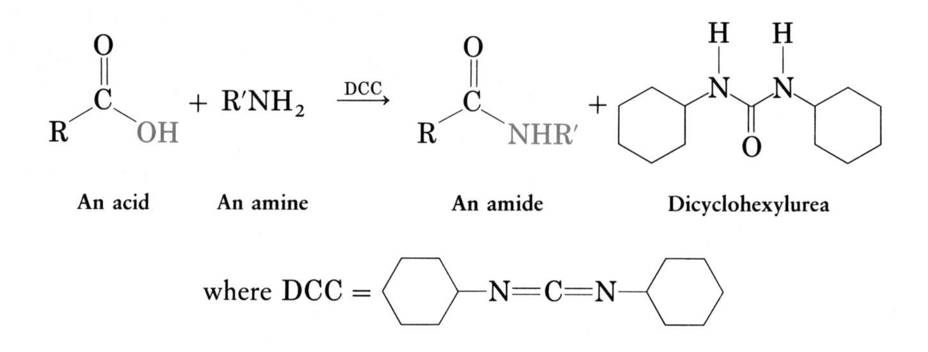

The five steps needed to synthesize H-Ala-Leu-OH are summarized:

1. Protect the amino group of alanine as the BOC derivative:

$$\text{H-Ala-OH} + (\text{BOC})_2\text{O} \longrightarrow \text{BOC-Ala-OH}$$

2. Protect the carboxyl group of leucine as the methyl ester:

$$\text{H-Leu-OH} + \text{CH}_3\text{OH} \longrightarrow \text{H-Leu-OCH}_3$$

3. Couple the two protected amino acids using DCC:

$$\text{BOC-Ala-OH} + \text{H-Leu-OCH}_3 \xrightarrow{\text{DCC}} \text{BOC-Ala-Leu-OCH}_3$$

4. Remove the BOC protecting group by acid treatment:

$$\text{BOC-Ala-Leu-OCH}_3 \xrightarrow{\text{CF}_3\text{COOH}} \text{H-Ala-Leu-OCH}_3$$

5. Remove the methyl ester protecting group by base treatment:

$$\text{H-Ala-Leu-OCH}_3 \xrightarrow[\text{H}_2\text{O}]{\text{NaOH}} \text{H-Ala-Leu-OH}$$

These steps can be repeated to add one amino acid at a time to the growing chain or to link two peptide chains together. Many remarkable achievements in peptide synthesis have been reported, including a complete synthesis of human insulin. Insulin, whose structure is shown in Figure 15.5, is composed of two chains totaling 51 amino acids and linked by cysteine disulfide bridges. Its structure was determined by Frederick Sanger, who received the 1958 Nobel prize for his work.

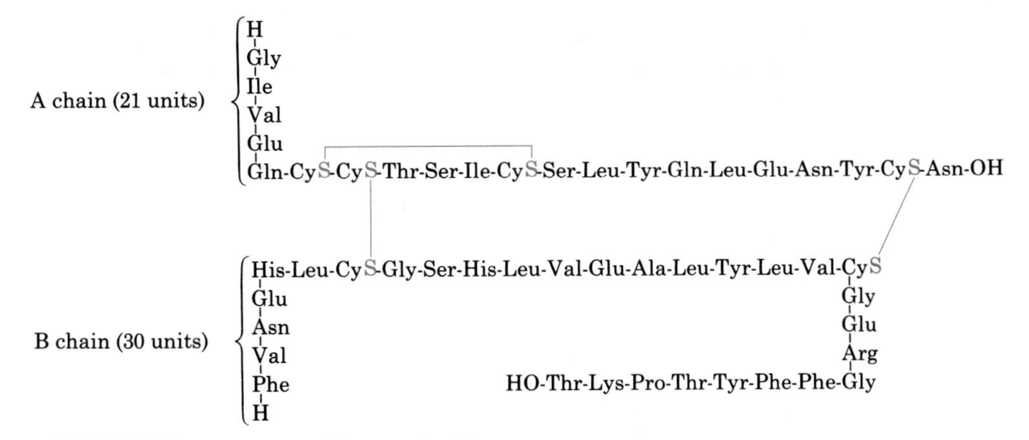

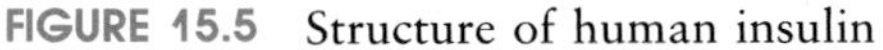
FIGURE 15.5 Structure of human insulin

PRACTICE PROBLEM 15.6 Write equations for the reaction of methionine with (a) CH_3OH, HCl and (b) di-*tert*-butyl dicarbonate.

SOLUTION

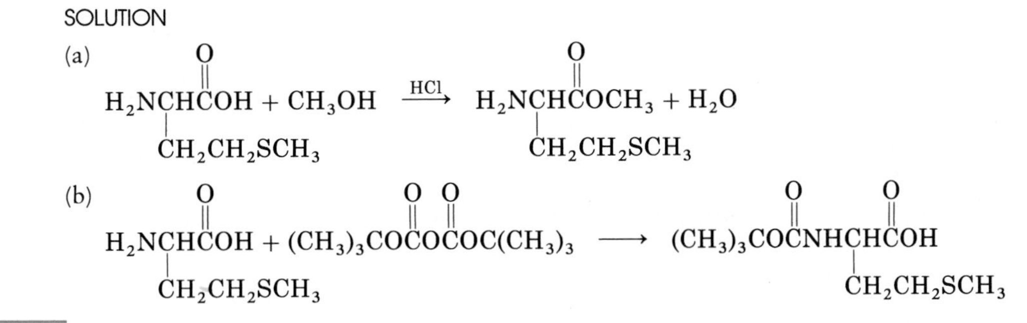

PROBLEM 15.15 Write the chemical structures of the intermediates in the five-step synthesis of H-Leu-Ala-OH from alanine and leucine.

PROBLEM 15.16 Show all of the steps involved in the synthesis of the tripeptide H-Val-Phe-Gly-OH.

15.9 CLASSIFICATION OF PROTEINS

simple protein a protein composed entirely of amino acids

conjugated protein a protein composed of an amino-acid part and a non-amino-acid part

Proteins are classified into two major types according to their composition. **Simple proteins,** such as blood serum albumin, are those that yield only amino acids and no other compounds on hydrolysis. **Conjugated proteins,** which are much more common than simple proteins, yield other compounds in addition to amino acids on hydrolysis. As shown in Table 15.2, conjugated proteins can be classified according to the chemical nature of the nonamino acid portion.

TABLE 15.2 Some conjugated proteins

Name	*Composition*
Glycoproteins	Proteins bonded to a carbohydrate. Cell membranes have a glycoprotein coating.
Lipoproteins	Proteins bonded to fats and oils (lipids). These proteins transport cholesterol and other fats through the body.
Metalloproteins	Proteins bonded to a metal ion. The enzyme cytochrome oxidase, necessary for biological energy production, is an example.
Nucleoproteins	Proteins bonded to RNA (ribonucleic acid). These are found in cell ribosomes.
Phosphoproteins	Proteins bonded to a phosphate group. Milk casein, which stores nutrients for growing embryos, is an example.

fibrous protein a tough, insoluble protein used in nature for structural purposes

globular protein a spherical, water-soluble protein found primarily inside cells

Another way to classify proteins is as either *fibrous* or *globular* according to their three-dimensional shape. **Fibrous proteins,** such as collagen and keratin, consist of polypeptide chains arranged side by side in long filaments. Because these proteins are tough and insoluble in water, they're used in nature for such structural materials as tendons, hoofs, horns, and muscles. **Globular proteins,** by contrast, are usually coiled into compact, nearly spherical shapes. These proteins are generally soluble in water and are mobile within cells. Most of the 2000 or so known enzymes, as well as hormonal and transport proteins, are globular. Table 15.3 lists some common examples of both fibrous and globular proteins.

TABLE 15.3 Some common fibrous and globular proteins

Name	*Occurrence and use*
Fibrous proteins (insoluble)	
Collagens	Found in animal hide, tendons, and other connective tissues
Elastins	Found in blood vessels, ligaments, and other tissues that must be able to stretch
Fibrinogen	Found in blood; necessary for blood clotting
Keratins	Found in skin, wool, feathers, hooves, silk, fingernails
Myosins	Found in muscle tissue
Globular proteins (soluble)	
Hemoglobin	Protein involved in oxygen transport
Immunoglobulins	Proteins involved in immune response
Insulin	Regulatory hormone for controlling glucose metabolism
Ribonuclease	Enzyme controlling RNA synthesis

Yet a third way to classify proteins is according to function. As shown in Table 15.4, there is an extraordinary diversity to the biological roles of proteins.

TABLE 15.4 **Some biological functions of proteins**

Type	*Function and example*
Enzymes	Proteins such as chymotrypsin that act as biological catalysts
Hormones	Proteins such as insulin that regulate body processes
Protective proteins	Proteins such as antibodies that fight infection
Storage proteins	Proteins such as casein that store nutrients
Structural proteins	Proteins such as keratin, elastin, and collagen that form an organism's structure
Transport proteins	Proteins such as hemoglobin that transport oxygen and other substances through the body

15.10 PROTEIN STRUCTURE

primary structure the amino acid sequence of a protein

secondary structure the specific way in which segments of a protein chain are oriented into a regular pattern

tertiary structure the specific way in which a protein molecule is oriented into an overall three-dimensional shape

quaternary structure the way in which several protein molecules aggregate to yield a larger structure

Proteins are so large that the word *structure* takes on a broader meaning when applied to such immense molecules than it does with other organic compounds. In fact, chemists speak of four different levels of structure when describing proteins. At its simplest, protein structure is the sequence in which amino acid residues are bound together. Called the **primary structure** of a protein, this is the most fundamental structural level.

There is much more to protein structure than amino acid sequence though. The chemical properties of a protein are also dependent on higher levels of structure: that is, on exactly how the peptide backbone is folded to give the molecule a specific three-dimensional shape. Thus, the term **secondary structure** refers to the way in which *segments* of the peptide backbone are oriented into a regular pattern; **tertiary structure** refers to the way in which the *entire* protein molecule is coiled into an overall three-dimensional shape; and **quaternary structure** refers to the way in which several protein molecules come together to yield large aggregate structures. Let's look at three examples—α-keratin (fibrous), fibroin (fibrous), and myoglobin (globular)—to see how higher structure affects a protein's properties.

α-Keratin

α helix a common kind of secondary structure in which a protein chain coils into a spiral

α-Keratin is the fibrous structural protein found in wool, hair, nails, and feathers. Studies show that α-keratin is coiled into a right-handed helical secondary structure, as illustrated in Figure 15.6. This **α helix** is stabilized by hydrogen bonding between amide N–H groups and other amide carbonyl groups four residues away. Although the strength of a single hydrogen bond (about 5 kcal/mol) is only about 5% the strength of a C–C or C–H covalent bond, the large number of hydrogen bonds made possible by helical winding imparts a great deal of stability to the α-helical structure. Each coil of the helix (the *repeat distance*) contains 3.6 amino acid residues, with a distance between coils of 5.4 Å.

Other evidence suggests that the α-keratins of wool and hair also have a quaternary structure. The individual helical strands are themselves coiled about one another in stiff bundles to form a superhelix that accounts for the thread-like

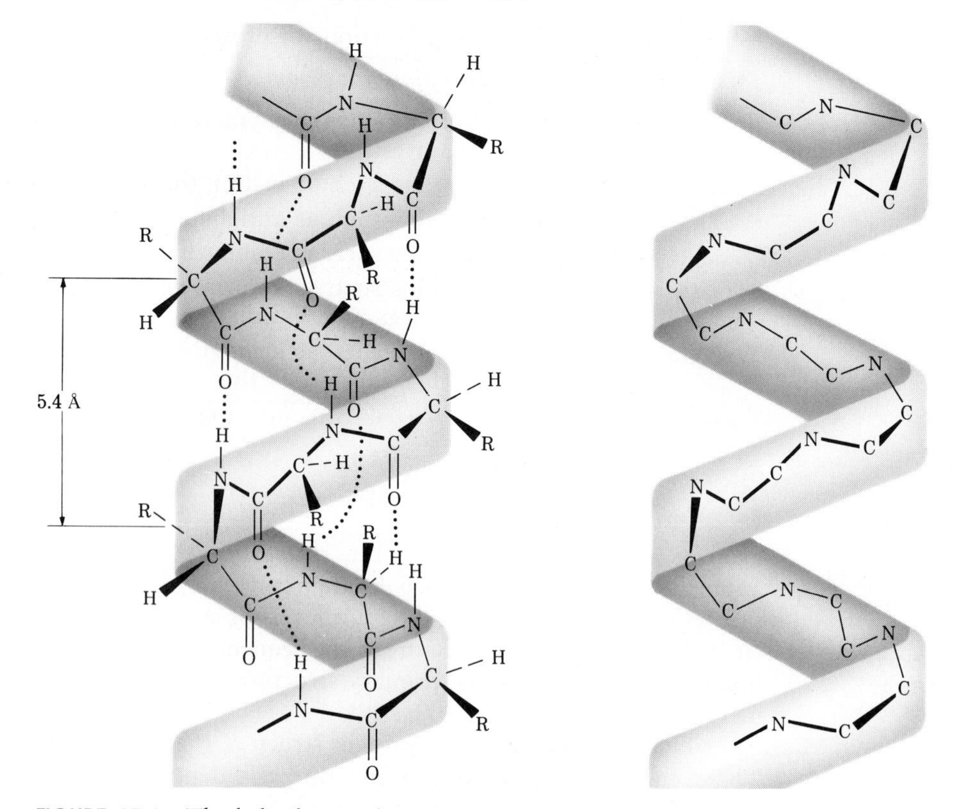

FIGURE 15.6 The helical secondary structure of α-keratin

properties and strength of these proteins. Although α-keratin is the best example of an almost entirely helical protein, most globular proteins contain α-helical *segments*. Both hemoglobin and myoglobin, for example, contain many short helical sections in their chains.

Fibroin

β pleated sheet a common kind of secondary structure in which a protein chain folds back on itself so that two sections of the chain run parallel

Fibroin, the fibrous protein found in silk, has a secondary structure known as a **β pleated sheet.** In this pleated-sheet structure, polypeptide chains line up in a parallel arrangement held together by hydrogen bonds between chains (Figure 15.7). Although not as common as the α helix, small pleated-sheet regions are often found in proteins where sections of peptide chains double back on themselves.

Myoglobin

Myoglobin is a small globular protein containing 153 amino acid residues in a single chain. A relative of hemoglobin, myoglobin is found in the skeletal muscles of sea mammals where it stores oxygen needed to sustain the animals during long dives. X-ray evidence has shown that myoglobin consists of eight straight segments, each of which adopts an α helical secondary structure. These helical sections are connected by bends to form a compact, nearly spherical, tertiary structure (Figure 15.8).

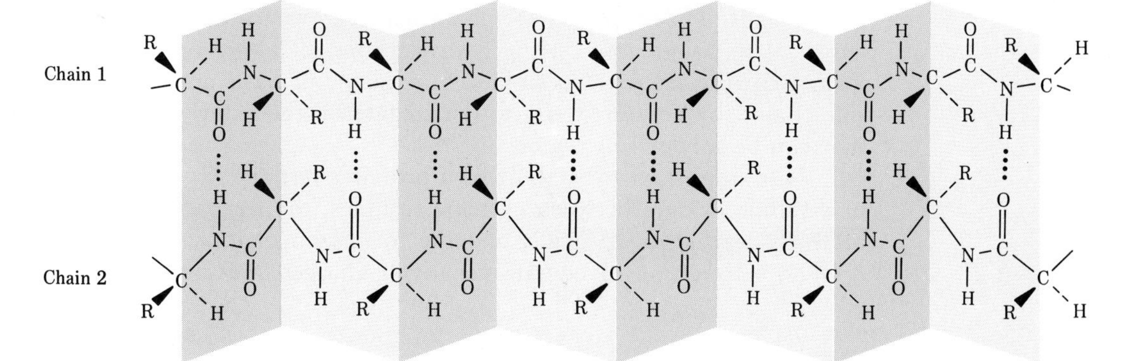

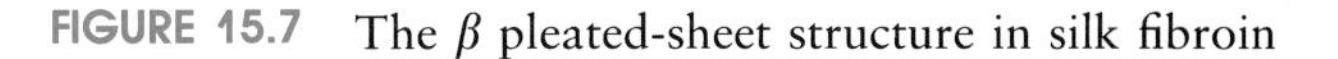
FIGURE 15.7 The β pleated-sheet structure in silk fibroin

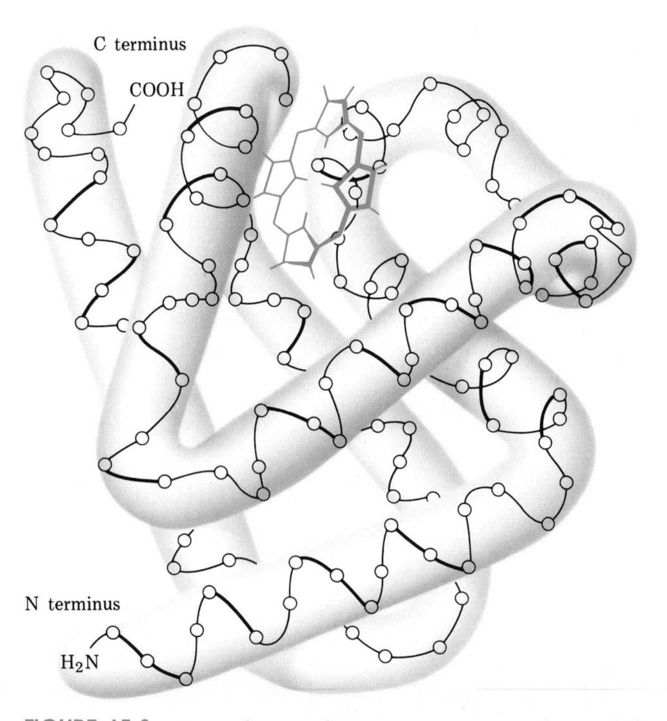

FIGURE 15.8 Secondary and tertiary structure of myoglobin

Why does myoglobin adopt the shape it does? The forces that determine the tertiary structure of myoglobin and other globular proteins are the same simple forces that act on all molecules, regardless of size. By bending and twisting in exactly the right way, myoglobin achieves maximum stability. Although the bends appear irregular and the three-dimensional structure appears random, this isn't the case. All myoglobin molecules adopt this same shape because it's more stable than any other.

The most important forces stabilizing a protein's tertiary structure are the hydrophobic (water-repelling) interactions of hydrocarbon side chains on neutral amino acids. Those amino acids with neutral, nonpolar side chains have a strong tendency to congregate on the hydrocarbon-like interior of a protein molecule, away from the aqueous medium. Those acidic or basic amino acids with charged side chains, by contrast, tend to congregate on the exterior of the protein where they can be solvated by water.

Also important for stabilizing a protein's tertiary structure are the formation of disulfide bridges between cysteine residues, the formation of hydrogen bonds between nearby amino acids, and the formation of ionic attractions, called *salt bridges,* between positively and negatively charged sites on the protein.

15.11 ENZYMES

enzyme a large globular protein that acts as a catalyst for a specific biological reaction

Enzymes are large proteins that act as catalysts for biological reactions. Unlike many of the simple catalysts that chemists use in the laboratory, enzymes are usually specific in their action. Often, in fact, an enzyme can catalyze only a single reaction of a single compound, called the enzyme's *substrate.* For example, the enzyme amylase found in the human digestive tract catalyzes only the hydrolysis of starch to yield glucose; cellulose and other polysaccharides are untouched by amylase.

Different enzymes have different specificities. Some, such as amylase, are specific for a single substrate, but others operate on a range of substrates. Papain, for instance, a globular protein of 212 amino acids isolated from papaya fruit, catalyzes the hydrolysis of many kinds of peptide bonds. In fact, it's this ability to hydrolyze peptide bonds that makes papain useful as a meat tenderizer and a contact-lens cleaner.

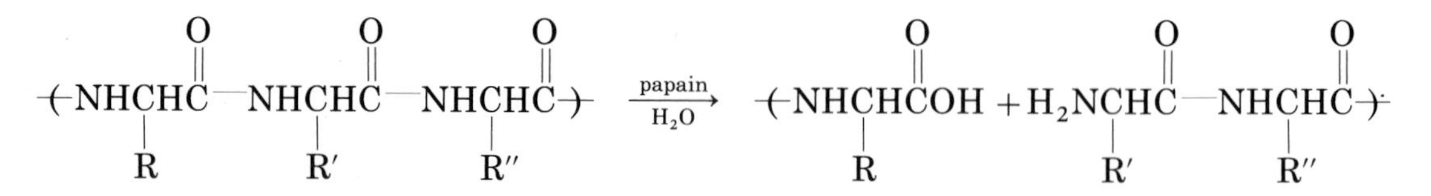

Like all catalysts, enzymes don't affect the equilibrium constant of a reaction and can't bring about chemical changes that are otherwise unfavorable. Enzymes act only to lower the activation energy for reaction, thereby making the reaction take place faster or at lower temperatures. Starch and water, for example, react very slowly in the absence of a catalyst because the activation energy is too high. When amylase is present, though, the energy barrier is lowered, and the hydrolysis reaction occurs rapidly.

15.12 STRUCTURE AND CLASSIFICATION OF ENZYMES

All of the 2000 or so known enzymes are globular proteins with a primary structure defined by amino acid sequence, a secondary structure defined by α helical or β pleated-sheet regions, and a tertiary structure defined by overall shape. In addition

cofactor
a small nonprotein part of an enzyme necessary for biological activity

apoenzyme
the protein part of an enzyme that needs a cofactor for biological activity

holoenzyme
the combination of apoenzyme and cofactor

coenzyme
a small organic molecule that acts as an enzyme cofactor

vitamin
a small organic molecule that must be obtained in the diet and that is required for proper growth

to the protein part, most enzymes also have small nonprotein parts called **cofactors.** The protein part in such enzymes is called an **apoenzyme,** and the combination of apoenzyme plus cofactor is called a **holoenzyme.** Only holoenzymes have biological activity; neither cofactor nor apoenzyme catalyze reactions by themselves.

$$\text{Holoenzyme} = \text{cofactor} + \text{apoenzyme}$$

Cofactors can be either inorganic ions such as Zn^{2+} or small organic molecules, called **coenzymes.** The requirement of many enzymes for inorganic cofactors is the main reason for our dietary need of trace minerals. Iron, zinc, copper, manganese, and numerous other metal ions are all essential minerals that act as enzyme cofactors, though the exact biological role isn't known in all cases.

A variety of organic molecules act as coenzymes. Many, though not all, coenzymes are **vitamins,** small organic molecules that must be obtained in the diet and that are required in trace amounts for proper growth. Table 15.5 lists the 13 known vitamins required in the human diet and their enzyme functions.

TABLE 15.5 Vitamins and their enzyme functions

Vitamin	*Enzyme function*	*Deficiency symptom*
Water-Soluble Vitamins		
Ascorbic acid (vitamin C)	Hydroxylases	Bleeding gums, bruising
Thiamin (vitamin B_1)	Reductases	Fatigue, depression
Riboflavin (vitamin B_2)	Reductases	Cracked lips, scaly skin
Pyridoxine (vitamin B_6)	Aminotransferases	Anemia, irritability
Niacin	Reductases	Dermatitis, dementia
Folic acid (vitamin M)	Methyltransferases	Megaloblastic anemia
Vitamin B_{12}	Isomerases	Megaloblastic anemia, neurodegeneration
Pantothenic acid	Acyltransferases	Weight loss, irritability
Biotin (vitamin H)	Carboxylases	Dermatitis, anorexia, depression
Fat-Soluble Vitamins		
Vitamin A	Visual system	Night blindness, dry skin
Vitamin D	Calcium metabolism	Rickets, osteomalacia
Vitamin E	Antioxidant	Hemolysis of red blood cells
Vitamin K	Blood clotting	Hemorrhage, delayed blood clotting

Enzymes are grouped into six main classes according to the kind of reaction they catalyze (Table 15.6). *Hydrolases* catalyze the hydrolysis of substrates; *isomerases* catalyze the isomerization of substrates; *ligases* catalyze the bonding

TABLE 15.6 Classification of enzymes

Main class	*Some subclasses*	*Type of reaction catalyzed*
Hydrolases	Lipases	Hydrolysis of an ester group
	Nucleases	Hydrolysis of a phosphate group
	Proteases	Hydrolysis of an amide group
Isomerases	Epimerases	Isomerization of chiral center
Ligases	Carboxylases	Addition of CO_2
	Synthetases	Formation of new bond
Lyases	Decarboxylases	Loss of CO_2
	Dehydrases	Loss of H_2O
Oxidoreductases	Dehydrogenases	Introduction of double bond by removal of H_2
	Oxidases	Oxidation
	Reductases	Reduction
Transferases	Kinases	Transfer of a phosphate group
	Transaminases	Transfer of an amino group

together of two substrates; *lyases* catalyze the breaking away of a small molecule such as H_2O from a substrate; *oxidoreductases* catalyze oxidations and reductions of substrate molecules; and *transferases* catalyze the transfer of a group from one substrate to another.

Although some enzymes, like papain and trypsin, have uninformative common names, the systematic name of an enzyme has two parts, ending with *-ase*. The first part identifies the enzyme's substrate, and the second part identifies its class. For example, *hexose kinase* is an enzyme that catalyzes the transfer of a phosphate group from adenosine triphosphate (ATP) to glucose.

PROBLEM 15.17 To what classes do these enzymes belong:

(a) Pyruvate decarboxylase (b) Chymotrypsin (c) Alcohol dehydrogenase

INTERLUDE

Protein and Nutrition

Dietary protein is needed by everyone, from weightlifters to infants. Children need large amounts of protein for proper growth, and adults need protein to replace what's lost each day by the body's normal biochemical reactions. Dietary protein is necessary because our bodies can synthesize only 10 of the 20 common amino acids from simple precursor molecules; the other 10 amino acids must be obtained from food by digestion of edible proteins. Table 15.7 shows the estimated amino acid requirements of an infant and an adult.

TABLE 15.7 **Estimated amino acid requirements**

	Daily requirement (mg/kg body weight)	
Amino acid	*Infant*	*Adult*
Arginine	?	?
Histidine	33	?
Isoleucine	83	12
Leucine	35	16
Lysine	99	12
Methionine + cysteine	49	10
Phenylalanine + tyrosine	141	16
Threonine	68	8
Tryptophan	21	3
Valine	92	14

Not all foods provide sufficient amounts of the 10 essential amino acids to meet our minimum daily needs. Most meat and dairy products are satisfactory, but many vegetable sources such as wheat and corn are *incomplete;* that is, many vegetable proteins contain too little of one or more essential amino acids to sustain the growth of laboratory animals. For example, wheat is low in lysine, and corn is low in both lysine and tryptophan.

Using an incomplete food as the sole source of protein can cause nutritional deficiencies, particularly in children. Vegetarians must therefore be careful to adopt a varied diet that provides proteins from several sources. Thus, legumes and nuts are useful for overcoming the deficiencies of wheat and grains. Some of the limiting amino acids found in various foods are listed in Table 15.8.

TABLE 15.8 **Limiting amino acids in some foods**

Food	*Limiting amino acid*
Wheat, grains	Lysine, threonine
Peas, beans, legumes	Methionine, tryptophan
Nuts, seeds	Lysine
Leafy green vegetables	Methionine

SUMMARY AND KEY WORDS

Proteins are large biomolecules consisting of **α-amino acid residues** linked together by amide bonds. Twenty amino acids are commonly found in proteins: All are α-amino acids, and all except glycine have stereochemistry similar to that of L-sugars.

Determining the structure of a large polypeptide or protein requires several steps. The identity and amount of each amino acid present in a peptide can be determined by **amino acid analysis.** The peptide is first hydrolyzed to its constituent α-amino acids, which are then separated and identified. Next, the peptide is **sequenced. Edman degradation** by treatment with phenyl isothiocyanate cleaves off one residue from the N terminus of the peptide and forms an easily identifiable derivative of that residue. A series of sequential Edman degradations sequences peptide chains up to 20 residues in length.

Peptide synthesis involves the use of selective **protecting groups.** An N-protected amino acid with a free carboxyl group is coupled using DCC to an O-protected amino acid with a free amino group. Amide formation occurs, the protecting groups are removed, and the sequence is repeated. Amines are usually protected as their *tert*-butoxycarbonyl (BOC) derivatives; acids are usually protected as esters.

Proteins are classified as either **globular** or **fibrous,** depending on their **secondary** and **tertiary structures.** Fibrous proteins such as α-keratin are tough and water insoluble; globular proteins such as myoglobin are water soluble and mobile within cells. Most of the 2000 or so known enzymes are globular proteins.

Enzymes are large globular proteins that act as biological catalysts. Like all catalysts, enzymes speed up the rate of a reaction without themselves being changed. They are classified into six groups according to the kind of reaction they catalyze: **oxidoreductases** catalyze oxidations and reductions; **transferases** catalyze transfers of groups; **hydrolases** catalyze hydrolysis; **isomerases** catalyze isomerizations; **lyases** catalyze bond breakages; and **ligases** catalyze bond formations.

In addition to their protein part, many enzymes contain **cofactors,** which can be either metal ions or small organic molecules. If the cofactor is an organic molecule, it is called a **coenzyme.** The combination of protein (**apoenzyme**) plus coenzyme is called a **holoenzyme.** Often, the coenzyme is a **vitamin,** a small molecule that must be obtained in the diet and that is required in trace amounts for proper growth.

ADDITIONAL PROBLEMS

15.18 Although only *S* amino acids occur in proteins, several *R* amino acids are found elsewhere in nature. For example, (*R*)-serine is found in earthworms, and (*R*)-alanine is found in insect larvae. Draw Fischer projections of (*R*)-serine and (*R*)-alanine.

15.19 Draw a Fischer projection of *S*-proline, the only secondary amino acid.

15.20 Define these terms.

(a) Amphoteric (b) Isoelectric point (c) Peptide
(d) N terminus (e) C terminus (f) Zwitterion

15.21 Using the three-letter code names for each amino acid, write the structures of the peptides containing the following amino acids.

(a) Val, Leu, Ser (b) Ser, Leu_2, Pro

15.22 Draw these amino acids in their zwitterionic forms.

(a) Serine (b) Tyrosine (c) Threonine

15.23 Draw structures of the predominant forms of lysine and aspartic acid at pH 3.0 and pH 9.7.

15.24 At what pH would you carry out an electrophoresis experiment if you wanted to separate a mixture of histidine, serine, and glutamic acid? Explain.

15.25 Predict the product of the reaction of valine with these reagents.

(a) CH_3CH_2OH, H^+ (b) NaOH, H_2O
(c) di-*tert*-Butyl dicarbonate

15.26 Write out full structures for these peptides and indicate the positions of the amide bonds.

(a) H-Val-Phe-Cys-OH (b) H-Glu-Pro-Ile-Leu-OH

15.27 The amino acid threonine, (2*S*,3*R*)-2-amino-3-hydroxybutanoic acid, has two chiral centers and a stereochemistry similar to that of the four-carbon sugar D-threose. Draw a Fischer projection of threonine.

15.28 Draw the Fischer projection of a diastereomer of threonine (Problem 15.27).

15.29 The amino acid analysis data in Figure 15.3 indicate that proline is not easily detected by reaction with ninhydrin. Suggest a reason for this.

15.30 Cytochrome *c*, an enzyme found in the cells of all aerobic organisms, plays a role in respiration. Elemental analysis of cytochrome *c* reveals it to contain 0.43% iron. What is the minimum molecular weight of this enzyme?

15.31 Draw the structure of the phenylthiohydantoin product you would expect to obtain from Edman degradation of these peptides.

(a) H-Val-Leu-Gly-OH (b) H-Ala-Pro-Phe-OH

15.32 Arginine, which contains a *guanidino* group in its side chain, is the most basic of the 20 common amino acids. How can you account for this basicity? (Hint: Use resonance structures to see how the protonated guanidino group is stabilized.)

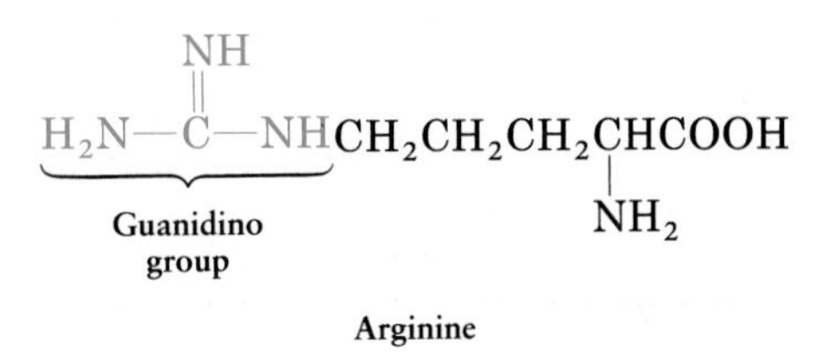

Arginine

15.33 Show the steps involved in a synthesis of H-Phe-Ala-Val-OH.

15.34 When unprotected α-amino acids are treated with dicyclohexylcarbodiimide (DCC), 2,5-diketopiperazines result. Explain.

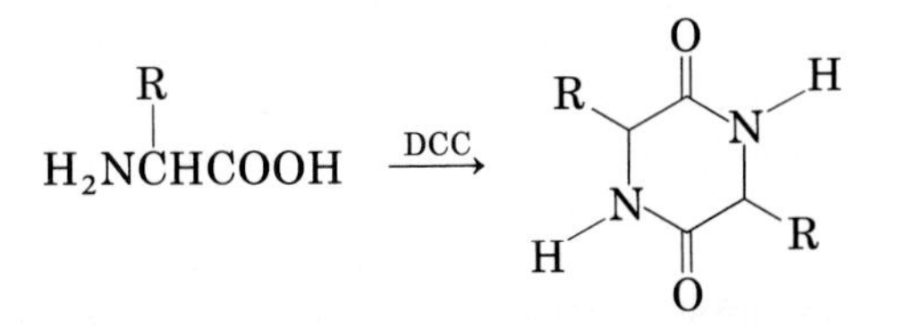

A 2,5-diketopiperazine

15.35 Which amide bonds in the following polypeptide are cleaved by trypsin? By chymotrypsin?

H-Phe-Leu-Met-Lys-Tyr-Asp-Gly-Gly-Arg-Val-Ile-Pro-Tyr-OH

15.36 Look up the structure of human insulin (Figure 15.5) and indicate where in each chain the molecule is cleaved by trypsin and by chymotrypsin.

15.37 A heptapeptide shows the composition Asp, Gly, Leu, Phe, Pro_2, Val on amino acid analysis. Edman degradation shows glycine to be the N-terminal group. Acidic hydrolysis gives the following fragments:

H-Val-Pro-Leu-OH, Gly, H-Gly-Asp-Phe-Pro-OH, H-Phe-Pro-Val-OH

Propose a structure for the starting heptapeptide.

15.38 Give the amino acid sequence of hexapeptides that produce these fragments on partial acid hydrolysis:

(a) Arg, Gly, Ile, Leu, Pro, Val gives H-Pro-Leu-Gly-OH, H-Arg-Pro-OH, H-Gly-Ile-Val-OH
(b) Asp, Leu, Met, Trp, Val_2 gives H-Val-Leu-OH, H-Val-Met-Trp-OH, H-Trp-Asp-Val-OH

15.39 How can you account for the fact that proline is never encountered in a protein α helix? The α-helical segments of myoglobin and other proteins stop when a proline residue is encountered in the chain.

15.40 Draw as many resonance forms as you can for the purple anion obtained by reaction of ninhydrin with an amino acid.

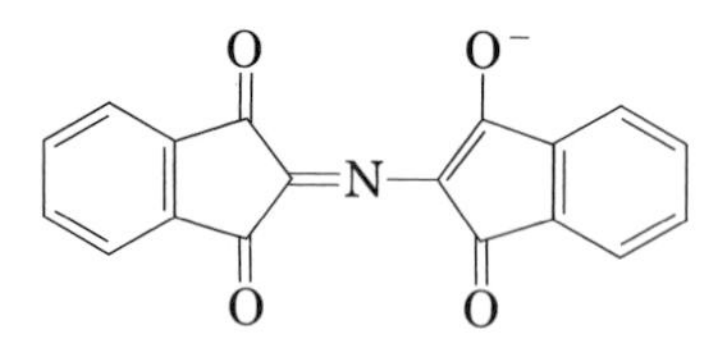

15.41 A nonapeptide gives the following fragments when cleaved by chymotrypsin and by trypsin.

Trypsin cleavage: H-Val-Val-Pro-Tyr-Leu-Arg-OH and H-Ser-Ile-Arg-OH
Chymotrypsin cleavage: H-Leu-Arg-OH and H-Ser-Ile-Arg-Val-Val-Pro-Tyr-OH

What is the structure of the nonapeptide?

15.42 Oxytocin, a nonapeptide hormone secreted by the pituitary gland, stimulates uterine contraction and lactation during childbirth. Its sequence was determined from the following evidence:

1. Oxytocin is a cyclic peptide containing a disulfide bridge between two cysteine residues.
2. When the disulfide bridge is reduced, oxytocin has the constitution Asn, Cys_2, Gln, Gly, Ile, Leu, Pro, Tyr.
3. Partial hydrolysis of reduced oxytocin yields seven fragments:
   H-Asp-Cys-OH H-Ile-Glu-OH H-Cys-Tyr-OH
   H-Leu-Gly-OH H-Tyr-Ile-Glu-OH H-Glu-Asp-Cys-OH
   H-Cys-Pro-Leu-OH
4. Gly is the C-terminal group.

5. Both Glu and Asp are present as their side-chain amides (Gln and Asn) rather than as free side-chain acids.

On the basis of this evidence, what is the amino acid sequence of reduced oxytocin? What is the structure of oxytocin?

15.43 *Aspartame*, a nonnutritive sweetener marketed under the trade name NutraSweet, is the methyl ester of a simple dipeptide, H-Asp-Phe-OCH_3.

(a) Draw the full structure of aspartame.
(b) The isoelectric point of aspartame is 5.9. Draw the principal structure present in aqueous solution at this pH.
(c) Draw the principal form of aspartame present at physiological pH 7.6.
(d) Show the products of hydrolysis on treatment of aspartame with H_3O^+.

CHAPTER 16

Biomolecules: Lipids and Nucleic Acids

In the previous two chapters, we've discussed the organic chemistry of carbohydrates and proteins, two of the four major classes of biomolecules. Let's now look at the two remaining classes: *lipids* and *nucleic acids*. Though chemically quite different from one another, all four classes are essential for life.

16.1 LIPIDS

lipid
a naturally occurring substance isolated from plants or animals by extraction with a nonpolar organic solvent

complex lipid
a lipid that has an ester linkage and can be hydrolyzed

Simple lipid
a lipid that has no ester linkage and can't be hydrolized

Lipids are naturally occurring organic substances isolated from cells and tissues by extraction with nonpolar organic solvents. Since they usually have large hydrocarbon portions in their structures, lipids are insoluble in water but soluble in organic solvents. Note that this definition differs from those used for carbohydrates and proteins in that lipids are defined by physical property (solubility) rather than by structure.

Lipids are further classified into two general types. **Complex lipids,** such as fats and waxes, contain ester linkages and can be hydrolyzed to yield smaller molecules. **Simple lipids,** such as cholesterol and other steroids, don't have ester linkages and can't be hydrolyzed.

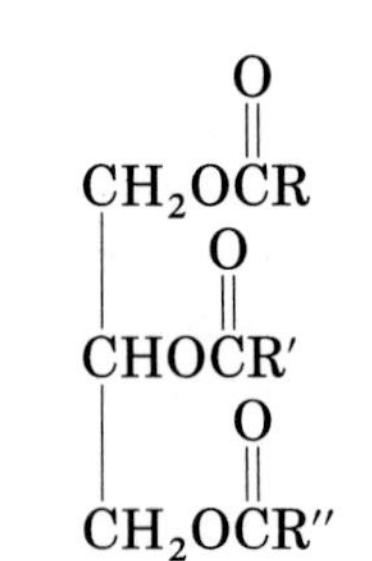

Animal fat, a complex lipid
(R, R', R'' = C_{11}–C_{19} chains)

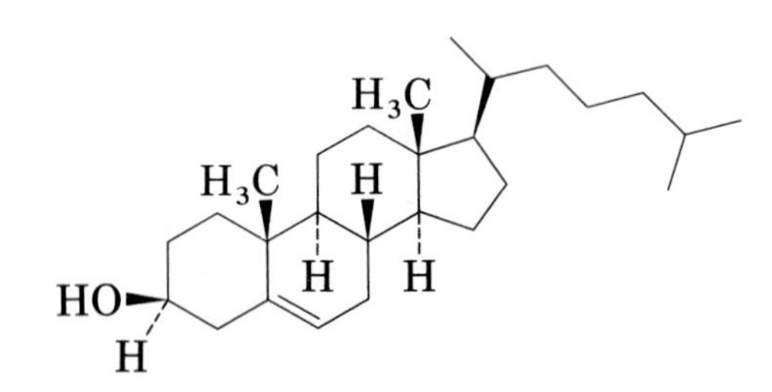

Cholesterol, a simple lipid

PROBLEM 16.1 Beeswax contains, among other things, a lipid with the structure

$$CH_3(CH_2)_{20}COO(CH_2)_{27}CH_3.$$

What products would you obtain by reaction of this lipid with aqueous NaOH followed by acidification?

16.2 FATS AND OILS

triacylglycerol a triester of glycerol with three fatty acids; an animal fat or vegetable oil

Animal fats and vegetable oils are the most widely occurring lipids. Although they look different—animal fats such as butter and lard are solids whereas vegetable oils such as corn oil and peanut oil are liquids—their structures are closely related. Chemically, fats and oil are **triacylglycerols** (also called *triglycerides*), triesters of glycerol with three long-chain carboxylic acids. Thus, hydrolysis of a fat or oil with aqueous sodium hydroxide yields glycerol and three fatty acids:

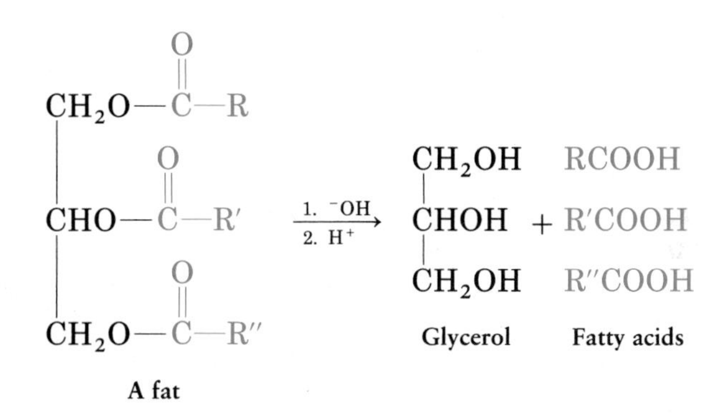

The fatty acids obtained by hydrolysis of triacylglycerols are generally unbranched and contain an even number of carbon atoms between 12 and 20. If double bonds are present, they usually have Z (cis) geometry. The three fatty acids of a specific molecule need not be the same, and a fat or oil from a given source is likely to be a complex mixture of many different triacylglycerols. Table 16.1 lists some of the commonly occurring fatty acids, and Table 16.2 lists the approximate composition of fats and oils from various sources.

The data in Table 16.1 show that unsaturated fatty acids generally have lower melting points than their saturated counterparts, a trend that's also true for triacylglycerols. Since vegetable oils generally have a higher proportion of unsaturated to saturated fatty acids than animal fats (Table 16.2), they have lower melting points. This behavior is due to the fact that saturated fats have a uniform shape that allows them to pack together easily in a crystal. Carbon–carbon double bonds in unsaturated vegetable oils, however, introduce bends and kinks into the hydrocarbon chains, making crystal formation difficult. The more double bonds there

TABLE 16.1 **Structures of some common fatty acids**

Name	*Carbons*	*Structure*	*Melting point (°C)*
Saturated			
Lauric	12	$CH_3(CH_2)_{10}COOH$	44
Myristic	14	$CH_3(CH_2)_{12}COOH$	58
Palmitic	16	$CH_3(CH_2)_{14}COOH$	63
Stearic	18	$CH_3(CH_2)_{16}COOH$	70
Arachidic	20	$CH_3(CH_2)_{18}COOH$	75
Unsaturated			
Palmitoleic	16	$CH_3(CH_2)_5CH{=}CH(CH_2)_7COOH$ (cis)	32
Oleic	18	$CH_3(CH_2)_7CH{=}CH(CH_2)_7COOH$ (cis)	4
Ricinoleic	18	$CH_3(CH_2)_5CH(OH)CH_2CH{=}CH(CH_2)_7COOH$ (cis)	5
Linoleic	18	$CH_3(CH_2)_4CH{=}CHCH_2CH{=}CH(CH_2)_7COOH$ (cis,cis)	−5
Arachidonic	20	$CH_3(CH_2)_4(CH{=}CHCH_2)_4CH_2CH_2COOH$ (all cis)	−50

TABLE 16.2 **Approximate fatty acid composition of some common fats and oils**

	Saturated fatty acids (%)				*Unsaturated fatty acids (%)*		
Source	C_{12} *Lauric*	C_{14} *Myristic*	C_{16} *Palmitic*	C_{18} *Stearic*	C_{18} *Oleic*	C_{18} *Ricinoleic*	C_{18} *Linoleic*
Animal fat							
Lard	—	1	25	15	50	—	6
Butter	2	10	25	10	25	—	5
Human fat	1	3	25	8	46	—	10
Whale blubber	—	8	12	3	35	—	10
Vegetable oil							
Coconut	50	18	8	2	6	—	1
Corn	—	1	10	4	35	—	45
Olive	—	1	5	5	80	—	7
Peanut	—	—	7	5	60	—	20
Linseed	—	—	5	3	20	—	20
Castor bean	—	—	—	1	8	85	4

are, the harder it is for the molecule to crystallize, and the lower the melting point of the oil. Figure 16.1 illustrates this effect with molecular models.

The carbon–carbon double bonds in vegetable oils can be reduced by catalytic hydrogenation (Section 4.6) to produce saturated solid or semisolid fats. Margarine and solid cooking fats such as Crisco are produced by hydrogenating soybean, peanut, or cottonseed oil until the right consistency is obtained.

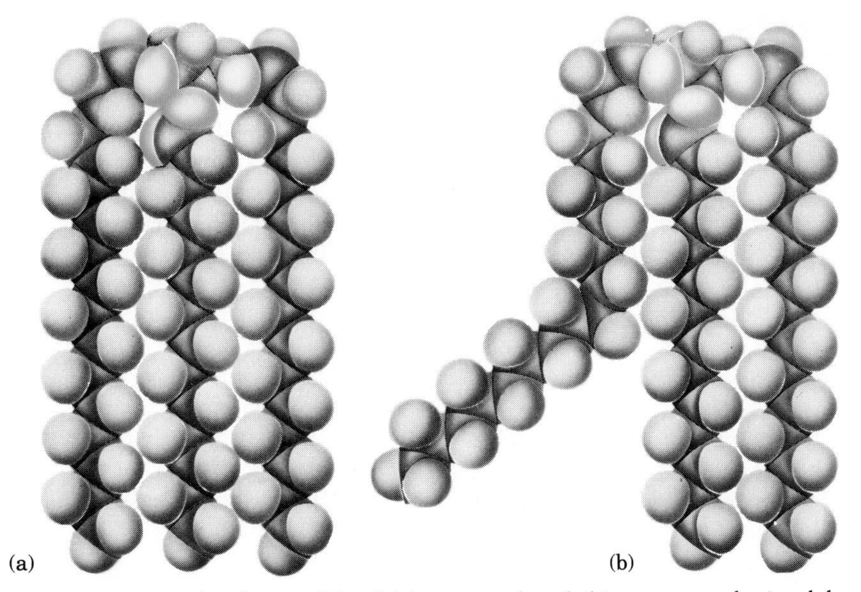

FIGURE 16.1 Molecular models of (a) saturated and (b) unsaturated triacylglycerols. The unsaturated acyl group in (b) prevents the molecule from adopting a regular shape and crystallizing easily.

PRACTICE PROBLEM 16.1 Draw the structure of glyceryl tripalmitate, a typical fat molecule.

SOLUTION Glyceryl tripalmitate is the triester of glycerol with three molecules of palmitic acid, $CH_3(CH_2)_{14}COOH$:

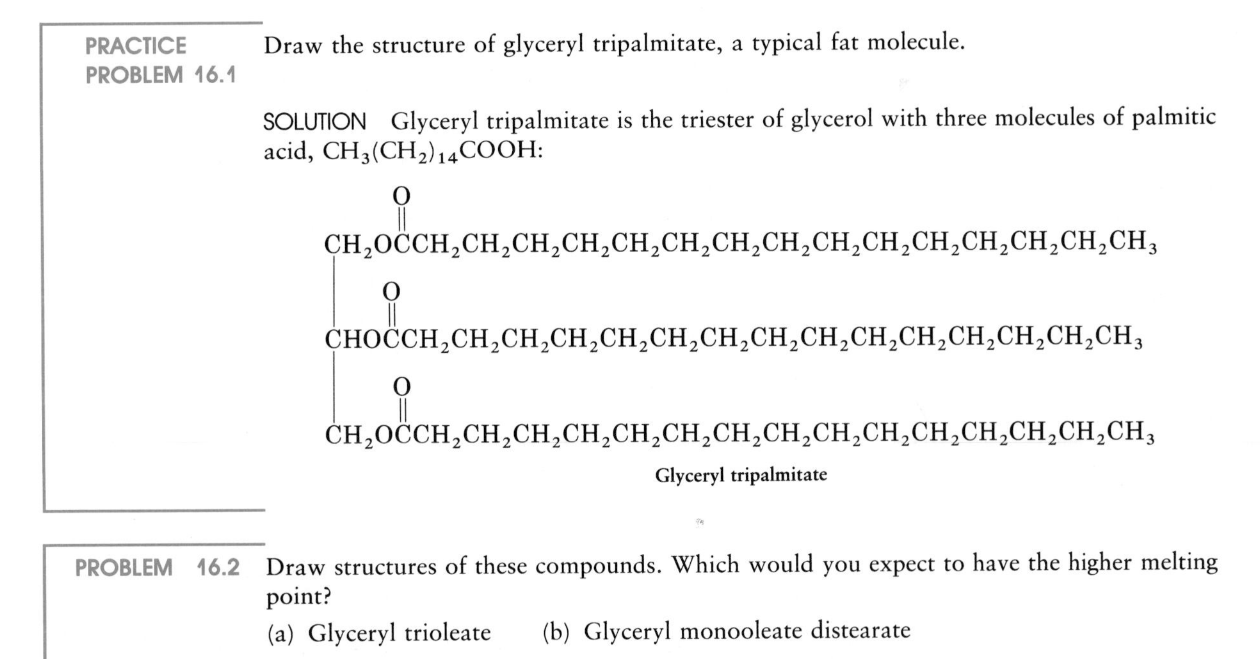

Glyceryl tripalmitate

PROBLEM 16.2 Draw structures of these compounds. Which would you expect to have the higher melting point?

(a) Glyceryl trioleate (b) Glyceryl monooleate distearate

PROBLEM 16.3 Fats and oils can be either optically active or optically inactive depending on their structures. Draw the structure of an optically active fat that gives 2 equiv palmitic acid and 1 equiv stearic acid on hydrolysis. Draw the structure of an optically inactive fat that gives the same products on hydrolysis.

16.3 SOAPS

Soap has been known since at least 600 BC, when the Phoenicians reportedly prepared a curdy material by boiling goat fat with extracts of wood ash. The cleansing properties of soap weren't generally recognized, however, and the use of soap didn't become widespread until the eighteenth century. Chemically, soap is a mixture of the sodium or potassium salts of long-chain fatty acids produced by hydrolysis (*saponification*) of animal fat with alkali.

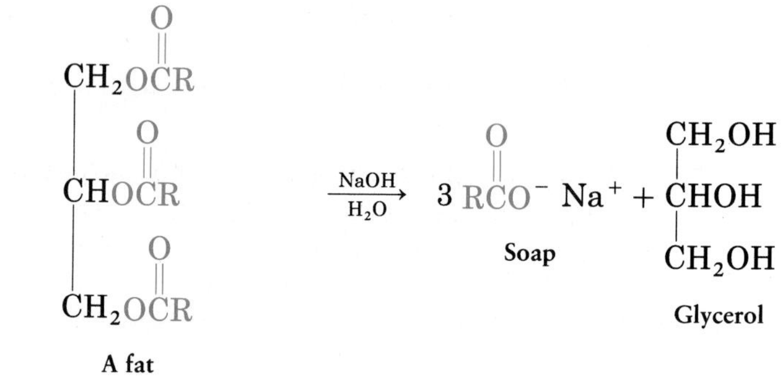

A fat
(R = C_{15}–C_{17} aliphatic chains)

Crude soap curds, which contain glycerol and excess alkali as well as soap, are purified by boiling with water and adding NaCl to precipitate the pure sodium carboxylate salts. The smooth soap that precipitates is dried, perfumed, and pressed into bars. Dyes are added for colored soaps, antiseptics are added for medicated soaps, pumice is added for scouring soaps, and air is blown in for soaps that float. Regardless of these extra treatments and regardless of price, all soaps are basically the same.

hydrophilic attracted to water (and repelled by hydrocarbons)

hydrophobic repelled by water (and attracted to hydrocarbons)

micelle a spherical cluster formed by aggregation of soap molecules in water

Soaps act as cleansers because the two ends of a soap molecule are so different. The sodium-salt end of the long-chain molecule is ionic and therefore **hydrophilic** (water loving); it tries to dissolve in water. The long aliphatic chain portion of the molecule, however, is **hydrophobic** (water fearing); it tries to avoid water and dissolve in grease. The net effect of these two opposing tendencies is that soaps are attracted to both grease and water.

When soaps are dispersed in water, the long hydrocarbon tails cluster together into a hydrophobic ball, while the ionic heads on the surface of the cluster stick out into the water layer. These spherical clusters, called **micelles**, are shown schematically in Figure 16.2. Grease and oil droplets are made soluble in water when they become coated by the hydrophobic, nonpolar tails of soap molecules in the center of micelles. Once solubilized, the grease and dirt can be rinsed away.

Although soaps make life much more pleasant than it would otherwise be, they also have drawbacks. In hard water, which contains metal ions, soluble sodium carboxylates are converted into insoluble calcium and magnesium salts, leaving the familiar ring of scum around bathtubs and the gray tinge on clothes. These problems have been circumvented by synthesizing a class of detergents based on

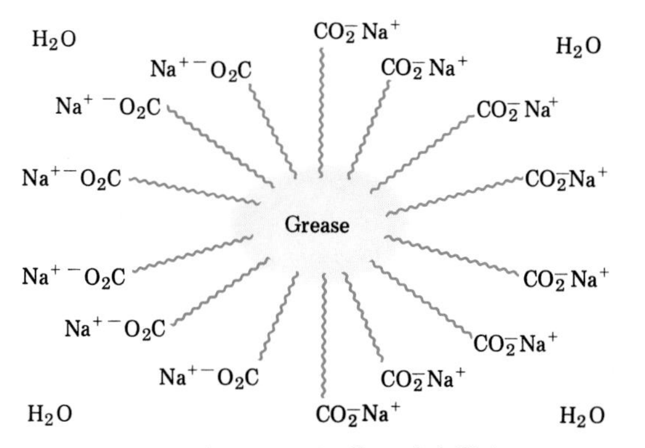

FIGURE 16.2 A soap micelle solubilizing a grease particle in water

salts of long-chain alkylbenzenesulfonic acids. The principle of synthetic detergents is identical to that of soaps: The alkylbenzene end of the molecule is lipophilic and attracts grease, but the sulfonate salt end is ionic and is attracted to water. Unlike soaps, though, sulfonate detergents don't form insoluble metal salts in hard water and don't leave an unpleasant scum.

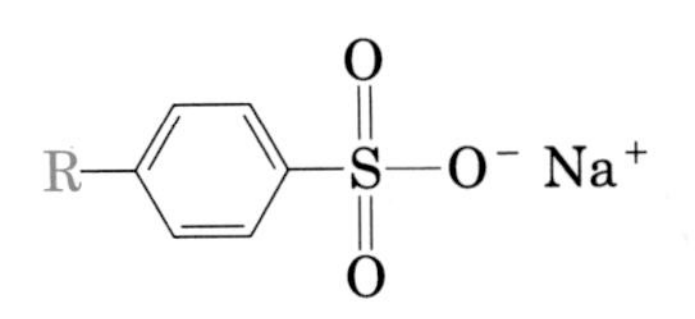

A synthetic detergent

PROBLEM 16.4 Draw the structure of magnesium oleate, one of the components of bathtub scum.

PROBLEM 16.5 Formulate the saponification reaction of glyceryl monopalmitate dioleate with aqueous NaOH.

16.4 PHOSPHOLIPIDS

phospholipid
a lipid that contains an ester link to phosphoric acid, H_3PO_4

phosphoglyceride
a phospholipid in which glycerol has ester links to two fatty acids and to phosphoric acid

Phospholipids are esters of phosphoric acid, H_3PO_4. Most phospholipids are closely related to fats, containing a glycerol backbone linked by ester bonds to two fatty acids and one phosphoric acid. Although the fatty acids in these **phosphoglycerides** can be any of the C_{12}–C_{20} units normally present in fats, the acyl group at carbon 1 is usually saturated, and that at carbon 2 is usually unsaturated. The phosphate group at carbon 3 is also bonded by a separate ester link to an amino alcohol such as choline, $HOCH_2CH_2\overset{+}{N}(CH_3)_3$, or ethanolamine, $HOCH_2CH_2NH_2$. The most important phosphoglycerides are the *lecithins* and the *cephalins*. Note that these compounds are chiral and that they have an L (or *R*) configuration at carbon 2.

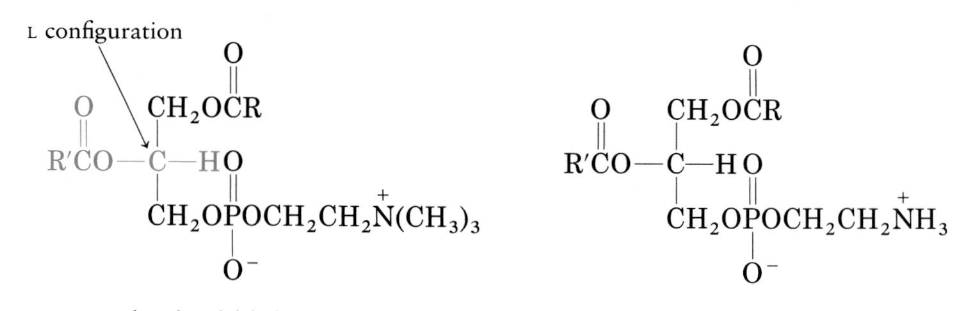

Found widely in plant and animal tissues, phosphoglycerides are the major lipid component of cell membranes (approximately 40%). Like soaps, phosphoglycerides have a long, nonpolar hydrocarbon tail bound to a polar ionic head (the phosphate group). Cell membranes are composed mostly of phosphoglycerides oriented into a bilayer about 50 Å thick. As shown in Figure 16.3, the hydrophobic tails aggregate in the center of the bilayer in much the same way that soap tails aggregate into the center of a micelle (Section 16.3). The bilayer thus forms an effective barrier to the passage of ions and other components into and out of the cell.

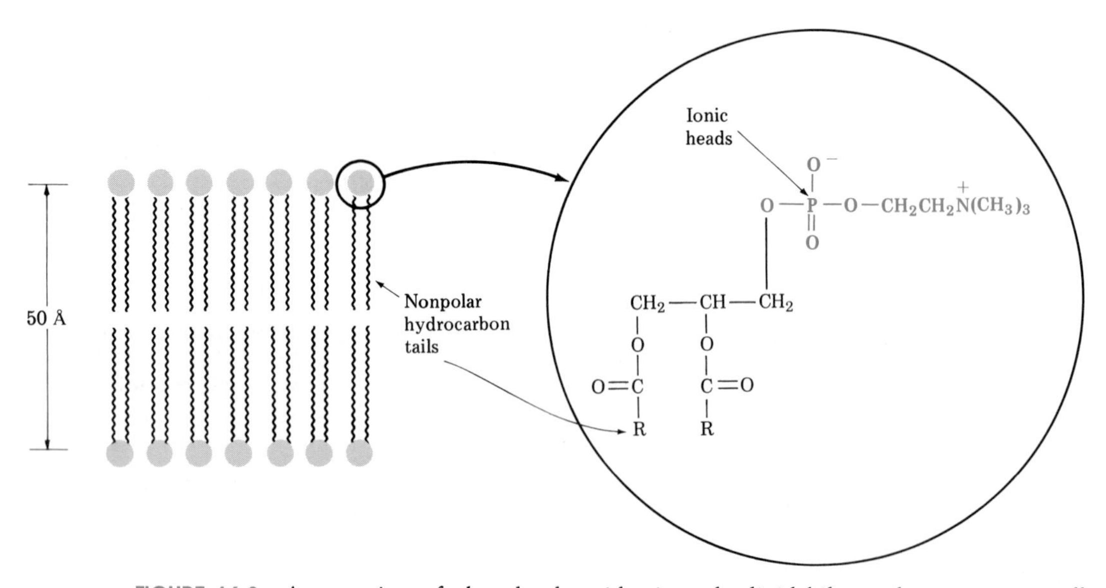

FIGURE 16.3 Aggregation of phosphoglycerides into the lipid bilayer that composes cell membranes

sphingolipid a phospholipid based on a sphingosine backbone rather than on glycerol

The second major group of phospholipids are the **sphingolipids.** These complex lipids, which have *sphingosine* or a related dihydroxyamine as their backbones, are constituents of plant and animal cell membranes. They're particularly abundant in brain and nerve tissue, where *sphingomyelins* are a major constituent of the coating around nerve fibers.

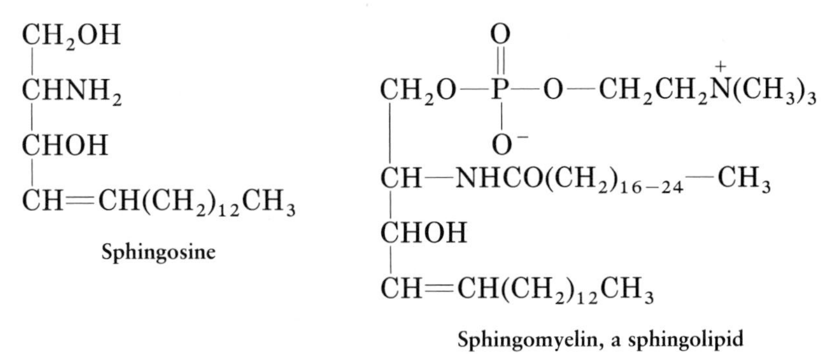

16.5 STEROIDS

steroid a lipid whose structure is based on a common tetracyclic ring system

In addition to fats, phospholipids, and terpenes, the lipid extracts of plants and animals also contain steroids. A **steroid** is an organic molecule whose structure is based on the tetracyclic ring system shown below. The four rings are designated A, B, C, and D, beginning at the lower left, and the carbon atoms are numbered beginning in the A ring.

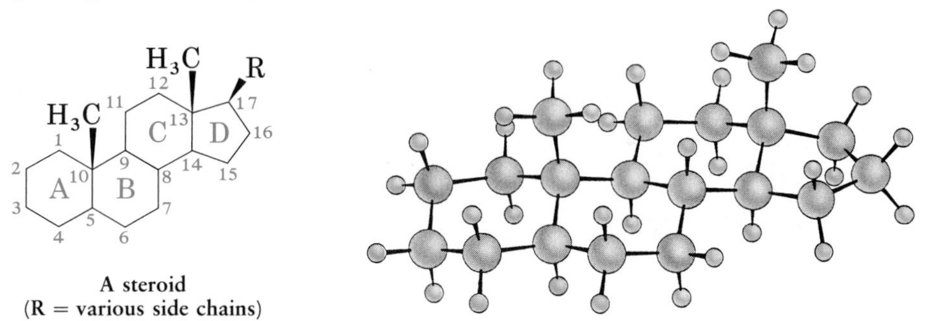

A steroid
(R = various side chains)

The steroid carbon skeleton has four rings joined so that the three six-membered rings (rings A, B, and C) adopt strain-free chair conformations. Unlike simple cyclohexane rings, though, steroids are constrained by their rigid conformation and can't undergo chair–chair interconversions by ring-flip (Section 2.11).

hormone a chemical messenger secreted by a specific gland and carried through the bloodstream to affect a target tissue

In humans, most steroids function as **hormones,** chemical messengers that are secreted by glands and are carried through the bloodstream to target tissues. There are two main classes of steroid hormones: the *sex hormones,* which control maturation and reproduction, and the *adrenocortical hormones,* which regulate a variety of metabolic processes.

Sex Hormones

androgen a male steroid sex hormone

Testosterone and *androsterone* are the two most important male sex hormones, or **androgens.** Androgens are responsible for the development of male secondary sex characteristics during puberty and for promoting tissue and muscle growth. Both are synthesized in the testes from cholesterol.

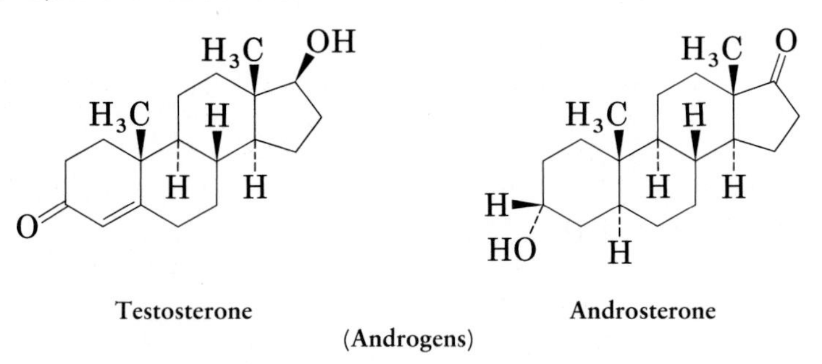

Testosterone (Androgens) Androsterone

Estrone and *estradiol* are the two most important female sex hormones, or **estrogens.** Synthesized in the ovaries from testosterone, estrogenic hormones are responsible for the development of female secondary sex characteristics and for regulation of the menstrual cycle. Note that both have a benzene-like aromatic A ring. In addition, another kind of sex hormone called a *progestin* is essential for preparing the uterus for implantation of a fertilized ovum during pregnancy. *Progesterone* is the most important progestin.

estrogen
a female steroid sex homone

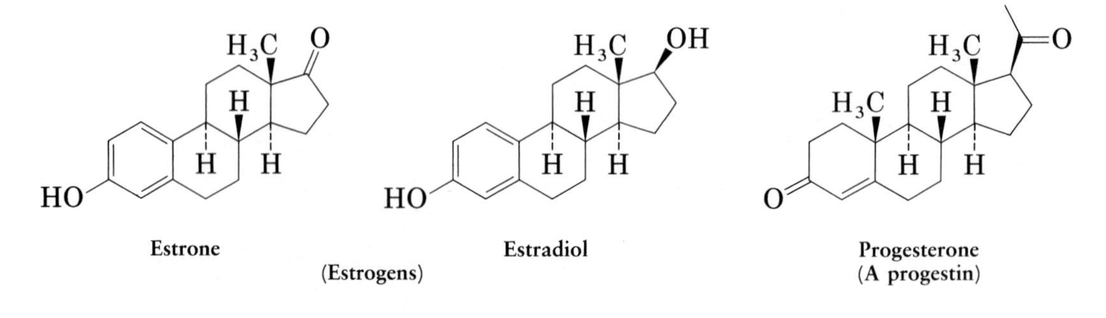

Estrone (Estrogens) Estradiol

Progesterone (A progestin)

Adrenocortical Hormones

Adrenocortical steroids are secreted by the adrenal glands, small organs located near the upper end of each kidney. There are two types of adrenocortical steroids, called *mineralocorticoids* and *glucocorticoids*. Mineralocorticoids such as *aldosterone* control tissue swelling by regulating cellular salt balance between Na^+ and K^+. Glucocorticoids such as *hydrocortisone* are involved in the regulation of glucose metabolism and in the control of inflammation. Glucocorticoid ointments are widely used to bring down the swelling from exposure to poison oak or poison ivy.

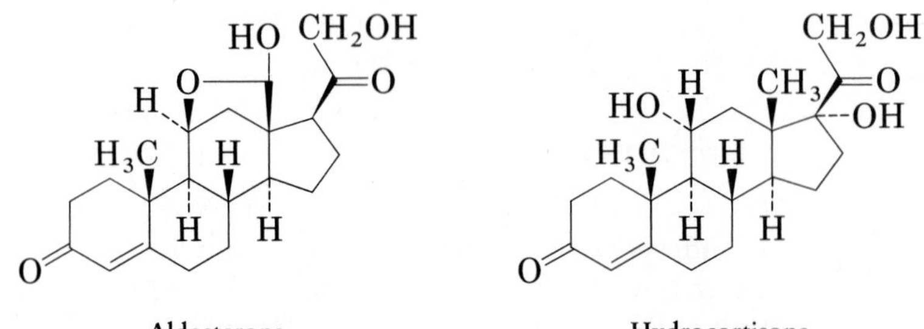

Aldosterone (A mineralocorticoid)

Hydrocortisone (A glucocorticoid)

Synthetic Steroids

In addition to the many hundreds of steroids isolated from plants and animals, thousands more have been synthesized in pharmaceutical laboratories in the search for new drugs. The idea is to start with a natural hormone, carry out a chemical modification of the structure, and then see what biological properties the modified steroid has.

Among the best-known synthetic steroids are oral contraceptive and anabolic agents. Most birth-control pills are a mixture of two compounds, a synthetic estrogen such as *ethynylestradiol* and a synthetic progestin such as *norethindrone*. Anabolic steroids such as stanozolol, detected in several athletes during the 1988 Olympics, are synthetic androgens that mimic the tissue-building effects of natural testosterone.

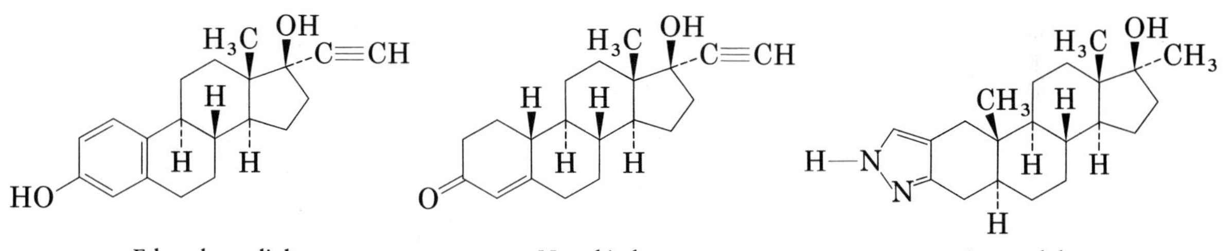

Ethynylestradiol (A synthetic estrogen) — Norethindrone (A synthetic progestin) — Stanozolol (An anabolic agent)

PROBLEM 16.6 Look at the structure of cholesterol shown in on the first page of this chapter and tell whether the hydroxyl group is axial or equatorial.

PROBLEM 16.7 Look at the structure of progesterone and identify all of the functional groups in the molecule.

PROBLEM 16.8 Look at the structures of estradiol and ethynylestradiol and point out the differences. What common structural feature do they share that makes both estrogens?

16.6 NUCLEIC ACIDS

nucleotide a building block for the construction of nucleic acids consisting of phosphoric acid, a pentose sugar, and a heterocyclic amine base

nucleoside the hydrolysis product of a nucleotide, consisting of a pentose sugar bonded to a heterocyclic amine base

The nucleic acids, deoxyribonucleic acid (DNA) and ribonucleic acid (RNA), are the chemical carriers of a cell's genetic information. Coded in a cell's DNA is all the information that determines the nature of the cell, controls cell growth and division, and directs biosynthesis of the enzymes and other proteins required for all cellular functions.

Just as proteins are polymers made of amino acid units, nucleic acids are polymers made up of individual building blocks called **nucleotides** linked together to form a long chain. Each nucleotide is composed of a **nucleoside** plus phosphoric acid, H_3PO_4, and each nucleoside is further composed of a simple aldopentose sugar plus a heterocyclic amine base (Section 12.7).

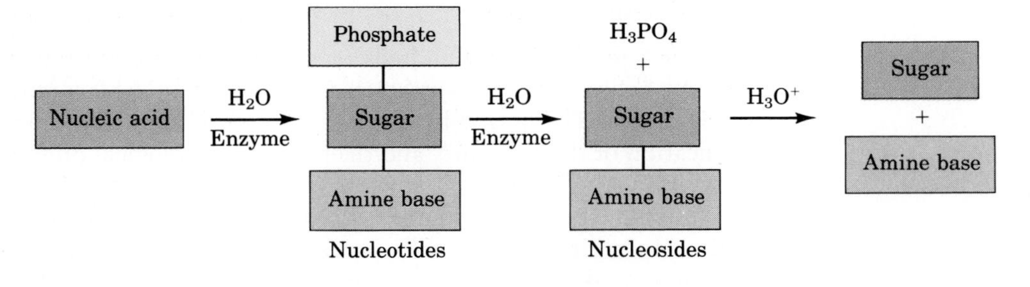

The sugar component in RNA is ribose, and the sugar in DNA is 2-deoxyribose (*2-deoxy* means that oxygen is missing from C2 of ribose.)

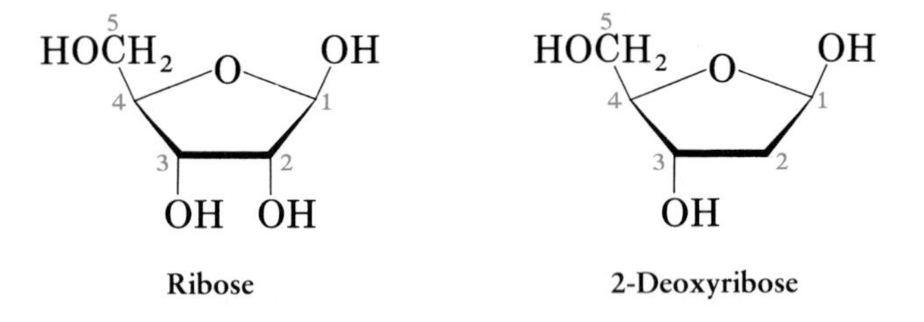

There are four different heterocyclic amine bases in DNA. Two are substituted *purines* (adenine and guanine), and two are substituted *pyrimidines* (cytosine and thymine). Adenine, guanine, and cytosine also occur in RNA, but thymine is replaced in RNA by a different pyrimidine base called uracil.

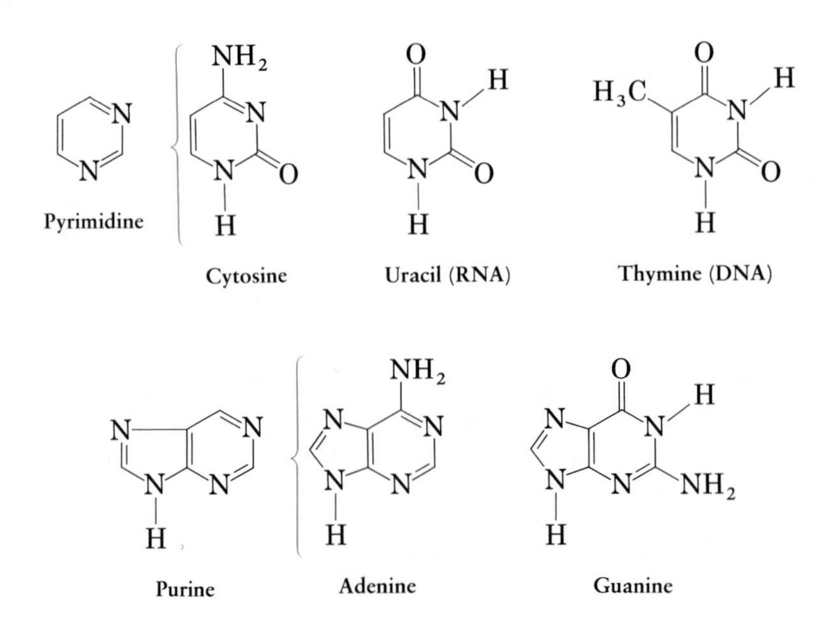

In both DNA and RNA, the heterocyclic amine base is bonded to C1′ of the sugar, and the phosphoric acid is bonded by a phosphate ester linkage to the C5′ sugar position. Thus, nucleosides and nucleotides have the general structure shown in Figure 16.4. (In discussions of RNA and DNA, numbers with a prime superscript refer to positions on the sugar component of a nucleotide; numbers without a prime refer to positions on the heterocyclic amine base.) The complete structures of all four deoxyribonucleotides and all four ribonucleotides are shown in Figure 16.5.

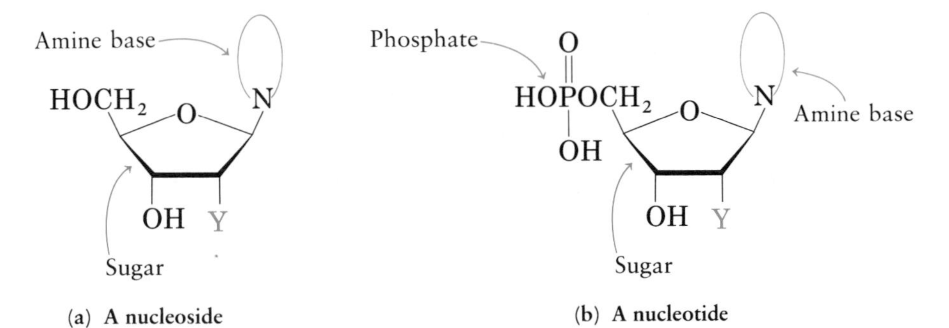

FIGURE 16.4 General structures of (a) a nucleoside and (b) a nucleotide. When Y = H, the sugar is deoxyribose; when Y = OH, the sugar is ribose.

Though chemically similar, DNA and RNA are different in size and have different roles in the cell. Molecules of DNA are enormous, with molecular weights of up to 150 billion and lengths of up to 12 cm; they are found mostly in the nucleus of the cell. Molecules of RNA, by contrast, are much smaller (as low as 35,000 mol wt) and are found mostly outside the cell nucleus. Let's consider the two kinds of nucleic acids separately, beginning with DNA.

16.7 STRUCTURE OF DNA

3′ end the end of a nucleic acid chain that has a free sugar hydroxyl group

5′ end the end of a nucleic acid chain that has a phosphoric acid unit

Nucleotides join together in DNA by forming a phosphate ester bond between the 5′-phosphate component of one nucleotide and the 3′-hydroxyl on the sugar component of another nucleotide (Figure 16.6). Regardless of how long the chain is, though, one end of the nucleic acid polymer has a free hydroxyl at C3′ (called the **3′ end**), and the other end has a phosphoric acid residue at C5 (the **5′ end**).

Just as the exact structure of a protein depends on the sequence in which individual amino acids are connected, the exact structure of a nucleic acid depends on the sequence of individual nucleotides. To carry the analogy further, just as a protein has a polyamide backbone with different side chains attached to it, a nucleic acid has an alternating sugar–phosphate backbone with different amine base side chains attached at regular intervals.

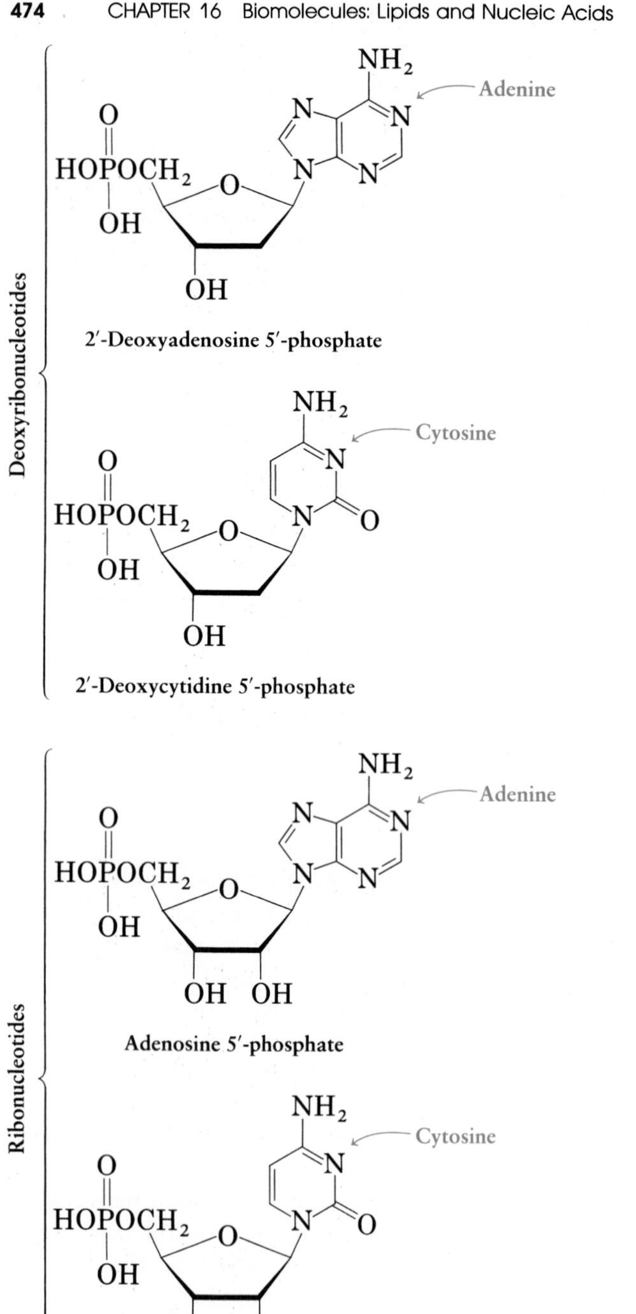

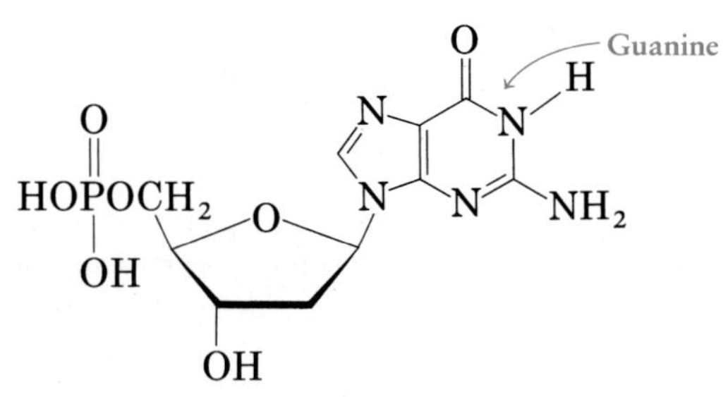

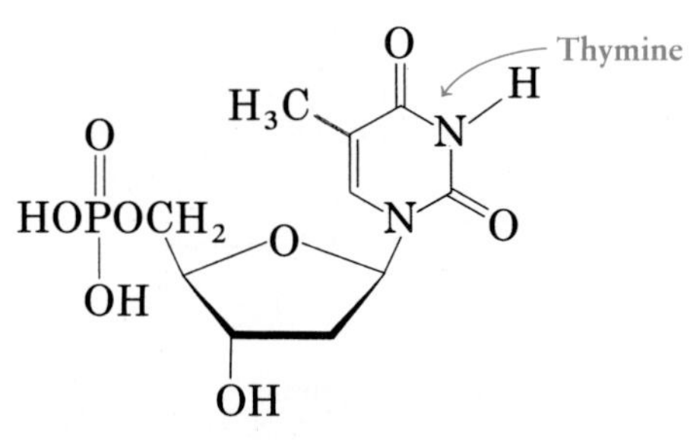

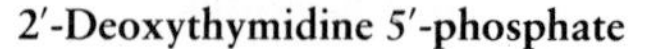

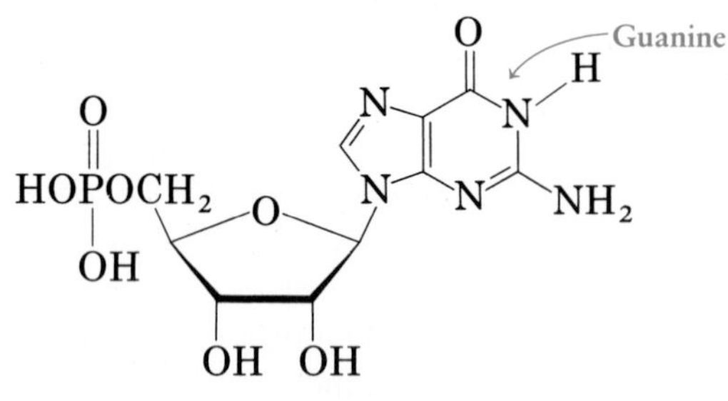

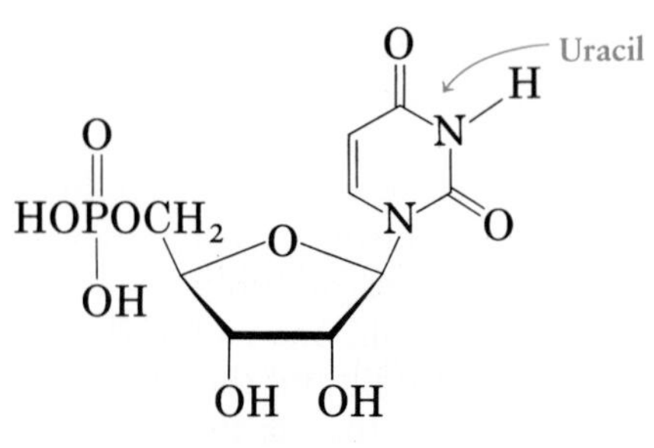

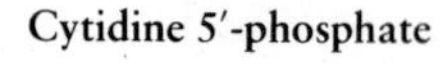

FIGURE 16.5 Structures of the four deoxyribonucleotides and four ribonucleotides

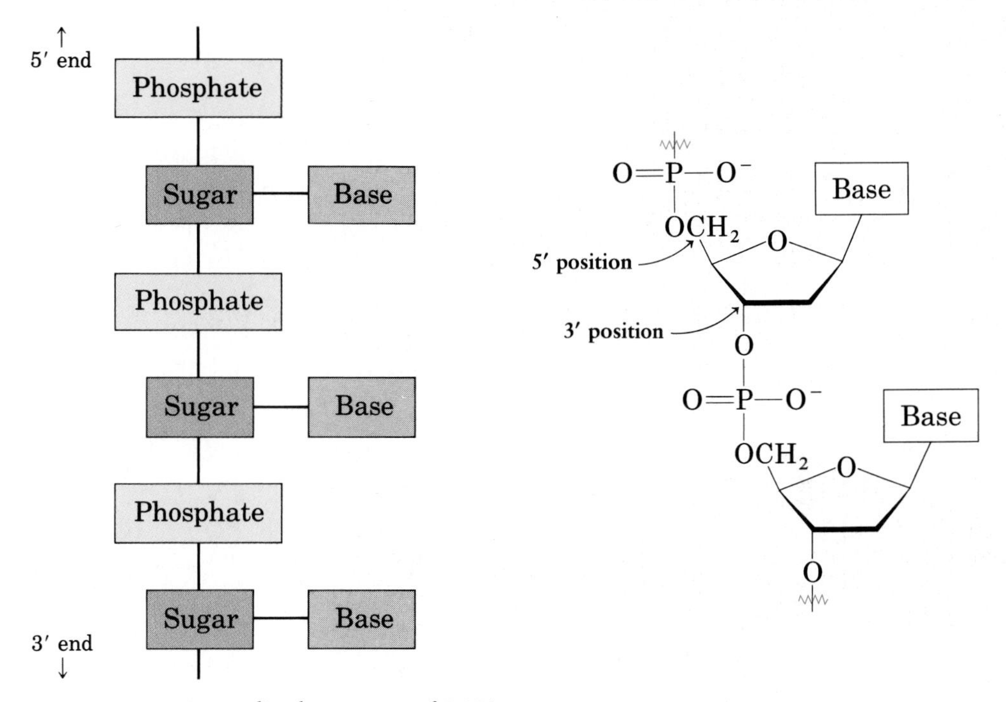

FIGURE 16.6 Generalized structure of DNA

A protein

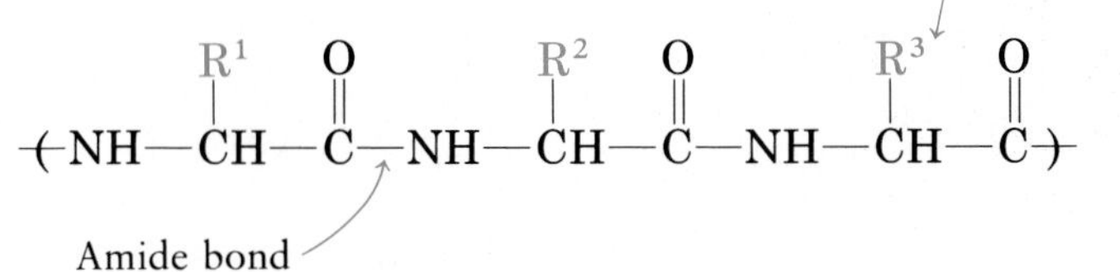

A nucleic acid

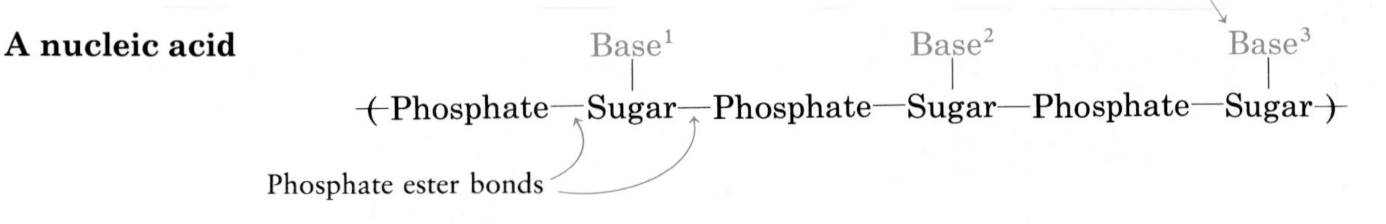

The sequence of nucleotides is described by starting at the 5′ end and identifying the bases in order of occurrence. Rather than write the full name of each nucleotide, though, it's easier to use abbreviations: A for adenosine, T for thymine, G for guanosine, and C for cytidine. Thus, a typical sequence might be written as -T-A-G-G-C-T-.

PRACTICE PROBLEM 16.2 Draw the full structure of the DNA dinucleotide C-T.

SOLUTION

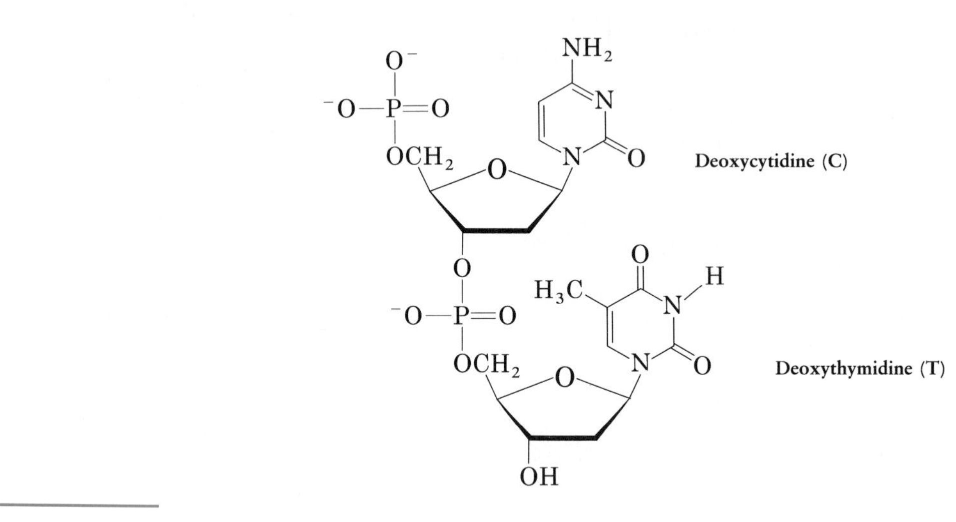

PROBLEM 16.9 Draw the full structure of the DNA dinucleotide A-G.

PROBLEM 16.10 Draw the full structure of the RNA dinucleotide U-A.

16.8 BASE PAIRING IN DNA: THE WATSON–CRICK MODEL

Samples of DNA isolated from different tissues of the same species have the same proportions of heterocyclic bases, but samples from different species can have greatly different proportions of bases. For example, human DNA contains about 30% each of adenine and thymine and about 20% each of guanine and cytosine. The bacterium *Clostridium perfringens,* however, contains about 37% each of adenine and thymine and only 13% each of guanine and cytosine. Note that in both of these examples, the bases occur in pairs. Adenine and thymine are usually present in equal amounts, as are guanine and cytosine. Why should this be?

In 1953, James Watson and Francis Crick made their now classic proposal for the secondary structure of DNA. According to the Watson–Crick model, DNA consists of two polynucleotide strands coiled around each other in a *double helix.* The two strands run in opposite directions and are held together by hydrogen bonds between specific pairs of bases. Adenine (A) and thymine (T) form strong hydrogen bonds to each other but not to G or C. Similarly, guanine (G) and cytosine (C) form strong hydrogen bonds to each other but not to A or T.

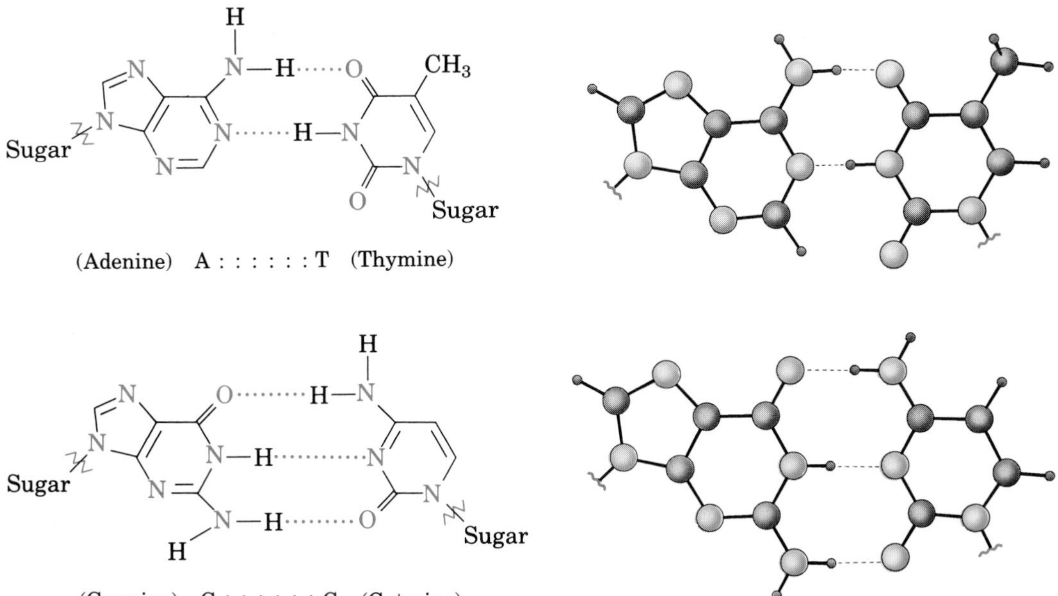

The two strands of the DNA double helix aren't identical; rather, they're complementary. Whenever a G base occurs in one strand, a C base occurs opposite it in the other strand. When an A base occurs in one strand, a T base occurs in the other strand. This complementary pairing of bases explains why A and T, and G and C, are always found in equal amounts. Figure 16.7 illustrates this base pairing, showing how the two complementary strands coil into the double helix. X-ray measurements show that the DNA double helix is 20 Å wide, that there are exactly 10 base pairs in each full turn, and that each turn is 34 Å in height.

A helpful mnemonic device to remember the pairing of the four DNA bases is the simple phrase "pure silver taxi":

pure	silver	taxi
Pur	Ag	TC
The purine bases,	A and G,	pair with T and C

PRACTICE PROBLEM 16.3 What sequence of bases on one strand of DNA is complementary to the sequence T-A-T-G-C-A-T on another strand?

SOLUTION Remembering that A and G (silver) form complementary pairs with T and C (taxi) respectively, we go through the given sequence replacing A by T, G by C, T by A, and C by G.

Original:	T-A-T-G-C-A-T
Complement:	A-T-A-C-G-T-A

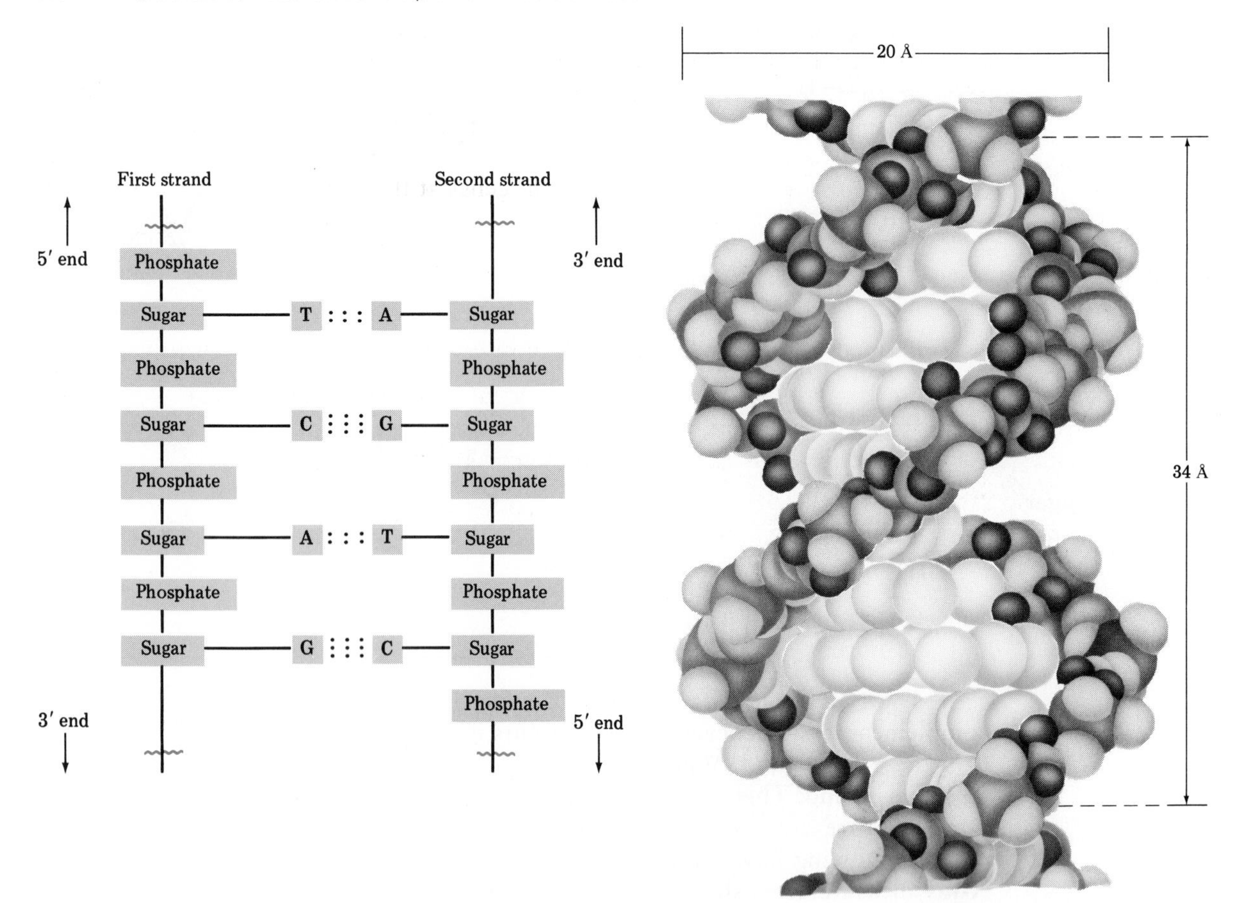

FIGURE 16.7 Complementarity of base pairing in the DNA double helix. The sugar–phosphate backbone of DNA is shown in gray; the atoms of the amine bases are shown in color and lie inside the helix; the small black atoms are hydrogen.

PROBLEM 16.11 What sequence of bases on one strand of DNA is complementary to the following sequence on another strand?

G-G-C-T-A-A-T-C-C-G-T

16.9 NUCLEIC ACIDS AND HEREDITY

A DNA molecule is the chemical repository of an organism's genetic information, which is stored as a sequence of deoxyribonucleotides strung together in the DNA chain. For the information to be preserved and passed on to future generations, a mechanism must exist for copying DNA. For the information to be used, a mechanism must exist for decoding the DNA message and for implementing the instructions it contains.

What Crick has termed the central dogma of molecular genetics says that the function of DNA is to store information and pass it on to RNA. The function of RNA, in turn, is to read, decode, and use the information received from DNA to make proteins. Each of the thousands of individual genes on a chromosome contains the instructions necessary to make a specific protein needed for a specific biological purpose. By decoding the right genes at the right time, an organism uses genetic information to synthesize the thousands of proteins necessary for smooth functioning.

$$\text{DNA} \longrightarrow \text{RNA} \longrightarrow \text{proteins}$$

replication the process by which DNA is copied

transcription the process by which RNA is made from DNA

translation the process by which proteins are made from RNA

Three fundamental processes take place in the transfer of genetic information:

Replication is the process by which identical copies of DNA are made, forming additional molecules and preserving genetic information.

Transcription is the process by which information in the DNA is read and decoded by RNA.

Translation is the process by which RNA uses the information to build proteins.

16.10 REPLICATION OF DNA

Replication of DNA is an enzyme-catalyzed process that begins by a partial unwinding of the double helix. As the DNA strands separate and bases are exposed, new nucleotides line up on each strand in an exactly complementary manner, A to T and C to G, and two new strands begin to grow. Each new strand is complementary to its old template strand, and two new identical DNA double helices are produced (Figure 16.8).

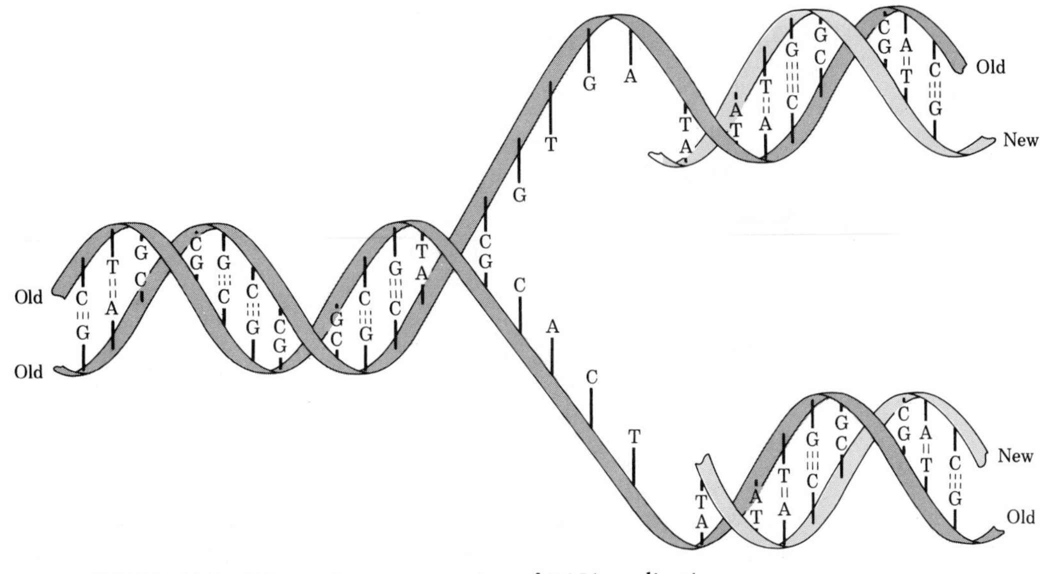

FIGURE 16.8 Schematic representation of DNA replication

Crick probably described the process best when he used the analogy of the two DNA strands fitting together like a hand in a glove. The hand and glove separate; a new hand forms inside the glove; and a new glove forms around the hand. Two identical copies now exist where only one existed before.

The process by which the individual nucleotides are joined to create new DNA strands is complex, involving many steps and different enzymes. Addition of new nucleotide units to the growing chain, a step catalyzed by the enzyme DNA polymerase, has been shown to occur by addition of a 5′-mononucleotide triphosphate to the free 3′-hydroxyl group of the growing chain, as indicated in Figure 16.9.

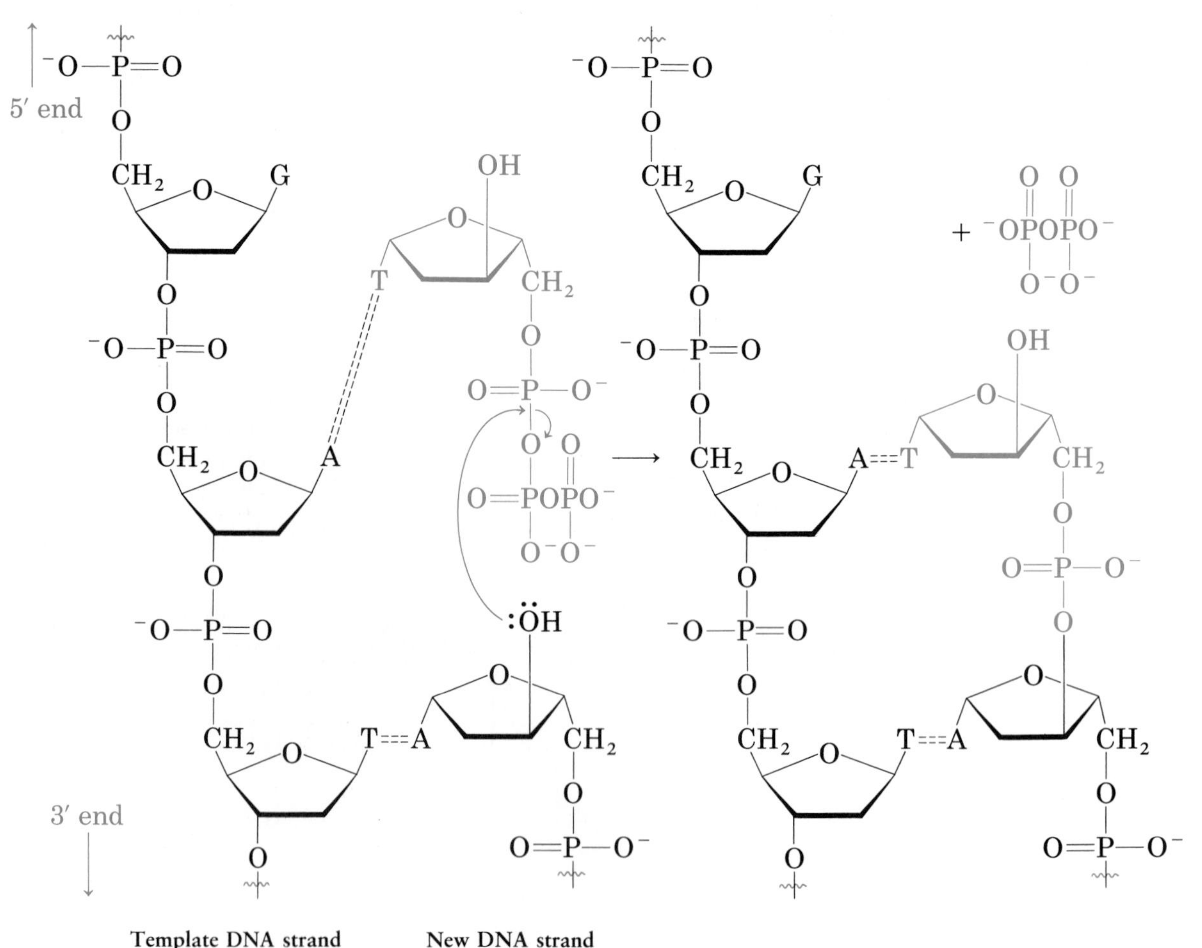

FIGURE 16.9 Addition of a new nucleotide to a growing DNA strand

It's difficult to conceive of the magnitude of the replication process. The nucleus of a human cell contains 46 chromosomes (23 pairs), each of which consists of one very large DNA molecule. Each chromosome, in turn, is made up of several thousand DNA segments called *genes,* and the sum of all genes in a human cell (the *genome*) is estimated to be approximately three billion base pairs. Regardless of the size of these massive molecules, the base sequence is faithfully copied during replication, with an error occurring only about once each 10–100 billion bases.

16.11 STRUCTURE AND SYNTHESIS OF RNA: TRANSCRIPTION

RNA is structurally similar to DNA. Both are sugar–phosphate polymers and both have heterocyclic bases attached. The only differences are that RNA contains ribose rather than 2-deoxyribose and contains uracil rather than thymine. Uracil in RNA forms strong hydrogen bonds to its complementary base, adenine, just as thymine does in DNA.

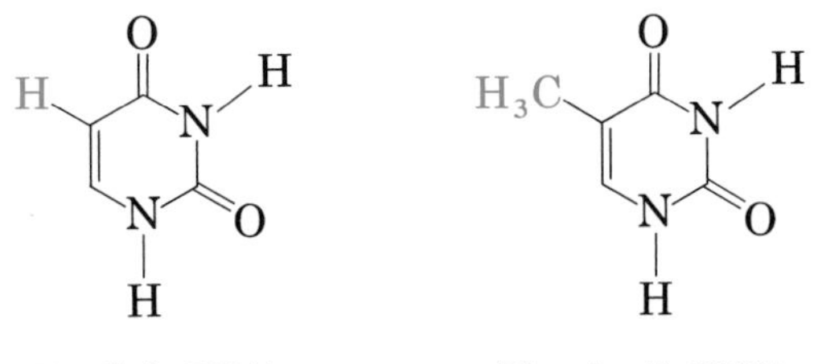

Uracil (in RNA) Thymine (in DNA)

There are three major kinds of ribonucleic acid, each of which serves a specific function.

messenger RNA (mRNA) the kind of RNA that decodes DNA and carries the message out of the cell nucleus

ribosomal RNA (rRNA) the kind of RNA used to construct ribosomes, where protein synthesis occurs

transfer RNA (tRNA) the kind of RNA that transports specific amino acids to ribosomes for protein synthesis

Messenger RNA (mRNA) carries genetic messages from DNA to *ribosomes,* small granular particles in the cell that act as "protein factories."

Ribosomal RNA (rRNA) is a structural component of ribosomes.

Transfer RNA (tRNA) transports specific amino acids to the ribosomes where they are joined together to make proteins.

All three kinds of RNA are much smaller molecules than DNA, and all occur as single polyribonucleotide strands rather than as double helices.

Molecules of RNA are synthesized in the nucleus of the cell by transcription of DNA. A small portion of the DNA double helix unwinds, and one of the two complementary DNA strands acts as a template for complementary ribonucleotides to line up. Bond formation then occurs in the $5' \rightarrow 3'$ sense as with DNA replication. Unlike DNA replication, though, the completed RNA molecule does not remain in a double helix with DNA but separates and migrates from the cell nucleus. The DNA then rewinds to its stable double-helix conformation (Figure 16.10).

PRACTICE PROBLEM 16.4

What RNA base sequence is complementary to the following DNA base sequence?

T-A-A-G-C-C-G-T-G

SOLUTION Go through the sequence replacing A by U, G by C, T by A, and C by G.

Original DNA: T-A-A-G-C-C-G-T-G

Complementary RNA: A-U-U-C-G-G-C-A-C

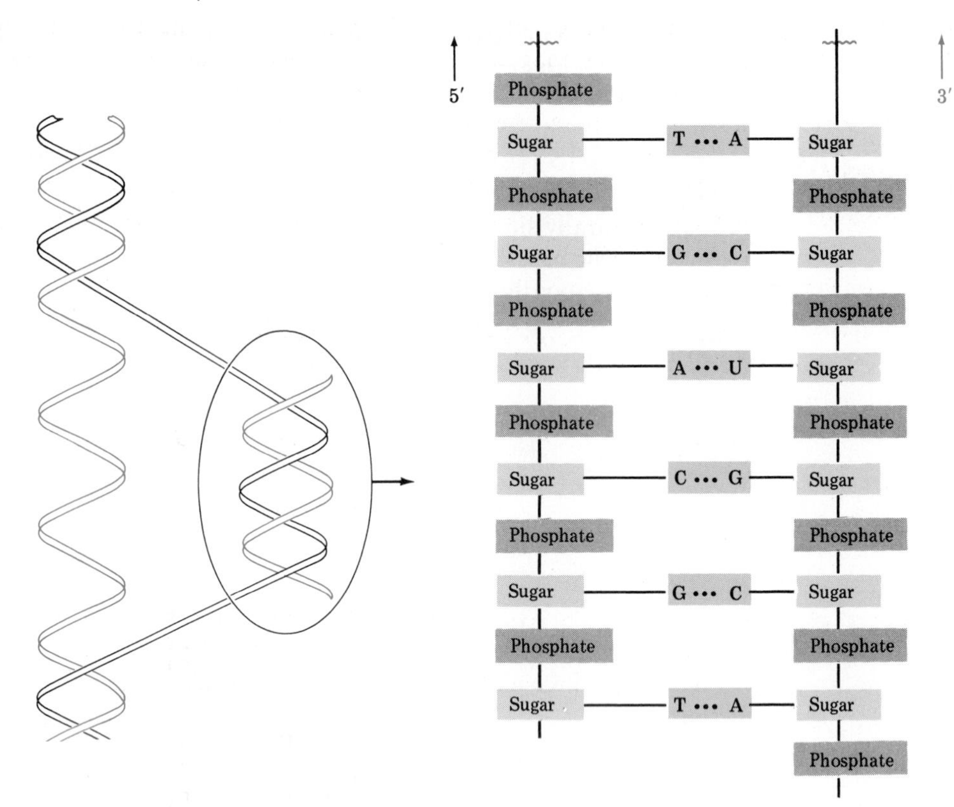

FIGURE 16.10 Synthesis of RNA using a DNA segment as template

PROBLEM 16.12 Show how uracil can form strong hydrogen bonds to adenine.

PROBLEM 16.13 What RNA base sequence is complementary to the following DNA base sequence?

G-A-T-T-A-C-C-G-T-A

PROBLEM 16.14 From what DNA base sequence was the following RNA sequence transcribed?

U-U-C-G-C-A-G-A-G-U

16.12 RNA AND PROTEIN BIOSYNTHESIS: TRANSLATION

Once the information in DNA has been transcribed into RNA, the information is used to synthesize proteins. The mechanics of protein biosynthesis are directed by messenger RNA and take place on ribosomes. The specific ribonucleotide sequence in mRNA acts like a long coded sentence to specify the order in which different amino acid residues are to be joined. Thus, each of the estimated 100,000 proteins in the human body is synthesized from a different mRNA that has been transcribed from a specific gene segment on DNA.

codon
a sequence of three ribonucleotides on mRNA that codes for incorporation of a specific amino acid into a protein sequence

Each "word" or **codon** along the mRNA chain consists of a series of three ribonucleotides that is specific for a given amino acid. For example, the series cytosine-uracil-guanine (C-U-G) on mRNA is a codon directing incorporation of the amino acid leucine into the growing protein. Similarly, guanine-adenine-uracil (G-A-U) codes for aspartic acid. Of the $4^3 = 64$ possible triads of the four bases in RNA, 61 code for specific amino acids (most amino acids are specified by more than one codon). In addition, 3 of the 64 codons specify chain termination. Table 16.3 shows the meaning of each codon.

TABLE 16.3 Codon assignments of base triads

First base (5′ end)	*Second base*	*Third base (3′ end)*			
		U	*C*	*A*	*G*
U	U	Phe	Phe	Leu	Leu
	C	Ser	Ser	Ser	Ser
	A	Tyr	Tyr	Stop	Stop
	G	Cys	Cys	Stop	Trp
C	U	Leu	Leu	Leu	Leu
	C	Pro	Pro	Pro	Pro
	A	His	His	Gln	Gln
	G	Arg	Arg	Arg	Arg
A	U	Ile	Ile	Ile	Met
	C	Thr	Thr	Thr	Thr
	A	Asn	Asn	Lys	Lys
	G	Ser	Ser	Arg	Arg
G	U	Val	Val	Val	Val
	C	Ala	Ala	Ala	Ala
	A	Asp	Asp	Glu	Glu
	G	Gly	Gly	Gly	Gly

The code expressed in mRNA is read by transfer RNA in the process called translation. There are at least 60 different tRNAs, one for each of the codons in Table 16.3. Each specific tRNA acts as a carrier to bring a specific amino acid into place so that it may be transferred to the growing protein chain. A typical tRNA is roughly the shape of a cloverleaf, as shown in Figure 16.11. It consists of about 70–100 ribonucleotides and is bonded to a specific amino acid by an ester linkage through the free 3′-hydroxyl on ribose at the 3′ end of the tRNA. Each tRNA also contains in its structure a segment called an **anticodon,** a sequence of three ribonucleotides complementary to the codon sequence. For example, the codon sequence C-U-G present on mRNA would be "read" by a leucine-bearing tRNA having the complementary anticodon sequence G-A-C.

anticodon
a sequence of three ribonucleotides on tRNA that is complementary to a codon on mRNA

As each successive codon on mRNA is read, different tRNAs bring the correct amino acids into position for enzyme-mediated transfer to the growing peptide. When synthesis of the proper protein is completed, a "stop" codon signals the

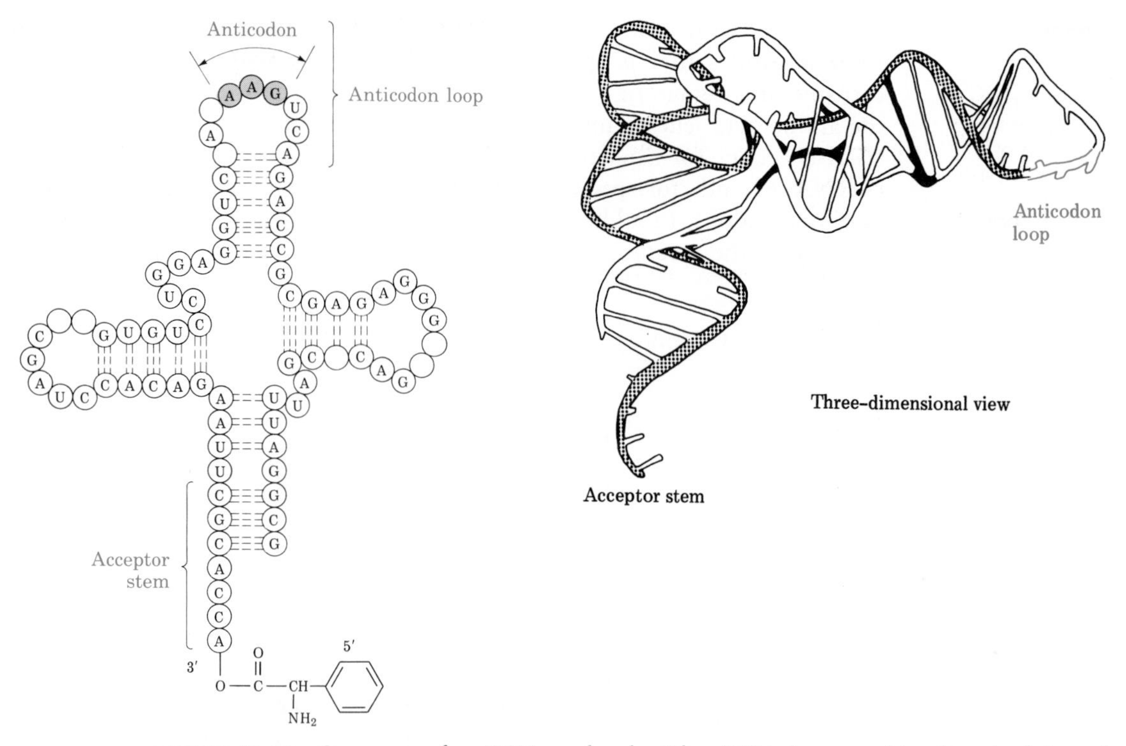

FIGURE 16.11 Structure of a tRNA molecule. The tRNA is a roughly cloverleaf-shaped molecule containing an anticodon triplet on one "leaf" and a covalently attached amino acid residue at its 3′ end. The example shown is a yeast tRNA that codes for phenylalanine. (The nucleotides that aren't specifically identified are chemically modified analogs of the four normal nucleotides.)

end, and the protein is released from the ribosome. The entire process of protein biosynthesis is illustrated schematically in Figure 16.12.

PRACTICE PROBLEM 16.5 List a codon sequence for valine.

SOLUTION According to Table 16.3, there are four codons for valine: G-U-U, G-U-C, G-U-A, G-U-G

PRACTICE PROBLEM 16.6 What amino acid sequence is coded for by the mRNA base sequence AUC-GGU?

SOLUTION Table 16.3 indicates that AUC codes for isoleucine and that GGU codes for glycine. Thus, AUC-GGU codes for H-Ile-Gly-OH.

PROBLEM 16.15 List codon sequences for these amino acids.

(a) Ala (b) Phe (c) Leu (d) Tyr

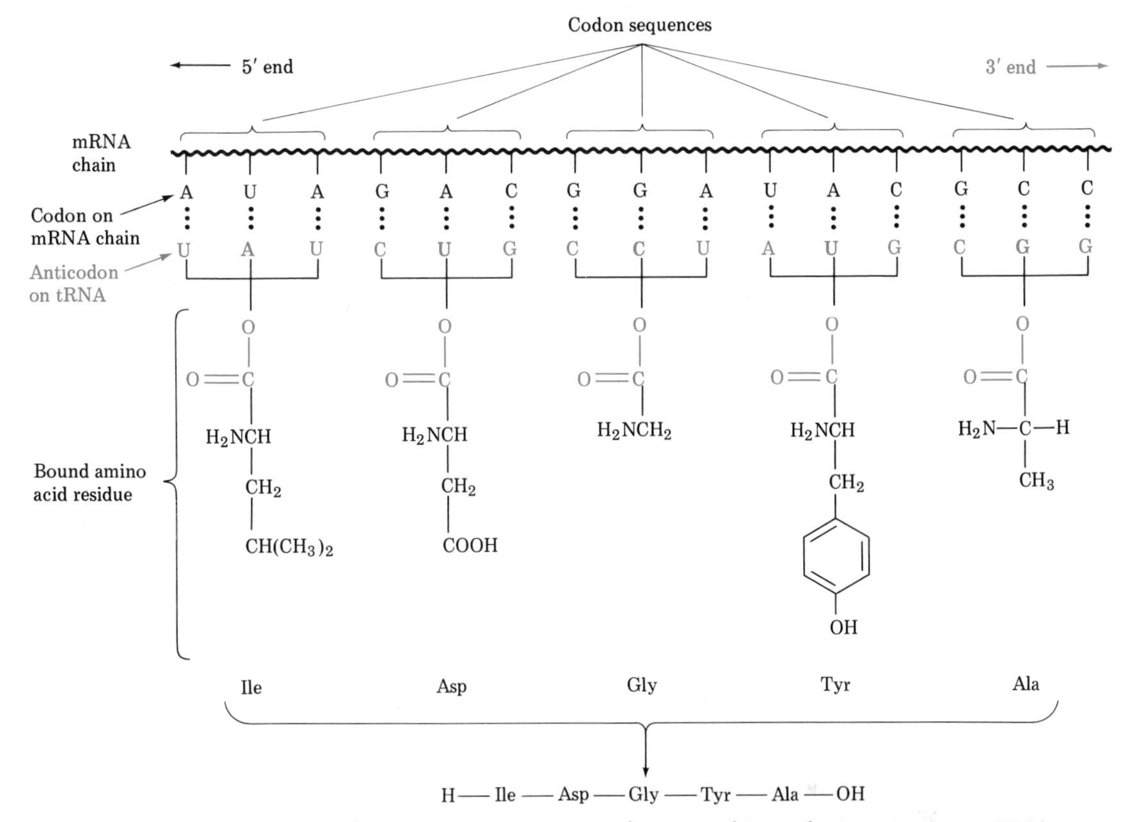

FIGURE 16.12 Schematic representation of protein biosynthesis. Messenger RNA, containing codon base sequences, is read by tRNA, containing complementary anticodon base sequences. Transfer RNA assembles proper amino acids into position for incorporation into the peptide.

PROBLEM 16.16 What amino acid sequence is coded for by the following mRNA base sequence?

CUU-AUG-GCU-UGG-CCC-UAA

PROBLEM 16.17 What anticodon sequences of tRNAs are coded for by the mRNA in Problem 16.16?

PROBLEM 16.18 What was the base sequence in the original DNA strand on which the mRNA sequence in Problem 16.16 was made?

16.13 SEQUENCING DNA

One of the greatest scientific revolutions in history is now under way in molecular biology as scientists are learning how to manipulate and harness the genetic machinery of organisms. None of the extraordinary advances of the past decade would have been possible, however, were it not for the discovery in 1977 of a method for sequencing immense DNA chains to find the messages therein. Most

DNA sequencing is carried out by a remarkably efficient and powerful method developed by Allan Maxam and Walter Gilbert. There are five steps:

Step 1. Since molecules of DNA are so enormous—some molecules of human DNA contain up to 250 million base pairs—the first problem in DNA sequencing is to find a method for cleaving the DNA chain at specific points to produce smaller, more manageable pieces. This problem has been solved by the use of enzymes called **restriction endonucleases.** Each different restriction enzyme, of which more than 200 are available, cleaves a DNA molecule between two nucleotides at a well-defined point along the chain where a specific base sequence occurs.

restriction endonuclease an enzyme that is able to cut a DNA strand at a specific base sequence in the chain

By cleavage of large DNA molecules with a given restriction enzyme, many different and well-defined segments of manageable length (100–200 nucleotides) are produced. For example, the restriction enzyme Alu I cleaves the linkage between G and C in the four-base sequence AG-CT. If the original DNA molecule is cut with another restriction enzyme, other segments are produced whose sequences partially overlap those produced by the first enzyme. Sequencing of all the segments, followed by identification of the overlapping sections, then allows complete DNA structure determination.

Step 2. After restriction enzymes have cleaved DNA into smaller pieces (restriction fragments), the various double-stranded fragments are isolated, and each is radioactively tagged by enzymatically incorporating a labeled ^{32}P phosphate group onto the 5′-hydroxyl of the terminal nucleotide. The fragments are then separated into two strands by heating, and the strands are isolated. Imagine, for example, that we now have a single-stranded DNA fragment with the following partial structure:

(5′ end) ^{32}P-G-A-T-C-A-G-C-G-A-T--- (3′ end)

Step 3. The labeled DNA strand is subjected to four parallel sets of chemical reactions under conditions that cause:

a. splitting of the DNA chain next to A,
b. splitting of the DNA chain next to G,
c. splitting of the DNA chain next to C, and
d. splitting of the DNA chain next to *both* T and C.

Mild reaction conditions are chosen so that *only a few of the many possible splittings occur in each reaction.* In our example, the pieces shown in Table 16.4 would be produced.

Chemically, the A and G cleavages are accomplished by treatment of a restriction fragment with dimethyl sulfate [$(CH_3O)_2SO_2$]. Deoxyadenosine (A) is methylated at N3 (S_N2 reaction), and deoxyguanosine (G) is methylated at N7, but T and C aren't affected. Treatment of the methylated DNA with an aqueous solution of the secondary amine piperidine then brings about destruction of the methylated nucleotides and opening of the DNA chain at both the 3′ and 5′ positions next to the methylated bases. By working carefully, it's possible to find reaction conditions that are selective for cleavage either at A or at G.

TABLE 16.4 **Splitting of a DNA fragment under four sets of conditions**

Cleavage conditions	*Labeled DNA pieces produced*
Original DNA fragment	^{32}P-G-A-T-C-A-G-C-G-A-T-
Next to A	^{32}P-G ^{32}P-G-A-T-C ^{32}P-G-A-T-C-A-G-C-G + larger pieces
Next to G	^{32}P-G-A-T-C-A ^{32}P-G-A-T-C-A-G-C + larger pieces
Next to C	^{32}P-G-A-T ^{32}P-G-A-T-C-A-G + larger pieces
Next to C + T	^{32}P-G-A ^{32}P-G-A-T ^{32}P-G-A-T-C-A-G ^{32}P-G-A-T-C-A-G-C-G-A + larger pieces

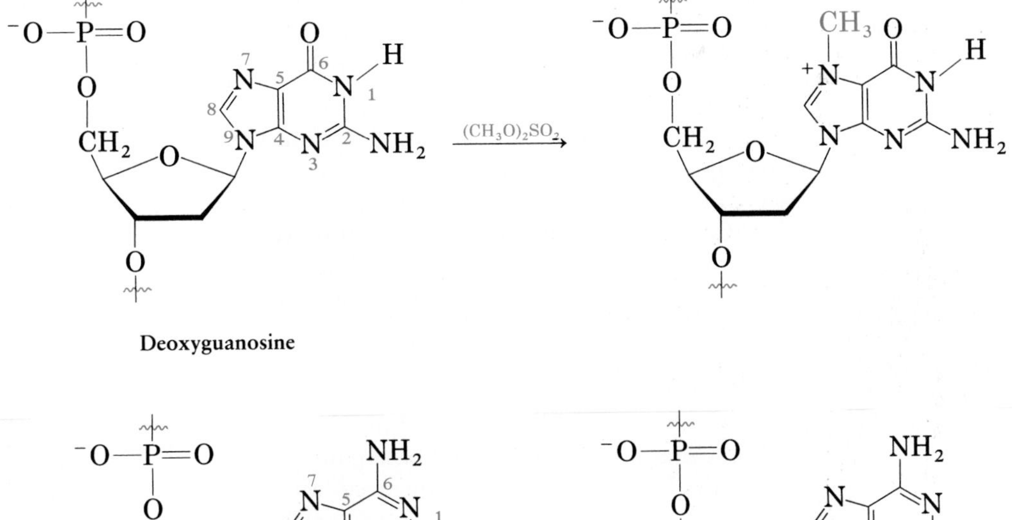

Breaking the DNA chain next to C and T is accomplished by treatment of DNA with hydrazine followed by heating with aqueous piperidine. Although conditions that are selective for cleavage next to T haven't been found, selective cleavage next to C is accomplished by carrying out the hydrazine reaction in 2 M NaCl.

Step 4. Product mixtures from the four cleavage reactions are separated by electrophoresis. When a mixture is placed at one end of a strip of buffered polyacrylamide gel and a voltage is applied to the two ends of the strip, electrically charged substances move along the gel. Each piece moves at a rate that depends on the number of negatively charged phosphate groups (that is, the number of nucleotides) it contains. Smaller fragments move rapidly and larger pieces move more slowly. The technique is so sensitive that up to 250 DNA pieces differing in size by only one nucleotide can be separated.

Once separated, the locations of each DNA fragment are detected by exposing the gel to a photographic plate. Each radioactive end piece containing a ^{32}P label appears as a dark band on the photographic plate, but nonradioactive pieces from the middle of the chain aren't visualized. The gel electrophoresis pattern shown in Figure 16.13 would be obtained in our hypothetical example.

Step 5. The DNA sequence is read directly from the gel. The band that appears farthest from the origin is the terminal mononucleotide (the smallest piece) and can't be identified. Since the terminal mononucleotide appears in the A column, though, it must have been produced by splitting *next to* an A. Thus, the *second* nucleotide in the DNA fragment is an A.

The second farthest band from the origin is a dinucleotide that appears only in the T + C column and is produced by splitting next to the third nucleotide, which must therefore be a T or C. Since this piece doesn't appear in the C column, though, the third nucleotide isn't a C and must therefore be a T. The third farthest band appears in both C and T + C columns, meaning that the fourth nucleotide is a C.

Continuing in this manner, the entire sequence of the DNA fragment is read from the gel simply by noting in what column the successively larger labeled polynucleotide pieces appear. Once read, the entire sequence can be checked by determining the sequence of the complementary strand. The identity of the 5′-terminal nucleotide can be determined by sequencing an overlapping segment produced by cleavage with another restriction enzyme.

The Maxam–Gilbert method of DNA sequencing is so efficient that a trained person can sequence up to 2000 base pairs per day. DNA strands of up to 170,000 base pairs have already been sequenced, and work is now under way to sequence the entire human genome with 3,000,000,000 base pairs. At least 10 years and several billion dollars will be needed.

PROBLEM 16.19 Show the labeled products you would obtain if the following DNA segment were subjected to each of the four cleavage reactions:

^{32}P-A-A-C-A-T-G-G-C-G-C-T-T-A-T-G-A-C-G-A

PROBLEM 16.20 Sketch what you would expect the gel electrophoresis pattern to look like if the DNA segment in Problem 16.19 were sequenced.

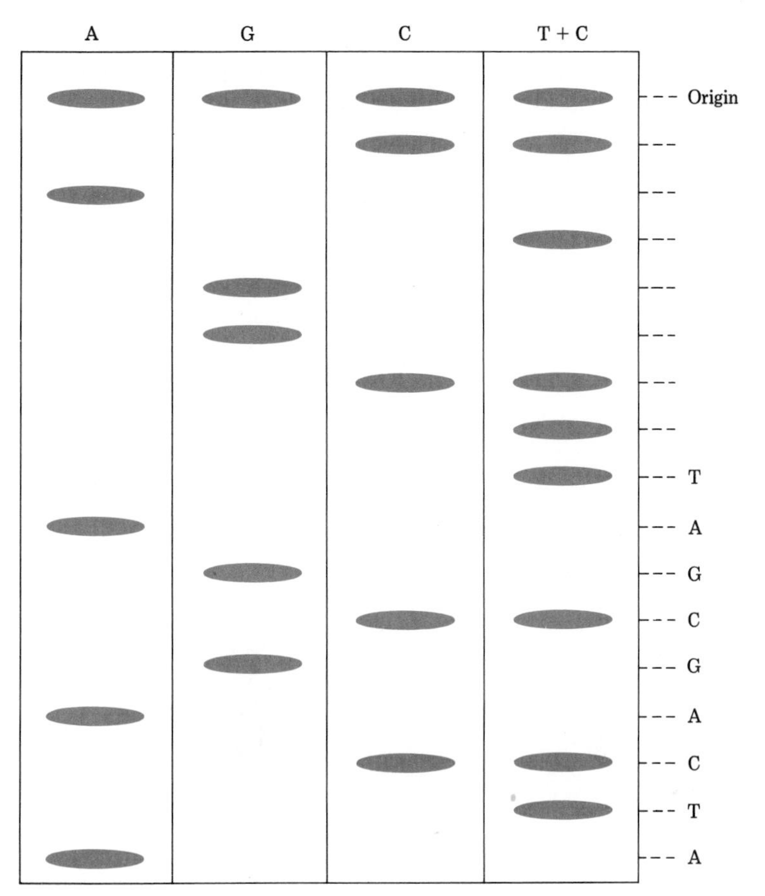

FIGURE 16.13 Representation of a gel electrophoresis pattern. The products of the four cleavage experiments are placed at the top of the gel, and a voltage is applied between top and bottom. Smaller products migrate along the gel at a faster rate and appear at the bottom. The DNA sequence is read by visualizing the radioactive spots.

PROBLEM 16.21 Finish assigning the sequence to the gel electrophoresis pattern shown in Figure 16.13.

INTERLUDE

Cholesterol and Heart Disease

What are the facts about the relationship between cholesterol and heart disease? It's well established that a diet rich in saturated animal fats often leads to an increase in blood-serum cholesterol, at least in sedentary, overweight people.

Conversely, a diet lower in saturated fats and higher in polyunsaturated fats (PUFAs) leads to a lower serum cholesterol level. Studies have shown that a serum cholesterol level greater than 300 mg/dL (a normal value is 150–240 mg/dL) is weakly correlated with an increased incidence of *atherosclerosis,* a form of heart disease in which cholesterol deposits build up on the inner walls of coronary arteries, blocking the flow of blood to the heart muscles.

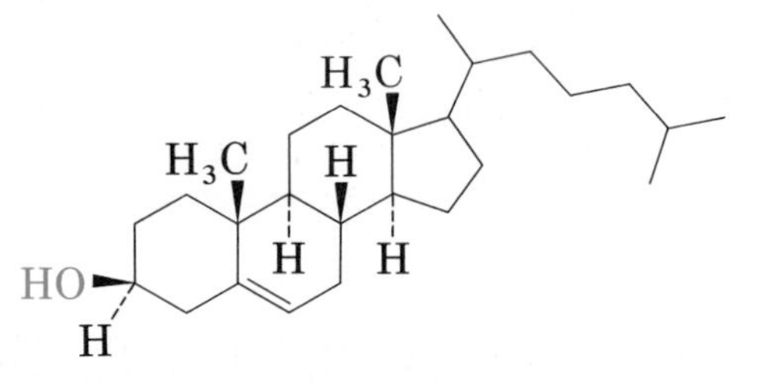

Cholesterol

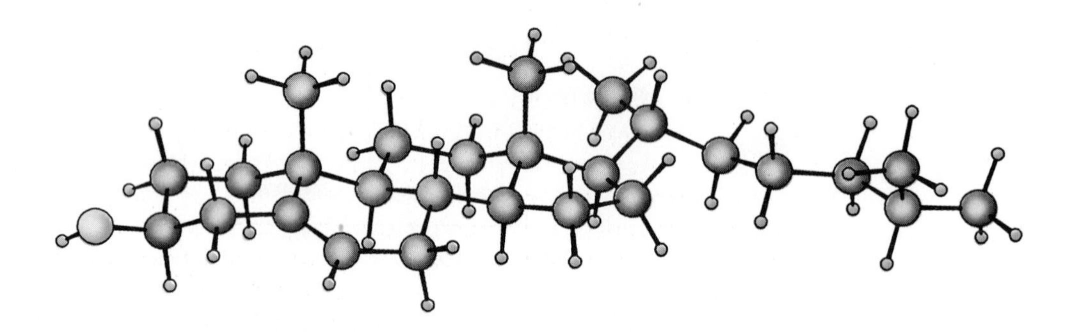

A better indication of a person's risk of heart disease comes from a measurement of blood lipoprotein levels. Lipoproteins are complex molecules with both lipid and protein parts that transport lipids through the body. They can be divided into four types according to density, as shown in Table 16.5. People with a high serum level of high-density lipoproteins (HDLs) seem to have a decreased risk of heart disease. As a rule of thumb, a person's risk drops about 25% for each increase of 5 mg/dL in HDL concentration. Normal values are about 45 mg/dL for men and 55 mg/dL for women, perhaps explaining why women are generally less susceptible than men to heart disease.

TABLE 16.5 Serum lipoproteins

Name	*Density (g/mL)*	*% Lipid*	*% Protein*
Chylomicrons	<0.94	98	2
VLDL (very-low-density lipoproteins)	0.940–1.006	90	10
LDL (low-density lipoproteins)	1.006–1.063	75	25
HDL (high-density lipoproteins)	1.063–1.210	60	40

Chylomicrons and VLDL act primarily as carriers of triglycerides from the intestines to peripheral tissues, whereas LDL and HDL act as carriers of cholesterol to and from the liver. Present evidence suggests that LDL transports cholesterol as its fatty-acid ester *to* peripheral tissues, whereas HDL removes cholesterol as its stearate ester *from* dying cells and transports it back to the liver. If LDL delivers more cholesterol than is needed, and if not enough HDL is present to remove it, the excess is deposited in arteries. The higher the HDL level, the less the likelihood of deposits and the lower the risk of heart disease.

Not surprisingly, the most important factor in gaining high HDL levels is a generally healthy lifestyle. Obesity, smoking, and lack of exercise lead to low HDL levels, whereas regular exercise and a sensible diet lead to high HDL levels. Distance runners, in particular, have HDL levels nearly 50% higher than the general population.

SUMMARY AND KEY WORDS

Lipids and nucleic acids, along with carbohydrates and proteins, comprise the four major classes of biomolecules. **Lipids** are the naturally occurring materials isolated from cells by extraction with organic solvents. **Animal fats** and **vegetable oils** are the most widely occurring lipids. Both fats and oils are **triacylglycerols,** triesters of glycerol with long-chain fatty acids. **Phosphoglycerides** such as **lecithin** and **cephalin** are closely related to fats. The glycerol backbone in these molecules is esterified to two fatty acids and to one phosphate ester. **Sphingolipids,** another major class of phospholipids, have an amino alcohol such as sphingosine for their backbone.

Steroids are plant and animal lipids with a characteristic tetracyclic carbon skeleton. Steroids occur widely in body tissue and have many different kinds of physiological activity. Among the more important kinds of steroids are the sex hormones (**androgens** and **estrogens**) and the **adrenocortical** hormones.

The nucleic acids, **DNA** (**deoxyribonucleic acid**) and **RNA** (**ribonucleic acid**), are biological polymers that act as chemical carriers of an organism's genetic information. Enzyme-catalyzed hydrolysis yields **nucleotides,** which in turn consist of a **purine** or **pyrimidine** heterocyclic amine base linked to C1′ of a simple pentose (ribose in RNA and 2-deoxyribose in DNA), with the sugar in turn linked through its C5′ hydroxyl to a phosphate group. Nucleotides are joined by ester links between the phosphate of one nucleotide and the 3′ hydroxyl on the sugar of another nucleotide.

Molecules of DNA consist of two polynucleotide strands held together by hydrogen bonds between heterocyclic bases on the different strands and coiled into a **double-helix** conformation. Adenine and thymine form hydrogen bonds to each other, as do cytosine and guanine. The two strands of DNA are complementary rather than identical.

Three main processes take place in deciphering the genetic information in DNA:

Replication of DNA is the process by which identical DNA copies are made and genetic information is preserved. This occurs when the DNA double helix unwinds, complementary deoxyribonucleotides line up in order, and two new DNA molecules are produced.

Transcription is the process by which RNA is produced to carry the genetic information from the nucleus to the ribosomes. This occurs when a segment of the DNA double helix unwinds, and complementary ribonucleotides line up to produce **messenger RNA** (mRNA).

Translation is the process by which mRNA directs protein synthesis. Each mRNA has segments called **codons** along its chain. These codons are ribonucleotide triads that are recognized by small amino-acid-carrying molecules of **transfer RNA** (tRNA) that then deliver the appropriate amino acids needed for protein synthesis.

Sequencing of DNA fragments is done by the Maxam–Gilbert method in which chemical reactions are carried out to cause specific cleavages of the DNA chain, followed by separation of the pieces by electrophoresis. The DNA sequence is read directly from the electrophoresis pattern.

ADDITIONAL PROBLEMS

16.22 Write representative structures for the following.

(a) A fat (b) A vegetable oil (c) A steroid

16.23 Write the structures of these molecules.

(a) Sodium stearate (b) Ethyl linoleate (c) Glyceryl palmitodioleate

16.24 Show the products you would expect to obtain from reaction of glyceryl trioleate with the following:

(a) Excess Br_2 in CCl_4 (b) H_2/Pd (c) NaOH, H_2O
(d) O_3, then Zn, CH_3COOH (e) $LiAlH_4$, then H_3O^+

16.25 How would you convert oleic acid into these substances?

(a) Methyl oleate (b) Methyl stearate
(c) Nonanal (d) Nonanedioic acid

16.26 Eleostearic acid, $C_{18}H_{30}O_2$, is a rare fatty acid found in tung oil. On ozonolysis followed by treatment with zinc, eleostearic acid yields one part pentanal, two parts glyoxal (OHC–CHO), and one part 9-oxononanoic acid [$OHC(CH_2)_7COOH$]. Propose a structure for eleostearic acid.

16.27 Stearolic acid, $C_{18}H_{32}O_2$, yields oleic acid on catalytic hydrogenation over the Lindlar catalyst. Propose a structure for stearolic acid.

16.28 Define these terms.

(a) Steroid (b) DNA (c) Base pair
(d) Codon (e) Lipid (f) Transcription

16.29 What DNA sequence is complementary to the following sequence?

G-A-A-G-T-T-C-A-T-G-C

16.30 Give codons for these amino acids.

(a) Ile (b) Asp (c) Thr

16.31 Draw the complete structure of the ribonucleotide codon UAC. For what amino acid does this sequence code?

16.32 Draw the complete structure of the deoxyribonucleotide sequence from which the mRNA codon in Problem 16.31 was transcribed.

16.33 What amino acids do the following ribonucleotide codons code for?

(a) AAU (b) GAG (c) UCC (d) CAU (e) ACC

16.34 From what DNA sequences were each of the mRNA codons in Problem 16.33 transcribed?

16.35 What anticodon sequences of tRNAs are coded for by each of the codons in Problem 16.33?

16.36 Give an mRNA sequence that codes for synthesis of met-enkephalin:

H-Tyr-Gly-Gly-Phe-Met-OH

16.37 What amino acid sequence is coded for by the following mRNA sequence?

CUA-GAC-CGU-UCC-AAG-UGA

16.38 What anticodon sequences of tRNAs are coded for by the mRNA in Problem 16.37? What was the base sequence in the original DNA strand on which this mRNA was made? What was the base sequence in the DNA strand *complementary* to that from which this mRNA was made?

16.39 Look up the structure of angiotensin II (Figure 15.2) and give an mRNA sequence that codes for its synthesis.

16.40 Diethylstilbestrol (DES) exhibits estradiol-like activity even though it is structurally unrelated to steroids. Once used widely as an additive in animal feed, DES has been implicated as a causative agent in several types of cancers. Look up the structure of estradiol (Section 16.5) and show how DES can be drawn so that it is sterically similar to estradiol.

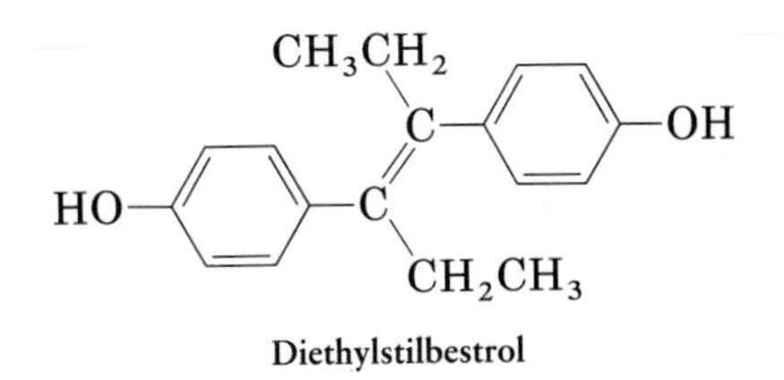

Diethylstilbestrol

16.41 How many chiral centers are present in estradiol (Problem 16.40)? Label each of them.

16.42 What products would you obtain from reaction of estradiol (Problem 16.40) with these reagents?

(a) NaOH, then CH_3I (b) CH_3COCl, pyridine (c) Br_2 (one equivalent)

16.43 Nandrolone is an anabolic steroid sometimes taken by athletes to build muscle mass. Compare the structures of nandrolone and testosterone and point out their structural similarities.

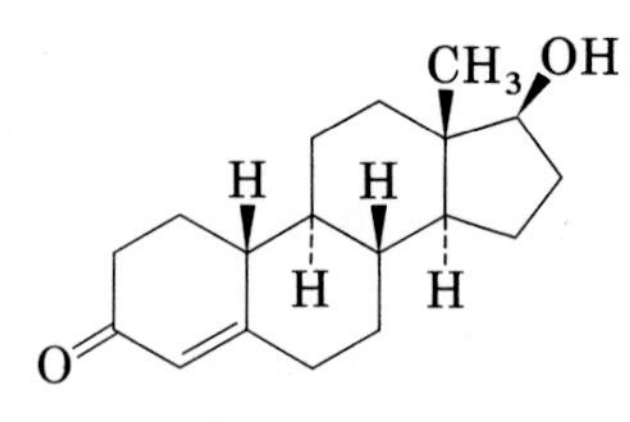

Nandrolone
(an anabolic steroid)

APPENDIX

Nomenclature of Polyfunctional Organic Compounds

Judging from the number of incorrect names that appear in the chemical literature, it's probably safe to say that relatively few practicing organic chemists are fully conversant with the rules of organic nomenclature. Simple hydrocarbons and monofunctional compounds present few problems; the basic rules governing the naming of such compounds are logical and easy to understand. However, problems are often encountered with polyfunctional compounds. Whereas most chemists could correctly identify hydrocarbon **1** as 3-ethyl-2,5-dimethylheptane, rather few could correctly identify polyfunctional compound **2**. Should we consider **2** as an ether? As an ethyl ester? As a ketone? As an alkene? It is, of course, all four, but it has only one correct name: ethyl 3-(4-methoxy-2-oxo-3-cyclohexenyl)propanoate.

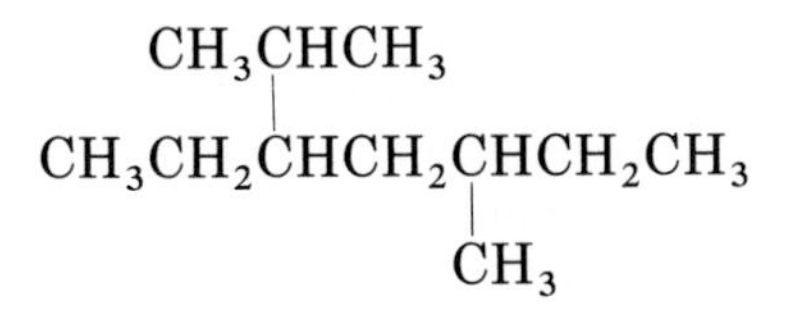

1. 3-Ethyl-2,5-dimethylheptane

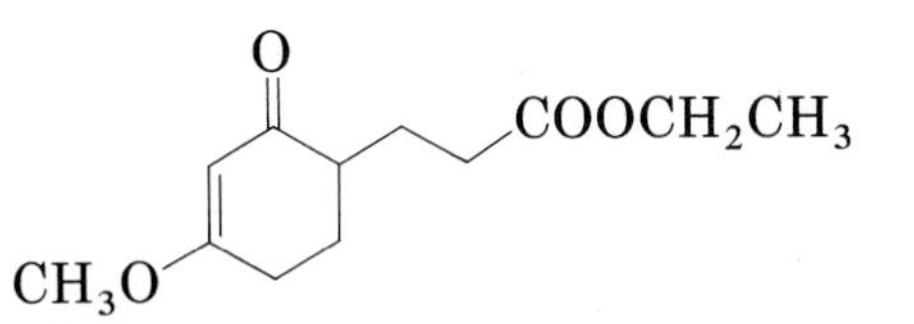

2. Ethyl 3-(4-methoxy-2-oxo-3-cyclohexenyl)propanoate

Naming polyfunctional organic compounds is really not much harder than naming monofunctional compounds. All that is required is a prior knowledge of monofunctional compound nomenclature and rigid application of a set of additional rules. In the following discussion, it's assumed that you have a good command of the rules of monofunctional compound nomenclature that are given throughout the text as each new functional group is introduced. A list of where these rules can be found is shown in Table A.1.

The name of a polyfunctional organic molecule has four parts:

1. **Suffix:** the part that identifies the principal functional group class to which the molecule belongs
2. **Parent:** the part that identifies the size of the main chain or ring

TABLE A.1 Where to find nomenclature rules for simple functional groups

Functional group	*Text section*	*Functional group*	*Text section*
Acid anhydrides	10.1	Amines	12.1
Acid halides	10.1	Aromatic compounds	5.1
Alcohols	8.1	Carboxylic acids	10.1
Aldehydes	9.3	Cycloalkanes	2.7
Alkanes	2.3	Esters	10.1
Alkenes	3.1	Ethers	8.1
Alkyl halides	7.1	Ketones	9.3
Alkynes	4.13	Nitriles	10.1
Amides	10.1		

3. **Substituent prefixes:** parts that identify what substituents are located on the main chain or ring
4. **Locants:** numbers that tell where substituents are located on the main chain or ring

To arrive at the correct name for a complex molecule, the above four name parts must be identified and then expressed in the proper order and format. Let's look at the four parts.

The Suffix: Functional Group Precedence

A polyfunctional organic molecule may contain many different kinds of functional groups, but for nomenclature purposes, we must choose just one suffix. It is not correct to use two suffixes; thus, keto ester **3** must be named either as a ketone with an *-one* suffix or as an ester with an *-oate* suffix, but can't be named as an *-onoate*. Similarly, amino alcohol **4** must be named either as an alcohol (*-ol*) or as an amine (*-amine*), but can't properly be named as an *-olamine*. The only exception to this rule is in naming compounds that have double or triple bonds. For example, the compound $H_2C{=}CHCH_2COOH$ is 3-butenoic acid, and $HC{\equiv}CCH_2CH_2CH_2CH_2OH$ is 5-hexyn-1-ol.

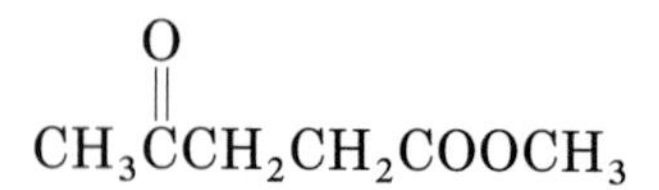

3. Named as an ester with a keto (oxo) substituent
Methyl 4-oxopentanoate

$$CH_3\overset{OH}{\overset{|}{C}}HCH_2CH_2CH_2NH_2$$

4. Named as an alcohol with an amino substituent
5-Amino-2-pentanol

How do we choose which suffix to use? Functional groups are divided into two classes, **principal groups** and **subordinate groups,** as shown in Table A.2. Principal groups are those that may be cited either as prefixes or as suffixes, whereas subordinate groups are those that may be cited only as prefixes. Within the principal groups, an order of precedence has been established. The proper suffix for a given compound is determined by identifying all the functional groups present and then choosing the principal group of highest priority. For example, Table A.2

TABLE A.2 Classification of functional groups for purposes of nomenclature[a]

Functional group class	*Structure*	*Name when used as suffix*	*Name when used as prefix*
Principal groups			
Carboxylic acids	—COOH	-oic acid -carboxylic acid	carboxy
Carboxylic anhydrides	$-\overset{O}{\overset{\Vert}{C}}-O-\overset{O}{\overset{\Vert}{C}}-$	-oic anhydride -carboxylic anhydride	
Carboxylic esters	—COOR	-oate -carboxylate	alkoxycarbonyl
Acyl halides	—COCl	-oyl halide -carbonyl halide	halocarbonyl (haloformyl)
Amides	$-CONH_2$	-amide -carboxamide	amido
Nitriles	—C≡N	-nitrile -carbonitrile	cyano
Aldehydes	—CHO	-al -carbaldehyde	formyl
	=O		oxo (either aldehyde or ketone)
Ketones	=O	-one	oxo
Alcohols	—OH	-ol	hydroxy
Phenols	—OH	-ol	hydroxy
Thiols	—SH	-thiol	mercapto, sulfhydryl
Amines	$-NH_2$	-amine	amino
Imines	=NH	-imine	imino
Alkenes	C=C	-ene	
Alkynes	C≡C	-yne	
Alkanes	C—C	-ane	
Subordinate groups			
Ethers	—OR		alkoxy
Sulfides	—SR		alkylthio
Halides	—F, —Cl, —Br, —I		halo
Nitro	$-NO_2$		nitro
Azides	N=N=N		azido
Diazo	=N=N		diazo

[a] The principal functional groups are listed in order of decreasing priority, but the subordinate functional groups have no established priority order. Principal functional groups may be cited either as prefixes or as suffixes; subordinate functional groups may be cited only as prefixes.

indicates that keto ester **3** must be named as an ester rather than as a ketone, since an ester functional group is higher in priority than a ketone. Similarly, amino alcohol **4** must be named as an alcohol rather than as an amine. The correct name of **3** is methyl 4-oxopentanoate, and the correct name of **4** is 5-amino-2-pentanol. Further examples are shown below.

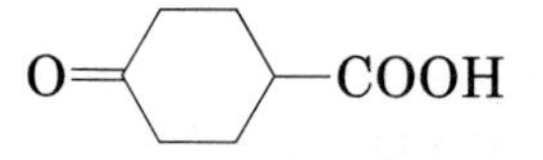

5. Named as a cyclohexanecarboxylic acid with an oxo substituent
4-Oxocyclohexanecarboxylic acid

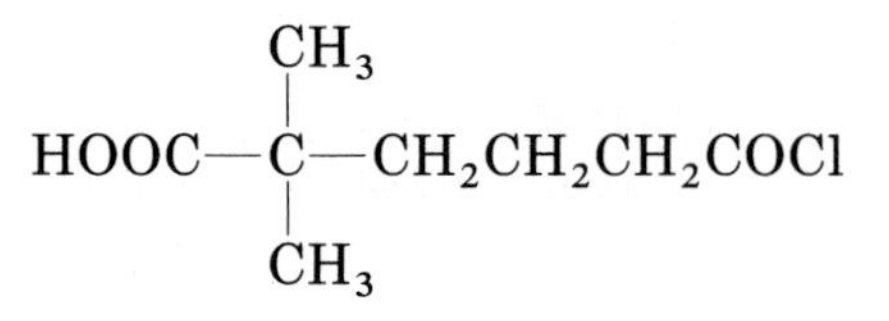

6. Named as a carboxylic acid with a chlorocarbonyl substituent
5-Chlorocarbonyl-2,2-dimethylpentanoic acid

$$\begin{array}{l} \quad\;\; CHO \\ \quad\;\; | \\ CH_3CHCH_2CH_2CH_2COOCH_3 \end{array}$$

7. Named as an ester with an oxo substituent
Methyl 5-methyl-6-oxohexanoate

The Parent: Selecting the Main Chain or Ring

The parent or base name of a polyfunctional organic compound is usually quite easy to identify. If the group of highest priority is part of an open chain, we simply select the longest chain that contains the largest number of principal functional groups. If the highest-priority group is attached to a ring, we use the name of that ring system as the parent. For example, compounds **8** and **9** are isomeric aldehydo acids, and both must be named as acids rather than as aldehydes according to Table A.2. The longest chain in compound **8** has seven carbons, and the substance is therefore named 6-methyl-7-oxoheptanoic acid. Compound **9** also has a chain of seven carbons, but the longest chain that contains both of the principal functional groups has only three carbons. The correct name of this compound is 3-oxo-2-pentylpropanoic acid.

$$\begin{array}{l} \quad\;\; CHO \\ \quad\;\; | \\ CH_3CHCH_2CH_2CH_2CH_2COOH \end{array}$$

8. Named as a substituted heptanoic acid
6-Methyl-7-oxoheptanoic acid

$$\begin{array}{r} CHO \quad\;\; \\ | \quad\;\;\;\; \\ CH_3CH_2CH_2CH_2CH_2CHCOOH \end{array}$$

9. Named as a substituted propanoic acid
3-Oxo-2-pentylpropanoic acid

Similar rules apply for compounds **10–13**, which contain rings. Compounds **10** and **11** are isomeric keto nitriles, and both must be named as nitriles according to Table A.2. Substance **10** is named as a benzonitrile since the –CN functional group is a substituent on the aromatic ring, but substance **11** is named as an acetonitrile since the –CN functional group is on an open chain. The correct names are 2-acetyl-4-methylbenzonitrile (**10**) and (2-acetyl-4-methylphenyl)-acetonitrile (**11**). Compounds **12** and **13** are both keto acids and must be named as acids. The correct names are 3-(2-oxocyclohexyl)propanoic acid (**12**) and 2-(3-oxopropyl)cyclohexanecarboxylic acid (**13**).

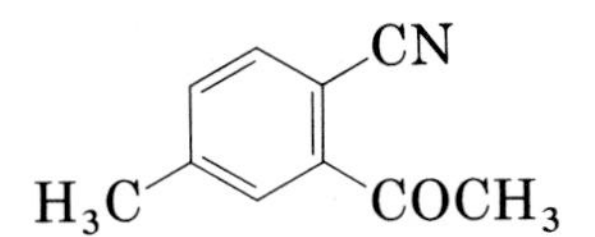

10. Named as a substituted benzonitrile
2-Acetyl-4-methylbenzonitrile

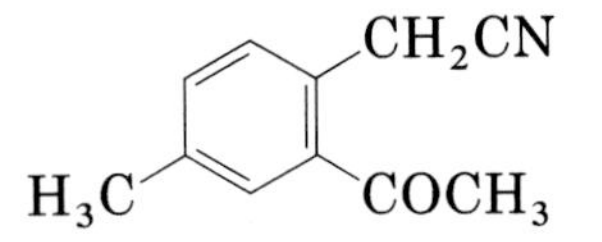

11. Named as a substituted acetonitrile
(2-Acetyl-4-methylphenyl)acetonitrile

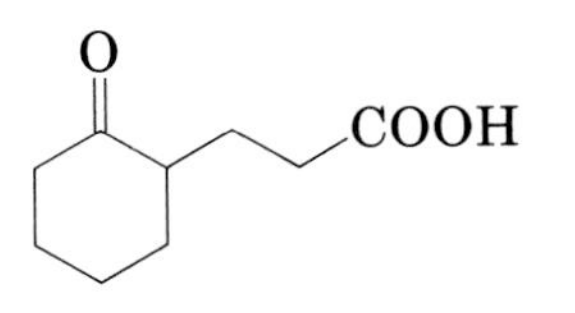

12. Named as a carboxylic acid
3-(2-Oxocyclohexyl)propanoic acid

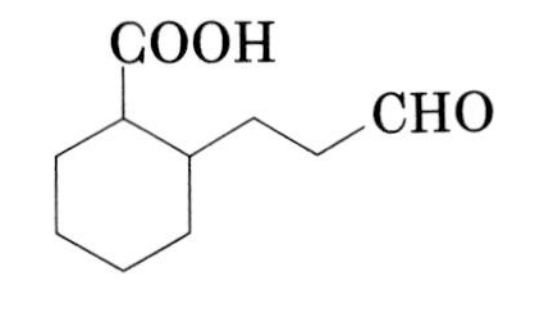

13. Named as a carboxylic acid
2-(3-Oxopropyl)cyclohexanecarboxylic acid

The Prefixes and Locants

With the suffix and parent name established, the next step is to identify and number all substituents on the parent chain or ring. These substituents include all alkyl groups and all functional groups other than the one cited in the suffix. For example, compound **14** contains three different functional groups (carboxyl, keto, and double bond). Since the carboxyl group is highest in priority, and since the longest chain containing the functional groups is seven carbons long, **14** is a heptenoic acid. In addition, the main chain has an oxo (keto) substituent and three methyl groups. Numbering from the end nearer the highest-priority functional group, we find that **14** is 2,5,5-trimethyl-4-oxo-2-heptenoic acid. Note that the final "e" of heptene is deleted in the word "heptenoic." This deletion only occurs when the name would have two adjacent vowels (thus, "heptenoic" has the final "e" deleted, but "heptenenitrile" retains the "e"). Look back at some of the other compounds we have considered to see other examples of how prefixes and locants are assigned.

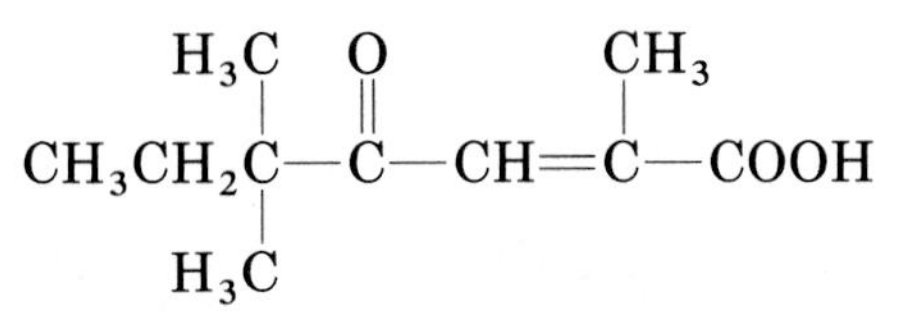

14. Named as a heptenoic acid
2,5,5-Trimethyl-4-oxo-2-heptenoic acid

Writing the Name

Once the name parts have been established, the entire name is written out. Several additional rules apply:

1. **Order of prefixes:** When the substituents have been identified, the main chain has been numbered, and the proper multipliers such as *di-* and *tri-* have been assigned, the name is written with the substituents listed in alphabetical, rather than numerical order. Multipliers such as *di-* and *tri-* are not used for alphabetization purposes, but the prefixes *iso-* and *tert-* are used.

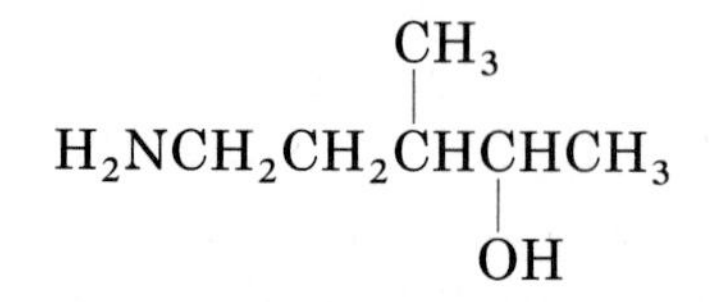

15. 5-Amino-3-methyl-2-pentanol (*NOT* 3-methyl-5-amino-2-pentanol)

2. **Use of hyphens; single- and multiple-word names:** The general rule is to determine whether the principal functional group is itself an element or compound. If it is either an element or a compound, then the name is written as a single word; if it is not, then the name is written as separate words. For example, methylbenzene (one word) is correct because the parent, benzene, is itself a compound. Diethyl ether, however, is written as two words because the parent, ether, is a class name rather than a compound name. Some further examples are shown below.

$$H_3C—Mg—CH_3$$

16. Dimethylmagnesium (one word, since magnesium is an element)

$$CH_3CHBrCOOH$$

17. 2-Bromopropanoic acid (two words, since "acid" is not a compound)

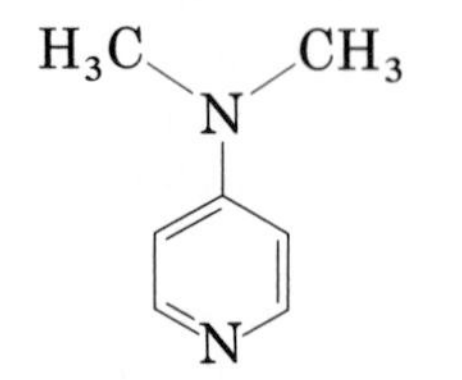

18. 4-(Dimethylamino)pyridine (one word, since pyridine is a compound)

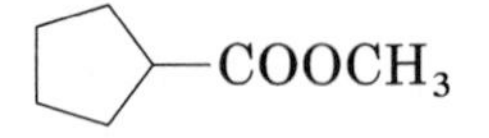

19. Methyl cyclopentanecarboxylate

3. **Parentheses:** Parentheses are used to denote complex substituents when ambiguity would otherwise arise. For example, chloromethylbenzene has two substituents on a benzene ring, but (chloromethyl)benzene has only one complex substituent. Note that the expression in parentheses is not set off by hyphens from the rest of the name.

Cl — (benzene ring) — CH_3

20. *p*-Chloromethylbenzene (two substituents)

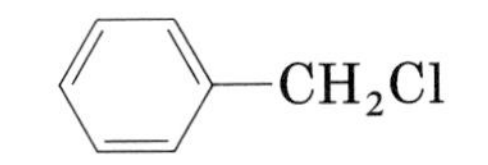

21. (Chloromethyl)benzene (one complex substituent)

$$\begin{array}{c} CH_3CHCH_2CH_3 \\ | \\ HOOC{-}CHCH_2CH_2COOH \end{array}$$

22. 2-(1-Methylpropyl)pentanedioic acid (The 1-methylpropyl group is a complex substituent on C2 of the main chain.)

Additional Reading

Further explanations of the rules of organic nomenclature can be found in the following references:

1. O. T. Benfey, "The Names and Structures of Organic Compounds," Wiley, New York, 1966.
2. J. H. Fletcher, O. C. Dermer, and R. B. Fox, "Nomenclature of Organic Compounds: Principles and Practice," Advances in Chemistry Series No. 126, American Chemical Society, Washington, D.C., 1974.
3. International Union of Pure and Applied Chemistry, "Nomenclature of Organic Chemistry, Sections A, B, C, D, E, F, and H," Pergamon Press, Oxford, 1979.
4. J. G. Traynham, "Organic Nomenclature: A Programmed Introduction," Prentice-Hall, Englewood Cliffs, N.J., 1985.

Index

References in color refer to boldface entries in the text where terms are defined.

A

D

E

F

N

O

P

Q

R

PERIODIC CHART

Atomic number	11
Name	Sodium
Symbol	**Na**
Atomic weight	22.98977

[a]Mass number of most stable or best-known isotope

[b]Mass of the isotope of longest half-life

← Transition elements →

Period	Group IA	IIA	IIIB	IVB	VB	VIB	VIIB	VIII	VIII
1	1 Hydrogen **H** 1.0079								
2	3 Lithium **Li** 6.941	4 Beryllium **Be** 9.01218							
3	11 Sodium **Na** 22.98977	12 Magnesium **Mg** 24.305							
4	19 Potassium **K** 39.098	20 Calcium **Ca** 40.08	21 Scandium **Sc** 44.9559	22 Titanium **Ti** 47.90	23 Vanadium **V** 50.9414	24 Chromium **Cr** 51.996	25 Manganese **Mn** 54.9380	26 Iron **Fe** 55.847	27 Cobalt **Co** 58.9332
5	37 Rubidium **Rb** 85.4678	38 Strontium **Sr** 87.62	39 Yttrium **Y** 88.9059	40 Zirconium **Zr** 91.22	41 Niobium **Nb** 92.9064	42 Molybdenum **Mo** 95.94	43 Technetium **Tc** 98.9062[b]	44 Ruthenium **Ru** 101.07	45 Rhodium **Rh** 102.9055
6	55 Cesium **Cs** 132.9054	56 Barium **Ba** 137.34	* 57 Lanthanum **La** 138.9055	72 Hafnium **Hf** 178.49	73 Tantalum **Ta** 180.9479	74 Wolfram (Tungsten) **W** 183.85	75 Rhenium **Re** 186.2	76 Osmium **Os** 190.2	77 Iridium **Ir** 192.22
7	87 Francium **Fr** (223)[a]	88 Radium **Ra** 226.0254[b]	** 89 Actinium **Ac** (227)[a]	104 Unnilquadium **Unq** (261)[a]	105 Unnilpentium **Unp** (262)[a]	106 Unnilhexium **Unh** (263)[a]			

* Lanthanide series 6	58 Cerium **Ce** 140.12	59 Praseodymium **Pr** 140.9077	60 Neodymium **Nd** 144.24	61 Promethium **Pm** (145)[a]	62 Samarium **Sm** 150.4
** Actinide series 7	90 Thorium **Th** 232.0381[b]	91 Protactinium **Pa** 231.0359[b]	92 Uranium **U** 238.029	93 Neptunium **Np** 237.0482	94 Plutonium **Pu** (242)[a]